RNA Technologies

For further volumes:
http://www.springer.com/series/8619

Volker A. Erdmann • Wojciech T. Markiewicz •
Jan Barciszewski
Editors

Chemical Biology of Nucleic Acids

Fundamentals and Clinical Applications

Springer

Editors
Volker A. Erdmann
Institute of Chemistry/Biochemistry
Free University Berlin
Berlin, Germany

Wojciech T. Markiewicz
Jan Barciszewski
Institute of Bioorganic Chemistry
Polish Academy of Sciences
Poznan, Poland

ISBN 978-3-642-54451-4 ISBN 978-3-642-54452-1 (eBook)
DOI 10.1007/978-3-642-54452-1
Springer Heidelberg New York Dordrecht London

Library of Congress Control Number: 2014937792

Printed on acid-free paper

Springer is part of Springer Science+Business Media (www.springer.com)

Preface

This year the RNA technologies are celebrating two very important anniversaries. The first one is the 60th anniversary of the founding of the RNA Tie Club and the other one is the 40th anniversary of the successful crystallization and structural determination of the phenylalanine-specific tRNA from yeast.

The 60th anniversary of the RNA Tie Club is indeed a very important date to remember! The 20 members of this club, among them Sydney Brenner, Erwin Chargaff, Francis Crick, Max Delbrück, Paul Doty, George Gamow, Leslie Orgel, Alexander Rich, Gunther Stent, and James Watson, had realized to the surprise of everyone, only 1 year after Watson and Crick had determined the double helical structure for DNA (Watson and Crick 1953), that there seems to be much more to RNA molecules than anticipated.

Indeed, Watson and Crick had surprised everyone with the statement in their DNA paper concerning RNA molecules: "It is impossible to build this structure with a ribose sugar in place of the deoxyribose, as the extra oxygen atom would make too close a van der Waals contact" (Watson and Crick 1953). It was therefore not expected when 1 year later the RNA Tie Club was founded with the goal to "to solve the riddle of the RNA structure and to understand how it built proteins".

The members of the RNA Tie Club had selected the guide slogan "Do or die, or don't try". Thus, very intensive RNA structural and functional studies were initiated, in which Alexander Rich played a key role. Indeed, Rich could demonstrate in the following years that RNA may form Watson and Crick base pairs, that an RNA strand could base pair with a DNA strand and that even triple strand nucleic acid structures were possible (Rich personal communication; Rich 2009). With the determination of the three-dimensional tRNA structure 40 years ago by the research groups from Alexander Rich (Kim et al. 1973) and Aaron Klug (Robertus et al. 1974), the unforeseen diversity of RNA structural potentials became apparent for the first time, and it can be safe to assume that we are still lacking the complete knowledge of all structural possibilities of the RNA molecules.

Parallel to the structural activities, studies were initiated in which the chemistry of nucleic acids was developed, so that oligonucleotides could be chemically synthesized. It was the knowledge of the chemical synthesis of nucleic acids which turned out to be the key to unlock the secrets of the genetic code.

Clearly these early RNA studies required the significant involvement of nucleic acid chemistry, which later on with the rapid developments of molecular biology and molecular genetics seemed to be less and less important. But now in the last few years there has been a unique revival of the employment of nucleic acid chemical methods in the natural sciences, so that we are currently speaking of the new field of chemical biology.

We, as the editors of this volume in the *RNA Technologies Series*, are very happy to present to the reader 29 of the world leading research groups in the area of chemical biology. You will see that the exiting research carried out by these groups will introduce us to new ideas how chemistry can add new elements to the areas of nucleic acids in biotechnologies, nanotechnologies, and, very importantly, in the areas of diagnostics and therapy in the field of molecular medicine.

This new volume of *RNA Technologies* starts with a contribution devoted to the newly defined field named *pre-biology*. The majority of papers address in a new way the fundamental questions: how the early genetic code was developed, how stable and prone to isomerisation RNAs are, and how stable nucleobases in RNAs are. The search for alternative and modified nucleobase systems has both a sense of basic questions and also applicational aspects in the fields of new approaches in therapies and technologies. In a similar context one can view the questions discussed in papers dealing with structural aspects of nucleic acids. The coped knowledge is very rich, yet it seems that still many possible structural features of nucleic acids are to be discovered. For example, the dynamic structures of nucleic acids and their analogs, with their ingenious modifications in the sugar moiety, which for example, determine their biological functions in processes such as DNA replication. But also very interestingly, their extra- and intracellular transport are the subject of several papers. This includes, for example, the important G-quadruplex motifs as potential therapeutic targets and still double-stranded nucleic acids as important molecular tools.

The development of technologies suitable to modify the level of gene expression is remarkable as well. The possibility of exon skipping by chemically modified RNAs brings again a new element to RNA functions. These technologies include editing of therapeutic genes, using modified riboswitches and approaches that are based on modification of Cap regions of the mRNAs. Not overlooked should be the developments of new powerful approaches to detect nucleic acids and the applications of appropriately modified RNAs by electron paramagnetic resonance (EPR) spectroscopy. One very interesting contribution, as a representative of these techniques, is also presented in this book.

And finally, we would like to bring the readers' attention to new approaches to study structure and interactions of nucleic acids and other biomolecules in an environment that is offered by ionic liquids. This field widens the range of observations of the most important biological molecules and might be considered as a step towards studying their structure and interactions under the conditions of molecular crowding, which is otherwise hardly accessible for detailed analysis.

The order of the chapters in the book could perhaps also be a different one. But this is more a personal opinion, because the areas covered in the field of chemical biology are so diverse that it should tempt the reader to just go through the book and read the chapters which interest him at that time the most.

In summary, we hope that this new volume of *RNA Technologies* will be of interest for chemists, biochemists, and life scientists and that it will not only stimulate their research but also our future research.

Berlin, Germany
Poznan, Poland
January, 2014

Volker A. Erdmann
Wojciech T. Markiewicz
Jan Barciszewski

References

Kim SH, Quigley GJ, Suddath FL et al (1973) Three-dimensional structure of yeast phenylalanine transfer RNA: folding of the polynucleotide chain. Science 179:285–288

Rich A (2009) The era of RNA awakening: structural biology of RNA in the early. Q Rev Biophys 42:117–137

Robertus JD, Ladner JE, Finch JT et al (1974) Structure of yeast phenylalanine tRNA at 3 Å resolution. Nature 250:546–551

Watson JD, Crick FH (1953) Molecular structure of nucleic acids; a structure for deoxyribose nucleic acid. Nature 171:737–738

Contents

RNA as Major Components in Chemical Evolvable Systems 1
Peter Strazewski

How the Early Genetic Code Was Established?: Inference from the Analysis of Extant Animal Mitochondrial Decoding Systems 25
Kimitsuna Watanabe and Shin-ichi Yokobori

Isomerization of RNA Phosphodiester Linkages 41
Harri Lönnberg

Effects of Ionic Liquid and Liposomes on the Structure, Stability, and Function of Nucleic Acids 57
Naoki Sugimoto

Oxidative Damage on RNA Nucleobases 75
Pascal A. Küpfer and Christian J. Leumann

Use of FRET to Study Dynamics of DNA Replication 95
Philip Nevin and Penny J. Beuning

Design, Characterization, and Application of Imidazopyridopyrimidine:Naphthyridine Base-Pairing Motifs Consisting of Four Hydrogen Bonds 113
Noriaki Minakawa and Akira Matsuda

Creation of Unnatural Base Pair Systems Toward New DNA/RNA Biotechnologies 131
Michiko Kimoto and Ichiro Hirao

Flexible Nucleobase Analogues: Novel Tools for Exploring Nucleic Acids 149
Sarah C. Zimmermann and Katherine L. Seley-Radtke

Sequence-Selective Recognition of Double-Stranded RNA 167
Eriks Rozners

Determining Transient Nucleic Acid Structures by NMR 181
Jeetender Chugh

Diastereomer-Specific Repertoire of 7′*R*- or 7′*S*-Me-Carba-*L*ocked *N*ucleic *A*cids (cLNAs) in Antisense Oligo/RNA Duplexes and Engineering of Physico-chemical and Enzymological Properties 199
Qing Li, Oleksandr Plashkevych, Ram Shankar Upadhayaya, Sachin Gangadhar Deshpande, Andras Földesi, and Jyoti Chattopadhyaya

Challenges and Opportunities for Oligonucleotide-Based Therapeutics by Antisense and RNA Interference Mechanisms 227
Ramon Eritja, Montserrat Terrazas, Santiago Grijalvo, Anna Aviñó, Adele Alagia, Sónia Pérez-Rentero, and Juan Carlos Morales

Progress in Chemically Modified Nucleic Acid Aptamers 243
Masayasu Kuwahara

Aptamers as Molecular Smugglers 271
Eileen Magbanua and Ulrich Hahn

Biochemical Aspects of Subcellular RNA Transport and Localization 293
Diana Bauermeister, Maike Claußen, and Tomas Pieler

Small Size, Big Impact: Bacterial Functional Nucleic Acids and Their Applications 309
Wendy W.K. Mok, Simon A. McManus, and Yingfu Li

Towards Defined DNA and RNA Delivery Vehicles Using Nucleic Acid Nanotechnology 325
Anders Hauge Okholm, David Schaffert, and Jørgen Kjems

Targeted Editing of Therapeutic Genes Using DNA-Based Transcriptional Activators: Scope and Challenges 347
Ganesh N. Pandian and Hiroshi Sugiyama

Interaction of DNA Intramolecular Structures with Their Complementary Strands: A Thermodynamic Approach for the Control of Gene Expression 367
Irine Khutsishvili, Sarah E. Johnson, Calliste Reiling, Iztok Prislan, Hui-Ting Lee, and Luis A. Marky

Site-Directed Spin Labeling of RNA for Distance Measurements by EPR 385
Joachim W. Engels, Christian Grünewald, and Lena Wicke

Chemo-enzymatic Strategies to Modify RNA in vitro or in Living Cells 409
Daniela Schulz and Andrea Rentmeister

Metal Dependence of Ligand Binding and Heavy-Atom Derivatization of Evolutionarily Distinct $PreQ_1$ Riboswitches 423
Joseph E. Wedekind, Joseph A. Liberman, Jermaine L. Jenkins, and Mohammad Salim

DNA G-Quadruplexes and I-Motifs in Therapeutics and Diagnostics 441
Yogini P. Bhavsar-Jog, Samantha M. Reilly, and Randy M. Wadkins

Peptides Targeting G-Quadruplex Structures 459
Kenji Usui and Arisa Okada

Synthesis of Site-Specifically Modified Long-mer RNAs 477
Darko Balke, Jennifer Frommer, Nico Rublack, Danilo Springstubbe, Bettina Appel, and Sabine Müller

Synthesis and Exon-Skipping Activity of Chemically Modified RNAs 497
Yoshiaki Masaki, Takeshi Yamada, Hisao Saneyoshi, Akihiro Ohkubo, Kohji Seio, and Mitsuo Sekine

mRNA and snRNA Cap Analogs: Synthesis and Applications 511
Janusz Stepinski and Edward Darzynkiewicz

Innovative Chemistry for Synthesis of Regular RNA, 5′-Triphosphate RNA, or 5′-Capped RNA 563
Yann Thillier, François Morvan, Jean-Jacques Vasseur, and Françoise Debart

Index 591

RNA as Major Components in Chemical Evolvable Systems

Peter Strazewski

Contents

1 Introduction: Synthetic Living Cells from an Experimental Bottom-Up Approach 2
1.1 What I Cannot Create, I Do Not Understand 2
1.2 Hierarchical Reduction, Metaphors, Prudent Compromises and Dogma 13 4
2 First Steps: Preparations for the Real Thing 5
2.1 Translate to Replicate, Rather than an RNA World 5
2.2 How to Compartment RNA? 7
2.3 A Hydrophobic Peptide Anchors RNA to Giant Lipidic Vesicles 10
2.4 How to Generate Plausibly Prebiotic Lipophilic "Random" Peptide Libraries? 16
3 Perspectives: Evolvable Systems Chemistry 18
3.1 Lipids and Glycans 19
3.2 Peptidyl RNA Esters 20
3.3 Why RNA? 21
References 22

Abstract A milestone for the origin of life was the onset of a sustained functional interplay between nucleic acids, peptides, lipids and sugars, to form chemical dynamic off-equilibrium systems. The bilayer surface of lipidic vesicles served for the anchoring and enrichment of macromolecules that could grow in size would interact with one another one thousand fold more frequently than unbound in solution and could sooner or later be internalised into the interior of the vesicles. Within a flow of exogenous high-energy compounds, internalised, possibly surface-bound, hereditary molecules like RNA could then grow in population size and length more rapidly, safer and more reliably. When "fed" with activated and activating monomers over a long enough time period, such systems would persist,

---

P. Strazewski (✉)
Institut de Chimie et Biochimie Moléculaires et Supramoléculaires (Unité Mixte de Recherche 5246), Université Claude Bernard Lyon 1, 69622 Villeurbanne Cedex, France

Université de Lyon, Lyon, France
e-mail: strazewski@univ-lyon1.fr

V.A. Erdmann et al. (eds.), *Chemical Biology of Nucleic Acids*, RNA Technologies,
DOI 10.1007/978-3-642-54452-1_1, © Springer-Verlag Berlin Heidelberg 2014

acquire partial control over their environment and eventually produce replicating protocells capable of autonomously producing high-energy compounds on their own.

We are outlining here how hydrophobic interactions between peptides and lipids can drive supramolecular multicomponent systems including RNA to assemble into models of functionalised protocells that carry potentially inheritable genetic information. Mixing chemically activated macromolecular libraries of synthetic biomolecules with an evolvable population of lipidic giant vesicles is an experimental bottom-up approach being rooted in a growing community of the emergent research area "Systems Chemistry". The aim of such experimentation is to initiate lifelike behaviour from inanimate chemical systems. The impact of having at one's disposal chemical protocols for the off-equilibrium search for synthetic peptides that transport synthetic RNA into giant lipidic vesicles (or cells) is difficult to overestimate. On a fundamental level, the impact of being able to prepare "living synthetic cells" would be quite breathtaking.

Keywords Systems chemistry • Evolvability • Off-equilibrium • Synthetic cells • Origin of life • Synthetic biology

1 Introduction: Synthetic Living Cells from an Experimental Bottom-Up Approach

1.1 What I Cannot Create, I Do Not Understand

To understand life is more than to define it, and to define life is presently a highly questionable endeavour (Pross 2012; Ellington 2012). I should like to advocate here my own fundamental conviction—and means of circumventing the definition problem, or postponing it until we know better—which is best summarised by Richard Feynman's famous "What I cannot create, I do not understand" (Hawking 2001). We can ultimately only understand what life is, once we are able to (re) create it ourselves—experimentally from clearly inanimate to clearly animate. It may be that we would then witness a life form somewhat different from our own. Possibly and quite likely, after repeating the experiments, we would perceive even more divergence. And yet we could watch "synthetic lives" emerge and evolve in complexity. We could deduce from these observations our comprehension of Life that appeared on Earth some 3.5–3.9 thousand million years ago. Besides, we could convince many more contemporaries that life emerges without any "external help" (panspermia, God, etc.), any other help than our own at least. If we humans can do it (create life from the inanimate), Nature—unlike building refrigerators, cars, and aeroplanes—can do it without us.

There is a *caveat* that is recurrently brought up during discussions on what Nature can do and what humans cannot. Time is a crucial factor. Nature, many say, disposed of "millions of years" to produce life, we scientists do not. My immediate, somewhat agitated answer to this is: "How do we know that Nature needed such a long time to produce the first animate matter? I don't, I cannot remember, I am too young to have witnessed the origin of life." There can be no proof for the million-years assumption, it must stay an unscientific statement, it is not falsifiable. We may quite safely assume that the evolution of our ecosphere has taken millions of years, not of life per se.

For the origin, we can reflect upon a reasonable minimal timeframe that must have been at disposal at this particular place and time when, now and again, chemical conditions on primordial Earth were such that, for limited periods of time, populations of compartmented matter appeared, consumed exogenous material and free energy, grew in population size, changed its immediate environment by excreting "waste", and so on. Would we be present and watch it happening, many of us were to call it "alive, animate, behaving as if it had a purpose, teleonomic"; until, for some lack or other (side reactions), the phenomenon would have vanished, degraded, fallen into its thermodynamic pithole, dematerialised, "died". Such events might have taken place many times in many places during the same or different stages of primordial Earth. They need neither have been homogenous (similar from case to case), nor have they led to any continuity (monotony) other than vanishing and reappearing for the same repeated physico-chemical reasons, and in the absence of any "systemic memory" being based on replication and inheritance. A single appear-and-vanish cycle could hardly endure and evolve over millions of years. The chemical persistence of the organic molecules—let us avoid here the term "lifetime"—needed for keeping the system away from thermodynamic equilibrium is much shorter. We are not considering the persistence of the possibly simple and thermodynamically quite stable building blocks (purine bases, amino acids), but of supramolecular assemblies based on hydrogen bonds and other chemically weak interactions.

A number of highly respectable colleagues would not quite describe the above phenomenon as animate. One axiomatic requisite for matter to "behave alive" must be a replication-based "systemic memory", thus including a templating device that would allow for heritage of much of the acquired "systemic knowledge". Also mandatory is the ability of the system to produce endogenously some of its needed high-energy compounds, to replace the exogenous compounds and to become efficient in energy gathering and thus more independent from the environment.

1.2 *Hierarchical Reduction, Metaphors, Prudent Compromises and Dogma 13*

So how to experiment on the creation of synthetic cells? My methodological conviction is in full agreement with what Addy Pross describes as "hierarchical reduction", the idea being that "phenomena at one hierarchical level can [only] be explained using concepts taken from a lower hierarchical level" (Pross 2012, p. 53): You can only explain biogenesis by using the concepts of the chemical and physical sciences; staying at the biological level of complexity will make it impossible to ultimately comprehend life's emergence, and even more so, to re-initiate life from the entirely inanimate. This means, for example, that we need to find chemical conditions for replication to actually emerge and evolve, rather than to start off with a designed replicating system right away.

My other methodological *credo* is the following. In our attempts to set up experiments that are aiming at the observation of lifelike behaviour in initially inanimate chemical systems, we should better avoid creating metaphors of life, i.e. mere images, prone to be mistaken for a system that is truly alive for a period of time. If a colleague constructed a chemical system that contained a compartmented and truly exponential replicator, whose function depended on a designed model metabolism, the system persisted unchanged unless it grew in population size as long as sufficient energy was provided, my short answer would likely be: "Ceci n'est pas une pipe" (Magritte 1929). My longer answer: An essential attribute of any living chemical system, irrespective of composition, size, complexity or location, must be the evolvability of its diversity in time. If the above chemical system represented a certain stage and complexity of life and was kept far off its thermodynamic equilibrium during that period of time, but without ever changing into any other slightly different persistent off-equilibrium system (only into thermodynamic degradation products), then it were not "living" after all. It would be just a metaphor, an image, if you like; perhaps a beautiful metaphor or photographic image of life. Metaphors can be cited, used for insightful descriptions, learning and teaching. Images can be framed, be carried around, and be useful for remembering. But, on their own, metaphors or images are never able to "walk out of the beaker" (act autonomously, as if on purpose), let alone to learn how to play the violin (evolve). In this sense, refrigerators, cars and aeroplanes are metaphors for, respectively, warm-blooded animals that live in hot areas (desert hedgehogs, dingos, humans), fast running animals (cheetahs, hares, horses, humans) and actively flying animals (bats, birds, dragonflies, mosquitos). Briefly, we cannot aim at a certain wanted and observed defined result that would well represent a living system, instead of actually being one. Almost by definition, we cannot know precisely what the experimental (successful!) outcome should be or look like, because the successful outcome must inherently diverge in time and variety space (Pross 2012, p. 171).

The construction of synthetic living cells has become a declared goal of a growing community of chemists (von Kiedrowski et al. 2010; Lynn et al. 2012).

Attempts to experimentally reconstruct presumably ancient macromolecular self-organisations must be a valid approach to follow this goal. A deeper knowledge about the spontaneous supramolecular organisation of macromolecules on primordial Earth may eventually serve to comprehend the emergence of dynamic complex chemical systems of any composition, natural, semi-natural or wholly synthetic. What we can do initially in first experiments is to opt for a "prudent compromise" and stay reasonably close to the molecular composition of the building blocks of current biological systems. A new kind of in vitro selection experiment is required, being held as far off thermodynamic equilibrium as experimentally feasible, and facing a plethora of dynamic, ill-defined selection barriers (instead of only one or two that appear to be chemically useful—ligand binding, for instance).

Let us follow *Dogma 13* (von Trier et al. 1995): (1) No biologically evolved information system, such as the universal genetic code, should be implicit in the starting conditions. (2) The emergence of a *de novo* genetic code is ultimately sought for. (3) The highest degree of information to embark on in the initial chemical system can be (stereo)chemically specific catalysts, "specific catalysis" being a form of chemical information (Pross 2012, pp. 150–153). (4) Prohibited is the use of any biologically evolved products, such as known enzymes or biosynthetic metabolites, i.e. secondary metabolites, cholesterol, sphingolipids, (5) except for those that are likely to have formed abiotically: certain amino acids, hydroxy acids, long-chain "fatty" acids, long-chain alcohols, carbohydrates, polyols, simple amines, N-heterocycles, nucleosides, nucleotides—and the oligomers and conjugates of all listed. (6) If we start out homochiral, this already would be a prudent compromise. Once the compromise experiments grow to be reasonably successful, one should go back and restart heterochirally (Hein and Blackmond 2012).

2 First Steps: Preparations for the Real Thing

2.1 Translate to Replicate, Rather than an RNA World

The long-term goal is to find "feeding" conditions for an energy-consuming chemical system to become dynamic, more precisely, dynamically sustained over a prolonged period of time during which a more or less constant turnover of its constituents is maintained. Its macromolecular constituents are synthetic nucleic acids, synthetic peptides and carbohydrates, all compartmented at or into lipidic vesicles that ultimately shall be replicating to give more vesicles. At least a part of the contained nucleic acids should replicate as well, i.e. multiply their nucleotide sequences and redistribute them into the daughter vesicles. In order to achieve this, and in view of the methodological *credos* described in Sect. 1.2, the best way is to provide conditions that seem favourable, if not compulsory, for the spontaneous formation of some kind of translation system between the nucleic acids and the peptides. The highest imaginable success would be the discovery of a de novo genetic code.

The decades-long research on enzyme-free autocatalytic or cross-catalytic nucleic acid amplification, particularly the search for truly exponential replicators (growing exponentially in population size), is pointing towards the nonexistence of any primordial "RNA World". The foremost argument in favour of RNA self-replication in a primordial RNA World is Nature alleged to be "parsimonious" (Cech 2012)—economical, sparing, cost-conscious. Therefore, one follows, Nature first generated an animate system where function (catalysis) and replication were combined in not more than one molecular class (RNA). This is not convincing anymore. Nature is not necessarily parsimonious. Nature does "whatever works" (Allen 2009) by letting evolvable "autonomous agents" explore the "adjacent possible" (Kauffman 2000).

If we agree upon an inhomogeneous messiness of the spontaneously available inanimate environment prevailing on primordial Earth, as opposed to the selected conditions maintained in a clean scientific laboratory, then an animate "single-molecular-class world" of any kind was hardly adjacent. Whatever was adjacently possible, for becoming the Cradle of Life out of a mixture and flow of then available prebiotic compounds, was very likely shouldered by several, if not many, quite different molecular classes at a time; by those (lipidic amphiphiles) that get spontaneously organised in functionalisable macromolecular assemblies bearing a hollow interior, those (carbohydrates) that are modular and offer rigid polar three-dimensional fragments that replace solvation water molecules at low entropic cost, those (amino acids, peptides and proteins) that are highly modular and extremely variety rich in their catalytic and structural properties and, last but of course not at all least, those (nucleic acids) that are perfectly water soluble at virtually any molecular length. Irrespective of the nucleic acid's actual backbone conformation, their two-dimensional (flat) hydrogen donor–acceptor patterns can be "read" as templates at relatively low entropic cost. These features make them ideal for the copying of any nucleotide sequence through twice loss-free doubling of the amounts of the molecules, which eventually results in exponential population growth. Nonetheless, the discovery of exponential self-replicators or cross-replicators is stubbornly withstanding any general resolution, in spite of long and intense research efforts on RNA and a number of related nucleic acids. The reason for this is largely known in principle—both product inhibition and side reactions cause over generations a steady loss in doubling efficiency.

However, any non-nucleotidic yet easily evolvable molecular device, being able to perform loss-free nucleic acid doubling, may carry out (catalyse) exponential nucleic acid replication. Modern Nature knows well how to loss-free double nucleic acids through the use of protein enzymes (polymerases), activating compounds (ATP) and chelating ions (Mg^{2+}). It seems a much longer way to evolve, from a mixture of different compound classes, an animate system that essentially relies on a "single performer" (as in an RNA World). From a strictly evolutionary point of view, where the initial condition is some mixture of abiotic molecules submerged in a flow of various higher-energy prebiotic compounds (*vide infra*), we can reasonably and convincingly ask: "Why bother using uniquely RNA for all functions including loss-free doubling, if a mixture of three to four available compound classes can do a better job?" Whatever works better at the moment will happen first! Whatever happens first will persist until it is outcompeted.

This is where translation comes in. This is the ultimate reason for evolving a reliable and recurrent production of proteins, for which a genetic code is prerequisite. Any genetic code will couple easily templating nucleic acids (RNA) with their loss-free doubling (replication) through protein enzymes. The reproducible production of useful proteins, as opposed to any peptides or proteins being available by mere chance, must have emerged within the relatively short time of a single appear-and-vanish cycle mentioned earlier. A chemical system unable to reliably replicate all its macromolecular constituents, thus remaining heavily dependent on a steady exogenous supply, but able to produce proteins that catalyse loss-free RNA replication, is the bottleneck where Nature was squeezing through. Whether to name such a chemical system capable of evolving a genetic code "alive" or "not yet alive" is academic. But, according to Pross' principle of hierarchical reduction, the emergence of a genetic code operating in an inanimate chemical system must precede the upheaval in complexity to a fully replicating chemical system. Wise experiments will aim at the emergence and evolution of a translation system implying proteins, instead of the replication of nucleic acids without proteins. A genetic code is based on actually two recognition principles, or else it will not work. Today's primary genetic code is the recognition and correct distinction of codons by anticodons, a pure templating quality between messenger RNA and transfer RNA. It is carried out by an ancient ribozyme, the ribosome. Today's secondary genetic code is the recognition and correct distinction of amino acids and "cognate" transfer RNAs by proteins. Aminoacyl transfer RNA synthetases (ARS) are the enzymes that guarantee for a correctly translated code. Both primordial functional fragments of ribosomes and of ARS may have co-evolved.

Experiments to recreate translation are the key for a chemical system to become animate. The questions now become: If we know where Nature was heading for ("translate to replicate"), where did She begin with? What were the pathways that allowed for stepping close to this adjacent possible? What physico-chemical feature—not yet a full-blown Darwinian advantage—would keep proteins and RNA in close contact to one another ?

2.2 *How to Compartment RNA?*

What follows is a summary of several scientific mostly published collaborations (Lamy et al. 2008; Le Chevalier Isaad et al. 2014). The making and description of the research results concerning RNA (Sect. 2.3) stem from the coauthorship with Pasquale Stano at the *Universitá degli Studi di Roma Tre* in Italy.

Natural liposomes are compartments ideal for the growth and harbouring of information carriers, useful metabolites and for their own reproduction. The bilayer surface of lipidic vesicles may have served since long for the anchoring and enrichment of macromolecules which sooner or later have been internalised and retained in the interior of the vesicles. Surface-bound internalised hereditary molecules could then grow in population size and length more rapidly, safer and more reliably. An obvious challenge for the deliberate creation of dynamic off-equilibrium systems that are expected to show "signs of life"—i.e. sustained population growth of certain, not all initially present individuals—is the concentration and compartmentation of nucleic acids at and into lipidic vesicles. This can be achieved in two ways: initially premixed and dried lipids and nucleic acids, or their precursors, may become trapped within vesicles as they form through hydration, or else the nucleic acids are compartmented through the use of helper compounds such as artificial cationic lipids and/or lipidic anchors, including fatty acid glycerol esters, terpenes, tocopherol and cholesterol (cf. Le Chevalier Isaad et al. 2014).

Nucleotide trapping in nascent lipid vesicles may be effective when possible, but is a quite crude and unselective process. Given the large differences between nucleic acids and lipids in atomic composition, charge distribution and solubility, the spontaneous concomitant and co-localised de novo enrichment of both compound classes is a chemically disputable issue for the discussion of biogenesis and appears somewhat forced when predisposed in designed experiments. Any acquired further attraction between nucleic acids and lipids, *viz.* any mediator compound bearing intermediate properties, will smoothen the mechanism of compartmentation of nucleic acids and thus may increase the selectivity for particularly useful nucleic acids to be compartmented.

Moreover, for a complex chemical system to become dynamically sustained, appropriate "feeding" conditions will not suffice (import of high-energy compounds and precursors, export of low-energy compounds as waste); it is the evolvability of a chemical system that is being primarily sought for. A priori there seems nothing particularly evolvable when nucleotides are being trapped within nascent lipid vesicles. Replicating primordial cells (natural protocells) are therefore expected to have emerged from chemical systems that were exploring the affinity between natural nucleic acids and natural lipidic vesicles by means of a molecular anchoring device. Useful devices should be sufficiently abundant, composed of prebiotically plausible precursors, and offer finely tunable properties without being highly evolved, thus enhancing the potential for the system to become self-evolvable (Iordache 2012). Lipidic anchoring compounds do bear some potential of evolvability but, in the absence of enzymes that catalyse the formation and breaking of carbon–carbon bonds, far less so than the evolvability of the highly modular compound class of peptides, where the hydrolysis and formation of carbon–nitrogen bonds seem more amenable to abiotic catalysis.

In prebiotic times, "activated" (particularly energy-rich) compounds, i.e. mineral or organic anhydrides of some sort, were likely available during geologically active periods on Earth (*vide infra*). Ultimately, it is owing to such anhydrides that the abiotic formation of organic macromolecules of the oligomer type may be explained. Saccharides, peptides, nucleic acids, lipids and fats are all products that form upon dehydration of their monomeric precursors or fragments thereof: aldehydes (or ketones), amino acids (or amines), phosphoric acids, fatty acids and alcohols or polyols react to give, respectively, acetals, amides and esters. However, the chemical selectivity of those abiotic condensing anhydrides is very limited. Consequently, different chemical families of "mixed" (chimeric) condensation products, such as peptidic nucleic acids or lipidic sugars, were being combinatorially produced and intermingled as well. The assembly and propagation of protocells could only be driven by quite simple and straightforward physical forces such as solubility, amphiphilicity, osmotic, pH or ionic strength gradients. Chimeric amphiphilic macromolecules are therefore likely to have played a crucial role in the assembly, organisation and development of such early protocellular structures.

Here we shall argue that lipophilic, hydrophobic peptides, composed of prebiotically available α-amino acids, are likely to have functioned as natural and readily evolvable anchoring devices for nucleic acids, for example RNA. Such peptides can be reversibly linked to RNA by means of prebiotically plausible condensing agents, without the requirement for specific catalysts such as ribozymes. When amphiphilic peptidyl RNAs are attracted to lipidic vesicles, barely soluble otherwise aggregated or precipitated hydrophobic peptides are being mobilised and thus made more readily available for chemical systems, while highly soluble RNAs are being immobilised and thus saved from rapid dilution out of the system (Strazewski 2009). A fraction of those randomly generated amphiphilic peptidyl RNA conjugates are expected to possess sufficient transmembrane properties to internalise into the interior volume of lipidic vesicles by virtue of any feasible membrane-crossing mechanism (Briers et al. 2012a, b). Once inside, the peptide anchors may hydrolyse and the RNA may grow in length and population size in the presence of chemically activated nucleotides.

This utility should initiate a process that enhances the sustainability of the functionalised vesicles, owing to a better import–export control across lipid membranes—reminiscent of modern transmembrane proteins. A logical consequence on a longer term could be the emergence, inside the vesicles, of a translation system and genetic code as we know it from modern organisms. Once experimental conditions are found that allow for an in vitro evolution of chemical systems that start with amino acids, peptides, nucleotides, sugars and lipids, but no predisposed genetically encoded enzymes, be they ribozymic or proteic, the de novo emergence of synthetic living cells should be in close reach.

2.3 A Hydrophobic Peptide Anchors RNA to Giant Lipidic Vesicles

2.3.1 A Non-dynamic Model System

Our model system investigates the interaction between an amphiphilic informational chimeric molecule, obtained from firmly joining a simple hydrophobic peptide with RNA, and giant vesicles (GVs) composed of natural phospholipids and synthetic glycolipids. We used glycolipids in some of the GVs due to their potential protective effect under desiccation conditions (Harland et al. 2009) and also for their phase patterning properties in lipid membranes (Subramanian et al. 2013). Spontaneous or glycan-induced demixing of fluid phases of lipid membranes has an important bearing on the formation of lipid rafts which, in evolved cellular systems, are involved in the spatial organisation of membrane-anchored proteins. The separation of fluid membrane phases on GVs might therefore be equally important for their interaction with externally added synthetic macromolecules, like peptidyl RNA, being anchored to the vesicle's membranes. One might expect anchored peptidyl RNA to enrich and concentrate even more at phase-separated fluid vesicle surfaces. The main goal was to see whether an amphiphilic peptide–RNA conjugate that might have been formed under prebiotically plausible conditions, and bearing a polyanionic RNA "head" and a hydrophobic peptide "tail" of comparable sizes, would stably interact with vesicles of a variety of lipid compositions, to spontaneously enrich at the water–lipid interface, to perhaps get across the membrane and to eventually become encapsulated.

2.3.2 Lipid Vesicles as Hosts

GVs are micrometre-sized (like cells); thus, membrane interactions on GVs can be directly and conveniently visualised by light microscopy. We considered the "natural swelling" method, i.e. the slow and undisturbed "gentle" hydration of a dry film of phospho- and glycolipid mixtures, to be most appropriate in mimicking the formation conditions of lipidic GVs as potential protocells during early stages of molecular evolution. The chemical composition of the GVs was varied (Fig. 1a), in order to form disorderly liquid monophasic membranes at 25 °C using 1-palmitoyl-2-oleoyl- or 1,2-dioleoyl-*sn*-glycero-3-phosphatidylcholine (POPC or, respectively, DOPC) or biphasic membranes containing higher melting 1,2-dipalmitoyl-*sn*-glycero-3-phosphatidylcholine (DPPC) or cholesterol (CHOL) or very fluid oleate, *viz.* DOPC/DPPC 1:4, DOPC/DPPC/CHOL 1:3:1 or POPC/oleate 85:15, where spontaneous lipid demixing into liquid–crystalline and gel phases would be evidenced by fluorescent probes. To assemble glycolipidic vesicles, 1,2-dipalmitoyl-*sn*-glycero-3-glycosides in either α- or β-anomeric forms (Fig. 1b), *viz.* α-D-mannopyranoside (DP-α-Man) or β-D-glucopyranoside

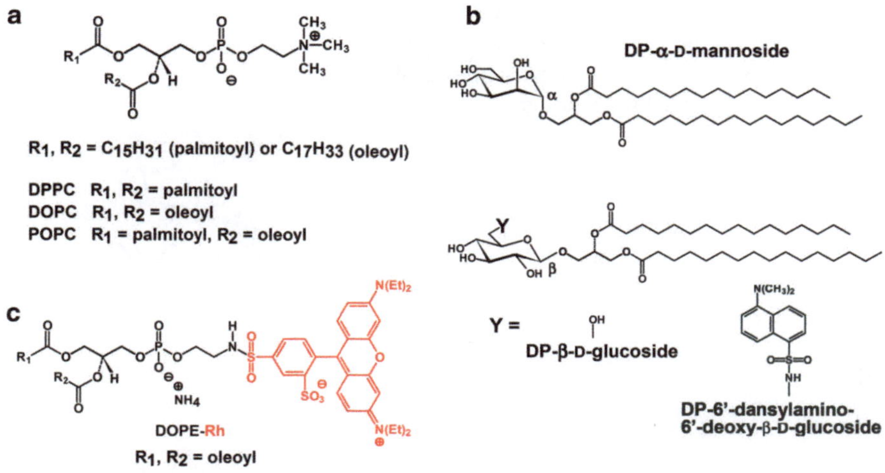

Fig. 1 Phospholipids (**a**, **c**) and glycolipids (**b**) used for naturally swelled GVs

(DP-β-Glc), were admixed with POPC. Their insertion and compatibility with POPC bilayers were assessed by imaging giant POPC vesicles that contained 5 mol % turquoise-fluorescent 1,2-dipalmitoyl-*sn*-glycero-3-(6′-dansylamino-6′-deoxy)-β-D-glucopyranoside, DP-β-(Dns)Glc (Fig. 1b, Y = dansyl). In many experiments the lipid membranes were made visible by admixing 0.01–0.2 mol % red-fluorescent 1,2-dioleoyl-*sn*-glycero-3-phosphatidylethanolamine-*N*-lissamine rhodamine B sulfonate (DOPE-Rh, Fig. 1c).

2.3.3 Peptidyl RNA: An Informational Amphiphile

The peptide sequence of the amphiphilic peptidyl RNA that we chose to begin with was a lipophilic 20-mer containing 6 leucines (L) and 14 alanines (A), cf. Fig. 2a. The predicted properties of the model icosapeptide amide H_3N^+–$LA_3LA_2LA_3LA_2LA_3LA$–$CONH_2$ at pH 7.0, $T = 298$ K and $I = 0.15$ M (salt) were propensities of 62.5 % α-helix, ≤66.9 % coiled-coil, ≤85.7 % β-sheet aggregations, and a lipid–water interfacial total hydrophobic moment 3.81 was calculated for the α-helix (central vector in Fig. 2e). Our icosamer "peptidyl-NHdA" (H_3N^+–$LA_3LA_2LA_3LA_2LA_3LA$-3′-amino-3′-deoxyadenosine), bearing a cationic N-terminus and being C-terminally linked through an amide bond to adenosine, showed circular dichroism (CD) spectra at both 25 and 4 °C being dominated by a single strong negative ellipticity below 200 nm wavelength. No signature of an α-helix was found. The infrared (IR) analysis of the peptide's amide I region confirmed that the icosamer's conformation was 84–88 % disordered and 12–16 % β-sheeted, rising to 22 (max. 33) % β in the presence of POPC at a molar lipid-to-peptide (L/P) ratio 40:1. Peptidyl-NHdA was simply too short to persistently favour any defined ordered conformation.

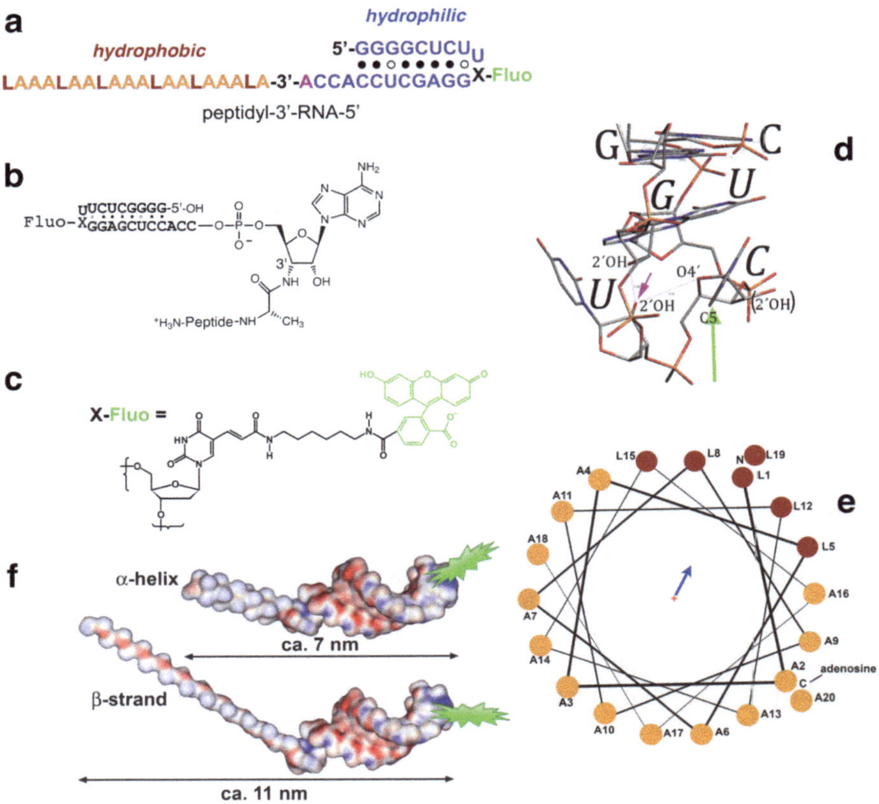

Fig. 2 Peptidyl-RNA(FAM) = amphiphilic (**a**), hydrolytically 3′-stabilised (**b**), peptidyl RNA labelled with fluorescein (**c**), RNA C<u>UUCG</u>G tetraloop NMR structure (**d**), icosapeptide $LA_3LA_2LA_3LA_2LA_3LA$ in a helical wheel model (**e**), and Ala_{20} peptide models 3′-connected to RNA hairpin NMR structure (**f**)

The RNA part of the peptidyl RNA was a 22-meric hairpin stem-loop molecule that was mimicking the aminoacyl stem of transfer RNA (Fig. 2a). In the present work the connection between peptide and RNA was a hydrolytically stable amide bond being isosteric with the natural but hydrolytically labile ester function usually found in peptidyl transfer RNA. This replacement of oxygen with nitrogen at the 3′-terminal position (Fig. 2b) was needed to reliably observe the constructs in clear fluorescence images over a long enough time period (a few hours, up to several days). The stem-loop folding of the RNA hairpin, closed by UUCG, is fully stable at least up to 60 °C (T_m = 87.6 °C, cf. Biała and Strazewski 2002). Based on this RNA hairpin's NMR solution structure (model 4 of Protein Data Bank code 1IKD), where the unpaired cytosine of the tetraloop sequence UUCG appeared unstacked—as opposed to UUCG tetraloop X-ray structure 1F7Y—from the uracil of the neighbouring non-canonical $G_{syn}{\cdot}U$ pair (Fig. 2d), the peptidyl RNA was made specifically visible through the replacement of the loop cytidine residue by a thymidine residue lacking 2′-OH (shown in brackets in Fig. 2d) and labelled with the green-fluorescent 6-fluoresceinyl carboxamido (FAM) fluorophore (Fig. 2c; the vertical arrow in Fig. 2d depicts the point of attachment C5 of the FAM spacer to the loop).

Highly amphiphilic peptidyl RNAs, composed of a large polar "head" (the RNA hairpin) and a single lipophilic "tail" (the peptide), can be considered as particularly large detergent molecules (Fig. 2f). The RNA head of this conjugate was about 5.0 nm long × 2.0–2.2 nm thick, which is typical for RNA in an A conformation. Depending on the peptide's conformation, the peptide tail could be 6.7–2.8 nm long × 0.60–0.85 nm thick, resulting in a head-to-tail thickness ratio 3.3–3.6 for a β-strand and 2.3–2.5 for an α-helix. The truncated cone dimension (as expressed by the head/tail thickness ratio) of the peptidyl RNA bearing a disordered "random coil" peptide is somewhere in between these extreme values (head/tail approximately 2.5–3.3). This peptidyl RNA is expected to form micelles in water at very low concentrations. When deposited on a glass surface, it self-aggregates in multilamellar spheroids and can spread into multilamellar supported bilayers (Terenzi et al. 2003; Coleman et al. 2006; Strazewski 2014).

2.3.4 Anchoring, Partitioning and (Im)Permeability

We first asked whether amphiphilic peptidyl RNA (9.19 kDa) could stably interact with a monophasic lipidic membrane. To this aim we incubated GVs with green-fluorescent "peptidyl-RNA(FAM)" and observed directly its localisation by confocal laser scanning fluorescence microscopy. When biphasic membranes were used, an intriguing question was whether peptidyl RNA displayed preferential binding to membranes of a specific physical state (liquid crystalline or gel phase) which, if true, could lead to an even more enhanced local concentration of RNA at the lipid surface. Upon binding, a possible outcome would be an alteration of membrane permeability. To test this possibility, we have carried out calcein (0.62 kDa) permeability tests across the lipid membrane. Conventional liposomes (140 nm average diameter) served for proton permeability assays. For all lipid compositions, control experiments were carried out by using (1) oligonucleotides "RNA(FAM)" of the same sequence as in the peptidyl RNA, yet being in part slightly 5′-truncated (5–7 kDa), all having FAM and all devoid of the 3′-peptide icosamer, and (2) the unlabeled icosameric peptide "peptidyl-NHdA" (1.94 kDa).

2.3.5 Results

Confocal Fluorescence Microscopy on Naturally Swelled GVs

Natural swelling produced a large variety of GVs in the 5–30 μm diameter range that were predominantly spherical or spheroidal, multilamellar and multivesicular, time and again of more complex topology and sometimes but rarely unilamellar (GUVs). GVs composed of POPC presented their membranes as a homogeneous phase. Heterogeneous zones on the membranes, as indicated by a differential two-dimensional enrichment of rhodamine, were clearly visible in most fluorescence images of mixed-lipid GVs prepared from DOPC/DPPC/DOPE-Rh 1:4:0.01 % or DOPC/DPPC/CHOL/DOPE-Rh 1:3:1:0.2 %. Cohydrated glycolipids were tolerated up to 20 mol % without being visibly detrimental to the shape and stability of the vesicles. This was confirmed by dynamic light scattering experiments on large unilamellar vesicles prepared from POPC (POPC LUVs); their average hydrodynamic radius remained in the range of 50–80 nm between 0 and 25 mol % glycolipids.

Confocal Fluorescence Microscopy on GVs Incubated with Peptidyl RNA

When a solution of peptidyl-RNA(FAM) was added to a population of naturally swelled POPC GVs, the green fluorescence immediately appeared enriched at their outer lipid membranes. At an initial molar lipid-to-peptidyl RNA (L/P) ratio 150:1, all unlabeled GVs in the sample—visible in confocal fluorescence images as black holes surrounded by a slightly green background—showed green rings due to the anchoring and enrichment of peptidyl-RNA(FAM) at their surface (Le Chevalier Isaad et al. 2014). No evident differences could be discerned in the images of anchoring experiments on pure POPC GVs when compared to those that contained 15 mol % glycolipids in their membranes. The green rings appeared within the mixing time of a few seconds, persisted for at least 18 h and, when 1-μm thin "confocal slices" were taken from the same objects at many different depths and cumulated, green and hollow spheroidal surfaces could be visualised in pseudo-3D images. At a molar L/P ratio 300:1 still a large majority of the vesicles showed green rings whereas only exceptionally few green rings materialised at L/P 1,500:1, yet those were very clearly visible. Typically, the GVs remained impermeable for calcein at all tested L/P ratios, added calcein would not visibly displace anchored peptidyl-RNA(FAM) from the vesicles, and peptidyl-RNA(FAM) was recurringly found anchored to filaments that protruded from numerous GVs. Where green filaments appeared in green-red labelled experiments, underlayed faintly red filaments were often discernible as well.

Essentially the same observations were made after incubation experiments on naturally swelled mixed-lipid DOPC/DPPC vesicles, with one more clearly visible feature. Anchored peptidyl-RNA(FAM) did spontaneously demix in highly enriched and rather depleted zones, but merely over the outermost membrane of every mixed-lipid GV. Sometimes the green peptidyl-RNA(FAM) and red rhodamine lipid (DOPE-Rh) labels co-localised over a large part of the vesicle's surface, but, more often than not, the green patterns were visibly distinct from those demixed red lipid patterns spreading underneath the inner lipid membranes not being accessed by peptidyl RNA. No green rings were found after incubating the vesicles with RNA(FAM) at L/RNA 300:1, nor with calcein, under otherwise identical conditions.

Light Scattering, Turbidity and Permeability Assays on LUVs

The incubation of unilamellar POPC LUVs with pept-NHdA (L/P 200:1) gave no measurable rise to changes in neither size distribution nor turbidity of the LUV suspensions. The incubation of pyranine-encapsulated LUVs with RNA(FAM) had no significant effect above the negative controls. In contrast, peptidyl RNA and pept-NHdA caused a prominent increase in proton permeation, comparable to the one of pore-forming gramicidin A. The proton transport-enhancing effect of the peptide was clearly dose dependent. Taken together, the permeability assays showed that peptidyl RNA anchoring was not detrimental to membrane stability, as evidenced by the failure of evident calcein permeation. The observation on proton permeability suggested a locally specific lipid–peptide architecture that could have functioned as a passive carrier of hydronium ions and, as matrix-assisted laser desorption ionisation (MALDI) mass spectrometry (MS) on peptidyl RNA suggested, probably also alkali metal ions. Our MALDI MS studies revealed that, despite exhaustive desalting of the peptidyl RNA samples, multi-sodium aggregates were much more abundant than in MALDI MS spectra of the RNA without the peptide.

2.3.6 Data Analysis and Discussion

The free energy and mode of partitioning of peptidyl-RNA(FAM) between the water and lipid phases were quantified by Pasquale Stano through a detailed analysis of green ring versus green background intensities in many fluorescence micrographs. The 21 negative charges on the RNA were revealed to be not detrimental to the partitioning of the conjugate between water and the lipid membranes; on the contrary, the value calculated for the peptide sequence was a good estimate. According to the empirical Wimley and White (1996) peptide hydrophobicity scale, the icosapeptide amide H_3N^+–$LA_3LA_2LA_3LA_2LA_3LA$–$CONH_2$ would partition from water with $\Delta G°(25\ °C) = -3.15$ kcal/mol in favour of the lipid membranes, as calculated by *Membrane Protein Explorer* (MPEx). The experimental free energy of partitioning of the peptidyl RNA was determined from fluorescence micrographs to be $\Delta G° = -3.4$ to -4.0 kcal/mol, being equivalent to $K = 300{:}1$ to 780:1 for excess anchored over unanchored peptidyl-RNA (FAM) in the presence of 0.1 mM lipids at 25 °C.

Quantitatively very similar CD and IR spectra of the icosapeptide pept-NHdA have been measured irrespective of the absence or presence of a 40-fold molar excess POPC. The main IR carbonyl stretching band at about 1,667 cm^{-1} was, in accord with the CD analysis, due to weakly or not hydrated/H-bonded disordered peptide coils originating from intermolecular hydrophobic peptide–peptide aggregates that were prevailing in water as micelles or more complex aggregates. In addition, a far from normal, highly discontinuous intensity distribution of the anchoring density of peptidyl-RNA(FAM) on otherwise unlabeled GVs was derived from a quantitative analysis of the green ring profile intensities of a number of representative fluorescence micrographs. This is consistent with the presence of aggregated sheets of peptidyl RNA being anchored at the outer water–lipid interface of the vesicles and cooperatively stabilised by peptide–peptide interactions, which appeared to remain intact upon anchoring. In conclusion, a favourable water-to-lipid partitioning of amphiphilic peptidyl RNA piloted its anchoring from peptidyl RNA micelles to the lipid vesicle membranes, but such lipid anchoring did not disintegrate the peptide–peptide associations, and green fluorescence intensities of mostly mono- but also bi- and tri-layered sheets of lipid-anchored peptidyl RNA became visible on the GVs.

2.3.7 Quod Nulla Capra Abligurrit ![1]

We have demonstrated that an amphiphilic peptidyl RNA bearing a hydrophobic, apart from the amino terminus charge-neutral Leu_6Ala_{14} peptide, being devoid of any particularly exceptional transmembrane association properties or pronounced folding characteristics, anchored by virtue of its simple lipophilicity spontaneously, rapidly and efficiently to the accessible bilayer membranes of naturally swelled lipidic vesicles, being composed of zwitterionic phospholipids and charge-neutral glycolipids, and stayed there. This designed peptide did not provoke any crossing of the attached RNA, or of calcein, through the essentially intact bilayer membranes.

2.4 How to Generate Plausibly Prebiotic Lipophilic "Random" Peptide Libraries?

2.4.1 Natural and Designed Cell Penetrating Peptides and Pore-Forming Proteins

Cell penetrating peptides (CPP) are short, sometimes hydrophilic peptides that gain access to the intracellular milieu. They have aroused great interest in both academic and applied research. First, cellular internalisation of CPPs often involves the crossing of a biological membrane, thus challenging the view of the nonpermeability of these structures to large hydrophilic molecules. Secondly, CPPs can drive the internalisation of hydrophilic cargoes into cells, a rate-limiting step in the development of many therapeutic substances. Interestingly, the two mostly used CPPs, TAT and Penetratin peptides, are derived from natural proteins,

[1] Old Swiss saying: *Das schläggt kei Geiss ewägg*, something like: no goat can lick this off !

HIV Tat protein and *Antennapedia* homeoprotein, respectively. The identification of the Penetratin peptide RQIKIWFQNRRMKWKK-NH_2 (Arg-Gln-Ile-Lys-Ile-Trp-Phe-Gln-Asn-Arg-Arg-Met-Lys-Trp-Lys-Lys-NH_2) is associated with the study of its parental protein (Dupont et al. 2011; Lamazière et al. 2008; Derossi et al. 1998).

These CPPs, and a great many of other natural transmembrane, transfecting, permeabilising or fusiogenic peptides, such as Melittin, Magainin or the transmembrane fragments of pore-forming proteins like Aquaporin or Acanthaporin, may serve to study their lipid membrane-crossing mechanisms, which may be quite varied (Nyholm et al. 2007). Such studies are inspiring the design of simpler transmembrane, haemolytic or fusiogenic peptides, such as the GALA/KALA peptide family (Plank et al. 1994; Midoux et al. 2003; Kol et al. 2003; Li et al. 2004), short lipid-like self-assembling peptides (Zhang 2012) or de novo designed His–Glu containing hexameric standalone pore-forming peptides (Zaccai et al. 2011). The variety of preferred amino acids (basic, acidic, neutral polar), primary sequences (hydrophobic–hydrophilic periodicities, amphipathic surface areas) and their associated tertiary or quaternary structures is quite overwhelming and a great deal of work has been sacrificed, based on known CPPs, for the design of peptide–nucleic acid conjugates that are able to cross cellular membranes one way or another (Järver et al. 2011).

It must be noted right here, however, that our goal is not finding yet another, better CPP being able to efficiently transport any desired RNA molecule into the interior of lipidic GVs or natural cells, not by design, nor with the help of evolved natural sequences (genetic information), cf. Sects. 1.2 and 3. Hence, all knowledge gained from the above-described work on natural and designed CPPs is being used with purposefully retentive naïvety. So how to generate plausibly prebiotic lipophilic "random" peptide libraries that might bear the potential of offering membrane-crossing properties ? We are deciding on combinations of a restricted choice of amino acids for peptide libraries to be tested, for example, which and how much of acidic, basic or neutral side chains will be combined in one batch.

2.4.2 Potentially Functional "Random" Co-oligopeptides

Prebiotic α-amino acids are thought to have evolved into today's proteins because copolymers derived from these residues offer catalytically competent side chains within an evolvable dynamic scaffold. The manner in which peptides of main chain lengths of 30 residues bearing different side chain functional groups might have formed spontaneously from chemically activated monomeric precursors is an issue that has attracted considerable attention (Luisi 2006). An effective enrichment of stably folded peptides bearing a few catalytically competent side chains, such as histidine or glutamate, could result from the low expected solubility of the vast majority of peptides formed from simple, more or less lipophilic amino acids, such as glycine, alanine, valine or leucine.

The synthesis of libraries of de novo oligopeptides from a mixture of different amino acids under chemically simple conditions has thus become an important challenge. The central difficulty in making such "random" oligopeptides lies in the fact that monomeric precursors of amino acids bearing potentially useful side-chain functional groups may polymerise at very different rates. Based on the principles of copolymerisation (Elias 1997), the amino acid distribution within a copolymer will not necessarily reflect the composition of the initial solution of the monomeric precursors. When rate differences between precursors reach several orders of magnitude, batch copolymerisation of the mixture results in the exclusive formation of homopolymers rather than the desired copolymer.

Whereas co-oligopeptides from non-aqueous solutions containing mixtures of *N*-carboxyanhydrides (NCAs = Leuch anhydrides) that polymerise at similar rates and homopolymers from simple amino acids such as oligoalanines sometimes longer than 30-mers have been obtained, the longest homopolymers from aqueous amino acid NCA solutions were often shorter than 10-mers. Aqueous in situ homopolymerisations from anionic Glu, Asp and phospho-serine NCAs produced 9- to 18-mers (Hill and Orgel 1996), but the longest reported co-oligopeptides synthesised from aqueous mixtures of different amino acid NCAs were pentamers (Taillades et al. 1999; Pascal et al. 2005). In order to generate copolymers requisite for the potential evolution of a protocellular function, it was thus necessary to overcome the difficulties inherent in the large differences in polymerisation rates of key amino acids under given aqueous conditions.

We elaborated, as a first important breakthrough (Lamy et al. 2008), the formation of "random" peptide libraries prepared from an aqueous mixture of pure amino acids, glycine, L-glutamate, or L-histidine, and dry carbonyl diimidazole (CDI), through the in situ formation of Leuch anhydrides (Fig. 3). Mass spectrometry allowed us to safely attribute peptide compositions and lengths up to 30-mers. Amino acids such as histidine, glutamate and serine are catalytically competent, whereas glycine, alanine, valine, leucine and similar amino acids can be considered as good spacers and folders. The synthetic protocol that we elaborated for the above amino acids is easily transferable to a choice of other similar amino acids that would produce peptide libraries of sufficient length (15–30 amino acid residues) and compositional variety (2–4 different amino acids) expected to bear the potential for transmembrane properties needed to transport RNA across membranes being composed of a mixture of lipids.

3 Perspectives: Evolvable Systems Chemistry

We aim at exploring through "experimental evolutionary reconstruction"—based on rules described in *Dogma 13*, Sect. 1.2—the process of how Nature learns to take advantage of physico-chemical properties of any available peptide that might spontaneously assemble from simple α-amino acids, that themselves might form spontaneously, and be reversibly linked to RNA that may form spontaneously

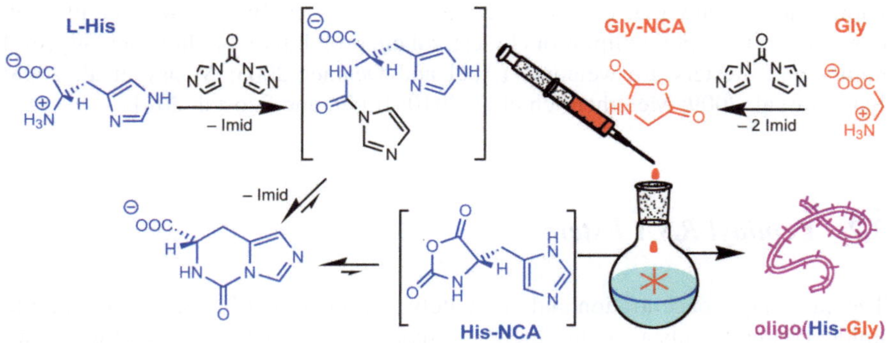

Fig. 3 The aqueous in situ formation of amino acid NCAs using a pulsed addition mode leads to combinatorial libraries of His-Gly co-oligopeptides of up to 29 residue length and controllable average His–Gly ratios

(Benner et al. 2012; Lehman 2013; Attwater et al. 2013), with the effect of compartmenting RNA molecules into the interior of dynamic lipidic vesicles that self-organise spontaneously, ideally, from spontaneously formed fatty acids and glycerol (Rushdi and Simoneit 2006), which is only a final goal so far (we are first using commercial lipids). We are planning to experimentally exploit the pathway from a moderately complex chemical system towards organised complexity, adaptability, eventually to evolvability, with the goal of giving a proof of concept that such chemical systems might become self-evolvable.

3.1 *Lipids and Glycans*

We shall profit very much from published work on mixed-lipid vesicles with separate (demixed) fluid phases, especially those composed of zwitterionic and anionic lipid mixtures (Keller et al. 1992; van Klompenburg et al. 1997; Speelmans et al. 1997; Kooijman et al. 2005, 2007; Raja et al. 2007), on the effect of glycans that are attached to lipidic membranes (DeMarco 2012), as well as on the currently described analytical methods for characterising the effects of macromolecules that have been brought into contact with lipidic GV populations (e.g. Lamazière et al. 2008). This will inspire us for the choice of mixed lipids to be utilised for the preparation of GVs, in order to enhance the chances of the import of, both, small dehydrating compounds, such as CDI, carbodiimides or other, geologically plausible "anhydrides", and charged macromolecules such as peptidyl RNA. Thereby, we shall never lose out of focus the desire to grow and divide our GV populations, best through the in situ formation of lipids or lipid-like ester compounds from added glycerol, fatty acids and small dehydrating compounds. In other words, not only prebiotically plausible amino acids and "anhydrides" shall be used, but also prebiotically plausible lipids, albeit of naturally evolved homochirality to begin with, which means that choline esters are about as evolved as we initially get. Compounds like sphingolipids or cholesterol are prohibitive, but fatty acid glyceryl mono- and diesters are welcome (Apel and Deamer 2005; Mansy et al. 2008; Maurer et al. 2009; Meierhenrich et al. 2010; Budin and Szostak 2011).

3.2 Peptidyl RNA Esters

The gathering, concentration and, ultimately, invagination of genetic material into lipidic compartments are considered a prerequisite for the emergence of life as we know it. Our anchoring experiments took a start on relatively low structural and supramolecular raffinesse and exploited prebiotically plausible interactions between a simple hydrophobic peptide, an unconspicuous RNA hairpin, and almost any mixture of phospho- and glycolipids forming highly polydisperse and polytopic lipid vesicles. The anchoring results offer a complement to any in situ entrapment of nucleic acids while lipid vesicles are forming; they are giving good grounds for the experimental in vitro reconstruction and evolution of synthetic model protocells. Locally very high concentrations of peptide–nucleic acid conjugates being anchored to fluid lipid membranes, in particular those phase-separated membranes being composed of mixed lipids, are favouring otherwise improbable interconjugate reactions that may be exploited by the harbouring vesicles and thus be apt to Darwinian selection (Szathmáry 2006).

Obviously useful properties of vesicles being decorated with conjugates would be the eventual gain of import–export control of nucleic acids through increasingly functional transmembrane peptides. It is also conceivable that nucleic acid-driven peptide synthesis could emerge from such surface vicinity phenomena (Strazewski 2009). The peptidyl RNA conjugate used so far was designed, for technical reasons, to remain reasonably stable toward the hydrolysis of the connection between peptide and RNA. Much shorter lived conjugates are needed to assemble chemically reversible, thus, in vitro selectable dynamic protocell model systems. Such peptide–nucleic acid esters are limited to estimated average half-lives in the order of minutes to hours at ambient temperatures and close to neutral pH values. Peptide–RNA ester libraries can be prepared from the addition of RNA, or RNA libraries, to a mixture of freshly prepared co-oligopeptide libraries. A constant or recurrent supply of fresh "anhydride" must compensate for the hydrolytic cleavage back to peptides and RNA. A fraction of those randomly generated amphiphilic peptide–RNA conjugates are expected to possess sufficient transmembrane properties to internalise into the interior volume of lipid vesicles should they be present in the pool. Once inside, the RNA may grow in length and population size in the presence of chemically activated nucleotides (Mansy et al. 2008; Luther et al. 1998; Deck et al. 2011).

3.3 Why RNA?

More than two decades of research on a plethora of different “nucleic acids”, i.e. synthetic or semi-synthetic structural analogues of DNA and RNA, have convincingly shown that an astonishing number of such oligomers or polymers can be used for templating and thus be potentially exploitable for faithful nucleic acid doubling, be it through polymerisation, ligation or a combination of both. This is good news and bad. The base-pairing potential of such nucleic acids may be promiscuously high. It makes us keep asking again and again: “Why was RNA chosen for inheritable replication, why not one of the other nucleic acid candidates?” “Because RNA replicates best !” is hardly a good answer anymore. “Because RNA basepairs best” is even worse an answer. “Because natural proteins can polymerise them best” is probably wrong as well (Yu et al. 2012). Proteins are just too good in adapting to any kind of useful (exploitable) molecular recognition and catalysis.

However, RNA molecules are expected to outclass other nucleic acids in synthesising peptides from chemically activated amino acids by virtue of their uniquely unhindered vicinal 2′,3′-*cis*-diol terminus, a feature that still is the basis of modern ribosomal protein biosynthesis (Acosta-Silva et al. 2012). No other furanosyl nucleotide offers two free *cis*-hydroxy groups at a terminus of the oligomer. The 3′-terminus of RNA is catalytically highly competent; it carries out conformationally defined internal movements by spontaneously switching between (pseudo) equatorial–axial and axial–equatorial at low energetic costs—pyranosyls are too rigid for this. As for lyxose, either one of its secondary hydroxyls is sterically too hindered in α- or β-lyxofuranosyl nucleotides. So the working hypothesis is: “Because RNA makes proteins best!” Which emphasises again the importance of finding chemical systems that will kick-start translation off its yet unknown grounds.

The vision: Suitably “nurtured” vesicle populations of appropriate lipidic composition are expected to evolve at the same time protoforms of (ribozymic) ribosomes, (proteic) aminoacyl transfer RNA synthetases, and nucleoside triphosphate synthetases, thus develop simultaneously a persistent, if primitive, primary and secondary genetic code. The components needed for a persistent “systemic memory”, i.e. a genetic replication machinery based on primordial nucleic acid polymerases, fatty acid synthetases and ribonucleotide deoxygenases, will follow rather than precede translation. This means that the experimental testing of evolvable chemical systems must be held under “full feeding conditions” for long enough a time period. They will permanently or at least recurrently provide the system with small to medium size molecular precursors and chemical energy to allow for the emergence of the first compartmented and functional proteic enzymes. I guess *Dogma 13* should help to achieve this within a scientist’s lifetime.

Acknowledgements This work would never have been possible without the devoted action of a number of colleagues and collaborators, in particular, Oliver Botta, Ewa Biała, Silvia Terenzi,

Nhat Quang Nguyen Trung, Carole Lamy, Peter Goekjian, Denis Bouchu, Alexandra Le Chevalier Isaad, Krishnakumar KS, Pasquale Stano, Paolo Carrara, René Buchet, Dominique Lafont, and Florian Albrieux. Discussions over the years with Günter von Kiedrowski, Eörs Szathmáry, Pier Luigi Luisi, Albert Eschenmoser, Hans Kuhn, Stuart Kauffman, and Addy Pross have moulded on my way of thinking about *le monde vivant* into what it has become now; my deepest thanks to all these scientists. The financial support from the 6th European Framework Programme in Synthetic Biology for the research programme "Synthcells" (contract no. 043359) and from the European Coordination of Science and Technology Action on "Systems Chemistry" (COST Action CM0703) is gratefully acknowledged.

References

Acosta-Silva C, Bertran J, Branchadell V et al (2012) Quantum-mechanical study on the mechanism of peptide bond formation in the ribosome. J Am Chem Soc 134:5817–5831

Allen W (2009) Whatever works, comedy, 92 minutes. Sony Pictures Classics, USA

Apel CL, Deamer DW (2005) The formation of glycerol monodecanoate by a dehydration condensation reaction: increasing the chemical complexity of amphiphiles on the early Earth. Orig Life Evol Biosph 35:323–332

Attwater J, Wochner A, Holliger P (2013) In-ice evolution of RNA I polymerase ribozyme activity. Nat Chem 5:1011–1018

Benner SA, Kim H-J, Carrigan MA (2012) Asphalt, water, and the prebiotic synthesis of ribose, ribonucleosides, and RNA. Acc Chem Res 45:2025–2034

Biała E, Strazewski P (2002) Internally mismatched RNA: pH and solvent dependence of the thermal unfolding of $tRNA^{Ala}$ acceptor stem microhairpins. J Am Chem Soc 124:3540–3545

Briers Y, Stäubli T, Schmid MC et al (2012a) Intracellular vesicles as reproduction elements in cell wall-deficient L-form bacteria. PLoS One 7:e38514

Briers Y, Walde P, Schuppler M et al (2012b) How did bacterial ancestors reproduce? Lessons from L-form cells and giant lipid vesicles. Bioesssays 34:1078–1084

Budin I, Szostak JW (2011) Physical effects underlying the transition from primitive to modern cell membranes. Proc Natl Acad Sci U S A 108:5249–5525

Cech T (2012) The RNA world's in context. Cold Spring Harb Perspect Biol 4:a006742

Coleman A, Lazar A, Terenzi S et al (2006) Observation of the formation of supported bilayers by amphiphilic peptidyl-RNA. Chem Commun (1):63–65

Deck C, Jauker M, Richert C (2011) Efficient enzyme-free copying of all four nucleobases templated by immobilized RNA. Nat Chem 3:603–608

DeMarco ML (2012) Three-dimensional structure of glycolipids in biological membranes. Biochemistry 51:5725–5732

Derossi D, Chassaing G, Prochiantz A (1998) Trojan peptides: the penetratin system for intracellular delivery. Trends Cell Biol 8:84–87

Dupont E, Prochiantz A, Joliot A (2011) Penetratin story: an overview. Methods Enzymol 683:21–29

Elias HG (1997) An introduction to polymer science. VCH, Weinheim

Ellington A (2012) Origins for everyone. Evol Educ Outreach 5:361–366

Harland CW, Botyanszki Z, Rabuka D, Bertozzi CR, Parthasarathy R (2009) Synthetic trehalose glycolipids confer desiccation resistance to supported lipid monolayers. Langmuir 25:5193–5198

Hawking S (2001) The universe in a nutshell. Bantam, London

Hein JE, Blackmond DG (2012) On the origin of single chirality of amino acids and sugars in biogenesis. Acc Chem Res 45:2045–2054

Hill AR Jr, Orgel L (1996) Oligomerization of negatively-charged amino acids by carbonyldiimidazole. Orig Life Evol Biosph 26:539–545

Iordache O (2012) Self-evolvable systems. Understanding complex systems. Springer, Heidelberg. ISBN 978-3-642-28881-4

Järver P, Coursindel T, Andaloussi SEL et al (2011) Peptide-mediated cell and in vivo delivery of antisense oligonucleotides and siRNA. Mol Ther Nucleic Acids 1:e27

Kauffman S (2000) Investigations. Oxford University Press, New York

Keller RCA, Killian JA, de Kruijff B (1992) Anionic phospholipids are essential for α-helix formation of the signal peptide of prePhoE upon interaction with phospholipid vesicles. Biochemistry 31:1672–1677

Kol MA, van Laak AN, Rijkers DT et al (2003) Phospholipid flop induced by transmembrane peptides in model membranes is modulated by lipid composition. Biochemistry 42:231–237

Kooijman EE, Carter KM, van Laar EG et al (2005) What makes the bioactive lipids phosphatidic acid and lysophosphatidic acid so special? Biochemistry 44:17007–17015

Kooijman EE, Tieleman DP, Testerink C et al (2007) An electrostatic/hydrogen bond switch as the basis for the specific interaction of phosphatidic acid with proteins. J Biol Chem 282:11356–11364

Lamazière A, Wolf C, Lambert O et al (2008) The homeodomain derived peptide Penetratin induces curvature of fluid membrane domains. PLoS One 3:e1938

Lamy C, Lemoine J, Bouchu D et al (2008) Glutamate-glycine and histidine-glycine co-oligopeptides: batch co-oligomerization versus pulsed addition of N-carboxyanhydrides. Chembiochem 9:710–713

Le Chevalier Isaad A, Carrara P, Stano P et al (2014) A hydrophobic peptide anchors RNA to giant lipidic vesicles. Org Biomol Chem (in press)

Lehman N (2013) Cold-hearted RNA heats up life. Nat Chem 5:987–989

Li W, Nicol F, Szoka FC Jr (2004) GALA: a designed synthetic pH-responsive amphipathic peptide with applications in drug and gene delivery. Adv Drug Deliv Rev 56:967–985, and references therein

Luisi PL (2006) The emergence of life. From chemical origins to synthetic biology. Cambridge University Press, Cambridge

Luther A, Brandsch R, von Kiedrowski G (1998) Surface-promoted replication and exponential amplification of DNA analogues. Nature 396:245–248

Lynn D, Burrows C, Goodwin J et al (2012) Origins of chemical evolution. Acc Chem Res 45:2023–2024

Magritte R (1929) La Trahison des images, huile sur toile, 59 × 65 cm. Art Institute of Chicago, Los Angeles

Mansy SS, Schrum JP, Krishnamurthy M et al (2008) Template-directed synthesis of a genetic polymer in a model protocell. Nature 454:122–125

Maurer SE, Deamer DW, Boncella JM et al (2009) Chemical evolution of amphiphiles: glycerol monoacyl derivatives stabilize plausible prebiotic membranes. Astrobiology 9:979–987

Meierhenrich UJ, Filippi J-J, Meinert C et al (2010) On the origin of primitive cells: from nutrient intake to elongation of encapsulated nucleotides. Angew Chem Int Ed 49:3738–3750

Midoux P, LeCam E, Coulaud D et al (2003) Histidine containing peptides and polypeptides as nucleic acid vectors. In: Luo D, Saltzman WC (eds) Synthetic DNA delivery systems. Kluwer Academic/Plenum, London, pp 23–44

Nyholm TMK, Özdirekcan S, Killian JA (2007) How protein transmembrane segments sense the lipid environment. Biochemistry 46:1457–1465

Pascal R, Boiteau L, Commeyras A (2005) From the prebiotic synthesis of alpha-amino acids towards a primitive translation apparatus for the synthesis of peptides. Top Curr Chem 259:69–122

Plank P, Oberhauser B, Mechtler K et al (1994) The influence of endosome-disruptive peptides on gene transfer using synthetic virus-like gene transfer systems. J Biol Chem 269:12918–12924

Pross A (2012) What is life? How chemistry becomes biology. Oxford University Press, Oxford

Raja M, Spelbrink RE, de Kruijff B et al (2007) Phosphatidic acid plays a special role in stabilizing and folding of the tetrameric potassium channel KcsA. FEBS Lett 581:5715–5722
Rushdi A, Simoneit BRT (2006) Abiotic condensation synthesis of glyceride lipids and wax esters under simulated hydrothermal conditions. Orig Life Evol Biosph 36:93–108
Speelmans G, Staffhorst RWHM, de Kruijff B (1997) The anionic phospholipid-mediated membrane interaction of the anti-cancer drug doxorubicin is enhanced by phosphatidylethanolamine compared to other zwitterionic phospholipids. Biochemistry 36:8657–8662
Strazewski P (2009) Adding to Hans Kuhn's thesis on the emergence of the genetic apparatus: of the Darwinian advantage to be neither too soluble, nor too insoluble, neither too solid, nor completely liquid. Colloids Surf B Biointerfaces 74:419–425
Strazewski P (2014) Amphiphilic peptidyl-RNA. In: Stulz E, Clever GH (eds) DNA in supramolecular chemistry and nanotechnology. Wiley, New York
Subramanian AB, Guidotti G, Manoharan VN et al (2013) Glycans pattern the phase behaviour of lipid membranes. Nat Mat 12:128–133
Szathmáry E (2006) The origin of replicators and reproducers. Phil Trans R Soc Lond B Biol Sci 361:1761–1776
Taillades J, Cottet H, Garrel L et al (1999) N-carbamoyl amino acid solid-gas nitrosation by NO/NOx: a new route to oligopeptides via alpha-amino acid N-carboxyanhydride. Prebiotic implications. J Mol Evol 48:638–645
Terenzi S, Biała E, Nguyen-Trung NQ et al (2003) Amphiphilic 3′-peptidyl-RNA conjugates. Angew Chem Int Ed 42:2909–2912
van Klompenburg W, Nilsson IM, von Heijne G et al (1997) Anionic phospholipids are determinants of membrane protein topology. EMBO J 16:4261–4266
von Kiedrowski G, Herdewijn P, Otto S (2010) Welcome home system chemists! J Syst Chem 1:1
von Trier L, Vinterberg T, Levring K et al (1995) Dogme 95, Denmark
Wimley WC, White SH (1996) Experimentally determined hydrophobicity scale for proteins at membrane interfaces. Nat Struct Biol 3:842–848
Yu H, Zang S, Chaput JC (2012) Darwinian evolution of an alternative genetic system provides support for TNA as an RNA progenitor. Nat Chem 4:183–187
Zaccai NR, Chi B, Thomson AR et al (2011) A de novo peptide hexamer with a mutable channel. Nat Chem Biol 7:935–941
Zhang S (2012) Lipid-like self-assembling peptides. Acc Chem Res 45:2142–2150

How the Early Genetic Code Was Established?: Inference from the Analysis of Extant Animal Mitochondrial Decoding Systems

Kimitsuna Watanabe and Shin-ichi Yokobori

Contents

1 Introduction ... 26
2 Comparison of Genetic Apparatus in Mitochondrial and Non-mitochondrial Decoding Systems ... 27
3 Characteristics of Mitochondrial Genetic Code Systems ... 28
4 The Codon–Anticodon Relationship in Animal Mitochondria ... 31
5 How Was the Mitochondrial Decoding System Generated? ... 32
6 The Early Genetic Code System ... 33
7 From the Early Decoding System to the Universal Decoding System ... 35
8 Concluding Remarks and Perspective ... 36
References ... 38

Abstract Mitochondria are intracellular organelles in eukaryotic cells that have their own genome and translational apparatus. The vertebrate mitochondrial decoding system is thought to be the simplest among all extant living systems and to have originated by retrogression from the universal decoding system, induced mainly by genome economisation and directional mutation pressure during mitochondrial evolution. Thus, it is reasonable to speculate that the vertebrate genetic code table is a typical model for the early genetic code table.

In some metazoan mitochondrial decoding systems, it was found that unmodified anticodons of tRNA have the potential to base-pair with cognate codons

K. Watanabe (✉)
Department of Applied Life Sciences, School of Life Sciences, Tokyo University of Pharmacy and Life Sciences, 1432-1 Horinouchi, Hachioji, Tokyo 192-0392, Japan

Department of Biotechnology, Graduate School of Agricultural and Life Sciences, The University of Tokyo, 1-1-1 Yayoi, Bunkyo-ku, Tokyo 113-8657, Japan
e-mail: kim-wata@toyaku.ac.jp; kwatanab@ft.catv.ne.jp

S. Yokobori
Department of Applied Life Sciences, School of Life Sciences, Tokyo University of Pharmacy and Life Sciences, 1432-1 Horinouchi, Hachioji, Tokyo 192-0392, Japan

V.A. Erdmann et al. (eds.), *Chemical Biology of Nucleic Acids*, RNA Technologies,
DOI 10.1007/978-3-642-54452-1_2,

and retain flexibility at the wobble pair, suggesting that the early decoding system consisted of unmodified RNA prior to the emergence of RNA-modifying systems. Competition would likely have occurred between G_{34}-tRNA having GNN anticodons and U_{34}-tRNA having UNN anticodons (or C_{34}-tRNA having CNN anticodons for AUR codons) in their binding to the ribosomal A site, which would have resulted in the discrimination between NNY codons and NNR codons in the two-codon sets. Thus, the early genetic code table would have been established in such a way that eight family box codons were deciphered by U_{34}-tRNA, and eight NNY and six NNR codons in the two-codon sets were deciphered by G_{34}-tRNA and U_{34}-tRNA (or C_{34}-tRNA), respectively.

This review describes the characteristics of an early decoding system inferred from the genetic code of present vertebrate mitochondria, and how the present universal decoding system may have originated from the early decoding system during evolutionary history.

Keywords Genetic code • Mitochondria • tRNA anticodon • Family box • 2-Codon set

1 Introduction

The genetic code is the critical principle underlying the cellular interpretation of genetic information and is considered common to all living systems (the universal genetic code; Osawa 1995). However, some non-universal codons are utilised by animal mitochondria (Barrell et al. 1979; Anderson et al. 1981) and free-living organisms such as *Mycoplasma* spp. (Yamao et al. 1985). It is thus hypothesised that the genetic code is changeable and species dependent (Osawa et al. 1992). Here, the early genetic code preceding the universal genetic code is deduced through inspection of the decoding systems of numerous metazoan mitochondria (Watanabe 2010).

Mitochondria are proposed to have originated from aerobic bacteria engulfed by ancestral host bacteria (Margulis 1970). Molecular phylogenetic analysis of ribosomal RNA sequences and genes housed in the mitochondrial genome indicates that the mitochondrial ancestor is α-proteobacterial (Andersson et al. 1998). The ancestral host cell for mitochondria is thought to be a cell wall-less archaeon such as thermoplasmas (Margulis 1993). The mitochondrial genome houses very many fewer genes than bacterial genomes; for example, there are 37 human mitochondrial genes, while most bacteria have more than 1,000 genes. Correspondingly, most of the proteins that function within the mitochondria are encoded by the nuclear genome and are synthesised in the cytoplasm, indicating large-scale gene transfer from the mitochondrial to the nuclear genome (Lang et al. 1977). The mitochondrial genome size is thus diminished, varies depending on animal phylum, and is only 20,000 bases in metazoan mitochondria (NCBI DB for Organelle genome). This diminution is called "genome economisation". Genome economisation is often accompanied by AT pressure, in which the base composition of DNA is biased towards AT richness by evolutionary mutation pressure. AT pressure would have been in effect during the evolutionary transition from α-proteobacteria, via proto-mitochondria, to the present metazoan mitochondrial state.

On the basis of the simplest anticodon composition of vertebrate mitochondria among various extant decoding systems, Jukes (1983) proposed that the mitochondrial genetic code was brought about by retrogression from the universal genetic code to the early genetic code through genomic economisation and accompanying AT pressure. Osawa (1995) proposed a simple early genetic code and its possible development into the present universal code; the simple code consists of 23 tRNA species for 61 sense codons and is based on the vertebrate mitochondrial genetic code.

Here, the evolution of mitochondrial genetic codes is re-examined using the most recent experimental results on mitochondrial tRNAs from various animal phyla. An early genetic code table and concomitant decoding systems are inferred, assuming retrogression from the present systems possessing the universal genetic code.

2 Comparison of Genetic Apparatus in Mitochondrial and Non-mitochondrial Decoding Systems

Table 1 shows the comparison of translation apparatus of vertebrate mitochondria as a representative of mitochondria, of *Mycoplasma capricolum* as a representative of free-living organisms possessing the simplest decoding system, and of *Escherichia coli* as a representative of organisms possessing the universal genetic code. Although the ribosomal components of *M. capricolum* and *E. coli* are almost the same, those of vertebrate mitochondria differ, having rRNA about half the size of *E. coli* rRNA and approximately 50 % more ribosomal proteins than *E. coli*. The functional domains composing the peptidyltransferase centre in the large ribosomal subunit (Suzuki et al. 2001a) and the decoding centre in the small ribosomal subunit (Suzuki et al. 2001b) are well conserved between vertebrate mitochondria and *E. coli*.

The number of tRNA species varies, with 22 in vertebrate mitochondria, 29 in *M. capricolum* and 48 in *E. coli*. One tRNA species exists per family box and two-codon set in vertebrate mitochondria, but multiple tRNA species exist in *M. capricolum* and *E. coli*. *M. capricolum* and *E. coli* possess two different species of $tRNA^{Met}$, one for initiation and one for elongation; however, vertebrate mitochondria possess a single species of $tRNA^{Met}$ that performs both functions (Takemoto et al. 2009). *E. coli* uses the universal genetic code, but

Table 1 Comparison of translation apparatus from vertebrate mitochondria and the free-living organisms *Mycoplasma capricolum* and *Escherichia coli*

	Mitochondria	Free-living organisms	
	Vertebrate mitochondria	*Mycoplasma capricolum*	*Escherichia coli*
	Non-universal		Universal
Genetic code	UGA = Trp, AUA = Met, AGA/G = Stop	UGA = Trp	
	60 sense codons	62 sense codons	61 sense codons
	4 stop codons	2 stop codons	3 stop codons
tRNA	22 species	29 species	42–48 species[a]
$tRNA^{Met}$	1 species	2 species	2 species
Ribosome	55S (28S + 39S)	70S (30S + 50S)	70S (30S + 50S)
rRNA	2,513 nt.	4,527 nt.	4,566 nt.
	12S + 16S	5S + 16S + 23S	5S + 16S + 23S
rProtein	78 species	51 species	54 species

[a]There are 42 different anticodons and a total of 48 tRNAs with different body sequences in *E. coli* (Näsvall et al. 2007)

M. capricolum and vertebrate mitochondria use non-universal genetic codes. In *M. capricolum*, the universal UGA stop codon becomes a Trp codon (Yamao et al. 1985), and in vertebrate mitochondria the AGA/AGG Arg codons become stop codons and the AUA Ile codon changes to a Met codon (Barrell et al. 1979; Anderson et al. 1981). The vertebrate mitochondrial system is the simplest genetic code system among all extant organisms so far examined (Suzuki et al. 2011a).

3 Characteristics of Mitochondrial Genetic Code Systems

The most prominent characteristic of animal mitochondrial genetic code systems is that they use non-universal genetic codes: (1) UGA, Stop becomes Trp in all animal mitochondria (Barrell et al. 1979); (2) AUA, Ile becomes Met in most metazoan mitochondria, excepting Echinodermata, Hemichordata, Platyhelminthes, Cnidaria, Ctenophora, Placozoa, and Porifera (Watanabe 2010); (3) AGR, Arg becomes Ser in most invertebrate mitochondria (Himeno et al. 1987), Arg becomes Gly in ascidian mitochondria (Yokobori et al. 1993), and Arg becomes Stop in vertebrate mitochondria (Anderson et al. 1981); (4) AAA, Lys becomes Asn in Echinodermata (Himeno et al. 1987) and Platyhelminth mitochondria (Telford et al. 2000); and (5) UAA, Stop becomes Tyr in a nematode, *Radopholus similis* (Jacob et al. 2009) and a planarian (Bessho et al. 1992) mitochondria. In the last case, no structural study on the corresponding tRNA was performed, and no release factor (RF) relevant to the UAA codon was identified, so discussion on the codon–anticodon relationship was not possible in this instance.

Modified nucleosides in the tRNA anticodons are associated with mitochondrial codon changes in all cases (Watanabe and Yokobori 2011): (1) The anticodon wobble position (the 34th) of $tRNA^{Trp}$ changes to 5-carboxymethyluridine ($cmnm^5U$) in nematode mitochondria (Sakurai et al. 2005) and to 5-taurinomethyluridine (τm^5U) in ascidian (Suzuki et al. 2011b), molluscan (Ohira et al. 2013), and vertebrate (Suzuki et al. 2002, 2011a) mitochondria; (2) The anticodon wobble position of $tRNA^{Met}$ changes to 5-formylcytidine (f^5C) in vertebrate (Moriya et al. 1994), arthropod (Tomita et al. 1999a), and nematode mitochondria (Watanabe et al. 1994) and to τm^5U in ascidian mitochondria (Suzuki et al. 2011b); (3) The anticodon wobble nucleoside of $tRNA^{Gly}$ for AGR codons becomes τm^5U in ascidian mitochondria (Suzuki et al. 2011b). The anticodon wobble nucleoside of $tRNA^{Ser}$ becomes m^7G in most invertebrate mitochondria (Matsuyama et al. 1998; Tomita et al. 1998) and unmodified U in nematode mitochondria (Watanabe et al. 1994); (4) The starfish mitochondrial $tRNA^{Asn}$ that deciphers AAA as Asn possesses the anticodon GΨU (Ψ, pseudouridine) (Tomita et al. 1999b); this is unusual in that it is a change at the second anticodon position. In vitro *E. coli* translation experiments indicate that ${tRNA^{Asn}}_{G\Psi U}$ has an approximately twofold higher translational efficiency than ${tRNA^{Asn}}_{GUU}$ (Tomita et al. 1999b). In other invertebrate mitochondria such as *Drosophila melanogaster*, $tRNA^{Lys}$ interprets AAR as Lys and has the anticodon CUU (Tomita et al. 1999a). Therefore, ${tRNA^{Lys}}_{CUU}$ should read AAR codons in a manner similar to the case of squid mitochondrial ${tRNA^{Met}}_{CAU}$ described subsequently.

The previously mentioned five nucleoside species ($cmnm^5U$, τm^5U, m^7G, Q, and f^5C) or seven species including the 2-thiolated forms of $cmnm^5U$ ($cmnm^5s^2U$) and τm^5U (τm^5s^2U) are all the modified nucleosides found in the anticodon wobble position of metazoan mitochondrial tRNAs. As shown in Table 2, $cmnm^5(s^2)U$, $\tau m^5(s^2)U$ and f^5C permit base pairing only with purine nucleosides (R_3) excluding base pairing with pyrimidine nucleosides (Y_3) of the cognate codons, probably through fixing their conformations in the C3′-endo form (Yokoyama and Nishimura 1995; Takemoto et al. 2009). This was proved experimentally using *E. coli* and bovine mitochondrial in vitro translation systems (Takai et al. 1994; Kurata et al. 2008; Takemoto et al. 2009). Queosine (Q, a G derivative), which forms base pairing only with pyrimidines, is sometimes found in the anticodon wobble position of *D. melanogaster* mitochondrial ${tRNA^{Asn}}_{GUU}$, but not in the anticodon wobble position of starfish mitochondrial ${tRNA^{Asn}}_{G\Psi U}$ (Tomita et al. 1999a, b). This indicates that *D. melanogaster* mitochondrial ${tRNA^{Asn}}_{G/QUU}$ restricts codon recognition to AAY codons, but starfish mitochondrial ${tRNA^{Asn}}_{G\Psi U}$ can decode not only AAY codons but also the AAA codon. It is likely that the m^7G located at the wobble position of $tRNA^{Ser}$ of most invertebrate mitochondria allows decoding of all four family box codons (Matsuyama et al. 1998; Tomita et al. 1998), although this speculation has not been verified experimentally (Watanabe and Yokobori 2011).

Nucleoside modifications often provide mechanisms for variable codon recognition; however, unmodified anticodon nucleosides have also been shown to decode cognate codons in some metazoan mitochondria (Table 2). For example, U_{34} was

Table 2 Mitochondrial pairing possibilities between the anticodon first position (N_{34}, wobble nucleotide) of tRNA and the codon third position (N_3)

Pairing rule				
	N_{34} of tRNA		N_3 of codon	Existence
Unmodified	U	⟺	U, C, A, G	All tRNAs belonging to family box of almost all animals
	A	⟺	U, C, A, G	Nematode $tRNA^{Arg}{}_{ACG}$
	G	⟺	U, C, A	Insect $tRNA^{Ser}{}_{GCU}$ for AGY/A codons
	C	⟺	A, G	Squid $tRNA^{Met}{}_{CAU}$, Insect $tRNA^{Lys}{}_{CUU}$
Modified	$cmnm^5(s^2)U$	⟺	A, G	Nematode $tRNA^{Trp}$
	$\tau m^5(s^2)U$	⟺	A, G	Ascidian $tRNA^{Trp}$, vertebrate $tRNA^{Trp}$, squid $tRNA^{Trp}$
				Ascidian $tRNA^{Met}$, ascidian $tRNA^{Gly}{}_{U*CU}$, vertebrate $tRNA^{Leu}{}_{U*AA}$
				Vertebrate tRNAs for NNR codons in the third column of codon table
	m^7G	⟺	U, C, A, G	Most invertebrate $tRNA^{Ser}{}_{GCU}$ for AGN codons
	Q	⟺	U, C	Vertebrate tRNAs for NNY codons in the third column of codon table
	f^5C	⟺	A, G	Vertebrate $tRNA^{Met}$, arthropod $tRNA^{Met}$, nematode $tRNA^{Met}$

shown experimentally to base-pair with all four nucleosides at the third codon position (Suzuki et al. 2011a). Unmodified U_{34} can adopt flexible conformations of either the C3′-endo or C2′-endo form, and this allows base pairing with pyrimidines in addition to purines. Unmodified A_{34} has a similar role in nematode mitochondrial $tRNA^{Arg}{}_{ACG}$ (Watanabe et al. 1997). Unmodified G is thought to be able to base-pair with A as well as C and U, because the AGG codon is absent in some metazoan mitochondria, such as those from *D. melanogaster*, and the anticodon wobble position of the relevant $tRNA^{Ser}{}_{GCU}$ is unmodified G (Tomita et al. 1999a). Therefore, the unmodified GCU anticodon of $tRNA^{Ser}{}_{GCU}$ should read the three codons AGU, AGC, and AGA as Ser. Recently, it was found that squid (*Loligo bleekeri*) mitochondrial $tRNA^{Met}$ possesses unmodified C at the anticodon wobble position (Ohira et al. 2013), suggesting that, in this case, the AUA codon is deciphered by the CAU anticodon via noncanonical A_3–C_{34} pairing. The possibility of such base-pair formation had previously been reported in the case of *D. melanogaster* $tRNA^{Lys}{}_{CUU}$ deciphering AAR codons (Tomita et al. 1999a).

Two mitochondrial RFs, mtRF1a and mtRF1, have been identified to date; mtRF1a recognises UAG and UAA codons (Soleimanpour-Lichaei et al. 2007), and mtRF1, which arose from a duplication of mtRF1a, probably binds the ribosome when the A site is devoid of mRNA (Huynen et al. 2012). An RF that recognised the UGA codon would have been lost in the transition process from pre-mitochondrial ancestral bacteria (α-proteobacteria) to the present mitochondria (Andersson et al. 1998). There has been no report identifying any RF able to recognise AGA/AGG codons directly. Instead, mtRF1a recognising UAR codons in vertebrate mitochondria (Soleimanpour-Lichaei et al. 2007) are thought to be used for translation termination of the AGR codons, as these would be changed to the UAG codon by -1 frame-shifting (Temperley et al. 2010). However, this hypothesis may not be generally applicable as the necessary U may not always be located prior to the AGR codons in the genome.

4 The Codon–Anticodon Relationship in Animal Mitochondria

As shown in previous sections, six non-universal codons (the UAA case is not discussed here) have been identified in animal mitochondria, and these vary over the course of animal evolution. As already mentioned, we have elucidated that the anticodons of tRNAs deciphering these non-universal codons ($tRNA^{Trp}$ for UGA, $tRNA^{Met}$ for AUA, $tRNA^{Asn}$ for AAA, and $tRNA^{Ser}$ and $tRNA^{Gly}$ for AGR) are all modified. However, besides these modified nucleosides involved in decoding non-universal codons, it has been found that unmodified nucleosides also exist at the wobble position of tRNA in the codon–anticodon interaction in some of the mitochondrial translation systems; unmodified U_{34} exists in $tRNA^{Ser}$ deciphering AGR codons in nematode (*Ascaris suum*) mitochondria (Watanabe et al. 1994). Apart from non-universal codon decoding, all tRNAs for the family box codons possess unmodified U_{34}. Unmodified U_{34} is also known to be present in the non-mitochondrial free-living organism systems such as *M. capricolum* (Inagaki et al. 1995) and decodes all four codons in the family boxes. Using an in vitro translation system of *M. capricolum*, it was experimentally verified that threonyl-$tRNA^{Thr}{}_{UGU}$ and alanyl-$tRNA^{Ala}{}_{UGC}$, both possessing unmodified U in the anticodon wobble position, could translate all four ACN and GCN codons, respectively (Andachi et al. 1989; Osawa 1995; Inagaki et al. 1995). An in vivo experiment in a plastid using knockout mutants for the $tRNA^{Gly}$ with G_{34}-containing anticodon in the GGN family box showed that the remaining $tRNA^{Gly}$ with U_{34}-containing anticodon was sufficient for survival of the plastid (Rogaiski et al. 2008), which also indicates that U_{34}-$tRNA^{Gly}$ can decipher all the GGN codons. In addition, unmodified A_{34} is present in the nematode (*Ascaris suum*) mitochondrial $tRNA^{Arg}$ deciphering GCN codons (Watanabe et al. 1997), in budding yeast (*Saccharomyces cerevisiae*) mitochondrial $tRNA^{Arg}{}_{ACU}$ deciphering CGN codons (Sibler et al. 1986), and in *Mycoplasma* spp. $tRNA^{Thr}{}_{AGU}$ deciphering ACN codons (Andachi et al. 1987). The translation activity of threonyl-$tRNA^{Thr}{}_{AGU}$ of *M. capricolum* toward all four ACN codons was experimentally verified (Inagaki et al. 1995). G_{34}-tRNA^{Ser} (G_{34} unmodified) deciphers three codons (AGU/AGC/AGA) in *D. melanogaster* mitochondria (Tomita et al. 1999a), C_{34}-tRNA^{Met} (C_{34} unmodified) deciphers AUR codons in squid (*Loligo bleekeri*) (Ohira et al. 2013), and C_{34}-tRNA^{Lys} (C_{34} unmodified) deciphers AAR codons in *D. melanogaster* (Tomita et al. 1999a). These data indicate that unmodified U_{34} (or, rarely, A_{34}) at the anticodon wobble position (N_{34}) can potentially pair with all four nucleosides at the codon third position (N_3), unmodified G_{34} can potentially pair with U_3, C_3, and A_3, and unmodified C_{34} can potentially pair with A_3 and G_3. Taken together, this strongly suggests that all codons were deciphered by unmodified tRNA in the early genetic code system; this would be prior to the emergence of the universal genetic code and RNA-modifying systems, the existence of which is difficult to imagine in the beginning of the genetic code systems.

Table 2 summarises the pairing patterns between anticodon first nucleoside (N_{34}) and codon third nucleoside (N_3) so far elucidated in animal mitochondrial translation systems.

5 How Was the Mitochondrial Decoding System Generated?

The mitochondrial decoding system is putatively derived from retrogression of the universal decoding system. This retrogression was mainly induced by genome economisation and directional mutation pressure, accompanying various events such as genetic code changes, reduction of tRNA species, and loss of modification enzymes.

Two main hypotheses have been proposed to explain genetic code changes: the "codon capture" hypothesis assuming an "unassigned codon" (Osawa and Jukes 1989) and the "ambiguous intermediate" hypothesis, which assumes two different tRNAs recognising a single codon (Schultz and Yarus 1994). The codon capture hypothesis successfully explains a codon change of the UGA Trp codon to a stop codon in *Mollicutes* (Osawa et al. 1992), but there has been little experimental evidence in support of the ambiguous intermediate hypothesis. Simulation studies on codon reassignment (Sengupta et al. 2007) suggest that the pathway of codon reassignment in favour of either "codon capture" or "ambiguous codon" depends on initial conditions such as genome size and number of codons involved. For example, as mitochondria possess a small genome size and a limited number of involved codons, the "codon capture" hypothesis can account for all the codon changes concerning the non-universal codons found in animal mitochondria (Watanabe and Yokobori 2011).

If codon change and the concomitant reduction of the number of tRNA anticodon species are regarded as the first event in the retrogression process of the mitochondrial decoding system, the second event is the removal of modification enzymes. The following events are proposed during the process of altering the universal genetic code to the vertebrate mitochondrial code. After removal of U_{34}-modification enzymes for U*NN anticodons of tRNA in family boxes during evolution, the resultant UNN anticodons (U unmodified) would become translatable for all four codons; this is observed in the present genetic codes for vertebrate mitochondria and in *Mycoplasma* spp. The GNN anticodons and CNN anticodons that had translated NNY and NNG codons would thus become unnecessary and be eliminated. In two-codon sets, U*NN anticodons for NNY codons and GNN or QNN anticodons for NNR codons would be retained because NNY and NNR codons code for different amino acids. In addition, $tRNA^{Arg}$ for AGR codons would be eliminated, probably by genome economisation. Once mtRF1a (if it recognised UAG codon generated by -1 frame-shifting of the AGR codons) or a new RF (recognising AGR codons directly) emerged, it would result in the change of AGR to a stop codon. This combination of events would finally result in 22 anticodon species and retention of modified nucleosides, as observed at present in vertebrate mitochondria.

6 The Early Genetic Code System

Vertebrate mitochondria use 60 sense codons and four stop codons and have a total of 22 tRNA species, including a single tRNA for eight family boxes and 14 - two-codon sets. One single $tRNA^{Met}$ functions as both initiator and elongator. The vertebrate mitochondrial genetic code system is the simplest genetic code system observed to date among extant organisms and can be used as a model of the early genetic code (Table 3). As noted previously, RNAs engaged in the early genetic code system would likely have been unmodified. All the family box codons are deciphered by U_{34}-tRNA having UNN anticodons. The two-codon sets are divided into two categories; the upper half (NNY codons) are read by GNN anticodons and the lower half (NNR codons) are read by UNN anticodons or CNN anticodons for AUR. How then would NNY codons have been distinguished from NNR codons in the early translation system? As illustrated in Fig. 1, U_{34}-tRNA has the potential to decipher any of the four possible codons, G_{34}-tRNA has a potential to decipher NNY and NNA codons, and C_{34}-tRNA has a potential to decipher NNY codons. Competition would likely have occurred between G_{34}-tRNA and U_{34}-tRNA (or C_{34}-tRNA for AUR codons) in their binding to the ribosomal A site. To separate the upper half and lower half codons in the two-codon sets, the binding affinity of G_{34}-tRNA with NNY codons at the ribosomal A site must be stronger than that of U_{34}-tRNA with NNY codons in the upper half, whereas the binding affinity of U_{34}-tRNA (or C_{34}-tRNA) with NNR codons must be stronger than that of G_{34}-tRNA with NNA codons in the lower half. Specifically, the base pairing between G_{34} and U_3 and between G_{34} and C_3 must be stronger than that between U_{34} and U_3 and between U_{34} and C_3. Likewise, the base pairing between U_{34} (or C_{34}) and A_3 must be stronger than that between G_{34} and A_3. These speculations are reasonable as R–Y pairing should be stronger than Y–Y or R–R pairing, and this is experimentally

Table 3 An early genetic code table proposed in this study

Codon	Anticodon	Amino acid	Codon	Anticodon	Amino acid	Codon	Anticodon	Amino acid	Codon	Anticodon	Amino acid
UUU	GAA	Phe	**UCU**	**UGA**	**Ser**	UAU	GUA[*3]	Tyr	UGU	GCA	Cys
UUC			**UCC**			UAC			UGC		
UUA	UAA[*1]	Leu	**UCA**			UAA		Stop	*UGA*	*UCA*[*1]	*Trp*
UUG			**UCG**			UAG			*UGG*		
CUU	**UAG**	**Leu**	**CCU**	**UGG**	**Pro**	CAU	GUG[*3]	His	**CGU**	**UCG**	**Arg**
CUC			**CCC**			CAC			**CGC**		
CUA			**CCA**			CAA	UUG[*4]	Gln	**CGA**		
CUG			**CCG**			CAG			**CGG**		
AUU	GAU	Ile	**ACU**	**UGU**	**Thr**	AAU	GUU[*3]	Asn	AGU	GCU	Ser
AUC			**ACC**			AAC			AGC		
AUA	*CAU*[*2]	*Met*	**ACA**			AAA	UUU[*4]	Lys	*AGA*		*Stop*
AUG			**ACG**			AAG			*AGG*		
GUU	**UAC**	**Val**	**GCU**	**UGC**	**Ala**	GAU	GUC[*3]	Asp	**GGU**	**UCC**	**Gly**
GUC			**GCC**			GAC			**GGC**		
GUA			**GCA**			GAA	UUC[*4]	Glu	**GGA**		
GUG			**GCG**			GAG			**GGG**		

Family box codons and specific codons related to codon change are shown by bold and italic emphases, respectively. This code is same as that in vertebrate mitochondria, with the exception that the following nucleosides are modified in vertebrate mitochondria: *1, 5-taurinomethyluridine (τm^5U); *2, 5-formylcytidine (f^5C); *3, queosine; *4, 5-taurinomethyl-2-thiouridine (τm^5s^2U)

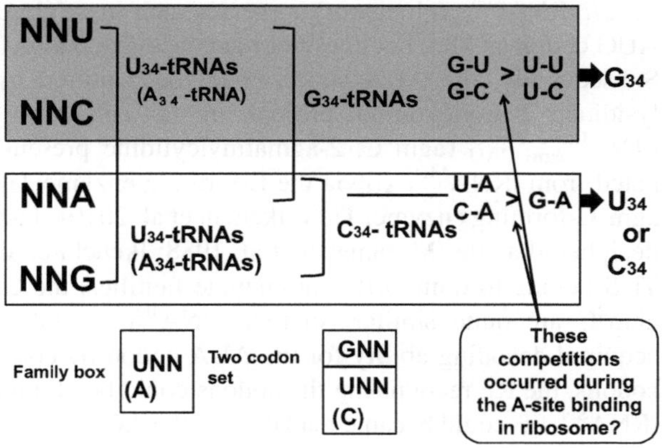

Fig. 1 Decoding scheme of two-codon set codons by competition between U_{34}-tRNA and G_{34}-tRNA (or C_{34}-tRNA for AUR codons) in the early genetic code system

verifiable. In summary, it is proposed that the early genetic code table would have consisted of eight family boxes, 14 two-codon sets, and four stop codons and would have been translated by 22 tRNA species. It remains to be experimentally determined whether all 64 codons were used in the early genetic code. UUR/CUN codons for Leu, CGN/AGR codons for Arg, and UCN/AGY codons for Ser are redundant, so the removal of one or several redundant codons would not diminish the number of different amino acids (20 species) available for protein synthesis. One of these redundancies is exploited in vertebrate mitochondria, where AGR codons become stop codons.

7 From the Early Decoding System to the Universal Decoding System

It is conceivable that the early decoding system could have evolved to the present universal decoding system by almost the reverse process of the evolution of the mitochondrial decoding system (Osawa 1995; Osawa et al. 1992). The process would, therefore, involve genome enlargement due to gene duplication, concomitant GC pressure, emergence of new tRNA genes accompanied by nucleoside modification, and the resulting reassignment of three kinds of codon: UGA from Trp to stop codon, AUA from Met to Ile, and AGR from stop codon to Arg. The progression of these reassignment processes is proposed as follows. For the UGA codon, the anticodon wobble position of $tRNA^{Trp}$ may have changed from U to C by GC pressure, so that the resultant CCA anticodon became readable only by the UGG codon, and the UGA codon became unassigned. An RF recognising the UGA codon (e.g. RF2 in *E. coli*) could then emerge and capture the UGA codon, transforming it into a stop codon. Acetylation enzyme modification of C_{34} of $tRNA^{Met}{}_{CAU}$ to ac^4C_{34} would lead the resulting $tRNA^{Met}{}_{ac^4CAU}$ to become readable only for the AUG codon, and the AUA codon would become unassigned. In fact,

E. coli $tRNA^{Met}{}_{m}$ (elongator) possesses such an ac^4CAU anticodon, which reads the AUG codon as Met, but does not read the AUA codon (Stern and Schulman 1978). Subsequently, the AUA codon could be captured by $tRNA^{Ile}{}_{LAU}$ (L (or k^2C), lysidine, 2-lysylcytidine present in *E. coli* and *Mycoplasma* $tRNA^{Ile}$) or $tRNA^{Ile}{}_{agm^2CAU}$ (agm^2C, 2-agmatinylcytidine present in Archaea), which originated from $tRNA^{Met}{}_{CAU}$ via the L-forming enzyme TilS (Soma et al. 2003) or the agm^2C-forming enzyme, TiaS (Ikeuchi et al. 2010). Thus, the AUA codon would be deciphered as Ile (Muramatsu et al. 1988; Ikeuchi et al. 2010). Although TilS and TiaS belong to quite different enzyme families, the chemical structures of L and agm^2C are quite similar, so that $tRNA^{Ile}{}_{LAU}$ and $tRNA^{Ile}{}_{agm^2CAU}$ could have acquired decoding ability for the AUA codon by convergent evolution. For AGR codons, the RF recognising the codons could be eliminated from the genome, then $tRNA^{Ser}{}_{GCU}$ could became readable for the AGA codon in addition to AGY codons, and the AGG codon would become unassigned. Genome enlargement would allow emergence of $tRNA^{Arg}{}_{UCU}$, which could capture the AGA codon as well as the AGG codon as Arg, because $tRNA^{Arg}{}_{UCU}$ would prevail over $tRNA^{Ser}{}_{GCU}$ in the binding of AGA codon, as discussed previously.

During this process, the number of tRNAs involved in the translation system would increase, probably initially to the number used in the *Mycoplasma* system (28 species), and finally to the number in the *E. coli* system (42–48 species). This would occur in parallel with nucleoside modifications, particularly G_{34}-tRNA modification to Q_{34}-tRNA corresponding to NNY codons in the upper half of two-codon sets, and modification of the wobble U of U_{34}-tRNA to various U derivatives, such as mnm^5U, mnm^5s^2U, $cmnm^5U$, $cmnm^5U_m$, or $cmnm^5s^2U$ in the lower half of two-codon sets, as in *Mycoplasma* and *E. coli* systems. Modification of G_{34} and U_{34} would be necessary for strict discrimination of UUY and UUR codon boxes by G_{34}-tRNA and U_{34}-tRNA.

To summarise, Fig. 2 shows a phylogenetic tree of three domains of life, mitochondrial origination from α-proteobacteria, and the emergence of universal and non-universal genetic codes. It should be emphasised that early decoding was performed by unmodified RNA, and the universal decoding system emerged after the development of RNA modification.

8 Concluding Remarks and Perspective

In this review, the following concepts were newly deduced from the study of metazoan mitochondrial decoding systems.

1. The change of AGR codons during the retrogression process was found to follow a route from Arg (universal code) to stop codons (vertebrate mitochondria) via Ser codons (most invertebrate mitochondria) (Gly codons in *Cnidarian* mitochondria seem to be a sub-route). This concept is useful for considering the

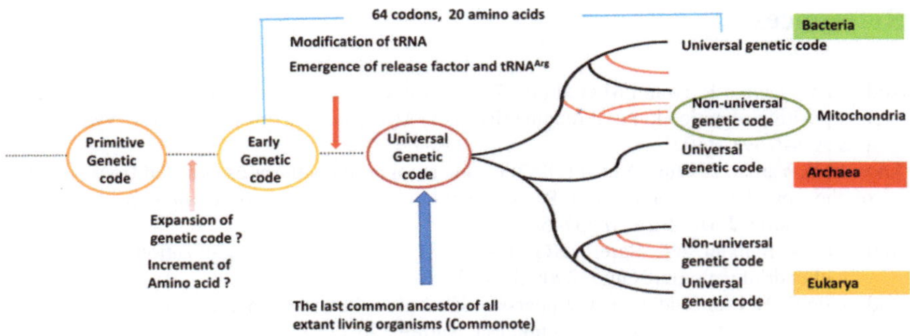

Fig. 2 Three domains of life indicating the position of mitochondria in the evolutionary process. The assignment of 20 amino acids to 64 codons would have been conserved from the early genetic code system. The early genetic system utilised unmodified RNA and developed into the universal code system after the emergence of RNA modification

evolutional processes by which AGR codon-use changed from the early code to the present universal code.

2. Examples from mitochondria indicate that any of the unmodified nucleosides at the anticodon wobble position of tRNA are available for the decoding of the corresponding codon at the ribosomal A site. It is thus possible to infer that unmodified tRNA had deciphered all codons in the early genetic code system, prior to the emergence of the universal genetic code and RNA-modifying systems.
3. The most plausible explanation for the discrimination of the NNY codon from the NNR codon frame in the two-codon sets is that the discrimination is fulfilled by competition between G_{34}-tRNA and U_{34}-tRNA (or C_{34}-tRNA for AUR codons) in their binding to the ribosomal A site. Thus, the early genetic code table would have been established in such a way that eight family box codons were deciphered by U_{34}-tRNA, and eight NNY and six NNR codons in the two-codon sets were deciphered by G_{34}-tRNA and U_{34}-tRNA (or C_{34}-tRNA in part), respectively. This genetic code table is the simplest among the extant living organisms, consisting of eight family boxes and 14 two-codon sets, decoded by 22 species of tRNA.

Studies of the mitochondrial genetic code and decoding apparatus from different animal phyla, alongside data from organisms using the universal code, have produced new insights into the evolution of non-universal genetic systems. Examination of such evolutionary traces from the mitochondrial genetic code systems of additional animal phyla would be useful to allow further elucidation of the early decoding system and the origins of the genetic code in the RNA world.

Acknowledgements We thank Dr. Tsutomu Suzuki of the University of Tokyo for supplying us with analytical data of mitochondrial tRNAs and his wholehearted support of this study. We also thank Dr. Akira Muto, Hirosaki University, for *Mycoplasma* ribosome data, and Dr. Takeo Suzuki, University of Tokyo, for *E. coli* tRNA data.

References

Andachi Y, Yamao F, Iwami M et al (1987) Occurrence of unmodified adenine and uracil at the first position of anticodon in threonine tRNAs in *Mycoplasma capricolum*. Proc Natl Acad Sci U S A 84:7398–7402

Andachi Y, Yamao F, Muto A et al (1989) Codon recognition patterns as deduced from sequences of the complete set of transfer RNA species in *Mycoplasma capricolum*: resemblance to mitochondria. J Mol Biol 209:37–54

Anderson S, Bankier AT, Barerll BG et al (1981) Sequence and organization of the human mitochondrial genome. Nature 290:457–465

Andersson SGE, Zomorodipour A, Andersson JO et al (1998) The genome sequence of *Rickettsia prowazekii* and the origin of mitochondria. Nature 396:133–143

Barrell BG, Bankier AT, Drouin J (1979) A different genetic code in human mitochondria. Nature 282:189–194

Bessho Y, Ohama T, Osawa S (1992) Planarian mitochondria II. The unique genetic code as deduced from cytochrome c oxydase subunit I gene sequence. J Mol Evol 34:331–335

Himeno H, Masaki H, Kawai T et al (1987) Unusual genetic codes and a novel gene structure for $tRNA^{Ser}{}_{AGY}$ in starfish mitochondria. Gene 56:219–230

Huynen MA, Duarte I, Chrzanowska-Lightowlers ZMA et al (2012) Structure based hypothesis of a mitochondrial ribosome rescue mechanism. Biol Direct 7:14

Ikeuchi Y, Kimura S, Numata T et al (2010) Agmatine-containing cytidine in tRNA anticodon essential for AUA decoding in Archaea. Nat Chem Biol 6:277–282

Inagaki Y, Kojima A, Bessho Y et al (1995) Translation of synonymous codons in family boxes by *Mycoplasma capricolum* tRNAs with unmodified uridine or adenosine at the first anticodon position. J Mol Biol 251:486–492

Jacob JEM, Vanholme B, Van Leeuwen T et al (2009) A unique genetic code change in the mitochondrial genome of the parasitic nematode *Radopholus similis*. BMC Res Notes 2:192

Jukes TH (1983) Evolution of the amino acid code: inferences from mitochondrial codes. J Mol Evol 19:219–225

Kurata S, Weixlbaumer A, Ohtsuki T et al (2008) Modified uridines with C5-methylene substituents at the first position of the tRNA anticodon stabilize U-G wobble pairing during decoding. J Biol Chem 283:18801–18811

Lang BF, Burger G, O'Kelly CJ et al (1977) An ancestral mitochondrial DNA resembling a eubacterial genome in miniature. Nature 387:493–497

Margulis L (1970) Origin of eukaryotic cells. Yale University Press, New Haven, CT

Margulis L (1993) Symbiosis in cell evolution: microbial communities in the Archean and Proterozoic eons, 2nd edn. W. H. Freeman, New York

Matsuyama S, Ueda T, Crain PF et al (1998) A novel wobble rule found in starfish mitochondria. Presence of 7-methylguanosine at the anticodon wobble position expands decoding capability of tRNA. J Biol Chem 273:3363–3368

Moriya J, Yokogawa T, Wakita K et al (1994) A novel modified nucleoside found at the first position of the anticodon of methionine tRNA from bovine liver mitochondria. Biochemistry 33:2234–2239

Muramatsu T, Nishikawa K, Nemoto F et al (1988) Codon and amino acid specificities of a transfer RNA are both converted by a single post-transcriptional modification. Nature 294:187–188

Näsvall SJ, Chen P, Björk GR (2007) The wobble hypothesis revisited: uridine-5-oxyacetic acid is critical for reading of G-ending codons. RNA 13:1–14

NCBI DB for Organelle genome. http://www.ncbi.nlm.gov/genomes/OrganelleResource,cgi?taxid=33208

Ohira T, Suzuki T, Miyauchi K et al (2013) Decoding mechanism of non-universal genetic codes in *Loligo bleekeri* mitochondria. J Biol Chem 288:7645–7652

Osawa S (1995) Evolution of the genetic code. Oxford University Press, Tokyo

Osawa S, Jukes TH (1989) Codon reassignment (codon capture) in evolution. J Mol Evol 28:271–278

Osawa S, Jukes TH, Watanabe K et al (1992) Recent evidence for evolution of the genetic code. Microbiol Rev 56:229–264

Rogaiski M, Karcher D, Bock R (2008) Superwobbling facilitates translation with reduced tRNA sets. Nat Struct Mol Biol 15:192–198

Sakurai M, Ohtsuki T, Suzuki T et al (2005) Unusual usage of wobble modifications in mitochondrial tRNAs of the nematode *Ascaris suum*. FEBS Lett 579:2767–2772

Schultz DW, Yarus M (1994) Transfer RNA mutation and the malleability of the genetic code. J Mol Biol 235:1377–1380

Sengupta S, Yang X, Higgs PG (2007) The mechanisms of codon reassignments in mitochondrial genetic codes. J Mol Evol 64:662–688

Sibler AP, Dirheimer G, Martin RP (1986) Codon reading patterns in *Saccharomyces cerevisiae* mitochondria based on sequences of mitochondrial tRNAs. FEBS Lett 194:131–138

Soleimanpour-Lichaei HR, Kühl I, Gaisne M et al (2007) mtRF1a is a human mitochondrial translation release factor decoding the major termination codons UAA and UAG. Mol Cell 27:745–757

Soma A, Ikeuchi Y, Kanemasa S et al (2003) An RNA-modifying enzyme that governs both the codon and amino acid specificities of isoleucine tRNA. Mol Cell 12:689–698

Stern L, Schulman LH (1978) The role of the minor base N^4-acetylcytidine in the function of the *Escherichia coli* non-initiator methionine transfer RNA. J Biol Chem 253:6132–6139

Suzuki T, Terasaki M, Takemoto-Hori C et al (2001a) Structural compensation for the deficit of rRNA with proteins in the mammalian mitochondrial ribosome. J Biol Chem 276:21724–21736

Suzuki T, Terasaki M, Takemoto-Hori C et al (2001b) Proteomic analysis of the mammalian mitochondrial ribosome. J Biol Chem 276:33181–33195

Suzuki T, Suzuki T, Wada T et al (2002) Taurine as a constituent of mitochondrial tRNAs: new insights into the function of taurine and human mitochondrial diseases. EMBO J 21:6581–6589

Suzuki T, Nagao A, Suzuki T (2011a) Human mitochondrial tRNAs: biogenesis, function, structural aspects, and diseases. Annu Rev Genet 45:299–329

Suzuki T, Miyauchi K, Suzuki T et al (2011b) Taurine-containing uridine modifications in tRNA anticodons are required to decipher non-universal genetic codes in ascidian mitochondria. J Biol Chem 286:35494–35498

Takai K, Horie N, Yamaizumi Z et al (1994) Recognition of UUN codons by two leucine tRNA species from *Escherichia coli*. FEBS Lett 344:31–34

Takemoto C, Spremulli LL, Benkowski LA et al (2009) Unconventional decoding of the AUA codon as methionine by mitochondrial $tRNA^{Met}$ with the anticodon f^5CAU as revealed with a mitochondrial *in vitro* translation system. Nucleic Acids Res 37:1616–1627

Telford MJ, Herniou EA, Russel RB et al (2000) Changes in mitochondrial genetic codes as phylogenetic characters: two examples from the flatworms. Proc Natl Acad Sci U S A 97:11359–11364

Temperley R, Richter R, Dennerlein S et al (2010) Hungry codons promote frameshifting in human mitochondrial ribosomes. Science 327:301

Tomita K, Ueda T, Watanabe K (1998) 7-methylguanosine at the anticodon wobble position of squid mitochondrial $tRNA^{Ser}$GCU: molecular basis for assignment of AGA/AGG codons as serine in invertebrate mitochondria. Biochim Biophys Acta 1399:78–82

Tomita K, Ueda T, Ishiwa S, Crain PF, McCloskey JA, Watanabe K (1999a) Codon reading patterns in *Drosophila melanogaster* mitochondria based on their tRNA sequences: a unique wobble rule in animal mitochondria. Nucleic Acids Res 27:4291–4297

Tomita K, Ueda T, Watanabe K (1999b) The presence of pseudouridine in the anticodon alters the genetic code: a possible mechanism for assignment of the AAA lysine codon as asparagine in echinoderm mitochondria. Nucleic Acids Res 27:1683–1689

Watanabe K (2010) Unique features of animal mitochondrial translation systems. The non-universal genetic code, unusual features of the translational apparatus and their relevance to human mitochondrial diseases. Proc Jpn Acad Ser B Phys Biol Sci 86:11–39

Watanabe K, Yokobori S (2011) tRNA modification and genetic code variations in animal mitochondria. J Nucleic Acids 2011:Article ID 623095

Watanabe Y, Tsurui H, Ueda T et al (1994) Primary and higher order structures of nematode (*Ascaris suum*) mitochondrial tRNAs lacking either the T or D stem. J Biol Chem 269:22902–22906

Watanabe Y, Tsurui H, Ueda T et al (1997) Primary sequence of mitochondrial $tRNA^{Arg}$ of a nematode *Ascaris suum*: occurrence of unmodified adenosine at the first position of the anticodon. Biochim Biophys Acta 1350:119–122

Yamao F, Muto A, Kawauchi Y et al (1985) UGA is read as tryptophan in *Mycoplasma capricolum*. Proc Natl Acad Sci U S A 82:2306–2309

Yokobori S, Ueda T, Watanabe K (1993) Codons AGA and AGG are read as glycine in ascidian mitochondria. J Mol Evol 36:1–8

Yokoyama S, Nishimura S (1995) Modified nucleosides and codon recognition. In: Söll D, RajBhandary UL (eds) tRNA: structure, biosynthesis and function. ASM, Washington, DC, pp 207–223

Isomerization of RNA Phosphodiester Linkages

Harri Lönnberg

Contents

1 Introduction 42
2 Isomerization Under Acidic Conditions via a Neutral Oxyphosphorane Intermediate 42
3 pH-Independent Isomerization via a Monoanionic Phosphorane Intermediate 46
4 Isomerization via Dianionic Phosphorane: Effect of Intramolecular Hydrogen Bonding . 48
5 Buffer-Catalyzed Isomerization 51
6 Metal Ion-Promoted Isomerization 52
7 Effect of Thio Substitution and Alkylation of a Non-bridging Oxygen on Isomerization 52
8 Epilogue 53
References 54

Abstract RNA phosphodiester linkages undergo two transesterification reactions catalyzed by Brönsted acids or bases, viz., cyclization to a 2′,3′-cyclic phosphate with concomitant release of the 5′-linked nucleoside and isomerization to a 2′,5′-linkage. While these processes are approximately as fast under acidic conditions, the cyclization gradually takes over on going to alkaline solutions. At physiological pH, the isomerization still successfully competes with the chain cleavage. Nevertheless, protein enzymes, ribozymes, and metal ion complexes with some exceptions catalyze only the cleavage reaction. The mechanistic reasons for the dissimilar pH dependence and susceptibility to catalysis are discussed.

Keywords RNA • Phosphodiesters • Isomerization • Kinetics • Mechanisms

---

H. Lönnberg (✉)
Department of Chemistry, University of Turku, 20014, Turku, Finland
e-mail: harlon@utu.fi

V.A. Erdmann et al. (eds.), *Chemical Biology of Nucleic Acids*, RNA Technologies,
DOI 10.1007/978-3-642-54452-1_3, © Springer-Verlag Berlin Heidelberg 2014

1 Introduction

The cleavage of RNA phosphodiester bonds by protein enzymes (Raines 1998) and small ribozymes (Cochrane and Strobel 2008), as well as by small molecular entities (Lönnberg 2011a) and Brönsted acids and bases (Perreault and Anslyn 1997; Oivanen et al. 1998), has been the subject of exceptionally extensive studies, owing to the obvious biological importance of this reaction. The overall course of the reaction is in all these cases the same: displacement of the 5′-linked nucleoside by intramolecular attack of the neighboring 2′-OH. The cyclic diester obtained is then rapidly hydrolyzed to 2′- and 3′-phosphomonoesters. Only large ribozymes behave differently. They utilize an intermolecular nucleophile for displacement of the 3′-linked nucleoside (Lönnberg 2011b).

On using Brönsted acid as a catalyst, another intramolecular transesterification, viz. conversion to a 2′,5′-diester, competes with formation of the 2′,3′-cyclic phosphate (Järvinen et al. 1991). Both reactions in all likelihood proceed via a common pentacoordinated oxyphosphorane intermediate, which may decompose into two alternative manners, by departure of either the 5′- or 3′-oxygen (Fig. 1). Interestingly, the pH-rate profiles of these two reactions are very different. At pH < 3, both reactions are hydronium ion catalyzed and isomerization is slightly faster than cleavage. Isomerization, however, turns pH independent at pH 4, while the cleavage experiences a change from hydronium ion to hydroxide ion-catalyzed reaction over a narrow pH range around pH 5. Consequently, isomerization is faster than cleavage up to pH 7 ($T = 90$ °C), in particular at pH 4–6, but cleavage takes over on passing pH 7. At high pH, cleavage then is the only reaction detected.

In contrast to Brönsted acids, protein enzymes, ribozymes, and most of catalytically active metal ion complexes promote only the cleavage reaction. In other words, they somehow favor the exocyclic fission of the P-O5′ over the endocyclic fission of the P-O3′ bond. The factors that influence the stability of the phosphorane intermediate and its partition to the cleavage and isomerization products are discussed in the following. Detailed understanding of these factors may help mechanistic chemists to sharpen the picture on details of enzyme action and synthetic chemists to avoid the harmful isomerization during the assembly of oligoribonucleotides.

2 Isomerization Under Acidic Conditions via a Neutral Oxyphosphorane Intermediate

As mentioned above, the cleavage and isomerization of RNA 3′,5′-phosphodiester linkages are both susceptible to acid catalysis, the isomerization being slightly faster than the cleavage (Järvinen et al. 1991; Kuusela and Lönnberg 1994). Both reactions in all likelihood proceed via a common oxyphosphorane intermediate obtained by an attack of the neighboring 2′-hydroxy group on the phosphorus atom.

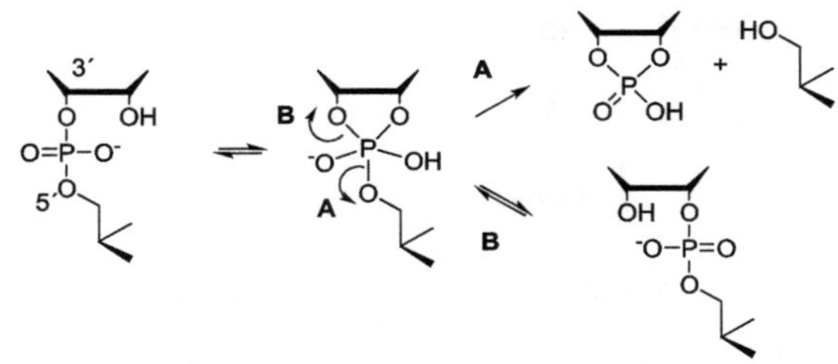

Fig. 1 Overall course of the cleavage (A) and isomerization (B) of RNA phosphodiester linkages

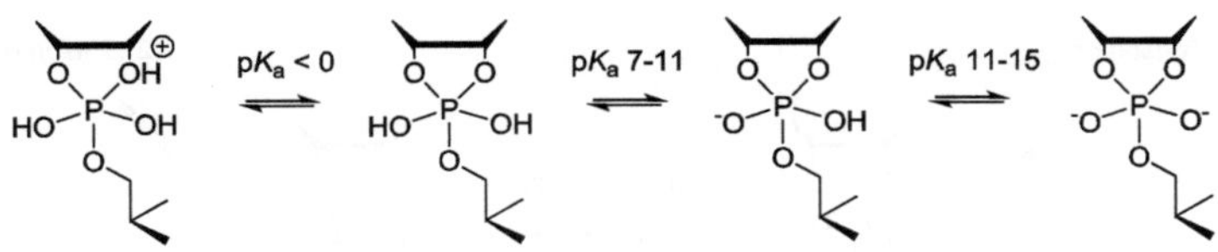

Fig. 2 Protolytic equilibria of oxyphosphorane intermediates

Depending on pH, a neutral (monoprotic) or monocationic(diprotic) phosphodiester linkage is attacked. The phosphorane intermediate is, hence, initially obtained as either a neutral (diprotic) or monocationic (triprotic) species (Fig. 2). At pH > 0, the monocationic species undergoes a rapid thermodynamically favored deprotonation to the neutral phosphorane. This is the stable ionic form of oxyphosphorane under acidic conditions. The accurate acidity constants of oxyphosphoranes are unknown, but the estimates for its first and second pK_a value fall in the range 7–11 and 11–15, respectively (Kluger et al. 1969; Guthrie 1977; Perreault and Anslyn 1997; Davies et al. 2002; Lopez et al. 2002; Range et al. 2004).

Oxyphosphorane intermediates have a structure of a trigonal bipyramid that contains two apical oxygens with longer and three equatorial oxygens with shorter bond lengths to phosphorus (Westheimer 1968; Range et al. 2004; Swamy and Kumar 2006; van Bochove et al. 2006). According to the well-known Westheimer's rules, nucleophiles enter and leave the phosphorane only via apical positions. Electron-rich oxygens prefer equatorial and electron-deficient apical positions. In case two of the oxygen atoms are members of a five-membered ring, one of them must be apical and the other equatorial. The attacking 2′-oxygen, hence, initially takes an apical position and forces the 3′-oxygen to an equatorial position (**1a** and **2a** in Fig. 3). To enable isomerization, the phosphorane must undergo a conformational rearrangement known as Berry pseudorotation (Stephen 1960): the apical ligands take equatorial and two of the equatorial ligands apical positions, one of the equatorial ligands remaining equatorial. This process brings the 3′-oxygen to apical position (**1b** and **2b** in Fig. 3), which is a prerequisite for the cleavage of the P-O3′ bond. The 5′-linked nucleoside may evidently take either an apical (in **1a**) or equatorial position (in **2a**) upon formation of the phosphorane intermediate. The apical position is probably favored, since theoretical calculations suggest a methoxy ligand to be more apicophilic than a hydroxy ligand (Range et al. 2004;

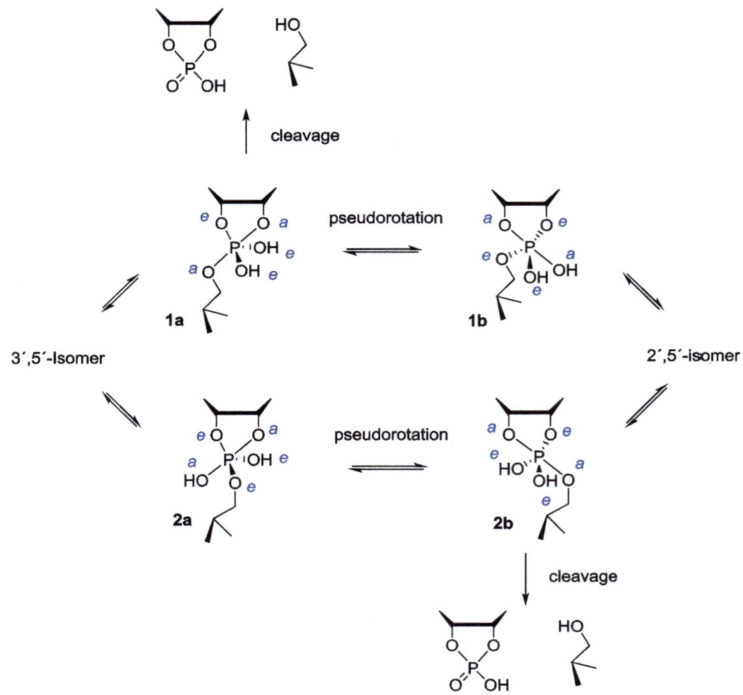

Fig. 3 Pseudorotation of oxyphosphoranes

Lopez et al. 2005). The phosphodiester linkage may, hence, be cleaved regardless of pseudorotation, while isomerization is realized only through pseudorotation of a reasonable stable intermediate.

The fact that isomerization of RNA phosphodiester bonds successfully competes with their cleavage at $pH < 6$ inevitably shows that neutral (dihydric) phosphoranes (**1a**, **b**; **2a**, **b**) are sufficiently stable to pseudorotate, as also suggested by ab initio gas phase calculations (Uchimaru et al. 1996; Zhou and Taira 1998) and DFT calculations with polarizable continuum models (Lopez et al. 2004) applied to the hydrolysis of methyl ethylene phosphate. These calculations additionally show that the pseudorotation barrier is at least 10 kcal mol^{-1} lower than the barrier for the exo- or endo-cyclic fission of the neutral phosphorane. In other words, various phosphorane structures are rapidly equilibrated before any of the rate-limiting PO bond fissions yielding either a 2′,5′-diester, 3′,5′-diester, or 2′,3′-cyclic phosphate takes place, and hence, according to Curtin–Hammett principle, the product distribution is not influenced by the pseudorotation process. Heavy atom kinetic isotope effects lend additional support for this interpretation. The primary ^{18}O effect for the hydrolysis of uridine 3′-(3-nitrobenzyl phosphate) is 1.0019 ± 0.0007 at pH 2.5, consistent with loss of bond order with the leaving benzyloxy group at the transition state, while the secondary isotope effect resulting from ^{18}O substitution of the non-bridging phosphodiesteroxygens is inverse (0.9885), probably owing to formation of a neutral phosphorane having additional OH bond compared to the starting material (Gerratana et al. 2000). Another piece of evidence for the crucial role of pseudorotating phosphorane as an intermediate of the isomerization reaction is offered by the pioneering studies of Buchwald et al. (1984). 2-[(*R*)-^{16}O,^{17}O,^{18}O] Phospho-(*S*)-propane-1,2-diol was shown to be isomerized by retention of configuration around phosphorus. Later, studies with R_P- and S_P-phosphorothioate analogues of 3′,5′-UpU have verified that the isomerization of ribonucleoside 3′-phosphodiesters also proceeds by retention of configuration (Oivanen et al. 1995).

The shape of the pH-rate profile reveals that the acid-catalyzed isomerization proceeds, depending on pH, via a monocationic (dihydric) or neutral (monohydric) phosphodiester (Järvinen et al. 1991). The former reaction, giving a monocationic (trihydric) oxyphosphorane, predominates at pH < 2 (Fig. 4). Thermodynamically favored deprotonation to neutral (dihydric) phosphorane evidently takes place, followed by rapid equilibration between various phosphorane structures. As discussed above, the pseudorotation barriers are considerably lower than the barriers for fission of the P–O bonds. Accordingly, any of the P–O bonds may adopt an apical position and undergo hydronium ion-catalyzed cleavage. In other words, three proton transfers take place, *viz.*, protonation of the starting material, deprotonation of the phosphorane obtained, and re-protonation of the phosphorane after pseudorotation. Since formation and/or cleavage of a P–O bond is rate limiting, the transition state is always monocationic, and hence, the reaction is first order in [H$^+$] at pH $<$ pK_a of the phosphodiester bond and second order in [H$^+$] at pH $>$ pK_a. With the phosphorane derived from 3′,5′-UpU, the relative rate constants for the formation of 2′,5′-diester, 3′,5′-diester, and 2′,3′-cyclic diester are 1.1, 1.0, and 1.3 (Järvinen et al. 1991). Exo- and endocyclic PO bonds are, hence, cleaved approximately as readily when the leaving group is alcohol, not alkoxide ion.

At pH 2–4, neutral (dihydric) phosphorane is obtained by an uncatalyzed attack of the 2′-OH on neutral (monohydric) phosphodiester linkage. This means that the reaction proceeds by proton transfer from the attacking OH to phosphorane non-bridging oxygen concerted with the P–O bond formation. On the basis of microscopic reversibility, the breakdown of the phosphorane then proceeds by proton transfer from an equatorial OH ligand to the departing oxygen. Now the endocyclic fission is slightly more favored than the exocyclic one, the relative rate constants for the formation of 2′,5′-diester, 3′,5′-diester, and 2′,3′ cyclic diester being 1.2, 1.0, and 0.4. Quantum chemical calculations in gas phase (Uchimaru et al. 1996) have shown that within a neutral phosphorane, the endocyclic PO bonds are inherently weaker than the exocyclic one. In fact the two- to three-fold difference in the cleavage rate is much less than what one might expect on the basis of these calculations.

The isomerization via neutral phosphorane is expectedly rather insensitive to the polar nature of the esterified alcohol. The reactions via a monocationic and neutral phosphodiester exhibit β values -0.23 ± 0.04 and -0.23 ± 0.11, respectively, on

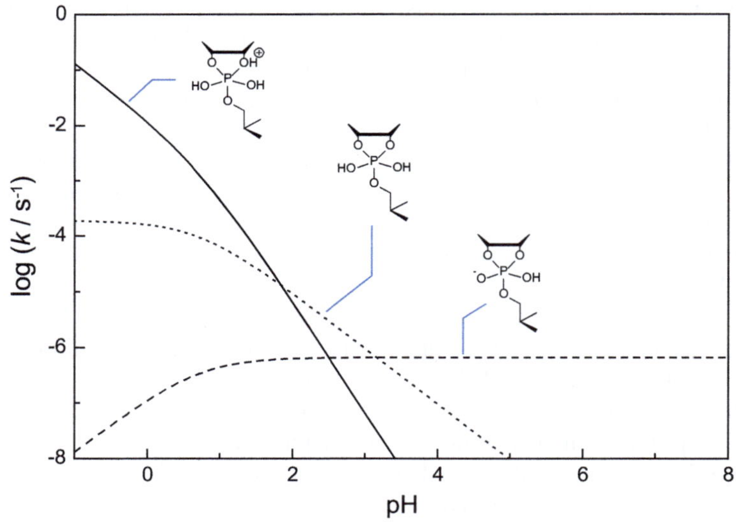

Fig. 4 Logarithmic first-order rate constants for the isomerization of the monocationic (*continous line*), neutral (*dotted line*), and monoanionic (*dashed line*) 3′,5′-phosphodiester bond to a 2′,5′-bond at 90 °C (Järvinen et al. 1991)

using ethyl, 2-ethoxyethyl, 2,2-dichloroethyl, and 2,2,2-trichloroethyl phosphates of uridine as model compounds (Kosonen et al. 1998b).

3 pH-Independent Isomerization via a Monoanionic Phosphorane Intermediate

The isomerization turns pH independent at pH > 4 (Järvinen et al. 1991). Evidently, the attack of 2′-OH on monoanionic phosphodiester is accompanied by concerted transfer of a proton from the attacking hydroxyl group to a non-bridging oxygen ligand of the developing monoanionic (monohydric) phosphorane intermediate (Lopez et al. 2006). Whether or not the proton transfer is mediated by a water molecule still appears to be an open question (Gerratana et al. 2000; Boero et al. 2002; Yang and Cui 2009). Anyway, a monoanionic phosphorane intermediate is obtained. This species evidently has a finite lifetime and it may well be sufficiently long-lived to undergo pseudorotation. DFT calculations with PMC, COSMO, and SM5 solvation models (Lopez et al. 2004, 2005) suggest that a monoanionic phosphorane exists and the barrier for its pseudorotation is even lower than that of the pseudorotation of neutral phosphorane. However, the barriers for the PO bond cleavage are also lower than with neutral phosphorane, and hence, the lifetime of monoanionic phosphorane is shorter than that of neutral phosphorane. With methyl ethylene phosphate, the barriers for the exo- and endocyclic cleavage are 14.4 and 3.24 kcal mol^{-1} (Lopez et al. 2004). In other words, the endocyclic fission is approximately as fast as pseudorotation and much faster than the exocyclic fission. Pseudorotation may at least partially limit the rate of isomerization. As indicated in Fig. 2, monoanionic (monohydric) phosphorane may undergo thermodynamically allowed protonation to a neutral diprotic form at pH < 7 but, as discussed above, pseudorotation without kinetically invisible protonation/deprotonation process is also plausible. Pseudorotation as a monoanion receives some support from kinetic isotope effects. The secondary effect resulting from ^{18}O substitution of the non-bridging phosphate oxygens is at pH 5.5 close to unity, in contrast to the inverse effect observed for the acid-catalyzed reactions proceeding via a neutral phosphorane (Gerratana et al. 2000). This has been taken as an indication of different ionic forms of the transition states.

It is worth noting that isomerization turns pH independent around pH 3.5, i.e., from one to two pH units earlier than the cleavage reaction (Järvinen et al. 1991; Kosonen et al. 1998b). Since both reactions proceed via a common phosphorane intermediate, breakdown of the monoanionic phosphorane favors endocyclic fission much more clearly than its neutral counterpart. Studies with uridine 3′-phosphotriesters have shed light to the situation (Mikkola et al. 2002). Let us consider the mixed isopropyl 2,2-difluoroethyl ester (**3**) as a model of UpU. The hydroxide ion-catalyzed isomerization and cleavage of this compound proceed via a monoanionic phosphorane, analogously to pH-independent isomerization and cleavage of UpU (Fig. 5). The pK_a values of 2,2-difluoroethanol and isopropanol are 13 and 17, respectively (Ballinger and Long 1960). Accordingly, 2,2-difluoroethoxide ion is a much better leaving group than isopropoxide ion and comparable to the 5′-oxyanion of nucleosides as a leaving group. Isopropyl group may be envisaged to take the role of proton within the monoanionic phosphorane. With **3**, the isomerization turns hydroxide ion catalyzed on passing pH 3. Evidently the 2′-OH is deprotonated in a rapid pre-equilibrium step, and the monoanionic phosphorane obtained by the attack of the 2′-oxyanion undergoes pseudorotation and subsequent endocyclic PO3′ fission that completes the isomerization. The 2,2-difluroethoxide group in all likelihood adopts apical position upon formation of the monoanionic phosphorane and may, hence, depart without pseudorotation. Still this reaction, the exocyclic cleavage of alkoxide ion, is two orders of magnitudes slower than the departure of 3′-oxyanion via pseudorotation and endocyclic cleavage (Kosonen et al. 1999).

The migration of an anionic alkylphosphate group between the 2′- and 3′-hydroxy functions is markedly insensitive to the polar nature of the group, the β value being -0.03 ± 0.01 (Kosonen et al. 1998b). In striking contrast to 3′-alkylphosphates, 3′-arylphosphates are isomerized only at pH < 2, and even then the reaction is slow compared to the cleavage (Oivanen and Lönnberg 1991). Evidently the 2′/3′-oxygens cannot compete as a leaving group with the aryloxy group.

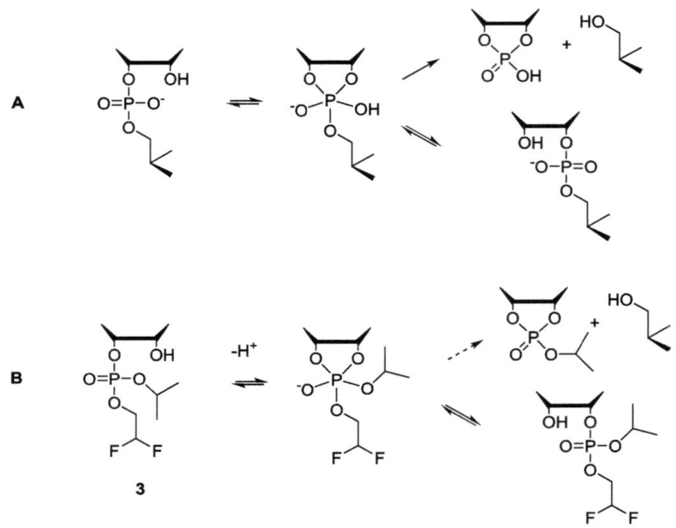

Fig. 5 Reactions of ribonucleoside 3′-phosphodiester (**A**) and 3′-phosphotriester (**B**) via a monoanionic phosphorane intermediate. Tetra-*O*-alkylphosphorane as a model of monohydric tri-*O*-alkylphosphorane

4 Isomerization via Dianionic Phosphorane: Effect of Intramolecular Hydrogen Bonding

Hydroxide ion-catalyzed cleavage of uridine 3′-phosphodiesters shows a nonlinear Brønsted graph, log k vs. pK_a^{lg}, with a convex break around the pK_a of the 2′-OH, which may be taken as compelling evidence for a stepwise mechanism via an at least marginally stable dianionic phosphorane intermediate (Lönnberg et al. 2004). Nevertheless, hydroxide ion-catalyzed migration has never been reported. DFT calculations lend support for the generally accepted assumption that dianionic phosphorane, if exists, cannot pseudorotate, since the barrier for the transfer of a negatively charged oxygen from an equatorial to apical position is too high, of the order of 10 kcal mol^{-1} (Lopez et al. 2005). Evidently this species is even too short-lived to pseudorotate via kinetically invisible temporary protonation of one of the non-bridging oxygens. Studies with a 5′-*C*-linked phosphonate analogues of ApA (**4a**, **b** in Fig. 6), however, suggest that stabilization of the dianionic phosphorane by intramolecular hydrogen bonding may allow isomerization. Usually replacement of the 5′-*O* with 5′-*C* has only a minor effect on the rate of isomerization. The phosphonate analogue of GpU (**5**), for example, exhibits a pH-rate profile similar to that of UpU, the phosphonate migration being twice as fast as the phosphate migration with UpU (Oivanen et al. 1993). In striking contrast, the pH-independent isomerization of **4a** and **4b** turns hydroxide ion catalyzed at pH 10 and levels again off to pH independence at pH > 12 ($T = 90$ °C). The isomerization, hence, seems to proceed via a dianionic intermediate at pH > 10 (Lönnberg et al. 2006).

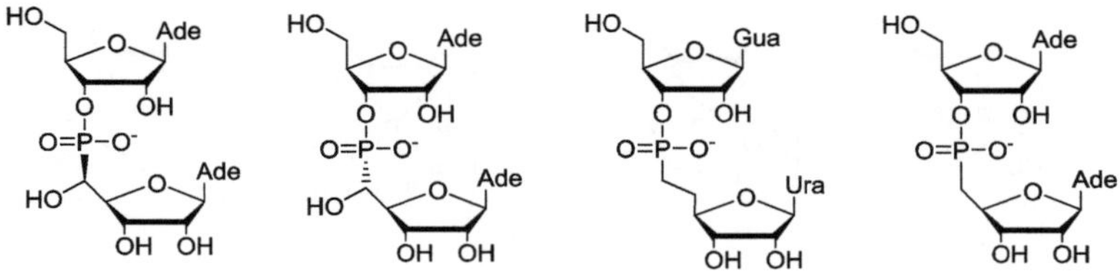

4a 4b 5 6

Fig. 6 Structures of some phosphonate analogues of dinucleoside-3′,5′-monophosphates

The 5′-OH of the 5′-*C*-linked adenosine evidently is responsible for this abnormal behavior of **4a**, **b**, since their 5′-deoxy analogue (**6**) does not exhibit a similar hydroxide ion-catalyzed reaction, but the isomerization rate remains pH independent at high alkalinities. Intramolecular hydrogen bonding between the 5′-OH and phosphorane oxyanion seems to stabilize the dianionic phosphorane to such an extent that pseudorotation takes place (Fig. 7a). Studies on isomerization of phosphorothioate triester **7** and its methylated analogue **8** (Fig. 8) lend some additional evidence for the importance of stabilization of anionic phosphorane intermediate (although in this case monoanionic one) by intramolecular hydrogen bonding (Lönnberg et al. 2007). Triester **7** undergoes hydroxide ion-catalyzed isomerization one order of magnitude faster than **8**, most likely due to stabilization of the intermediate (Fig. 7b).

On the basis of the foregoing conclusions, hydrogen-bond donors in the vicinity of a phosphodiester linkage expectedly affect the rate and equilibrium of the phosphate migration between the 2′- and 3′-OH. This subject has recently been studied by using an analogue of uridylyl-3′,5′-thymidine bearing two aminomethyl groups at C4′ of the uridine moiety (**9** in Fig. 9) as a model (Lain et al. 2013). At, pH 3–5, i.e., when the amino groups are fully protonated, the isomerization is pH independent, as with UpU. Interestingly, the migration from 3′-O to 2′-O is 50 times as fast as with UpU, whereas the reverse reaction is not enhanced compared to 2′,5′-UpU, and, hence, the thermodynamic equilibrium is strongly biased on the side of the 2′,5′-isomer (Fig. 10). Owing to shorter distance, the protonated 4′-aminomethyl substituents reduce the electron density at the 3′-phosphate inductively, electrostatically, and possibly by hydrogen bonding more markedly than at the 2′-phosphate. For the same reason, nucleophilicity of the 3′-OH is reduced compared to 2′-OH. Migration from 3′-O to 2′-O through initial formation of intermediate $\mathbf{I^1}$ is, hence, accelerated compared to UpU, while the influence on the rate of the reverse reaction by initial formation of $\mathbf{I^2}$ remains low. Equilibration of $\mathbf{I^1}$and $\mathbf{I^2}$ evidently is fast.

At pH 7–9, i.e., when the amino groups are largely deprotonated, the isomerization is approximately as fast as at pH 3–5, but the equilibrium is now biased on the side of the 3′,5′-isomer. The reactions most likely proceed *via* a minor zwitterionic tautomer having the attacking sugar hydroxy group ionized and one of the amino groups protonated. The attack of the 2′- or 3′-oxyanion on the monoanionic phosphodiester linkage gives a dianionic phosphorane that should be too unstable to

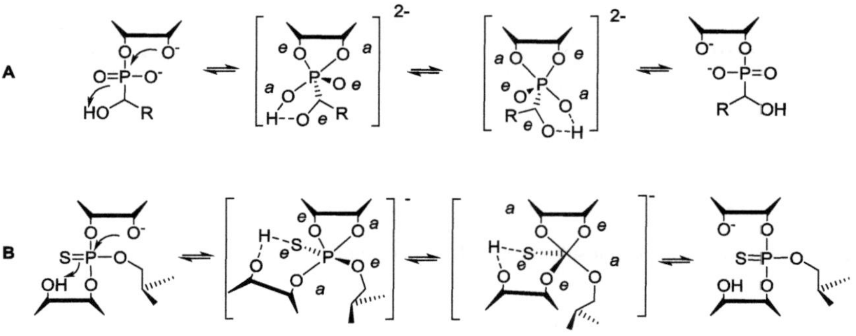

Fig. 7 Stabilization of dianionic (**a**) and monoanionic (**b**) phosphorane intermediates by intramolecular hydrogen bonding

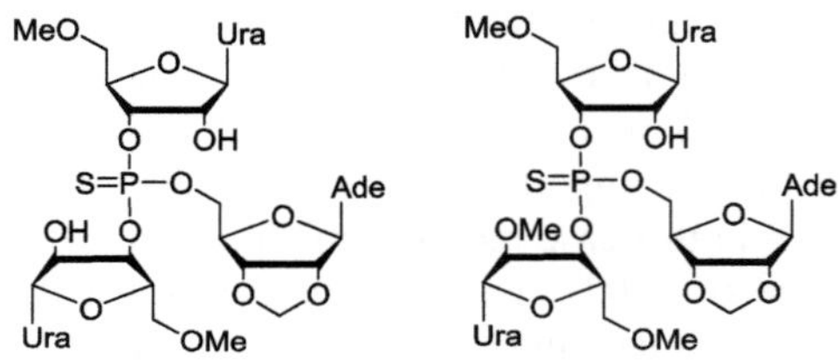

7 **8**

Fig. 8 Trinucleoside-3′,3′,5′-phosphorothioates with 2′-OH and 2′-OMe ribonucleoside as the departing 3′-linked nucleoside

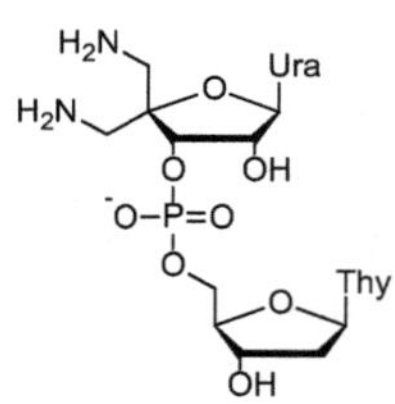

9

Fig. 9 A model compound for studies aimed at elucidating the nonbonding and H-bonding effects of amino groups on the isomerization of RNA phosphodiester linkages

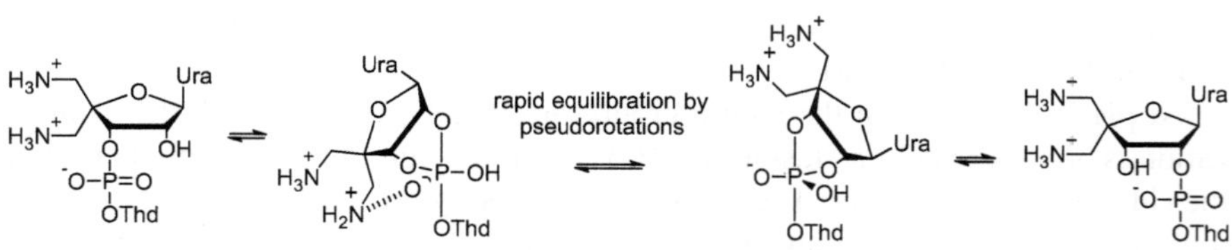

I[1] I[2]

Fig. 10 Isomerization of bis(aminomethyl) analogue **9** of UpT at pH 3–5

pseudorotate. Consistent with this expectation, the 3′,5′-isomer seems to undergo only cleavage to a 2′,3′-cyclic phosphate. The 2′,5′-isomer, however, is isomerized approximately as fast as it is cleaved to cyclic phosphate, suggesting that stabilization of the dianionic phosphorane by intramolecular hydrogen bonding plays a role.

5 Buffer-Catalyzed Isomerization

While the cleavage of RNA phosphodiester bonds is susceptible to general acid catalysis, the situation is less clear with the isomerization reaction. A modest general acid catalysis by imidazolium ion has been reported for the isomerization of UpU (Breslow et al. 1996; Beckmann et al. 1998). For example, upon increasing the concentration of an imidazole buffer ([B]/[BH$^+$] = 3:7, I = 1.0 M, T = 80.0 °C) from 0.1 to 0.7 M, the isomerization of UpU was accelerated by a factor of 3. By contrast, increasing the concentration of a somewhat more basic morpholine buffer even retarded this reaction, although the influence was again rather modest: on going from 0.2 to 0.8 M buffer ([B]/[BH$^+$] = 9:1), a 40 % rate retardation took place when the ionic strength was kept constant with NaCl (Breslow et al. 1996).

Studies with 3′-isopropyl and 2-ethoxyethyl phosphates of uridine have not provided convincing evidence for buffer-catalyzed isomerization, either (Mäki et al. 1999). When the ionic strength was adjusted to 1.0 M with NaCl and the concentration of imidazole buffer was varied from zero to 0.5 M, the rate constant for the mutual isomerization of the 2′- and 3′-isopropyl esters was increased by only 44 % and 16 % at the buffer ratio [ImH$^+$]/[Im] 4:1 and 1:4, respectively. With the 2-ethoxyethyl esters, the corresponding rate increments were even more modest, 23 and 0 %, and with uridine 2′/3′-methylphosphonates 25 and 9 %.

In summary, the buffer catalysis of the isomerization reaction is very modest at pH > 5, the buffer acid being the catalytically active species. The most plausible mechanism is sequential specific acid general base catalysis: a non-bridging phosphoryl oxygen is protonated in a rapid pre-equilibrium step and the buffer base then deprotonates the attacking hydroxy function concerted with formation of the pentacoordinated intermediate. Partially rate-limiting pseudorotation may explain the observed inefficiency of the buffer catalysis.

Isomerization in more acidic carboxylic acid buffers has been studied with uridine 2′/3′-methylphosphonates (Mäki et al. 1999). The rate of isomerization is proportional to [H$^+$][HA], where HA stands for the buffer acid. The reaction most likely is subject to sequential specific acid general acid catalysis. Consistent with these findings, the isomerization of UpU has been reported to be in acetic acid buffers second order in the concentration of AcOH without any contribution of AcO$^-$ (Breslow et al. 1996). Evidently rapid initial protonation of one of the non-bridging phosphoryl oxygens is followed by proton transfer from the buffer acid to the remaining non-bridging oxygen concerted with rate-limiting attack of the 2′/3′-hydroxy function on phosphorus. The general acid catalysis is now

Fig. 11 A dinuclear Zn^{2+} complex that efficiently cleaves RNA phosphodiester bonds and additionally accelerates the isomerization

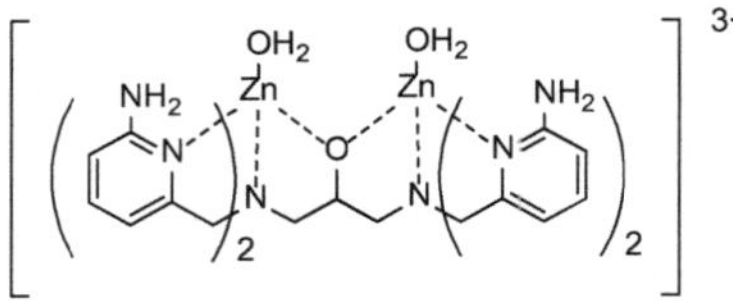

10

indisputable. At [HA] = [A$^-$] = 0.4 M (I = 1.0 M with NaCl), the rate of the buffer-catalyzed reaction is with acetic, formic, and chloroacetic acid 0.5-, 1.7-, and 5.3-fold, respectively, compared to the buffer-independent rate at the same pH (Mäki et al. 1999).

6 Metal Ion-Promoted Isomerization

Metal ions and their complexes do not usually catalyze the 2′/3′-isomerization of ribonucleoside 3′-phosphodiesters, only their cleavage. Dinuclear Zn^{2+} complex **10** (Fig. 11) that, apart from lanthanide complexes, evidently is the most efficient metal-based cleaving agent of RNA so far reported catalyzes also the isomerization, although less efficiently than the cleavage (Linjalahti et al. 2008).

A 500-fold rate acceleration has been observed at 1 mM concentration of **10** (Korhonen et al. 2012). It has been suggested that binding of **10** to both non-bridging oxygens of the phosphodiester linkage sufficiently stabilizes the dianionic phosphorane to allow it to pseudorotate. Migration of an alkylphosphate group is only moderately sensitive to the polar nature of the alkyl group, the β value being −0.5. The kinetic solvent deuterium isotope effect, k_H/k_D, is 1.5.

Another example of metal ion-catalyzed isomerization is offered by the phosphoromonothioate analogue of UpU. Zn^{2+} has been reported to accelerate at 5 mM concentration the isomerization of this compound by a factor of 6.4 (Ora et al. 1998). Evidently binding of thiophilic Zn^{2+} to the non-bridging sulfur stabilizes the monoanionic phosphorane intermediate, and, hence, the isomerization is enhanced, although much less than the cleavage which experiences a 400-fold acceleration.

7 Effect of Thio Substitution and Alkylation of a Non-bridging Oxygen on Isomerization

As mentioned above, mutual isomerization of 2′,5′- and 3′,5′-phosphoromonothiates proceeds by retention of configuration around phosphorus (Oivanen et al. 1995). According to DFT calculations with various solvation models, mono- or di-thio substitution does not markedly alter the pseudorotation barriers of neutral or monoanionic phosphoranes (Lopez et al. 2004). Desulfurization, however, offers an additional route for the breakdown of the phosphorane intermediate, and, hence, the migration of phosphorothioate and phosphorodithioate groups is under neutral and acidic conditions considerably slower than the phosphate migration, the thio effects, k_O/k_S, ranging from 4 to 9 at pH 2–8 (Oivanen et al. 1995; Ora et al. 2000). In more acidic solutions, the thio effect can be even much larger.

In contrast to thio substitution, alkylation of a non-bridging oxygen dramatically accelerates the phosphate migration. The half-life is around 40 h at pH 2 at 25 °C, but since the reaction is first order in hydroxide ion concentration at pH $>$ 3, the half-life is of the order of seconds under neutral conditions (Kosonen et al. 1994). The reaction is not markedly susceptible to buffer catalysis (Kosonen et al. 1998a). Even in cases where apparent buffer catalysis is observed, the buffer-dependent proportion of the observed rate constant is always less than 50 %. Increasing electronegativity of the alkyl group modestly accelerates the isomerization, the β value being -0.57 ± 0.06 (Kosonen et al. 1999). Under acidic conditions, hydrolysis to a diester competes with the migration. The β value for the polar influence of the esterified alcohol is -0.44 ± 0.05.

8 Epilogue

In summary, a sufficiently stable phosphorane intermediate is a prerequisite for mutual isomerization of ribonucleoside 2′- and 3′-phosphodiesters. Protonation is the most straightforward way to stabilize the phosphorane, and hence, acidic conditions favor isomerization compared to cleavage. Good hydrogen-bond donors in the vicinity of non-bridging oxygens may also provide sufficient stabilization, or even electrostatic stabilization may facilitate the isomerization. Metal ions, in spite of their positive charge, are able to promote isomerization only when the binding is exceptionally tight, a situation that may be best achieved by double Lewis acid activation with dinuclear metal ion complexes. The fact that protein enzymes or ribozymes do not enhance isomerization implies that upon formation of a pentacoordinated intermediate or transition state, the cleavage of the exocyclic P–O5′ bond becomes much more favored than the cleavage of the endocyclic P–O3′ bond. Either interaction with the catalyst prevents pseudorotation by freezing the O2′- and O5′-oxygens to an apical position, or proton transfer to the departing O5′ is much more efficient than to departing O3′. Finally, it is interesting to note that even most of small molecular catalysts studied so far enhance only the cleavage reaction (Lönnberg 2011a), which may refer to the importance of the leaving group protonation as a source of preferred exocyclic cleavage.

References

Ballinger P, Long FA (1960) Acid ionization constants of alcohols. II. Acidities of some substituted methanols and related compounds. J Am Chem Soc 82:795–798

Beckmann C, Kirby AJ, Kuusela S et al (1998) Mechanisms of catalysis by imidazole buffers of the hydrolysis and isomerization of RNA models. J Chem Soc Perkin Trans 2:573–581

Boero M, Terakura K, Tateno M (2002) Catalytic role of metal ion in the selection of competing reaction paths: a first principles molecular dynamics study of the enzymatic reaction in ribozyme. J Am Chem Soc 124:8949–8957

Breslow R, Dong SD, Webb Y et al (1996) Further studies on the buffer catalyzed cleavage and isomerization of uridylyluridine. Medium and ionic strength effect on catalysis by morpholine, imidazole, and acetate buffers help clarify the mechanisms involved and their relationship to the mechanism used by the enzyme ribonuclease and by a ribonuclease mimic. J Am Chem Soc 118:6588–6600

Buchwald SL, Pliura DH, Knowles JR (1984) Stereochemical evidence for pseudorotation in the reaction of a phosphoric monoester. J Am Chem Soc 106:4916–4922

Cochrane JC, Strobel SA (2008) Catalytic strategies of self-cleaving ribozymes. Acc Chem Res 41:1027–1035

Davies JE, Doltsinis NL, Kirby AJ et al (2002) Estimating pKa values for pentaoxyphosphoranes. J Am Chem Soc 124:6594–6599

Gerratana B, Sowa GA, Cleland WW (2000) Characterization of the transition-state structures and mechanisms for the isomerization and cleavage reactions of uridine 3′-*m*-nitrobenzyl phosphate. J Am Chem Soc 122:12615–12621

Guthrie JP (1977) Hydration and dehydration of phosphoric acid derivatives: free energies of formation of the pentacoordinate intermediates for phosphate ester hydrolysis and of monomeric metaphosphate. J Am Chem Soc 99:3991–4001

Järvinen P, Oivanen M, Lönnberg H (1991) Interconversion and phosphoester hydrolysis of 2′,5′- and 3′,5′-dinucleoside monophosphates: kinetics and mechanisms. J Org Chem 56:5396–5401

Kluger R, Covitz F, Dennis E et al (1969) pH-Product and pH-rate profiles for the hydrolysis of methyl ethylene phosphate. Rate-limiting pseudorotation. J Am Chem Soc 91:6066–6072

Korhonen H, Mikkola S, Williams NH (2012) The mechanism of cleavage and isomerization of RNA promoted by an efficient dinuclear Zn^{2+} complex. Chem Eur J 18:659–670

Kosonen M, Oivanen M, Lönnberg H (1994) Hydrolysis and interconversion of the dimethyl esters of 5′-O-Methyluridine 2′- and 3′-monophosphates: kinetics and mechanism. J Org Chem 59:3704–3708

Kosonen M, Hakala K, Lönnberg H (1998a) Hydrolysis and intramolecular transesterification of ribonucleoside3′-phosphotriesters: the effect of alkyl groups on the general and specific acid–base-catalyzed reactions of 5′-*O*-pivaloyluridin-3′-yl dialkyl phosphates. J Chem Soc Perkin Trans 2:663–670

Kosonen M, Yousefi-Salakdeh E, Strömberg R et al (1998b) pH- and buffer-independent cleavage and mutual isomerization of uridine 2′- and 3′-alkyl phosphodiesters: implications for the buffer catalyzed cleavage of RNA. J Chem Soc Perkin Trans 2:1589–1595

Kosonen M, Seppänen R, Wichmann O et al (1999) Hydrolysis and intramolecular transesterification of ribonucleoside3-phosphotriesters: comparison of structural effects in the reactions of asymmetric and symmetric dialkyl esters of 5′-*O*-pivaloyl-3′-uridylic acid. J Chem Soc Perkin Trans 2:2433–2439

Kuusela S, Lönnberg H (1994) Hydrolysis and isomerization of the internucleosidic phosphodiester bonds of polyuridylic acid: kinetics and mechanism. J Chem Soc Perkin Trans 2:2109–2113

Lain L, Lönnberg H, Lönnberg T (2013) Intramolecular participation of amino groups in the cleavage and isomerization of ribonucleoside 3-phosphodiesters: the role of stabilization of the phosphorane intermediate. Chem Eur J 19:12424–12434

Linjalahti H, Feng G, Mareque-Rivas JC et al (2008) Cleavage and isomerization of UpU promoted by dinuclear metal ion complexes. J Am Chem Soc 130:4232–4233
Lönnberg H (2011a) Cleavage of RNA phosphodiester bonds by small molecular entities: mechanistic insight. Org Biomol Chem 9:1687–1703
Lönnberg T (2011b) Understanding catalysis of phosphate-transfer reactions by the large ribozymes. Chem Eur J 17:7140–7153
Lönnberg H, Strömberg R, Williams A (2004) Compelling evidence for a stepwise mechanism of the alkaline cyclisation of uridine 3′-phosphate esters. Org Biomol Chem 2:2165–2167
Lönnberg T, Kralikova S, Rosenberg I et al (2006) Kinetics and mechanisms for the isomerization of internucleosidic 3′-O-P-CH_2-5′ and 3′-O-P-CH(OH)-5′ linkages to their 2′,5′-counterparts. Collect Czech Chem Commun 71:859–870
Lönnberg T, Ora M, Virtanen S et al (2007) Thio effects on the departure of the 3′-linked ribonucleoside from diribonucleoside 3′,3′-phosphorodithioate diesters and triribonucleoside3′,3′,5′-phosphoromonothioate triesters: implications for ribozyme catalysis. Chem Eur J 13:4614–4627
Lopez X, Schaefer M, Dejaegere A et al (2002) Theoretical evaluation of pKa in phosphoranes: implications for phosphate ester hydrolysis. J Am Chem Soc 124:5010–5018
Lopez CS, Faza ON, Gregersen BA et al (2004) Pseudorotation of natural and chemically modified biological phosphoranes: implications for RNA hydrolysis. Chem Phys Chem 5:1045–1049
Lopez CS, Faza ON, de Lera AR et al (2005) Pseudorotation barriers of biological oxyphosphoranes: a challenge for simulations of ribozyme catalysis. Chem Eur J 11:2081–2093
Lopez X, Dejaegere A, Leclerc F et al (2006) Nucleophilic attack on phosphate diesters: a density functional study of in-line reactivity in dianionic, monoanionic and neutral system. J Phys Chem B 110:11525–11539
Mäki E, Oivanen M, Poijärvi P et al (1999) Buffer-catalyzed interconversion of ribonucleoside 2′/3′-methylphosphonates and 2′/3′-alkylphosphates. J Chem Soc Perkin Trans 2:2493–2499
Mikkola S, Kosonen M, Lönnberg H (2002) Cleavage and isomerization of RNA phosphodiester bonds: nucleoside phosphotriesters and chimeric ribo/2′-*O*-methylribo oligonucleotides as tools for mechanistic studies. Curr Org Chem 6:523–538
Oivanen M, Lönnberg H (1991) Kinetics and mechanisms of the hydrolysis of the 2-chlorophenyl ester of uridine 3′-monophosphate: comparison with the methyl ester. Acta Chem Scand 45:968–971
Oivanen M, Mikhailov SN, Padyukova NS et al (1993) Kinetics of mutual isomerization of the phosphonate analogs of dinucleoside 2′,5′- and 3′,5′-monophosphates in aqueous solution. J Org Chem 58:1617–1619
Oivanen M, Ora M, Almer H et al (1995) Hydrolytic reactions of the diastereomeric phosphoromonothioate analogs of uridylyl(3′,5′)uridine: kinetics and mechanism for desulfurization, phosphoester hydrolysis and transesterification to the 2′,5′-isomers. J Org Chem 60:5620–5627
Oivanen M, Kuusela S, Lönnberg H (1998) Kinetics and mechanisms for the cleavage and isomerization of the phosphodiester bonds of RNA by Brønsted acids and bases. Chem Rev 98:961–990
Ora M, Peltomäki M, Oivanen M et al (1998) Metal-ion-promoted cleavage, isomerization, and desulfurization of the diastereomeric phosphoromonothioate analogues of uridylyl(3′,5′)uridine. J Org Chem 63:2939–2947
Ora M, Järvi J, Oivanen M et al (2000) Hydrolytic reactions of the phosphorodithioate analogue of uridylyl(3′,5′)uridine: kinetics and mechanisms for the cleavage, desulfurization, and isomerization of the internucleosidic linkage. J Org Chem 65:2651–2657
Perreault DM, Anslyn EV (1997) Unifying the current data on the mechanism of cleavage-transesterification of RNA. Angew Chem Int Ed Engl 36:432–450
Raines RT (1998) Ribonuclease A. Chem Rev 98:1045–1066

Range K, McGrath MJ, Lopez DM et al (2004) The structure and stability of biological metaphosphate, phosphate, and phosphorane compounds in the gas phase and in solution. J Am Chem Soc 126:1654–1665

Stephen BR (1960) Correlation of rates of intramolecular tunneling processes, with application to some group V compounds. J Chem Phys 32:933–938

Swamy KCK, Kumar NS (2006) New features in pentacoordinate phosphorus chemistry. Acc Chem Res 39:324–333

Uchimaru T, Uebayasi M, Hirose T et al (1996) Electrostatic interactions that determine the rate of pseudorotation processes in oxyphosphorane intermediates: implications with respect to the roles of metal ions in the enzymatic cleavage of RNA. J Org Chem 61:1599–1608

van Bochove MA, Swart M, Bickelhaupt FM (2006) Nucleophilic substitution at phosphorus ($S_N2@P$): disappearance and reappearance of reaction barriers. J Am Chem Soc 128:10738–10744

Westheimer FH (1968) Pseudo-rotation in the hydrolysis of phosphate esters. Acc Chem Res 1:70–78

Yang Y, Cui Q (2009) Does water relay play an important role in phosphoryl transfer reactions? Insights from theoretical study of a model reaction in water and *tert*-butanol. J Phys Chem B 113:4930–4939

Zhou D-M, Taira K (1998) The hydrolysis of RNA: from theoretical calculations to the hammerhead ribozyme-mediated cleavage of RNA. Chem Rev 98:991–1026

Effects of Ionic Liquid and Liposomes on the Structure, Stability, and Function of Nucleic Acids

Naoki Sugimoto

Contents

1 Introduction 58
2 Nucleic Acids with and on Liposomes 59
3 Nucleic Acids in a Hydrated Ionic Liquid 65
4 New Sensing System for DNA Sequences 68
5 Perspectives 70
References 71

Abstract The influences of homogenous molecular crowding on the structures and thermodynamics of various nucleic acid sequences in aqueous solution have been reported. Little is known, however, about the behaviors of nucleic acids in heterogeneous conditions such as in the presence of liposomes and ionic liquids (ILs). In this chapter, I discuss the effects of liposome and ionic liquid on the structure, stability, and function of nucleic acids. A novel DNA sequence sensor based on unique interactions between the hydrated IL of choline dihydrogen phosphate (choline dhp) and the groove of the DNA triplex is described. The instability of Hoogsteen base pairs at neutral pH has been a bottleneck in the progress of DNA nanobiotechnology, but the use of ILs will allow development of environment-friendly DNA sensors. The system in this study provides a promising example of large stabilization of Hoogsteen base pairs by choline dhp, allowing sensing of a target duplex at neutral pH.

Keywords DNA • RNA • Liposome • Ionic liquid • Biosensor

N. Sugimoto (✉)
Frontier Institute for Biomolecular Engineering Research (FIBER), Konan University, 7-1-20 Minatojima-Minamimachi, Chuo-ku, Kobe 650-0047, Japan

Faculty of Frontiers of Innovative Research in Science and Technology (FIRST), Konan University, 7-1-20 Minatojima-Minamimachi, Chuo-ku, Kobe 650-0047, Japan
e-mail: sugimoto@konan-u.ac.jp

V.A. Erdmann et al. (eds.), *Chemical Biology of Nucleic Acids*, RNA Technologies,
DOI 10.1007/978-3-642-54452-1_4,

1 Introduction

Nucleic acids play pivotal roles not only in genetic information storage but also in various biological functions. Therefore, the structure, stability, and function of DNA and RNA in living cells are of interest in the medical, pharmaceutical, biological, chemical, biophysical, and material sciences fields. The stability of nucleic acid structures depends on the balance of interactions, such as base pairing and electrostatic interactions, that are sensitive to the environment. In particular, hydration impacts the structure and function of nucleic acids (Miyoshi and Sugimoto 2008; Nakano et al. 2009), and the so-called molecular crowding conditions related to osmotic pressure in living cells influence nucleic acid structure and stability (Zimmerman and Minton 1993). Cell experiences a variety of environmental stresses that reduce or increase the amount of intracellular water. To overcome water stresses, organisms accumulate low molecular weight compounds called osmolytes or osmoprotectants that allow cellular processes to continue even under stress conditions (Yancey 2005; Gilles 1997). We have shown systematically and quantitatively that the neutral osmolyte polyethylene glycol (PEG) affects the thermodynamic stability of nucleic acid structures (Nakano et al. 2004; Miyoshi et al. 2006b, 2009). It is also known that Watson–Crick and Hoogsteen base pair formations are associated with hydration and dehydration, respectively (Miyoshi et al. 2009).

Both homogenous and heterogeneous conditions are found in environments where biomolecules, such as nucleic acids and proteins. Biomembranes of organelles separate certain biomolecules from the rest of the cellular environment (Lodish et al. 2008). In eukaryotic cells, the nuclear membrane is a highly organized double lipid bilayer membrane that surrounds the nuclear genome. Nuclear membranes serve not only as physical boundaries but also may be involved in chromatin function and gene expression (Reddy et al. 2008). These membranes are also physically linked with the endoplasmic reticulum and contain various proteins. As demonstrated by the fact that membrane proteins are over half of recently validated pharmaceutical targets (Katzen et al. 2009), regulation of their expression is extremely important.

DNAs can undergo structural transitions in response to environmental stimuli (Ohmichi et al. 2005; Miyoshi et al. 2006a, 2007; Liu and Lu 2007; Seeman 2003; Venkatesan et al. 2008; Taton et al. 2000). Researchers have taken advantage of the switching properties of DNA that depend on stimuli to develop DNA-based materials such as sensors (Ohmichi et al. 2005; Liu and Lu 2007), logic devices (Miyoshi et al. 2006a), circuits (Seeman 2003), and drugs (Balasubramanian et al. 2011). We have developed logic devices that respond to input molecules (e.g., pH and cations) that monitor the change in DNA structure from G-quadruplexes to duplexes, and these are of considerable interest in DNA computing (Miyoshi et al. 2006a, 2007).

In this chapter, I discuss the effect of liposomes on the structure and stability of DNA and RNA G-quadruplexes and the effect of ionic liquid (IL) on the stabilities of Watson–Crick and Hoogsteen base pairs. Quantitative thermodynamic analyses show that A-T base pairs are more stable than G-C base pairs in the hydrated IL because of specific interactions between DNA bases and choline. Moreover, a novel DNA sequence sensor based on unique interactions between a hydrated IL of choline dihydrogen phosphate (choline dhp) and the groove of the DNA triplex is described. The presence of ILs alters the relative stabilities of Hoogsteen and Watson–Crick base pairs at neutral pH. The instability of triplex structures at neutral pH has been a bottleneck in the progress of DNA nanobiotechnology. The system in this study provides a promising example of large stabilization of Hoogsteen base pairs by choline dhp, allowing sensing of a target duplex at neutral pH, that will allow development of environmentally friendly DNA sensors.

2 Nucleic Acids with and on Liposomes

The G-quadruplex structures that can be formed by DNA and RNA are attracting wide attention not only due to their biotechnological applications but also because they are important structures in biological reactions such as telomere elongation, transcription, and translation (Belotserkovskii et al. 2010; Bugaut and Balasubramanian 2012; Lam et al. 2013). The structural diversity of the G-quadruplexes is associated with the environmental conditions and sequence (Gray et al. 2012; Heddi and Phan 2011; Phan et al. 2006; Trajkovski et al. 2012; Mukundan and Phan 2013). In living cells, a variety of organelles, biomolecules, and other soluble and insoluble components result in a heterogeneous and crowded environment. To better predict the behavior of the noncanonical structures of nucleic acids such as the G-quadruplexes in cells, the structure and stability of the nucleic acids under conditions of molecular crowding induced by both non-interacting (Nakano et al. 2004; Miyoshi et al. 2006b, 2009; Pramanik et al. 2011a) and interacting cosolutes (Pramanik et al. 2011b) have been studied. The intracellular environment is also associated with micro-phases and micro-domains. We recently investigated the importance of heterogeneous confined media in cell nucleus using reverse micelles and found that a significant fraction of the telomeric region of genomic DNA adopts noncanonical tetraplex structures under these conditions (Pramanik et al. 2012).

The biomembranes of organelles separate certain biomolecules from the rest of the cellular environment (Lodish et al. 2008). The phospholipid bilayer is the basic building block of biomembranes. Phospholipid molecules generally consist of hydrophobic tails (two fatty acids) and a hydrophilic head group (Lodish et al. 2008). Liposomes are simple artificial systems that mimic biomembranes. Liposomes have been used to study the dynamics and structural features of many cellular processes (Kunimoto et al. 2003). For example, it has been demonstrated that a membrane surface can regulate transcriptional activity of genomic DNA by inducing conformational transitions (Tsuji and Yoshikawa 2010). Moreover, the efficiency of in vitro gene expression is enhanced in the presence of liposomes (Noireaux and Libchaber 2004). We recently showed that the formation of a

G-quadruplex in an open reading frame (ORF) suppresses translation of the mRNA into protein (Endoh et al. 2013). Although most proteins are translated on free ribosomes in the cytoplasm, translation of membrane proteins occurs on ribosomes bound to the membrane surface (membrane-bound ribosome) (Lodish et al. 2008). The structures of mRNAs that encode membrane proteins might be affected by this proximity to the membrane, and this might affect the efficiency of translation. Therefore, detailed studies of the G-quadruplex formation in the presence of a biomembrane-mimicking surface are required to elucidate how biomembranes affect the biological functions of these nucleic acids at the molecular level.

We investigated the effect of liposomes on the structure and stability of DNA and RNA G-quadruplexes. The sequences of the oligonucleotides studied are those of the human telomeric DNA sequence 5′-TAG$_n$TTAG$_n$TTAG$_n$TTAG$_n$-3′ (G_n: $n = 2$, 3, and 4) (Fig. 1a). We used zwitterionic 1-palmitoyl-2-oleoyl-*sn*-glycero-3-phosphocholine (POPC) to form liposomes, because POPC is a component of eukaryotic biomembranes (Fig. 1b). Since cholesterol-based anchoring molecules are also found in eukaryotic membranes and these molecules can be incorporated into lipid membranes without disrupting the bilayer structure, we modified the 5′ end of certain oligonucleotides with different numbers of the cholesteryl-triethyleneglycol (TEG) spacers (mcG$_n$: $m = 1$, 2, 3, and 4) to induce binding of the DNAs to the liposome surface (Fig. 1c). We also used a cholesterol-modified RNA oligonucleotide with a TEG spacer, 5′-cholesteryl-TEG-UAG$_3$UUAG$_3$UUAG$_3$UUAG$_3$-3′ (1crG$_3$).

The structure of oligonucleotides was studied by circular dichroism (CD) spectroscopy in both the absence and presence of liposome. We used a buffer containing 10 mM sodium phosphate (pH 7.0), 10 mM NaCl, and 1 mM Na_2EDTA throughout this work. The CD spectra of 20 μM of all G_n ($n = 2$, 3, and 4) oligonucleotides in the absence of liposome had a positive peak around 295 nm and a negative peak around 260 nm, indicating formation of a basket-type, anti-parallel G-quadruplex structure. The cholesterol-tagged 1cG$_n$ ($n = 2$, 3, and 4) oligonucleotides also formed antiparallel G-quadruplex structures in the absence of liposome. In the presence of 2 mM POPC, the CD signatures of G_ns and 1cG$_n$s were unaltered compared to those in the absence of liposome, indicating that the liposomes studied here induced no structural changes in the oligonucleotides.

We also evaluated the effect of liposomes on the G-quadruplex stability using CD spectroscopy. For 1cG$_3$, successive increase of temperature from 0 to 90 °C resulted in decrements and increments of the positive and negative peak intensities, respectively, with clear isodichroic points in the absence and presence of 2 mM POPC. This finding indicates that 1cG$_3$ undergoes a two-state transition under these conditions. The CD peak intensities of oligonucleotides with different numbers of guanines were recorded as a function of temperature to investigate the stability of G-quadruplexes in different conditions. The thermal stability for G_3 was the same in the absence of POPC and in the presence of 0.5, 1, and 2 mM POPC (the melting temperature, T_m, was 43.5 °C). In contrast, the thermal stability of the cholesterol-conjugated 1cG$_3$ decreased as the concentration of POPC increased (Table 1). Structures formed by 1cG$_2$ and 1cG$_4$ were also destabilized in the presence of

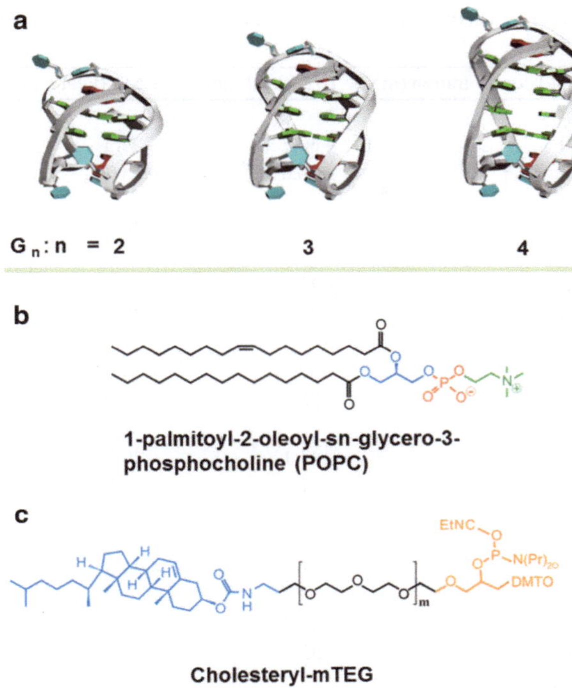

Fig. 1 (**a**) Schematic representation of G_n ($n = 2$, 3, and 4). (**b**) Chemical structure of POPC. (**c**) Chemical structure of cholesteryl-mTEG ($m = 1$, 2, 3, and 4)

POPC. These data suggest that the thermal stability of these DNA G-quadruplexes depends on location of the oligonucleotide with respect to the liposome surface.

To gain insight into how the lipid–water interface alters the stabilities of the G-quadruplexes, we extracted thermodynamic parameters from melting curves as described previously (Nakano et al. 2004; Miyoshi et al. 2006b, 2009; Pramanik et al. 2011a). The calculated thermodynamic parameters for the formation of G-quadruplexes in each POPC concentration are shown in Table 1. The value of $\Delta G°_{25}$ (free energy change at 25 °C) for $1cG_3$ increased from -3.1 to -1.2 kcal mol^{-1} as the POPC concentration increased from 0 to 2 mM. The values of $\Delta H°$ and $T\Delta S°$ for formation of the G-quadruplex also increased as the POPC concentration increased from 0 to 2 mM (Table 1). Thus, the $1cG_3$ G-quadruplex was destabilized by the liposome due to an unfavorable enthalpic contribution. Like the DNA G-quadruplex, the RNA G-quadruplex ($1crG_3$) was also destabilized in the presence of POPC liposome due to an unfavorable enthalpic contribution (Table 1). Structures formed by $1cG_2$ and $1cG_4$ were also destabilized in the presence of liposome, again due to unfavorable enthalpic contributions. We reported recently that choline ions bind to bases of single-stranded DNA, especially unpaired guanines, and decrease duplex stability with an unfavorable enthalpic contribution (Tateishi-Karimata and Sugimoto 2012). It is likely that the choline group at the lipid interface bound to the guanine bases of $1cG_n$ resulting in the decreased G-quadruplex stability.

Table 1 Thermodynamic parameters for formations of the G-quadruplex structures of mcG_n and $1crG_3$[a]

Lipid concentration (mM)	$\Delta H°$ (kcal mol^{-1})	$T\Delta S°$ (kcal mol^{-1})	$\Delta G°_{25}$ (kcal mol^{-1})	T_m (°C)
1cG₂				
0.0	−30.7	−29.3	−1.4	38.5
2.0	−26.6	−25.7	−0.9	33.0
1cG₃				
0.0	−40.0	−36.9	−3.1	50.0
0.5	−30.7	−28.4	−2.3	48.5
1.0	−23.3	−21.9	−1.4	43.9
2.0	−22.1	−20.9	−1.2	40.5
1cG₄				
0.0	−62.1	−54.7	−7.4	65.5
2.0	−38.1	−33.9	−4.2	59.5
2cG₃				
0.0	−43.1	−40.2	−2.9	46.0
2.0	−27.8	−26.4	−1.4	40.5
3cG₃				
0.0	−40.2	−37.9	−2.3	42.0
2.0	−31.7	−30.3	−1.4	38.0
4cG₃				
0.0	−43.3	−41.1	−2.2	40.0
2.0	−33.1	−31.8	−1.3	36.0
1crG₃				
0.0	−45.8	−40.6	−5.2	62.5
2.0	−23.5	−22.0	−1.5	54.5

[a]All experiments were carried out in a buffer containing 10 mM sodium phosphate (pH 7.0), 10 mM NaCl, and 1 mM Na_2EDTA with or without POPC. Melting temperatures were measured at a total strand concentration of 20 μM

To confirm the effect of the number of guanines on the G-quadruplex stabilities in the presence of liposome, we evaluated the extent of change of thermodynamic parameters ($\Delta X = (X$ of G-quadruplex formation in the presence of liposome) − (X of G-quadruplex formation in the absence of the liposome); X: $\Delta H°$, $T\Delta S°$, and $\Delta G°_{25}$). In the presence of 2.0 mM liposome, the $\Delta\Delta G°_{25}$ values were estimated to be +0.5, +1.9, and +3.2 kcal mol^{-1} for $1cG_2$, $1cG_3$, and $1cG_4$, respectively (Fig. 2). The destabilization was correlated with the number of guanines in the oligonucleotide. Thus, we suggest that the G-quadruplexes were destabilized at the liposome surface due to the binding of choline to the guanine bases. To determine the importance of proximity to the liposome surface on the G-quadruplex stability, we incorporated different numbers of TEG spacers between the oligonucleotide and the cholesterol group. Oligonucleotides $2cG_3$, $3cG_3$, and $4cG_3$ have the same G-quadruplex forming sequence but differ in the spacer length. In the presence of liposome, all were destabilized compared to stability in the absence of liposome. With increasing spacer length, however, the values of the net free energy change ($\Delta\Delta G°_{25}$) decreased (Fig. 3). This is due to the fact that additional spacer length moves the G-quadruplex structure away from the liposome

Fig. 2 (**a**) $\Delta\Delta H°$, (**b**) $\Delta(T\Delta S°)$, and (**c**) $\Delta\Delta G°_{25}$ in the presence of 2.0 mM POPC liposome for $1cG_2$, $1cG_3$, and $1cG_4$

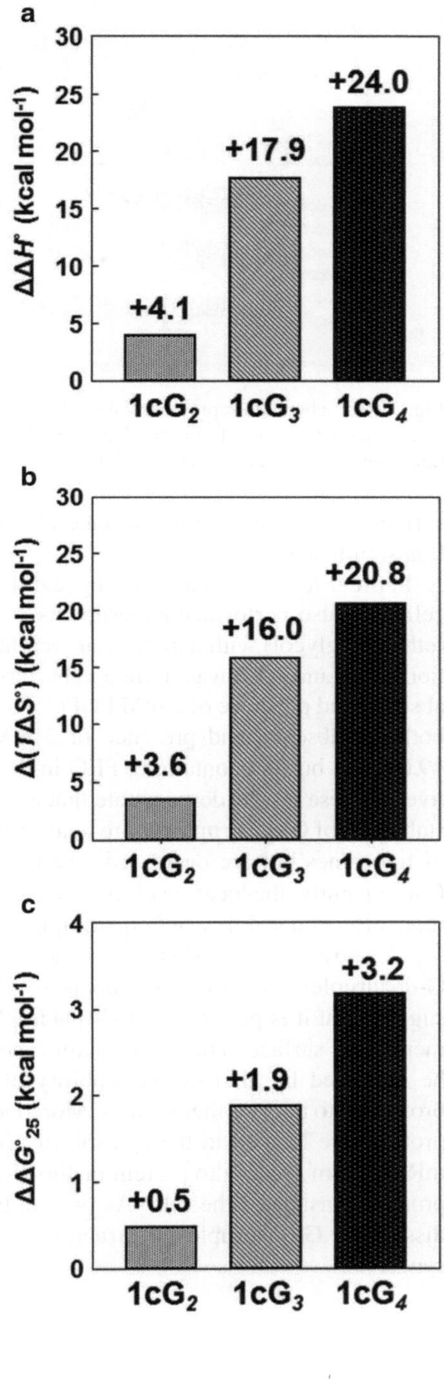

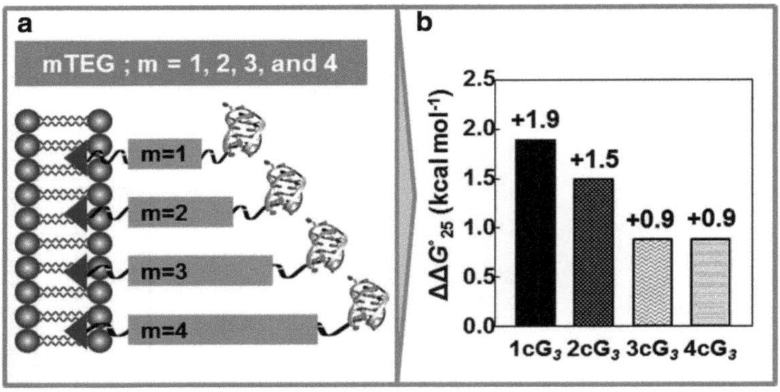

Fig. 3 (**a**) Schematic representation of the location of G-quadruplex structures at the liposome surface as a function of spacer (TEG) length. (**b**) $\Delta\Delta G°_{25}$ in the presence of 2.0 mM POPC liposome for $1cG_3$, $2cG_3$, $3cG_3$, and $4cG_3$

surface and the membrane surface effects on the G-quadruplex stability become less significant.

In order to mimic the crowding and the presence of biomembranes found inside cells, we also performed experiments in the presence of a crowding reagent poly (ethylene glycol) with a molecular weight of 200 (PEG 200). G-quadruplex formation by G_3 and $1cG_3$ was studied in buffers containing 1 M PEG 200 (20 wt%) in the absence and presence of 2 mM POPC. For G_3, T_ms of the transitions were 51.5 °C in both the absence and presence of 2 mM POPC. For $1cG_3$, T_ms were 52.5 and 49.0 °C in buffers containing PEG in the absence and presence of POPC, respectively. These results demonstrate that even in the presence of PEG 200, the thermal stabilities of G-quadruplexes are unaltered when they are in solution in the presence of liposomes but are decreased when they were located at the liposome surface. Consequently, the location of the nucleic acid with respect to membrane surface has more effect on stability of G-quadruplex structure than does the crowded condition.

We showed previously that histone H3 tail and nuclear confinement can stabilize G-quadruplex structures (Pramanik et al. 2011b, 2012). Based on our data, we suggest that it is possible that G-quadruplexes unfold when they are located at the membrane surface. Thus, chromatin function and levels of gene expression could be regulated by altering the stability of DNA G-quadruplexes by varying their proximity to nuclear membranes. Moreover, when the mRNAs encoding membrane proteins are located in the cytosol, the G-quadruplex structure would prevent the mRNA from binding to protein factors. In contrast, during the course of membrane protein translation, the mRNAs proximity to the membrane surface may serve to dissociate G-quadruplexes structures and increase the efficiency of protein synthesis.

3 Nucleic Acids in a Hydrated Ionic Liquid

Ionic liquids (ILs) are used as green solvents in the field of nanotechnology and nanobiotechnology as they provide favorable environments for a wide range of chemical reactions (Welton 1999; Seddon 2003; Fujita et al. 2007; Fujita and Ohno 2010). Biomolecules, especially nucleic acids, are not sufficiently stable in an aqueous solution for practical use in nano-machines (Fujita and Ohno 2012). Nucleic acids are susceptible to degradation by contaminating nucleases and to unproductive aggregation (Sasaki et al. 2007). Furthermore, aqueous solutions are useless in small-volume solution technologies, since water immediately vaporizes when used in small volumes under open air conditions or at high temperature (Armand et al. 2009). Consequently, atmospheric biosensing technology requires nonaqueous solutions. Certain remarkable features of ILs make these liquids attractive alternatives to water. For example, a representative IL, choline dihydrogen phosphate (choline dhp), dissolved in a small amount of water (~20 wt%) ensures long-term stability of biomolecules like DNA and has a negligible vapor pressure (Vijayaraghavan et al. 2010). Thus, ILs should be suitable reaction media in small-volume wet devices. Certain structural properties of biomolecules are drastically changed due to interactions with ILs, and it is possible to take advantage of this effect of ILs. For example, we recently proved that A-T base pairs are more stable than the G-C base pairs in choline dhp due to specific stabilizing interactions between choline and bases in both grooves (Tateishi-Karimata and Sugimoto 2012). This result suggests that one can take advantage of the unique interactions between choline dhp and DNA to develop DNA devices.

One of the most convenient methods for controlling DNA stability is through changes in cosolute and cosolvent conditions (Nakano et al. 2004; Miyoshi et al. 2005, 2006b). The formation of Hoogsteen base pairs is stabilized by the addition of cosolutes such as polyethylene glycol and glycerol, although these base pairs are still less stable than Watson–Crick base pairs in these cosolutes (Goobes and Minsky 2001; Miyoshi et al. 2006b, 2009). A triplex consisting of T-A*T base triplets (the dash and asterisk indicate the Watson–Crick and Hoogsteen base pairs, respectively) is stabilized in a deep-eutectic solvent containing a choline ion that is similar to an IL (Mamajanov et al. 2010). It has been also suggested that molecules that bind to grooves of A-T-rich duplexes will bind to the major and minor groove in a triplex, stabilizing the triplex structure (Arya 2011; Willis and Arya 2010; Arya et al. 2001; Xi et al. 2010). Therefore, we hypothesized that the presence of a hydrated IL would stabilize Hoogsteen base pairs in a triplex.

Here, we investigated the formation of DNA triplexes in the hydrated IL of choline dhp using the thermodynamic analysis. We designed and synthesized oligonucleotides to form three intermolecular DNA triplexes (Ts1, Ts2, and Ts3) with different G*C pairs formed by the third strand (Fig. 4). Three intermolecular double-stranded DNAs (Ds1, Ds2, and Ds3) of the same sequences of Watson–Crick base pairs as in Ts1, Ts2, and Ts3 were also prepared (Fig. 4). We measured normalized UV melting curves at 260 nm for 30 μM triplexes and also at 295 nm,

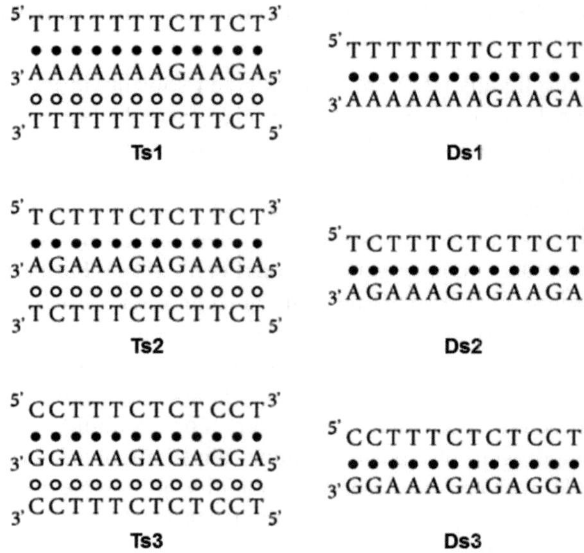

Fig. 4 Sequences and schematic structures of triplexes (Ts1, Ts2, and Ts3) and duplexes (Ds1, Ds2, and Ds3). *Filled* and *open circles* indicate Watson–Crick and Hoogsteen base pairs, respectively

because the dissociation of Hoogsteen base pairs can be specifically monitored at this wavelength (Miyoshi et al. 2009). When a melting curve for triplexes has two sigmoidal melting transitions, one corresponds to melting of the duplex and the other to dissociation of the triplex strand. If two transitions occur at 260 nm, the one with the same midpoint as that of the melting transition observed at 295 nm is assumed to correspond to the dissociation of Hoogsteen base pairs ($T_{m\text{-}H}$). A transition observed only at 260 nm corresponds to dissociation of Watson–Crick base pairs ($T_{m\text{-}W}$). If UV melting curves at 260 nm and 295 nm have single transitions with very similar T_ms, this indicates that the Hoogsteen and Watson–Crick base pairs dissociate at the same temperature ($T_{m\text{-}H}$&W).

The melting curves of Ts1 in 4 M NaCl showed single sigmoidal melting transitions with almost the same T_ms at 260 nm and 295 nm, suggesting that Watson–Crick and Hoogsteen base pairs dissociated at the same time. The melting curves for Ts2 in 4 M NaCl at 260 nm had two sigmoidal melting transitions. The lower temperature transition corresponded to the dissociation of Hoogsteen base pairs, because the T_m value was similar to that of the transition at 295 nm, whereas the higher temperature transition corresponded to the dissociation of Watson–Crick base pairs. Ts3 in 4 M NaCl solution showed one sigmoidal melting transition at 260 nm but no transitions at 295 nm, indicating that under these conditions third strand binding via Hoogsteen base pairing did not occur. Moreover, in the NaCl solution, the T_m values of Ds3 were almost identical to the T_m values for Ts3 in 4 M NaCl solution at 260 nm (Table 2). This indicates that the transitions for Ts3 and Ds3 correspond to dissociations of duplex. The $T_{m\text{-}W}$ values for Ds1, Ds2, and Ds3 increased with increasing G-C base pair content, because the G-C base pairs are more stable than the A-T base pairs in this solution. In contrast, the $T_{m\text{-}H}$ value of

Table 2 Melting temperatures at 260 nm for DNA triplexes and duplexes in 4 M NaCl and in 4 M choline dhp[a]

	T_m(°C)[b]						
	NaCl			Choline dhp		NaCl	Choline dhp
Sequence	$T_{m\text{-}H}$	$T_{m\text{-}W}$	$T_{m\text{-}H}$&W	$T_{m\text{-}H}$&W		$T_{m\text{-}W}$	$T_{m\text{-}W}$
Ts1 [2][c]			39.4	55.5	Ds1 [2][d]	43.8	51.4
Ts2 [4]	14.5	48.1		49.3	Ds2 [4]	47.5	43.2
Ts3 [6]	n.d.[e]	51.6		37.3	Ds3 [6]	51.2	40.2

[a]All experiments were carried out in buffers containing 50 mM Tris (pH 7.0), 1 mM Na_2EDTA, and either 4 M NaCl or 4 M choline dhp
[b]Total DNA strand concentration of triplexes and duplexes were 30 and 20 μM, respectively
[c]The number of G*C Hoogsteen base pairs is shown in parentheses
[d]The number of G-C Watson–Crick base pairs is shown in parentheses
[e]n.d. indicates that the T_m was too low to be determined

Ts3, which had the most G*C base pairs, was too low to be determined because G*C base pairs are unstable at neutral pH in aqueous solution (Plum et al. 1990).

In contrast, UV melting curves for Ts1, Ts2, and Ts3 in 4 M (80 wt%) choline dhp solution showed each clear sigmoidal melting transition with almost the same T_ms at 260 nm and 295 nm (Table 2). In the hydrated IL, therefore, Ts1, Ts2, and Ts3 formed the triplexes even at pH 7.0, and the stability of Hoogsteen base pairs was comparable to that of the Watson–Crick base pairs. The extent of stability differences in NaCl vs. choline dhp depended on G-C base pair content and likely resulted from specific interactions between DNA bases and choline ions (Tateishi-Karimata and Sugimoto 2012). The $T_{m\text{-}H}$&W values for Ts1, Ts2, and Ts3 in the choline dhp solution were higher than the $T_{m\text{-}W}$ values for Ds1, Ds2, and Ds3 (Table 2), respectively, although Ts3 did not form triplexes in the NaCl solution, suggesting that the duplex structures were also stabilized by the binding of third strand. These results show that choline dhp stabilized the formation of Hoogsteen base pairs independently of sequence, although the stability of Watson–Crick base pairs was destabilized or stabilized by choline dhp depending on G-C content (Tateishi-Karimata and Sugimoto 2012). The stabilization of Hoogsteen base pairs in choline dhp is very significant as compared to stabilization observed in previous studies with triplex-forming oligonucleotides with DNA backbone modifications (Chen et al. 2008; Mergny et al. 1991; Sugimoto et al. 2001). Moreover, we evaluated melting of Ts1, Ts2, and Ts3 in 4 M choline chloride at pH 7.0, which is not a hydrated IL. The T_m values of all the triplexes increased in the choline chloride solution relative to those in the NaCl solution, although the choline dhp stabilized the triplex structure more significantly than choline chloride. In both cases, the stabilization is due to an interaction between choline ions and DNA atoms.

4 New Sensing System for DNA Sequences

Systems for sensing particular DNA sequences, including single nucleotide polymorphisms (SNPs), are important in the fields of medicine and nanobiotechnology (Shen et al. 2013; Krieg et al. 2004; Erdogan et al. 2001; Gao et al. 2013; Gresham et al. 2006; Kawai et al. 2010). Traditional methods for sensing DNA sequences, like molecular beacons, DNA microarrays, and in situ hybridization, are based on the formation of A-T and G-C Watson–Crick base pairs between target sequence and probe DNA (Drummond et al. 2003; Lee et al. 2010; Schena et al. 1995; Francois et al. 1988). In general, duplexes of fully matched A-T and G-C base pairs are much more stable than those with mismatched base pairs; therefore, an optimally designed probe DNA can discriminate completely complementary target sequence from one that contains mismatches due to differences in thermodynamic stability. Certain mismatches, such as the G-T mismatch (Ebel et al. 1992; Leonard et al. 1990), are quite thermodynamically stable, and the stability depends on where the mismatch is located within the duplex (Leonard et al. 1990). In fact, about 30 % of SNPs result in G-T mismatches between target and probe (Gu et al. 1991; Mokry et al. 2010), and these mismatches cannot be readily discriminated from perfectly complementary strands using conventional probe technologies. The formation of thermodynamically stable mismatched duplexes is thus a bottleneck to development of sensing systems for DNA sequences.

In a triple helix, a third strand, also called the triplex-forming oligonucleotide, binds with sequence specificity via A*T and G*C base pairs in the major groove of a Watson–Crick base-paired DNA duplex (Chen et al. 2008; Mergny et al. 1991; Sugimoto et al. 2001). If there is a mutation of an A-T to a G-C base pair in the duplex, as is observed with an SNP, the Hoogsteen pair will be also mismatched. Thus, the triplex is a promising recognition motif for sequence-specific sensing of double-stranded DNA targets (Grossmann et al. 2007; Patterson et al. 2010). Use of triplex sensors simplifies the detection protocol as traditional methods of DNA sequence detection require the generation of single-stranded DNA prior to analysis. However, in an aqueous buffer at neutral pH, triplex formation is stable only at certain sequence motifs (Plum et al. 1990; Wu et al. 2001). In polypurine tracts, for example, found in HIV-1 provirus and triplet repeat diseases, cytosines in the third strand must be protonated at N3 ($pK_a = 4.5$) to form C-G*C$^+$ base triplets (Patterson et al. 2010). Therefore, formation of C-G*C$^+$ base triplets is pH dependent and disfavored at neutral pH, although the formation of T-A*T base triplets is not pH dependent (Sugimoto et al. 2001). Triplexes of mixed G and A sequences are not stable at neutral pH (Sugimoto et al. 2001). Moreover, even under acidic conditions, Hoogsteen base pairs are less stable than Watson–Crick pairs. Despite considerable efforts over the past decade, the instability of triplexes continues to limit their biological applications.

To take advantage of the large stabilization of triplex formation by choline dhp to develop the new sensing systems of the double-stranded DNA, we designed a 29-mer DNA molecular beacon (Fig. 5a) (Bonnet et al. 1999). This molecular

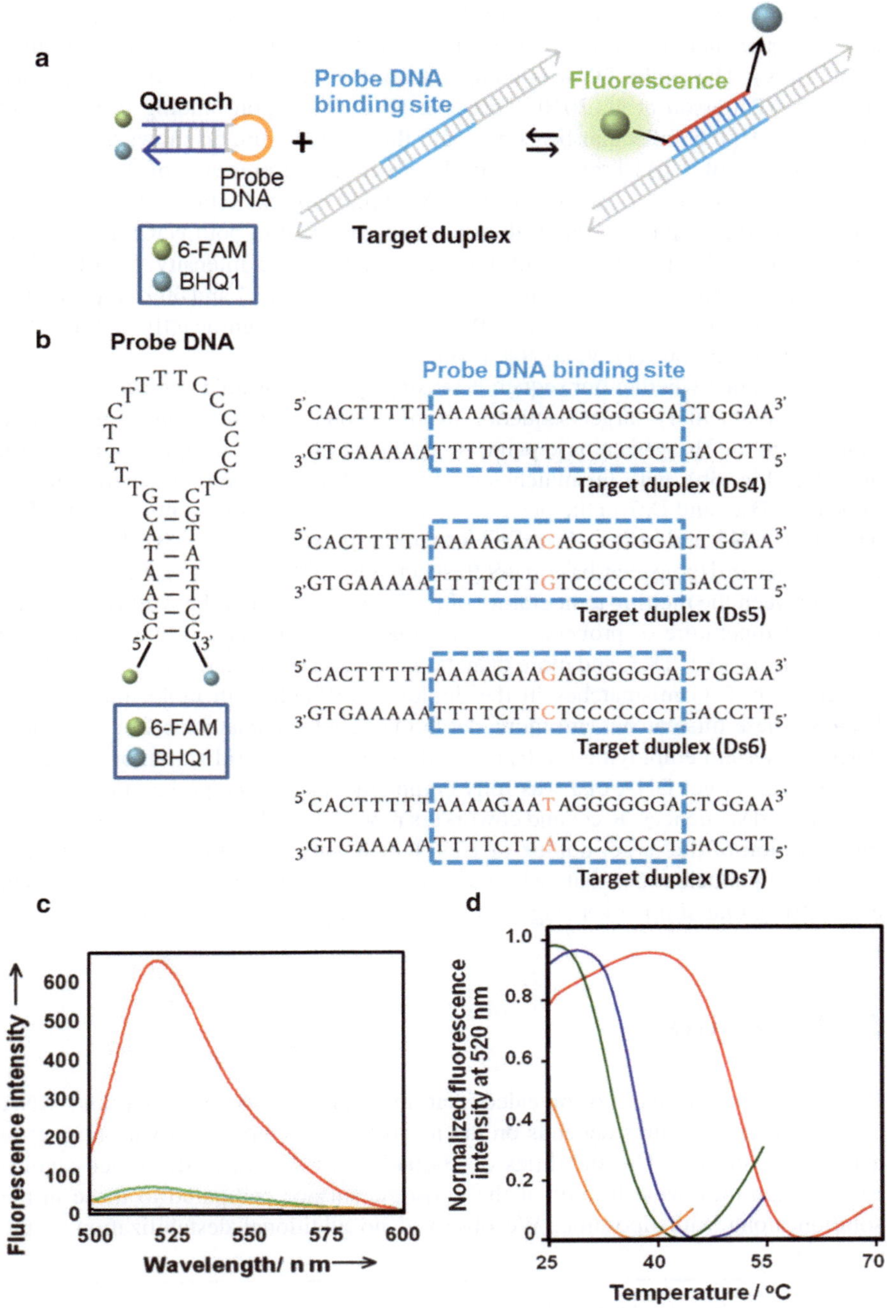

Fig. 5 (**a**) Schematic illustration of molecular beacon detection system. (**b**) Sequences of the DNAs used in this study. Probe DNA and target duplexes Ds4, Ds5, Ds6, and Ds7 form DNA triplexes. (**c**) Emission spectra recorded at 25 °C for 2 μM probe DNA in NaCl (*black*) and in choline dhp (*yellow*) and for 2 μM probe DNA and 2 μM Ds4 in NaCl (*green*) and in choline dhp

beacon can form a hairpin to bring a 5′ fluorophore (6-carboxylfluorescein, 6-FAM) and a 3′ quencher (Black Hole Quencher 1, BHQ1) into close proximity (Bonnet et al. 1999). The probe DNA is complementary to a conserved HIV-1 sequence (Fig. 5b) (Patterson et al. 2010; Turner et al. 2008). Upon binding to the target duplex (Ds4), the molecular beacon should fluoresce. Figure 5c shows the fluorescence spectra of 2 μM probe DNA in the absence and presence of 2 μM Ds4 at 25 °C. The fluorescence intensity of 6-FAM at 520 nm (excitation wavelength was 494 nm) in the absence of target duplex was very small in both NaCl and choline dhp solutions. In the presence of Ds4, the fluorescence intensity of 6-FAM at 520 nm in 4 M NaCl and 4 M choline dhp solutions were 45.7 and 609, respectively. Thus, the interaction of the probe DNA with Ds4 was significantly stabilized in choline dhp solution relative to NaCl solution at pH 7.

To determine whether our sequence sensing system was able to discriminate the fully complementary target sequence from a sequence with a single nucleotide mutation in the Hoogsteen base pairs, we designed three DNA duplexes, Ds5, Ds6, and Ds7 (Fig. 5b), with mismatches relative to Ds4. The thermal denaturation of Ds4, Ds5, Ds6, and Ds7 in the presence of the DNA probe was examined (Fig. 5d). The probe DNA and Ds5, Ds6, or Ds7 form the triplexes with the T*C, T*G, or T*T mismatches in Hoogsteen base pairs, respectively. The values of ΔT_m, the difference between the melting temperature of perfectly matched probe with Ds4 and the melting temperature of probe with the mismatched sequences, in the choline dhp solution were 13.1, 18.3, and more than 25 °C (Fig. 5d), respectively. In general, the detection of T–G mismatches in the duplex is difficult with molecular beacons because these mismatches are more stable than other mismatches. Our sequence-detection system employing the IL resulted in selective hybridization of molecular beacon with target DNA that was fully complementary to the probe but not to the mismatched sequences. Ricci and coworkers reported an electrochemical sensor for the detection of the target duplex of HIV-1 PCR amplicons that employs triplex-forming oligonucleotides that can detect their target at as low as 10 nM (Patterson et al. 2010). Our simpler sensing system has a comparable sensitivity.

5 Perspectives

Our experimental findings revealed that thermal stability of DNA and RNA G-quadruplex structures depends on the location of the molecules with respect to a liposome surface. The stabilities of G-quadruplex structures decreased significantly when they were located at the liposome surface compared to those in the solution along with liposome. We observed no additional destabilization in the

Fig. 5 (continued) (*red*) solutions. (**d**) Normalized fluorescence melting curves for DNA triplexes of probe DNA and Ds4 (*red*), Ds5 (*blue*), Ds6 (*green*), and Ds7 (*orange*) in solution containing 50 mM Tris (pH 7.0), 1 mM Na_2EDTA, and 4 M choline dhp

presence of PEG. Consequently, G-quadruplex stability appears to be regulated by membrane surfaces but not by crowding. Such destabilization of G-quadruplex structures originates from the unfavorable binding of the phospholipid head group to the guanine bases of DNA and RNA.

We have also demonstrated a direct and sequence-specific sensing system for double-stranded DNA enabled by a significant stabilization of Hoogsteen base pairs in choline dhp solution. Our new sensing system can specifically detect target duplex at concentrations as low as 30 nM. Hydrated ILs, such as choline dhp used in our study, are nonvolatile, green solvents and good media for nanotechnology. Although the sensitivity of our system must be improved in order for it to find utility in diagnostic applications, our system surpasses previously reported sensing systems in both sequence specificity and simplicity. The DNA duplex sensor described based on the unique interactions between ILs and DNAs brings development of environmentally friendly, low-volume DNA sensors closer to reality.

Acknowledgments The author is grateful to the colleagues named in the cited papers released from my laboratory and institute (FIBER), especially Drs. Tateishi-Karimata and Pramanik, and for technical assistance from Mses. Yamaguchi and Yoshioka. This work was supported in part by Grants-in-Aid for Scientific Research, the "Strategic Research Foundation at Private Universities" (2009–2014) from the Ministry of Education, Culture, Sports, Science and Technology (MEXT), Japan, and the Hirao Taro Foundation of the Konan University Association for Academic Research.

References

Armand M, Endres F, MacFarlane DR et al (2009) Ionic-liquid materials for the electrochemical challenges of the future. Nat Mater 8:621–629

Arya DP (2011) New approaches toward recognition of nucleic acid triple helices. Acc Chem Res 44:134–146

Arya DP, Coffee RL Jr, Willis B et al (2001) Aminoglycoside-nucleic acid interactions: remarkable stabilization of DNA and RNA triple helices by neomycin. J Am Chem Soc 123:5385–5395

Balasubramanian S, Hurley LH, Neidle S (2011) Targeting G-quadruplexes in gene promoters: a novel anticancer strategy? Nat Rev 10:261–275

Belotserkovskii BP, Liu R, Tornaletti S et al (2010) Mechanisms and implications of transcription blockage by guanine-rich DNA sequences. Proc Natl Acad Sci U S A 107:12816–12821

Bonnet G, Tyagi S, Libchaber A et al (1999) Thermodynamic basis of the enhanced specificity of structured DNA probes. Proc Natl Acad Sci U S A 96:6171–6176

Bugaut A, Balasubramanian S (2012) 5′-UTR RNA G-quadruplexes: translation regulation and targeting. Nucleic Acids Res 40:4727–4741

Chen H, Meena, McLaughlin LW (2008) A Janus-Wedge DNA triplex with A-W1-T and G-W2-C base triplets. J Am Chem Soc 130:13190–13191

Drummond TG, Hill MG, Barton JK (2003) Electrochemical DNA sensors. Nat Biotechnol 21:1192–1199

Ebel S, Lane AN, Brown T (1992) Very stable mismatch duplexes: structural and thermodynamic studies on tandem G.A mismatches in DNA. Biochemistry 31:12083–12086

Endoh T, Kawasaki Y, Sugimoto N (2013) Suppression of gene expression by G-quadruplexes in open reading frames depends on G-quadruplex stability. Angew Chem Int Ed 52:5522–5526

Erdogan F, Kirchner R, Mann W et al (2001) Detection of mitochondrial single nucleotide polymorphisms using a primer elongation reaction on oligonucleotide microarrays. Nucleic Acids Res 29:E36

Francois JC, Saison-Behmoaras T, Helene C (1988) Sequence-specific recognition of the major groove of DNA by oligodeoxynucleotides via triple helix formation. Footprinting studies. Nucleic Acids Res 16:11431–11440

Fujita K, Ohno H (2010) Enzymatic activity and thermal stability of metallo proteins in hydrated ionic liquids. Biopolymers 93:1093–1099

Fujita K, Ohno H (2012) Stable G-quadruplex structure in a hydrated ion pair: cholinium cation and dihydrogen phosphate anion. Chem Commun (Camb) 48:5751–5753

Fujita K, MacFarlane DR, Forsyth M et al (2007) Solubility and stability of cytochrome c in hydrated ionic liquids: effect of oxo acid residues and kosmotropicity. Biomacromolecules 8:2080–2086

Gao Z, Shen W, Deng H et al (2013) Detection of single-nucleotide polymorphisms based on the formation of an electron-transfer impeding layer on an electrode surface. Chem Commun (Camb) 49:370–372

Gilles R (1997) "Compensatory" organic osmolytes in high osmolarity and dehydration stresses: history and perspectives. Comp Biochem Physiol 117:279–290

Goobes R, Minsky A (2001) Thermodynamic aspects of triplex DNA formation in crowded environments. J Am Chem Soc 123:12692–12693

Gray RD, Buscaglia R, Chaires JB (2012) Populated intermediates in the thermal unfolding of the human telomeric quadruplex. J Am Chem Soc 134:16834–16844

Gresham D, Ruderfer DM, Pratt SC et al (2006) Genome-wide detection of polymorphisms at nucleotide resolution with a single DNA microarray. Science 311:1932–1936

Grossmann TN, Roglin L, Seitz O (2007) Triplex molecular beacons as modular probes for DNA detection. Angew Chem Int Ed 46:5223–5225

Gu XF, Lee JS, Delfau MH et al (1991) PCR detection of a G/T polymorphism at exon 10 of the porphobilinogen deaminase gene (PBG-D). Nucleic Acids Res 19:1966

Heddi B, Phan AT (2011) Structure of human telomeric DNA in crowded solution. J Am Chem Soc 133:9824–9833

Katzen F, Peterson TC, Kudlicki W (2009) Membrane protein expression: no cells required. Trends Biotechnol 27:455–460

Kawai K, Kodera H, Majima T (2010) Photocatalytic formation of I-I bonds using DNA which enables detection of single nucleotide polymorphisms. J Am Chem Soc 132:14216–14220

Krieg A, Laib S, Ruckstuhl T et al (2004) Fast detection of single nucleotide polymorphisms (SNPs) by primer elongation with monitoring of supercritical-angle fluorescence. Chembiochem 5:1680–1685

Kunimoto S, Murofushi W, Yamatsu I et al (2003) Cholesteryl glucoside-induced protection against gastric ulcer. Cell Struct Funct 28:179–186

Lam EY, Beraldi D, Tannahill D et al (2013) G-quadruplex structures are stable and detectable in human genomic DNA. Nat Commun 4:1796

Lee JB, Campolongo MJ, Kahn JS et al (2010) DNA-based nanostructures for molecular sensing. Nanoscale 2:188–197

Leonard GA, Booth ED, Brown T (1990) Structural and thermodynamic studies on the adenine. guanine mismatch in B-DNA. Nucleic Acids Res 18:5617–5623

Liu J, Lu Y (2007) Rational design of "turn-on" allosteric DNAzyme catalytic beacons for aqueous mercury ions with ultrahigh sensitivity and selectivity. Angew Chem Int Ed 46:7587–7590

Lodish H, Berk A, Kaiser C et al (2008) Molecular cell biology, 6th edn. W. H. Freeman, New York

Mamajanov I, Engelhart AE, Bean HD et al (2010) DNA and RNA in anhydrous media: duplex, triplex, and G-quadruplex secondary structures in a deep eutectic solvent. Angew Chem Int Ed 49:6310–6314

Mergny JL, Sun JS, Rougee M et al (1991) Sequence specificity in triple-helix formation: experimental and theoretical studies of the effect of mismatches on triplex stability. Biochemistry 30:9791–9798

Miyoshi D, Sugimoto N (2008) Molecular crowding effects on structure and stability of DNA. Biochimie 90:1040–1051

Miyoshi D, Karimata H, Sugimoto N (2005) Drastic effect of a single base difference between human and tetrahymena telomere sequences on their structures under molecular crowding conditions. Angew Chem Int Ed 44:3740–3744

Miyoshi D, Inoue M, Sugimoto N (2006a) DNA logic gates based on structural polymorphism of telomere DNA molecules responding to chemical input signals. Angew Chem Int Ed 45:7716–7719

Miyoshi D, Karimata H, Sugimoto N (2006b) Hydration regulates thermodynamics of G-quadruplex formation under molecular crowding conditions. J Am Chem Soc 128:7957–7963

Miyoshi D, Karimata H, Wang ZM et al (2007) Artificial G-wire switch with 2,2′-bipyridine units responsive to divalent metal ions. J Am Chem Soc 129:5919–5925

Miyoshi D, Nakamura K, Tateishi-Karimata H et al (2009) Hydration of Watson-Crick base pairs and dehydration of Hoogsteen base pairs inducing structural polymorphism under molecular crowding conditions. J Am Chem Soc 131:3522–3531

Mokry M, Feitsma H, Nijman IJ et al (2010) Accurate SNP and mutation detection by targeted custom microarray-based genomic enrichment of short-fragment sequencing libraries. Nucleic Acids Res 38:e116

Mukundan VT, Phan AT (2013) Bulges in G-quadruplexes: broadening the definition of G-quadruplex-forming sequences. J Am Chem Soc 135:5017–5028

Nakano S, Tateishi-Karimata H, Ohmichi T et al (2004) The effect of molecular crowding with nucleotide length and cosolute structure on DNA duplex stability. J Am Chem Soc 126:14330–14331

Nakano S, Tateishi-Karimata H, Kitagawa Y et al (2009) Facilitation of RNA enzyme activity in the molecular crowding media of cosolutes. J Am Chem Soc 131:16881–16888

Noireaux V, Libchaber A (2004) A vesicle bioreactor as a step toward an artificial cell assembly. Proc Natl Acad Sci U S A 101:17669–17674

Ohmichi T, Kawamoto Y, Wu P et al (2005) DNA-based biosensor for monitoring pH in vitro and in living cells. Biochemistry 44:7125–7130

Patterson A, Caprio F, Vallee-Belisle A et al (2010) Using triplex-forming oligonucleotide probes for the reagentless, electrochemical detection of double-stranded DNA. Anal Chem 82:9109–9115

Phan AT, Kuryavyi V, Patel DJ (2006) DNA architecture: from G to Z. Curr Opin Struct Biol 16:288–298

Plum GE, Park YW, Singleton SF et al (1990) Thermodynamic characterization of the stability and the melting behavior of a DNA triplex: a spectroscopic and calorimetric study. Proc Natl Acad Sci U S A 87:9436–9440

Pramanik S, Nagatoishi S, Saxena S et al (2011a) Conformational flexibility influences degree of hydration of nucleic acid hybrids. J Phys Chem B 115:13862–13872

Pramanik S, Nakamura K, Usui K et al (2011b) Thermodynamic stability of Hoogsteen and Watson-Crick base pairs in the presence of histone H3-mimicking peptide. Chem Commun (Camb) 47:2790–2792

Pramanik S, Nagatoishi S, Sugimoto N (2012) DNA tetraplex structure formation from human telomeric repeat motif (TTAGGG):(CCCTAA) in nanocavity water pools of reverse micelles. Chem Commun (Camb) 48:4815–4817

Reddy KL, Zullo JM, Bertolino E et al (2008) Transcriptional repression mediated by repositioning of genes to the nuclear lamina. Nature 452:243–247
Sasaki Y, Miyoshi D, Sugimoto N (2007) Regulation of DNA nucleases by molecular crowding. Nucleic Acids Res 35:4086–4093
Schena M, Shalon D, Davis RW et al (1995) Quantitative monitoring of gene expression patterns with a complementary DNA microarray. Science 270:467–470
Seddon KR (2003) Ionic liquids: a taste of the future. Nat Mater 2:363–365
Seeman NC (2003) DNA in a material world. Nature 421:427–431
Shen W, Deng H, Ren Y et al (2013) An electronic sensor array for label-free detection of single-nucleotide polymorphisms. Biosens Bioelectron 43:165–172
Sugimoto N, Wu P, Hara H et al (2001) pH and cation effects on the properties of parallel pyrimidine motif DNA triplexes. Biochemistry 40:9396–9405
Tateishi-Karimata H, Sugimoto N (2012) A-T base pairs are more stable than G-C base pairs in a hydrated ionic liquid. Angew Chem Int Ed 51:1416–1419
Taton TA, Mirkin CA, Letsinger RL (2000) Scanometric DNA array detection with nanoparticle probes. Science (New York) 289:1757–1760
Trajkovski M, da Silva MW, Plavec J (2012) Unique structural features of interconverting monomeric and dimeric G-quadruplexes adopted by a sequence from the intron of the N-myc gene. J Am Chem Soc 134:4132–4141
Tsuji A, Yoshikawa K (2010) ON-OFF switching of transcriptional activity of large DNA through a conformational transition in cooperation with phospholipid membrane. J Am Chem Soc 132:12464–12471
Turner KB, Brinson RG, Yi-Brunozzi HY et al (2008) Structural probing of the HIV-1 polypurine tract RNA:DNA hybrid using classic nucleic acid ligands. Nucleic Acids Res 36:2799–2810
Venkatesan N, Seo YJ, Kim BH (2008) Quencher-free molecular beacons: a new strategy in fluorescence based nucleic acid analysis. Chem Soc Rev 37:648–663
Vijayaraghavan R, Izgorodin A, Ganesh V et al (2010) Long-term structural and chemical stability of DNA in hydrated ionic liquids. Angew Chem Int Ed 49:1631–1633
Welton T (1999) Room-temperature ionic liquids. Solvents for synthesis and catalysis. Chem Rev 99:2071–2084
Willis B, Arya DP (2010) Triple recognition of B-DNA by a neomycin-Hoechst 33258-pyrene conjugate. Biochemistry 49:452–469
Wu P, Hara H, Kawamoto Y et al (2001) Effect of cytosine protonation and cation on thermodynamic properties of parallel DNA triplex family. Nucleic Acids Res Suppl 1:39–40
Xi H, Kumar S, Dosen-Micovic L et al (2010) Calorimetric and spectroscopic studies of aminoglycoside binding to AT-rich DNA triple helices. Biochimie 92:514–529
Yancey PH (2005) Organic osmolytes as compatible, metabolic and counteracting cytoprotectants in high osmolarity and other stresses. J Exp Biol 208:2819–2830
Zimmerman SB, Minton AP (1993) Macromolecular crowding: biochemical, biophysical, and physiological consequences. Annu Rev Biophys Biomol Struct 22:27–65

Oxidative Damage on RNA Nucleobases

Pascal A. Küpfer and Christian J. Leumann

Contents

1 Introduction 76
2 Definition of Reactive Oxygen and Nitrogen Species and Their Properties 77
3 Reactivity of ROS and RNS Versus Nucleic Acids 79
4 Oxidative Lesions in RNA 80
4.1 Base-Pairing Properties of Lesion Containing Duplexes 81
4.2 Mutagenicity of Oxidatively Damaged Nucleobases 84
5 Consequences of Oxidatively Damaged RNA 86
6 Fate of Oxidized RNA 87
7 Redox Chemistry of Oxidized Nucleosides 88
8 Conclusions and Perspective 89
References 89

Abstract Oxidatively damaged RNA has recently gathered more attention and has been closely related to different neurodegenerative diseases. The principles of oxidative stress and its influence on nucleic acids are reported. In contrast to DNA oxidative lesions of RNA have been scarcely described in the literature so far. These known stable RNA base modifications which arise under oxidative stress are reviewed here with regard to their biophysical properties and their potential mutagenicity. Furthermore the possible mechanisms of how cells deal with oxidized RNA are discussed. Posttranscriptional RNA modifications and the oxidation of RNA as an early event in several neurodegenerative diseases are not in the scope of this review.

Keywords RNA oxidation • Reactive oxygen species • Reverse transcription • 8-Oxo-guanosine • Modified RNA

P.A. Küpfer • C.J. Leumann (✉)
Department of Chemistry and Biochemistry, University of Bern, Freiestrasse 3, 3012 Bern, Switzerland
e-mail: kuepfer@dcb.unibe.ch; leumann@dcb.unibe.ch

V.A. Erdmann et al. (eds.), *Chemical Biology of Nucleic Acids*, RNA Technologies,
DOI 10.1007/978-3-642-54452-1_5,

1 Introduction

Oxidation of DNA has been known for over 40 years and considerable efforts have been made to describe the generation, metabolism, and repair of such lesions. The progress in this field is a recurrent topic in review articles (Cadet et al. 2010; Barciszewski et al. 1999). Unlike in DNA, RNA oxidative damage has not been in the focus of research until the past decade. This might be due to the assumed transient nature of RNA. Nonetheless it has been shown that most mRNAs have an average half-life of ~7 h in mouse cells and ~10 h in human cells, attributing the small percentages of mRNA with considerable lower half-lives to transcription factor mRNAs (Yang et al. 2003; Sharova et al. 2009). Moreover mRNA just accounts for a few percent of total RNA in cells, whereas the majority of RNA includes stable tRNA and rRNA which have considerably lower turnovers: the half-lives in vitro are estimated to be in the range of several hours to days for tRNA and several days for rRNA (Defoiche et al. 2009).

In the recent years the central dogma of RNA being "the messenger in the middle" between DNA and proteins had to be expanded by the disclosure of the activities of non-protein-coding RNAs (ncRNA): while over 85 % of our genome is transcribed to RNA, the known protein-coding gene exons only account for less than 3 % of the human genome (Hangauer et al. 2013). Such ncRNAs have been linked to diverse regulatory mechanisms (Nelson and Keller 2007).

Overall these data imply that oxidative damage to RNA could in principle have a strong and probably so far underestimated impact on different cell mechanisms. Recently, oxidatively damaged RNA has been closely linked to aging, a still growing number of neurodegenerative and other diseases (Poulsen et al. 2012). The data suggest that RNA oxidation is an early event in disease rather than the effect of cell decay (Shan et al. 2007; Chang et al. 2008). The progress in this field has been thoroughly reviewed elsewhere (Nunomura et al. 2009; Poulsen et al. 2012).

It is a long known fact that RNA is hydrolytically less stable than DNA due to the fact that the 2′-hydroxyl group of the ribose undergoes cyclophosphate formation with the 3′-*O*-phosphate under basic conditions which leads to strand cleavage. Compared to DNA, purified RNA is known to have a greater chemical oxidative stability (Thorp 2000). Nevertheless RNA is considered to be more prone to oxidation due to its higher abundance in cells, its extensive subcellular distribution (vicinity to mitochondria), and lesser association with protective proteins (Li et al. 2006). Indeed it has been found that in mitochondrial DNA there is a 16-fold higher abundance of 8-oxo-dG compared to nuclear DNA in rat liver. This has been explained by the spatial vicinity of the oxygen metabolism, the absence of histones in mitochondria, and the relatively inefficient DNA repair (Richter et al. 1988). Moreover it was shown recently that in H_2O_2-treated *E. coli* cells, levels of 8-oxo-rG in rRNA were the same or even higher than that for non-ribosomal RNA (Liu et al. 2012). According to the authors the results are consistent with the observation of rRNA oxidation in vivo and suggest that complex structures and association with proteins do not protect rRNA from oxidation.

It has been found that oxidative stress causes greater damage to RNA than DNA in different mammalian cells: the basic levels of 8-oxo-rG in RNA increase up to the fivefold after oxidative stress and are 10–25 times higher than for 8-oxo-dG in DNA (Fiala et al. 1989; Shen et al. 2000; Hofer et al. 2005, 2006).

2 Definition of Reactive Oxygen and Nitrogen Species and Their Properties

Reactive oxygen and nitrogen species are collective terms for either oxygen or nitrogen containing molecules, radical or nonradical, that act as oxidizing agents or that can be easily converted into radicals.

Reactive oxygen and nitrogen species (ROS, RNS) are either produced endogenously as natural by-products of the normal cellular metabolism of oxygen and nitrogen or from exogenous sources like tobacco, drugs, xenobiotics, and ionizing radiation. Under oxidative stress, which implies an imbalance between the generation of ROS and the system's capability of neutralizing ROS or subsequent damage, these reactive intermediates may damage a wide spectrum of vital cell compounds like nucleic acids, amino acids, and unsaturated fatty acids.

With regard to nucleic acid damage the main reactive oxygen species include hydrogen peroxide (H_2O_2), singlet oxygen ($^1O_2$), and the radical species superoxide ($^{\bullet}O_2^-$) and hydroxyl radical ($^{\bullet}OH$). Reactive nitrogen species include peroxynitrite ($ONOO^-$), nitric oxide ($^{\bullet}NO$), and nitrogen dioxide ($^{\bullet}NO_2$). Endogenous sources of ROS mainly include the mitochondrial respiratory chain, NADPH oxidases (NOX), and 5-lipoxygenases (LOX) among others which mainly form the primary ROS $^{\bullet}O_2^-$ (Nathan and Cunningham-Bussel 2013). Mitochondria are among the most reactive organelles in the cell and consume nearly 90 % of the total oxygen content (Bolisetty and Jaimes 2013). Overall seven sites of superoxide production are known (Brand 2010). In the mitochondrial respiratory chain up to 5 % of the electrons flowing through the electron transport chain can be abstracted mainly at the levels of complex I (NADH/CoQ reductase) and complex III (ubiquinol/cytochrome c reductase) to form $^{\bullet}O_2^-$ (Novo and Parola 2008; Lenaz 2012). Superoxide, having a quite short half-life and low reactivity, is readily converted to hydrogen peroxide by superoxide dismutase (SOD), as depicted in Fig. 1. The further inactivation of H_2O_2 is catalyzed by a series of enzymes including glutathione peroxidase (GPX), peroxiredoxins (PRX), thioredoxin, or catalase in peroxisomes. Compared to $^{\bullet}O_2^-$, hydrogen peroxide is electrophobic and can easily diffuse across membranes. It was found that the concentration of H_2O_2 in mitochondria is about 100 times higher compared to $^{\bullet}O_2^-$ (Cadenas and Davies 2000). In the presence of divalent metal ions H_2O_2 can further react in Fenton-type reactions to yield the highly reactive hydroxyl radical.

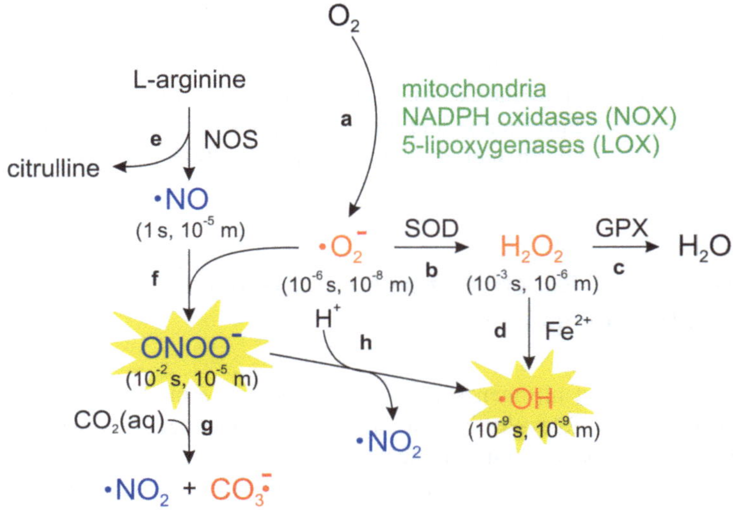

Fig. 1 Main ROS (*red*) and RNS (*blue*) formed in cells. Superoxide is generated mainly in mitochondria or by NOX and LOX (a) and is further dismuted by superoxide dismutase (SOD) to form hydrogen peroxide (b). H_2O_2 is then further metabolized to water by glutathione peroxidases (GPX) or other enzymes (c). Reactive hydroxyl radicals are produced by Fenton-type reactions of H_2O_2 in the presence of Fe^{2+} (d). Nitric oxide is formed in the conversion of L-arginine to citrulline by nitric oxide synthases (NOS) (e) and readily forms peroxynitrite upon reaction with superoxide (f). Reaction of peroxynitrite with carbon dioxide yields nitrite and carbonate radicals (g), while peroxynitrous acid generates hydroxyl and nitrite radicals (h). Magnitudes of half-life times and the respective mobilities of the different species are given in parentheses

Nitric oxide, which is formed through the conversion of L-arginine to citrulline by nitric oxide synthases (NOS), is a highly diffusible and relatively stable RNS. Upon reaction with $^{•}O_2^-$ it forms the highly reactive peroxynitrite ($ONOO^-$) at diffusion-controlled rates (Beckman et al. 1990; Radi et al. 2001). Peroxynitrite is a short-lived species and it can as well in its protonated form participate directly in oxidation reactions. Peroxynitrous acid itself immediately undergoes homolytic fission and forms a hydroxyl radical and nitrogen dioxide (Fig. 1, h) or a nitrate anion. Upon reaction with carbon dioxide, peroxynitrite forms [$ONOOCO_2^-$] that decays to give either a carbonate radical and nitrogen dioxide (Fig. 1, g) or nitrate and carbon dioxide (Radi et al. 2001).

The magnitudes of half-life times and the respective mobilities of ROS (Pryor 1986; Karuppanapandian et al. 2011) and RNS (Pacher et al. 2007; Szabó et al. 2007) are depicted in Fig. 1.

3 Reactivity of ROS and RNS Versus Nucleic Acids

While H_2O_2 and $^{\bullet}O_2^-$ have been shown to be unreactive themselves with respect to nucleic acids (Lesko et al. 1980; Brawn and Fridovich 1981), $^{\bullet}OH$ efficiently reacts at diffusion-controlled rates near the site of its formation with all four nucleobases and the (deoxy-)ribose sugar. Hydroxyl radical footprinting has been used to cleave nucleic acids to the nucleotide level and for revealing RNA folding pathways (Tullius and Dombroski 1985; Tullius and Greenbaum 2005). Mechanisms for the oxidation on the sugar have been widely explored for DNA (Pogozelski and Tullius 1998; Evans et al. 2004; Dedon 2007), whereas for RNA the existing data are limited to a few examples. Hydroxyl radicals can in principle react with sugar units by the abstraction of H-atoms of any of the five carbon atoms present. It was found that oxidation at the C1′-center of the ribose sugar is slower than for the 2′-deoxyribose sugar, presumably due to the polar effect of the 2′-hydroxyl group in RNA as the authors suggest (Neyhart et al. 1995). A similar behavior was found for the generation of C4′ radicals in model nucleotides: fragmentation of the C3′–O3′ bond by formation of a C3′ radical was shown to be far slower in the presence of a 2′-*O* substituent on the sugar moiety (Crich and Mo 1997). It has been further shown that iron-bleomycin (FeBLM), which efficiently cleaves DNA by the generation of C4′ radicals, is far less reactive but more sequence selective with RNA as the substrate (Hecht 1994). The data suggest, although not entirely conclusive, that the ribose unit seems to be more stable against hydroxyl radical attack compared to deoxyribose.

Upon reaction with RNA, peroxynitrite was found to form 8-oxo-rG and 8-nitro-rG (Masuda et al. 2002). Peroxynitrite seems to react with RNA according to different mechanisms: it was found that bicarbonate caused a dose-dependent increase in the formation of 8-nitro-rG whereas it produced no apparent effect on 8-oxo-rG formation. While 8-nitro-dG in DNA was found to depurinate rapidly (Yermilov et al. 1995), in RNA it was shown that 8-nitro-rG was far more stable (Masuda et al. 2002). Data on reactions of peroxynitrite with DNA suggest two different mechanisms: an increasing bicarbonate concentration seems to favor the formation of $^{\bullet}NO_2$ (Fig. 1, g) and a low pH seems to induce the formation of hydroxyl radicals which can lead to strand breaks (Fig. 1, h) (Yermilov et al. 1996). Moreover it has been shown that peroxynitrite reacts mainly with dG nucleosides to form 8-oxo-dG and 8-nitro-dG while dA, dC, and dT showed only minimal reactivity (Burney et al. 1999).

Peroxynitrite has also been shown to induce strand cleavage in DNA (Salgo et al. 1995; Kennedy et al. 1997; Szabó and Ohshima 1997; Niles et al. 2006). According data for RNA cleavage is still missing, but it can be suggested that RNA cleavage happens to a lower extent. Overall the data suggest that the main oxidative damage seems to be coming from H_2O_2-generated $^{\bullet}OH$ and peroxynitrite itself, as well from $ONOO^-$-generated $^{\bullet}OH$ because of the high diffusibility of H_2O_2 and peroxynitrite. Other forms of ROS hardly get into the vicinity of nucleic acids because of their limited half-life and their lacking moving capability.

Singlet oxygen, another strong oxidant generated by light (photooxidative stress), has so far only been shown to be able to oxidize rG (Schneider et al. 1993). Overall the data show that the reactivity of RNA under the influence of ROS and RNS is comparable to DNA with respect to base modifications. On the other hand, RNA might be less vulnerable against sugar backbone oxidation and subsequent strand cleavage compared to DNA.

4 Oxidative Lesions in RNA

The four main RNA oxidative lesions (Fig. 2) have been isolated from *Torula* yeast RNA (Yanagawa et al. 1990, 1992). Two other base lesions that are generated indirectly by oxidative stress through lipid peroxidation products are 1,N^6-ethenoadenosine (ε-rA) and 3,N^4-ethenocytidine (ε-rC) (El Ghissassi et al. 1995). All of these lesions have subsequently been synthesized and chemically introduced into RNA: 8-oxo-7,8-dihydroguanosine (Kim et al. 1998; Koga et al. 2013), 8-oxo-7,8-dihydroadenosine (Kim et al. 2002), 5-hydroxyuridine (Cui et al. 2009), 5-hydroxycytidine (Küpfer and Leumann 2011), and ε-rA and ε-rC (Srivastava et al. 1994; Calabretta and Leumann 2013).

These oxidative modifications described for RNA only represent a small fraction of the lesions found in DNA (Cadet et al. 2010). For pyrimidines potentially interesting oxidative lesions include 5,6-dihydroxy-5,6-dihydropyrimidines (pyrimidine glycols) and 5-hydroxymethyl- and 5-formylpyrimidines (hm^5C, hm^5U, f^5C and f^5U). While pyrimidine glycols are known precursors for 5-HO-rC and 5-HO-rU, 5-hydroxymethyl- and 5-formylpyrimidines could be potential oxidation products of 5-methylcytidine and 5-methyluridine, which are found in tRNA and mRNA (for m^5C) or rRNA and tmRNA (for m^5U) (Cantara et al. 2011). For 8-oxo-purines, the low redox potentials make them especially vulnerable to undergo further oxidation which leads to 4,6-diamino-5-formamidopyrimidine (FapyA) and 2,6-diamino-4-hydroxy-5-formamidopyrimidine nucleosides (FapyG) (Fig. 2) among a number of other derivatives for G including imidazolone, oxazolone, and spiroiminodihydantoin derivatives (Cadet et al. 2010). For FapyG it could be shown that the *N*-glycosidic bond of the isolated ribonucleoside is remarkably stable against anomerization and cleavage (Burgdorf and Carell 2002). Furthermore FapyG and FapyA were just recently quantified among other oxidative lesions in the analysis of RNA damage in Alzheimer's disease patients and it can therefore be assumed that they are also present in the pool of oxidatively damaged RNA lesions (Bradley-Whitman and Lovell 2013). Also it is known from DNA that lipid peroxidation products lead to a wide variety of substituted etheno-, propano-, as well as malondialdehyde adducts. So far there is no data available on the hybridization properties or possible mutagenicity of these lesions in RNA.

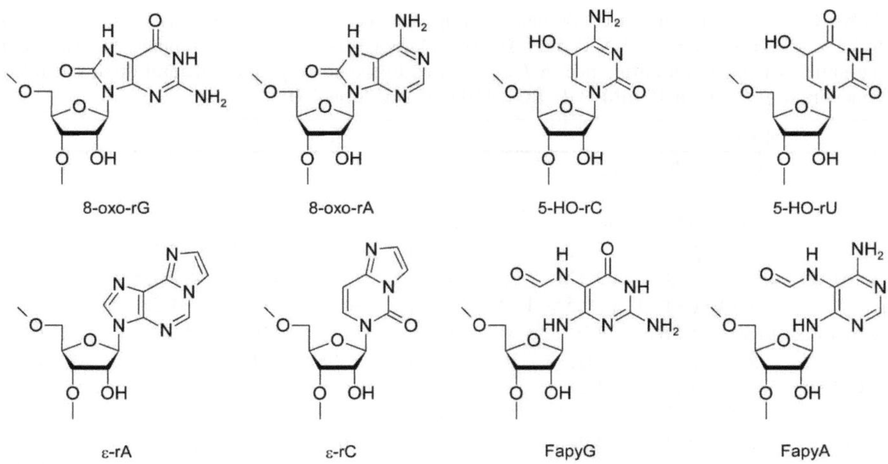

Fig. 2 Main stable oxidative lesions found in RNA: 8-oxo-7,8-dihydroguanosine (8-oxo-rG), 8-oxo-7,8-dihydroadenosine (8-oxo-rA), 5-hydroxycytidine (5-HO-rC), 5-hydroxyuridine (5-HO-rU), 1,N^6-ethenoadenosine (ε-rA), and 3,N^4-ethenocytidine (ε-rC). Further oxidation products of 8-oxo-purines: 2,6-diamino-4-hydroxy-5-formamidopyrimidine (FapyG) and 4,6-diamino-5-formamidopyrimidine (FapyA)

4.1 Base-Pairing Properties of Lesion Containing Duplexes

The influence of the oxidized bases on duplex stability was elucidated by UV melting experiments with RNA duplexes and RNA/DNA hybrids. The results of these T_m measurements in the same sequence context are summarized in Table 1.

5-Hydroxypyrimidines lead to no notable destabilization in neither of the duplexes compared to the natural cases and all formed base pairs are in accordance with Watson and Crick. An important fact to note is that in the case of 5-hydroxycytidine a biphasic melting behavior was observed (Küpfer and Leumann 2011).[1]

8-Substituted purine nucleosides have long been known to adopt the *syn* conformation (Uesugi and Ikehara 1977). For 8-oxo-rG it has been shown that it exists mainly in the neutral 6,8-diketo form with a pK_a of 8.5 for the 6-enolate-8-keto form and a pK_a of 11.2 for the 6,8-dienolate form (Cho et al. 1990). In contrast to 8-oxo-rG, the pK_a of 8.7 for the 8-enolate form of 8-oxo-rA is quite low (Cho and Evans 1991). The different tautomeric forms of the oxidized nucleosides with their respective pK_a values are depicted in Fig. 3.

[1] The melting curves for the matched case 5-HO-rC/G in RNA duplexes and RNA/DNA heteroduplexes show two distinct changes in hyperchromicity: the higher T_m is close to the matched case whereas the lower T_m is close to the mismatched case. This phenomenon is not yet fully understood and still under investigation.

Table 1 T_m data (°C, 260 nm) of duplexes of oligoribonucleotide 5′-AUGCU**X**AGUCGA-3′ with 5′-UCGACU**Y**AGCAU-3′ (RNA) or 5′-TCGACT**Y**AGCAT-3′ (DNA) as complement; values in parentheses indicate the difference in T_m compared to the corresponding natural duplex; duplex concentration: 2 μM in 10 mM NaH_2PO_4, 150 mM NaCl, pH 7.0

	RNA				DNA			
Y=	rA	rC	rG	rU	dA	dC	dG	dT
X=								
8-oxo-rA	41.8 (−3.1)	37.8 (−8.8)	46.5 (−0.2)	50.5 (−7.1)	25.8 (−3.8)	20.8 (−13.8)	25.8 (−4.1)	35.2 (−10.1)
5-OH-rC	47.4 (−0.4)	44.0 (−1.6)	65.2, 47.7 (−0.5, −17.5)	44.6 (−1.4)	27.2 (−1.2)	25.4 (−1.3)	48.6, 25.9 (−1.3, −24)	25.4 (−1.2)
8-oxo-rG	54.9 (+9.9)	55.6 (−8.4)	52.0 (+5)	43.7 (−9.3)	37.6 (+6.6)	40.0 (−11)	31.3 (0)	29.0 (−9.6)
5-OH-rU	58.8 (−0.4)	43.8 (−0.4)	54.2 (−0.4)	45.9 (−0.4)	42.6 (+0.3)	23.6 (0)	32.6 (+0.6)	27.6 (−0.6)
ε-rA	43.9 (−1)	40.6 (−6)	44.9 (−1.8)	42.0 (−15.6)	28.6 (−1)	23.6 (−11)	29.3 (−0.6)	25.4 (−19.9)
ε-rC	43.9 (−3)	39.3 (−3.7)	46.6 (−18)	41.9 (−2)	25.2 (−2)	20.8 (−3.7)	25.3 (−23.3)	24.3 (−2.9)

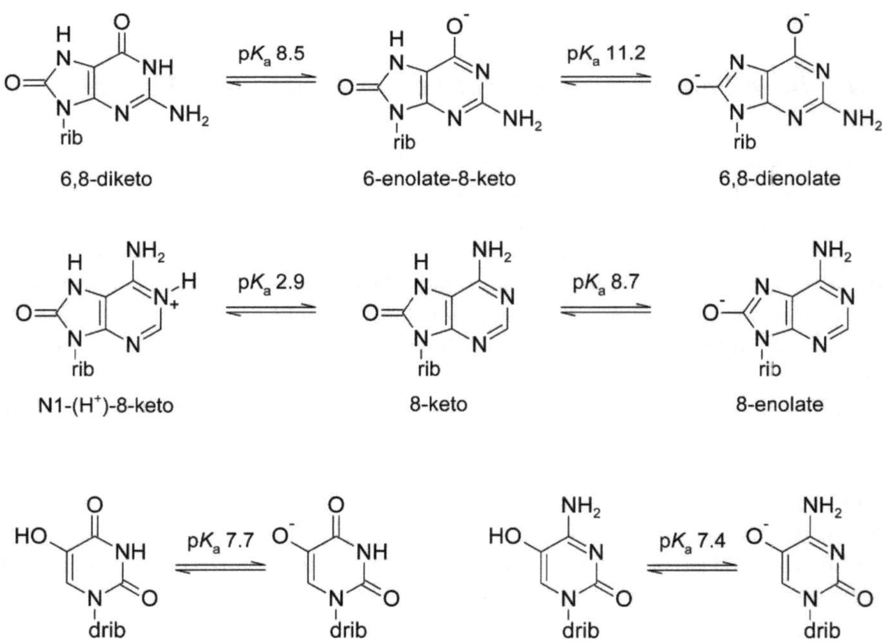

Fig. 3 pK_a values for the different tautomeric forms of 8-oxo-rG, 8-oxo-rA, 5-HO-rU, and 5-HO-rC. Values for the 8-oxo-purines are given for the ribonucleosides and values for the 5-hydroxypyrimidines are given for the deoxynucleosides

For 8-oxoguanosine the duplexes are destabilized in the case where 8-oxo-rG pairs to pyrimidine bases. When 8-oxo-rG is paired against purine bases the duplexes gain in stability compared to the natural mismatch duplexes: for 8-oxo-rG/dG the duplex has the same stability as the mismatch duplex with rG/dG; for 8-oxo-rG pairing against rG, rA, and dA the duplexes gain considerably in stability. Especially in the case of 8-oxo-rG/rA the duplex stability equals almost the stability of the 8-oxo-rG/rC duplex. These results have just recently been confirmed showing that the T_m of 8-oxorG/rA is even slightly higher than for the natural case (Koga et al. 2013).[2] This remarkable stabilization for 8-oxo-rG pairing rG and rA could be explained by data measured previously in DNA: it has been found that the Hoogsteen face of 8-oxo-dG pairs with dA forming a *syn*-8-oxo-dG/*anti*-dA base pair (Kouchakdjian et al. 1991; Gannett and Sura 1993). Such a *syn*-8-oxo-rG/*anti*-rA base pair would be expected to add to the overall duplex stability compared to an *anti*-rG/*anti*-rA mismatch. Furthermore it has been shown that rG can pair with the Hoogsteen face of rA forming an *anti*-rG/*syn*-rA base pair (Pan et al. 1999). It has been suggested that in a similar manner this could also give rise to an *anti*-8-oxo-rG/*syn*-rG base pair (Koga et al. 2013). The data suggest that 8-oxo-rG easily switches between *anti* and *syn* conformations to comfort the opposing base (Fig. 4).

For 8-oxoadenosine moderate to strong destabilization of the duplexes was found with the strongest decrease for the pairing to pyrimidine bases in both RNA/RNA and RNA/DNA duplexes. In DNA 8-oxo-dA was shown to pair dT according to Watson and Crick (Guschlbauer et al. 1991). In analogy to 8-oxo-dG it has also been found that 8-oxo-dA can pair to dG via its Hoogsteen face forming a *syn*-8-oxo-dA/*anti*-dG base pair (Leonard et al. 1992). Later on base pairs of *syn*-8-oxo-dA with dC and dA have been postulated (Kamiya et al. 1995). Nevertheless 8-oxoadenosine still discriminates the binding partners in favor of the matched base.

These findings suggest that all oxidized bases are capable of forming stable base pairs but with a lower discrimination of mismatches. This might subsequently lead to the synthesis of modified proteins via the interaction of oxidatively damaged mRNA with tRNA as was suggested earlier (Kong and Lin 2010).

Etheno-adenosine and -cytidine lesions show major destabilization in RNA duplexes and RNA/DNA heteroduplexes because of the loss of their ability to form standard Watson–Crick hydrogen bonds. Duplexes formed with a G opposite the lesion are slightly more stable than the others, most likely because of stronger base stacking interactions of the etheno bases (Calabretta and Leumann 2013).

[2] The published findings confirm our earlier observations that the previously used 6*O*,7*N*-bis (dimethylcarbamyl) protected phosphoramidite of 8-oxo-rG is not fully deprotectable once incorporated into RNA. The authors here used a 6*O*,7*N*-bis(diphenylcarbamyl)-protected while we used an 6*O*,7*N*-unprotected phosphoramidite of 8-oxo-rG (unpublished results). Both phosphoramidites were fully deprotectable as confirmed by mass spectrometry. The T_m values measured by Koga et al. correspond nicely to our findings (Table 1).

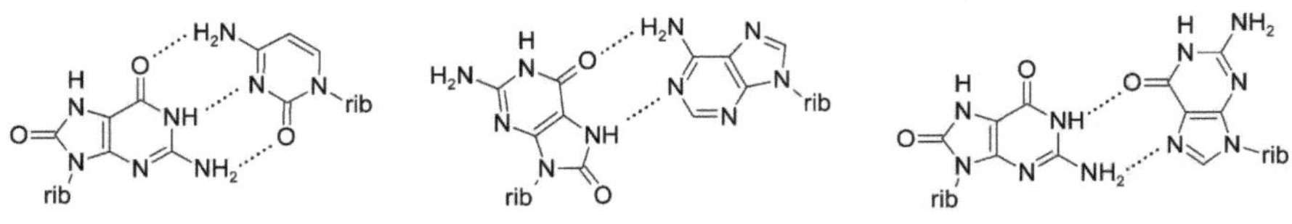

Fig. 4 Base-pairing modes for 8-oxo-rG: canonical 8-oxo-rG/rC base pair according to Watson and Crick (*left*), 8-oxo-rG/A base pair where 8-oxo-rG pairs with its Hoogsteen side (*middle*) and the 8-oxo-rG/G base pair where the Watson–Crick side of 8-oxo-rG pairs to the Hoogsteen side of rG

4.2 Mutagenicity of Oxidatively Damaged Nucleobases

The potential mutagenicity of oxidized RNA bases was elucidated by reverse transcription assays. In these assays a DNA primer is elongated on a modified RNA template using reverse transcriptases and the efficiency of the incorporation of natural 2′-*O*-deoxynucleoside monophosphates (dNMP) is determined.

4.2.1 5-Hydroxypyrimidines

For a 5-HO-rU containing RNA template it was found that reverse transcriptases[3] incorporated preferentially dAMP opposite the lesion. Notable misincorporation of dGMP was dependent on the reverse transcriptase used: MMLV-RT incorporated up to one-third, Superscript™II-RT up to one-fourth, and AMV only trace amounts of dGMP compared to incorporated dAMP (Cui et al. 2009).

In the case of a 5-HO-rC containing template there was found substantial misincorporation of dAMP besides the expected incorporation of dGMP: HIV1-RT incorporated dAMP almost as efficiently as dGMP while for the AMV-RT and MMLV-RT this process is two- to threefold less efficient. Interestingly, HIV1-RT also incorporated dTMP with about half the efficiency of dGMP (Küpfer and Leumann 2011). In DNA 5-HO-dU and 5-HO-dC have also been tested for potential mutagenicity during DNA replication and it was found that 5-HO-dU and 5-HO-dC cause C/T transition mutations (Kreutzer and Essigmann 1998; Suen et al. 1999). The mutagenicity has been associated with tautomeric variability of the oxidized base: while the amino form of 5-hydroxycytidine pairs to guanosine, the imino form would be able to pair to adenosine (Suen et al. 1999; La Francois et al. 2000).

---

[3] AMV-RT: avian myeloblastosis virus reverse transcriptase; MMLV-RT: moloney murine leukemia virus reverse transcriptase; HIV1-RT: human immunodeficiency virus type 1 reverse transcriptase; RAV2-RT: Rous-associated virus-2 reverse transcriptase.

4.2.2 8-Oxo-7,8-Dihydropurines

Reverse transcription assays with 8-oxo-rG containing templates were performed using MMLV-RT and AMV-RT (Kim et al. 1999) and HIV-RT and RAV2-RT (Kim et al. 2004). While MMLV-RT seemed to incorporate preferentially dTMP opposite 8-oxo-rG, AMV inserted the correct dCMP residue. Using HIV-1 RT and RAV2-RT, almost no dAMP and dGMP were incorporated opposite 8-oxo-rG. - HIV-1 RT incorporated almost exclusively dCMP opposite 8-oxo-rG. Using RAV2-RT, dCMP was only incorporated 1.5-fold of dTMP opposite 8-oxo-rG. These results have to be looked at carefully since there is evidence that the authors might have used the not fully deprotected 8-oxo-rG-DMC derivative (see Footnote 2). It was recently shown that duplexes of 8-oxo-rG-DMC containing RNA with DNA have considerably lower T_m values compared to 8-oxo-rG containing heteroduplexes and clearly discriminate in favor of dC opposite 8-oxo-rG-DMC (Koga et al. 2013).

Similar assays with 8-oxo-rA containing RNA were performed using AMV-RT, MMLV-RT, and RAV2-RT (Kim et al. 2002). All reverse transcriptases incorporated preferentially dTMP opposite the lesion. AMV-RT incorporated also dGMP and dAMP to a three- and fourfold lesser extent and MMLV- and RAV2-RT incorporated dGMP to a fivefold lesser extent compared to dTMP.

In DNA 8-oxo-dG is known to lead to G/T transversions in vitro (Wood et al. 1990) and in vivo (Moriya et al. 1991; Cheng et al. 1992). DNA polymerases incorporate dAMP and dCMP differently opposite 8-oxo-dG: DNA pol α, pol δ, and pol III preferentially incorporate dAMP, whereas DNA pol β and pol I incorporate dCMP (Wang et al. 1998). Replication of 8-oxo-dA containing DNA templates revealed that exclusively dTMP is inserted opposite the lesion by bacterial DNA polymerases pol I (Klenow fragment) and Taq DNA pol (Guschlbauer et al. 1991). Nonetheless, 8-oxo-dA is found to be mutagenic in mammalian cells where DNA pol α is found to misinsert dGMP and recombinant rat DNA pol β can misinsert dAMP and dGMP (Shibutani et al. 1993; Kamiya et al. 1995).

4.2.3 ε-Adenosine and ε-Cytidine

For reverse transcription assays it was found that only the error-prone HIV-1 RT was able to bypass the lesions. In case of ε-rA containing RNA templates HIV-1 RT incorporated deoxynucleoside triphosphates in the order dAMP $>$ dGMP $\gg$ dCMP and dTMP. For ε-rC containing templates the incorporation order observed was dAMP $>$ dTMP $\gg$ dCMP $>$ dGMP. In both cases full-length cDNA was obtained in the presence of all deoxynucleoside triphosphates (Calabretta and Leumann 2013).

The mutagenicity of ε-dA and ε-dC has been shown to be strongly mutagenic when tested in mammalian cell lines, whereas in *E. coli* minor mutagenic properties were observed. In mammalian cells predominantly A/G transitions and A/T transversions were found for ε-dA and C/T transitions for ε-dC (Pandya and Moriya 1996; Levine et al. 2000). In *E. coli* only minor mutagenic properties were found for ε-dA (Basu et al. 1993; Pandya and Moriya 1996) and only low frequent C/A transversions for ε-dC (Moriya et al. 1994).

4.2.4 Oxidative Damage to the Nucleotide Pool

Oxidative damage to the nucleotide pool can be induced by direct oxidation of ribonucleotides or by degradation of oxidatively damaged RNA. Leukocytes that were put under oxidative stress revealed increased levels of 8-oxo-rG in the nucleotide pool (Shen et al. 2000). Quality control of RNA synthesis is thus crucial and mechanisms have evolved to reduce incorporation of oxidatively damaged ribonucleotides. It was shown that 8-oxo-rGTP was incorporated into RNA by *E. coli* polymerase at a rate of ~10 % of that of rGTP (Taddei et al. 1997). To prevent the incorporation of 8-oxo-rGTP, MutT protein in *E. coli* degrades the di- and triphosphates of 8-oxo-rG to the monophosphate 8-oxo-rGMP (Ito et al. 2005). The reutilization of 8-oxo-rGMP is prevented since it was found that guanylate kinase does not phosphorylate 8-oxo-rGMP to 8-oxo-rGDP and therefore no 8-oxo-rGTP can be synthesized (Sekiguchi et al. 2013). A similar pathway has been disclosed for human cells. Here 8-oxo-rGTP is incorporated by human RNA polymerase II at a rate of 2 % of that of rGTP (Hayakawa et al. 1999). MTH1 protein has been shown to hydrolyze 8-oxo-rGTP to 8-oxo-rGMP. Additionally it could be shown that MTH5 and NUDT5 hydrolyze 8-oxo-rGDP to 8-oxo-rGMP (Ishibashi et al. 2005). Normally GTP is synthesized from GDP through phosphorylation by nucleotide diphosphate kinase (ND kinase) and GDP is itself produced through phosphorylation of GMP by guanylate kinase (GK). While ND kinase has been found to phosphorylate 8-oxo-rGDP as well, GK is not able to phosphorylate 8-oxo-rGMP, thus preventing the synthesis and incorporation of 8-oxo-rGTP into RNA (Hayakawa et al. 1999).

5 Consequences of Oxidatively Damaged RNA

During the last few years increasing evidence emerged that oxidatively damaged RNA causes severe effects on cellular function. Messenger RNA oxidation in vitro was shown to cause reduction in protein expression and ribosome stalling on the transcripts (Shan et al. 2007). In another study the authors found that oxidized mRNA induces translation errors (Tanaka et al. 2007). It was also shown that nitric oxide and peroxynitrite accelerate mutation of RNA viruses in vitro and in vivo, probably via the formation of 8-oxo-rG and 8-nitro-rG (Akaike et al. 2000).

Preliminary studies on oxidatively damaged ribosomes have shown that if the catalytic adenine, A2451 in the peptidyl transferase center (PTC) of the 23S ribosome subunit of *E. coli*, is replaced by an 8-oxo-adenine unit, peptide synthesis is abolished (Polacek, unpublished results). This is quite remarkable since it has been shown earlier that several base modifications on A2451 only have a minor influence on the activity of the PTC as long as the 2′-hydroxyl group of the ribose is present (Erlacher and Polacek 2008).

Ongoing studies on the efficiency of translation on site specifically damaged mRNA show that in rabbit reticulocyte lysate ribosome stalling as well as read-through can be observed. Whether oxidized mRNA leads to mutations in the synthesized proteins is yet unclear (Leumann, unpublished results).

6 Fate of Oxidized RNA

High levels of oxidized RNA are not tolerated by cells. Several studies have shown that levels of oxidized RNA dropped after removal of oxidative stress inducing conditions (Shen et al. 2000; Kajitani et al. 2006). Recently it could be shown that H_2O_2-induced 8-oxo-rG levels in *E. coli* cells dropped rapidly after switching to a H_2O_2-free medium (Liu et al. 2012). These results suggest that oxidatively damaged RNA is actively removed from total RNA. In theory this might happen either by repair of the oxidized lesions or by degradation. So far there is no strong evidence supporting the existence of RNA repair mechanisms. The only repair mechanism reported for RNA is the dealkylation of alkylated RNA by DNA demethylases (Aas et al. 2003). Therefore it seems more likely that oxidatively damaged RNA is removed by degradation.

Removal of damaged RNA can be achieved by ribonucleases. Polynucleotide phosphorylase (PNPase) is a 3′–5′ exoribonuclease that is important for the degradation of RNA. In *E. coli* it has been shown that PNPase specifically binds 8-oxo-rG containing RNA with a higher affinity than undamaged RNA (Hayakawa et al. 2001). Undamaged RNA has been shown to be degraded efficiently while 8-oxo-rG containing RNA is tightly bound to PNPase and thus protected from nuclease attack. Human PNPase is located mainly in mitochondria and also specifically binds 8-oxo-rG containing RNA. Under induced oxidative stress, levels of hPNPase decreased rapidly while the levels of other proteins were left unchanged (Hayakawa and Sekiguchi 2006). The authors speculate that the PNP might thus play an important role in maintaining the high fidelity of translation by sequestering the oxidatively damaged RNA molecules and possibly direct them to degradation. In a further study it has been shown that PNPase-deficient cells are hypersensitive to H_2O_2 and less viable than wild-type cells (Wu et al. 2009). Introduction of a plasmid-borne PNP gene restores the viability against oxidative stress. Furthermore overexpression of hPNPase in HeLa cells reduces 8-oxo-rG in RNA and improves cell viability against H_2O_2 treatment (Wu and Li 2008).

Another protein that shows a similar binding capacity for 8-oxo-rG containing RNA is the mammalian Y box-binding protein 1 (YB-1 protein). YB-1 has multiple regulatory activities and was originally identified as a transcription factor which binds to the Y box. YB-1 is a major core protein of messenger ribonucleoprotein (mRNP) and has a weak RNA binding capacity. It was found that YB-1 selectively binds 8-oxo-rG containing RNA to form stable complexes. Cells of *E. coli* expressing YB-1 were found to acquire resistance against oxidative stress. These data imply that YB-1 protein may be involved in sequestering oxidatively damaged RNA from normal cellular processes (Hayakawa et al. 2002).

Oxidized nucleobases are known to have low one-electron redox potentials and are therefore prone to undergo further oxidation processes (Yanagawa et al. 1992). In DNA 8-oxo-dG and 8-nitro-dG showed a higher reactivity with $ONOO^-$ than dG itself (Burney et al. 1999).

That such ongoing oxidation induces abasic sites could be shown by the reaction of RNA treated under Fenton conditions with the aldehyde reactive probe (ARP) (Tanaka et al. 2011a, b). The ARP has been shown earlier to react specifically with the aldehyde functions of abasic sites and formylcytidine but no other oxidative lesion, e.g., 8-oxo-rA, 8-oxo-rG, or pyrimidine glycols (Ide et al. 1993). RNA abasic sites have been shown to be remarkably more stable than DNA abasic sites (Küpfer and Leumann 2007). That such abasic sites can be substrates for enzymatic degradation has been shown by the disclosure of a new class of enzymes: the ribosomal RNA apurinic site-specific lyase (RALyase) (Ogasawara et al. 1999; Ito et al. 2002). These enzymes exhibited a very high specificity for the cleavage of the abasic site in the sarcin–ricin loop of intact ribosomes. On the contrary, RALyase failed to cleave DNA abasic sites as well as protein-free abasic rRNA. Thus, rigorous proof for broader RNA-specific lyase activity is still missing, as no other members of this class of enzymes could be identified so far.

7 Redox Chemistry of Oxidized Nucleosides

The one-electron redox potentials of the four main oxidative lesions have been reported to be far lower than for the standard nucleosides: 8-oxo-rG: 0.58 V (1.29 V for rG), 8-oxo-rA: 0.92 V (1.42 V for rA), 5-HO-rC: 0.62 V (1.6 V for dC), and 5-HO-rU: 0.64 V (1.7 V for dT) versus NHE (Yanagawa et al. 1992; Steenken and Jovanovic 1997); for 8-oxo-dG: 0.74 (Steenken et al. 2000). So far two examples of exploiting the low redox potentials of hydroxynucleosides have been published.

For oxidized RNA nucleosides it could be shown that they catalyze the oxidoreduction of NADH and $K_3Fe(CN)_6$ (Yanagawa et al. 1992). In this system 5-HO-rC showed by far the highest activity followed by 5-HO-rU and 8-oxo-rG which showed roughly half and a quarter of the reactivity compared to 5-HO-rC. The authors conclude that these hydroxyribonucleosides might also serve as redox cofactors at active sites of ribozymes with redox activity. So far this was not experimentally proven.

8-Oxo-dG, incorporated in proximity to a cyclobutane pyrimidine dimer (CPD), can mimic the function of a flavin in photorepair. The study supports a photolyase-type mechanism in which in a first step the excited state of 8-oxo-dG transfers an electron to the CPD effecting cleavage of the cyclobutane and in a second step the back electron transfer regenerates 8-oxo-dG and the repaired pyrimidines (Nguyen and Burrows 2011, 2012b). The free tri-acetylated 8-oxo-rG and ribofuranosyluric acid nucleosides also showed enhanced repair of T=T and U=U CPDs (Nguyen and Burrows 2012a).

The possible role of oxidized RNA nucleosides as redox catalysts in a "RNA world" is not known. Their use in ribozymes might be a future prospect.

8 Conclusions and Perspective

For DNA a large number of oxidative lesions, their metabolism, and distinct repair mechanisms have been reported in the past. As reviewed here, only half a dozen lesions that arise under oxidative stress have been described, along with their biophysical properties for RNA. Nevertheless, in accordance with DNA it can be assumed that more lesions exist which are still lacking a detailed description.

While 85 % of the genomic DNA is transcribed into RNA, the function of 97 % of this RNA is still largely unknown. Thus, oxidatively damaged RNA cannot only interfere with the translation machinery but possibly also with a large number of mostly unknown regulation mechanisms of ncRNA. The number of such long intergenic noncoding RNAs (lincRNA) is still growing and their functions are yet elusive (Hangauer et al. 2013).

The data show that RNA is at least as prone to oxidation as DNA due to its spatial vicinity to the sources of ROS and RNS and the (so far) absence of a distinct repair machinery. Moreover the amount of RNA in a cell is considerably higher than that of DNA. Within this context, RNA cannot be viewed only as a transient messenger anymore: average lifetimes of RNA are probably long enough to have an impact on different cell mechanisms.

Oxidatively damaged RNA has been linked over the past decade to a still increasing number of neurodegenerative and other diseases (Poulsen et al. 2012). The data show that the oxidative damage is rather an early event in the development of such diseases than a consequence of cell decay. A deeper understanding of the biophysical properties of oxidative lesions in RNA would undoubtedly help to understand the more complex biological impact and mechanisms of such diseases.

Acknowledgments The authors wish to thank Dr. Alessandro Calabretta for the contribution of T_m data for Table 1.

References

Aas PA, Otterlei M, Falnes PO et al (2003) Human and bacterial oxidative demethylases repair alkylation damage in both RNA and DNA. Nature 421:859–863

Akaike T, Fujii S, Kato A et al (2000) Viral mutation accelerated by nitric oxide production during infection *in vivo*. FASEB J 14:1447–1454

Barciszewski J, Barciszewska M, Siboska G et al (1999) Some unusual nucleic acid bases are products of hydroxyl radical oxidation of DNA and RNA. Mol Biol Rep 26:231–238

Basu AK, Wood ML, Niedernhofer LJ et al (1993) Mutagenic and genotoxic effects of three vinyl chloride-induced DNA lesions: 1, N6-ethenoadenine, 3, N4-ethenocytosine, and 4-amino-5-(imidazol-2-yl)imidazole. Biochemistry 32:12793–12801

Beckman JS, Beckman TW, Chen J et al (1990) Apparent hydroxyl radical production by peroxynitrite: implications for endothelial injury from nitric oxide and superoxide. Proc Natl Acad Sci U S A 87:1620–1624

Bolisetty S, Jaimes E (2013) Mitochondria and reactive oxygen species: physiology and pathophysiology. Int J Mol Sci 14:6306–6344

Bradley-Whitman M, Lovell M (2013) Increased oxidative damage in RNA in Alzheimer's disease progression. J Anal Bioanal Tech. doi:10.4172/2155-9872.s2-004

Brand MD (2010) The sites and topology of mitochondrial superoxide production. Exp Gerontol 45:466–472

Brawn K, Fridovich I (1981) DNA strand scission by enzymically generated oxygen radicals. Arch Biochem Biophys 206:414–419

Burgdorf LT, Carell T (2002) Synthesis, stability, and conformation of the formamidopyrimidine G DNA lesion. Chemistry 8:293–301

Burney S, Niles JC, Dedon PC et al (1999) DNA damage in deoxynucleosides and oligonucleotides treated with peroxynitrite. Chem Res Toxicol 12:513–520

Cadenas E, Davies KJA (2000) Mitochondrial free radical generation, oxidative stress, and aging. Free Radic Biol Med 29:222–230

Cadet J, Douki T, Ravanat J-L (2010) Oxidatively generated base damage to cellular DNA. Free Radic Biol Med 49:9–21

Calabretta A, Leumann CJ (2013) Base pairing and miscoding properties of 1, N6-ethenoadenine- and 3, N4-ethenocytosine-containing RNA oligonucleotides. Biochemistry 52:1990–1997

Cantara WA, Crain PF, Rozenski J et al (2011) The RNA modification database, RNAMDB: 2011 update. Nucleic Acids Res 39(suppl 1):D195–D201

Chang Y, Kong Q, Shan X et al (2008) Messenger RNA oxidation occurs early in disease pathogenesis and promotes motor neuron degeneration in ALS. PLoS One 3:e2849

Cheng KC, Cahill DS, Kasai H et al (1992) 8-Hydroxyguanine, an abundant form of oxidative DNA damage, causes G→T and A→C substitutions. J Biol Chem 267:166–172

Cho BP, Evans FE (1991) Structure of oxidatively damaged nucleic acid adducts. 3. Tautomerism, ionization and protonation of 8-hydroxyadenosine studied by ^{15}N NMR spectroscopy. Nucleic Acids Res 19:1041–1046

Cho BP, Kadlubar FF, Culp SJ et al (1990) ^{15}N nuclear magnetic resonance studies on the tautomerism of 8-hydroxy-2′-deoxyguanosine, 8-hydroxyguanosine, and other C8-substituted guanine nucleosides. Chem Res Toxicol 3:445–452

Crich D, Mo X-S (1997) Nucleotide C3′,4′-radical cations and the effect of a 2′-oxygen substituent. The DNA/RNA paradox. J Am Chem Soc 119:249–250

Cui S, Kim Y-H, Jin C-H et al (2009) Synthesis and base pairing properties of DNA-RNA heteroduplex containing 5-hydroxyuridine. BMB Rep 42:373–379

Dedon PC (2007) The chemical toxicology of 2-deoxyribose oxidation in DNA. Chem Res Toxicol 21:206–219

Defoiche J, Zhang Y, Lagneaux L et al (2009) Measurement of ribosomal RNA turnover *in vivo* by use of deuterium-labeled glucose. Clin Chem 55:1824–1833

El Ghissassi F, Barbin A, Nair J et al (1995) Formation of 1, N6-ethenoadenine and 3, -N4-ethenocytosine by lipid peroxidation products and nucleic acid bases. Chem Res Toxicol 8:278–283

Erlacher MD, Polacek N (2008) Ribosomal catalysis: the evolution of mechanistic concepts for peptide bond formation and peptidyl-tRNA hydrolysis. RNA Biol 5:5–12

Evans MD, Dizdaroglu M, Cooke MS (2004) Oxidative DNA damage and disease: induction, repair and significance. Mutat Res 567:1–61

Fiala ES, Conaway CC, Mathis JE (1989) Oxidative DNA and RNA damage in the livers of Sprague-Dawley rats treated with the hepatocarcinogen 2-nitropropane. Cancer Res 49:5518–5522

Gannett PM, Sura TP (1993) Base pairing of 8-oxoguanosine and 8-oxo-2′-deoxyguanosine with 2′-deoxyadenosine, 2′-deoxycytosine, 2′-deoxyguanosine, and thymidine. Chem Res Toxicol 6:690–700

Guschlbauer W, Duplaa A-M, Guy A et al (1991) Structure and *in vitro* replication of DNA templates containing 7,8-dihydro-8-oxoadenine. Nucleic Acids Res 19:1753–1758

Hangauer MJ, Vaughn IW, McManus MT (2013) Pervasive transcription of the human genome produces thousands of previously unidentified long intergenic noncoding RNAs. PLoS Genet 9:e1003569

Hayakawa H, Sekiguchi M (2006) Human polynucleotide phosphorylase protein in response to oxidative stress. Biochemistry 45:6749–6755

Hayakawa H, Hofer A, Thelander L et al (1999) Metabolic fate of oxidized guanine ribonucleotides in mammalian cells. Biochemistry 38:3610–3614

Hayakawa H, Kuwano M, Sekiguchi M (2001) Specific binding of 8-oxoguanine-containing RNA to polynucleotide phosphorylase protein. Biochemistry 40:9977–9982

Hayakawa H, Uchiumi T, Fukuda T et al (2002) Binding capacity of human YB-1 protein for RNA containing 8-oxoguanine. Biochemistry 41:12739–12744

Hecht SM (1994) RNA degradation by bleomycin, a naturally occurring bioconjugate. Bioconjug Chem 5:513–526

Hofer T, Badouard C, Bajak E et al (2005) Hydrogen peroxide causes greater oxidation in cellular RNA than in DNA. Biol Chem 386:333

Hofer T, Seo Arnold Y, Prudencio M et al (2006) A method to determine RNA and DNA oxidation simultaneously by HPLC-ECD: greater RNA than DNA oxidation in rat liver after doxorubicin administration. Biol Chem 387:103–111

Ide H, Akamatsu K, Kimura Y et al (1993) Synthesis and damage specificity of a novel probe for the detection of abasic sites in DNA. Biochemistry 32:8276–8283

Ishibashi T, Hayakawa H, Ito R et al (2005) Mammalian enzymes for preventing transcriptional errors caused by oxidative damage. Nucleic Acids Res 33:3779–3784

Ito Y, Ozawa A, Sawasaki T et al (2002) OsRALyase1, a putative F-Box protein identified in rice, *Oryza sativa*, with enzyme activity identical to that of wheat RALyase. Biosci Biotechnol Biochem 66:2727–2731

Ito R, Hayakawa H, Sekiguchi M et al (2005) Multiple enzyme activities of *Escherichia coli* MutT protein for sanitization of DNA and RNA precursor pools. Biochemistry 44:6670–6674

Kajitani K, Yamaguchi H, Dan Y et al (2006) MTH1, an oxidized purine nucleoside triphosphatase, suppresses the accumulation of oxidative damage of nucleic acids in the hippocampal microglia during kainate-induced excitotoxicity. J Neurosci 26:1688–1698

Kamiya H, Miura H, Murata-Kamiya N et al (1995) 8-Hydroxyadenine (7, 8-dihydro-8-oxoadenine) induces misincorporation in *in vitro* DNA synthesis and mutations in NIH 3T3 cells. Nucleic Acids Res 23:2893–2899

Karuppanapandian T, Moon J-C, Kim C et al (2011) Reactive oxygen species in plants: their generation, signal transduction, and scavenging mechanisms. Aust J Crop Sci 5:709–725

Kennedy LJ, Moore K, Caulfield JL et al (1997) Quantitation of 8-oxoguanine and strand breaks produced by four oxidizing agents. Chem Res Toxicol 10:386–392

Kim SK, Yokoyama S, Takaku H et al (1998) Oligoribonucleotides containing 8-oxo-7,8-dihydroguanosine and 8-oxo-7,8-dihydro-2′-O-methylguanosine: synthesis and base pairing properties. Bioorg Med Chem Lett 8:939–944

Kim SK, Kim JY, Yokoyama S et al (1999) Misreading of RNA templates containing 8-oxo-7,8-dihydroguanosine or 8-oxo-2′-O-methylguanosine in cDNA synthesis by reverse transcriptases. Nucleosides Nucleotides 18:1335–1337

Kim SK, Kim JY, Baek AK et al (2002) Base pairing properties of 8-oxo-7,8-dihydroadenosine in cDNA synthesis by reverse transcriptases. Bioorg Med Chem Lett 12:1977–1980

Kim SK, Lee SH, Kwon O-S et al (2004) DNA-RNA heteroduplex containing 8-oxo-7,8-dihydroguanosine: base pairing, structures, and thermodynamic stability. J Biochem Mol Biol 37:657–662

Koga Y, Taniguchi Y, Sasaki S (2013) Synthesis of the oligoribonucleotides incorporating 8-oxo-guanosine and evaluation of their base pairing properties. Nucleosides Nucleotides Nucleic Acids 32:124–136

Kong Q, Lin C-L (2010) Oxidative damage to RNA: mechanisms, consequences, and diseases. Cell Mol Life Sci 67:1817–1829

Kouchakdjian M, Bodepudi V, Shibutani S et al (1991) NMR structural studies of the ionizing radiation adduct 7-hydro-8-oxodeoxyguanosine (8-oxo-7H-dG) opposite deoxyadenosine in a DNA duplex. 8-Oxo-7H-dG(syn):dA(anti) alignment at lesion site. Biochemistry 30:1403–1412

Kreutzer DA, Essigmann JM (1998) Oxidized, deaminated cytosines are a source of C→T transitions *in vivo*. Proc Natl Acad Sci U S A 95:3578–3582

Küpfer PA, Leumann CJ (2007) The chemical stability of abasic RNA compared to abasic DNA. Nucleic Acids Res 35:58–68

Küpfer PA, Leumann CJ (2011) Synthesis, base pairing properties and trans-lesion synthesis by reverse transcriptases of oligoribonucleotides containing the oxidatively damaged base 5-hydroxycytidine. Nucleic Acids Res 39:9422–9432

La Francois CJ, Jang YH, Cagin T et al (2000) Conformation and proton configuration of pyrimidine deoxynucleoside oxidation damage products in water. Chem Res Toxicol 13:462–470

Lenaz G (2012) Mitochondria and reactive oxygen species. Which role in physiology and pathology? In: Scatena R, Bottoni P, Giardina B (eds) Advances in mitochondrial medicine, vol 942, Advances in experimental medicine and biology. Springer, Dordrecht, pp 93–136

Leonard GA, Guy A, Brown T et al (1992) Conformation of guanine:8-oxoadenine base pairs in the crystal structure of d(CGCGAATT(O8A)GCG). Biochemistry 31:8415–8420

Lesko SA, Lorentzen RJ, Ts'o POP (1980) Role of superoxide in deoxyribonucleic acid strand scission. Biochemistry 19:3023–3028

Levine RL, Yang I-Y, Hossain M et al (2000) Mutagenesis induced by a single 1, -N6-ethenodeoxyadenosine adduct in human cells. Cancer Res 60:4098–4104

Li Z, Wu J, Deleo CJ (2006) RNA damage and surveillance under oxidative stress. IUBMB Life 58:581–588

Liu M, Gong X, Alluri Ravi K, Wu J, Sablo T, Li Z (2012) Characterization of RNA damage under oxidative stress in *Escherichia coli*. Biol Chem 393:123–132

Masuda M, Nishino H, Ohshima H (2002) Formation of 8-nitroguanosine in cellular RNA as a biomarker of exposure to reactive nitrogen species. Chem Biol Interact 139:187–197

Moriya M, Ou C, Bodepudi V et al (1991) Site-specific mutagenesis using a gapped duplex vector: a study of translesion synthesis past 8-oxodeoxyguanosine in *E. coli*. Mutat Res 254:281–288

Moriya M, Zhang W, Johnson F et al (1994) Mutagenic potency of exocyclic DNA adducts: marked differences between *Escherichia coli* and simian kidney cells. Proc Natl Acad Sci U S A 91:11899–11903

Nathan C, Cunningham-Bussel A (2013) Beyond oxidative stress: an immunologist's guide to reactive oxygen species. Nat Rev Immunol 13:349–361

Nelson PT, Keller JN (2007) RNA in brain disease: no longer just "the messenger in the middle". J Neuropathol Exp Neurol 66:461–468

Neyhart GA, Cheng C-C, Thorp HH (1995) Kinetics and mechanism of the oxidation of sugars and nucleotides by oxoruthenium(IV): model studies for predicting cleavage patterns in polymeric DNA and RNA. J Am Chem Soc 117:1463–1471

Nguyen KV, Burrows CJ (2011) A prebiotic role for 8-oxoguanosine as a flavin mimic in pyrimidine dimer photorepair. J Am Chem Soc 133:14586–14589

Nguyen KV, Burrows CJ (2012a) Photorepair of cyclobutane pyrimidine dimers by 8-oxopurine nucleosides. J Phys Org Chem 25:574–577

Nguyen KV, Burrows CJ (2012b) Whence flavins? Redox-active ribonucleotides link metabolism and genome repair to the RNA world. Acc Chem Res 45:2151–2159

Niles JC, Wishnok JS, Tannenbaum SR (2006) Peroxynitrite-induced oxidation and nitration products of guanine and 8-oxoguanine: structures and mechanisms of product formation. Nitric Oxide 14:109–121

Novo E, Parola M (2008) Redox mechanisms in hepatic chronic wound healing and fibrogenesis. Fibrogenesis Tissue Repair 1:5

Nunomura A, Hofer T, Moreira PI et al (2009) RNA oxidation in Alzheimer disease and related neurodegenerative disorders. Acta Neuropathol (Berl) 118:151–166

Ogasawara T, Sawasaki T, Morishita R et al (1999) A new class of enzyme acting on damaged ribosomes: ribosomal RNA apurinic site specific lyase found in wheat germ. EMBO J 18:6522–6531

Pacher P, Beckman JS, Liaudet L (2007) Nitric oxide and peroxynitrite in health and disease. Physiol Rev 87:315–424

Pan B, Mitra SN, Sundaralingam M (1999) Crystal structure of an RNA 16-mer duplex r $(GCAGAGUUAAAUCUGC)_2$ with nonadjacent G(syn):A^+(anti) mispairs. Biochemistry 38:2826–2831

Pandya GA, Moriya M (1996) 1, N6-ethenodeoxyadenosine, a DNA adduct highly mutagenic in mammalian cells. Biochemistry 35:11487–11492

Pogozelski WK, Tullius TD (1998) Oxidative strand scission of nucleic acids: routes initiated by hydrogen abstraction from the sugar moiety. Chem Rev 98:1089–1108

Poulsen HE, Specht E, Broedbaek K et al (2012) RNA modifications by oxidation: a novel disease mechanism? Free Radic Biol Med 52:1353–1361

Pryor WA (1986) Oxy-radicals and related species: their formation, lifetimes, and reactions. Annu Rev Physiol 48:657–667

Radi R, Peluffo G, Alvarez MN et al (2001) Unraveling peroxynitrite formation in biological systems. Free Radic Biol Med 30:463–488

Richter C, Park JW, Ames BN (1988) Normal oxidative damage to mitochondrial and nuclear DNA is extensive. Proc Natl Acad Sci U S A 85:6465–6467

Salgo MG, Bermudez E, Squadrito GL et al (1995) Peroxynitrite causes DNA damage and oxidation of thiols in rat thymocytes. Arch Biochem Biophys 322:500–505

Schneider JE, Phillips JR, Pye Q et al (1993) Methylene blue and rose bengal photoinactivation of RNA bacteriophages: comparative studies of 8-oxoguanine formation in isolated RNA. Arch Biochem Biophys 301:91–97

Sekiguchi T, Ito R, Hayakawa H et al (2013) Elimination and utilization of oxidized guanine nucleotides in the synthesis of RNA and its precursors. J Biol Chem 288:8128–8135

Shan X, Chang Y, Lin CL (2007) Messenger RNA oxidation is an early event preceding cell death and causes reduced protein expression. FASEB J 21:2753–2764

Sharova LV, Sharov AA, Nedorezov T et al (2009) Database for mRNA half-life of 19977 genes obtained by DNA microarray analysis of pluripotent and differentiating mouse embryonic stem cells. DNA Res 16:45–58

Shen Z, Wu W, Hazen SL (2000) Activated leukocytes oxidatively damage DNA, RNA, and the nucleotide pool through halide-dependent formation of hydroxyl radical. Biochemistry 39:5474–5482

Shibutani S, Bodepudi V, Johnson F et al (1993) Translesional synthesis on DNA templates containing 8-oxo-7,8-dihydrodeoxyadenosine. Biochemistry 32:4615–4621

Srivastava SC, Raza SK, Misra R (1994) 1, N6-etheno deoxy and ribo adenosine and 3, N4-etheno deoxy and ribo cytidine phosphoramidites. Strongly fluorescent structures for selective introduction in defined sequence DNA and RNA molecules. Nucleic Acids Res 22:1296–1304

Steenken S, Jovanovic SV (1997) How easily oxidizable is DNA? One-electron reduction potentials of adenosine and guanosine radicals in aqueous solution. J Am Chem Soc 119:617–618

Steenken S, Jovanovic SV, Bietti M, Bernhard K (2000) The trap depth (in DNA) of 8-oxo-7,8-dihydro-2′-deoxyguanosine as derived from electron-transfer equilibria in aqueous solution. J Am Chem Soc 122:2373–2374
Suen W, Spiro TG, Sowers LC et al (1999) Identification by UV resonance Raman spectroscopy of an imino tautomer of 5-hydroxy-2′-deoxycytidine, a powerful base analog transition mutagen with a much higher unfavored tautomer frequency than that of the natural residue 2′-deoxycytidine. Proc Natl Acad Sci U S A 96:4500–4505
Szabó C, Ohshima H (1997) DNA damage induced by peroxynitrite: subsequent biological effects. Nitric Oxide 1:373–385
Szabó C, Ischiropoulos H, Radi R (2007) Peroxynitrite: biochemistry, pathophysiology and development of therapeutics. Nat Rev Drug Discov 6:662–680
Taddei F, Hayakawa H, Bouton M-F et al (1997) Counteraction by MutT protein of transcriptional errors caused by oxidative damage. Science 278:128–130
Tanaka M, Chock PB, Stadtman ER (2007) Oxidized messenger RNA induces translation errors. Proc Natl Acad Sci U S A 104:66–71
Tanaka M, Han S, Küpfer PA et al (2011a) An assay for RNA oxidation induced abasic sites using the aldehyde reactive probe. Free Radic Res 45:237–247
Tanaka M, Han S, Küpfer PA et al (2011b) Quantification of oxidized levels of specific RNA species using an aldehyde reactive probe. Anal Biochem 417:142–148
Thorp HH (2000) The importance of being r: greater oxidative stability of RNA compared with DNA. Chem Biol 7:R33–R36
Tullius T, Dombroski B (1985) Iron(II) EDTA used to measure the helical twist along any DNA molecule. Science 230:679–681
Tullius TD, Greenbaum JA (2005) Mapping nucleic acid structure by hydroxyl radical cleavage. Curr Opin Chem Biol 9:127–134
Uesugi S, Ikehara M (1977) Carbon-13 magnetic resonance spectra of 8-substituted purine nucleosides. Characteristic shifts for the syn conformation. J Am Chem Soc 99:3250–3253
Wang D, Kreutzer DA, Essigmann JM (1998) Mutagenicity and repair of oxidative DNA damage: insights from studies using defined lesions. Mutat Res 400:99–115
Wood ML, Dizdaroglu M, Gajewski E et al (1990) Mechanistic studies of ionizing radiation and oxidative mutagenesis: genetic effects of a single 8-hydroxyguanine (7-hydro-8-oxoguanine) residue inserted at a unique site in a viral genome. Biochemistry 29:7024–7032
Wu J, Li Z (2008) Human polynucleotide phosphorylase reduces oxidative RNA damage and protects HeLa cell against oxidative stress. Biochem Biophys Res Commun 372:288–292
Wu J, Jiang Z, Liu M et al (2009) Polynucleotide phosphorylase protects *Escherichia coli* against oxidative stress. Biochemistry 48:2012–2020
Yanagawa H, Ogawa Y, Ueno M et al (1990) A novel minimum ribozyme with oxidoreduction activity. Biochemistry 29:10585–10589
Yanagawa H, Ogawa Y, Ueno M (1992) Redox ribonucleosides. Isolation and characterization of 5-hydroxyuridine, 8-hydroxyguanosine, and 8-hydroxyadenosine from *Torula* yeast RNA. J Biol Chem 267:13320–13326
Yang E, van Nimwegen E, Zavolan M et al (2003) Decay rates of human mRNAs: correlation with functional characteristics and sequence attributes. Genome Res 13:1863–1872
Yermilov V, Rubio J, Ohshima H (1995) Formation of 8-nitroguanine in DNA treated with peroxynitrite *in vitro* and its rapid removal from DNA by depurination. FEBS Lett 376:207–210
Yermilov V, Yoshie Y, Rubio J et al (1996) Effects of carbon dioxide/bicarbonate on induction of DNA single-strand breaks and formation of 8-nitroguanine, 8-oxoguanine and base-propenal mediated by peroxynitrite. FEBS Lett 399:67–70

Use of FRET to Study Dynamics of DNA Replication

Philip Nevin and Penny J. Beuning

Contents

1 Introduction 96
2 Förster Resonance Energy Transfer 98
3 Conformational Changes in DNA Polymerases During Substrate Binding and Nucleotide Incorporation 98
4 Replisome Assembly 104
5 Conclusions 107
References 107

Abstract The DNA replisome is a large protein complex that accurately and efficiently copies an entire genome in every cell cycle. Numerous dynamic interactions must be regulated in order to control this process. Cellular integrity depends on the correct incorporation of each nucleotide and incorrectly incorporated bases must be removed by proofreading to prevent mutations. Lagging strand DNA polymerases and processivity clamps must be continuously recycled. When DNA replication forks encounter damaged DNA, translesion synthesis DNA polymerases replace replicative polymerases to allow bypass of the DNA damage. Förster resonance energy transfer (FRET) has provided key insights into the dynamics of these processes, including how DNA polymerases translocate on DNA, how DNA switches between polymerization and proofreading modes, and the numerous

P. Nevin
Department of Chemistry and Chemical Biology, Northeastern University, 360 Huntington Ave, Boston, MA 02115, USA
e-mail: nevin.p@husky.neu.edu

P.J. Beuning (✉)
Department of Chemistry and Chemical Biology, Northeastern University, 360 Huntington Ave, Boston, MA 02115, USA

Center for Interdisciplinary Research on Complex Systems, Northeastern University, Boston, MA, USA
e-mail: p.beuning@neu.edu

V.A. Erdmann et al. (eds.), *Chemical Biology of Nucleic Acids*, RNA Technologies,
DOI 10.1007/978-3-642-54452-1_6,

protein–protein and protein–nucleic acid interactions that regulate replication processes. This chapter discusses advances that have been achieved through the application of FRET, including the use of fluorescent nucleotide analogs, to a number of DNA replication processes.

Keywords DNA polymerase • DNA damage • Translesion synthesis • Processivity clamp • Replisome assembly

1 Introduction

DNA replication, the process by which organisms copy their DNA, is the basis for biological heredity and is one of the most fundamental aspects of life. The enzymes responsible for carrying out DNA replication, DNA polymerases, do so by selecting a 2′-deoxyribonucleoside-5′-triphosphate (dNTP) that is complementary to the templating base on a template DNA strand and adding it to the 3′ terminus of a primer DNA strand. Normal chromosomal DNA replication involves several accessory proteins: a DNA helicase unwinds the DNA double helix to create a single-stranded DNA (ssDNA) substrate; ssDNA-binding proteins prevent ssDNA from re-annealing or forming other secondary structures and protect the DNA from chemical and enzymatic attack; and a sliding clamp, which is loaded onto DNA by the clamp loader complex, binds both the polymerase and DNA and slides along the DNA helix (O'Donnell et al. 2013).

In general, DNA polymerases are able to insert the correct nucleotide opposite the complementary templating base with amazing specificity, making only one error in every 10^6–10^8 nucleotides incorporated (Kunkel 2004), far surpassing what would be expected from only the energetics of base paring (Kunkel and Bebenek 2000). This accuracy is essential for maintaining the fidelity of the genome as errors made by DNA polymerases may lead to mutations, genomic instability, and cell death. The study of DNA replication mechanisms and how DNA polymerases facilitate mutagenesis is of great interest since DNA replication defects are linked to antibiotic resistance (Napolitano et al. 2000; Miller et al. 2004), cellular tolerance to cancer chemotherapy agents (Bassett et al. 2004; Albertella et al. 2005; Chen et al. 2006; Ho et al. 2011), as well as cancer and other human diseases (Pages and Fuchs 2002; Loeb and Monnat 2008; Lange et al. 2011).

Structural and kinetic data are available for many different types of DNA polymerases with different functions, rates of incorporation, and error frequencies. A minimal model of nucleotide incorporation has been developed for several DNA polymerases (Fig. 1) (Kuchta et al. 1987, 1988; Dahlberg and Benkovic 1991; Eger et al. 1991; Patel et al. 1991; Wong et al. 1991; Capson et al. 1992; Kati et al. 1992; Washington et al. 2001). In this model, the DNA polymerase (Pol) first binds the primer–template DNA substrate (p/t), followed by binding of dNTP to give a

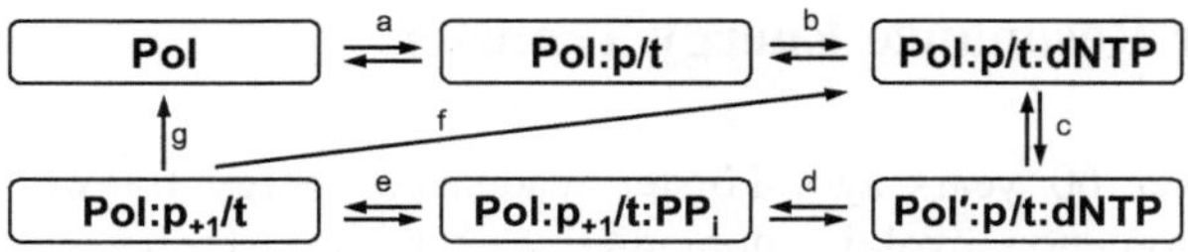

Fig. 1 A minimal model of nucleotide incorporation (Kuchta et al. 1987, 1988; Dahlberg and Benkovic 1991; Eger et al. 1991; Patel et al. 1991; Wong et al. 1991; Capson et al. 1992; Kati et al. 1992; Washington et al. 2001; Rothwell et al. 2005): the DNA polymerase (Pol) first binds the primer–template DNA substrate (p/t) (a), followed by binding of dNTP to give a ternary Pol:p/t: dNTP complex (b). Next, a conformational change in the Pol takes place to form Pol′ (c), followed by phosphodiester bond formation (d) and release of pyrophosphate (e). Finally, in the case of processive DNA synthesis, which is the case for most replicative DNA polymerases, the extended DNA is translocated (f) to allow for incorporation of the next nucleotide. In the case of distributive synthesis, the extended DNA is instead released (g)

ternary Pol:p/t:dNTP complex. Structural studies of this ternary complex have demonstrated the steric complementarity between Pol active sites and Watson–Crick base pairs as well as hydrogen bonding between the minor groove side of the nascent base pair and the enzyme as contributing to fidelity (Doublie et al. 1998; Kiefer et al. 1998; Li et al. 1998; Pelletier et al. 1994; Franklin et al. 2001). The next step in the minimal model is a conformational change in the Pol, followed by phosphodiester bond formation and release of pyrophosphate. Finally, in the case of processive DNA synthesis, which is the case for most replicative DNA polymerases, the extended DNA is translocated to allow for incorporation of the next nucleotide. In the case of distributive synthesis, the extended DNA is instead released.

Although replicative DNA polymerases are highly accurate and processive, they lack the ability to copy damaged DNA. DNA damage is ubiquitous and is normally repaired by DNA repair pathways (Friedberg et al. 2006). However, lesions in the DNA that escape repair can disrupt DNA replication and lead to mutagenesis, chromosomal instability, and cell death. Specialized DNA polymerases, most notably members of the Y family, can copy damaged DNA in a process termed translesion synthesis (TLS) (Friedberg et al. 2006). Y-family DNA polymerases have been found in all domains of life (Ohmori et al. 2001) and are characterized by the ability to carry out TLS, low fidelity when copying undamaged DNA, low processivity, and the lack of intrinsic proofreading (Yang 2005; Pata 2010). These DNA polymerases are typically part of DNA damage tolerance pathways and rescue disrupted replication forks by copying DNA damage that blocks replicative DNA polymerases. Because Y-family DNA polymerases are error prone, they rescue replication at a potentially mutagenic cost. Although Y-family DNA polymerases have relatively open active sites, recently it has become clear that there is a degree of specificity in the lesion bypass properties of Y-family DNA polymerases (Prakash et al. 2005; Walsh et al. 2011).

2 Förster Resonance Energy Transfer

Described over 60 years ago, Förster resonance energy transfer (FRET) is a physical phenomenon involving nonradiative energy transfer between two chromophores (Förster 1948). A donor chromophore, initially in its electronic excited state, may transfer energy to an acceptor chromophore via dipole–dipole interaction without the emission of a photon. Consequently, the acceptor becomes excited and the donor fluorescence is quenched. If the acceptor is a fluorophore it may return to the ground state by fluorescence emission. The efficiency of the energy transfer is proportional to the inverse sixth power of the distance between the donor and acceptor (Stryer and Haugland 1967). The FRET efficiency also depends on the overlap of the donor emission spectrum and the acceptor absorption spectrum and the mutual orientation of their transition dipole moments. The Förster radius, the distance at which FRET is 50 % efficient, is in the range of 10–100 Å for most donor–acceptor pairs. This limited range of distances over which FRET occurs is the basis for utilizing FRET as a spectroscopic ruler to provide information about the structure and dynamics of proteins or nucleic acids. FRET can be observed in a number of ways, including a decrease in the fluorescence quantum yield of the donor, a shortening of the donor fluorescence lifetime, and an increased fluorescence emission from the acceptor.

3 Conformational Changes in DNA Polymerases During Substrate Binding and Nucleotide Incorporation

Crystallographic studies have captured DNA polymerases from the different families (A, B, C, D, X, Y, and RT), which share a right-hand fold with characteristic thumb, palm, and fingers domains (Fig. 2) (Ollis et al. 1985), in various states during the catalytic cycle. These structural snapshots have revealed conformational changes as the DNA polymerase sequentially binds DNA and nucleotides. Generally, comparison of binary (Pol:p/t) and ternary (Pol:p/t:dNTP) complexes reveals that nucleotide binding to the binary Pol:DNA complex induces a conformational change involving an open-to-closed transition of the fingers domain (Fig. 2) in A-, B-, and some RT- and X-family DNA polymerases (Li et al. 1998; Doublie et al. 1998; Franklin et al. 2001; Johnson et al. 2003; Huang et al. 1998; Pelletier et al. 1996). This conformational change is characterized by the inward rotation of the fingers domain resulting in the positioning of key residues in the active site and formation of the binding site for the nascent base pair. For many DNA polymerases, it is believed that this fingers-closing step is rate limiting and contributes most to dNTP discrimination (Kuchta et al. 1987, 1988; Patel et al. 1991; Wong et al. 1991). However, the identity of the rate-limiting step has been under debate and for various systems there is evidence that fingers-closing is not rate limiting (Vande Berg et al. 2001; Zhong et al. 1997; Kim et al. 2003; Purohit et al. 2003; Shah et al. 2003).

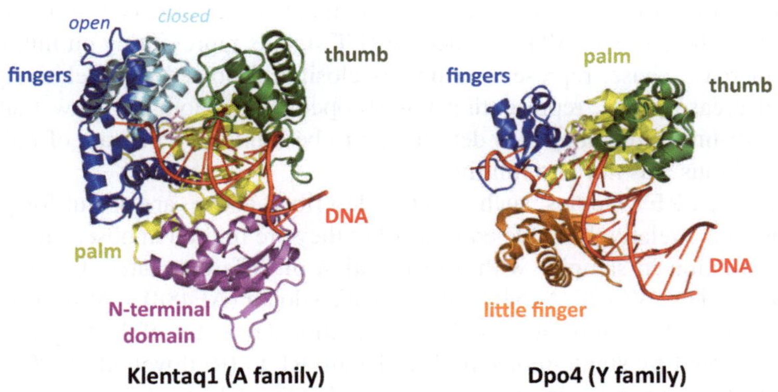

Fig. 2 Comparison of crystal structures of ternary complexes of Klentaq1 (*open*, PDB: 3SZ2; *closed*, PDB: 1QTM) and Dpo4 (PDB: 1JX4). The change in position of the fingers domain is shown in *cyan* in the closed Klentaq1 complex. The polymerase core adopts the right-hand fold characteristic of DNA polymerases with the palm (*yellow*), fingers (*blue*), and thumb (*green*) domains. Klentaq1 also contains a catalytically inactive N-terminal domain (*purple*), while Dpo4 contains a little finger domain (*orange*). The p/t DNA is shown in *red* and the next correct nucleotide is shown in *pink*

A FRET system was developed to determine the rate of the fingers-closing step in Klentaq1 (Rothwell et al. 2005), an N-terminal truncated variant of *Thermus aquaticus* DNA polymerase I and a well-characterized A-family DNA polymerase. Klentaq1 consists of a catalytically inactive N-terminal domain, corresponding to the 3′–5′ exonuclease domain in full-length DNA polymerase I, and a C-terminal polymerase domain, which adopts the characteristic right-hand fold with palm, thumb, and fingers domains (Fig. 2). The active site is located within the palm domain while the thumb domain binds DNA, and the fingers domain contains the dNTP-binding site. Using Klentaq1 labeled with an acceptor fluorophore (Alexa Fluor 594; AF594) attached to an engineered cysteine residue on the fingers domain (V649C) and a primer labeled with a donor fluorophore (AF488) at the sixth base from the 3′ terminus, FRET was observed due to formation of the binary Pol:p/t complex and addition of the correct dNTP resulted in an increase in FRET efficiency due to the closure of the fingers domain (Rothwell et al. 2005). In contrast, addition of the incorrect dNTP had no effect on the FRET efficiency. Stopped-flow measurements provided evidence that the fingers-closing step is relatively fast and not rate limiting for Klentaq1.

Although this FRET system was able to report directly on the closure of the fingers domain in Klentaq1, it requires fluorophore-labeled DNA to be bound to the polymerase and therefore is unable to monitor structural changes in the apoenzyme or the binary Pol:dNTP complex. To overcome these limitations, an intramolecular FRET system was developed to probe the movements of the fingers domain of Klentaq1 during nucleotide binding and after catalysis (Allen et al. 2008). Using the structures of the open and closed forms of Klentaq1 (Li et al. 1998) two residues were mutated to cysteine (V649C/E825C) and reacted sequentially with donor and acceptor fluorophores. Stopped-flow measurements of the binary Pol:p/t complex mixed with the correct dNTP produced FRET signals representing an initial rapid FRET increase phase, representing fingers-closing, followed by a relatively slow FRET decrease phase, representing fingers-opening. Although a slow transition, this rate of fingers-opening was determined to be limited by the rate of polymerization and thus was not rate limiting.

Ensemble FRET studies, such as those described above, are useful for probing changes with a relatively stable end state, but they are limited in observing changes in conformation in samples with a distribution of different states. Using another Klentaq1 variant (V649C/A454C) labeled with a donor (AF488) at position 649 and an acceptor (AF647) at position 454, conformational changes of the fingers domain were monitored by single-molecule FRET (smFRET) (Rothwell et al. 2013). The distribution of donor–acceptor distances revealed that Klentaq1 was in equilibrium between three distinct conformational states, termed open (O), nucleotide-binding (NB), and closed (C), whose level of occupancy was dependent on the nature of the bound substrate. The apoenzyme and binary Pol:p/t complex favored the O conformation. Only binding of the correct dNTP resulted in a stable closed conformation. Monitoring of time-dependent changes in the FRET distributions showed that for Klentaq1 molecules alone and binary Pol:p/t complexes, both the NB and C states were able to transition to the O state reversibly. For both Pol:p/t:dTTP$_{(\text{correct})}$ or Pol:p/t:dATP$_{(\text{incorrect})}$ ternary complexes, direct transitions from the NB to the C state were observed. This direct transition between the NB and C states was not observed in the apoenzyme, Pol:p/t, or Pol:dNTP complexes, suggesting that p/t and dNTP are both required for transition to a closed ternary complex via the NB conformation.

The rate of fingers-closing in *E. coli* DNA polymerase I (Klenow fragment; KF) was probed by attaching an acceptor (AF555) to the fingers domain (S751C) and using a p/t substrate containing the fluorescent base analog 1,3-diaza-2-oxophenothiazine (tC) adjacent to the primer 3′ terminus as donor (Stengel et al. 2007). The FRET efficiency increased following binding of the correct dNTP to the KF:p/t complex due to closure of the fingers domain over the active site. The rate of the nucleotide-induced fingers-closing, determined by stopped-flow FRET measurements, was highly similar to the rate of correct nucleotide incorporation, measured in rapid quench-flow experiments, indicating that fingers-closing was the rate-limiting step in nucleotide incorporation for the labeled KF-DNA system, although the fingers-closing occurred faster in unlabeled KF.

In a similar study, the rate of fingers-closing was probed by attaching the donor 5-(((2-iodoacetyl)amino)ethyl)aminonaphthalene-1-sulfonic acid (IAEDANS) to the fingers domain of KF (L744C) and using a Dabcyl acceptor (quencher) attached to a p/t substrate (Joyce et al. 2008). The rate of fingers-closing, measured by stopped-flow FRET, was higher than the rate of nucleotide incorporation measured in chemical quench experiments and was therefore not rate limiting. Moreover, fingers-closing was blocked by rNTPs or mispaired dNTPs. To further probe the conformational changes that precede phosphoryl transfer, smFRET studies were performed using a KF variant with fluorophores attached to L744C in the fingers domain and K550C in the thumb domain (Santoso et al. 2010). Monitoring FRET in the apoenzyme showed that KF samples the open and closed conformations in the millisecond time scale in the absence of substrates. Moreover, unique FRET efficiencies were observed in ternary complexes with incorrect dNTPs or complementary rNTPs, suggesting the existence of additional conformational states that are relevant for discrimination against these incorrect substrates. Another smFRET study using immobilized donor-labeled DNA and acceptor-labeled KF revealed that the fingers domain of KF can adopt a third, intermediate conformation (ajar) in addition to the open and closed conformations (Berezhna et al. 2012). As for Klentaq1, the binary KF:p/t complex usually adopted the open conformation and only the correct dNTP shifted the equilibrium toward the closed conformation. Furthermore, subpopulations of binary (KF:p/t) and ternary ($KF:p/t:dNTP_{(incorrect)}$) complexes were identified with DNA bound at the 3′–5′ exonuclease (exo) site, which is separated from the pol active site by 35 Å (Freemont et al. 1988). A similar smFRET study using an intramolecular FRET system also demonstrated a third, partially closed conformation of KF (Hohlbein et al. 2013).

Several smFRET studies have used immobilized p/t DNA labeled with Cy3 (donor) and KF labeled with Cy5 (acceptor) at the palm domain residue C907 to investigate polymerase interactions with DNA in real time (Christian et al. 2009; Markiewicz et al. 2012; Vrtis et al. 2013). These experiments revealed that primer binding to the exo site only takes place with mismatched primer termini (Markiewicz et al. 2012) or with p/t substrates containing adducts (2-aminofluorene (AF) or *N*-acetyl-2-aminofluorene (AAF)) at the templating base (Vrtis et al. 2013). Moreover, an adduct- or mismatch-induced intermediate FRET state was identified, which may be relevant for proofreading. In a similar study using AF488-labeled (donor) p/t and KF labeled with AF594 (acceptor) at the thumb domain (K550C), smFRET trajectories provided evidence that DNA can switch between the pol and exo site in both directions without dissociating from KF (Lamichhane et al. 2013). The rate for this intramolecular transfer, determined by dwell-time analysis, was higher than the rate for intermolecular transfer, where DNA dissociates from one site before rebinding the other site.

The transition of DNA between the pol and exo sites in the replicative DNA polymerase from bacteriophage RB69, a model B-family DNA polymerase, was investigated using a two-nucleobase FRET pair consisting of tC (donor) at the 3′ primer terminus (position N) and tC_{nitro} (acceptor) at the N-7 position of the primer strand (Xia et al. 2013). The $tC\text{-}tC_{nitro}$ pair has a well-defined orientation in duplex DNA, which provides the orientation factor in FRET efficiency calculations (Preus et al. 2010). For p/t DNA containing an A/A mismatch at N-1, the FRET efficiency was lower than for a fully complementary p/t DNA due to unwinding of DNA at the p/t junction. Binding of RB69 Pol resulted in a FRET decrease for both matched and mismatched p/t DNA substrates. However, the change was more dramatic for the mismatched p/t, suggesting that the primer terminus was transferred from the pol site to the exo site. When highly conserved L415 in the nucleotide-binding pocket was replaced by Ala or Gly, binding of RB69 Pol to both p/t substrates resulted in similar FRET efficiencies, suggesting that no transfer to the exo site occurred and that the ability of these RB69 Pol variants to differentiate fully complementary p/t substrates from p/t substrates containing a mismatch at the p/t junction is abolished. Partitioning between the pol and exo sites was not significantly compromised with the L415M variant, which produced FRET efficiencies similar to wild-type RB69 Pol with both p/t substrates (Xia et al. 2013).

FRET analyses have provided insights into initial dNTP selection, proofreading, and other aspects of catalysis by these model DNA polymerases. However, it has become apparent that the paradigms for these model DNA polymerases do not universally apply to specialized DNA polymerases of the Y family. Y-family DNA polymerases are characterized by small thumb and fingers domains relative to replicative DNA polymerases (Fig. 2) (Ling et al. 2001; Zhou et al. 2001; Silvian et al. 2001; Yang 2005; Pata 2010) and specifically lack the α-helix in the fingers domain, known as the O-helix in KF, that interacts with dNTP and ensures the fidelity of base pair formation (Kaushik et al. 1996; Ogawa et al. 2001; Beard and Wilson 2003). Y-family DNA polymerases therefore have relatively open active sites that allow them to accommodate lesions in the templating DNA strand that would block replicative DNA polymerases (Ling et al. 2001; Pata 2010). In addition to the characteristic palm, thumb, and finger domains, each Y-family DNA polymerase possesses a little finger (LF) domain, also referred to as the polymerase associated domain (PAD) in eukaryotic polymerases, unique to the Y family (Ling et al. 2001; Pata 2010), that plays an important role in defining the specific lesion bypass properties of each Y-family DNA polymerase (Boudsocq et al. 2004; Wilson et al. 2013).

In contrast to many replicative DNA polymerases, no large-scale movements of the fingers domain have been observed for Y-family DNA polymerases (Pata 2010). Only subtle repositioning of certain active site residues are observed when comparing the binary and ternary complex crystal structures of the Y-family DNA polymerases Dpo4 from *Sulfolobus solfataricus* (Wong et al. 2008) and Dbh from *Sulfolobus acidocaldarius* (Wilson and Pata 2008). To probe the motions of specific domains relative to the polymerase-bound DNA substrate, a FRET system consisting of a primer with a covalently attached donor on the ninth base from the primer 3′-terminus and an acceptor linked to an engineered cysteine residue in Dpo4 was employed (Xu et al. 2009). Steady-state measurements revealed a decrease in FRET upon addition of the correct dNTP to the binary Dpo4:p/t complex for all labeled domain variants (fingers, N70C; palm, S112C and N130C; thumb, K172C and 207C; LF, R267C and K329C). Similar decreases in FRET were observed when dideoxy-terminated (non-extendable) primers were used, suggesting that the change in FRET efficiency represents a pre-catalytic conformational change in Dpo4. Interestingly, the opposite was observed for KF (Stengel et al. 2007; Joyce et al. 2008) or Klentaq1 (Rothwell et al. 2005), where binding of the correct dNTP resulted in an increase in FRET due to the fingers-closing. For the labeled N70C fingers variant as well as the S112C and N130C palm variants, stopped-flow measurements produced three phases in the FRET signals. A rapid decrease in FRET was followed by a rapid increase in FRET, which was then followed by a final slow decrease in FRET. These three phases were proposed to represent a rapid DNA translocation event during the first phase, closure of the fingers and palm domains to "grip" DNA during the second phase, and "reopening" of the fingers and palm domains during the third phase, respectively. The initial DNA translocation probably occurred through the rotation of the DNA duplex that can be inferred from comparison of binary and ternary crystal structures of Dpo4 (Wong et al. 2008). FRET experiments with the remaining Dpo4 variants (thumb, K172C and S207C; LF, R267C and K329C; fingers, E49C) revealed only two phases, although with varying directionality. The thumb domain variants exhibited "gripping–reopening" transitions, while the labeled LF variants and the E49C fingers variant exhibited the reverse motions. Using a non-extendable DNA substrate, the "gripping" phase was absent for all Dpo4 variants, suggesting that this phase represents either a post-catalytic transition or the chemistry step. Analysis of the temperature dependence of the "gripping" and "reopening" phases demonstrated that the activation energies for both conformational changes were lower than the previously determined activation energy for the rate-limiting step preceding phosphodiester bond formation (Fiala et al. 2008). Thus, neither of these phases in the FRET signals ("gripping" or "reopening") is related to the rate-limiting step.

The motions of the fingers domain relative to the LF domain in Dpo4 were probed using a Dpo4 variant containing an intrinsic tryptophan donor in the LF domain (Y274W) and a 7-diethylamino-3-(4′-maleimidylphenyl)-4-methyl-coumarin (CPM) acceptor attached to an engineered residue in the fingers domain (K26C or N70C) (Xu et al. 2009). For both variants, stopped-flow fluorescence showed an initial rapid FRET decrease phase followed by a relatively slow FRET increase phase. Determination of rate constants suggested that these phases correlated to the "opening" and "gripping" phases observed in the experiments using acceptor-labeled DNA discussed above. These observations showed that the LF and fingers domains move away from each other before catalysis and move closer together after phosphoryl transfer.

Evidence from crystallographic studies suggests that Dpo4 makes contacts with a DNA substrate containing an 8-oxo-7,8-dihydro-2′-deoxyguanosine (8-oxoG) lesion, the major DNA lesion formed by oxidative damage in vivo (Cadet et al. 2003), that are not possible with an undamaged DNA substrate (Rechkoblit et al. 2006, 2009; Zang et al. 2006). Several Y-family DNA polymerases, including yeast and human DNA polymerase η (Haracska et al. 2000) and Dpo4 (Rechkoblit et al. 2006), are able to perform error-free bypass of 8-oxoG. The FRET system described above, consisting of labeled Dpo4 domain variants and 3′-labeled primer, was used to probe conformational changes that occur in Dpo4 during dCTP incorporation opposite 8-oxoG (Maxwell et al. 2012). Labeled S112C (palm), S207C (thumb), and K329C (LF) variants of Dpo4 exhibited FRET changes similar to those observed during correct dNTP incorporation opposite undamaged DNA. In contrast, the FRET signal for labeled N70C (fingers) contained an initial rapid decrease in FRET followed by a slow increase in FRET, suggesting that the fingers domain moves away from DNA prior to catalysis and reverses its motion after 8-oxoG bypass, which is quite different than what was observed with undamaged

DNA. The conformational changes in Dpo4 during 8-oxoG bypass that were inferred from these FRET studies are a "gripping–reopening" motion by the thumb and palm domains and an "opening–closing" motion by the fingers and LF domains. In comparison, the fingers domain undergoes a "closing–opening" transition during correct dNTP incorporation on undamaged DNA. Only the LF domain undergoes similar conformational transitions during correct dNTP incorporation on undamaged DNA, insertion of a nucleotide opposite 8-oxoG, and extension beyond a base pair containing 8-oxoG. The LF domain is unique to Y-family DNA polymerases and its conformation may play a role in specificity of TLS (Boudsocq et al. 2004; Wilson et al. 2013).

4 Replisome Assembly

A number of components of the replisome must be assembled in order for DNA replication to occur. In particular, processive DNA replication relies on recruitment of the processivity factor, which is usually a sliding clamp that encircles DNA, to DNA at primer–template junctions. The ring-shaped clamps need to be opened and loaded onto DNA in an ATP-dependent process carried out by clamp loader complexes (Kelch et al. 2012). Several studies have utilized FRET to study interactions or conformational changes during clamp loading and DNA polymerase holoenzyme assembly in different systems, including bacteriophage T4 (Trakselis et al. 2001; Xi et al. 2005a, b; Zhang et al. 2005; Smiley et al. 2006; Perumal et al. 2013), *E. coli* (Goedken et al. 2004, 2005), yeast (Zhuang et al. 2006), and humans (Hedglin et al. 2013).

Conformational changes in the yeast sliding clamp, proliferating cell nuclear antigen (PCNA), were probed using a donor–acceptor pair consisting of an intrinsic tryptophan donor (F185W) and IAEDANS (acceptor) attached to K107C in the subunit interface of the trimeric PCNA ring (Zhuang et al. 2006). Steady-state FRET measurements showed that both ATP and ATPγS, a non-hydrolyzable analog of ATP, facilitated PCNA ring opening by the clamp loader replication factor C (RFC). Only the presence of ATP produced a DNA-induced FRET decrease, representing ATP hydrolysis-induced release of PCNA. Using a different FRET system consisting of 5′-Cy3-labeled (donor) primer and Cy5-labeled (acceptor) human PCNA at N107C, it was shown by stopped-flow measurements that human RFC loads PCNA onto a p/t junction in an ATP-dependent manner and remains transiently bound to DNA (Hedglin et al. 2013). Moreover, PCNA was unloaded by RFC in the absence of DNA polymerase, while in the presence of DNA polymerase δ, loaded PCNA was captured from DNA-bound RFC, which subsequently dissociated.

To monitor interactions between the *E. coli* γ/τ clamp loader complex and the sliding clamp (β clamp), a donor was attached to D106C in the δ subunit of the γ/τ complex and an acceptor to D253C in the β clamp (Goedken et al. 2005). Steady-state measurements produced ATP- and Mg^{2+}-dependent FRET increases, confirming the requirements for these cofactors for the association of the β clamp and the clamp loader. Subsequent addition of p/t DNA resulted in a FRET decrease, presumably due to DNA-stimulated ATP hydrolysis and clamp release. Addition of excess ADP completely reversed the FRET signal, suggesting that the clamp loader and the clamp are reversibly bound. Several mutations were made in the γ subunit of the clamp loader to investigate the function of conserved residues located within the inner surface of the clamp loader. Examination of the DNA-dependent FRET decrease showed that clamp loaders with mutations near the clamp-interacting and central helices (R98E, K100E, R105E, S132A, R133A, and R133E) were unable to carry out DNA-dependent clamp release, whereas mutations near the SRC helix (K161A and K161E), which is important for ATP hydrolysis, had only modest effects on this ability, even though all of the variants showed fairly modest defects in binding to the β clamp. However, these variants all showed reduced binding to DNA. The clamp loader variants with control mutations (S33A and R80E) of residues that are expected to be on the outer surface showed no change in DNA-dependent clamp release. This FRET system was then used to show that release of the β clamp by clamp loader was most efficient with DNA containing 5′ overhangs and recessed 3′ ends, which mimic primer termini.

The bacteriophage T4 DNA polymerase holoenzyme consists of gp43 (DNA polymerase) and gp45 (sliding clamp), which is loaded onto DNA by gp44/62 (clamp loader). To probe opening and closing of gp45 during the formation of the holoenzyme, a coumarin (CPM) acceptor was attached to a cysteine residue in the interface of the trimeric gp45 ring (S158C, T168C, or V163C), across from an intrinsic tryptophan donor (W92) in the adjacent subunit (Trakselis et al. 2001). Stopped-flow measurements and interdomain distance calculations were consistent with a conformation of gp45 that is initially partially open and opens further in the presence of gp44/62 and ATP, closes slightly upon the addition of DNA, and then closes further with the addition of gp43 polymerase. The final state featured a partially open subunit interface that would allow the C-terminal tail of gp43 to be accommodated in the subunit interface of the gp45 clamp, which is important for processive replication.

Diffusion of the β clamp on DNA was studied by smFRET using donor-labeled (AF488) β clamp at C180 loaded onto primed circular M13ssDNA (Laurence et al. 2008). By using 5′-acceptor-labeled (AF647) primers of varying length and restricting the motion of the β clamp to the double-stranded DNA region by using *E. coli* ssDNA-binding protein (SSB) or its human homolog RPA, free diffusion of the β clamp on DNA was observed. The diffusion constant for the β clamp on DNA was ~10^{-14} m^2/s, at least three orders of magnitude less than that for diffusion in water. In the presence of SSB, but not RPA, the β clamp remained stationary at the p/t junction (the primer 3′ terminus where the clamp was loaded). Detailed measurements of the diffusion of human PCNA with a fluorescence system gave two rates consistent with two different sliding modes on DNA, a slower one in which PCNA tracks the DNA groove and a faster mode in which PCNA slides along DNA but does not track the groove (Kochaniak et al. 2009).

In addition to DNA polymerases, the β clamp binds other proteins acting on DNA in replication or repair. Because the β clamp is a homodimer, it is possible that it can bind two protein partners simultaneously. To test this hypothesis, *E. coli* Y-family DNA polymerase DinB (Pol IV) was labeled with a donor (Oregon Green 488) and DNA polymerase III α was labeled with an acceptor (Alexa Red 555) (Indiani et al. 2005). Steady-state FRET was observed only in the presence of the dimeric β clamp, suggesting that both DinB and α were bound to the same β clamp molecule.

The effect of the *E. coli* SOS response proteins UmuD and UmuD′ on the DNA polymerase III α-β complex was probed using donor-labeled (AF488) β clamp and acceptor-labeled (AF647) α subunit (Silva et al. 2012). Assembly of the α-β complex resulted in FRET and steady-state measurements revealed a UmuD-induced FRET decrease. In contrast, the cleaved form UmuD′ had no effect on the FRET efficiency, indicating that the α-β complex is selectively disrupted by UmuD after DNA damage.

The diffusion of SSB on ssDNA and its interaction with the RecA protein, involved in the SOS response and recombinational DNA repair, was studied using three-color smFRET (Roy et al. 2009). These experiments utilized immobilized ssDNA of 130 nucleotides with a Cy5-acceptor and a Cy5.5-acceptor attached to each end and a Cy3-donor placed in the middle of the DNA. Both dye pairs displayed transient high FRET states as wrapping of the ssDNA by SSB brings the middle closer to either end, indicating that SSB is able to migrate at least 65 nucleotides to either end on ssDNA. A similar setup was used to demonstrate that SSB is displaced by formation of the RecA filament on ssDNA. Addition of SSB to an ssDNA construct containing a Cy3-donor and a Cy5-acceptor on either side of a hairpin structure partially converted a single high FRET state to several lower FRET states, indicating that DNA hairpins were melted due to SSB diffusion. Formation of the RecA filament on a similar hairpin construct occurred 40-fold faster in the presence of SSB. Whereas RecA displaces SSB from ssDNA, SSB facilitates RecA formation on ssDNA that contains secondary structure.

While analysis of individual components of replisomes is informative, FRET experiments can also shed light on the assembly of replisome complexes. The order of assembly of labeled T4 replication proteins on DNA substrates has been investigated in a number of FRET studies (Smiley et al. 2006; Perumal et al. 2013; Xi et al. 2005a, b; Zhang et al. 2005). In one study, N terminally donor-labeled (AF488) gp43 and acceptor-labeled (AF555) gp45-V163C were used together with immobilized p/t substrates, clamp loader, and Mg^{2+}·ATP to demonstrate, by single-molecule fluorescence microscopy, that the T4 holoenzyme can be assembled through four different pathways: with either clamp and clamp loader binding first or both binding simultaneously, followed by the DNA polymerase; or DNA polymerase binding to DNA first, followed by the clamp loader–clamp complex (Smiley et al. 2006). In all cases, Mg^{2+}·ATP was required for functional holoenzyme assembly.

5 Conclusions

Stopped-flow, bulk, and single-molecule FRET techniques have proven to be powerful tools for investigating conformational changes in DNA polymerases as well as the dynamics of protein–protein interactions. These studies have provided important insights into the mechanisms of DNA polymerases as well as the pathways of replisome assembly. Rational placement of donors and acceptors is crucial and will be facilitated by high-resolution structures of the proteins of interest. Further development of flexible labeling strategies will enable additional FRET experiments with replisomes and lead to better understanding of the mechanisms of DNA replication.

References

Albertella MR, Green CM, Lehmann AR et al (2005) A role for polymerase η in the cellular tolerance to cisplatin-induced damage. Cancer Res 65:9799–9806

Allen WJ, Rothwell PJ, Waksman G (2008) An intramolecular FRET system monitors fingers subdomain opening in Klentaq1. Protein Sci 17:401–408

Bassett E, King NM, Bryant MF et al (2004) The role of DNA polymerase η in translesion synthesis past platinum-DNA adducts in human fibroblasts. Cancer Res 64:6469–6475

Beard WA, Wilson SH (2003) Structural insights into the origins of DNA polymerase fidelity. Structure 11:489–496

Berezhna SY, Gill JP, Lamichhane R et al (2012) Single-molecule Förster resonance energy transfer reveals an innate fidelity checkpoint in DNA polymerase I. J Am Chem Soc 134:11261–11268

Boudsocq F, Kokoska RJ, Plosky BS et al (2004) Investigating the role of the little finger domain of Y-family DNA polymerases in low fidelity synthesis and translesion replication. J Biol Chem 279:32932–32940

Cadet J, Douki T, Gasparutto D et al (2003) Oxidative damage to DNA: formation, measurement and biochemical features. Mutat Res 531:5–23

Capson TL, Peliska JA, Kaboord BF et al (1992) Kinetic characterization of the polymerase and exonuclease activities of the gene 43 protein of bacteriophage T4. Biochemistry 31:10984–10994

Chen YW, Cleaver JE, Hanaoka F et al (2006) A novel role of DNA polymerase eta in modulating cellular sensitivity to chemotherapeutic agents. Mol Cancer Res 4:257–265

Christian TD, Romano LJ, Rueda D (2009) Single-molecule measurements of synthesis by DNA polymerase with base-pair resolution. Proc Natl Acad Sci U S A 106:21109–21114

Dahlberg ME, Benkovic SJ (1991) Kinetic mechanism of DNA polymerase I (Klenow fragment): identification of a second conformational change and evaluation of the internal equilibrium constant. Biochemistry 30:4835–4843

Doublie S, Tabor S, Long AM et al (1998) Crystal structure of a bacteriophage T7 DNA replication complex at 2.2 Å resolution. Nature 391:251–258

Eger BT, Kuchta RD, Carroll SS et al (1991) Mechanism of DNA replication fidelity for three mutants of DNA polymerase I: Klenow fragment KF(exo+), KF(polA5), and KF(exo-). Biochemistry 30:1441–1448

Fiala KA, Sherrer SM, Brown JA et al (2008) Mechanistic consequences of temperature on DNA polymerization catalyzed by a Y-family DNA polymerase. Nucleic Acids Res 36:1990–2001

Förster T (1948) Zwischenmolekulare Energiewanderung und Fluoreszenz. Ann Phys 2:55–75

Franklin MC, Wang J, Steitz TA (2001) Structure of the replicating complex of a pol alpha family DNA polymerase. Cell 105:657–667

Freemont PS, Friedman JM, Beese LS et al (1988) Cocrystal structure of an editing complex of Klenow fragment with DNA. Proc Natl Acad Sci U S A 85:8924–8928

Friedberg EC, Walker GC, Siede W et al (2006) DNA repair and mutagenesis, 2nd edn. ASM, Washington, DC

Goedken ER, Levitus M, Johnson A et al (2004) Fluorescence measurements on the *E. coli* DNA polymerase clamp loader: implications for conformational changes during ATP and clamp binding. J Mol Biol 336:1047–1059

Goedken ER, Kazmirski SL, Bowman GD et al (2005) Mapping the interaction of DNA with the *Escherichia coli* DNA polymerase clamp loader complex. Nat Struct Mol Biol 12:183–190

Haracska L, Yu SL, Johnson RE et al (2000) Efficient and accurate replication in the presence of 7,8-dihydro-8-oxoguanine by DNA polymerase eta. Nat Genet 25:458–461

Hedglin M, Perumal SK, Hu Z et al (2013) Stepwise assembly of the human replicative polymerase holoenzyme. Elife 2:e00278

Ho TV, Guainazzi A, Derkunt SB et al (2011) Structure-dependent bypass of DNA interstrand crosslinks by translesion synthesis polymerases. Nucleic Acids Res 39:7455–7464

Hohlbein J, Aigrain L, Craggs TD et al (2013) Conformational landscapes of DNA polymerase I and mutator derivatives establish fidelity checkpoints for nucleotide insertion. Nat Commun 4:2131

Huang H, Chopra R, Verdine GL et al (1998) Structure of a covalently trapped catalytic complex of HIV-1 reverse transcriptase: implications for drug resistance. Science 282:1669–1675

Indiani C, McInerney P, Georgescu R et al (2005) A sliding-clamp toolbelt binds high- and low-fidelity DNA polymerases simultaneously. Mol Cell 19:805–815

Johnson SJ, Taylor JS, Beese LS (2003) Processive DNA synthesis observed in a polymerase crystal suggests a mechanism for the prevention of frameshift mutations. Proc Natl Acad Sci U S A 100:3895–3900

Joyce CM, Potapova O, Delucia AM et al (2008) Fingers-closing and other rapid conformational changes in DNA polymerase I (Klenow fragment) and their role in nucleotide selectivity. Biochemistry 47:6103–6116

Kati WM, Johnson KA, Jerva LF et al (1992) Mechanism and fidelity of HIV reverse transcriptase. J Biol Chem 267:25988–25997

Kaushik N, Pandey VN, Modak MJ (1996) Significance of the O-helix residues of *Escherichia coli* DNA polymerase I in DNA synthesis: dynamics of the dNTP binding pocket. Biochemistry 35:7256–7266

Kelch BA, Makino DL, O'Donnell M et al (2012) Clamp loader ATPases and the evolution of DNA replication machinery. BMC Biol 10:34

Kiefer JR, Mao C, Braman JC et al (1998) Visualizing DNA replication in a catalytically active *Bacillus* DNA polymerase crystal. Nature 391:304–307

Kim SJ, Beard WA, Harvey J et al (2003) Rapid segmental and subdomain motions of DNA polymerase beta. J Biol Chem 278:5072–5081

Kochaniak AB, Habuchi S, Loparo JJ et al (2009) Proliferating cell nuclear antigen uses two distinct modes to move along DNA. J Biol Chem 284:17700–17710

Kuchta RD, Mizrahi V, Benkovic PA et al (1987) Kinetic mechanism of DNA polymerase I (Klenow). Biochemistry 26:8410–8417

Kuchta RD, Benkovic P, Benkovic SJ (1988) Kinetic mechanism whereby DNA polymerase I (Klenow) replicates DNA with high fidelity. Biochemistry 27:6716–6725

Kunkel TA (2004) DNA replication fidelity. J Biol Chem 279:16895–16898

Kunkel TA, Bebenek K (2000) DNA replication fidelity. Annu Rev Biochem 69:497–529

Lamichhane R, Berezhna SY, Gill JP et al (2013) Dynamics of site switching in DNA polymerase. J Am Chem Soc 135:4735–4742

Lange SS, Takata K, Wood RD (2011) DNA polymerases and cancer. Nat Rev Cancer 11:96–110

Laurence TA, Kwon Y, Johnson A et al (2008) Motion of a DNA sliding clamp observed by single molecule fluorescence spectroscopy. J Biol Chem 283:22895–22906

Li Y, Korolev S, Waksman G (1998) Crystal structures of open and closed forms of binary and ternary complexes of the large fragment of *Thermus aquaticus* DNA polymerase I: structural basis for nucleotide incorporation. EMBO J 17:7514–7525

Ling H, Boudsocq F, Woodgate R et al (2001) Crystal structure of a Y-family DNA polymerase in action: a mechanism for error-prone and lesion-bypass replication. Cell 107:91–102

Loeb LA, Monnat RJ Jr (2008) DNA polymerases and human disease. Nat Rev Genet 9:594–604

Markiewicz RP, Vrtis KB, Rueda D et al (2012) Single-molecule microscopy reveals new insights into nucleotide selection by DNA polymerase I. Nucleic Acids Res 40:7975–7984

Maxwell BA, Xu C, Suo Z (2012) DNA lesion alters global conformational dynamics of Y-family DNA polymerase during catalysis. J Biol Chem 287:13040–13047

Miller C, Thomsen LE, Gaggero C et al (2004) SOS response induction by beta-lactams and bacterial defense against antibiotic lethality. Science 305:1629–1631

Napolitano R, Janel-Bintz R, Wagner J et al (2000) All three SOS-inducible DNA polymerases (Pol II, Pol IV and Pol V) are involved in induced mutagenesis. EMBO J 19:6259–6265

O'Donnell M, Langston L, Stillman B (2013) Principles and concepts of DNA replication in bacteria, archaea, and eukarya. Cold Spring Harb Perspect Biol 5:a010108

Ogawa M, Tosaka A, Ito Y et al (2001) Enhanced ribonucleotide incorporation by an O-helix mutant of *Thermus aquaticus* DNA polymerase I. Mutat Res 485:197–207

Ohmori H, Friedberg EC, Fuchs RP et al (2001) The Y-family of DNA polymerases. Mol Cell 8:7–8

Ollis DL, Brick P, Hamlin R et al (1985) Structure of large fragment of *Escherichia coli* DNA polymerase I complexed with dTMP. Nature 313:762–766

Pages V, Fuchs RP (2002) How DNA lesions are turned into mutations within cells? Oncogene 21:8957–8966

Pata JD (2010) Structural diversity of the Y-family DNA polymerases. Biochim Biophys Acta 1804:1124–1135

Patel SS, Wong I, Johnson KA (1991) Pre-steady-state kinetic analysis of processive DNA replication including complete characterization of an exonuclease-deficient mutant. Biochemistry 30:511–525

Pelletier H, Sawaya MR, Kumar A et al (1994) Structures of ternary complexes of rat DNA polymerase beta, a DNA template-primer, and ddCTP. Science 264:1891–1903

Pelletier H, Sawaya MR, Wolfle W et al (1996) Crystal structures of human DNA polymerase beta complexed with DNA: implications for catalytic mechanism, processivity, and fidelity. Biochemistry 35:12742–12761

Perumal SK, Ren WH, Lee TH et al (2013) How a holoenzyme for DNA replication is formed. Proc Natl Acad Sci U S A 110:99–104

Prakash S, Johnson RE, Prakash L (2005) Eukaryotic translesion synthesis DNA polymerases: specificity of structure and function. Annu Rev Biochem 74:317–353

Preus S, Kilsa K, Wilhelmsson LM et al (2010) Photophysical and structural properties of the fluorescent nucleobase analogues of the tricyclic cytosine (tC) family. Phys Chem Chem Phys 12:8881–8892

Purohit V, Grindley ND, Joyce CM (2003) Use of 2-aminopurine fluorescence to examine conformational changes during nucleotide incorporation by DNA polymerase I (Klenow fragment). Biochemistry 42:10200–10211

Rechkoblit O, Malinina L, Cheng Y et al (2006) Stepwise translocation of Dpo4 polymerase during error-free bypass of an oxoG lesion. PLoS Biol 4:e11

Rechkoblit O, Malinina L, Cheng Y et al (2009) Impact of conformational heterogeneity of OxoG lesions and their pairing partners on bypass fidelity by Y family polymerases. Structure 17:725–736

Rothwell PJ, Mitaksov V, Waksman G (2005) Motions of the fingers subdomain of klentaq1 are fast and not rate limiting: implications for the molecular basis of fidelity in DNA polymerases. Mol Cell 19:345–355

Rothwell PJ, Allen WJ, Sisamakis E et al (2013) dNTP-dependent conformational transitions in the fingers subdomain of Klentaq1 DNA polymerase: insights into the role of the "nucleotide-binding" state. J Biol Chem 288:13575–13591

Roy R, Kozlov AG, Lohman TM et al (2009) SSB protein diffusion on single-stranded DNA stimulates RecA filament formation. Nature 461:1092–1097

Santoso Y, Joyce CM, Potapova O et al (2010) Conformational transitions in DNA polymerase I revealed by single-molecule FRET. Proc Natl Acad Sci U S A 107:715–720

Shah AM, Maitra M, Sweasy JB (2003) Variants of DNA polymerase Beta extend mispaired DNA due to increased affinity for nucleotide substrate. Biochemistry 42:10709–10717

Silva MC, Nevin P, Ronayne EA et al (2012) Selective disruption of the DNA polymerase III alpha-beta complex by the *umuD* gene products. Nucleic Acids Res 40:5511–5522

Silvian LF, Toth EA, Pham P et al (2001) Crystal structure of a DinB family error-prone DNA polymerase from *Sulfolobus solfataricus*. Nat Struct Biol 8:984–989

Smiley RD, Zhuang Z, Benkovic SJ et al (2006) Single-molecule investigation of the T4 bacteriophage DNA polymerase holoenzyme: multiple pathways of holoenzyme formation. Biochemistry 45:7990–7997

Stengel G, Gill JP, Sandin P et al (2007) Conformational dynamics of DNA polymerase probed with a novel fluorescent DNA base analogue. Biochemistry 46:12289–12297

Stryer L, Haugland RP (1967) Energy transfer: a spectroscopic ruler. Proc Natl Acad Sci U S A 58:719–726

Trakselis MA, Alley SC, Abel-Santos E et al (2001) Creating a dynamic picture of the sliding clamp during T4 DNA polymerase holoenzyme assembly by using fluorescence resonance energy transfer. Proc Natl Acad Sci U S A 98:8368–8375

Vande Berg BJ, Beard WA, Wilson SH (2001) DNA structure and aspartate 276 influence nucleotide binding to human DNA polymerase beta. Implication for the identity of the rate-limiting conformational change. J Biol Chem 276:3408–3416

Vrtis KB, Markiewicz RP, Romano LJ et al (2013) Carcinogenic adducts induce distinct DNA polymerase binding orientations. Nucleic Acids Res 41:7843–7853

Walsh JM, Hawver LA, Beuning PJ (2011) *Escherichia coli* Y family DNA polymerases. Front Biosci 16:3164–3182

Washington MT, Prakash L, Prakash S (2001) Yeast DNA polymerase eta utilizes an induced-fit mechanism of nucleotide incorporation. Cell 107:917–927

Wilson RC, Pata JD (2008) Structural insights into the generation of single-base deletions by the Y family DNA polymerase dbh. Mol Cell 29:767–779

Wilson RC, Jackson MA, Pata JD (2013) Y-family polymerase conformation is a major determinant of fidelity and translesion specificity. Structure 21:20–31

Wong I, Patel SS, Johnson KA (1991) An induced-fit kinetic mechanism for DNA replication fidelity: direct measurement by single-turnover kinetics. Biochemistry 30:526–537

Wong JH, Fiala KA, Suo Z et al (2008) Snapshots of a Y-family DNA polymerase in replication: substrate-induced conformational transitions and implications for fidelity of Dpo4. J Mol Biol 379:317–330

Xi J, Zhang ZQ, Zhuang ZH et al (2005a) Interaction between the T4 helicase loading protein (gp59) and the DNA polymerase (gp43): unlocking of the gp59-gp43-DNA complex to initiate assembly of a fully functional replisome. Biochemistry 44:7747–7756

Xi J, Zhuang Z, Zhang Z et al (2005b) Interaction between the T4 helicase-loading protein (gp59) and the DNA polymerase (gp43): a locking mechanism to delay replication during replisome assembly. Biochemistry 44:2305–2318

Xia S, Wood M, Bradley MJ et al (2013) Alteration in the cavity size adjacent to the active site of RB69 DNA polymerase changes its conformational dynamics. Nucleic Acids Res 41:9077–9089

Xu C, Maxwell BA, Brown JA et al (2009) Global conformational dynamics of a Y-family DNA polymerase during catalysis. PLoS Biol 7:e1000225

Yang W (2005) Portraits of a Y-family DNA polymerase. FEBS Lett 579:868–872

Zang H, Irimia A, Choi JY et al (2006) Efficient and high fidelity incorporation of dCTP opposite 7,8-dihydro-8-oxodeoxyguanosine by *Sulfolobus solfataricus* DNA polymerase Dpo4. J Biol Chem 281:2358–2372

Zhang Z, Spiering MM, Trakselis MA et al (2005) Assembly of the bacteriophage T4 primosome: single-molecule and ensemble studies. Proc Natl Acad Sci U S A 102:3254–3259

Zhong X, Patel SS, Werneburg BG et al (1997) DNA polymerase beta: multiple conformational changes in the mechanism of catalysis. Biochemistry 36:11891–11900

Zhou BL, Pata JD, Steitz TA (2001) Crystal structure of a DinB lesion bypass DNA polymerase catalytic fragment reveals a classic polymerase catalytic domain. Mol Cell 8:427–437

Zhuang Z, Yoder BL, Burgers PM et al (2006) The structure of a ring-opened proliferating cell nuclear antigen-replication factor C complex revealed by fluorescence energy transfer. Proc Natl Acad Sci U S A 103:2546–2551

Design, Characterization, and Application of Imidazopyridopyrimidine:Naphthyridine Base-Pairing Motifs Consisting of Four Hydrogen Bonds

Noriaki Minakawa and Akira Matsuda

Contents

1 Introduction 114
2 Design, Synthesis, and Evaluation of Novel Base-Pairing Motifs Consisting of Four H-Bonds 117
 2.1 **Im:Im** Base Pairs (First Generation) 117
 2.2 **Im:Na** Base Pairs (Second Generation) 121
3 Application of **ImO^{N}:NaN^{O}** Pair to Develop a Thermally Stabilized Decoy Molecule 125
4 Conclusion and Perspective 127
References 128

Abstract Beginning with a simple question, i.e., how is a DNA duplex thermally stabilized if the base pair has four hydrogen bonds, we designed a series of imidazo [5′,4′:4,5]pyrido[2,3-*d*]pyrimidine (**Im**) nucleosides and evaluated their base-pairing ability when incorporated into DNA duplexes. All possible **Im**:**Im** pairs thermally destabilized the duplexes, although the pair formed four hydrogen bonds between **Im** units. On the other hand, the resulting duplexes were highly stabilized when three pairs were consecutively incorporated. To stabilize the duplexes independent of the mode of incorporation of the pair(s), we designed a series of 1,8-naphthyridine (**Na**) *C*-nucleosides as complementary nucleobases of **Im**. The newly designed **Im**:**Na** base-pairing motifs, especially **ImN^{N}**:**NaO^{O}**, thermally stabilized the duplexes by nearly 10 °C more per pair than the A:T and 8 °C

N. Minakawa
Graduate School of Pharmaceutical Sciences, The University of Tokushima, Shomachi 1-78-1, Tokushima 770-8505, Japan
e-mail: minakawa@tokushima-u.ac.jp

A. Matsuda (✉)
Faculty of Pharmaceutical Sciences, Hokkaido University, Kita-12, Nishi-6, Kita-ku, Sapporo 060-0812, Japan
e-mail: matuda@pharm.hokudai.ac.jp

V.A. Erdmann et al. (eds.), *Chemical Biology of Nucleic Acids*, RNA Technologies, DOI 10.1007/978-3-642-54452-1_7,

more than the G:C pair, independent of the mode of incorporation. We describe herein the results concerning the properties of the base-pairing motifs, **Im**:**Im** and **Im**:**Na** pairs, as well as potential therapeutic application toward a thermally stabilized decoy molecule.

Keywords Four hydrogen bonds • Imidazo[5′,4′:4,5]pyrido[2,3-*d*]pyrimidine nucleoside • 1,8-naphthyridine *C*-nucleoside • Base pair • Thermal stability • Decoy

1 Introduction

Hydrogen bonding (H-bonding) is one of the most typical nonbonded molecular interactions. Therefore, it is important to understand the molecular recognition process in biology and in design of functional organic supramolecules through H-bonding. In the field of nucleic acid chemistry, for example, specific H-bonding between adenine (A) and thymine (T), and guanine (G) and cytosine (C), the so-called Watson–Crick base pairs (Watson and Crick 1953), plays a critical role in not only conserving and transmitting genetic information but also in duplex stability. Since the A:T and G:Cpairs were selected during the evolution of life starting from prebiotic era, these two base pairs should be ideal as a genetic polymer (Szathmary 2003). Since the early 1990s, chemists have expanded their efforts to development of artificial extra base pairs beyond the ideal Watson–Crick base pairs with the goal of developing biological, bioengineering, and therapeutic applications. If these alternative base pairs in the DNA duplex showed specificity in formation of the helical structure and enzymatic replication, they could potentially allow one to expand the genetic code, explore synthetic biology, and create new genetic systems. Thus far, a number of alternative base pairs have been developed, and successful applications have also been reported (Hirao and Kimoto 2012).

Figure 1 illustrated typical artificial base pairs that have been developed to date. Benner and his coworkers proposed a new base pair, **isoG**:**isoC** (Switzer et al. 1989) and investigated its enzymatic incorporation with the goal of extending “the genetic alphabet”. Starting with this pioneering work, they also designed other artificial base pairs with different H-bonding patterns, such as **X**:**κ**(Piccirilli et al. 1990) and **P**:**Z** (Yang et al. 2007). Different from Benner’s work, which was based on H-bonding patterns, Kool and his coworkers developed a novel base pair between **Z** and **F**as a steric isostere of the A:T pair (Switzer et al. 1989; Kool 2002; Kool and Sintim 2006). Although the **Z**:**F** pair has no typical H-bonding, replication studies using this pair revealed that the **Z** and **F** were equally replaceable with A and T, respectively. These results suggested the importance of the shape complementarity of the base pair as well as the H-bonding interaction. Hirao’s group developed new base pairs based on the aforementioned factors, that is, H-bonding interaction and shape complementarity. Thus, they designed base pairs between **s** and **y** (Hirao et al. 2002), and **s** and **z** (Hirao et al. 2004).

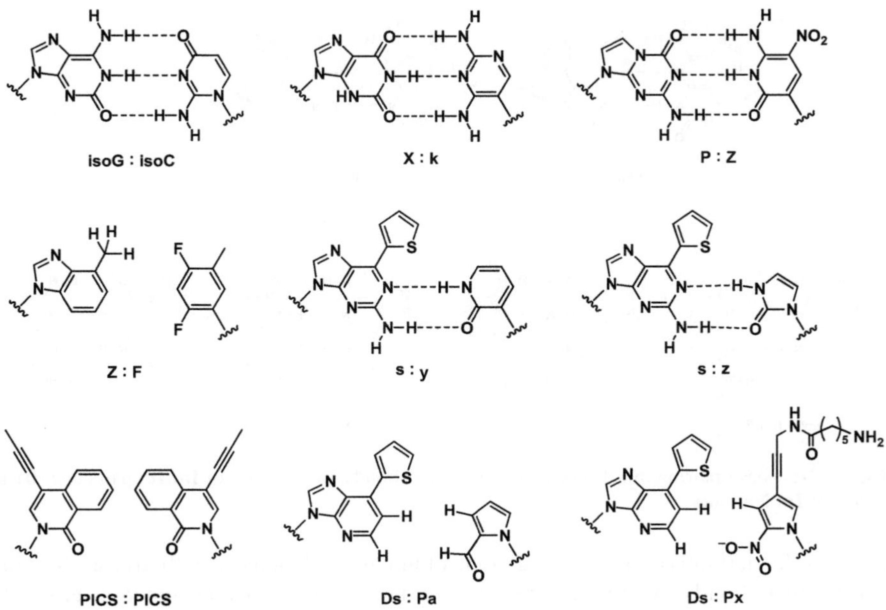

Fig. 1 Artificial base pairs developed to date

Romesberg and his coworkers reported hydrophobic bases without H-bonding interaction toward the development of an artificial extra base pair beyond the Watson–Crick base pairs. They developed a self-pair between **PICS** bases and showed that 5′-triphosphate of the **PICS** was effectively incorporated against **PICS** in template DNA (McMinn et al. 1999). Unlike the **Z:F** pair developed by Kool et al. (Switzer et al. 1989), the **PICS:PICS** pair does not seem to satisfy shape complementarity with a natural purine:pyrimidine base pair. Thus, their results indicated that an additional factor, that is, the hydrophobic interaction, may also contribute to the development of an artificial extra base pair. Idea of hydrophobic interaction as well as shape complementarity was well utilized by Hirao's group to develop base pairs consisting of **Ds** and **Pa** (Hirao et al. 2006) and **Ds** and **Px** (Kimoto et al. 2009). The resulting **Ds:Pa** and **Ds:Px** pairs showed high efficiency and selectivity even in PCR amplification and are currently utilized in efforts toward the expansion of the genetic alphabet and in vitro selection.

As described above, four research groups have been intensively studying the development of artificial extra base pairs beyond the Watson–Crick base pairs and attempting to utilize the resulting new base pair to develop new technologies. In contrast with the aforementioned investigations, we started our research project with two simple questions: (1) why were A:T pair possessing two H-bonds and G:C pair possessing three H-bonds chosen as a genetic alphabet in DNA during evolution and (2) how is a DNA duplex thermally stabilized if the base pair has more than three H-bonds? To answer these questions, we envisioned the preparation of

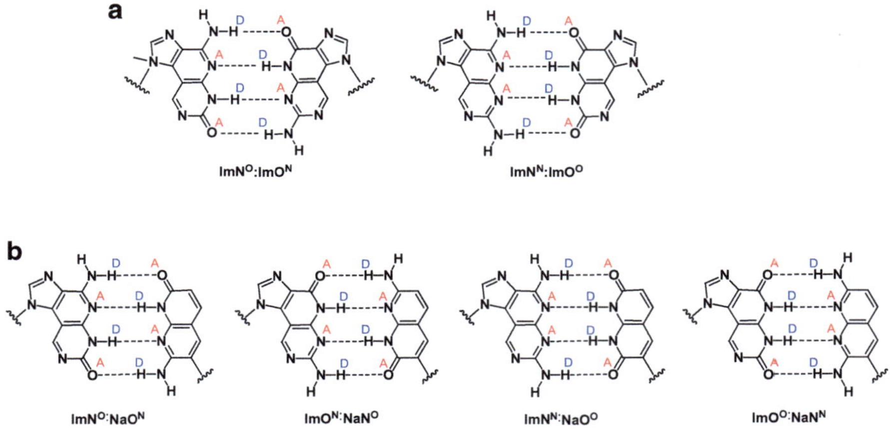

Fig. 2 New base-pairing motifs consisting of four H-bonds. (**a**) A series of **Im**:**Im** pairs and (**b**) a series of **Im**:**Na** pairs

nucleoside derivatives possessing noncanonical nucleobases with the ability to form four H-bonds. As a first generation of the base pairs, we designed a series of imidazo[5′,4′:4,5]pyrido[2,3-*d*]pyrimidine (**Im**) nucleosides, **ImN$^{\mathbf{O}}$**, **ImO$^{\mathbf{N}}$**, **ImN$^{\mathbf{N}}$**, and **ImO$^{\mathbf{O}}$** (Minakawa et al. 2003). As depicted in Fig. 2a, these nucleoside derivatives are expected to form the complementary base pairs, **ImN$^{\mathbf{O}}$**:**ImO$^{\mathbf{N}}$** and **ImN$^{\mathbf{N}}$**:**ImO$^{\mathbf{O}}$**, with four H-bonds when incorporated into oligodeoxynucleotides (ODNs). Contrary to our expectation, neither **ImN$^{\mathbf{O}}$**:**ImO$^{\mathbf{N}}$** nor **ImN$^{\mathbf{N}}$**:**ImO$^{\mathbf{O}}$** thermally stabilized a duplex relative to the natural A:T pair when one pair was incorporated into the duplex. On the other hand, when three pairs were consecutively incorporated, the resulting duplexes were highly stabilized. Since this phenomenon was thought to arise from the shape complementarity factor with a natural purine:pyrimidine base pair, for the second generation we designed 1,8-naphthyridine (**Na**) *C*-nucleosides, **NaN$^{\mathbf{O}}$**, **NaO$^{\mathbf{N}}$**, **NaN$^{\mathbf{N}}$**, and **NaO$^{\mathbf{O}}$**, as complementary units towards a series of **Im** units in the duplexes (Hikishima et al. 2005; Kuramoto et al. 2011). Since the resulting **ImN$^{\mathbf{O}}$**:**NaO$^{\mathbf{N}}$**, **ImO$^{\mathbf{N}}$**:**NaN$^{\mathbf{O}}$**, **ImN$^{\mathbf{N}}$**:**ImO$^{\mathbf{O}}$**, and **ImO$^{\mathbf{O}}$**:**NaN$^{\mathbf{N}}$** pairs satisfy shape complementarity (Fig. 2b), all pairs in the duplexes were thermally stabilized independent of the mode of incorporation. These results suggested the importance of the shape complementarity of the base pair as well as the H-bonding interaction in the thermal stability of the duplex. In addition, we also noticed that the stability of base pair is affected by the arrangement of H-bonds arising from a secondary interaction.

In this chapter, we would like to review our research project to develop new base-pairing motifs consisting of four H-bonds, that is, a series of **Im**:**Im** and **Im**:**Na** pairs and their behavior in the duplexes. In addition, a biological application to develop decoy molecules will also be described.

2 Design, Synthesis, and Evaluation of Novel Base-Pairing Motifs Consisting of Four H-Bonds

2.1 *Im:Im Base Pairs (First Generation)*

To append four H-bonding sites on the nucleobases, we first designed a series of **Im** nucleosides (Minakawa et al. 2003). These nucleosides have DADA (**ImN^{O}**), ADAD (**ImO^{N}**), DAAD (**ImN^{N}**), and ADDA (**ImO^{O}**) H-bonding patterns (D = donor, A = acceptor) in their structures, respectively, and complementary **ImN^{O}**:**ImO^{N}** and **ImN^{N}**:**ImO^{O}** pairs are expected to form with four H-bonds (Fig. 2a). The desired nucleosides were prepared through the Stille coupling reaction of a 5-iodoimidazole nucleoside with an appropriate 5-stannylpyrimidine derivative, followed by intramolecular cyclization. Prior to incorporation of the resulting nucleosides into ODNs, characterization of the H-bonding motifs of the tricyclic nucleosides was investigated with ^{1}H NMR studies. The silylated **Im** derivatives were used for ^{1}H NMR measurements in the nonpolar solvent $CDCl_3$, and the H-bonds were monitored to identify downfield shifts of their amino and amide proton signals. As an example, partial ^{1}H NMR spectra of **ImN^{O}**, an equimolar mixture of **ImN^{O}** and **ImO^{N}**, and **ImO^{N}** at −60 °C are shown in Fig. 3. As can be seen, the amide proton signal of **ImN^{O}** was observed at 13.85 ppm and shifted downfield with decreasing temperatures (appeared at 13.62 ppm at 30 °C). In addition, the amino protons were split into two sets of broad singlets, clearly observed at 10.43 and 6.62 ppm (white arrows). Therefore, the proton signals at 13.85 and 10.43 ppm would be involved in H-bonding, while that at 6.62 ppm is not involved in the H-bonding. Based on these results, **ImN^{O}** is considered to form the homodimer in an antiparallel direction with four H-bonds as illustrated in Fig. 3a. A similar homodimer formation was also observed in **ImO^{N}** (Fig. 3c). The ^{1}H NMR spectrum of the mixture of **ImN^{O}** and **ImO^{N}** was more complicated (Fig. 3b). However, the proton signals, except for those of the homodimers, were easily detected. The proton signals indicated by black arrows would correspond to a new structure formed at the low temperature, and these are assigned to the aglycon protons of the heterodimer, namely, **ImN^{O}**:**ImO^{N}**, with four H-bonds. Thus, the proton signals at 13.58 and 13.37 ppm are likely to be two sets of the amide protons. The amino protons involved in H-bonding can be seen at 10.24 and 9.49 ppm, while those at 6.78 and 5.68 ppm are assigned as non-H-bonded amino protons. In addition, four aromatic protons were also detected between 9.3 and 8.0 ppm.

Since formation of base pair with four H-bonds was confirmed at a nucleoside level, we next investigated base-pair formation based on thermal stability of DNA duplexes. Thus, two sets of complementary 17mers (ODNs I and II, and ODNs III and IV) were prepared to investigate the H-bonding abilities of the tricyclic nucleosides, where the **ImN^{O}**, **ImO^{N}**, **ImN^{N}**, or **ImO^{O}** units were incorporated in the X and Y positions. Thermal stability of duplexes formed by ODN I and complementary ODN II, which contained one molecule of the **ImN^{O}**, **ImO^{N}**,

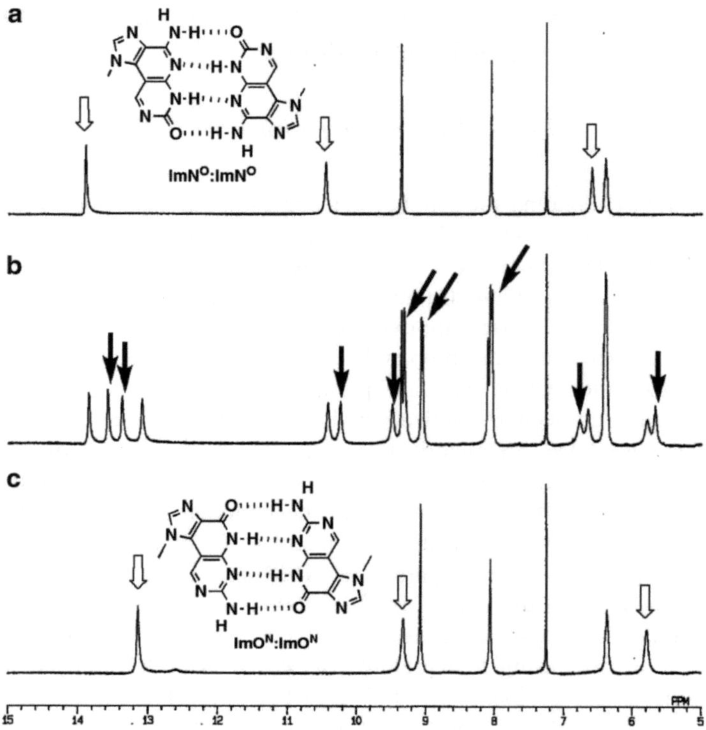

Fig. 3 Partial ^{1}H NMR spectra of (**a**) **ImN^O**, (**b**) a mixture of **ImN^O** and **ImO^N**, and (**c**) **ImO^N** at −60 °C. The amide and amino protons were indicated by *white* and *black arrows*

ImN^N, or **ImO^O** in their X or Y position, was first studied by thermal denaturation in a buffer of 10 mM sodium cacodylate (pH 7.0) containing 0.1 M NaCl. The resulting melting temperatures (T_ms) and ΔT_ms calculated based on the T_m of the duplex (X:Y = A:T, common to ODN I:ODN II and ODN III:ODN IV) are listed in Table 1. Contrary to our expectation, the possible pairs including the complementary **ImN^O:ImO^N** and **ImN^N:ImO^O** pairs were all less stable than natural A:T pair consisting of two H-bonds in ODN I:ODN II. On the other hand, when three molecules of the tricyclic nucleosides were consecutively incorporated into the center of each ODN (ODN III:ODN IV), the thermal stabilization of the duplexes due to the specific base pairings was observed. The best result was obtained in the case of the duplex containing the **ImN^O:ImO^N** pair (T_m = 84.0 °C). The value was between 18.2 and 23.5 °C higher than those of the duplexes containing three consecutive G:C (T_m = 65.8 °C) and A:T pairs (T_m = 60.5 °C), respectively. Consequently, it was found that the **ImN^O:ImO^N** pair stabilized the duplex by about +6 and +8 °C per modification compared with those of the duplexes containing the G:C and A:T pairs, respectively. Although the **ImN^N:ImO^O** pair is

Table 1 Hybridization data of a series of **Im:Im** pairs

ODN I; 5'-GCACCGAA**X**AAACCACG-3' ODN II; 3'-CGTGGCTT**Y**TTTGGTGC-5'				ODN III; 5'-GCACCGA**XXX**AACCACG-3' ODN IV; 3'-CGTGGCT**YYY**TTGGTGC-5'			
X	Y	T_m (°C)	ΔT_m (°C)	X	Y	T_m (°C)	ΔT_m (°C)
ImN^{O}	ImN^{O}	57.2	−3.3	**ImN^{O}**	ImN^{O}	70.0	+9.5
	ImO^{N}	**59.6**	**−0.9**		**ImO^{N}**	**84.0**	**+23.5**
	ImN^{N}	55.1	−5.4		ImN^{N}	46.3	−14.2
	ImO^{O}	54.0	−6.6		ImO^{O}	56.0	−4.5
ImO^{N}	**ImN^{O}**	**58.4**	**−2.1**	**ImO^{N}**	**ImN^{O}**	**83.8**	**+23.3**
	ImO^{N}	56.9	−3.6		ImO^{N}	65.8	+5.3
	ImN^{N}	55.4	−5.1		ImN^{N}	41.5	−19.0
	ImO^{O}	53.3	−7.2		ImO^{O}	57.5	−3.0
ImN^{N}	ImN^{O}	54.5	−6.0	**ImN^{N}**	ImN^{O}	49.4	−11.1
	ImO^{N}	56.1	−4.4		ImO^{N}	57.1	−8.8
	ImN^{N}	54.5	−6.0		ImN^{N}	55.7	−4.8
	ImO^{O}	**56.3**	**−4.2**		**ImO^{O}**	**70.6**	**+10.1**
ImO^{O}	ImN^{O}	54.0	−6.5	**ImO^{O}**	ImN^{O}	58.3	−2.2
	ImO^{N}	54.3	−6.2		ImO^{N}	62.3	+1.8
	ImN^{N}	**56.0**	**−4.5**		**ImN^{N}**	**67.1**	**+6.6**
	ImO^{O}	54.3	−6.2		ImO^{O}	53.5	−7.0
G	C	61.9	+1.4	G	C	65.8	+5.3
A	**T**	**60.5**	**–**	**A**	**T**	**60.5**	**–**
A	G	53.8	−6.7	A	G	47.2	−13.3
A	A	52.4	−8.1	A	A	43.1	−17.4

All measurements were carried out using 3 μM each of ODN in 10 mM sodium cacodylate (pH 7.0) containing 0.1 M NaCl. The ΔT_m values were obtained by subtracting data for the T_m possessing X:Y = A:T from each duplex. The complementary pairs were represented in bold

also expected to form a base pair with four H-bonds (Fig. 2a), the stabilization effect was much smaller than that of the **ImN^{O}:ImO^{N}** pair. A reasonable explanation for the above result would be that this base pair has fewer H-bonds than the expected four. A tautomeric form of **ImO^{O}**, represented as **tImO^{O}**, can be considered, and this could form a base pair with **ImN^{N}** by three H-bonds. The validity of this consideration will be verified in the next section (see Fig. 5).

As described above, it was shown that the H-bonding abilities of the **Im** bases are essential for the stability of the duplexes. However, the stabilizing effect seemed too high to conclude that it was due to the number of H-bonds since the duplexes were thermally stabilized to 1.4–1.8 °C per pair when the H-bonds increased to three (G:C pair) from two (A:T pair). Thus, this effect undoubtedly implies that the stacking interaction arising from extended aromatic surfaces would also be an important factor contributing to stability. For this reason, evaluation of the stacking abilities was examined. According to the method reported by Guckian et al. (1996), a series of duplexes, where Z was added at the end of each paired duplex (dangling end), were prepared, and the T_m values of the duplexes were

Table 2 Stacking abilities of **Im** and **Na** bases

5'-**Z**-CGCGCG-3'
3'-GCGCGC-**Z**-5'

Z	T_m (°C)	$\Delta T_m/Z$ (°C)	Z	T_m (°C)	$\Delta T_m/Z$ (°C)	Z	T_m (°C)	$\Delta T_m/Z$ (°C)
None	40.8							
ImN^O	58.8	+9.0	NaN^O	61.0	+10.1	A	54.9	+7.1
ImO^N	57.4	+8.3	NaO^N	56.7	+8.0	G	54.6	+6.9
ImN^N	57.0	+8.1	NaN^N	59.5	+9.4	C	49.1	+4.2
ImO^O	52.5	+5.9	NaO^O	56.9	+8.1	T	49.1	+4.2

All measurements were carried out using 2.5 μM each of ODN in 10 mM sodium cacodylate (pH 7.0) containing 1 M NaCl. The $\Delta T_m/Z$ values were obtained by 1/2 of subtracting data for the T_m (Z = none) from each duplex

determined (Table 2) (the stacking abilities of a series of **Na**, described in the next section, are also presented in Table 2). As a comparison, the T_m data for non-dangling and natural bases (A, G, C, or T) at the dangling end are also included. The two unpaired Ts added +4.2 °C per base of T_m relative to the self-complementary duplex (Z = none), and the Cs also added a similar stability. The unpaired purines (Z = A and G) showed higher stability relative to the pyrimidines arising from expansion of the aromatic surface. These results agreed with those reported by Guckian et al. Among the tested compounds, $\mathbf{ImN^O}$ showed the highest T_m increase over the unsubstituted duplex of 9.0 °C per base. Other **Im** bases, except for $\mathbf{ImO^O}$, also showed higher stacking ability than those of the natural purines and pyrimidines. From these results, it was revealed that the stacking abilities of the **Im** bases also contributed to increase duplex stability, and this would be enhanced by three consecutive introductions of these bases in the duplexes.

The question that has to be considered is why introduction of one molecule of the **Im:Im** base pairs slightly destabilized the duplexes, while three consecutive introductions of these base pairs, especially the $\mathbf{ImN^O{:}ImO^N}$ pair, markedly stabilized the duplexes? To elucidate these results, we further examined the sequence dependence of T_ms using the duplexes containing the G:$\mathbf{ImN^N}$ pair, which was considered as an alternative to the **Im:Im** pairs. Thus, setting the T_m of the duplex between ODNs V and VI as a control (68.3 °C), the sequence-dependent variation was compared by substituting the G:C pair with the G:$\mathbf{ImN^N}$ pair at various positions (ODN V:ODNsVII–XII). The sequences of ODNs and resulting T_ms are shown in Table 3 (W = $\mathbf{ImN^N}$). It was found that the duplexes became thermally less stable as the numbers of nonconsecutive isolated introductions of the G:$\mathbf{ImN^N}$ pairs increased (ODNs VII, VIII, IX, and X). On the other hand, the duplexes containing the consecutive G:$\mathbf{ImN^N}$ pairs (ODN V:ODNs XI and XII) were more stable than the control duplex. Through comparison of the results of ODN V:ODN IX (T_m = 63.0 °C) and ODN V:ODN XI (T_m = 72.8 °C), it was revealed that consecutive introduction of the preferable base pairs is essential to contribute to the thermal

Table 3 Hybridization data of G:**ImN**^{N} pairs

ODN V	3'-C G T G G C T G G G T T G G T G C-5'	T_m (°C)
ODN VI	5'-G **C** A **C C** G A **C C C** A A **C C** A **C** G-3'	68.3
ODN VII	- - - - - - - - **W** - - - - - - - -	67.2
ODN VIII	- - - - - - - **W** - **W** - - - - - - -	64.3
ODN IX	- - - **W** - - - - **W** - - - - **W** - - -	63.0
ODN X	- **W** - - **W** - - **W** - - - - **W** - - **W** -	55.7
ODNXI	- - - - - - - **W W W** - - - - - - -	72.8
ODN XII	- **W** - **W W** - - **W W W** - - **W W** - **W** -	75.7

All measurements were carried out using 3 μM each of ODN in 10 mM sodium cacodylate (pH 7.0) containing 0.1 M NaCl. Sequences are same as ODN VI except for W = **ImN**^{N}

stability of the duplex. The hypothetical duplex structures are illustrated in Fig. 4. The **ImN**^{O}:**ImO**^{N} (and G:**ImN**^{N}) base pairs form stable H-bonds in the duplex. However, the pairs would spread the width of the helix around the strand where they were introduced. It is known that the average intra-strand C1′–C1′ distance in a canonical Watson–Crick base pair is 10.5 (±0.2) Å, while the distances in purine (anti)–purine(anti) base pairs are typically much longer (Cote and Georgiadis 2001). For example, Privé et al. (1987) reported the C1′–C1′ distance in the G (anti)–A(anti) mismatch to be 12.5 Å in an X-ray structure. The intra-strand C1′–C1′ distance in **ImN**^{O}:**ImO**^{N} (and G:**ImN**^{N}) would be similar to the above-mentioned distance. Therefore, we speculated that thermal destabilization of the duplex occurs at both sides of these pairs. When one pair of the tricyclic nucleoside was incorporated into the duplex, the destabilization factor arising from disruption of the Watson–Crick base pairs next to the pair would be greater than the stabilization arising from the stable H-bonds. Consequently, the duplex would become less stable (Fig. 4a). This undesirable effect would be enhanced as the numbers of the nonconsecutive tricyclic nucleosides increased (Fig. 4b). On the other hand, when the tricyclic nucleosides were consecutively incorporated, the duplex would be thermally stabilized to a great extent since the base pairs between the tricyclic nucleosides have four H-bonds and a strong stacking ability with the adjacent **Im** bases. These stabilization factors would be superior to the conformational destabilization around the boundary of the base pairs (Fig. 4c).

2.2 *Im:Na Base Pairs (Second Generation)*

As described above, the **ImN**^{O}:**ImO**^{N} pair markedly stabilized the duplex arising from noncanonical four H-bonds and strong stacking ability. However, this base pair does not seem to satisfy the shape complementarity, since it behaves like a purine:purine pair. To develop the ideal base pairs beyond the Watson–Crick base pairs with four H-bonds, we next designed a series of **Na** *C*-nucleosides, **NaO**^{N}, **NaN**^{O}, **NaO**^{O}, and **NaN**^{N}, as the complementary units against a series of **Im** units

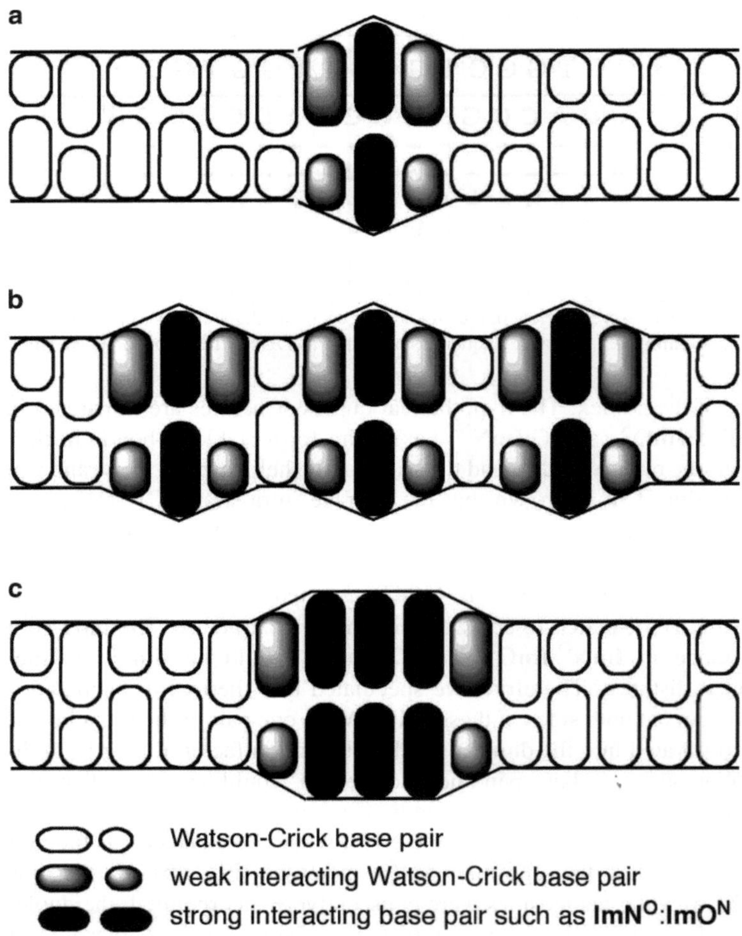

Fig. 4 Hypothetical duplex structures based on our consideration

(Hikishima et al. 2005; Kuramoto et al. 2011). These nucleosides can be recognized as expanded pyrimidine derivatives, and thus, the resulting **Im**:**Na** pairs (Fig. 2b), **ImN**$^{\mathbf{O}}$:**NaO**$^{\mathbf{N}}$, **ImO**$^{\mathbf{N}}$:**NaN**$^{\mathbf{O}}$, **ImN**$^{\mathbf{N}}$:**ImO**$^{\mathbf{O}}$, and **ImO**$^{\mathbf{O}}$:**NaN**$^{\mathbf{N}}$ are expected to stabilize the duplexes not only with four H-bonds and strong stacking ability but also to satisfy the shape complementarity. The desired nucleosides were prepared using the Heck reaction between a glycal and 1,8-naphthyridine derivative. After being converted into phosphoramidite units, three classes of complementary duplexes were synthesized to investigate the base-pairing properties of **Im**:**Na** pairs. As shown in Table 4, the first class consists of duplexes (a series of ODN I:ODN II) containing one X:Y pair at the center of the duplexes (with the noncanonical and natural bases in their X or Y positions). The second class is made up of duplexes

Table 4 Hybridization data of a series of **Im**:**Na** pairs

ODN I; 5'-GCACCGAAXAAACCACG-3' ODN II; 3'-CGTGGCTTYTTTGGTGC-5'				ODN XIII; 5'-GCXCCGAAXAAACCXCG-3' ODN XIV; 3'-CGXGGCTTYTTTGGYGC-5'				ODN III; 5'-GCACCGAXXXAACCACG-3' ODN IV; 3'-CGTGGCTYYYTTGGTGC-5'			
X	Y	T_m (°C)	ΔT_m (°C)	X	Y	T_m (°C)	ΔT_m (°C)	X	Y	T_m (°C)	ΔT_m (°C)
ImN^O^	**NaO^N^**	**56.1**	**+7.5**	**ImN^O^**	**NaO^N^**	**79.6**	**+31.0**	**ImN^O^**	**NaO^N^**	**80.1**	**+31.5**
ImO^N^		50.4	+1.8	ImO^N^		65.8	+17.2	ImO^N^		64.6	+16.0
ImN^N^		48.6	0	ImN^N^		58.3	+9.7	ImN^N^		55.5	+6.9
ImO^O^		49.3	+0.7	ImO^O^		57.5	+8.9	ImO^O^		56.5	+7.9
ImN^O^	NaN^O^	53.3	+4.7	ImN^O^	NaN^O^	66.7	+18.1	ImN^O^	NaN^O^	67.2	+18.6
ImO^N^		**56.4**	**+7.8**	**ImO^N^**		**81.4**	**+32.8**	**ImO^N^**		**79.0**	**+30.2**
ImN^N^		51.5	+2.9	ImN^N^		61.7	+13.1	ImN^N^		59.1	+10.5
ImO^O^		51.0	+2.4	ImO^O^		58.8	+10.2	ImO^O^		61.6	+13.0
ImN^O^	NaO^O^	54.6	+6.0	ImN^O^	NaO^O^	70.5	+21.9	ImN^O^	NaO^O^	76.8	+28.2
ImO^N^		53.1	+4.5	ImO^N^		72.7	+24.1	ImO^N^		70.7	+22.1
ImN^N^		**60.0**	**+11.4**	**ImN^N^**		**88.0**	**+39.4**	**ImN^N^**		**88.9**	**+40.3**
ImO^O^		51.8	+3.2	ImO^O^		65.8	+17.2	ImO^O^		69.7	+21.1
ImN^O^	NaN^N^	50.4	+1.8	ImN^O^	NaN^N^	64.7	+16.1	ImN^O^	NaN^N^	62.8	+14.2
ImO^N^		50.9	+2.3	ImO^N^		68.0	+19.4	ImO^N^		64.1	+15.5
ImN^N^		53.4	+4.8	ImN^N^		70.7	+22.1	ImN^N^		70.0	+21.4
ImO^O^		**56.5**	**+7.9**	**ImO^O^**		**80.5**	**+31.9**	**ImO^O^**		**81.3**	**+32.7**
G	C	49.9	+1.3	G	C	56.7	+8.1	G	C	55.2	+6.6
A	T	48.6	–	A	T	60.5	–	A	T	48.6	–

All measurements were carried out using 3 μM each of ODN in 10 mM sodium cacodylate (pH 7.0) containing 1 mM NaCl. The ΔT_m values were obtained by subtracting data for the T_m possessing X:Y = A:T from each duplex. The complementary pairs were represented in bold

(a series of ODN XIII:ODN XIV) having three nonconsecutive X:Y pairs, and the third class (a series of ODN III:ODN IV) is made up of three consecutive X:Y pairs. Measurement of the thermal stability was carried out in a buffer of 10 mM sodium cacodylate (pH 7.0) containing 1 mM NaCl. The resulting melting temperature T_ms and the ΔT_ms values calculated based on the T_m of the duplex (X:Y = A:T, common to all duplexes) are listed in Table 4. As expected, the complementary base pairs, **ImN^{O}:NaO^{N}**, **ImO^{N}:NaN^{O}**, **ImN^{N}:ImO^{O}**, and **ImO^{O}:NaN^{N}**, are all thermally stabilized the duplexes. Thus, when these pairs were incorporated into ODN I:ODN II in which the **Im**:**Im** base pairs destabilized the duplexes, the **ImO^{N}**:**NaN^{O}**, **ImN^{O}:NaO^{N}**, and **ImO^{O}:NaN^{N}** pairs stabilized the duplexes by +7.8 °C, +7.5 °C, and +7.9 °C respectively, relative to the duplex containing the A:T pair. The **ImN^{N}:NaO^{O}** pair stabilized the duplex by +11.4 °C, which was the highest among the four pairs. The preferable base-pairing motif by the **ImN^{N}:NaO^{O}** pair was emphasized in the series of ODN XIII:ODN XIV and ODN III:ODN IV. Both duplexes containing three **ImN^{N}:NaO^{O}** pairs were stabilized by ca. +40 °C, which was a dramatic improvement. From these results, it can be concluded that the newly designed **Im**:**Na** base-pairing motifs, especially **ImN^{N}:NaO^{O}**, thermally stabilized the duplexes by nearly 10 °C more per pair than the A:T pair and 8 °C more than the

G:C pair independent of the mode of incorporation of the base pair(s). This effect is presumably caused by the noncanonical four H-bonds and the enhanced stacking abilities of not only **Im** bases but also **Na** bases (Table 2) as well as the shape complementarity of the **Im**:**Na** pairs.

One may wonder why the **ImN^{N}**:**NaO^{O}** pair showed the highest T_m value relative to the other three pairs despite formation of base pairs with four H-bonds. As one explanation of this phenomenon, Jorgensen and Pranata suggested the importance of secondary interaction for the stability of the hydrogen-bonded complexes (Jorgensen and Pranata 1990). They revealed that a pair of 9-methylguanine:1-methylcytosine is much more thermodynamically stable than that of 1-methyluracil:2,6-diaminopyridine despite consisting of three H-bonds in both pairs. They explained that this comes from the secondary interaction arising from the arrangement of H-bonds, and the validity of their consideration is well demonstrated and evaluated for many complexes possessing a variety of H-bonding patterns (Murray and Zimmerman 1992; Beijer et al. 1998; Djurdjevic et al. 2007; Leigh et al. 2013). Our results can also be understood in view of their hypothesis. The **ImN^{O}**:**NaO^{N}** pair has a DADA:ADAD H-bonding pattern (Fig. 5). In this pair, six repulsive (−6) secondary interactions (represented by dotted lines) arising from D–D and A–A repulsion have to be considered together with four primary H-bonds. Accordingly, the overall strength of interaction of the **ImN^{O}**:**NaO^{N}** pair can be estimated as four primary H-bonds and six repulsive secondary interactions (−6) represented as "index". The **ImO^{N}**:**NaN^{O}** pair possessing an ADAD:DADA H-bonding pattern is expected to have the same overall strength of interaction. This estimation agrees well with the calculated T_m values of the **ImN^{O}**:**NaO^{N}** and **ImO^{N}**:**NaN^{O}** pairs (+7.5 °C vs. +7.8 °C). Since the **ImN^{N}**:**NaO^{O}** pair has a DAAD:ADDA H-bonding pattern, this pair is expected to have four repulsive (−4) and two attractive (+2) secondary interactions (represented by bold lines) together with four primary H-bonds. Accordingly, the overall strength of the interactions of the **ImN^{N}**:**NaO^{O}** pair can be estimated as four primary H-bonds and −2 of secondary interactions, which was, in fact, thermally more stable than the **ImN^{O}**:**NaO^{N}** and **ImO^{N}**:**NaN^{O}** pairs (+11.4 °C vs. +7.5 and +7.8 °C). For the **ImO^{O}**:**NaN^{N}** pair, if this pair has an ADDA:DAAD H-bonding pattern as illustrated in Fig. 2b, the overall strength of interaction of this pair can be estimated as four H-bonds and −2 of secondary interactions, which should be same as those of the **ImN^{N}**:**NaO^{O}** pair. However the thermal stability of this pair (+7.9 °C) was rather low compared with the **ImN^{N}**:**NaO^{O}** pair and almost the same as those of the **ImN^{O}**:**NaO^{N}** and **ImO^{N}**:**NaN^{O}** pairs. As described in the previous section, we suggested that the **ImO^{O}** exists as a tautomeric form, represented as **tImO^{O}**, possessing an ADAA H-bonding pattern. This being the case, then the base pair between **tImO^{O}** and **NaN^{N}** should have four repulsive (−4) and two attractive (+2) secondary interactions together with three primary H-bonds (Fig. 5). Therefore, the overall strength of interaction of the **tImO^{O}**:**NaN^{N}** pair can be estimated as three H-bonds and −2 of secondary interactions, which was thermally similar to those of four H-bonds and −6 of secondary interactions. The attached $-\Delta G^0$ data also support these

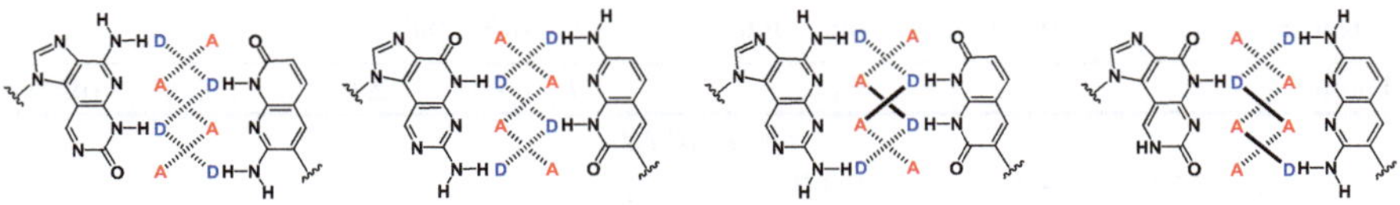

	ImN^O:NaO^N	ImO^N:NaN^O	ImN^N:NaO^O	$tImO^O$:NaN^N
H-bonds	4	4	4	3
index	–6	–6	–2	–2
ΔT_m	+7.5 °C	+7.8 °C	+11.4 °C	+7.9 °C
$-\Delta G^0$	17.64 kcal/mol	17.93 kcal/mol	21.64 kcal/mol	17.07 kcal/mol

Fig. 5 Consideration of thermal and thermodynamic stability of a series of **Im**:**Na** base pairs. The *dotted line* represented repulsive secondary interaction and the *bold line* represented attractive secondary interaction. The index represented the sum of secondary interactions

considerations. Thus, the **ImN**$^{\mathbf{N}}$:**NaO**$^{\mathbf{O}}$ pair, through a favorable contribution to the enthalpy of formation, was most thermodynamically stable, and the remaining pairs showed similar thermodynamic parameters.

3 Application of ImO^{N}:NaN^{O} Pair to Develop a Thermally Stabilized Decoy Molecule

As described above, a series of **Im**:**Na** base pairs markedly stabilized duplexes independent of the mode of incorporation of the pairs due to the noncanonical four H-bonds, the stacking effect of the expanded aromatic surfaces, and shape complementarity with the Watson–Crick base pairs. These successful results prompted us to utilize the **Im**:**Na** base pairs toward therapeutic applications. Since the **Im**:**Na** base pairs markedly stabilized DNA duplexes, application toward a decoy strategy seemed to be ideal. Because the decoy molecule can compete with a target sequence in promoter and enhancer for binding with transcription regulatory proteins, this molecule is expected to be a powerful tool for regulating gene expression (Cho-Chung et al. 1999). However, appending not only nuclease resistance but also thermal stability to the decoy molecule by chemical modification is required. The most common strategy employed is the replacement of an oxygen atom in the phosphodiester linkage with a sulfur atom to give phosphorothioate-modified decoy molecules (Bielinska et al. 1990). Although these molecules are highly resistant to nuclease digestion, thermal stability is reduced relative to the unmodified decoy molecule due to the stereochemistry created at the phosphate center (Stec et al. 1995). To append thermal stability, efforts have been dedicated to the preparation of a circular decoy molecule and to investigations of its effectiveness in vitro and in vivo (Ahn et al. 2002a, b; Nakane et al. 2008). As alternative approaches, modifications of the sticky ends of decoy molecules, for example, by peptide nucleic acids (PNAs) (Romanelli et al. 2001) and locked nucleic acids (LNAs) (Crinelli et al. 2002) have also been examined. It would therefore be worth

Table 5 Properties of decoy molecules containing **ImO^N:NaN^O** pairs

Duplex	T_m (°C)	ΔT_m (°C)	IC_{50} (nM)
Natural-NF			
5′-CGCTGGGGACTTTCCACGG-**3′** **3′**-GCGACCCCTGAAAGGTGCC-**5′**	55.5	–	22.5 ± 4.7
2-NF			
5′-ImO^{N}GCTGGGGACTTTCCACG**ImO^N-3′** **3′-NaN^O**CGACCCCTGAAAGGTGC**NaN^O-5′**	60.6	+5.1	20.1 ± 13.3
4-NF			
5′-ImO^{N}ImO^{N}CTGGGGACTTTCCAC**ImO^{N}ImO^N-3′** **3′-NaN^{O}NaN^O**GACCCCTGAAAGGTG**NaN^{O}NaN^O-5′**	76.1	+20.6	10.9 ± 4.8

The T_m measurement was carried out using 3 μM each of ODN in 10 mM sodium cacodylate (pH 7.0) containing 1 mM NaCl. The sequences of underlined are binding site of NF-κB

investigating whether duplexes possessing the **Im**:**Na** base pairs act as decoy molecules (Hikishima et al. 2006).

The duplexes used in this experiment are shown in Table 5. All duplexes include an NF-κB binding site (underlined), and **2-NF** contains one each of an **ImO^N:NaN^O** pair and **4-NF** contains two each of **ImO^N:NaN^O** pairs on its 3′- and 5′-ends. Characterization of the duplexes and their application to decoy molecules were examined by comparison with the duplex, **natural-NF**. The T_ms and the ΔT_ms values were calculated based on the T_m of **natural-NF**. The **2-NF** showed a T_m value of 60.6 °C which is +5.1 °C higher than that of **natural-NF**. This result indicated that the **ImO^N:NaN^O** pair stabilized the duplex by +2.5 °C per pair, which was much less than that of our previous case (ca +7.5 °C per pair). When the **ImO^N: NaN^O** pairs were consecutively incorporated, the resulting duplex, **4-NF**, was stabilized by +20.6 °C relative to **natural-NF**. Although the average stabilizing effect is calculated to be +5.1 °C per pair, one can see a drastic improvement of thermal stability between **2-NF** and **4-NF**. This indicates that **4-NF** was stabilized by +15.5 °C relative to **2-NF** and that the additional stabilizing effect is estimated to be +7.8 °C per pair, which is approximately the same as our previous data. These results indicated that, as well as H-bonds and shape complementarity, stacking ability with the adjacent nucleobases is an important stabilizing factor (Holmes et al. 2003). Thus, incorporation of one pair each of **ImO^N:NaN^O** on the sticky ends of the duplex is insufficient for thermal stabilization, and at least two consecutive incorporations of pairs are required.

Since the duplexes containing the **ImO^N:NaN^O** pairs on their sticky ends were thermally stabilized, we next turned our attention to thermally stabilized decoy molecules. To check the potency of each molecule, the binding affinity of the molecule to NF-κB (p50) was examined in a competition assay. The experiments were performed in a buffer solution with increasing concentrations of the unlabeled duplex, i.e., **2-NF** or **4-NF**, together with a fixed amount of the radiolabeled **natural-NF**. As a positive control, self-competition by unlabeled **natural-NF** was also examined. The calculated IC_{50} values of **2-NF** and **4-NF** were 20.1 ± 13.3 and 10.9 ± 4.8 nM, respectively. These values were equal to and

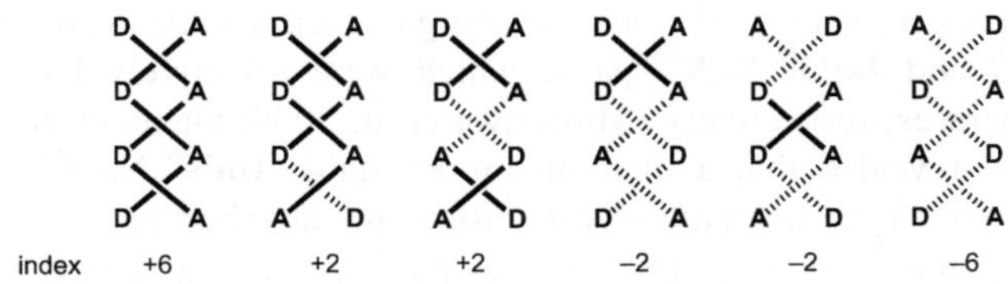

Fig. 6 Possible base-pairing motifs and their indexes consisting of four H-bonds

greater than that of **natural-NF** (22.5 ± 4.7 nM). These data indicate that the duplexes possessing **ImO^N:NaN^O** pairs on their sticky ends do not distort the helical structure and thus act as decoy molecules with high affinity to the NF-κB (p50) protein.

4 Conclusion and Perspective

During the past a decade, we developed a series of noncanonical base pairs consisting of four H-bonds. We designed two sets of base pairs, **ImN^O:ImO^N** and **ImN^N:ImO^O**, and investigated their effect for thermal stabilization of the duplex. Interestingly, incorporation of one base pair destabilized the duplexes, while three consecutive incorporation of the base pair, especially **ImN^O:ImO^N**, markedly stabilized the duplexes. This contradictory result can be explained as follows. The **ImN^O:ImO^N** pair stabilized the duplex due to four H-bonds and strong stacking ability with the adjacent bases. However, this base pair does not satisfy shape complementarity, since it behaves like a purine:purine pair. Accordingly, the stabilization effect of the **ImN^O:ImO^N** pair depends on the mode of incorporation. To satisfy shape complementarity, we developed a series of **Na** *C*-nucleosides as the complementary units against a series of **Im** units. As expected, the resulting **Im**:**Na** base pairs markedly stabilized the duplexes independent of the mode of incorporation of the pairs due to the noncanonical four H-bonds, the stacking effect of the expanded aromatic surfaces, and shape complementarity with the Watson–Crick base pairs. In addition, we also revealed the importance of secondary interaction arising from the arrangement of H-bonds for the stability of the duplexes. Among the possible H-bonding patterns consisting of four H-bonds (Fig. 6), the **ImN^O:NaO^N** and **ImO^N:NaN^O** pairs show the most unfavorable secondary interaction (index = −6). The **ImN^N:NaO^O** pair, which showed the best result in our study, is also an unfavorable base pair for the secondary interaction (index = −2). In our ongoing project, one objective we may pursue would be the design, synthesis, and evaluation of base pairs possessing favorable secondary interaction, such as DDDD:AAAA (index = +6) and AADD:DDAA (index = +2).

In addition to the fundamental characterization of the **Im**:**Na** base pairs, we demonstrated the utility of the **ImO^N:NaN^O** pair for developing a thermally stabilized decoy molecule. We also investigated enzymatic recognition of the **ImN^{O}:NaO^{N}** and **ImO^{N}:NaN^{O}** pairs, which were selectively incorporated by DNA polymerases, respectively (Minakawa et al. 2009; Ogata et al. 2009). Furthermore, we revealed that a thermally more stable **ImN^{N}:NaO^{O}** pair showed higher selectivity for enzymatic incorporation (unpublished results). Given these results, development of new **Im:Na** base pairs possessing DDDD:AAAA and AADD:DDAA H-bonding patterns should be an attractive challenge in our continuing research project. Further investigations would appear as articles, and the complete results would be summarized elsewhere as a review article.

Acknowledgments We thank all of our colleagues, especially Dr. N. Kojima, Dr. S. Hikishima, and Mr. K. Kuramoto, who contributed to our research described here. This work was supported by Grants-in-Aid for Scientific Research from the Ministry of Education, Culture, Sports, Science and Technology in Japan.

References

Ahn JD, Morishita R, Kaneda Y et al (2002a) Novel E2F decoy oligodeoxynucleotides inhibit *in vitro* vascular smooth muscle cell proliferation and *in vivo* neointimal hyperplasia. Gene Ther 9:1682–1692

Ahn JD, Morishita R, Kaneda Y et al (2002b) Inhibitory effects of novel AP-1 decoy oligodeoxynucleotides on vascular smooth muscle cell proliferation in vitro and neointimal formation in vivo. Circ Res 90:1325–1332

Beijer FH, Sijbesma RP, Kooijman H et al (1998) Strong dimerization of ureidopyrimidones via quadruple hydrogen bonding. J Am Chem Soc 120:6761–6769

Bielinska A, Shivdasani R, Zhang L et al (1990) Regulation of gene expression with double-stranded phosphorothioate oligonucleotides. Science 250:997–1000

Cho-Chung YS, Park YG, Lee YN (1999) Oligonucleotides as transcription factor decoys. Curr Opin Mol Ther 1:386–392

Cote ML, Georgiadis MM (2001) Structure of a pseudo-16-mer DNA with stacked guanines and two G-A mispairs complexed with the N-terminal fragment of Moloney murine leukemia virus reverse transcriptase. Acta Crystallogr D Biol Crystallogr 57:1238–1250

Crinelli R, Bianchi M, Gentilini L et al (2002) Design and characterization of decoy oligonucleotides containing locked nucleic acids. Nucleic Acids Res 30:2435–2443

Djurdjevic S, Leigh DA, McNab H et al (2007) Extremely strong and readily accessible AAA-DDD triple hydrogen bond complexes. J Am Chem Soc 129:476–477

Guckian KM, Schweitzer BA, Ren RXF et al (1996) Experimental measurement of aromatic stacking affinities in the context of duplex DNA. J Am Chem Soc 118:8182–8183

Hikishima S, Minakawa N, Kuramoto K et al (2005) Synthesis of 1,8-naphthyridine C-nucleosides and their base-pairing properties in oligodeoxynucleotides: thermally stable naphthyridine: imidazopyridopyrimidine base-pairing motifs. Angew Chem Int Ed Engl 44:596–598

Hikishima S, Minakawa N, Kuramoto K et al (2006) Synthesis and characterization of oligodeoxynucleotides containing naphthyridine:imidazopyridopyrimidine base pairs at their sticky ends. Application as thermally stabilized decoy molecules. Chembiochem 7:1970–1975

Hirao I, Kimoto M (2012) Unnatural base pair systems toward the expansion of the genetic alphabet in the central dogma. Proc Jpn Acad Ser B Phys Biol Sci 88:345–367

Hirao I, Ohtsuki T, Fujiwara T et al (2002) An unnatural base pair for incorporating amino acid analogs into proteins. Nat Biotechnol 20:177–182

Hirao I, Harada Y, Kimoto M et al (2004) A two-unnatural-base-pair system toward the expansion of the genetic code. J Am Chem Soc 126:13298–13305
Hirao I, Kimoto M, Mitsui T et al (2006) An unnatural hydrophobic base pair system: site-specific incorporation of nucleotide analogs into DNA and RNA. Nat Methods 3:729–735
Holmes SC, Arzumanov AA, Gait MJ (2003) Steric inhibition of human immunodeficiency virus type-1 Tat-dependent trans-activation in vitro and in cells by oligonucleotides containing 2′-O-methyl G-clamp ribonucleoside analogues. Nucleic Acids Res 31:2759–2768
Jorgensen WL, Pranata J (1990) Importance of secondary interactions in triply hydrogen bonded complexes: guanine-cytosine vs uracil-2,6-diaminopyridine. J Am Chem Soc 112:2008–2010
Kimoto M, Kawai R, Mitsui T et al (2009) An unnatural base pair system for efficient PCR amplification and functionalization of DNA molecules. Nucleic Acids Res 37:e14
Kool ET (2002) Replacing the nucleobases in DNA with designer molecules. Acc Chem Res 35:936–943
Kool ET, Sintim HO (2006) The difluorotoluene debate – a decade later. Chem Commun (Camb) 35:3665–3675
Kuramoto K, Tarashima N, Hirama Y et al (2011) New imidazopyridopyrimidine:naphthyridine base-pairing motif, ImN^N:NaO^O, consisting of a DAAD:ADDA hydrogen bonding pattern, markedly stabilize DNA duplexes. Chem Commun 47:10818–10820
Leigh DA, Robertson CC, Slawin AMZ et al (2013) AAAA-DDDD quadruple hydrogen-bond arrays featuring NH···N and CH···N hydrogen bonds. J Am Chem Soc 135:9939–9943
McMinn DL, Ogawa AK, Wu Y et al (1999) Efforts toward expansion of the genetic alphabet: DNA polymerase recognition of a highly stable, self-pairing hydrophobic base. J Am Chem Soc 121:11585–11586
Minakawa N, Kojima N, Hikishima S et al (2003) New base pairing motifs. The synthesis and thermal stability of oligodeoxynucleotides containing imidazopyridopyrimidine nucleosides with the ability to form four hydrogen bonds. J Am Chem Soc 125:9970–9982
Minakawa N, Ogata S, Takahashi M et al (2009) Selective recognition of unnatural imidazopyridopyrimidine:naphthyridine base pairs consisting of four hydrogen bonds by the Klenow fragment. J Am Chem Soc 131:1644–1645
Murray TJ, Zimmerman SC (1992) New triply hydrogen bonded complexes with highly variable stabilities. J Am Chem Soc 114:4010–4011
Nakane M, Ichikawa S, Matsuda A (2008) Triazole-linked dumbbell oligodeoxynucleotides with NF-κB binding ability as potential decoy molecules. J Org Chem 73:1842–1851
Ogata S, Takahashi M, Minakawa N et al (2009) Unnatural imidazopyridopyrimidine: naphthyridine base pairs: selective incorporation and extension reaction by Deep Vent (exo−) DNA polymerase. Nucleic Acids Res 37:5602–5609
Piccirilli JA, Benner SA, Krauch T et al (1990) Enzymatic incorporation of a new base pair into DNA and RNA extends the genetic alphabet. Nature 343:33–37
Privé G, Heinemann U, Chandrasegaran S et al (1987) Helix geometry, hydration, and G.A mismatch in a B-DNA decamer. Science 238:498–504
Romanelli A, Pedone C, Saviano M et al (2001) Molecular interactions with nuclear factor kappaB (NF-kappaB) transcription factors of a PNA-DNA chimera mimicking NF-kappaB binding sites. Eur J Biochem 268:6066–6075
Stec WJ, Grajkowski A, Kobylanska A et al (1995) Diastereomers of nucleoside 3′-O-(2-thio-1,3,2-oxathia(selena)phospholanes): building blocks for stereocontrolled synthesis of oligo (nucleoside phosphorothioate)s. J Am Chem Soc 117:12019–12029
Switzer C, Moroney SE, Benner SA (1989) Enzymatic incorporation of a new base pair into DNA and RNA. J Am Chem Soc 111:8322–8323
Szathmary E (2003) Why are there four letters in the genetic alphabet? Nat Rev Genet 4:995–1001
Watson JD, Crick FHC (1953) Molecular structure of nucleic acids: a structure for deoxyribosenucleic acid. Nature 171:737–738
Yang Z, Sismour AM, Sheng P et al (2007) Enzymatic incorporation of a third nucleobase pair. Nucleic Acids Res 35:4238–4249

Creation of Unnatural Base Pair Systems Toward New DNA/RNA Biotechnologies

Michiko Kimoto and Ichiro Hirao

Contents

1 Introduction 132
2 Replication System by Genetic Alphabet Expansion Using the Ds–Px Pair 135
 2.1 Sequencing Involving the Ds–Px Pair 135
 2.2 Replication Fidelity Involving the Ds–Px Pair 138
3 DNA Aptamer Generation 143
 3.1 SELEX Method Involving the Ds–Px Pair 143
 3.2 Aptamer Generation Targeting Proteins 145
4 Conclusion and Future Studies 146
References 147

Abstract The formation of A–T (U) and G–C base pairs is the fundamental rule of genetic information flow in the central dogma of living organisms on Earth. By the same token, the present genetic recombinant biotechnology relies on this base pair complementarity, but the use of only two types of base pairs restricts further technological advancement. Thus, the expansion of the genetic alphabet of DNA, using an artificial third base pair (unnatural base pair), has the potential to open the door to new biotechnologies. Polymerase reactions mediated by an unnatural base pair could site specifically incorporate extra functional components of interest within nucleic acids. To pursue this possibility, researchers have sought to create unnatural base pairs that function as a third base pair, along with the natural base pairs, in polymerase reactions. Recently, several unnatural base pairs with high efficiency and selectivity in in vitro replication and transcription have been created. Here, we describe the recent advancements in unnatural base pair development, focusing on our studies and applications to DNA aptamer selection.

M. Kimoto • I. Hirao (✉)
RIKEN Center for Life Science Technologies, 1-7-22 Suehiro-cho, Tsurumi-ku, Yokohama, Kanagawa 230-0045, Japan
e-mail: ihirao@riken.jp

V.A. Erdmann et al. (eds.), *Chemical Biology of Nucleic Acids*, RNA Technologies,
DOI 10.1007/978-3-642-54452-1_8,

Keywords Genetic alphabet expansion • Unnatural base pair • PCR • Aptamer • SELEX

1 Introduction

Nucleic acids are unique molecules that are amplifiable by polymerase reactions. They display a wide variety of functions, including ligands (aptamers) and catalysts (ribozymes). Thus, an evolutionary engineering method involving repetitive cycles of selection and amplification will provide the ability to artificially generate new functional nucleic acids. In 1990, in vitro selection methods (SELEX) were first reported for creating RNA aptamers that bind to target molecules, such as dyes and proteins (Ellington and Szostak 1990; Tuerk and Gold 1990), and a ribozyme that specifically cleaves single-stranded DNA (Robertson and Joyce 1990). Since then, an enormous number of aptamers and ribozymes have been generated, and the evolutionary engineering method has prompted the rapid advancement of DNA/RNA biotechnologies, including diagnostic and therapeutic applications of aptamers. However, only one aptamer, which binds to human vascular endothelial growth factor-165 (VEGF165), has been approved as a treatment for age-related macular degeneration (Ruckman et al. 1998; Lee et al. 2005). While there are several reasons for this slow progress in aptamer therapeutics, one significant task is to increase the selectivity and affinity of the aptamer to the target molecule, for the first screening in drug development.

One of the attractive methods for increasing nucleic acid functionality is the expansion of the genetic alphabet, by introducing artificial extra bases (unnatural bases) into nucleic acids. As compared with proteins comprising 20 standard amino acids, nucleic acids consist of nucleotides, bearing only four bases with similar chemical and physical properties. Thus, introducing unnatural bases with different properties into DNA and RNA might improve nucleic acid functionalities. The DNA fragments containing unnatural bases must function in replication and transcription, and thus the unnatural base pairs developed as a third base pair should be recognized in nucleic acid polymers by polymerases (Fig. 1).

The idea of the creation of unnatural base pairs was mentioned first by Alexander Rich (1962). He wrote "In general, the trend in macroscopic organic evolution is toward increased complexity," and proposed an unnatural base pair between isoguanine (**isoG**) and isocytosine (**isoC**). The **isoG–isoC** pair has different hydrogen bonding geometry from those of the natural A–T and G–C pairs (Fig. 2a–c). The nucleotide derivatives of **isoG** and **isoC** were chemically synthesized, and they examined polymerase reactions involving the **isoG–isoC** pair (Switzer et al. 1989). They also designed other variations of unnatural base pairs with different hydrogen bonding patterns (Piccirilli et al. 1990; Horlacher et al. 1995). Although these unnatural base pairs were less selective and efficient in replication and transcription

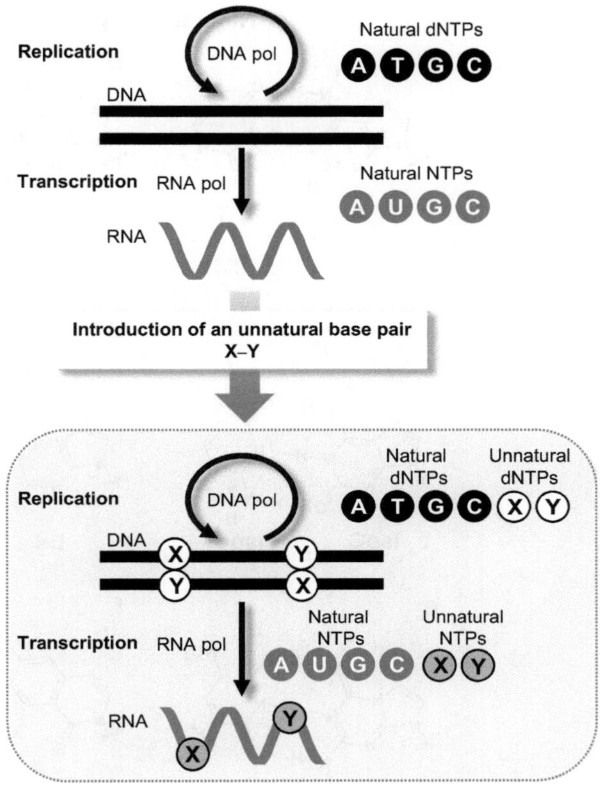

Fig. 1 Genetic alphabet expansion by introducing an artificial third base pair (unnatural base pair, X–Y) that works in replication and transcription. This unnatural base pair system allows the site-specific incorporation of the unnatural components into nucleic acids through polymerase reactions

relative to the natural base pairs, these initial experiments suggested the possibility of genetic alphabet expansion.

Since the first trials, several researchers have tackled the creation of unnatural base pairs that exhibit highly selective pairing complementarity in polymerase reactions (Kool 2000; Henry and Romesberg 2003; Benner 2004; Hirao 2006b; Bergstrom 2009; Kimoto et al. 2011; Hirao and Kimoto 2012). The triphosphate substrates of unnatural bases must be site specifically incorporated by polymerases into DNA or RNA, opposite their pairing partners in templates. In particular, replication requires extremely high selectivity of each cognate pairing of the natural and unnatural base pairs. This is because the replicated DNAs are subsequently used as templates, and thus even slight misincorporation (<1 %) accumulates through repetitive amplification. For example, if the selectivity of an unnatural base pairing is 99.0 % in each cycle of PCR, only ~74 % of the unnatural base pair (if the amplification efficiency = 100% per cycle) will theoretically exist in the amplified DNA after 30 cycles of PCR ($0.99^{30} = \sim 0.74$). However, if the selectivity is 99.9 %, then ~97 % of the unnatural base pair will survive after 30 cycles of PCR ($0.999^{30} = \sim 0.97$).

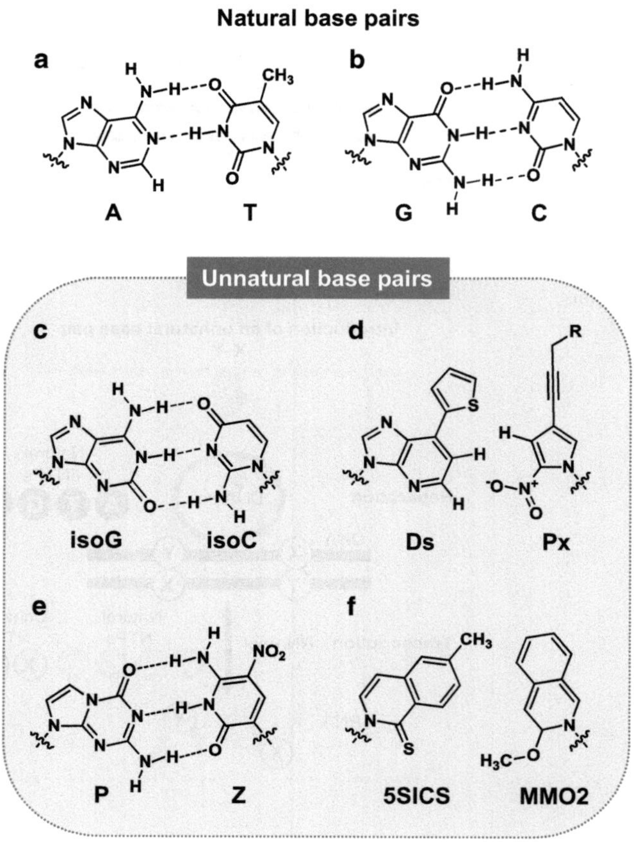

Fig. 2 Chemical structures of the natural and unnatural base pairs. (**a**) The natural A–T pair. (**b**) The natural G–C pair. (**c**) The unnatural **isoG–isoC** pair. (**d**) The unnatural **Ds–Px** pair developed by Hirao's group. (**e**) The unnatural **P–Z** pair developed by Benner's group. (**f**) The unnatural **5SICS–MMO2** pair developed by Romesberg's group

For 15 years, our group has been developing unnatural base pairs by thoroughly pursuing the shape complementarity between pairing bases (Hirao et al. 2002, 2006, 2012; Hirao and Kimoto 2012). As a consequence, we have developed several unnatural base pairs that function in replication, transcription, and/or translation. Among them, the hydrophobic base pair between 7-(2-thienyl)imidazo[4,5-*b*]pyridine (**Ds**) and 4-propynyl-2-nitropyrrole (**Px**) exhibits high fidelity and efficiency in replication (Fig. 2d) (Kimoto et al. 2009; Yamashige et al. 2012). The shape of the **Ds** base only fits well with that of the **Px** base. The **Ds** base is larger than the natural purine bases and the **Px** base is smaller than the natural pyrimidine bases, and thus mispairing with the natural bases is prevented. Furthermore, several other ideas were introduced into the design of these unnatural bases to increase the pairing fidelity. For example, the oxygen of the nitro group in **Px** electrostatically repels the 1-nitrogen of A, preventing the A–**Px** mispairing. The fidelity of the **Ds–Px** pair is around 99.9 % in PCR, using exonuclease-proficient Deep Vent DNA polymerase. Importantly, for practical uses, the misincorporation rates of these unnatural base substrates opposite the natural bases are very low (0.005 % per base per replication). DNA fragments containing the **Ds–Px** pair can be amplified 5×10^{27}-fold by PCR with 100 cycles in total, and more than 97 % of the **Ds–Px** pair survived in the amplified DNA (Yamashige et al. 2012).

Recently, other types of unnatural base pairs for PCR amplification have also been developed. The initial base pairs, such as the **isoG–isoC** pair, have been improved and an unnatural hydrogen-bonded base pair between 2-aminoimidazo [1,2-*a*]-1,3,5-triazin-4(8*H*)-one (**P**) and 6-amino-5-nitro-2(1*H*)-pyridone (**Z**) has been newly designed (Fig. 2e) (Yang et al. 2007, 2011). The **P–Z** pair exhibits 99.8 % of the unnatural base pair selectivity per replication and 0.2 % of the misincorporation opposite the natural bases per replication in PCR using Taq DNA polymerase. Meanwhile, many types of hydrophobic unnatural base pairs have been examined, and among them, **5SICS** (2-((2*R*,4*S*,5*R*)-4-hydroxy-5-(hydroxymethyl)tetrahydrofuran-2-yl)-6-methylisoquinoline-1(2*H*)-thione) and **MMO2** ((2*R*,3*S*,5*R*)-2-(hydroxymethyl)-5-(2-methoxy-4-methylphenyl)tetrahydrofuran-3-ol) pair reached 99.9 % selectivity in PCR using OneTaq (a mixture of Taq and exonuclease-proficient Deep Vent DNA polymerases) (Fig. 2f) (Malyshev et al. 2009, 2012). The misincorporation rate of the **5SICS–MMO2** pairs is 0.01–0.1 % per replication per base.

Here, we describe the characterization and application of unnatural base pairs, focusing on the replicable hydrophobic **Ds–Px** pair. Recently, we applied the **Ds–Px** pair to SELEX. We successfully generated unnatural base-containing DNA aptamers that specifically bind to target proteins with higher affinities, beyond those of the existing nucleic acid aptamers.

2 Replication System by Genetic Alphabet Expansion Using the Ds–Px Pair

2.1 *Sequencing Involving the Ds–Px Pair*

In the creation of unnatural base pairs, it is important to develop extremely precise methods to determine the sequences containing the unnatural bases and to assess the fidelity of the unnatural base pairings in replication. In this section, we describe our methods to determine the unnatural base pair fidelity in PCR involving the **Ds–Px** pair.

We developed an assessment method for the **Ds–Px** pairing fidelity in PCR amplification, by employing DNA sequencing involving the unnatural base pair, as summarized in Figs. 3 and 4 (Hirao et al. 2006, 2007; Kimoto et al. 2009; Yamashige et al. 2012). The assessment should include the confirmation of the

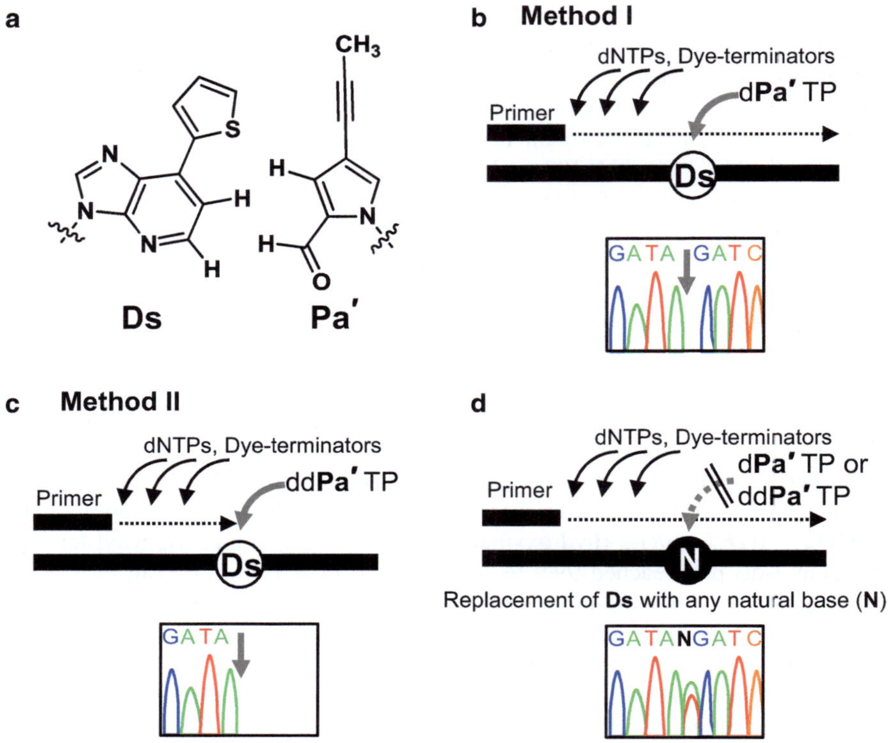

Fig. 3 Scheme of DNA sequencing methods for **Ds**-containing DNA fragments. (**a**) Chemical structures of the **Ds** base and its pairing partner, **Pa**′, used in the DNA sequencing instead of **Px**. (**b**) DNA sequencing supplemented with d**Pa**′TP (Method I) and an estimated sequencing peak pattern. (**c**) DNA sequencing supplemented with dd**Pa**′TP (Method II) and an estimated sequencing pattern. (**d**) Estimated DNA sequencing pattern for a DNA fragment, in which the **Ds** base was replaced with the natural bases (N) after PCR amplification

sequence and the determination of some parameters: the amplification fold of DNA fragments containing the unnatural bases, the selectivity of the unnatural base pairing, and the misincorporation rates of the unnatural base substrates opposite the natural bases in the template.

Our DNA sequencing method involving unnatural base pairs is based on the Sanger method, dideoxy dye termination, and can identify not only the sequence but also the content rate of the unnatural bases at a specific position in DNA fragments, by quantifying the sequence peaks. The retention rate of the **Ds–Px** pairing in PCR-amplified DNA fragments can be determined by comparing the unnatural base contents between the initial and amplified DNAs. Amplification folds are determined by gel analysis or real-time quantitative PCR. Using the retention rate and the amplification fold, the selectivity of the **Ds–Px** pair in PCR is determined.

We developed two sequencing methods for **Ds**-containing DNA fragments. The first method to determine the unnatural base positions is the conventional dideoxy

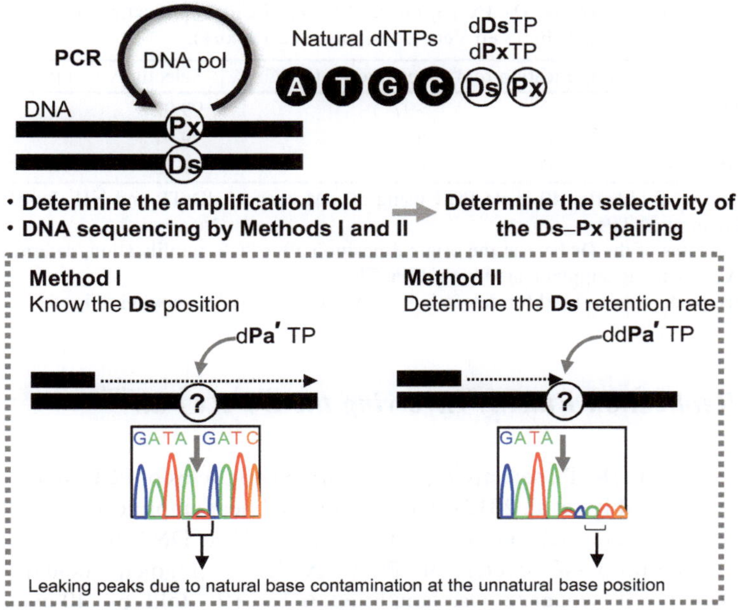

Fig. 4 Scheme for determining the **Ds–Px** pair selectivity in PCR amplification

dye-terminator method, modified by adding the deoxyribonucleoside triphosphate of an unnatural base, 4-propynylpyrrole-2-carbaldehyde (**Pa**′), which is another pairing partner of **Ds** (Hirao et al. 2006). As compared to d**Px**TP, d**Pa**′TP is favorably incorporated into DNA opposite **Ds** in the template when using Taq DNA polymerase. Since the dideoxy dye terminator of **Pa**′ is not added, the sequencing peak patterns have a gap at the **Ds** base position. The second method uses the dideoxyribonucleoside triphosphate of **Pa**′(dd**Pa**′TP), instead of d**Pa**′TP (Kimoto et al. 2009; Yamashige et al. 2012). Upon the dd**Pa**′ incorporation opposite **Ds**, the sequencing reaction stops, and there are no peaks following the **Ds** position. If natural bases contaminate the **Ds** position, then the following peaks are observed and their heights linearly reflect the contamination level. This second method is useful to assess the retention rate of the unnatural base pairs in the amplified DNAs by PCR, followed by the determination of the unnatural base pair selectivity. Practically, the retention rate of the unnatural base [F (%)] is calculated by comparing the leaking peak heights with those of the standard reference sequencing peak patterns. The selectivity[f (%)] of the **Ds–Px** pair in PCR is calculated by the following formula: $f = 100 \times (F/100)^{1/n}$, where n is the actual replication times [$= \log_2$(Amplification fold)].

Table 1 Selectivities of the **Ds–Px** pairing in 40-cycle PCR amplification in different DNA template sequence contexts by Deep Vent DNA polymerase (exo+)

DNA template	Amplification fold[a]	Retention rate[b] (F) [%]	Selectivity (f) [%/replication]
3′-TAT**Ds**CTA-5′	6.7×10^{10}	>97[c]	99.92
3′-GTG**Ds**CAT-5′	4.8×10^{10}	96	99.91
3′-CAA**Ds**GTA-5′	8.4×10^{9}	92[c]	99.75

[a]Amplification folds after 40-cycle PCR using 50 μM each of d**Ds**TP and NH_2-hx-d**Px**TP and 300 μM natural dNTPs

[b]Retention rates of the **Ds** base at the original position after PCR amplification, determined from the DNA sequencing supplemented with dd**Pa**′TP

[c]Sequence peak patterns used for determination of these values are shown in Fig. 5a

2.2 *Replication Fidelity Involving the Ds–Px Pair*

To determine the **Ds–Px** selectivity, we performed 40 cycles of PCR to increase the accuracy (Yamashige et al. 2012). To maintain the linear amplification of DNA, the process of 10-cycle PCR and dilution of the amplified DNA was repeated four times, to perform 40 cycles of PCR. A DNA fragment containing the 3′-TAT**Ds**CTA-5′ sequence was amplified 6.7×10^{10}-fold by 40-cycle PCR, using Deep Vent DNA polymerase (exo+), 50 μM d**Ds**TP and d**Px**TP, and 300 μM natural base dNTPs, and the selectivity of the **Ds–Px** pairing was 99.92 % per replication. As shown in Table 1 and Fig. 5a, the selectivities of the **Ds–Px** pairing ranged 99.75–99.92 % per replication, depending on the sequence contexts around the unnatural base in the DNA. Although purine–**Ds**–purine sequences, such as 3′-CAA**Ds**GTA-5′, are unfavorable, these sequences can also be amplified, and they survived during SELEX to generate **Ds**-containing DNA aptamers, as described later.

We also performed 100 cycles (10 cycles × 10 times) of PCR, using the Deep Vent and AccuPrime *Pfx* DNA polymerases (Yamashige et al. 2012). In the amplified DNA fragments (5×10^{27}-fold by Deep Vent and 2×10^{28}-fold by AccuPrime *Pfx*), the retentions were more than 97 % for Deep Vent DNA polymerase and 88 % for AccuPrime *Pfx* DNA polymerase (Fig. 5b). Thus, the selectivities of the **Ds–Px** pair, calculated from these retention rates and amplification folds, were more than 99.97 % and 99.9 % per replication for the Deep Vent and AccuPrime *Pfx* DNA polymerases, respectively.

Several functional groups can be attached to the **Px** base via the propyne moiety, and these modified **Px** substrates are also incorporated into DNA, opposite **Ds**, by PCR (Fig. 6). Within the favored sequence contexts, the selectivities of the unnatural base pairs between **Ds** and each modified **Px** are as high as 99.9 % per replication. In contrast, the selectivities in the unfavorable 3′-CAA**Ds**GTA-5′ sequence varied from 99.22 to 99.88 %, depending on the **Px** modification. Interestingly, as described below, the misincorporation rates of these pairs also varied.

The misincorporation rate of the unnatural base substrates opposite the natural bases is determined by using a DNA template consisting of only the natural bases

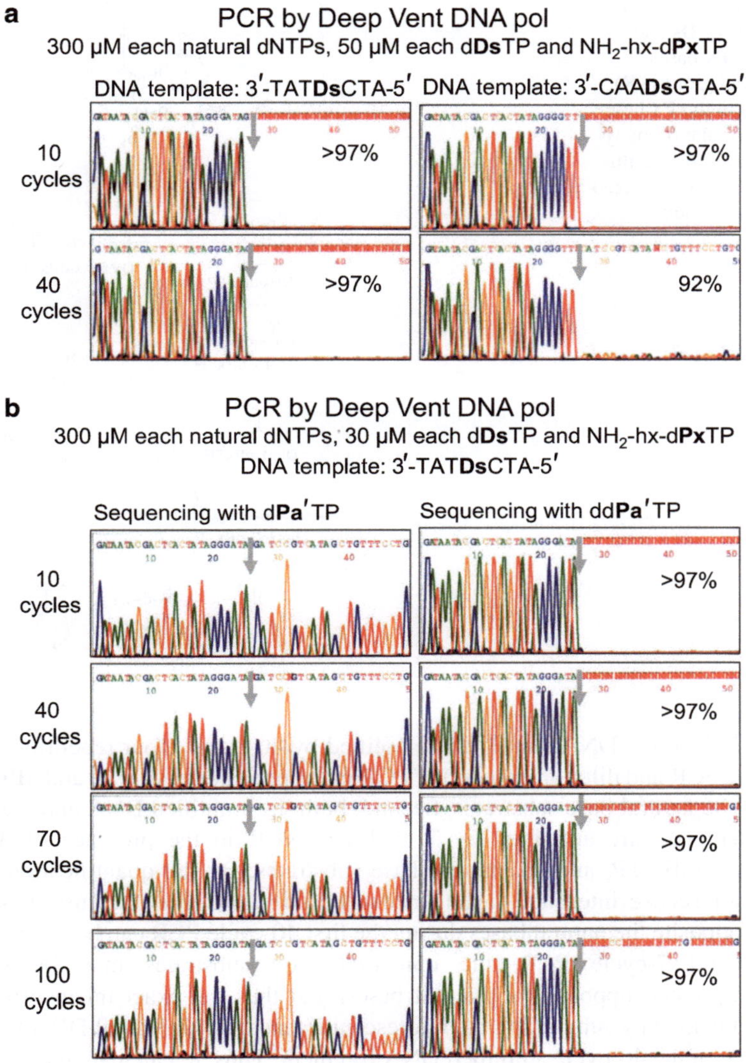

Fig. 5 Sequencing analysis of the PCR products after 40 or 100 cycles of PCR by Deep Vent DNA polymerase (exo+). Sequencing reactions were performed by using amplified products after 40 cycles (**a**) or 100 cycles (**b**) of PCR, in the presence of d**Pa**′TP (Method I) for the *left panels* in (**b**) or dd**Pa**′TP (Method II) for (**a**) and the *right panels* in (**b**). The *gray arrows* indicate the original unnatural base position. The percentages indicated in the panels for Method II are the retention rates of the **Ds–Px** pair. The sequence contexts around the unnatural base of the DNA template and the substrate concentrations used in PCR amplification are shown above the panels

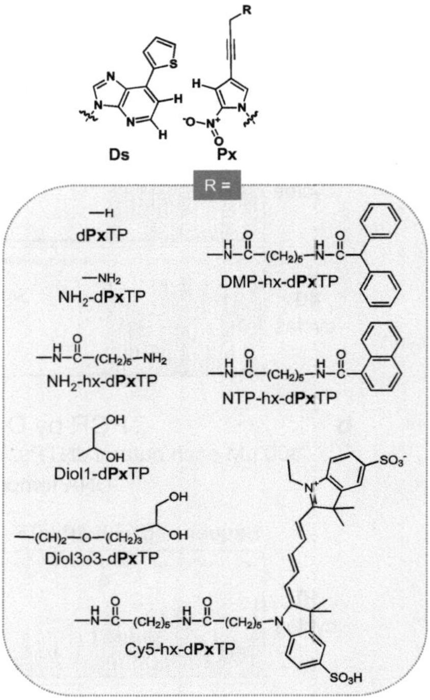

Fig. 6 The chemical structures of **Ds** and modified **Px** bases. Functional groups (**R**), attached to the **Px** base moiety via the propynyl linker, described in this chapter, are summarized in the enclosed box

(Fig. 7). First, the DNA template is amplified by 40 cycles (four repetitions of the 10-cycle PCR and dilution process) of PCR in the presence of d**Ds**TP and d**Px**TP, as well as the natural base substrates. Second, both the amplified DNA and the initial DNA template are amplified by 7 cycles of PCR in the presence of d**Ds**TP, Cy5-linked d**Px**TP, and the natural base substrates. By the quantification of the Cy5 fluorescence intensity, the misincorporation rate of the unnatural base substrates opposite the natural bases during the first 40-cycle PCR can be estimated. In the second 7-cycle PCR, the unnatural base substrates may be slightly misincorporated opposite the natural bases, and thus the exact misincorporation rate is obtained by subtracting the fluorescent intensity of the initial DNA template from that of the 40-cycle amplified DNA fragment. Thus, to calculate the misincorporation rate per base pair per replication, $m\%$, the ratios of the unnatural base misincorporation in the total PCR products ($M\%$) are quantified, at first, by comparison with the fluorescence intensity of dye-linked **Px** incorporation into the second PCR-amplified products. Then, the misincorporation rate is calculated by the following formula: $m = [1 - (1 - M/100n)^{1/L}] \times 100$, where n is the actual replication times [= $\log_2$(Amplification fold)], and L is the base pair length (bp) outside the primer hybridizing regions.

Table 2 shows the misincorporation rates of the unnatural base substrates opposite the natural bases in DNA templates during PCR amplifications by Deep Vent DNA polymerase (exo+), in the presence of 50 μM d**Ds**TP and d**Px**TP and

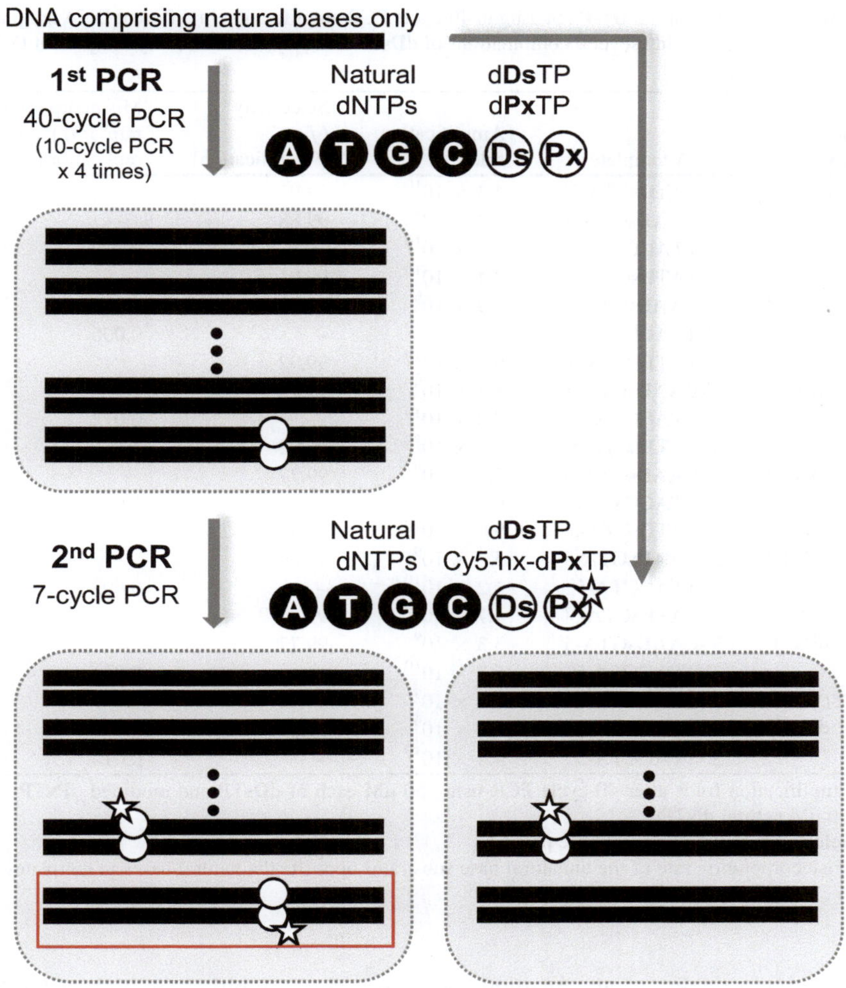

Fig. 7 Scheme for determining unnatural base misincorporation rates opposite natural bases in PCR amplification. The unnatural base misincorporation ratios in the 1st PCR product can be determined by comparing the fluorescence intensities of the 2nd PCR products for the 1st PCR product and DNA comprising natural bases only (without the 1st PCR amplification)

300 μM natural base dNTPs (Yamashige et al. 2012). The misincorporation rates of the **Ds** and modified **Px** pairs ranged from 0.003 to 0.019 % per base pair per replication. In general, increasing the concentrations of the unnatural substrates relative to those of the natural substrates (dNTPs) enhanced the **Ds**–**Px** selectivity (*f*%), whereas the undesired misincorporation rates (*m*%) might increase by competition with natural substrate selective incorporation and misincorporation.

Table 2 Fidelities of the **Ds–Px** pairing in 40-cycle PCR amplification in different DNA template sequence contexts using several combinations of **dDs**TP and modified **dPx**TP by Deep Vent DNA polymerase (exo+)

Modified **dPx**TP	DNA template	Amplification fold[a]	Selectivity[b] (f) [%/replication]	Misincorporation rate[c] (m) [%/bp/replication]
dPxTP	3′-TAT**Ds**CTA-5′	4.3×10^{10}	99.91	
	3′-CAA**Ds**GTA-5′	4.3×10^{9}	99.88	
	3′-TTACCTA-5′	4.2×10^{10}	–	0.019
NH_2-**dPx**TP	3′-TAT**Ds**CTA-5′	3.4×10^{10}	99.91	
	3′-CAA**Ds**GTA-5′	2.9×10^{9}	99.22	
	3′-TTACCTA-5′	5.0×10^{10}	–	0.006
NH_2-hx-**dPx**TP	3′-TAT**Ds**CTA-5′	6.7×10^{10}	99.92	
	3′-CAA**Ds**GTA-5′	8.4×10^{9}	99.75	
	3′-TTACCTA-5′	7.2×10^{10}	–	0.012
Diol1-**dPx**TP	3′-TAT**Ds**CTA-5′	6.2×10^{10}	99.92	
	3′-CAA**Ds**GTA-5′	7.6×10^{9}	99.77	
	3′-TTACCTA-5′	6.1×10^{10}	–	0.005
Diol3o3-**dPx**TP	3′-TAT**Ds**CTA-5′	9.0×10^{10}	99.92	
	3′-CAA**Ds**GTA-5′	5.3×10^{9}	99.64	
	3′-TTACCTA-5′	1.2×10^{10}	–	0.003
DMP-hx-**dPx**TP	3′-TAT**Ds**CTA-5′	1.7×10^{10}	99.91	
	3′-CAA**Ds**GTA-5′	3.3×10^{9}	99.77	
	3′-TTACCTA-5′	5.8×10^{10}	–	0.011
NTP-hx-**dPx**TP	3′-TAT**Ds**CTA-5′	6.9×10^{9}	99.91	
	3′-CAA**Ds**GTA-5′	3.3×10^{9}	99.87	
	3′-TTACCTA-5′	4.4×10^{10}	–	0.016

[a]Amplification folds after 40-cycle PCR using 50 μM each of **dDs**TP and modified **dPx**TP and 300 μM natural dNTPs
[b]Selectivity of the unnatural base pairing
[c]Misincorporation rate of the unnatural base substrates opposite the natural bases in templates

The best fidelity of an unnatural base pair is that exhibiting the highest **Ds–Px** selectivity (*f*%) and the lowest misincorporation rates (*m*%). We found that a dihydroxy derivative of **Px** (Diol-**Px**) reduced the misincorporation rates of the **Ds–Px** pairing. The combination of the unnatural base substrates of Diol1-d**Px**TP and d**Ds**TP in PCR by Deep Vent DNA polymerase showed the best **Ds–Px** fidelity: the **Ds–Px** pairing selectivity was 99.77–99.92 % per replication, and the misincorporation rate was as low as 0.005 % per base pair per replication. This misincorporation rate corresponds to an error rate of 5×10^{-5} error per base pair, which is close to the intrinsic error rate of the natural base mispairings by Deep Vent DNA polymerase (~2×10^{-5}).

3 DNA Aptamer Generation

3.1 SELEX Method Involving the Ds–Px Pair

The high-fidelity **Ds–Px** pair using the Diol1-**Px** base was applied to SELEX for DNA aptamer generation, to examine whether the genetic alphabet expansion could increase nucleic acid functionality (Fig. 8) (Kimoto et al. 2013b). Nucleic acids are highly hydrophilic, and even the natural bases with hydrogen-bonded residues are soluble in water. In fact, nucleic acid aptamers do not bind well to hydrophobic regions in target proteins. Therefore, we added the hydrophobic **Ds** base as a fifth base into the randomized region of DNA libraries for SELEX. However, we did not add the pairing partner **Px** into the libraries. The **Ds** base cannot pair with any bases, and it increased the structural diversity of the DNA fragments in the libraries. By SELEX using the **Ds**-containing library, we performed DNA aptamer generation for target proteins and found that the genetic alphabet expansion significantly augments the aptamer affinity.

The present limitation in the application of the genetic alphabet expansion to SELEX is the sequence determination of each clone, in the unnatural-base-containing DNA library that is ultimately isolated by several selection and amplification rounds. In the conventional SELEX methods, each sequence is determined either by a cloning–sequencing process by the transformation of the library-introduced plasmids into *E. coli* or by direct deep sequencing using a next-generation DNA sequencer. However, we currently have no means of both cloning into a cell and deep sequencing involving unnatural-base-containing DNA fragments, although these techniques are being developed. Thus, we created a new method using sublibraries containing unnatural bases at predetermined positions. Eventually, we found that this predetermined method is suitable for deep sequencing of the genetic alphabet expansion SELEX method.

We developed a new SELEX method using 22 different sublibraries, in which one to three **Ds** bases were embedded at different, specific positions in a random natural base region (43 bases) (Kimoto et al. 2013b). Each sublibrary has a unique recognition tag sequence of two or three natural bases between the forward primer and the random region, to identify the **Ds** positions after sequencing. For SELEX, these sublibraries are mixed and used as a library, in which one to three **Ds** bases are located at a variety of positions. The DNA fragments that bind to the target proteins are selected from the library and amplified by PCR using natural and unnatural (**Ds** and **Px**) base substrates and AccuPrime *Pfx* DNA polymerase.

After several rounds of selection and amplification, each sequence is determined by replacing the unnatural bases with the natural bases. For the replacement, the isolated library is amplified by PCR using the natural base and **Pa′** substrates, to replace the **Ds** bases with the natural bases. The replacement without the unnatural base substrates is inefficient due to the high fidelity of the **Ds–Px** pair, and thus we add d**Pa′**TP as an intermediate, instead of d**Ds**TP and d**Px**TP, to the PCR. Since **Pa′** efficiently pairs with **Ds** but slightly mispairs with A, the **Pa′** substrate is useful for

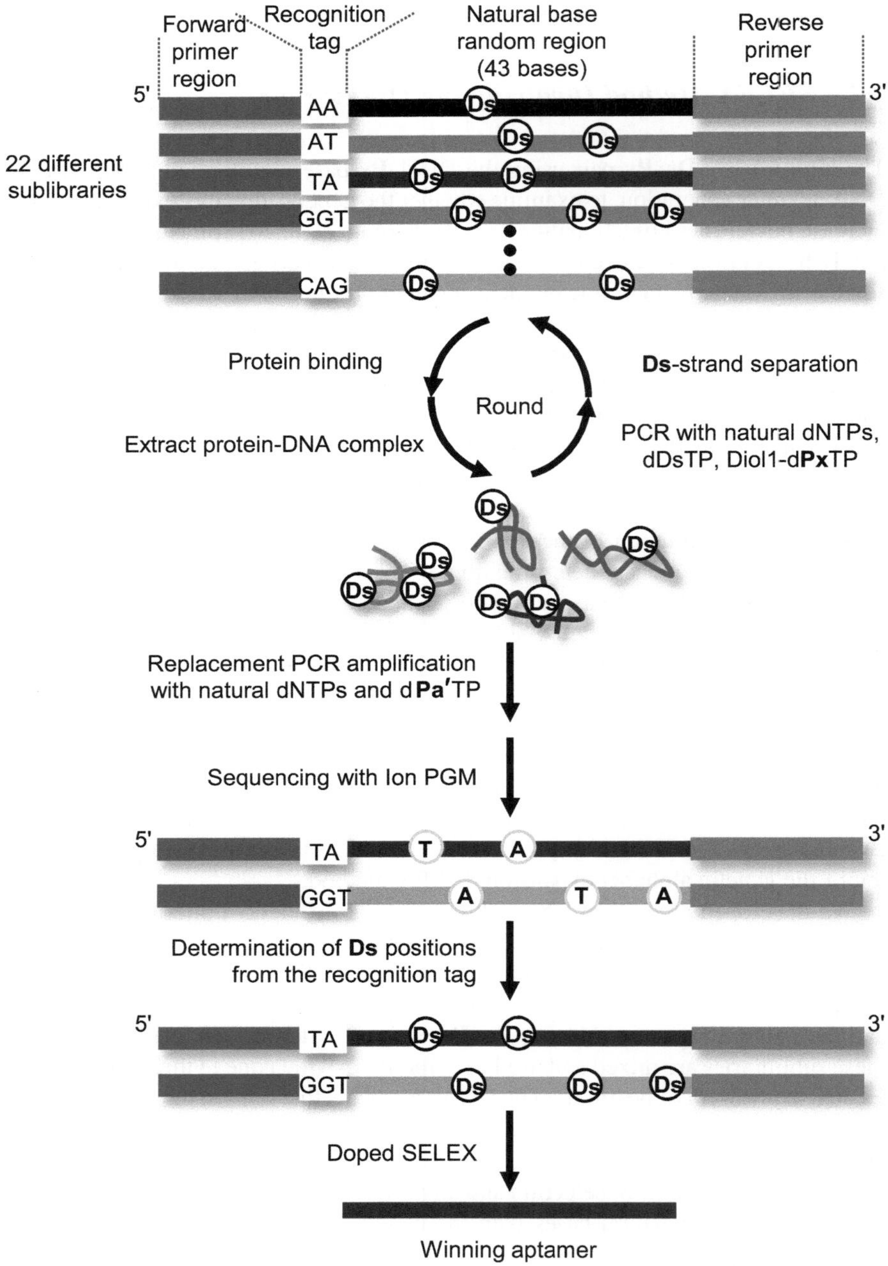

Fig. 8 Scheme of a SELEX procedure for DNA aptamer selection using libraries containing hydrophobic **Ds** bases. In the first selection, a mixture of 22 different chemically synthesized DNA sublibraries was used as the initial library. Each sublibrary contains oligonucleotides with a central

the replacement of the unnatural bases with the natural bases in PCR. Then, each sequence in the natural base library is determined by deep sequencing, using Ion Torrent PGM. From the tag sequence in each clone, the original **Ds** positions and the complete sequence of the aptamer can be determined.

3.2 Aptamer Generation Targeting Proteins

We demonstrated the genetic alphabet expansion SELEX for two target proteins, VEGF165 and interferon-γ (IFNγ) (Kimoto et al. 2013b). After seven rounds of selection and optimization of aptamer sequences, we obtained a 47-mer DNA aptamer containing two **Ds** bases for VEGF165 and a 49-mer DNA aptamer containing three **Ds** bases for IFNγ. The binding affinities were determined by surface plasmon resonance (SPR), and the K_D values of these anti-VEGF165 and anti-IFNγ aptamers were 0.65 and 38 pM, respectively. These affinities are more than 100-fold improved over the existing DNA aptamers with only natural bases (370 pM for VEGF165 and 3–28 nM for IFNγ) and the modified RNA aptamer (49–130 pM for VEGF165 as Macugen). The binding experiments using other proteins revealed the high specificity of these aptamers.

We examined the **Ds**→A mutant aptamers and confirmed that the affinities of these mutants were significantly reduced (347 pM for anti-VEGF165 aptamer and 7.21 nM for anti-IFNγ aptamer). In the forced mutation by PCR without the unnatural base substrates, the **Ds–Px** pair was mainly replaced with the A–T pair (and slightly by T–A). The crucial point of the genetic alphabet expansion SELEX using the **Ds**-predetermined library is the high fidelity of the unnatural base pairing in PCR amplification. If the replacement of unnatural to natural or natural to unnatural bases occurs in the library during SELEX with more than 150 cycles of PCR in total, we could have not determined the **Ds** positions in the sequences from each recognition tag. The high **Ds** dependency of the aptamer affinities also confirmed high fidelity of the **Ds–Px** pair in PCR amplification.

Since only a few unnatural bases significantly augmented the affinity and specificity of DNA aptamers, the **Ds**-predetermined library is more powerful for

Fig. 8 (continued) random region of 43 nucleotides with 1–3 **Ds** bases at predetermined positions, flanked by constant regions for forward and reverse primers. In each selection round, the single-stranded DNA library was mixed with a target protein, and the DNA fragments that bound to the target proteins were extracted, followed by asymmetric PCR amplification involving the **Ds–Px** pair for the next round of selection. After several rounds of selection, the enriched DNA library was PCR amplified in the presence of the natural dNTPs and the unnatural d**Pa′**TP as substrates, to replace the **Ds** bases in the DNA fragments with the natural bases (mainly A or T) via the **Ds–Pa′** pair. All amplified DNA fragments were sequenced with a next-generation DNA sequencer, Ion PGM. After binding analyses of the obtained DNA aptamers, the most potent aptamers for the target protein were optimized through doped in vitro selection and the "winning" aptamers were finally obtained

the genetic alphabet expansion SELEX than we first anticipated. By utilizing deep sequencing, we obtained tens of thousands of aptamer candidate sequence data from the selected library, after several rounds of selection. To find the winning aptamer sequence from the enormous number of sequences, the alignment by the tag sequences of each sublibrary is very useful as an initial screening. In addition, the number of base sequence combinations in a completely randomized library with five different bases is a lot larger than that with the four natural bases. All possible combinations with five different bases in the 43-base randomized region are greatly increased to 1.14×10^{30}, relative to those (7.74×10^{25}) with four natural bases. This increased complexity is unfavorable for SELEX because of the scale limitation, and thus the incorporation of up to three **Ds** bases into the natural base randomized region addresses this downside of the genetic alphabet expansion.

This SELEX experiment proved that the genetic alphabet expansion greatly increases the nucleic acid functionality. The high-affinity **Ds**-containing aptamers could be applied to diagnostics and therapeutics. In addition, other functional nucleic acids, such as catalysts, could be generated by the genetic alphabet expansion SELEX.

4 Conclusion and Future Studies

We have described the high-fidelity **Ds–Px** pair, which functions as a third base pair in PCR, and its application to DNA aptamer generation. Just 5 years ago, it was hard to imagine such an unnatural base pair surviving in amplified DNAs through more than 100 cycles of PCR. The high-fidelity unnatural base pair has enabled its application to an evolutionary engineering method, SELEX, to generate high-affinity DNA aptamers containing unnatural bases. Introducing only few unnatural bases significantly improved the aptamer affinity and selectivity to target proteins, indicating the high potential of the genetic alphabet expansion technology.

Several unnatural base pairs have been developed, not only for replication (Malyshev et al. 2012; Yang et al. 2011; Yamashige et al. 2012), but also for transcription and/or translation (Hirao et al. 2002; Hirao 2006a; Kimoto et al. 2004, 2007; Seo et al. 2009, 2011). For example, another hydrophobic base pair between **Ds** and pyrrole-2-carbaldehyde (**Pa**) is useful for in vitro transcription (Hirao et al. 2006; Morohashi et al. 2012; Kimoto et al. 2013a). The unnatural base substrates are both complementarily incorporated into RNA by T7 RNA polymerase. Functional groups, such as an amino group, biotin, and fluorescent dyes, can be attached to the **Pa** base via the propynyl linker, and these functional substrates are also incorporated into RNA opposite **Ds** in DNA templates. Since DNA templates containing **Ds** are chemically synthesized and amplified by PCR involving the **Ds–Px** pairing, large DNA templates can be prepared to introduce functional **Pa** bases into large RNA molecules (Kimoto et al. 2012).

In the future, genetic alphabet expansion will be applied to in vivo systems. In vivo studies might provide valuable information about life's mechanisms, through analyses of gene replication, repair, expression, and regulation in cells or organisms involving unnatural base pair systems. In addition, unnatural base pair recombinant techniques might increase the safety level of the containment of genetic recombinant products. Since the unnatural base substrates cannot be synthesized in cells and organisms, the recombinant creatures cannot proliferate without unnatural base nucleotide supplementation. The large-scale production of artificial proteins containing unnatural amino acids could also become possible, by using recombinant cells and organisms.

References

Benner SA (2004) Understanding nucleic acids using synthetic chemistry. Acc Chem Res 37:784–797

Bergstrom DE (2009) Unnatural nucleosides with unusual base pairing properties. Curr Protoc Nucleic Acid Chem Chapter 1:Unit 1.4

Ellington AD, Szostak JW (1990) In vitro selection of RNA molecules that bind specific ligands. Nature 346:818–822

Henry AA, Romesberg FE (2003) Beyond A, C, G and T: augmenting nature's alphabet. Curr Opin Chem Biol 7:727–733

Hirao I (2006a) Placing extra components into RNA by specific transcription using unnatural base pair systems. Biotechniques 40:711–715

Hirao I (2006b) Unnatural base pair systems for DNA/RNA-based biotechnology. Curr Opin Chem Biol 10:622–627

Hirao I, Kimoto M (2012) Unnatural base pair systems toward the expansion of the genetic alphabet in the central dogma. Proc Jpn Acad Ser B Phys Biol Sci 88:345–367

Hirao I, Ohtsuki T, Fujiwara T et al (2002) An unnatural base pair for incorporating amino acid analogs into proteins. Nat Biotechnol 20:177–182

Hirao I, Kimoto M, Mitsui T et al (2006) An unnatural hydrophobic base pair system: site-specific incorporation of nucleotide analogs into DNA and RNA. Nat Methods 3:729–735

Hirao I, Mitsui T, Kimoto M et al (2007) An efficient unnatural base pair for PCR amplification. J Am Chem Soc 129:15549–15555

Hirao I, Kimoto M, Yamashige R (2012) Natural versus artificial creation of base pairs in DNA: origin of nucleobases from the perspectives of unnatural base pair studies. Acc Chem Res 45:2055–2065

Horlacher J, Hottiger M, Podust VN et al (1995) Recognition by viral and cellular DNA polymerases of nucleosides bearing bases with nonstandard hydrogen bonding patterns. Proc Natl Acad Sci U S A 92:6329–6333

Kimoto M, Endo M, Mitsui T et al (2004) Site-specific incorporation of a photo-crosslinking component into RNA by T7 transcription mediated by unnatural base pairs. Chem Biol 11:47–55

Kimoto M, Mitsui T, Harada Y et al (2007) Fluorescent probing for RNA molecules by an unnatural base-pair system. Nucleic Acids Res 35:5360–5369

Kimoto M, Kawai R, Mitsui T et al (2009) An unnatural base pair system for efficient PCR amplification and functionalization of DNA molecules. Nucleic Acids Res 37:e14

Kimoto M, Cox RS 3rd, Hirao I (2011) Unnatural base pair systems for sensing and diagnostic applications. Expert Rev Mol Diagn 11:321–331

Kimoto M, Yamashige R, Yokoyama S et al (2012) PCR amplification and transcription for site-specific labeling of large RNA molecules by a two-unnatural-base-pair system. J Nucleic Acids 2012:230943

Kimoto M, Hikida Y, Hirao I (2013a) Site-specific functional labeling of nucleic acids by in vitro replication and transcription using unnatural base pair systems. Isr J Chem 53:450–468

Kimoto M, Yamashige R, Matsunaga K et al (2013b) Generation of high-affinity DNA aptamers using an expanded genetic alphabet. Nat Biotechnol 31:453–457

Kool ET (2000) Synthetically modified DNAs as substrates for polymerases. Curr Opin Chem Biol 4:602–608

Lee JH, Canny MD, De Erkenez A et al (2005) A therapeutic aptamer inhibits angiogenesis by specifically targeting the heparin binding domain of VEGF(165). Proc Natl Acad Sci U S A 102:18902–18907

Malyshev DA, Seo YJ, Ordoukhanian P et al (2009) PCR with an expanded genetic alphabet. J Am Chem Soc 131:14620–14621

Malyshev DA, Dhami K, Quach HT et al (2012) Efficient and sequence-independent replication of DNA containing a third base pair establishes a functional six-letter genetic alphabet. Proc Natl Acad Sci U S A 109:12005–12010

Morohashi N, Kimoto M, Sato A et al (2012) Site-specific incorporation of functional components into RNA by an unnatural base pair transcription system. Molecules 17:2855–2876

Piccirilli JA, Krauch T, Moroney SE et al (1990) Enzymatic incorporation of a new base pair into DNA and RNA extends the genetic alphabet. Nature 343:33–37

Rich A (1962) Problems of evolution and biochemical information transfer. In: Kasha M, Pullman B (eds) Horizons in biochemistry. Academic, New York, pp 103–126

Robertson DL, Joyce GF (1990) Selection in vitro of an RNA enzyme that specifically cleaves single-stranded DNA. Nature 344:467–468

Ruckman J, Green LS, Beeson J et al (1998) 2′-Fluoropyrimidine RNA-based aptamers to the 165-amino acid form of vascular endothelial growth factor (VEGF165). Inhibition of receptor binding and VEGF-induced vascular permeability through interactions requiring the exon 7-encoded domain. J Biol Chem 273:20556–20567

Seo YJ, Matsuda S, Romesberg FE (2009) Transcription of an expanded genetic alphabet. J Am Chem Soc 131:5046–5047

Seo YJ, Malyshev DA, Lavergne T et al (2011) Site-specific labeling of DNA and RNA using an efficiently replicated and transcribed class of unnatural base pairs. J Am Chem Soc 133:19878–19888

Switzer C, Moroney SE, Benner SA (1989) Enzymatic incorporation of a new base pair into DNA and RNA. J Am Chem Soc 111:8322–8323

Tuerk C, Gold L (1990) Systematic evolution of ligands by exponential enrichment: RNA ligands to bacteriophage T4 DNA polymerase. Science 249:505–510

Yamashige R, Kimoto M, Takezawa Y et al (2012) Highly specific unnatural base pair systems as a third base pair for PCR amplification. Nucleic Acids Res 40:2793–2806

Yang Z, Sismour AM, Sheng P et al (2007) Enzymatic incorporation of a third nucleobase pair. Nucleic Acids Res 35:4238–4249

Yang Z, Chen F, Alvarado JB et al (2011) Amplification, mutation, and sequencing of a six-letter synthetic genetic system. J Am Chem Soc 133:15105–15112

Flexible Nucleobase Analogues: Novel Tools for Exploring Nucleic Acids

Sarah C. Zimmermann and Katherine L. Seley-Radtke

Contents

1 Introduction 150
2 Nucleobase Modifications 150
3 Flexibility and Medicinal Chemistry 154
4 Flexible Nucleobases and Helix Stability 157
5 Flexible Nucleobases as Fluorescent Bioprobes 160
6 Summary 161
References 162

Abstract Flexible nucleosides such as Holý's sugar-modified acyclics have seen great utility in antiviral therapeutics. Another area of modification to the nucleoside scaffold that has only recently garnered attention involves imparting flexibility to the nucleobase instead of the sugar moiety. This flexibility can be achieved by splitting the purine base into its respective imidazole and pyrimidine components or by increasing the number of aromatic rings. These modifications also allow installation of additional substituents to the nucleobase scaffold that can potentially enhance biological activity and enzyme recognition. Nucleobase flexibility can also provide other advantages such as enhancing fluorescence properties that can be utilized for bioprobes as well as to assist in exploring the effects of the nucleobase on RNA and DNA helix stability.

Keywords Fleximers • Flexible nucleosides • Nucleobase • Acyclic • Nucleic acid

---

S.C. Zimmermann • K.L. Seley-Radtke (✉)
Department of Chemistry and Biochemistry, University of Maryland, Baltimore County, Baltimore, MD 21250, USA
e-mail: kseley@umbc.edu

V.A. Erdmann et al. (eds.), *Chemical Biology of Nucleic Acids*, RNA Technologies, DOI 10.1007/978-3-642-54452-1_9, © Springer-Verlag Berlin Heidelberg 2014

1 Introduction

Nucleoside and nucleotide analogues have long served as the cornerstone in the arsenal of currently available therapeutics exhibiting potent activity against a number of cancers, viruses, and bacterial and parasitic infections. To date more than 25 nucleoside/tide drugs have earned FDA approval (Jordheim et al. 2013). As a result, strategic modifications to the nucleoside scaffold can be used to increase their efficacy as therapeutics, but also to develop novel ways for investigating nucleic acid structure and function. Although varying the substituents on the sugar and the base has provided the majority of nucleoside analogues, other successful modifications have focused on altering the natural scaffold of the nucleoside. For example, as shown in Fig. 1, carbocyclic nucleosides such as Abacavir and Entecavir (Bisacchi et al. 1997; Boyle et al. 2012) and the acyclics such as Tenofovir are FDA-approved nucleosides with modified sugars (De Clercq 1988, 2004; Tuske et al. 2004; De Clercq and Holý 2005). Notably, the acyclic nucleosides have exhibited a wide range of biological activity against many viruses. Most significantly, Tenofovir currently serves as the primary component in a number of different highly active antiretroviral combination therapies (HAART) due to its ability to overcome resistance mechanisms in HIV reverse transcriptase (HIV-RT), a key enzyme in the HIV replication pathway. This has been attributed to its ability to "wiggle and jiggle" in the HIV RT binding site (Das et al. 2005; Tuske et al. 2004).

Modifications to the heterocyclic bases offer forth another route for installing advantageous properties. Base modifications have led to novel nucleosides that have found use in areas such as medicinal, computational, and biophysical chemistries as well as to increase our toolbox of heterocyclic chemistry techniques. Switching the positions of atoms such as those found in the pyrazolopyrimidines or removing nitrogens, as is typical with the deaza purines, are two approaches that have found success (Seley et al. 1997, 2003a; Yang et al. 2006). Another approach has been to build onto or insert atoms into the nucleobase framework. Less attention, however, has been paid to installing flexibility to the base, and it is that aspect that has provided the primary focus for this review.

2 Nucleobase Modifications

An early pioneer in the area of nucleobase modifications, Nelson Leonard constructed a series of "expanded" and "extended" purines and pyrimidines to investigate enzyme recognition (Agasimundin et al. 1985). The *linear* (**1**), *distal* (**2**), and *proximal* (**3**) benzoadenine analogues, shown in Fig. 2, exhibited increased conjugation due to the insertion of a benzene ring between the imidazole and pyrimidine rings of the purine scaffold (Leonard and Hiremath 1986). These analogues were initially developed as dimensional probes for investigating enzyme

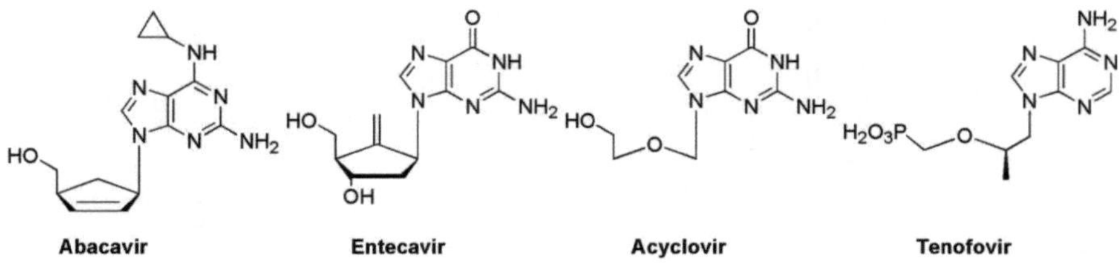

Fig. 1 Carbocyclic and acyclic nucleosides

binding sites, but also exhibited fluorescent properties (Leonard and Hiremath 1986; Spencer et al. 1974; Secrist et al. 1972a, b; Barrio et al. 1972). The Seley-Radtke and Tor groups have continued this work with their hetero-expanded purines (**4**) and extended pyrimidines (**5** and **6**) (Wauchope et al. 2010, 2012a; O'Daniel et al. 2008; Temburnikar et al. 2013, 2014; Tor et al. 2007). The use of heterocyclic spacers serves to increase polarizability and aromatic character as well as to provide a less drastic curvature to the expanded base, thereby decreasing the base pair distance in DNA.

In contrast to the inherently rigid expanded and extended bases, several groups have studied the effects of introducing flexibility to the nucleobase. Early reports from Miller et al. have suggested that flexible bases such as C5-substituted uracils (Fig. 3) may have been present in primordial times due to the ease of coupling amino acid side chains to uracil in prebiotic conditions (Robertson and Miller 1995).

More recently, the Seley-Radtke group introduced a series of compounds called "fleximers" whereby the purine ring has been split into its imidazole and pyrimidine components. This allows for rotation, while still retaining the elements essential for base pairing and molecular recognition (Seley et al. 2002, 2005). A number of different possibilities for connectivity exist; however, the Seley-Radtke studies have focused on **A** and **B** shown in Fig. 4, since these connectivities most closely resemble the purine scaffold.

When this concept was first explored, the fleximers were investigated using various computational methods. The potential energies for rotation about the carbon–carbon bond were studied for the adenine, guanosine, and inosine fleximers (Seley et al. 2002; Bardon and Wetmore 2005). A second study focused on the connectivity of the carbon–carbon bond between the C4 or C5 of the imidazole and the C5′ or C6′ of the pyrimidine, as shown in Fig. 4. Without the ribose moiety, a planar configuration of the base was observed with calculated energy barriers less than 40 kJ/mol. One exception was noted with the adenine fleximer possessing a C5–C5′ connectivity, where a nonplanar global minimum was observed due to the interaction of the imidazole and the amino of the pyrimidine (Bardon and Wetmore 2005). With the addition of a ribose to the imidazole, a nonplanar orientation was favored; however, this was dependent upon the connectivity of the fleximer base and various interactions between the base and the sugar. In that regard, the proximal

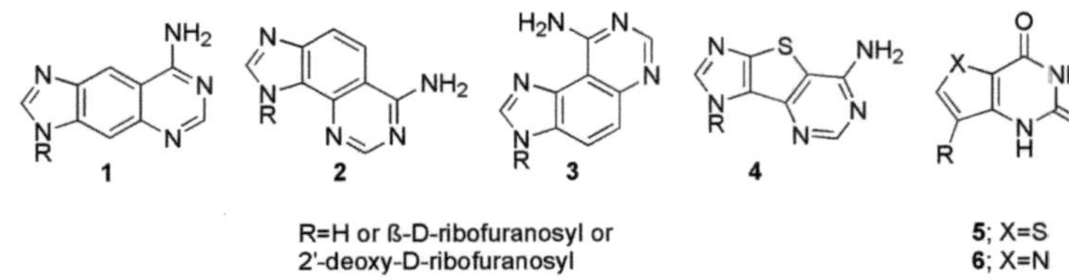

Fig. 2 *lin*- (**1**), *dist*- (**2**), and *prox*-benzoadenine (**3**) analogues developed by Nelson J. Leonard, and expanded analogues from the Seley-Radtke group (**4–6**)

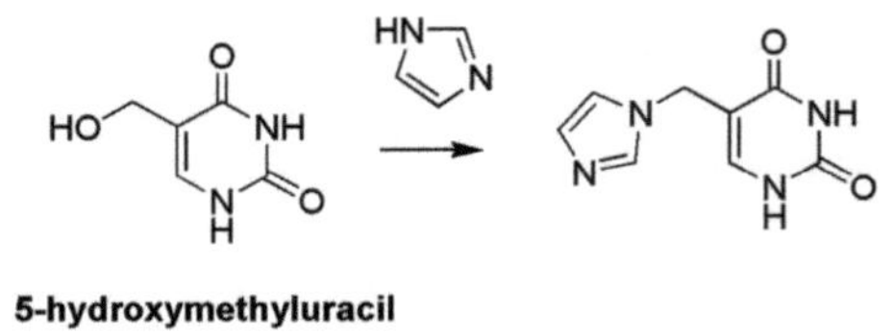

Fig. 3 Possible prebiotic synthesis of a 5-substituted uracil

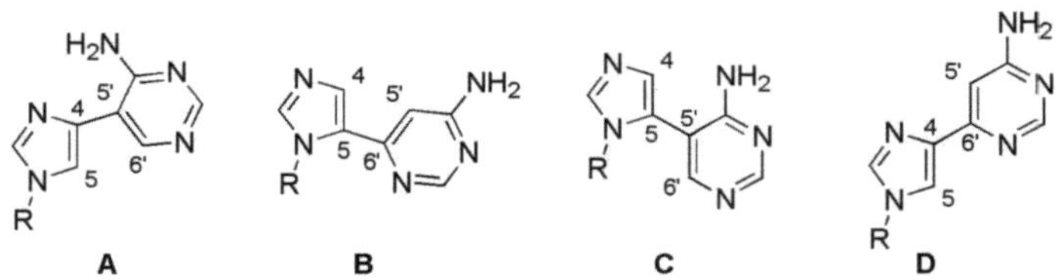

Fig. 4 Variations of base connectivity for the adenosine fleximers

fleximer nucleosides (**A**) appeared to exhibit a planar orientation for their bases, while the distal (**B**) analogues strongly preferred a biplanar mode.

Following the initial computational studies, Seley-Radtke et al. reported the synthesis for a series of ribose and 2′-deoxyribose fleximers (Seley et al. 2002; Wauchope et al. 2012b). Notably, the distal fleximers were initially synthesized from their tricyclic "expanded purine" counterparts (**9** shown in Fig. 5) by reducing the sulfur of the thiophene spacer ring with Raney Nickel (Seley et al. 2001). In contrast, the proximal fleximers were realized by various cross-coupling methodologies including Stille and Suzuki palladium-catalyzed coupling (Wauchope et al. 2012b). Organometallic coupling techniques offer forth many advantages since there are a wide variety of catalysts available that can be utilized for different conditions. For example, as shown in Fig. 6, Len et al. have developed a system of coupling 6-iodouridine (**10**) with various boronic acids using Na_2PdCl_4 at room temperature. These gave moderate to good yields depending on the nature of the boronic acid (Enderlin et al. 2013). Berteina-Raboin et al. have developed a microwave-assisted Suzuki-Miyaura cross-coupling reaction of unprotected 2′-deoxyuridine (**11**) that can be performed in water with standard palladium catalysis using $Pd(OAc)_2$ and triphenylphosphine (Fresneau et al. 2012).

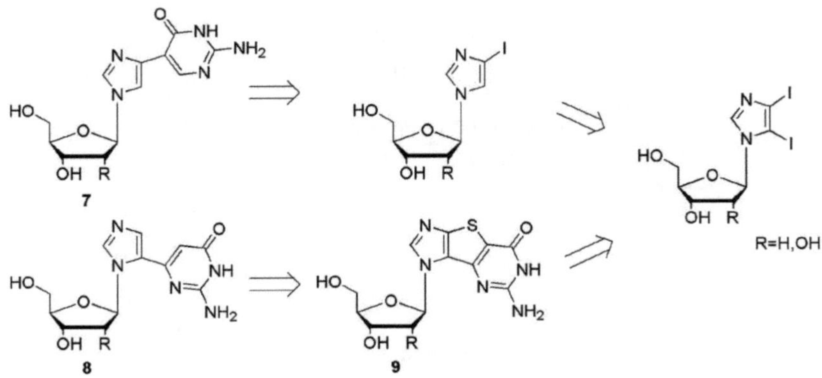

Fig. 5 Seley-Radtke's proximal (**7**) and distal fleximers (**8**)

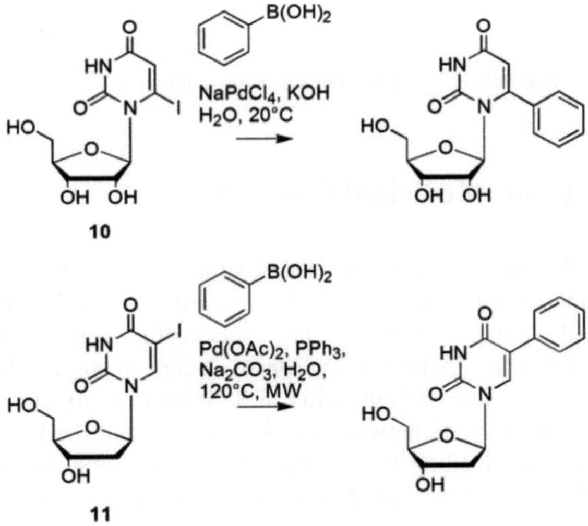

Fig. 6 Suzuki-Miyaura cross-coupling reactions

More recent versions of the fleximer approach have been reported. For example, Hudson's triazole fleximer (**12**, Fig. 7), where the imidazole is replaced with a triazole, was achieved using click chemistry (St. Amant et al. 2012). These triazole analogues led to further modifications by Koszytkowska-Stawińska using a Banert cascade reaction (Loren and Sharpless 2005) to form analogues with a 4,5-disubstituted-NH-1,2,3-triazoles (**13**) (Koszytkowska-Stawińska and Sas 2013). Importantly, the use of the Barnet cascade reaction allowed for a solvent-free "green chemistry" approach.

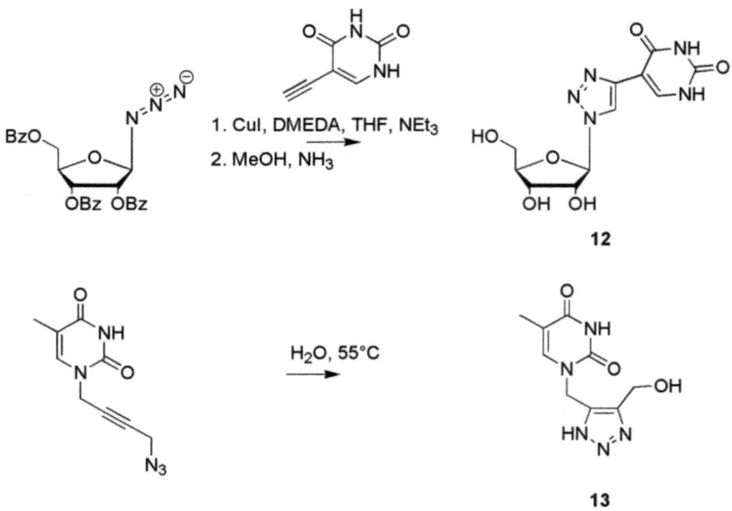

Fig. 7 Hudson and Koszythowska-Stawińska's triazole analogues

3 Flexibility and Medicinal Chemistry

Seley-Radtke's fleximers and other related flexible analogues have been studied for their potential therapeutic properties. In that regard, the distal guanosine fleximer was observed to inhibit an adenosine-metabolizing enzyme, S-adenosylhomocysteine hydrolase (SAHase) (Seley et al. 2003b). It has been speculated that this unexpected inhibition of SAHase is likely due to the existence of an intramolecular H-bond between the 5′-hydroxyl and the C2 carbonyl of the base (Fig. 8), thereby leading to an atypical syn conformation. This essentially results in an adenosine mimic since this syn conformation then places the amino group of the guanosine in the appropriate position in the enzyme binding site where the adenosine NH_2 would normally reside. The influence of the enzyme on the molecule was further investigated by NMR; in solution, an anti-confirmation was more energetically favorable (Polak et al. 2004), whereas in SAHase, the syn conformation was preferred.

Further studies with the triphosphate of the guanosine fleximer (Flex-GTP) revealed additional advantages for the fleximer approach. Notably, Flex-GTP was preferred as a substrate over the natural substrate guanosine triphosphate (GTP) by GTP-fucosepyrophosphorylase (GFPP). In addition, Flex-GTP was able to retain all activity when confronted with active site mutations. Both observations appear to be due to the ability of Flex-GTP to interact with secondary binding site amino acids not previously involved in the mechanism of action for GFPP (Quirk and Seley 2005a, b).

Gundersen et al. studied a series of 6,9-disubstituted purines (**14**, Fig. 9) which were found to be inhibitors of *Mycobacterium tuberculosis* (Braendvang

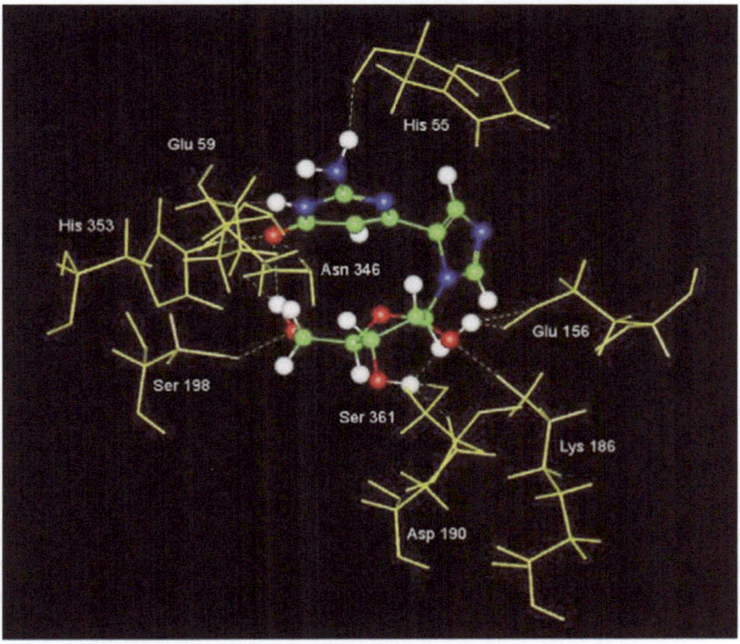

Fig. 8 Seley-Radtke's distal guanosine fleximer in SAHase

et al. 2009). To expand the structure activity relationship of these compounds a second series of fleximers was synthesized. The flexible purine-like bases were found to exhibit antimycobacterial activity with low mammalian toxicity (Read et al. 2010). This series included fleximers where the C4 of the imidazole was connected to the C5′ of the pyrimidine (**15**), and conversely the C5 of the imidazole connected to the C6′ of the pyrimidine (**16**), with better activity being noted for fleximer **16** (Read et al. 2010).

"Reverse" fleximers, whereby the connectivity of the base has been switched, feature an aryl substitution on the C5 of the pyrimidine (Fig. 10). These uridine nucleoside analogues have provided important leads such as 5-iodo-2′-deoxyuridine (IdUrd) and have been used clinically for the treatment of herpes. In addition, 5-(2-halogenovinyl)-2′-deoxyuridines have displayed strong activity against herpes simplex virus type 1 (HSV-1) and varicella-zoster virus (VZV). Using Stille coupling techniques, Herdewijn and colleagues constructed a series of 5-thien-2-yl (**17**, Fig. 10) and related 5-furan-2-yl substituted 2′-deoxyuridines (Wigerinck et al. 1991). The 5-bromothien-2-yl and 5-chlorothien-2-yl derivatives of deoxyuridine displayed the strongest activity against HSV-1 and VZV with potencies similar to Brivudine, likely due to phosphorylation by the viral thymidine kinase (TK) (Wigerinck et al. 1991). These compounds have a high affinity for TK as they exhibit similar binding energetics to thymidine, but the differences in affinity are likely due to the hydrophobicity of the C5 substitution or the differences

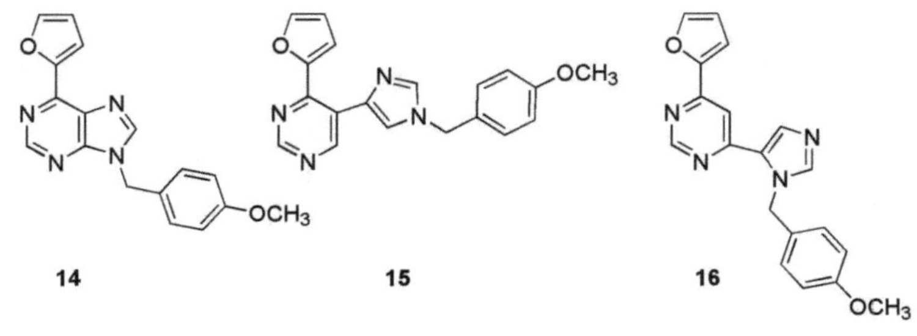

Fig. 9 Gunderson's antimycobacterial compounds

in their interaction with TK (De Winter and Herdewijn 1996). Surprisingly, when Delarue investigated the analogous monophosphates of **17** and the related aryl C5-substituted thymidines for activity with *Mycobacterium tuberculosis* thymidylate kinase, none of the compounds served as a substrate, but rather as moderate inhibitors (Haouz et al. 2003).

Carbocyclic reverse fleximers, where the furanose oxygen has been replaced with a methylene group, have also been evaluated (Fig. 10). Modest HSV-1 activity was observed for the Aristeromycin analogue (**18**) (Popescu et al. 1995); however, the truncated analogue (**19**), lacking the 5′-hydroxyl, failed to display any meaningful antiviral activity. They did, however, exhibit modest adenosine deaminase activity despite their pyrimidine character (Zimmermann et al. 2013). Similarly, Holý et al. studied C5-substituted pyrimidines such as **20** and found that the acyclic 5-phenyl uracils exhibited HSV-1 and HSV-2 activity, whereas the cytosine analogues were active against HCMV and VZV (Krecmerova et al. 2007). Another series of acyclics with interesting activity are Orr's C5-aryl analogues (**21**, Fig. 10) of 5-benzylacyclouridine (BAU). These compounds were screened against uridine phosphorylase (UrdPase), since inhibition of this enzyme is thought to enhance the therapeutic effect of nucleoside drugs such as 5-fluorouracil (Orr et al. 1995). Later analogues exhibited increased inhibition of UrdPase, with enhanced pharmacodynamic effects compared to those noted for BAU (Orr et al. 1997).

Interestingly, when the imidazole of a flexible nucleoside analogue was replaced with a triazole and the pyrimidine replaced with a benzene ring (**22**, Fig. 11) significant anticancer activity against $HepG_2$ cells was observed, possibly due to the conjugation between the aromatic ring systems (Yu et al. 2010). Related to this, Kim et al. studied a series of isoxazole and triazole analogues (**23**) that were synthesized from C5-ethynyl-2′-deoxyuridines using Sonogashira coupling and click chemistry. These compounds were evaluated against six types of cancer cell lines. The isoxazole compounds exhibited superior anticancer activity, although the triazoles displayed lower cytotoxicity (Lee et al. 2009).

Similarly, Agrofoglio's group synthesized a series of C5-disubstituted 1,2,3-triazolo deoxyuridines (**24**, Fig. 11) as potential antivirals. These compounds were realized using copper- or ruthenium-catalyzed azide–alkyne cycloaddition

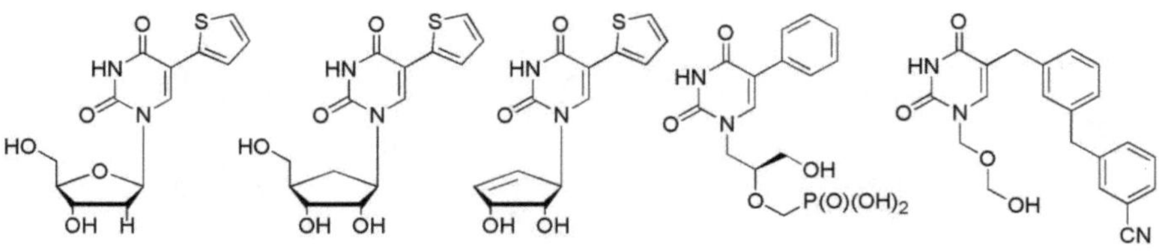

Fig. 10 C5-substituted uracils

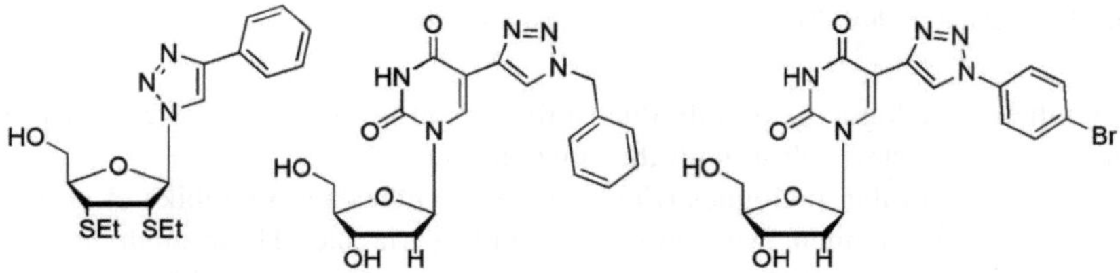

Fig. 11 Triazole containing nucleosides

conditions. The 1,4-disubstituted compounds displayed activity against VZV, whereas the 1,5-disubstituted compounds showed no antiviral activity (Montagu et al. 2011).

4 Flexible Nucleobases and Helix Stability

Modified nucleosides have also been utilized to study the structure and function of nucleic acids. Since DNA carries the genetic code, investigating DNA helix stability has been important due to its critical role in replication and gene expression (Kool 2001). Weisz et al. constructed a series of nonnatural imidazole/phenyl fleximers with different substituents on the phenyl ring in an effort to study their ability to bind to CG Watson–Crick base pairs (Fig. 12) (Wang et al. 2004). Triplex formation studies are of interest due to the ability of a single-stranded oligomer to bind to a double-stranded helix, which can be useful for gene regulation and repression of protein production. The binding to duplex DNA was studied using melting experiments (Wachowius et al. 2008). The 4-(3-*n*-butylureidophenyl)imidazole-2′-deoxy-nucleoside (**25**, Fig. 12) showed moderate binding affinity for all possible base pairings with little discrimination between bases (Wachowius et al. 2008).

Interestingly, the unsubstituted (**27**) and benzoylated aminophenyl (**26**) analogues (Fig. 12) did not exhibit the high-affinity binding observed with the

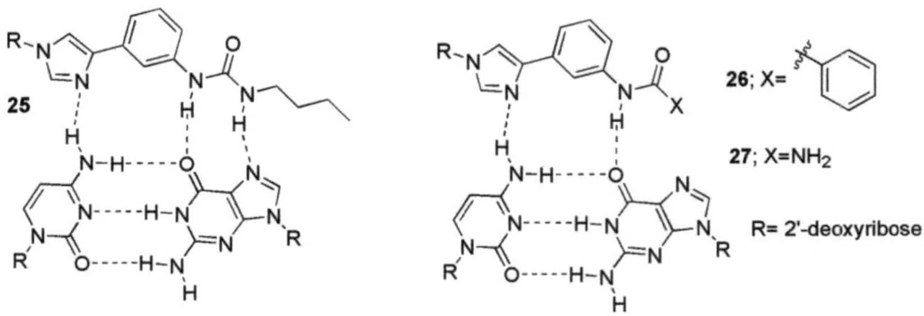

Fig. 12 Potential triplex CG base pairing of Weisz's flexible nucleosides

ureidophenyl analogues, possibly due to the amino group being a weak hydrogen bond donor, or a steric clash with the benzamido group.

A series of similar analogues (Fig. 13) were synthesized by Obika et al. where the imidazole component was substituted with a triazole. These analogues were synthesized by the use of copper(I)-catalyzed alkyne–azide cycloaddition reactions. As shown in Fig. 13, triplex formation was evaluated, with the 1-(4-ureidophenyl) triazole exhibiting the best affinity and selectivity for the CG base pairs (Hari et al. 2011). As a result, use of a triazole instead of imidazole ring may lead to better binding for triplex formation in select sequences.

Although not used to study triplex formation, Seela et al. investigated various alkynyl nucleobases of pyrimidines and purines (Fig. 14) where click chemistry was used to construct 1,2,3-triazole nucleosides with the pyrimidine and purine bases tethered to the C4 of the triazole with a methylene linker. These compounds were found to display antiviral activity against HCV and ssRNA viruses (YFV, DENV, and WNV), but unfortunately all compounds exhibited cellular toxicity within the same concentration range as the antiviral activity. Interestingly, when the adenine (**28c**) and uridine (**28b**) analogues were incorporated into DNA, they displayed abasic-like properties, with a destabilizing effect on DNA duplexes (Chittepu et al. 2008).

As shown in Fig. 15, a series of reverse fleximer uridine analogues were synthesized by Froehler using palladium-catalyzed Stille coupling with 2-pyridyl (**29**), 2-thienyl (**30**), 2-thiazoyl (**31**), or 2-imidazoyl (**32**) moieties. The nucleosides were then incorporated into short oligodeoxynucleotides (ODNs) and studied for their ability to form stable duplexes with complementary RNA strands (Gutierrez et al. 1994). The modified ODNs exhibited increased melting temperature (T_m) compared to the natural thymidine-containing ODNs, likely due to increased base stacking (Gutierrez et al. 1994). Moreover, the favorable interactions observed for the ODNs and RNA may lead to the development of new ODNs capable of sequence-specific inhibition of gene expression.

Rozners et al. studied the C5 substitution of an imidazole ring to uridine, as shown in Fig. 16, which resulted in a non-canonical base pair, known as a G-U wobble. This pairing resulted in a slight reduction in thermal stability and also altered the spine of hydration of RNA helices. This observation suggests the

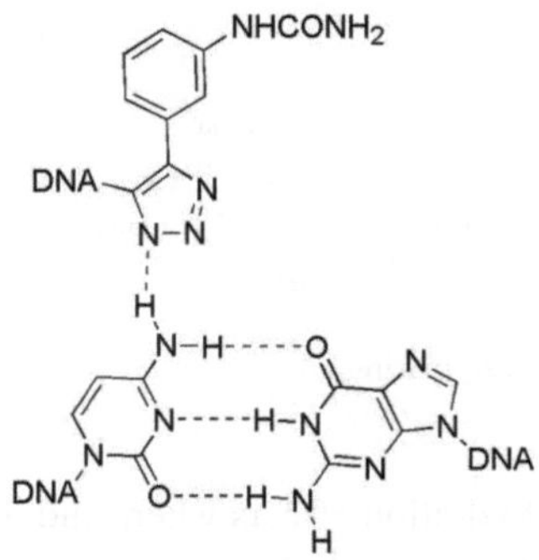

Fig. 13 Obika's triazole triplex formation

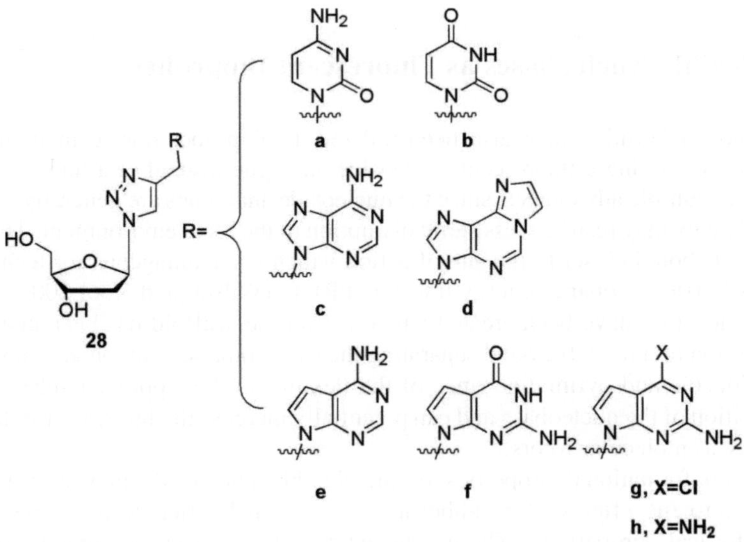

28

Fig. 14 Seela's methylene-linked triazoles

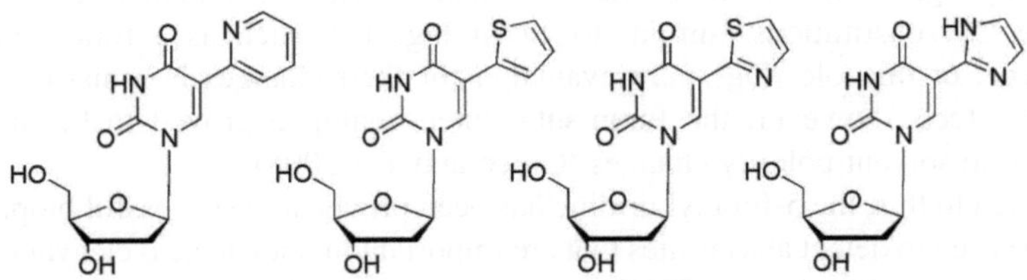

29 30 31 32

Fig. 15 Froehler's C5-substituted uridines

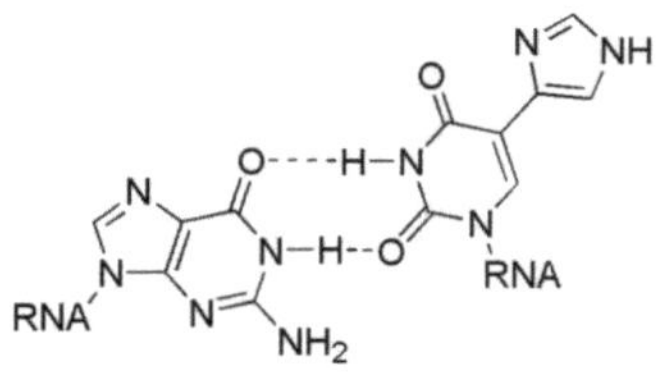

Fig. 16 Rozners' G-U wobble base pairing

importance of considering hydration effects when studying modified ODNs. These findings may lead to the improved design of aptamers and ribozymes (Rozners et al. 2005).

5 Flexible Nucleobases as Fluorescent Bioprobes

Modified nucleosides have also been utilized as bioprobes due to their inherent fluorescence. Using a fluorescent nucleoside analogue instead of a fluorescent tag can have multiple advantages, since the nucleoside analogues are generally smaller. As such, they may result in less steric disruption to the local environment. They can also be anchored closer to the site of action which is advantageous for techniques such as Förster resonance energy transfer (FRET) (Wilson and Kool 2006). Emissive nucleosides have been created where the purine scaffold has been altered by the addition of a rotatable bond separating the two aromatic components, similar to the imidazole and pyrimidine rings of the fleximers. This approach enhances the conjugation of the nucleobase and can potentially increase the fluorescence, leading to utility as molecular rotors.

The conformational properties of the flexible nucleoside may also provide insight into the effects of neighboring groups which often lead to changes in photophysical properties. These phenomena have been observed by Tor et al. while studying emissive properties. Since solvent polarity and viscosity play key roles, changes in Stokes shifts were observed for flexible analogues, while increasing rigidity was found to increase quantum yields (Sinkeldam et al. 2011). Uridine C5 substitutions (similar to **17** in Fig. 10), such as a furan, oxazole, thiophene, or thiazole ring, were evaluated for their changes in emission due to solvent effects; however, the furan-substituted analogue proved to be the most sensitive to solvent polarity changes (Greco and Tor 2005).

Related to this, the 5-furanyl uridine has been proven as a successful bioprobe as it can be used to detect abasic sites that are important in nucleic acid cleavage. It has also been incorporated using T7 RNA polymerase as well as into the A-site to study aminoglycoside antibiotics (Greco and Tor 2005; Srivatsan and Tor 2007). Increasing conjugation with a furan at the C5-position of pyrimidine and C8-position of purines yielded fluorescent nucleosides emitting in the visible spectrum. Notably, the uridine moiety was still able to maintain normal base pairing as well as its

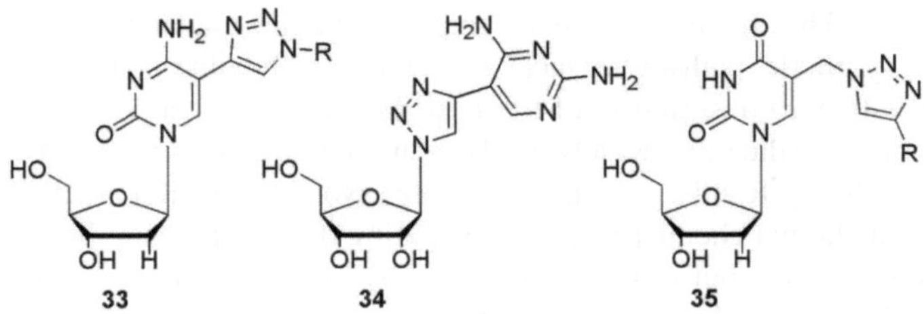

Fig. 17 Fluorescent triazole nucleosides

absorption and emission properties and thus could prove useful in various biophysical applications (Greco and Tor 2007).

As shown in Fig. 17, a series of N3 aryl-substituted triazolylnucleosides were synthesized by Hudson. The N3 aryl substitutions ("R" on **33**, Fig. 17) included a thiophene, phenyl, bithiophene, or fluorenone ring. The fleximers were synthesized by the derivatization of 5-ethynyl-2′-deoxycytidine using copper-catalyzed azide–alkyne Huisgen cycloadditions (CuAAC). The addition of the aryl substitutions resulted in blue fluorescent nucleosides that displayed strong solvatochromism (Dodd et al. 2010).

Related to these, a second series of fluorescent 1,2,3-triazole fleximers (**34**, Fig. 17) were then synthesized using the CuAAC reaction with a ribosyl azide and various 5-ethynylpyrimidines including uracil, cytosine, 4-aminopyrimidine, 2,4-diaminopyrimidine (**34**), as well as aryls such as fluorene and pyrene (St. Amant et al. 2012). These analogues displayed interesting conformational properties due to a preferred coplanar arrangement; however, only the diaminopyrimidine analogue **34** exhibited luminescent properties that could be useful for further studies employing it as a bioprobe (St. Amant et al. 2012).

Similarly, Engels et al. constructed a series of triazole analogues, such as **35**, with various aryl and alkyl substituents on the C4 of the triazole linked to the C5 by a methylene group. Notably, these were synthesized in high yields (74–90 %) using solvent-free microwave conditions (Krim et al. 2012). These compounds were later evaluated for antibacterial and antiviral activity, which unfortunately proved to be poor. They did, however, exhibit high spin labeling efficiency for EPR analytical techniques (Krim et al. 2012) and thus proved to be useful despite their lack of therapeutic activity.

6 Summary

As has been highlighted in this brief review, nucleosides can be strategically modified in a number of ways to enhance their properties and uses. While typically these modifications are made to the sugar and/or base, the spectrum of outcomes varies widely. This review has focused primarily on flexible nucleobase modifications. This flexibility can be imparted to the nucleobase scaffold by a variety of synthetic pathways, such as the click chemistry approach used by Hudson, the linear construction of the base as was employed for Seley-Radtke's distal fleximers, or the more widely used organometallic catalyzed couplings as was seen with the Seley-Radtke proximal fleximers. Notably, flexible nucleobases have seen use in medicinal chemistry and investigating biologically significant enzymes, as well as in studying helix stability and fluorescence properties. Advantages have also been observed with the "reverse" fleximers, where the C5 of the pyrimidine ring has been substituted with various aryl groups. Given their wide use and interesting properties, pursuit of additional nucleosides with innovative flexible base modifications may yield many new chemotherapeutic agents as well as provide new and valuable tools in the rapidly growing field of bioprobes.

References

Agasimundin YS, Oakes FT, Kostuba LJ et al (1985) Annelation of guanosine by reaction with methyl N-cyanomethanimidate and sodium methoxide to give a tricyclic, fluorescent analog of adenosine. J Org Chem 50:2468–2474

Bardon AB, Wetmore SD (2005) How flexible are fleximer nucleobases? A computational study. J Phys Chem 109:262–272

Barrio JR, Secrist JA 3rd, Leonard NJ (1972) A fluorescent analog of nicotinamide adenine dinucleotide. Proc Natl Acad Sci U S A 69:2039–2042

Bisacchi GS, Chao ST, Bachard C et al (1997) BMS-200475, a novel carbocyclic 2′-deoxyguanosine analog with potent and selective anti-hepatitis B virus activity in vitro. Bioorg Med Chem Lett 7:127–132

Boyle GA, Edlin CD, Li Y et al (2012) Enantioselective synthesis of the carbocyclic nucleoside (−)-abacavir. Org Biomol Chem 10:1870–1876

Braendvang M, Bakken V, Gundersen LL (2009) Synthesis, structure, and antimycobacterial activity of 6-[1(3H)-isobenzofuranylidenemethyl]purines and analogs. Bioorg Med Chem 17:6512–6516

Chittepu P, Sirivolu VR, Seela F (2008) Nucleosides and oligonucleotides containing 1,2,3-triazole residues with nucleobase tethers: synthesis via the azide-alkyne 'click' reaction. Bioorg Med Chem 16:8427–8439

Das K, Lewi PJ, Hughes SH et al (2005) Crystallography and the design of anti-AIDS drugs: conformational flexibility and positional adaptability are important in the design of non-nucleoside HIV-1 reverse transcriptase inhibitors. Prog Biophys Mol Biol 88:209–231

De Clercq E (1988) Recent advances in the search for selective antiviral agents. Adv Drug Res 17:1–59

De Clercq E (2004) Antivirals and antiviral strategies. Nat Rev Microbiol 2:704–720

De Clercq E, Holý A (2005) Acyclic nucleoside phosphonates: a key class of antiviral drugs. Nat Rev Drug Discov 4:928–940

De Winter H, Herdewijn P (1996) Understanding the binding of 5-substituted 2′-deoxyuridine substrates to thymidine kinase of herpes simplex virus type-1. J Med Chem 39:4727–4737

Dodd DW, Swanick KN, Price JT et al (2010) Blue fluorescent deoxycytidine analogues: convergent synthesis, solid-state and electronic structure, and solvatochromism. Org Biomol Chem 8:663–666

Enderlin G, Sartori G, Hervé G et al (2013) Synthesis of 6-aryluridines via Suzuki–Miyaura cross-coupling reaction at room temperature under aerobic ligand-free conditions in neat water. Tetrahedron Lett 54:3374–3377

Fresneau N, Hiebel M-A, Agrofoglio L et al (2012) Efficient synthesis of unprotected C-5-aryl/heteroaryl-2′-deoxyuridine via a Suzuki-Miyaura reaction in aqueous media. Molecules 17:14409–14417

Greco NJ, Tor Y (2005) Simple fluorescent pyrimidine analogues detect the presence of DNA abasic sites. J Am Chem Soc 127:10784–10785

Greco NJ, Tor Y (2007) Furan decorated nucleoside analogues as fluorescent probes: synthesis, photophysical evaluation and site-specific incorporation. Tetrahedron 63:3515–3527

Gutierrez AJ, Terhorst TJ, Matteucci MD et al (1994) 5-heteroaryl-2′-deoxyuridine analogs. Synthesis and incorporation into high-affinity oligonucleotides. J Am Chem Soc 116:5540–5544

Haouz A, Vanheusden V, Munier-Lehmann H et al (2003) Enzymatic and structural analysis of inhibitors designed against Mycobacterium tuberculosis thymidylate kinase. New insights into the phosphoryl transfer mechanism. J Biol Chem 278:4963–4971

Hari Y, Nakahara M, Pang J et al (2011) Synthesis and triplex-forming ability of oligonucleotides bearing 1-substituted 1H-1,2,3-triazole nucleobases. Bioorg Med Chem 19:1162–1166

Jordheim LP, Durantel D, Zoulim F et al (2013) Advances in the development of nucleoside and nucleotide analogues for cancer and viral diseases. Nat Rev Drug Discov 12:447–464

Kool ET (2001) Hydrogen bonding, base stacking, and steric effects in DNA replication. Annu Rev Biophys Biomol Struct 30:1–22

Koszytkowska-Stawińska M, Sas W (2013) Synthesis of novel NH-1,2,3-triazolo-nucleosides by the Banert cascade reaction. Tetrahedron 69:2619–2627

Krecmerova M, Holý A, Masojidkova M (2007) Pyrimidine acyclic nucleotide analogues with aromatic substituents in C-5 position. Collect Czech Chem Commun 72:927–951

Krim J, Grunewald C, Taourirte M et al (2012) Efficient microwave-assisted synthesis, antibacterial activity and high fluorescence of 5 benzimidazolyl-2′-deoxyuridines. Bioorg Med Chem 20:480–486

Lee YS, Park SM, Kim HM et al (2009) C5-Modified nucleosides exhibiting anticancer activity. Bioorg Med Chem Lett 19:4688–4691

Leonard NJ, Hiremath SP (1986) Dimensional probes of binding and activity. Tetrahedron 42:1917–1961

Loren JC, Sharpless KB (2005) The Banert cascade: a synthetic sequence to polyfunctional NH-1,2,3-triazoles. Synthesis 2005(EFirst):1514–1520

Montagu A, Roy V, Balzarini J et al (2011) Synthesis of new C5-(1-substituted-1,2,3-triazol-4 or 5-yl)-2′-deoxyuridines and their antiviral evaluation. Eur J Med Chem 46:778–786

O'Daniel PI, Jefferson M, Wiest O et al (2008) A computational study of expanded heterocyclic nucleosides in DNA. J Biomol Struct Dyn 26:283–292

Orr GF, Musso DL, Boswell GE et al (1995) Inhibition of uridine phosphorylase: synthesis and structure-activity relationships of aryl-substituted 5-benzyluracils and 1-[(2-hydroxyethoxy)methyl]-5-benzyluracils. J Med Chem 38:3850–3856

Orr GF, Musso DL, Kelley JL et al (1997) Inhibition of uridine phosphorylase. Synthesis and structure-activity relationships of aryl-substituted 1-((2-hydroxyethoxy)methyl)-5-(3-phenoxybenzyl)uracil. J Med Chem 40:1179–1185

Polak M, Seley KL, Plavec J (2004) Conformational properties of shape modified nucleosides – Fleximers. J Am Chem Soc 126:8159–8166

Popescu A, Hörnfeldt A-B, Gronowitz S et al (1995) Catalytic osmylation and antiviral activity of some garbocyclic 5-substituted uridine and cytidine analogues. Nucleosides Nucleotides 14:1639–1657

Quirk S, Seley KL (2005a) Identification of catalytic amino acids in the human GTP fucose pyrophosphorylase active site. Biochemistry 44:13172–13178

Quirk S, Seley KL (2005b) Substrate discrimination by the human GTP fucose pyrophosphorylase. Biochemistry 44:10854–10863

Read ML, Brændvang M, Miranda PO et al (2010) Synthesis and biological evaluation of pyrimidine analogs of antimycobacterial purines. Bioorg Med Chem 18:3885–3897

Robertson MP, Miller SL (1995) Prebiotic synthesis of 5-substituted uracils: a bridge between the RNA world and the DNA-protein world. Science 268:702–705

Rozners E, Smicius R, Uchiyama C (2005) Expanding functionality of RNA: synthesis and properties of RNA containing imidazole modified tandem G-U wobble base pairs. Chem Commun 46:5778–5780

Secrist JA 3rd, Barrio JR, Leonard NJ (1972a) A fluorescent modification of adenosine triphosphate with activity in enzyme systems: 1, N 6 -ethenoadenosine triphosphate. Science 175:646–647

Secrist JA 3rd, Barrio JR, Leonard NJ et al (1972b) Fluorescent modification of adenosine-containing coenzymes. Biological activities and spectroscopic properties. Biochemistry 11:3499–3506

Seley KL, Schneller SW, Rattendi D et al (1997) (+)-7-deaza-5′-noraristeromycin as an anti-trypanosomal agent. J Med Chem 40:622–624

Seley KL, Zhang L, Hagos A (2001) "Fleximers". Design and synthesis of two novel split nucleosides. Org Lett 3:3209–3210

Seley KL, Zhang L, Hagos A et al (2002) "Fleximers". Design and synthesis of a new class of novel shape-modified nucleosides. J Org Chem 67:3365–3373

Seley KL, O'Daniel PI, Salim S (2003a) Design and synthesis of a series of chlorinated 3-deazaadenine analogues. Nucleosides Nucleotides Nucleic Acids 22:2133–2144

Seley KL, Quirk S, Salim S et al (2003b) Unexpected inhibition of S-adenosyl-L-homocysteine hydrolase by a guanosine nucleoside. Bioorg Med Chem Lett 13:1985–1988

Seley KL, Salim S, Zhang L (2005) "Molecular chameleons". Design and synthesis of C-4-substituted imidazole fleximers. Org Lett 7:63–66

Sinkeldam RW, Wheat AJ, Boyaci H et al (2011) Emissive nucleosides as molecular rotors. ChemPhysChem 12:567–570

Spencer RD, Weber G, Tolman GL et al (1974) Species responsible for the fluorescence of 1: N6-ethenoadenosine. Eur J Biochem 45:425–429

Srivatsan SG, Tor Y (2007) Fluorescent pyrimidine ribonucleotide: synthesis, enzymatic incorporation, and utilization. J Am Chem Soc 129:2044–2053

St. Amant AH, Bean LA, Guthrie JP et al (2012) Click fleximers: a modular approach to purine base-expanded ribonucleoside analogues. Org Biomol Chem 10:6521–6525

Temburnikar K, Brace K, Seley-Radtke KL (2013) Synthesis of 2′-deoxy-9-deaza nucleosides using Heck methodology. J Org Chem 78:7305–7311

Temburnikar KW, Zimmermann SC, Kim NT et al (2014) Antiproliferative activities of halogenated thieno [3,2-d]pyrimidines. Bioorg Med Chem (in press)

Tor Y, Del Valle S, Jaramillo D et al (2007) Designing new isomorphic fluorescent nucleobase analogues: the thieno[3,2-d]pyrimidine core. Tetrahedron 63:3608–3614

Tuske S, Sarafianos SG, Clark AD Jr et al (2004) Structures of HIV-1 RT-DNA complexes before and after incorporation of the anti-AIDS drug tenofovir. Nat Struct Mol Biol 11:469–474

Wachowius F, Rettig M, Palm G et al (2008) Synthesis and DNA duplex recognition of a triplex-forming oligonucleotide with an ureido-substituted 4-phenylimidazole nucleoside. Tetrahedron Lett 49:7264–7267

Wang W, Purwanto MG, Weisz K (2004) CG base pair recognition by substituted phenylimidazole nucleosides. Org Biomol Chem 2:1194–1198

Wauchope OR, Tomney MJ, Pepper JL et al (2010) Tricyclic 2′–C-modified nucleosides as potential anti-HCV therapeutics. Org Lett 12:4466–4469

Wauchope OR, Johnson C, Krishnamoorthy P et al (2012a) Synthesis and biological evaluation of a series of thieno-expanded tricyclic purine 2′-deoxy nucleoside analogues. Bioorg Med Chem 20:3009–3015

Wauchope OR, Velasquez M, Seley-Radtke K (2012b) Synthetic routes to a series of proximal and distal 2′-deoxy fleximers. Synthesis 44:3496–3504
Wigerinck P, Pannecouque C, Snoeck R et al (1991) 5-(5-Bromothien-2-yl)-2′-deoxyuridine and 5-(5-chlorothien-2-yl)-2′-deoxyuridine are equipotent to (E)-5-(2-bromovinyl)-2-′-deoxyuridine in the inhibition of herpes simplex virus type I replication. J Med Chem 34:2383–2389
Wilson JN, Kool ET (2006) Fluorescent DNA base replacements: reporters and sensors for biological systems. Org Biomol Chem 4:4265–4274
Yang MM, Zhou J, Schneller SW (2006) The Mitsunobu reaction in preparing 3-deazapurine carbocyclic nucleosides. Tetrahedron 62:1295–1300
Yu JL, Wu QP, Zhang QS et al (2010) Synthesis and antitumor activity of novel 2′,3′-dideoxy-2′,3′-diethanethionucleosides bearing 1,2,3-triazole residues. Bioorg Med Chem Lett 20:240–243
Zimmermann SC, Sadler JM, O'Daniel PI et al (2013) "Reverse" carbocyclic fleximers: synthesis of a new class of adenosine deaminase inhibitors. Nucleosides Nucleotides Nucleic Acids 32:137–154

Sequence-Selective Recognition of Double-Stranded RNA

Eriks Rozners

Contents

1 Introduction 168
2 Affinity and Selectivity of dsRNA Recognition by PNA 169
3 Triple Helical Recognition of Pyrimidine Nucleobases 171
4 Triple Helical Recognition of dsRNA at Physiological Conditions 174
5 PNA–Lysine Conjugates: Improved RNA Binding and Cellular Uptake 175
6 Conclusions and Outlook: Recognition of Biologically Significant dsRNA 177
References 178

Abstract The important role that noncoding RNA plays in cell biology makes it an attractive target for molecular recognition. However, the discovery of small molecules that bind such double-stranded RNA (dsRNA) selectively and may serve as biochemical probes and potential drug leads has been relatively slow. The surface of dsRNA is relatively uniform, dominated by negatively charged phosphates, and presents little opportunity for the traditional shape-selective molecular recognition. On the other hand, hydrogen-bonded base pairing is the key feature of nucleic acids and, therefore, inherently the most effective way of sequence-selective recognition of RNA. This chapter reviews recent progress in developing oligonucleotide analogues for selective binding to dsRNA. The emphasis is on hydrogen bonding to nucleobases, such as triple helix formation between RNA and peptide nucleic acids (PNAs). PNA having cationic nucleobase modifications binds to dsRNA with low nanomolar affinity and excellent sequence selectivity under physiologically relevant conditions. PNA is uniquely selective ligand for dsRNA as binding to the same sequences of dsDNA is at least an order of magnitude weaker. Conjugation of PNA with short lysine peptides further enhances binding affinity and RNA over DNA selectivity without compromising sequence selectivity of the triple helix formation.

E. Rozners (✉)
Department of Chemistry, Binghamton University, SUNY, Binghamton, NY 13902, USA
e-mail: erozners@binghamton.edu

V.A. Erdmann et al. (eds.), *Chemical Biology of Nucleic Acids*, RNA Technologies,
DOI 10.1007/978-3-642-54452-1_10,

Preliminary results suggest that the modified PNAs can bind and recognize noncoding dsRNAs through the triple helix formation. The cationic nucleobase and backbone (Lys) modifications enhance the cellular uptake of PNA. Thus, modified PNAs may be promising compounds for detecting and interfering with the function of biologically relevant dsRNA in live cells.

Keywords Sequence-selective RNA recognition • Peptide nucleic acid • PNA • Triple helix • Modified nucleobases

1 Introduction

Historically, molecular recognition of double-stranded RNA (dsRNA) has received relatively little attention. For a long time, RNA was viewed merely as a middleman in transfer of genetic information from DNA to proteins. While the importance of some noncoding RNAs, such as ribosomal and transfer RNAs, was appreciated, the molecular recognition of these species was not a high priority research. However, since the discovery that RNA can catalyze chemical reactions, the number and variety of noncoding RNAs and the important roles they play in biology have been growing steadily. Most notable gene expression regulators are short interfering RNAs, microRNAs, riboswitches, and the RNA motifs involved in splicing machinery (Meister and Tuschl 2004; Mandal and Breaker 2004; Plasterk 2006). While less than 2 % of DNA encodes for functional proteins, almost 70 % is transcribed into RNA. The ability to selectively recognize, detect, and inhibit the function of such RNAs will be highly useful for both fundamental biology and practical applications in biotechnology and medicine. However, since most noncoding RNAs form well-organized double helical structures, molecular recognition of such species is a formidable challenge (Chow and Bogdan 1997; Sucheck and Wong 2000; Thomas and Hergenrother 2008; Guan and Disney 2012).

RNA helix has a relatively uniform and polar surface that presents little opportunity for hydrophobic shape-selective recognition. On the other hand, binding to RNA bulges and internal loops, which are the most common targets of small molecules, is frustrated by conformational flexibility of non-helical RNA. Most of the current RNA binding compounds rely on electrostatic interactions with the negatively charged phosphate backbone to boost the affinity at the expense of selectivity. For example, the cationic aminoglycoside antibiotics are very strong RNA binders; however, their applications are limited by high toxicity due to indiscriminate binding to a variety of RNA species. Thus, discovery of small molecules that bind *double helical RNA sequence selectively* remains a highly desirable yet challenging goal (Thomas and Hergenrother 2008; Guan and Disney 2012).

Hydrogen bonding to nucleobases is an inherently effective means of selective recognition of helical nucleic acids. This concept has been elegantly applied by Dervan and coworkers (2003) in targeting of DNA by minor groove-binding polyamides. In contrast to DNA, the shallow minor groove of RNA is not a good fit for Dervan's compounds (Chenoweth et al. 2013). Recently, we proposed that triple helical binding of neutral oligonucleotide analogs, such as peptide nucleic acids (PNA), might provide straightforward recognition of double-stranded RNA (dsRNA) in the major groove with substantially higher selectivity than that of traditional RNA binders.

Triple helices have been extensively studied in the DNA context (Fox and Brown 2005; Maher et al. 1989; Moser and Dervan 1987). However, compared to DNA, the major groove of an RNA helix is deep and narrow, which hinders binding of the third strand oligonucleotide. RNA triple helices are less well studied than the DNA counterparts. Relatively stable RNA triple helices are formed via parallel binding of a pyrimidine-rich third strand to a purine-rich strand of the double helix at mildly acidic conditions (Roberts and Crothers 1992; Han and Dervan 1993; Escude et al. 1993). The sequence selectivity derives from uridine recognition of A-U base pairs (U*A-U triplet) and protonated cytidine recognition of G-C base pairs (C^+*G-C triplet) via the Hoogsteen hydrogen bonding (Fig. 1). In theory, triple helix has great potential for sequence-selective recognition of dsRNA. In practice, applications of triple helices are limited by three problems: (1) *low affinity of the third strand oligonucleotide* for the double helix caused, at least in part, by electrostatic repulsion between the negatively charged phosphate backbones of duplex and the third strand; (2) the *requirement for long homopurine tracts*, as only U*A-U and C^+*G-C triplets can be used in recognition; and (3) *low stability* of triple helices *at physiologic pH* due to need for cytosine protonation (unfavorable due to pK_a of ~4.5) to form the C^+*G-C triplet. This chapter discusses recent progress in using chemically modified PNA to overcome the problems of triple helical recognition of dsRNA. The key advances have been nucleobase modifications, especially, introduction of a cationic nucleobase (2-aminopyridine), and backbone modifications by attachment of short lysine peptides that allowed nanomolar recognition of biologically relevant RNAs at physiological conditions. The cationic modifications enhanced cellular uptake of PNA, suggesting that PNA may be promising compounds for applications that require dsRNA recognition in live cells.

2 Affinity and Selectivity of dsRNA Recognition by PNA

PNA has the entire sugar-phosphate backbone replaced by a neutral *N*-(2-aminoethyl)glycine moiety (Fig. 1) that does not cause electrostatic repulsion when binding to negatively charged nucleic acids (Egholm et al. 1993; Nielsen et al. 1991). PNA binds to double-stranded DNA (dsDNA) via two competing binding modes, triple helix (PNA:DNA, 1:1) and strand invasion, where PNA

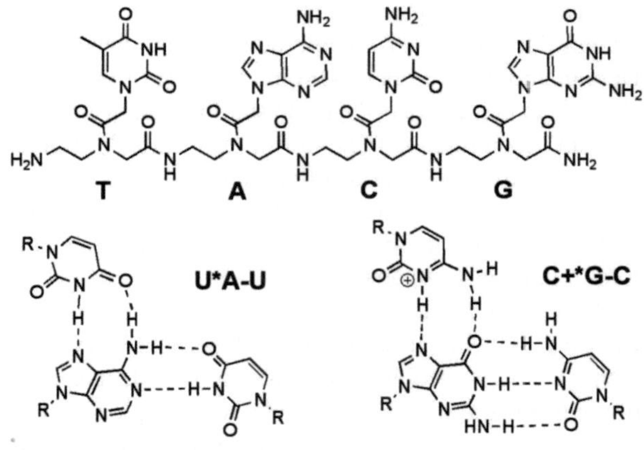

Fig. 1 Structure of PNA and common Hoogsteen base triplets in RNA

displaces one of the DNA strands, typically followed by a triplex formation (PNA: DNA, 2:1) (Hansen et al. 2009; Nielsen et al. 1991). The binding mode depends on sequence of PNA, salt concentration, DNA duplex stability, and PNA concentration (Wittung et al. 1997).

While PNA–dsDNA triple helices have been well studied, at the outset of our studies, we were surprised to discover that triple helix between PNA and dsRNA was virtually unexplored. There was only one report by Toulme and coworkers suggesting that PNA had low affinity for an RNA hairpin responsible for the ribosome gag-pol frameshift in HIV-1 (Aupeix et al. 1999). However, this result was not conclusive because the complex studied contained two adjacent mismatched T*C-G triplets and the affinity of PNA for the dsRNA was estimated only qualitatively. Thus, the ability of PNA to bind dsRNA via triple helix formation was unknown.

To probe if PNA could form a stable triple helix with dsRNA, we adopted the model system of short RNA hairpins used by Roberts and Crothers (Roberts and Crothers 1992) in one of the early studies on RNA triple helices. Using isothermal titration calorimetry (ITC), we found that a PNA 12-mer (**PNA1**, Fig. 2) had low nanomolar affinity ($K_a = 3.5 \times 10^8\ M^{-1}$, $K_d = 2.9$ nM) for the matched **HRP1** at pH 5.5 in 100 mM sodium acetate [conditions used by Roberts and Crothers (1992)] (Li et al. 2010). The PNA:RNA stoichiometry was 1:1, consistent with the expected triple helix. The affinity of **PNA1** for dsRNA was 30 times higher than affinity of an RNA third strand of the same sequence making the PNA–RNA complex more thermally stable than the original RNA duplex (Li et al. 2010).

As additional bonus, the binding rate of the neutral PNA was at least three times faster than that of the oligoribonucleotide third strand, as could be judged qualitatively from the inspection of ITC traces (Li et al. 2010). A sequence selectivity study revealed that the binding mode was triple helix formation and that PNA *as short as hexamers* (**PNA2**, Fig. 2) had high binding affinity and specificity ($K_a > 10^7\ M^{-1}$, $K_d < 100$ nM). When compared against the four hairpins (**HRP1**–**HRP4**, Fig. 2) having a variable base pair in the middle (bold in Fig. 2), only the

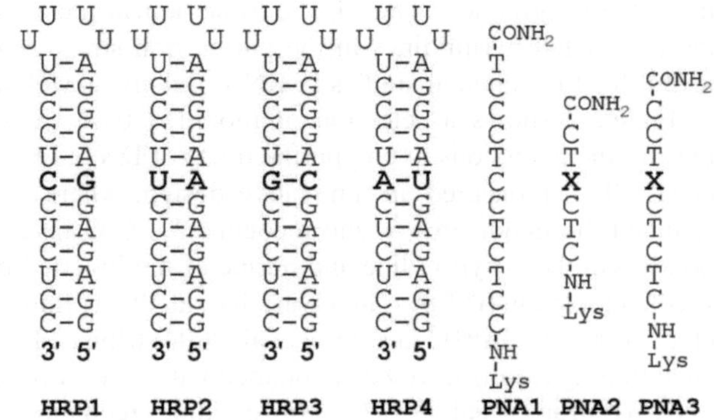

Fig. 2 Sequences of RNA hairpins and PNA ligands used to study PNA-dsRNA triple helix formation

two matched combinations involving C and T as the variable base X in **PNA2** had high affinity for their target hairpins, **HRP1** and **HRP2**, respectively (Li et al. 2010). All other mismatched combinations were significantly destabilized. Typically, the mismatches with **HRP1** were the most stable, but even these had $K_a < 10^6\ M^{-1}$, an order of magnitude lower than the matched complexes. Of particular importance for potential in vivo applications, PNA had an order of magnitude higher affinity for an RNA than for a DNA hairpin of the same sequence, which inside cells should minimize side effects due to competing binding to DNA targets. To the best of our knowledge, this was the first study to suggest a strong and sequence-selective triple helical binding of PNA to dsRNA (Li et al. 2010).

Although UV thermal melting and CD spectroscopy confirmed the ITC results, we found the spectroscopic methods less informative, mostly because of the high melting temperatures of PNA–RNA complexes (frequently >90 °C) and generally low resolution of UV and CD spectra. In our hands, ITC clearly emerged as the method of choice for studying binding of PNA to dsRNA and was used as the main tool in all of our follow-up studies discussed below.

3 Triple Helical Recognition of Pyrimidine Nucleobases

While the purine nucleosides are easily recognized by the standard Hoogsteen triplets (Fig. 1), recognition of pyrimidine nucleosides is a formidable challenge presumably because they present fewer hydrogen-bonding sites in the major groove. Despite extensive research, the requirement for polypurine tracts, i.e., long stretches of purine nucleosides on one strand of dsDNA, remains a major limitation for triple helix formation (Purwanto and Weisz 2003; Buchini and Leumann 2003; Robles et al. 2002). Biologically relevant dsRNAs typically do not contain long polypurine tracts. However, it is common to find eight and more contiguous purines interrupted by one or two pyrimidines in ribosomal RNAs (http://www.rna.ccbb.utexas.edu/) and miRNAs (Griffiths-Jones et al. 2008). It is conceivable that other noncoding RNAs contain similar sequences suitable for triple helix formation. Thus, if the triple helical recognition could be expanded to target isolated pyrimidines in the purine-rich strand, the approach could become useful for fundamental studies in RNA biology as well as practical applications.

Figure 3 shows a selection of modified base triplets designed to recognize pyrimidine inversions in polypurine tracts of DNA. Dervan and coworkers (Griffin et al. 1992) pioneered an innovative design, where a nucleobase analogue (**D₃**) spanned the major groove and recognized a C-G inversion by forming hydrogen bonds with both pyrimidine and purine of the inverted base pair. A similar design was later used in **S***T-A (Guianvarc'h et al. 2001; Ranasinghe et al. 2005; Rusling et al. 2006) and **Ind***C-G (Mertz et al. 2004) triplets. For recognition of thymidine, encouraging results have been obtained with heterocycle **E** (Fig. 3) incorporated in PNA (Eldrup et al. 1997). For cytidine recognition, various derivatives of pyrimidinone (e.g., **MP** and $^{\mathbf{Me}}$**P**) (Ranasinghe et al. 2005; Buchini and Leumann 2004) and cytosines modified by cationic (Semenyuk et al. 2010) ($\mathbf{C^{cat}}$) and intercalating (Gianolio and McLaughlin 1999) ($\mathbf{C^{int}}$) N-substituents have given promising results when incorporated in short DNA fragments. Although these analogs form only one hydrogen bond, they may engage in stabilizing bonding interactions with the C5-H of pyrimidine. Triple helix recognition of mixed sequences has also been demonstrated by switching the polarity of oligonucleotides to read the purine tracts on alternate DNA strands (Beal and Dervan 1992; Horne and Dervan 1990). Overall, the recognition of T-A and C-G inversions has been demonstrated in DNA with useful selectivity, but the stability of resulting triplets (Fig. 3) is typically lower than that of the standard Hoogsteen triplets.

While various modified heterocycles have been studied for DNA triplexes (Fig. 3), recognition of pyrimidine nucleosides in dsRNA had no precedent prior to our studies. We discovered that PNA nonamers (**PNA3**, Fig. 2) modified with 2-pyrimidinone **P** and 3-oxo-2,3-dihydropyridazine **E** (Fig. 4) recognized C-G and U-A inversions in **HRP3** and **HRP4**, respectively, with high affinity ($K_a = 0.4$ and $2.8 \times 10^7\ M^{-1}$) and sequence selectivity when placed at the variable position X in **PNA3** (Gupta et al. 2011). The use of **P** base in PNA had no precedent before our study, although derivatives of pyrimidinone had been used as oligonucleotide modifications for recognition of C-G inversions in DNA (Ranasinghe et al. 2005; Buchini and Leumann 2004). The **E** base was originally designed by Nielsen and coworkers (Eldrup et al. 1997) as a PNA modification to recognize T-A and dU-A inversions in DNA (Fig. 3).

The experiments described above were done in 100 mM sodium acetate buffer at pH 6.25 because at pH 5.5 **E**-modified **PNA3** had very strong but nonselective binding to all four hairpins **HRP1–HRP4**. This was in contrast to earlier studies by Nielsen and coworkers who reported that **E** was highly selective for T and dU in DNA at pH 7 (Eldrup et al. 1997). It was conceivable that the lower pH may have favored alternative tautomeric forms of **E** that formed triplets with other nucleobases. While **E**-modified **PNA3** did not bind to **HRP4** at pH 7, good affinity and selectivity were observed at pH 6.25.

In our design of pyrimidine recognition, extending the linker between **P** and PNA backbone by an extra CH_2 group (Fig. 4) enhanced the affinity and selectivity

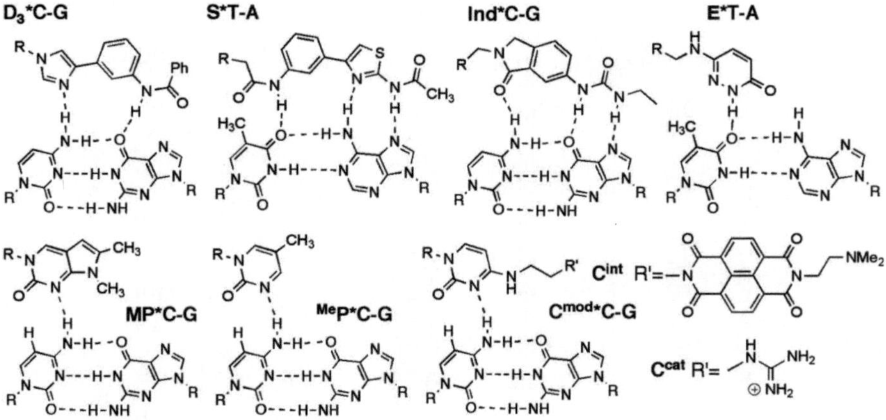

Fig. 3 Modified Hoogsteen base triplets for pyrimidine recognition in DNA

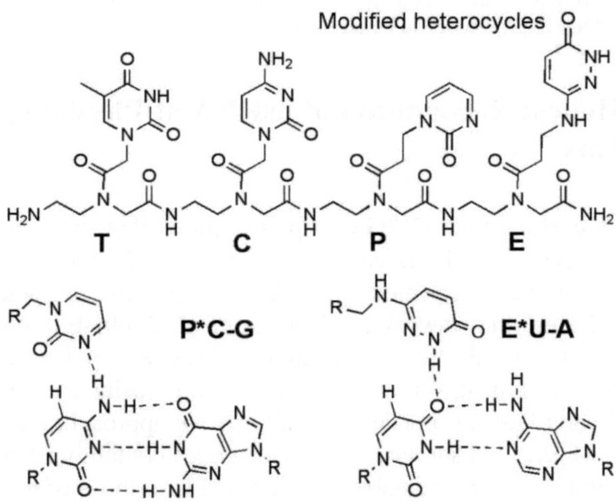

Fig. 4 Structure of nucleobase-modified PNA and modified base triplets for recognition of pyrimidines in RNA triple helices

of pyrimidine recognition (Gupta et al. 2011). The linker for **E** was originally designed by Nielsen and coworkers to be two atoms longer than in regular PNA. The thermodynamic stability of **P***C-G and **E***U-A triplets was similar to that of the standard Hoogsteen triplets allowing PNA nanomers to recognize RNA hairpins (Fig. 2) with mid-nanomolar affinity at pH 6.25 (Gupta et al. 2011).

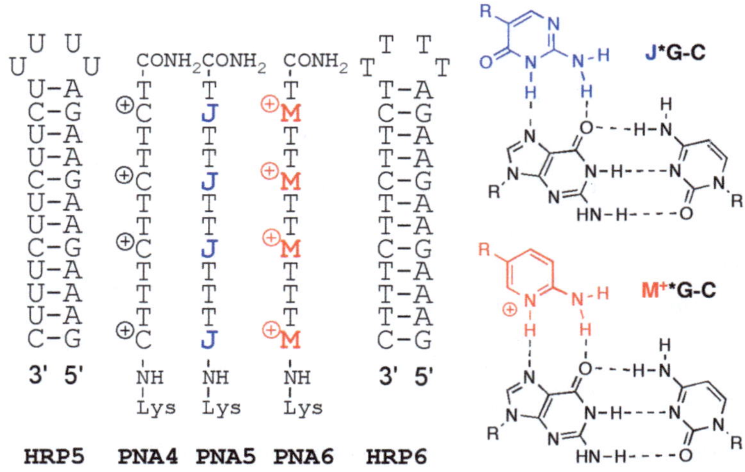

Fig. 5 Sequences of RNA and DNA hairpins, PNA ligands, and modified Hoogsteen triplets for recognition of dsRNA at physiological conditions

4 Triple Helical Recognition of dsRNA at Physiological Conditions

Formation of the Hoogsteen C^{+}*G-C triplet requires that cytosines of the third strand are protonated (Fig. 1). Because of the $pK_a = 4.5$, cytosine is hardly protonated at physiological conditions, which is a problem for potential in vivo applications. Nielsen and coworkers (Hansen et al. 2009; Egholm et al. 1995) used pseudoisocytosine (**J**, Fig. 5), originally designed by Kan and coworkers (Ono et al. 1991), as a neutral mimic of protonated cytosine to solve the problem in PNA. Povsic and Dervan pioneered an alternative approach, chemical modulation of cytosine's pK_a, and showed that triple helices containing 5-methylcytosine (more basic than C) were more stable at higher pH than those of unmodified DNA (Povsic and Dervan 1989). More recently, derivatives of 2-aminopyridine (**M**, $pK_a = 6.7$, Fig. 5) have been used to increase the stability of DNA triple helices at high pH (Rusling et al. 2005; Hildbrand et al. 1997; Cassidy et al. 1997; Chen and McLaughlin 2000). Before our studies, 2-aminopyridine or other more basic cytosine analogues had not been used in PNA.

In a recent study, we compared the potential of **J**- and **M**-modified PNAs (e.g., Fig. 5, **PNA5** and **PNA6**, respectively) to bind to dsRNA (**HRP5**) at physiological conditions (Zengeya et al. 2012). We found that **M**-modified PNAs were significantly more efficient RNA binders than **J**-modified PNAs at physiological conditions: 2 mM $MgCl_2$, 90 mM KCl, 10 mM NaCl, and 50 mM potassium phosphate, pH 7.4 and 37 °C.

While **PNA4** had good affinity, $K_a = 7.6 \times 10^6\ M^{-1}$ for **HRP5** in acetate buffer at pH 5.5, we could not observe any binding at pH 7 under physiologically relevant conditions. While substitution of C with **J** restored some binding affinity for **PNA5**, $K_a = 1.7 \times 10^6\ M^{-1}$, the **M**-modified **PNA6** was clearly superior with $K_a = 1.8 \times 10^7\ M^{-1}$ being even higher than the K_a of **PNA4** at pH 5.5 (Zengeya et al. 2012). Thus, while the neutral **J** provided the correct hydrogen-bonding scheme, the stability of the triplet was significantly higher when the positively charged **M** was used. Most remarkably, binding of **PNA6** to the matched DANN hairpin (having the same sequence as **HRP5**) was about two orders of magnitude weaker ($K_a = 3 \times 10^5$) than binding to the RNA hairpin **HRP5**. This result confirmed the unique selectivity of PNA for the recognition of the deep and narrow major groove of RNA over the wider major groove of DNA, as we had initially observed at pH 5.5.

Our results suggested that the positive charge on cytosine contributed significantly to stability of the Hoogsteen triplet, presumably via electrostatic attraction to the negatively charged nucleic acid. Thus, an ideal design for the recognition of G-C pairs would include correct hydrogen-bonding scheme and a positive charge on the heterocycle. To check that the cationic **M** had not reduced the sequence selectivity, we studied binding of **M**-modified **PNA7** and **PNA8** to the model hairpins **HRP1**–**HRP4** (Fig. 6) featuring a variable base pair (in bold) in the middle of helix. As expected, **PNA7** and **PNA8** formed stable 1:1 complexes with the matched **HRP1** and **HRP2** ($K_a = 2.0$ and $0.4 \times 10^7\ M^{-1}$), respectively (Zengeya et al. 2012). Confirming high sequence selectivity, we could observe only very weak interactions ($K_a < 10^4$) of **PNA7** and **PNA8** with other mismatched hairpins. The lower affinity of **PNA8** for its matched target **HRP2** confirmed that the charge contributed significantly to triplex stability. In all experiments with **M**-modified PNAs we observed the expected 1:1 PNA:RNA stoichiometry.

5 PNA–Lysine Conjugates: Improved RNA Binding and Cellular Uptake

Unmodified PNA does not readily cross cellular membranes. A promising approach to enhancing cellular uptake has been conjugation of PNA with cell penetrating peptides (CPP) that deliver the conjugate through endocytosis (Torres et al. 2012; Said Hassane et al. 2010; Shiraishi and Nielsen 2006). Recently, the groups of Corey (Hu et al. 2009; Hu and Corey 2007) and Gait (Fabani and Gait 2008; Abes et al. 2007; Turner et al. 2005) showed that conjugation of PNA with short lysine peptides enabled efficient delivery in various cancer cell lines. Four lysine residues achieved similar efficiency as R6-penetratin, a CPP optimized for cellular delivery of PNA (Abes et al. 2007). Using short lysine peptide instead of longer CPP significantly reduced the synthetic complexity required for PNA preparation and use in cell culture and enabled PNA delivery in mice (Wancewicz et al. 2011; Fabani et al. 2010). Addition of a terminal Cys residue further increased the cellular uptake of Cys-Lys-PNA-Lys_3 conjugate (Torres et al. 2012).

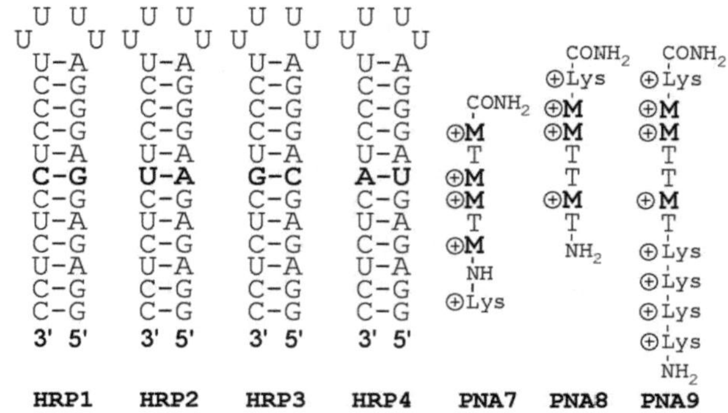

Fig. 6 Sequences of RNA hairpins and cationic nucleobase and backbone-modified PNAs

Our interest in testing short lysine peptides was further stimulated by the work of Strömberg and coworkers (Murtola et al. 2010) who reported that addition of cationic peptides to UV melting buffer enhanced the thermal stability of 2′-*O*-MeRNA/RNA duplexes, but had no effect on the stability of dsDNA. We hypothesized that the cationic peptides stabilized dsRNA by binding selectively in its deep and narrow major groove (but not in the wider major groove of dsDNA). Since the triplex forming PNAs also bind in the major groove, we further hypothesized that conjugation of short lysine peptides to PNA may increase the stability of PNA–RNA triple helices. If these hypotheses were true the resulting conjugates should be selective ligands for dsRNA while binding less strongly to dsDNA. Thus, conjugation of PNA with lysines would enhance the cellular uptake, affinity of PNA for dsRNA, and the unique RNA over DNA selectivity observed in our previous studies.

The above hypotheses were tested using **PNA9** (Fig. 6), a variant of **PNA8** carrying four additional D-lysines at the amino end. While **PNA8** had modest affinity ($K_a = 0.4 \times 10^7\ M^{-1}$) for the matched **HRP2**, the affinity of **PNA9** was in the low nanomolar range ($K_a = 16.5 \times 10^7\ M^{-1}$) (Muse et al. 2013). Consistent with Strömberg's results (Murtola et al. 2010) and our hypothesis, addition of four lysine residues to PNA enhanced the binding affinity about 40 times. Despite the highly cationic nature, **PNA9** had excellent sequence selectivity (Muse et al. 2013). Most importantly, we could not detect any binding of **PNA9** to the matched DNA hairpin having the same sequence as **HRP2**. These results strongly supported our hypotheses and were very encouraging for using D-lysines as affinity and selectivity enhancing agents for RNA recognition. Finally, the **M**- and Lys-modified PNA showed encouraging cellular uptake (Muse et al. 2013). Interestingly, both modifications were important for efficient uptake of modified PNAs, suggesting that the cationic nucleobases may contribute to crossing the cellular membrane, endosomal escape, or both.

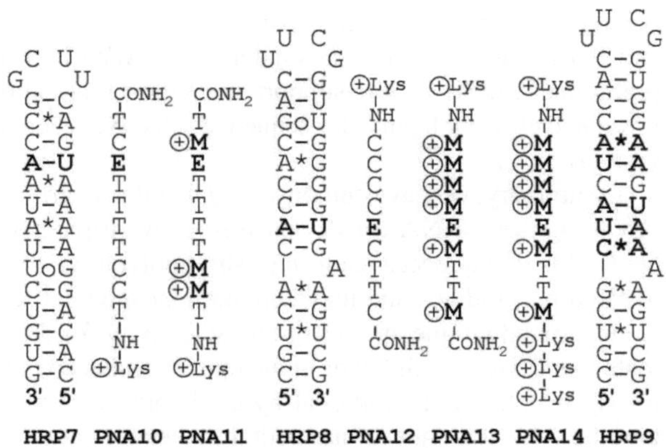

Fig. 7 Sequences of RNA hairpins modeling biologically important RNA and cationic nucleobase and backbone-modified PNAs

6 Conclusions and Outlook: Recognition of Biologically Significant dsRNA

M- and lysine-modified PNAs carrying multiple positive charges showed high affinity and excellent sequence selectivity in triple helical recognition of dsRNA. Conjugation with short lysine peptides strongly enhanced binding of PNA to dsRNA while no binding could be observed to matched dsDNA, which should eliminate nonspecific binding of PNA to nuclear dsDNA (Muse et al. 2013). The unique RNA selectivity will be important for potential applications in cells and other biological systems. The improved cellular uptake suggested that PNAs carrying cationic **M** and Lys modifications may be promising compounds for modulating the function of biologically important dsRNA in live cells.

To explore the potential of modified PNAs in recognition of biologically significant dsRNA sequences we studied two hairpins: **HRP7** modeling the sequence of pri-microRNA-215 and **HRP8** modeling the sequence of bacterial ribosomal A-site (Fig. 7). Under physiological conditions binding of **PNA10** to **HRP7** was relatively weak ($K_a \sim 1 \times 10^5$ M^{-1}), as expected because of the unfavorable C protonation. In contrast, strong binding was restored by substitution of C with **M** in **PNA11** ($K_a = 1.2 \times 10^7$ M^{-1}) (Zengeya et al. 2012).

Similar trend was observed for **HRP8**. While we could not observe any binding for the C-rich **PNA12** ($K_a < 10^3$ M^{-1}) at physiological conditions, the **M**-modified **PNA13** was binding strongly ($K_a = 3.7 \times 10^7$ M^{-1}) to **HPR8**. Addition of three lysines at the carboxyl end of PNA increased the binding affinity of **PNA14** to low nanomolar ($K_a = 1.3 \times 10^8$ M^{-1}). The PNA:RNA stoichiometry was 1:1, as expected for the proposed triple helical recognition. Interestingly, the presence of several noncanonical base pairs in **HRP7** and **HRP8** and a bulge in **HRP8** did not prevent strong binding of PNAs to the RNA hairpins. When testing against **HRP9** that models the human A-site sequence, we could not detect any binding of **PNA13** ($K_a < 10^3$ M^{-1}); **PNA14** had only very weak interaction ($K_a < 10^4$ M^{-1}) with **HRP9**. This result was not unexpected because **HRP8** and **HRP9** share only four common base pairs out of nine base pair recognition site; in other words, for **PNA13** and **PNA14** there are five mismatches in **HRP9** indicated in bold (Fig. 7). While preliminary, these results support the notion that cationic modified PNA may be novel and efficient ligands for sequence-selective recognition of dsRNA in biological systems.

In summary, we have demonstrated that PNA carrying appropriate modifications can recognize dsRNA via sequence-selective triple helix formation. The method is applicable to sequences featuring a short polypurine tract that may be interrupted by isolated pyrimidines and may contain noncanonical secondary structure. PNA has demonstrated unique and intriguing selectivity for dsRNA over dsDNA. We speculate that PNA may fit better in the narrow and deep major groove of dsRNA where it may benefit from additional hydrophobic interactions that contribute to strong binding. This notion is in line with the effect of cationic peptides on RNA duplexes observed by Strömberg and coworkers (Murtola et al. 2010). It is conceivable that the wider major groove of dsDNA does not provide optimal binding pocket for PNA and peptides. Promising preliminary results that the cationic **M** and Lys modifications enhance cellular uptake suggest that the modified PNA may be promising probes to study biologically important dsRNA species in live cells and may even become novel drug leads.

Acknowledgments The author thanks all the students and postdoctoral associates who contributed to PNA work in his laboratory. Their names appear in the references cited throughout the chapter. RNA projects in author's laboratory have been supported by Binghamton University and NIH (GM071461).

References

Abes S, Turner JJ, Ivanova GD et al (2007) Efficient splicing correction by PNA conjugation to an R6-penetratin delivery peptide. Nucleic Acids Res 35:4495–4502

Aupeix K, Le Tinevez R, Toulme JJ (1999) Binding of oligopyrimidines to the RNA hairpin responsible for the ribosome gag-pol frameshift in HIV-1. FEBS Lett 449:169–174

Beal PA, Dervan PB (1992) Recognition of double helical DNA by alternate strand triple helix formation. J Am Chem Soc 114:4976–4982

Buchini S, Leumann CJ (2003) Recent improvements in antigene technology. Curr Opin Chem Biol 7:717–726

Buchini S, Leumann CJ (2004) Stable and selective recognition of three base pairs in the parallel triple-helical DNA binding motif. Angew Chem Int Ed 43:3925–3928

Cassidy SA, Slickers P, Trent JO et al (1997) Recognition of GC base pairs by triplex forming oligonucleotides containing nucleosides derived from 2-aminopyridine. Nucleic Acids Res 25:4891–4898

Chen DL, McLaughlin LW (2000) Use of pKa differences to enhance the formation of base triplets involving C-G and G-C base pairs. J Org Chem 65:7468–7474

Chenoweth DM, Meier JL, Dervan PB (2013) Pyrrole-imidazole polyamides distinguish between double-helical DNA and RNA. Angew Chem Int Ed 52:415–418

Chow CS, Bogdan FM (1997) A structural basis for RNA-ligand interactions. Chem Rev 97:1489–1513

Dervan PB, Edelson BS (2003) Recognition of the DNA minor groove by pyrrole-imidazole polyamides. Curr Opin Struct Biol 13:284–299

Egholm M, Buchardt O, Christensen L et al (1993) PNA hybridizes to complementary oligonucleotides obeying the Watson-Crick hydrogen-bonding rules. Nature 365:566–568

Egholm M, Christensen L, Dueholm KL et al (1995) Efficient pH-independent sequence-specific DNA binding by pseudoisocytosine-containing bis-PNA. Nucleic Acids Res 23:217–222

Eldrup AB, Dahl O, Nielsen PE (1997) A novel peptide nucleic acid monomer for recognition of thymine in triple-helix structures. J Am Chem Soc 119:11116–11117

Escude C, Francois JC, Sun JS et al (1993) Stability of triple helixes containing RNA and DNA strands: experimental and molecular modeling studies. Nucleic Acids Res 21:5547–5553

Fabani MM, Gait MJ (2008) miR-122 targeting with LNA/2′-O-methyl oligonucleotide mixmers, peptide nucleic acids (PNA), and PNA-peptide conjugates. RNA 14:336–346

Fabani MM, Abreu-Goodger C, Williams D et al (2010) Efficient inhibition of miR-155 function in vivo by peptide nucleic acids. Nucleic Acids Res 38:4466–4475

Fox KR, Brown T (2005) An extra dimension in nucleic acid sequence recognition. Q Rev Biophys 38:311–320

Gianolio DA, McLaughlin LW (1999) Selective recognition of a dC-dG base pair by oligonucleotide-directed triplex formation using a dC residue tethering an intercalator. J Am Chem Soc 121:6334–6335

Griffin LC, Kiessling LL, Beal PA, Gillespie P, Dervan PB (1992) Recognition of all four base pairs of double-helical DNA by triple-helix formation: design of nonnatural deoxyribonucleosides for pyrimidine-purine base pair binding. J Am Chem Soc 114:7976–7982

Griffiths-Jones S, Saini HK, van Dongen S et al (2008) miRBase: tools for microRNA genomics. Nucleic Acids Res 36(Database Issue):D154–D158

Guan L, Disney MD (2012) Recent advances in developing small molecules targeting RNA. ACS Chem Biol 7:73–86

Guianvarc'h D, Benhida R, Fourrey J-L, Maurisse R, Sun J-S (2001) Incorporation of a novel nucleobase allows stable oligonucleotide-directed triple helix formation at the target sequence containing a purine-pyrimidine interruption. Chem Commun:1814–1815

Gupta P, Zengeya T, Rozners E (2011) Triple helical recognition of pyrimidine inversions in polypurine tracts of RNA by nucleobase-modified PNA. Chem Commun 47:11125–11127

Han H, Dervan PB (1993) Sequence-specific recognition of double helical RNA and RNA-DNA by triple helix formation. Proc Natl Acad Sci U S A 90:3806–3810

Hansen ME, Bentin T, Nielsen PE (2009) High-affinity triplex targeting of double stranded DNA using chemically modified peptide nucleic acid oligomers. Nucleic Acids Res 37:4498–4507

Hildbrand S, Blaser A, Parel SP et al (1997) 5-substituted 2-aminopyridine C-nucleosides as protonated cytidine equivalents: increasing efficiency and specificity in DNA triple-helix formation. J Am Chem Soc 119:5499–5511

Horne DA, Dervan PB (1990) Recognition of mixed-sequence duplex DNA by alternate-strand triple-helix formation. J Am Chem Soc 112:2435–2437

Hu J, Corey DR (2007) Inhibiting gene expression with peptide nucleic acid (PNA)-peptide conjugates that target chromosomal DNA. Biochemistry 46:7581–7589

Hu J, Matsui M, Gagnon KT et al (2009) Allele-specific silencing of mutant huntingtin and ataxin-3 genes by targeting expanded CAG repeats in mRNAs. Nat Biotechnol 27:478–484

Li M, Zengeya T, Rozners E (2010) Short peptide nucleic acids bind strongly to homopurine tract of double helical RNA at pH 5.5. J Am Chem Soc 132:8676–8681

Maher LJ III, Wold B, Dervan PB (1989) Inhibition of DNA binding proteins by oligonucleotide-directed triple helix formation. Science 245:725–730

Mandal M, Breaker RR (2004) Gene regulation by riboswitches. Nat Rev Mol Cell Biol 5:451–463

Meister G, Tuschl T (2004) Mechanisms of gene silencing by double-stranded RNA. Nature 431:343–349

Mertz E, Mattei S, Zimmerman SC (2004) Synthesis and duplex DNA recognition studies of oligonucleotides containing a ureido isoindolin-1-one homo-N-nucleoside. A comparison of host-guest and DNA recognition studies. Bioorg Med Chem 12:1517–1526

Moser HE, Dervan PB (1987) Sequence-specific cleavage of double helical DNA by triple helix formation. Science 238:645–650

Murtola M, Zaramella S, Yeheskiely E et al (2010) Cationic Peptides that increase the thermal stabilities of 2′-O-MeRNA/RNA duplexes but do not affect DNA/DNA melting. ChemBioChem 11:2606–2612

Muse O, Zengeya T, Mwaura J et al (2013) Sequence selective recognition of double-stranded RNA at physiologically relevant conditions using PNA-peptide conjugates. ACS Chem Biol 8 (8):1683–1686

Nielsen PE, Egholm M, Berg RH et al (1991) Sequence-selective recognition of DNA by strand displacement with a thymine-substituted polyamide. Science 254:1497–1500

Ono A, Ts'o POP, Kan LS (1991) Triplex formation of oligonucleotides containing 2′-O-methylpseudoisocytidine in substitution for 2′-deoxycytidine. J Am Chem Soc 113:4032–4033

Plasterk RHA (2006) Micro RNAs in animal development. Cell 124:877–881

Povsic TJ, Dervan PB (1989) Triple helix formation by oligonucleotides on DNA extended to the physiological pH range. J Am Chem Soc 111:3059–3061

Purwanto MGM, Weisz K (2003) Non-natural nucleosides for the specific recognition of Watson-Crick base pairs. Curr Org Chem 7:427–446

Ranasinghe RT, Rusling DA, Powers VEC, Fox KR, Brown T (2005) Recognition of CG inversions in DNA triple helices by methylated 3H-pyrrolo[2,3-d]pyrimidin-2(7H)-one nucleoside analogues. Chem Commun:2555–2557

Roberts RW, Crothers DM (1992) Stability and properties of double and triple helices: dramatic effects of RNA or DNA backbone composition. Science 258:1463–1466

Robles J, Grandas A, Pedroso E et al (2002) Nucleic acid triple helices: stability effects of nucleobase modifications. Curr Org Chem 6:1333–1368

Rusling DA, Powers VEC, Ranasinghe RT et al (2005) Four base recognition by triplex-forming oligonucleotides at physiological pH. Nucleic Acids Res 33:3025–3032

Rusling DA, Brown T, Fox KR (2006) DNA triple-helix formation at target sites containing duplex mismatches. Biophys Chem 123:134–140

Said Hassane F, Saleh AF, Abes R et al (2010) Cell penetrating peptides: overview and applications to the delivery of oligonucleotides. Cell Mol Life Sci 67:715–726

Semenyuk A, Darian E, Liu J et al (2010) Targeting of an interrupted polypurine: polypyrimidine sequence in mammalian cells by a triplex-forming oligonucleotide containing a novel base analogue. Biochemistry 49:7867–7878

Shiraishi T, Nielsen PE (2006) Enhanced delivery of cell-penetrating peptide-peptide nucleic acid conjugates by endosomal disruption. Nat Protoc 1:633–636

Sucheck SJ, Wong CH (2000) RNA as a target for small molecules. Curr Opin Chem Biol 4:678–686

Thomas JR, Hergenrother PJ (2008) Targeting RNA with small molecules. Chem Rev 108:1171–1224

Torres AG, Fabani MM, Vigorito E et al (2012) Chemical structure requirements and cellular targeting of microRNA-122 by peptide nucleic acids anti-miRs. Nucleic Acids Res 40:2152–2167

Turner JJ, Ivanova GD, Verbeure B et al (2005) Cell-penetrating peptide conjugates of peptide nucleic acids (PNA) as inhibitors of HIV-1 Tat-dependent trans-activation in cells. Nucleic Acids Res 33:6837–6849

Wancewicz EV, Maier MA, Siwkowski AM et al (2011) Peptide nucleic acids conjugated to short basic peptides show improved pharmacokinetics and antisense activity in adipose tissue. J Med Chem 53:3919–3926

Wittung P, Nielsen P, Norden B (1997) Extended DNA-recognition repertoire of peptide nucleic acid (PNA): PNA-dsDNA triplex formed with cytosine-rich homopyrimidine PNA. Biochemistry 36:7973–7979

Zengeya T, Gupta P, Rozners E (2012) Triple helical recognition of RNA Using 2-aminopyridine-modified PNA at physiologically relevant conditions. Angew Chem Int Ed 51:12593–12596

Determining Transient Nucleic Acid Structures by NMR

Jeetender Chugh

Contents

1 Introduction 182
2 The $R_{1\rho}$ NMR Experiment 184
  2.1 Understanding Exchange Broadening and Suppression 187
3 Sample Preparation 189
4 Data Acquisition 189
  4.1 Protocol 190
5 Data Analysis 193
6 Model Building 195
  6.1 Proving the Model 195
7 Concluding Remarks 196
References 196

Abstract Current structure determination techniques for bio-macromolecules including X-ray crystallography, nuclear magnetic resonance spectroscopy (NMR) and cryo-electron microscopy, focus on obtaining snapshots of the most stable conformer in the free-energy landscape. It is, however, imperative to fathom the conformations of other occupants of energy landscape as well that are transient in existence but are often the functionally active structures. NMR spectroscopy is a versatile technique that, in addition to providing static snapshots, has ability to measure various motional modes of a macromolecule. These motional modes connect distinct local minima in the free-energy landscape and help decipher additional conformational states that are accessible to a macromolecule. This book chapter gives a glimpse of the concept of existence of excited states for nucleic acids on free-energy landscape. Subsequently, this chapter focuses on using $R_{1\rho}$ relaxation dispersion NMR experiments to characterize the transient

J. Chugh (✉)
Department of Chemistry, Indian Institute of Science Education and Research (IISER), Dr. Homi Bhabha Road, Pune 411008, Maharashtra, India
e-mail: cjeet@iiserpune.ac.in

V.A. Erdmann et al. (eds.), *Chemical Biology of Nucleic Acids*, RNA Technologies, DOI 10.1007/978-3-642-54452-1_11,

structures of nucleic acids that are otherwise not possible with standard structure determination techniques. A step-by-step guide and basic principles—including NMR data acquisition, data fitting and statistical analysis, model building and mutational approaches to test the proposed models—will be discussed to first hypothesize and then confirm the existence of such unique transient structures in nucleic acids.

Keywords CPMG • Exchange broadening • Nucleic acid excited-state structures • Relaxation dispersion NMR spectroscopy • $R_{1\rho}$

1 Introduction

Atomic-resolution structural snapshots of bio-macromolecules provided by routine structure determination techniques including X-ray crystallography, NMR spectroscopy and cryo-electron microscopy are not sufficient enough to completely understand the functional diversity exhibited by these molecules (Dethoff et al. 2012a). The functional diversity of bio-macromolecules is embedded in their ability to acquire distinct conformations. This ability, i.e. the process of inter-conversion among several distinct conformations, is termed as 'molecular dynamics' and is responsible for increasing the conformational space of a macromolecule. The conformational space is not degenerate and is shown (Dill and Chan 1997) to have rugged free-energy landscape where bio-macromolecules with distinct conformations occupy distinct energy levels (Fig. 1). Proteins that function as enzymes have long been recognized as molecular machines that walk along this conformational space to carry out a particular biological process (Heine et al. 2001; Kuhlenkoetter et al. 2011; Granata et al. 2013; Schramm 2013). Similarly, DNA and RNA are also being recognized as parts of various molecular machines (Kuhlenkoetter et al. 2011; Nikolova et al. 2011; Dethoff et al. 2012a, b) that carry out several biological processes by undergoing conformational exchange due to molecular dynamics, either inherent or induced by a cellular cue.

Molecular dynamics in these nano-machines takes place at varying orders of timescale including picoseconds to milliseconds to seconds and so on (Schwartz and Schramm 2009; Wolfe and O'Brien 2009). A timescale in molecular dynamics—ms-to-μs motions—that is responsible to carry out many important biological processes has been explored in detail in proteins (Palmer 2009; Kay 2011) but is still poorly understood in nucleic acids. A very recent discovery of transient Hoogsteen base pairs in canonical duplex DNA has generated a new understanding for accessing functionally important but transient structures in nucleic acids with greater details (Nikolova et al. 2011, 2012). Transient structures occupy a slightly higher state in the free-energy landscape than that is occupied by equilibrium (or ground-state) structures. These transient structures are, therefore,

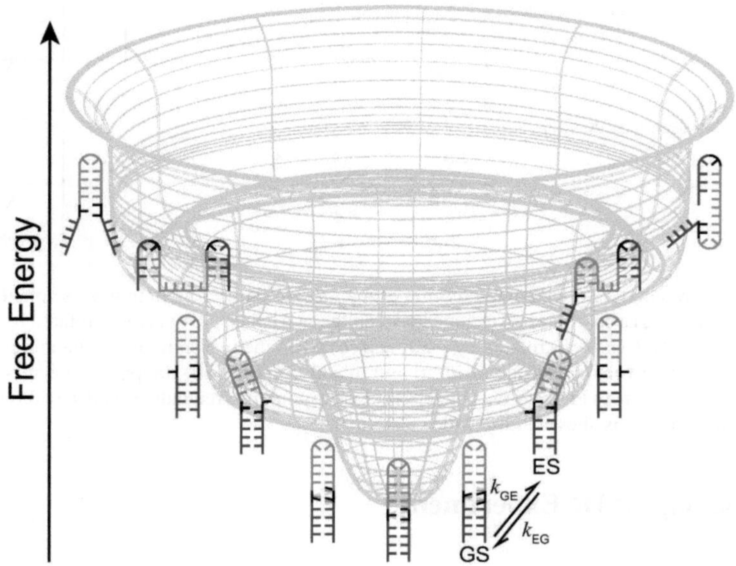

Fig. 1 A rugged and dynamic free-energy landscape is shown in *grey* containing schematic RNA conformations. Exchange process between a ground-state conformation (GS) and an excited-state conformation (ES) is *highlighted*. Three shades of *grey/black* are used to highlight how the same RNA sequence is inter-converting among distinct conformations

often referred to as excited-state structures or simply the excited states (ESs). Similar ESs have also been discovered in case of RNAs that are essential to perform various biological functions in a tightly regulated manner. For example, in ribosomal A-site RNA ES conformation sequesters two functionally important adenines A1492 and A1493 by flipping them inside the helix and makes them unavailable for decoding messenger RNA (mRNA) during the process of translation (Dethoff et al. 2012b).

Such transient conformations exist in too little abundance (typically with populations <5 %) and for too small a duration (typical lifetime < milliseconds) to carry out a detailed biophysical investigation. In such a situation, a low spin-lock field $R_{1\rho}$ NMR experiment comes handy and can measure molecular dynamics precisely within the window of above-described timescales (Massi et al. 2004; Palmer and Massi 2006; Bothe et al. 2011). A strategy to visualize ES of nucleic acids by NMR citing A-site ribosomal RNA as an example is described in the sections below.

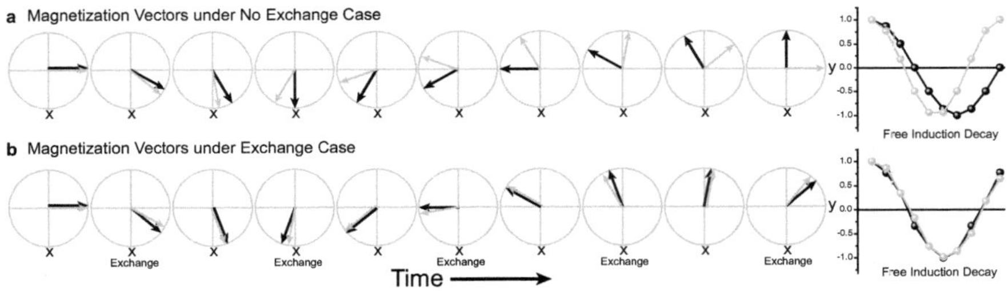

Fig. 2 A representative precessional motion of magnetization vectors in transverse plane as time is progressing in (**a**) a case of no exchange between the two spins, *grey vector* is rotated by $-40°$ in each step, and *black vector* is rotated by $-30°$ and (**b**) when the two spins are exchanging; angle of rotation of the two vectors is same as in '**a**' and in addition there is swapping of vectors (*grey* to *black* and *black* to *grey*) in every exchange step. Corresponding free induction decay (considering no relaxation decay) is shown at the end

2 The $R_{1\rho}$ NMR Experiment

Relaxation dispersion NMR spectroscopy is one of the few techniques that are used to site-specifically characterize slow motions in ms-to-μs timescales (Palmer 2004; Korzhnev et al. 2005). This technique manipulates NMR line broadening caused due to chemical exchange processes and can obtain the structural, thermodynamic and kinetic information of the exchange process. Exchange process may be understood by considering a nucleus residing in two conformations on the free-energy landscape—ground-state conformation (GS) and excited-state conformation (ES) (Fig. 1). In a typical NMR experiment during free evolution, each nucleus has a magnetization vector precessing around the *z*-axis (or the applied magnetic field direction) at its Larmor frequency (the frequency that translates into chemical shift). For a nucleus residing in two conformations, there will be two such vectors precessing at two different chemical shifts (Fig. 2). Consider that a detector is placed along the *y*-axis to capture intensity of the *y*-component of the precessing vector (a precessing vector at any given time has an *x*-component and a *y*-component) after a fixed time interval. In the absence of any exchange between the two conformations (Fig. 2a), intensity map of the two vectors in one full cycle of precession is shown at the far right end, called as free induction decay (*fid*). It is important to note that transverse relaxation decay has been excluded from the *fid* to keep the discussion simple. It is evident though, from the two non-overlapping *fid*s, that these two vectors have distinct precessional frequencies and hence distinct chemical shifts (Fig. 2a). In Sect. 2.1, a hypothetical scenario of two racing tracks is used to further explain the chemical exchange process and its suppression using two different NMR techniques.

In the presence of exchange between the two conformations, GS and ES $\left(\mathrm{GS} \underset{k_{\mathrm{EG}}}{\overset{k_{\mathrm{GE}}}{\rightleftharpoons}} \mathrm{ES}\right)$, at a rate ($k_{\mathrm{ex}} = k_{\mathrm{GE}} + k_{\mathrm{EG}}$) comparable to difference in their

NMR resonance frequencies ($\Delta\omega = \omega_G - \omega_E$), magnetization vectors of the two conformations stochastically swap their Larmor frequencies with each other (Fig. 2b). Since the process is stochastic and residence time of nucleus in a particular conformation can vary, it results in de-synchronization of the magnetization vectors' precessional motion. This causes fanning out of the magnetization vector resulting in broadening (R_{ex}) of the NMR line—in addition to that caused by the transverse relaxation process, R_2. In this particular example (Fig. 2b), the exchange rate and Larmor frequency of the two conformations have been chosen such that both the spins eventually have same chemical shift, as indicated by their overlapping *fid*s. The exchange broadening may also result in total disappearance of the NMR signal if the intensity is low to start with—typically in the case of ES conformations due to low populations. In such a situation, relaxation dispersion experiments come to rescue by suppressing the exchange contribution to the line broadening of the visible ground-state signal and quantifying the same. This suppression of exchange process is carried out either by application of a series of 180° pulses in a Carl–Purcell–Meiboom–Gill (CPMG) relaxation dispersion experiment or a constant radiofrequency (RF) field in a rotating frame $R_{1\rho}$ relaxation dispersion experiment ($R_{1\rho}$ experiment).

The exchange contribution is measured as a function of τ_{cp} in a CPMG experiment (Wang et al. 2001) and the power and frequency of radiofrequency field in $R_{1\rho}$ experiment (Massi et al. 2004). The resulting relaxation dispersion curve is typically fit to a two-state model yielding a set of parameters depending on if the exchange is slow ($k_{ex} \ll |\Delta\omega|$) to intermediate ($k_{ex} = \sim |\Delta\omega|$) or fast ($k_{ex} \gg |\Delta\omega|$) on the chemical shift time scale. Information about population (p_E), lifetime $\left(\tau_{ex} = {}^1/_{k_{ex}}\right)$ and chemical shift (ω_E) of the excited state can be obtained if the exchange is slow to intermediate; and only τ_{ex} and $\Phi = p_G p_E \Delta\omega^2$ can be determined in fast-exchanging systems and additional experiments are required to resolve ($\Delta\omega$) and (p_E), where $\Delta\omega = \omega_G - \omega_E$. The CPMG experiments have another disadvantage when compared with $R_{1\rho}$ experiment as in they only yield the absolute difference in chemical shift $|\omega_G - \omega_E|$ and additional experiments are needed to determine the sign of the $\Delta\omega$ (Baldwin and Kay 2012). In the above equations, k_{ex} is exchange rate constant; τ_{ex} is lifetime of the excited state; ω_G is chemical shift of GS; ω_E is chemical shift of ES; $\Delta\omega$ is chemical shift difference between GS and ES; p_G is population of GS; and p_E is population of ES.

Slower millisecond motions can be accessed by CPMG relaxation dispersion experiments and has been shown to work well for proteins (Vallurupalli et al. 2012); however extensive ^{13}C–^{13}C scalar coupling networks in the base and sugar moieties of nucleic acids severely complicate these experiments (Vallurupalli et al. 2007). The problem due to scalar couplings can, in principle, be solved by selectively labelling ^{13}C at distinct carbon positions and measuring data on multiple samples (Johnson and Hoogstraten 2008; Kloiber et al. 2011; Wunderlich et al. 2012). This, however, increases the cost of the study by several folds. Further, absence of sufficient number of ^{15}N–^{1}H probes in bases (especially in non-canonical regions of RNA where chances of occurrence of such slow motions

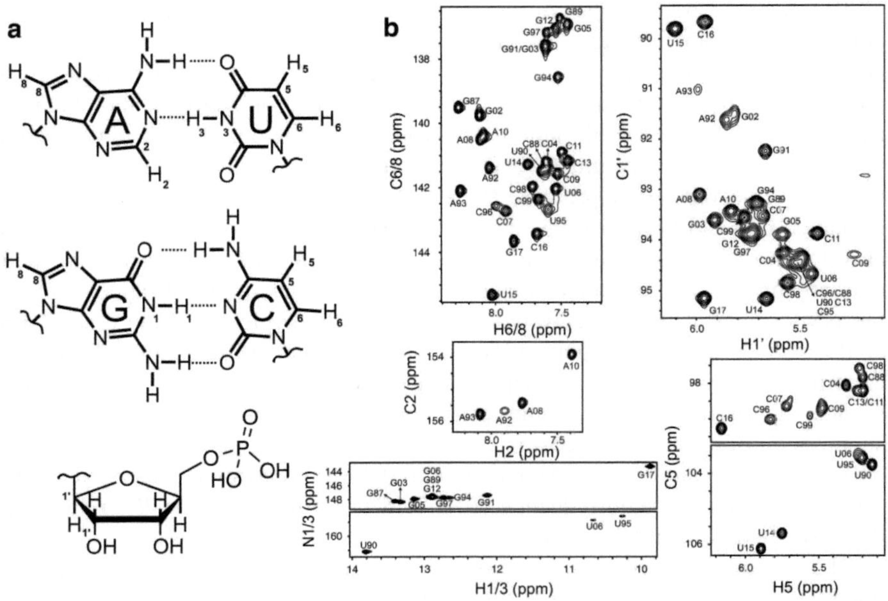

Fig. 3 (**a**) Protonated carbon and nitrogen atoms are numbered (N_1–H_1, N_3–H_3, C_2–H_2, C_5–H_5, C_6–H_6, C_8–H_8, $C_{1'}$–$H_{1'}$) to highlight the probes available for relaxation dispersion measurement in nucleic acids; figure also highlights the large number of C–C bonds (and hence scalar couplings) present in bases and ribose sugars and the scarcity of N–H bonds in canonical base pairs and total absence of them in ribose sugars, making CPMG measurement on nucleic acids challenging; (**b**) different regions of 2D [^{13}C, ^{1}H]-HSQC and [^{15}N, ^{1}H]-HSQC spectra of the A-site ribosomal RNA oligonucleotide of the marked probes; resonance assignments are marked on each peak

are highest) and total absence of ^{15}N–^{1}H probes in sugar moieties of DNA/RNA (Fig. 3a) makes the use of CPMG experiment on nucleic acids even more challenging.

$R_{1\rho}$ experiments are widely used, when compared with CPMG experiments, for the study of nucleic acids as spinlock fields can be used more selectively to eliminate or diminish the effect of unwanted ^{13}C–^{13}C scalar coupling interactions (Blad et al. 2005; Hansen et al. 2009). Additionally, $R_{1\rho}$ experiment can directly yield the magnitude as well as the sign of $\Delta\omega$ from the data measured on a single magnetic field, unlike CPMG experiments, where measurement at a minimum of two magnetic fields is required (Kovrigin et al. 2006). In an $R_{1\rho}$ experiment, line broadening is measured as a function of the spinlock power and frequency offset of a constant radiofrequency spinlock field. Early variants of $R_{1\rho}$ experiment employed effective RF fields typically in the range of 1–6 kHz thus limiting the sensitivity of exchange processes occurring on microsecond timescales (Blad et al. 2005). Very recently, range of accessible time scales has been extended to ~10 ms using low spinlock power in $R_{1\rho}$ experiment (Hansen et al. 2009). Appropriate use of ^{1}H decoupling and magnetization alignment schemes permits the use of significantly weaker RF fields (typically ~100 Hz) thus increasing the sensitivity to slower motions. Using these schemes, $R_{1\rho}$ experiment can now be measured for protonated carbons and protonated nitrogens in uniformly ^{13}C, ^{15}N-labelled and unlabelled (only for protonated carbon measurements) nucleic acids (Nikolova et al. 2011, 2012; Dethoff et al. 2012b).

2.1 Understanding Exchange Broadening and Suppression

To understand a no exchange situation, lets consider two concentric athletic tracks (namely walk-track and run-track) in a stadium that are rotating at two different speeds '*w*' and '*r*', where $r > w$ (Fig. 4a). Now consider a set of people asked to stand on walk-track for a certain length of time '*t*'. Also consider another set of people asked to stand on run-track for the same amount of time, with the same starting positions as walk-track people. Visualize that people do not have to walk or run to cover a distance, but it's the track itself that is rotating at two unique speeds. If we now plot distance travelled on '*y*' axis and time used to travel on '*x*' axis, it is very obvious to envisage that all the people on walk-track reach at a unique point and all the runners reach at another unique point, both defined by the respective track speeds (slope of the curve) (Fig. 4b). This situation is comparable to a nucleus residing in two conformations without undergoing exchange, and the NMR spectrum of that sample displays two distinct peaks for that nucleus (Fig. 4c).

In the exchange situation, walkers and runners are travelling again for the same duration, '*t*', but this time some of the people from walk-track decide to jump on the run-track and some of the runners decide to jump on walk-track—multiple times during the time '*t*'. Now walkers and runners would not be reaching at two unique points after time '*t*', but there will be a distribution of positions around pure walker and pure runner positions (Fig. 4d). The width of this distribution is directly correlated with the jump frequency. This distribution of position is what we know as NMR peak broadening due to exchange. If the population of walkers and runners is highly skewed, overall distribution of the two positions might merge and shift towards the higher population set, resulting in a single broad distribution (Fig. 4e).

Let us now consider the suppression of exchange using CPMG relaxation dispersion experiment (Fig. 4g). During time '*t*', if we blow a whistle and at that point both the tracks reverse their direction of rotation, it is intuitive to think that the overall effect of reversing the direction of rotation would be to decrease the width of the distribution of positions around pure walker and pure runner positions. If the whistle blowing frequency is increased comparable to exchange frequency of walkers and runners, there will not be any effective distribution, and all walkers and runners positions will match with either pure walker or pure runner position, depending on the population of the two sets. This is precisely what is done during a CPMG experiment where a series of 180° pulses are applied that results in reversing the order of precession (fast precessing vector becomes slow and slow precessing vector becomes fast) of the two magnetization vectors (corresponding to two

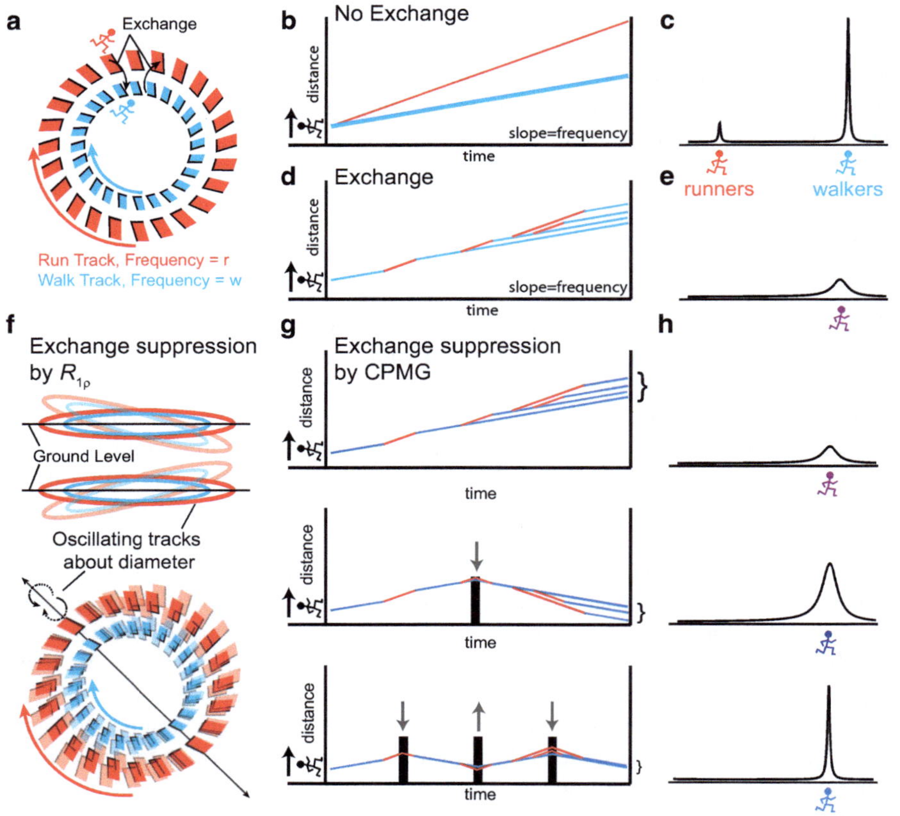

Fig. 4 Cartoon representation of exchange process and its suppression using $R_{1\rho}$ and CPMG relaxation dispersion experiments. *Red* and *blue tracks* are represented as run track and walk track, respectively, rotating at respective frequencies. (**a**) In a no exchange case, the two tracks are rotating at respective frequencies and people standing on the tracks reach at respective unique points at the end of time '*t*'. (**b**) These unique points are shown on the *y*-axis in the time–distance plot. (**c**) NMR spectrum of a nucleus populating two conformations without undergoing exchange. (**d**) In the case of exchange, frequent change of speed of people standing on the two tracks by jumping between the tracks results in a distribution of position at the end of time '*t*'. (**e**) Exchange broadened NMR spectrum of a nucleus populating two conformations. (**f**) Exchange suppression by $R_{1\rho}$ experiment where the two tracks are made to oscillate about the diameter such that tracks no longer can run parallel and jumping frequency of people from one track to the other is decreased. (**g**) Exchange suppression by CPMG experiment where reversing the direction of rotation of the tracks results in narrowing of the distribution of position in the time-distance plot. (**h**) Suppression of exchange results in narrowing of NMR line of the GS peak

conformations of a nuclei) every time a pulse is applied at a constant interval of τ_{cp}. When the pulse is applied at close enough intervals (shorter τ_{cp} delays), no sufficient time is given to magnetization vector to evolve and exchange its speed with the other vector and hence exchange contribution is suppressed.

In suppression of exchange using $R_{1\rho}$ relaxation dispersion experiment (Fig. 4f), during time 't', the two tracks are made to oscillate about their diameter at a certain frequency in such a way that half of the track (divided by the diameter) moves out of the plane of the earth and the other half goes below the earth. In the next cycle of oscillation, the half that was below the earth moves above the ground and vice versa. In this situation, the rotating tracks no longer run parallel for sufficient amount of time and the chances of people to jump from one track to the other are decreased. Jumping frequency further decreases as the oscillation frequency is increased and thereby suppressing the exchange. This is what is done in a typical $R_{1\rho}$ experiment where a resonant RF field is applied (consider it similar to oscillating track) and rotating transverse magnetization vector is locked along this RF field, suppressing the chemical exchange. A detailed mathematical treatment of the experiment is presented elsewhere (Massi et al. 2004).

Suppression of exchange, either by CPMG or by $R_{1\rho}$ experiment, is typically applied on a GS peak (or a peak corresponding to a conformation with more population) that is broadened by exchange. A gradual narrowing of NMR peak is seen as the parameters in the two experiments are varied (Fig. 4h).

3 Sample Preparation

The preparation and purification of ^{13}C, ^{15}N-labelled oligonucleotides (DNA and RNA) samples required to carry out above studies have been described in detail earlier (Zhang et al. 2006). Following sample requirements are to be met for these kinds of NMR studies: (1) typical size of the DNA/RNA construct should be between 12 and 60 nucleotides; and (2) typical concentration should be ~0.5 mM for a ^{13}C, ^{15}N-labelled DNA/RNA sample and ~4 mM for an unlabelled DNA/RNA sample. Phosphate buffer with a slightly acidic pH (~6.4) and low salt concentration (~25 mM) is a preferred buffer for such studies. Other buffers may also be explored.

4 Data Acquisition

Rotating frame $R_{1\rho}$ relaxation dispersion profiles on nucleic acids may be acquired using several pulse sequences available in the literature (Massi et al. 2004; Auer et al. 2009). Here, as an example, measurements using selective carbon experiment with a one-dimensional (1D) acquisition scheme is discussed (Hansen et al. 2009). This scheme extends the sensitivity to chemical exchange into millisecond timescale relative to conventional two-dimensional (2D) relaxation dispersion methods. Acquisition in 1D mode is particularly suited for nucleic acids, where chemical exchange is very often limited to a small number of nucleotides in non-canonical sites, e.g. loops and bulges, making it unnecessary to record full 2D experiments for the canonical helical regions.

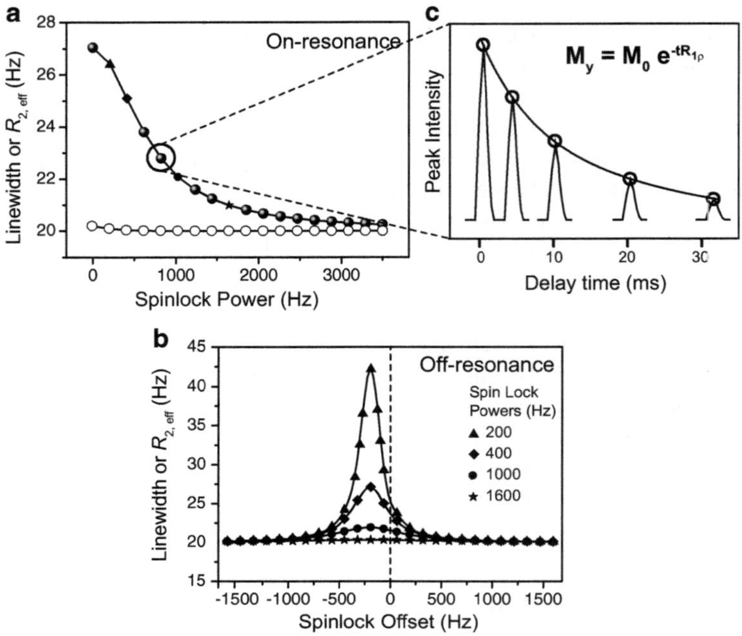

Fig. 5 (**a**) An example on-resonance $R_{1\rho}$ profile in the presence of exchange (*filled spheres*) and no-exchange (*open circles*), where points in *shapes* other than *filled spheres* are used for (**b**) off-resonance $R_{1\rho}$ measurement by varying the spinlock offset. Spinlock power corresponding to the points used for off-resonance measurement is mentioned in the graph. Each point in (**a**) is a $R_{1\rho}$ value that comes as a result of (**c**) mono-exponential fitting of the NMR peak intensities acquired as a function of relaxation delays—with equation in the *inset*

4.1 Protocol

1. First of all, a 2D heteronuclear single quantum coherence (HSQC) non-TROSY (Dethoff et al. 2012b) spectrum is acquired on the oligonucleotide sample (Fig. 3b). Two heteronuclei in the experiment could be either ^{13}C–^{1}H and/or ^{15}N–^{1}H, depending on what nuclei is to be probed. Hereafter, the case of carbon relaxation dispersion measurements is discussed and all the steps/principles can be directly applied as is for nitrogen relaxation dispersion measurements. It is assumed that the NMR peak assignments (knowledge of each correlation peak in the HSQC spectrum belonging to corresponding bond vector, see Fig. 3b) for the nucleic acid sample is already known. The NMR peak assignments for $C_{1'}H_{1'}$, $C_{2'}H_{2'}$, C_2H_2, C_4H_4, C_5H_5, C_6H_6 and C_8H_8 are minimally required.
2. A list of NMR peaks is prepared corresponding to nuclei in the nucleic acid sample to be probed, e.g. C_2H_2-A93, C_8H_8-A92 and ^{13}C and ^{1}H chemical shifts of these spins are carefully noted from the HSQC experiment acquired in step 1. These values are used in the $R_{1\rho}$ experiment.

3. It is also useful to measure longitudinal (R_1) and transverse (R_2) relaxation rates for the nuclei of interest (Hansen and Al-Hashimi 2007). These numbers help in deciding an appropriate length of relaxation delays in the $R_{1\rho}$ experiment. These measurements are, however, not mandatory.
4. The $R_{1\rho}$ experiment is acquired in two steps; first an on-resonance profile is measured where only the power of the spinlock field is varied (Fig. 5a). Typically, 10–12 spinlock powers with 3–4 repeats are used for each peak. Subsequently, an off-resonance profile is measured that is discussed later in this protocol (Fig. 5b). Each $R_{1\rho}$ data point (in both on-resonance and off-resonance experiments) in turn is measured by acquiring a set of experiments as a function of transverse relaxation delay (Fig. 5c).
5. ^{13}C and ^{1}H chemical shift values of the NMR peak under study are taken from the step 2 in the $R_{1\rho}$ pulse sequence parameter file and several other parameters used in the $R_{1\rho}$ pulse sequence including pulse lengths, water offset frequency, pulse powers, etc. are optimized.
6. For deciding the length and number of transverse relaxation delays in the experiment R_2 numbers from step 3 comes handy. As a rule of thumb, maximum delay is chosen so as to keep the decayed intensity as one-third of the intensity with zero ms delay. Also, the number of scans is chosen so as to keep the signal-to-noise (S/N) ratio of at least 20:1 for the longest delay peak intensity. Now, between zero ms and maximum delay, 2–4 additional delays are chosen that are spread apart in time as well as in NMR peak intensity. For error calculation 1–2 repeat delays are chosen. All the relaxation delays are input as a list and the experiment is recorded in an interleaved fashion. A typical $R_{1\rho}$ experiment using above scheme at a single-spinlock power takes anywhere from ~5 to ~20 min, depending upon S/N of the NMR peak chosen, number of delays and length of delays. It is advisable to run this experiment at this stage, process and check if one is getting a good exponential decay of peak intensities at various relaxation delays.
7. Although this pulse sequence contains a robust heat compensation block that ensures that same amount of heat is generated for all lengths and powers of spinlock field, it is suggested to confirm the temperature of the sample after a few runs of the experiments by measuring water line position or using a d_4-methanol sample (Findeisen et al. 2007) before and after the experiment. A necessary correction in the sample temperature can be applied accordingly.
8. An additional parameter in the pulse sequence called as 'zeta delay' is required to be added if a peak of interest has a nearby peak in carbon dimension (i.e. having similar ^{1}H chemical shift and nearby ^{13}C chemical shift). Zeta delay is used to suppress the nearby ^{13}C signal. Typically, peaks in the proximity of 50–100 Hz are taken into consideration. If there is a peak that is closer than 50 Hz in carbon dimension, one cannot practically measure the peak of interest, as it is difficult to suppress signal from a peak that close.
9. List of parameters that must be optimized in this pulse sequence (Bruker scheme), in addition to regular essential parameters:

(a) cnst30: ^{1}H chemical shift for the spin of interest (in ppm).
(b) o2: ^{13}C chemical shift for the spin of interest (in Hz).
(c) p11: $1/J$, where J is the one bond coupling constant between ^{13}C and ^{1}H for the spin of interest.
(d) sp1: power level for p11 pulse; optimized for efficient Hartman–Hahn matching condition between ^{1}H and ^{13}C.
(e) cnst12: spinlock power in Hz.
(f) pl23: spinlock power in db.
(g) d31: average of all transverse relaxation delays.
(h) d30: maximum transverse relaxation delay + 2 ms.
(i) nbl: number of transverse relaxation delays.
(j) vdlist: list of transverse relaxation delays.
(k) zgoptns: if zeta delay is being used, write −D_ZETADELAY; for an off-resonance $R_{1\rho}$ experiment, write −DC13_OFFSET; for both the above conditions, write both separated by a space.
(l) cnst10: value of zeta delay in Hz.
(m) cnst28: desired value of offset in Hz.
(n) p13 and p14 pulses for efficient water suppression.

10. Upon optimization of all the parameters and the required relaxation delays, the experiment is copied over to several numbers of experiments and spinlock power values are changed in each experiment ranging from ~100 to ~3,500 Hz (caution—maximum value of spinlock power that can be applied is defined by the NMR probe hardware parameters). One can prepare a list for convenience.
11. An example of on-resonance $R_{1\rho}$ profile for a spin of interest is shown in Fig. 5a (filled spheres). In a non-exchanging case, or if the exchange is out of the ms–μ s time scale, one gets a flat profile (Fig. 5a, open circles). Once we have an estimate of whether a particular spin is undergoing chemical exchange or not, off-resonance $R_{1\rho}$ profile is measured on exchanging spins.
12. To measure off-resonance $R_{1\rho}$ profile, a minimum of three spinlock powers are chosen from the on-resonance $R_{1\rho}$ plot that are spread apart in corresponding $R_{2,\mathrm{eff}}$ numbers (e.g. 200 Hz, 400 Hz, 1,000 Hz and 1,600 Hz in Fig. 5b). Now on-resonance $R_{1\rho}$ experiment with above identified spinlock powers is copied over to several new experiments and offset where these spinlock powers are applied is varied in each of these experiments to measure a complete off-resonance $R_{1\rho}$ profile. As a rule of thumb, maximum offset = ~ ±3 × spinlock power. For example, for a 200 Hz spinlock power, offset can be varied from −600 to +600 Hz in steps of 25–50 Hz. Few of the offsets are first measured to check the quality of the data followed by increased number of offsets to get better fitting of the data. In certain cases, where chemical shift difference ($\Delta\omega$) is large, on-resonance $R_{1\rho}$ profile may look flat, deceiving it for a no exchange case. However, off-resonance $R_{1\rho}$ experiment might be able to measure the exchange in these cases. Therefore, it is useful to measure off-resonance $R_{1\rho}$ profile even for peaks having flat on-resonance $R_{1\rho}$ profiles

with less number of offsets. A detailed number of offsets can then be measured on exchanging spins.

5 Data Analysis

Each experiment acquired above at a given spinlock power and offset is a set of interleaved experiments containing a given number of transverse relaxation delays. Each of these interleaved experiments is processed in a similar fashion as 1D NMR experiment to obtain the intensities of the peaks. In a typical $R_{1\rho}$ profile, on-resonance experiment is acquired with a set of 17 spinlock powers (13 + 4 repeats) and off-resonance experiment is acquired with a set of 3 spinlock powers each with ~20 offsets. This makes a total of 77 experiments each acquired with 6 relaxation delays making a grand total of 462 experiments for a single NMR peak. One can write scripts in NMRPipe (or other NMR processing software) to automate the processing, peak picking and peak-intensity tabulating process.

The intensities of the interleaved experiments (corresponding to a single spinlock power and spinlock offset) are then plotted against transverse relaxation delays and fit to mono-exponential decay to obtain the $R_{1\rho}$ numbers (Fig. 5c) using a Mathematica notebook (Spyracopoulos 2006) that uses the Monte-Carlo approach to fit the intensity data, outputs all the plots of intensity vs. relaxation delay in addition to $R_{1\rho}$ numbers and corresponding error in a text file. Further, this notebook also determines if the input experimental data is adequately fit for a mono-exponential decay.

In uniformly labelled samples, remote carbons in aromatic bases and adjacent carbons in ribose sugar rings may have sizeable homonuclear couplings (to the carbon being measured) and care must be taken to avoid the spins (from analysis) having efficient Hartman–Hahn matching condition. Computing the Hartman–Hahn transfer efficiency (A_{HAHA}) has been described elsewhere (Hansen et al. 2009) and must be calculated for all the spins being measured at all the spinlock powers and offsets. Data points with computed $A_{HAHA} > 1\ \%$ for the C–C couplings with a $J_{C\text{–}C}$ of 1 Hz and >0.1 % for C–C couplings with a $J_{C\text{–}C}$ of 10–40 Hz are removed from further analysis.

Above $R_{1\rho}$ numbers are then plotted against spinlock powers (in an on-resonance $R_{1\rho}$ experiment) and against spinlock offsets (in an off-resonance $R_{1\rho}$ experiment) and data is fitted using different models (e.g. two-state model/three-state model, etc.) (Palmer and Massi 2006). Data for different spins is first fit individually and fit parameters are analyzed to ascertain whether the data can be fitted using a global fit or not. Model selection is then carried out using an F-test that uses chi-square, applying the Levenberg–Marquardt minimization algorithm (Levenberg 1944; Marquardt 1963; Lourakis 2005), to determine the feasibility of a model (for example, individual fits versus a more complex model or shared-parameter/2-state or 3-state fits). In a complete (on-resonance + off-resonance) $R_{1\rho}$ profile, following parameters are obtained from the data fitting:

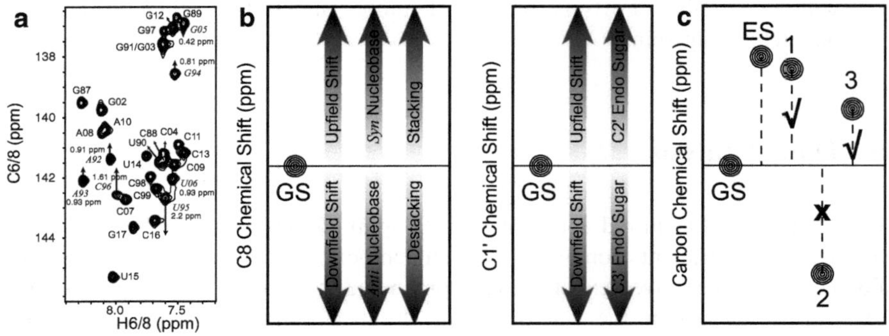

Fig. 6 (**a**) C_6H_6/C_8H_8 HSQC spectrum for A-site ribosomal RNA displaying the ground-state chemical shifts and excited-state chemical shifts. *Small arrows* mark the magnitude and direction of the excited-state chemical shift for the residues labelled in *italics* along with the $\Delta\omega$ values. (**b**) Sensitivity of C8 and C1′ chemical shift towards the base and sugar conformation. Ground-state (GS) chemical shift of a resonance is shown at the reference line and any deviation of the excited state (ES) from that line in ppm is shown to associate with conformational change as sensed by that very resonance. Greater the deviation from ground-state NMR chemical shift in a particular direction, greater is the change in corresponding conformation. For example, upfield chemical shift deviation of C8 in nucleic acids is associated with a change towards *syn* conformation and increased stacking. (**c**) Comparison of GS and ES carbon chemical shift with three different mutants; mutant 1 matches with ES, mutant 2 does not comply with ES and mutant 3 adopts similar conformation as ES but does not match completely

1. $\Delta\omega$: chemical shift difference from the major state chemical shift (values are mentioned for each spin measured in Fig. 6a)
2. Direction of the minor state chemical shift from the major state (shown as arrows in Fig. 6a)
3. R_1: longitudinal relaxation rate for the major/minor state (assuming the two states have same R_1 value)
4. R_2: transverse relaxation rate for the major/minor state (assuming the two states have same R_2 value)
5. k_{ex}: exchange rate constant for the reaction
6. p_E: population of minor state

Errors in these parameters are determined using standard Monte Carlo simulations and verified using Boot strapping approach for error analysis. It is also useful to convert the off-resonance $R_{1\rho}$ data into $R_2 + R_{ex}$ (or $R_{2,eff}$) and plot against spinlock offsets (Fig. 5b).

$$R_2 = \frac{R_{1\rho}}{\sin^2\theta} - \frac{R_1}{\tan^2\theta},$$

where

$$\tan\theta = \frac{\text{power}}{\text{offset}}.$$

This plot gives useful insights into the exchange process that is being observed. For example, the $R_2 + R_{ex}$ maxima is shifted from the on-resonance line by a value equal to the $\Delta\omega$; left shift is for negative $\Delta\omega$ and right shift for positive $\Delta\omega$—thus giving a physical sense to the sign of chemical shift difference.

6 Model Building

Once all the parameters are obtained using data fitting, it is required to fit these parameters into appropriate model. Chemical shifts of the excited-state conformation, obtained from above data analysis, are very sensitive reporters of the sugar pucker, *syn* versus *anti* glycosidic angles and the base stacking (Fig. 6b). A few thumb-rules include a downfield-shifted C1′ chemical shift of ES hints toward a pure C3′-endo (and upfield-shifted C1′ ES hints toward C2′-endo) sugar pucker characteristics of a helical conformation; a downfield-shifted base ES is assigned to a *syn* base and decreased stacking whereas a upfield-shifted base ES is assigned to a *anti* base and increased stacking conformation. Based on these carbon chemical shift signatures, a secondary structure model is built. For example, in the case of A-site ribosome, downfield-shifted base ES chemical shift of U95 and upfield-shifted base ES chemical shifts of A92, A93, G94 and C96 indicated transition towards a structural rearrangement leading to bulging out of U95 and formation of three consecutive non-canonical base pairs (Dethoff et al. 2012b).

The model is then looked up in distinct secondary structures predicted by a secondary structure prediction program, MC-Fold (Parisien and Major 2008), to confirm the feasibility of the model within a certain energy difference. For example, above A-site ES was found to be second most favorable structure (first structure is GS conformation) in MC-Fold secondary structure predictions.

6.1 *Proving the Model*

Model building is no good unless the model is proven experimentally. Mutate-and-chemical-shift-fingerprint (MCSF) strategy used recently (Dethoff et al. 2012b), works effectively for nucleic acids as base and sugar carbon chemical shifts have directional signatures, as described previously. In this strategy, a mutation or a chemical modification is introduced in the wild-type sequence to stabilize the candidate ES conformation (based on the above model building). Mutant's carbon chemical shift signatures are recorded using 2D [^{13}C,^{1}H]-HSQC experiments and assignment is obtained using standard NOESY experiments. Although it is a change of single atom (in case of chemical modifications) or a single residue (in case of mutation), but the HSQC spectrum can be quite unique as the mutant maps a completely different conformation. These carbon chemical shift signatures are then compared with the wild type's ES (or GS, depending upon if the mutant was designed to stabilize ES or GS) chemical shift signatures and thus the ES conformation is confirmed (Fig. 6c). Typically, two to three mutations are designed in distinct samples to confirm the ES conformation, to have multiple proofs of the conformation.

7 Concluding Remarks

Above described strategy is useful to look into structural transitions between the ground state and the excited state that are localized around the non-canonical base pairs. These structural transitions occur at rates much faster than large structural transitions that occur in RNA switches and are thus difficult to characterize otherwise. The ability to visualize such structures that may be functionally relevant has opened a spectrum of whole new possibilities to design new drugs using these new conformations as potential drug targets. For example, a docking study to design and capture such ES conformations for A-site ribosomal RNA may be an innovative way of designing new antibiotics.

Acknowledgments I acknowledge Indian Institute of Science Education and Research for constant support during the writing of this chapter. I thank Dr. Elizabeth Dethoff for help with Fig. 4. I thank Prof Anil Kumar, Prof R V Hosur, Dr Neel Sarovar Bhavesh, Dr Sai Krishnan and Dr Shilpy Sharma for useful comments and suggestions on the manuscript.

References

Auer R, Neudecker P, Muhandiram DR et al (2009) Measuring the signs of 1H(alpha) chemical shift differences between ground and excited protein states by off-resonance spin-lock R(1rho) NMR spectroscopy. J Am Chem Soc 131:10832–10833

Baldwin AJ, Kay LE (2012) Measurement of the signs of methyl 13C chemical shift differences between interconverting ground and excited protein states by R(1ρ): an application to αB-crystallin. J Biomol NMR 53:1–12

Blad H, Reiter NJ, Abildgaard F et al (2005) Dynamics and metal ion binding in the U6 RNA intramolecular stem-loop as analyzed by NMR. J Mol Biol 353:540–555

Bothe JR, Nikolova EN, Eichhorn CD et al (2011) Characterizing RNA dynamics at atomic resolution using solution-state NMR spectroscopy. Nat Methods 8:919–931

Dethoff EA, Chugh J, Mustoe AM et al (2012a) Functional complexity and regulation through RNA dynamics. Nature 482:322–330

Dethoff EA, Petzold K, Chugh J et al (2012b) Visualizing transient low-populated structures of RNA. Nature 491:724–730

Dill K, Chan H (1997) From Levinthal to pathways to funnels. Nat Struct Mol Biol 4:10–19

Findeisen M, Brand T, Berger S (2007) A 1H-NMR thermometer suitable for cryoprobes. Magn Reson Chem 45:175–178

Granata D, Camilloni C, Vendruscolo M et al (2013) Characterization of the free-energy landscapes of proteins by NMR-guided metadynamics. Proc Natl Acad Sci U S A 110:6817–6822

Hansen AL, Al-Hashimi HM (2007) Dynamics of large elongated RNA by NMR carbon relaxation. J Am Chem Soc 129:16072–16082

Hansen AL, Nikolova EN, Casiano-Negroni A et al (2009) Extending the range of microsecond-to-millisecond chemical exchange detected in labeled and unlabeled nucleic acids by selective carbon R(1rho) NMR spectroscopy. J Am Chem Soc 131:3818–3819

Heine A, Desantis G, Luz JG et al (2001) Observation of covalent intermediates in an enzyme mechanism at atomic resolution. Science 294:369–374

Johnson JE, Hoogstraten CG (2008) Extensive backbone dynamics in the GCAA RNA tetraloop analyzed using 13C NMR spin relaxation and specific isotope labeling. J Am Chem Soc 130: 16757–16769

Kay LE (2011) NMR studies of protein structure and dynamics. J Magn Reson 213:477–491

Kloiber K, Spitzer R, Tollinger M et al (2011) Probing RNA dynamics via longitudinal exchange and CPMG relaxation dispersion NMR spectroscopy using a sensitive 13C-methyl label. Nucleic Acids Res 39:4340–4351

Korzhnev DM, Orekhov VY, Kay LE (2005) Off-resonance R(1rho) NMR studies of exchange dynamics in proteins with low spin-lock fields: an application to a Fyn SH3 domain. J Am Chem Soc 127:713–721

Kovrigin E, Kempf J, Grey M et al (2006) Faithful estimation of dynamics parameters from CPMG relaxation dispersion measurements. J Magn Reson 180:93–104

Kuhlenkoetter S, Wintermeyer W, Rodnina MV (2011) Different substrate-dependent transition states in the active site of the ribosome. Nature 476:351–354

Levenberg K (1944) A method for the solution of certain non-linear problems in least squares. Q Appl Math 2:164–168

Lourakis M (2005) A brief description of the Levenberg-Marquardt algorithm implemented by levmar. Technical report

Marquardt DW (1963) An algorithm for the least-squares estimation of nonlinear parameters. SIAM J Appl Math 11:431–441

Massi F, Johnson E, Wang C et al (2004) NMR R1 rho rotating-frame relaxation with weak radio frequency fields. J Am Chem Soc 126:2247–2256

Nikolova EN, Kim E, Wise AA et al (2011) Transient Hoogsteen base pairs in canonical duplex DNA. Nature 470:498–502

Nikolova EN, Gottardo FL, Al-Hashimi HM (2012) Probing transient Hoogsteen hydrogen bonds in canonical duplex DNA using NMR relaxation dispersion and single-atom substitution. J Am Chem Soc 134:3667–3670

Palmer AG (2004) NMR characterization of the dynamics of biomacromolecules. Chem Rev 104: 3623–3640

Palmer AG (2009) A topical issue: NMR investigations of molecular dynamics. J Biomol NMR 45:1–4

Palmer AG, Massi F (2006) Characterization of the dynamics of biomacromolecules using rotating-frame spin relaxation NMR spectroscopy. Chem Rev 106:1700–1719

Parisien M, Major F (2008) The MC-Fold and MC-Sym pipeline infers RNA structure from sequence data. Nature 452:51–55

Schramm VL (2013) Transition states, analogues, and drug development. ACS Chem Biol 8:71–81

Schwartz SD, Schramm VL (2009) Enzymatic transition states and dynamic motion in barrier crossing. Nat Chem Biol 5:551–558

Spyracopoulos L (2006) A suite of Mathematica notebooks for the analysis of protein main chain 15N NMR relaxation data. J Biomol NMR 36:215–224
Vallurupalli P, Scott L, Williamson JR et al (2007) Strong coupling effects during X-pulse CPMG experiments recorded on heteronuclear ABX spin systems: artifacts and a simple solution. J Biomol NMR 38:41–46
Vallurupalli P, Bouvignies G, Kay LE (2012) Studying "invisible" excited protein states in slow exchange with a major state conformation. J Am Chem Soc 134:8148–8161
Wang C, Grey MJ, Palmer AG (2001) CPMG sequences with enhanced sensitivity to chemical exchange. J Biomol NMR 21:361–366
Wolfe AE, O'brien PJ (2009) Kinetic mechanism for the flipping and excision of 1, N(6)-ethenoadenine by human alkyladenine DNA glycosylase. Biochemistry 48:11357–11369
Wunderlich CH, Spitzer R, Santner T et al (2012) Synthesis of (6-13C)pyrimidine nucleotides as spin labels for RNA dynamics. J Am Chem Soc 134:7558–7569
Zhang Q, Sun X, Watt ED et al (2006) Resolving the motional modes that code for RNA adaptation. Science 311:653–656

Diastereomer-Specific Repertoire of 7′*R*- or 7′*S*-Me-Carba-*L*ocked *N*ucleic *A*cids (cLNAs) in Antisense Oligo/RNA Duplexes and Engineering of Physico-chemical and Enzymological Properties

Qing Li, Oleksandr Plashkevych, Ram Shankar Upadhayaya, Sachin Gangadhar Deshpande, Andras Földesi, and Jyoti Chattopadhyaya

Contents

1 Introduction 200
2 Synthesis of Diastereomerically Pure (7′*S*-Me or 7′*R*-Me)-cLNA-A, -G, -^{Me}C, and -T Nucleosides and Their Phosphoramidites 202
2.1 Key Free-Radical Ring Closure to cLNA-A, -G, -^{Me}C, and -T Nucleosides 203
2.2 Synthesis of Phosphoramidites of 7′*S*- and 7′*R*-Me-cLNA-A, -G, -^{Me}C, and -T 208
3 Binding Affinity of cLNA-A, -G, -^{Me}C, -T and LNA-A, -G, -C, -T Single Modified AONs Toward Complementary Native RNA and DNA and Thermal of Stability of the Resulting Duplexes 209
3.1 RNA Affinity of AONs Containing cLNA-A, -G, -^{Me}C, -T or LNA-A, -G, -C, -T Single Modification 211
3.2 Relative RNA Selectivity of cLNA-A-, -G-, -^{Me}C-, -T-Modified AONs in Comparison with LNA-A-, -G-, -C-, -T-Modified AONs 215
4 3′-Exonuclease Stability Studies of cLNA-A-, -G-, -MeC-, -T-Modified AONs 215
4.1 Relative 3′-Exonucleolytic Stabilities of cLNA-A-, -G-, -^{Me}C-, -T-Modified AONs 217
4.2 The Effect of Sequence Context of a Given AON on 3′-Exonucleolytic Stability . 217
5 Stability of cLNA-A-, -G-, -^{Me}C-, -T-Modified AONs in Human Blood Serum 219
6 Impact of cLNA Modifications on the Duplex Structure and Properties: Theoretical Study 220
7 Conclusions 222
References 224

Abstract Modified oligos (AON), with pure 2′, 4′-locked 7′*S*- or 7′*R*-Me-cLNA-A, -G, -^{Me}C, and -T (Upadhayaya et al., J Org Chem 76:4408–4431, 2011; Srivastava et al., J Am Chem Soc 129:8362–8379, 2007), show higher RNA affinity and RNA

Qing Li and Oleksandr Plashkevych contributed equally.

Q. Li • O. Plashkevych (✉) • R.S. Upadhayaya • S.G. Deshpande • A. Földesi • J. Chattopadhyaya (✉)
Chemical Biology Program, Department of Cell and Molecular Biology, Biomedical Center, Uppsala University, Box 581, 751 23 Uppsala, Sweden
e-mail: oleksandr@boc.uu.se; jyoti@boc.uu.se

V.A. Erdmann et al. (eds.), *Chemical Biology of Nucleic Acids*, RNA Technologies, DOI 10.1007/978-3-642-54452-1_12, © Springer-Verlag Berlin Heidelberg 2014

selectivity, highly improved exonuclease (SVPDE), and blood serum stabilities in comparison with native oligos as well as maintained or higher RNase H recruitment capability depending upon the modification site. The AON with 7′*S*-Me-cLNA-^{Me}C is found to be ~40 times more stable against SVPDE than 7′*S*- and 7′*R*-Me-cLNA-T-modified AONs, which are in turn much more stable than 7′*S*- and 7′*R*-Me-cLNA-A- and G-modified counterparts. The T_m increase of the duplexes is found to be dependent on the AON-sequence context, -modification site, and the cLNA diastereomer type used (7′*S or* 7′*R*). MD simulations of the AON/RNA duplexes with 7′*S*- or 7′*R*-Me-cLNA have demonstrated that the modifications have only small local effect on the duplex structures, stacking and Watson–Crick base pairing.

Keywords Nucleotides analogs • Antisense oligonucleotides • Thermal stability • Snake venom phosphodiesterase exonuclease stability • Blood serum stability • Antisense technology

Abbreviations

AONs	Antisense oligonucleotides
Carba-ENA (cENA)	2′-*C*,4′-*C*-Ethylene-bridged carbocyclic analogue of 2′-*O*,4′-*C*-ethylene-bridged nucleic acid (ENA)
Carba-LNA (cLNA)	Carbocyclic locked nucleic acid
LNA	Locked nucleic acid
SVPDE	Snake venom phosphodiesterase
T_m	Duplex melting temperature

1 Introduction

The potential of chemically modified oligonucleotides to act as efficient gene silencing agents through knock-down of mRNA target has been in development for decades. Generally, there are several main types of anti-mRNA strategies, involving antisense approach (Crooke 2004), ribozymes, DNA enzymes (Schubert and Kurreck 2004; Santoro and Joyce 1997), RNA interference (Novina and Sharp 2004), micro-RNAs (miRNA) (Bartel 2004), or triple helix-forming oligonucleotides (TFOs) (Buchini and Leumann 2003). Chemical modifications of antisense oligonucleotides (AONs) and siRNAs have been shown to increase their potency as RNA-silencing agent due to their high efficiency and accuracy to target specific mRNA (Peterson et al. 2002; Koshkin et al. 1998; Obika et al. 1997, 1998; Singh et al. 1997, 1998; Altmann et al. 1994; Rodriguez et al. 1993; Tarkoy and Leumann 1993). One particular class of compound, so-called locked nucleic acid (LNA, also known as BNA), has attracted extensive attention as potential RNA-silencing agent due to its ability to enhance unprecedentedly the affinity of LNA-modified DNA strand toward complementary RNA (Koshkin et al. 1998; Obika et al. 1997, 1998; Singh et al. 1997). LNA contains 2′,4′-linkage which is known to be able to lock the conformation of sugar moiety into a perfect *North*-type (*N*-type) conformation (Obika et al. 1997). Thereby upon incorporation into oligo-DNA, LNA can induce its neighboring natural nucleotides, up to 4–5 nt from the modification toward the 5′-end, to shift *S*outh–*N*orth sugar conformational equilibrium to the latter through conformational transmission (Thibaudeau et al. 2005; Acharya et al. 2002). Such conformational change is resulting in alteration of the local duplex structure from DNA/RNA-type to RNA/RNA A-type helix which contributes positively to the overall thermostability of the modified heteroduplexes. The thermal stability of LNA-modified DNA/RNA duplexes was found to be 2–5 °C/modification higher than that of the native counterpart (Peterson et al. 2002).

Excellent thermostability of LNA-modified oligos toward RNAs has stimulated synthesis of a number of *North*-type conformationally constrained nucleoside analogues (Zhou and Chattopadhyaya 2009) in which the 2′,4′-bridge has been altered to yield amino-LNA (Singh et al. 1998), 6′-substituted LNA (Seth et al. 2010a; Enderlin and Nielsen 2008), 2′,4′-BNA^{COC} (Hari et al. 2006, 2011), 2′,4′-BNA^{NC} (Prakash et al. 2010; Rahman et al. 2008), 2′-*O*,4′-*C*-ethylene-bridged nucleic acid (ENA) (Morita et al. 2002, 2003), aza-ENA (Honcharenko et al. 2007, 2008; Varghese et al. 2006), and carba-ENA (Kumar et al. 2009; Albæk et al. 2006). In addition, a new class of 1′,2′-bridged nucleosides such as 1′,2′-oxetane (Pradeepkumar et al. 2004) or 1′,2′-azetidine (Honcharenko et al. 2006) analogue has also been developed. Introduction of 1′,2′-oxetane (Pradeepkumar et al. 2004) or 1′,2′-azetidine (Honcharenko et al. 2006) modifications into oligonucleotides led to similar or moderately lower target RNA affinity relative to the native and LNA counterparts, whereas the exonuclease resistance and RNase H recruitment properties of AONs containing them were found to be relatively more favorable than those of the native (Honcharenko et al. 2006; Pradeepkumar et al. 2004).

Recently, the intramolecular free-radical cyclization of the C2′ radical to the C4′-tethered –CH=CH_2, –CH=NOR, or –C≡CH group had been applied to synthesize 2′-*C*,4′-*C*-bridged carba-LNA and carba-ENA, including C6′, C7′-substituted carba-LNAs, C6′, C8′-substituted carba-ENAs, and C6′, C7′-substituted α-L-carba-LNAs analogues (Li et al. 2010; Liu et al. 2010; Seth et al. 2010b; Zhou and Chattopadhyaya 2010; Xu et al. 2009; Zhou et al. 2008, 2009a, b; Srivastava et al. 2007). It had been shown that carba-LNA and carba-ENA derivatives enhance melting temperatures (T_m) of modified AON/RNA hybrids by 3–5 °C and 1–2 °C per modification, respectively, depending upon different modification sites, while α-L-carba-LNAs incorporation decreases T_m values of modified AON/RNA heteroduplexes by 2–3 °C/modification. It is worth to note that AONs containing these hydrophobic cLNAs, cENAs, and α-L-cLNAs had exhibited unprecedented nucleolytic and human blood serum stability without impairing RNase H recruitment capabilities of the hybrid AON/RNA duplexes compared to the native counterpart (Li et al. 2010; Liu et al. 2010; Seth et al. 2010b; Zhou and Chattopadhyaya 2010; Xu et al. 2009; Zhou et al. 2008, 2009a, b; Srivastava et al. 2007).

Remarkably prolonged lifetime of these carbocyclic nucleoside-modified AON analogues in human blood serum is highly desired pharmacokinetic property leading potentially to net reduction of the required dosage and, correspondingly, the toxicity in possible therapeutic application.

We have previously reported an unambiguous synthesis of 7′-Me-carba-LNA thymine (7′-Me-cLNA-T), which upon incorporation into AONs had modulated physico-chemical and biochemical properties of AONs and their duplexes with DNA and RNA (Srivastava et al. 2007). These molecules had demonstrated excellent biological properties and thus prompted us to synthesize the related 7′-Me-cLNA-^{5Me}C/A/G analogues and incorporate them into antisense oligos for biological evaluations. In our earlier report (Srivastava et al. 2007) on cLNA-T-modified AONs, we employed diastereomeric mixtures of 7′(*S*/*R*)-Me-cLNA-T, which did not allow us to address how the stereochemistry of C7′ center actually influences the biophysical and biochemical outcome. Recently (Upadhayaya et al. 2011) we had also incorporated mixtures of 7′(*S*/*R*)-Me-cLNA-^{5Me}C diastereoisomers as well as diastereomerically pure 7′*R*-Me-cLNA-A and -G isomers into antisense oligos for their physico-chemical and biological evaluation, which, however, gave us only limited insight into diastereospecific properties of these compounds (Upadhayaya et al. 2011).

To elucidate the stereospecific effects of *S*- and *R*-configured 7′-methyl group of cLNA-A/G/^{5Me}C/T on the thermostability, nuclease stability, and RNase H efficiency, we have now synthesized and separated pure diastereomer at the C7′ center (minor *vs.* major) (Fig. 1) of cLNA-A/G/^{5Me}C/T nucleotides. The nucleotides were derived from the key 5-*exo* free-radical cyclization step following the procedure developed in our previous works (Upadhayaya et al. 2011; Srivastava et al. 2007) and outlined in Figs. 6 and 7. Individual diastereomerically pure cLNA-type nucleosides have been incorporated at different positions of 20mer AON sequence and investigated for their diastereospecific RNA affinity and selectivity, 3′-exonucleolytic resistance, and human blood serum stability as well as for elicitation of *E. coli* RNase H1 activity against modified AON/RNA duplexes (not reported here).

2 Synthesis of Diastereomerically Pure (7′*S*-Me or 7′*R*-Me)-cLNA-A, -G, -^{Me}C, and -T Nucleosides and Their Phosphoramidites

In our previous work (Upadhayaya et al. 2011; Srivastava et al. 2007) diastereomerically pure 7′*R*-Me-cLNA-A and -G were prepared and used in oligo synthesis, whereas diastereomeric mixtures of 7′(*S*/*R*)-Me-cLNA-^{Me}C and -T were employed for this purpose. The 6-*endo* products, cENA-A, -G, -^{Me}C and -T, were also formed and isolated from the products of the key free-radical reaction step in minute amount (Upadhayaya et al. 2011; Srivastava et al. 2007). However, in the present

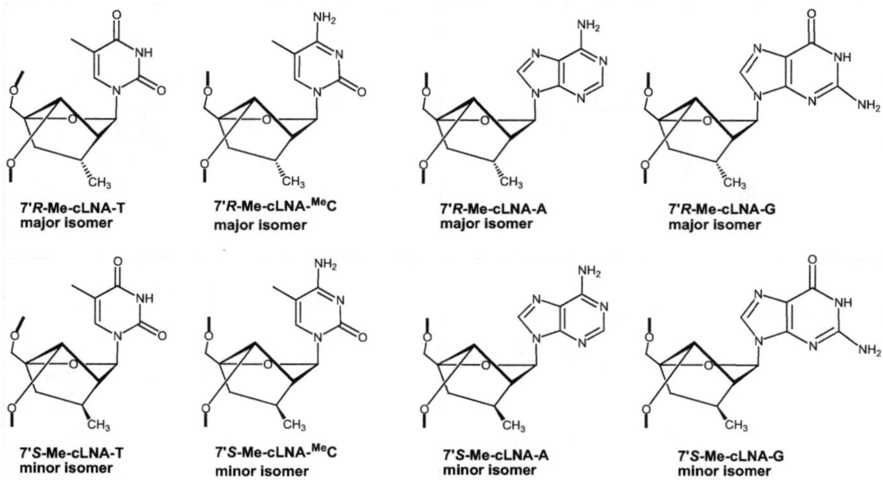

Fig. 1 Molecular structures of carbocyclic LNA (cLNA)-A/G/T/^{5Me}C analogues

study, all diastereomerically pure 7′*S*- and 7′*R*-Me-cLNA-A, -G, -^{Me}C, and -T nucleosides have been synthesized in large scale, separated by preparative HPLC (see Experimental Section and Figs. SII.11–24 in Supporting Information (SI) (Li et al. 2013) Parts IV and II) and incorporated into 20mer oligo-DNA sequence (AON) at different positions. Since bicyclic ring formation and stereochemistry of the ring closure products of 7′*R*- and 7′*S*-Me-cLNA-A, -G, and -^{Me}C as well as of cENA-A, -G, and -^{Me}C nucleosides had been previously discussed by us (Upadhayaya et al. 2011), the detailed 1D and 2D NMR characterization data are provided only for the free-radical ring closure products 7′*R*- and 7′*S*-Me-cLNA-T and 6-*endo*-cENA-T (cENA-T) nucleosides (Figs. SI.1–42 in SI (Li et al. 2013) Part I). Synthesis of correct cLNA- and LNA-modified AONs has been confirmed using MALDI-TOF mass spectrometry (Fig. 5).

2.1 Key Free-Radical Ring Closure to cLNA-A, -G, -^{Me}C, and -T Nucleosides

Free-radical precursors, 4′-*C*-allyl-2′-*O*-phenoxythiocarbonyl-*β*-D-ribofuranoside (compounds **1a–d**), were synthesized according to previously described procedure (Upadhayaya et al. 2011; Srivastava et al. 2007) to be used for 5-hexenyl free-radical cyclization to afford the bicyclic 7′*S*- and 7′*R*-Me-cLNA-A, -G, -^{Me}C, and -T analogues (**2a/b**, **3a/b**, **4a/b,** and **5a/b**), along with the trace amount of 6-*endo*-cENA-A, -G, -^{Me}C, and -T **2–5c** (Fig. 6).

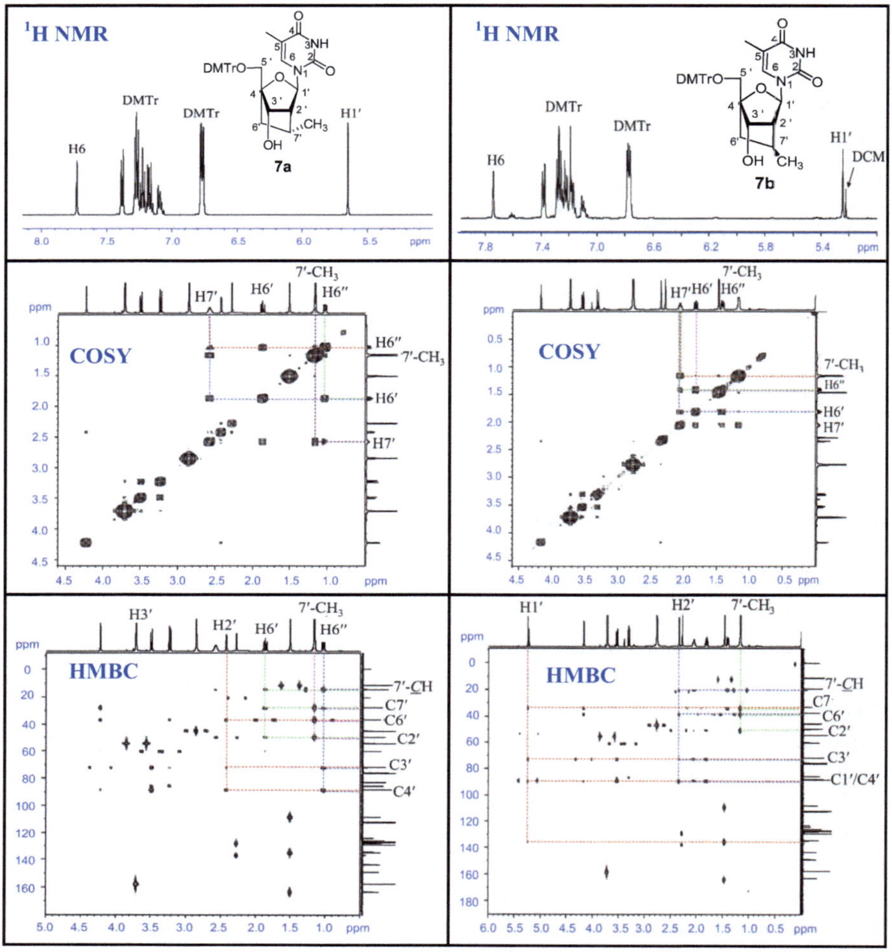

Fig. 2 NMR characterizations (^{1}H, COSY, HMBC) of (1*R*,3*R*,4*R*,5*R*,7*S*)-7-hydroxyl-1-(4,4′-dimethoxy-trityloxymethyl)-3-(thymin-1-yl)-2-oxa-bicyclo[2.2.1]heptane (**7a** in Fig. 7) and (1*R*,3*R*,4*R*,5*S*,7*S*)-7-hydroxyl-1-(4,4′-dimethoxytrityloxymethyl)-3-(thymin-1-yl)-2-oxa-bicyclo[2.2.1] heptane (**7b**). Other 1D and 2D NMR spectra can be found in SI Part I (Li et al. 2013)

Since our present study required pure 7′*R*- and 7′*S*-Me-cLNAs diastereomer of all carba-nucleosides (Upadhayaya et al. 2011) in a relatively large amount to prepare the modified oligos for biochemical studies, we synthesized the free-radical precursors **1a–d** in corresponding amounts of 3.5–15.3 mmol (see Experimental Section in SI (Li et al. 2013) Part IV for details). These precursors were treated then

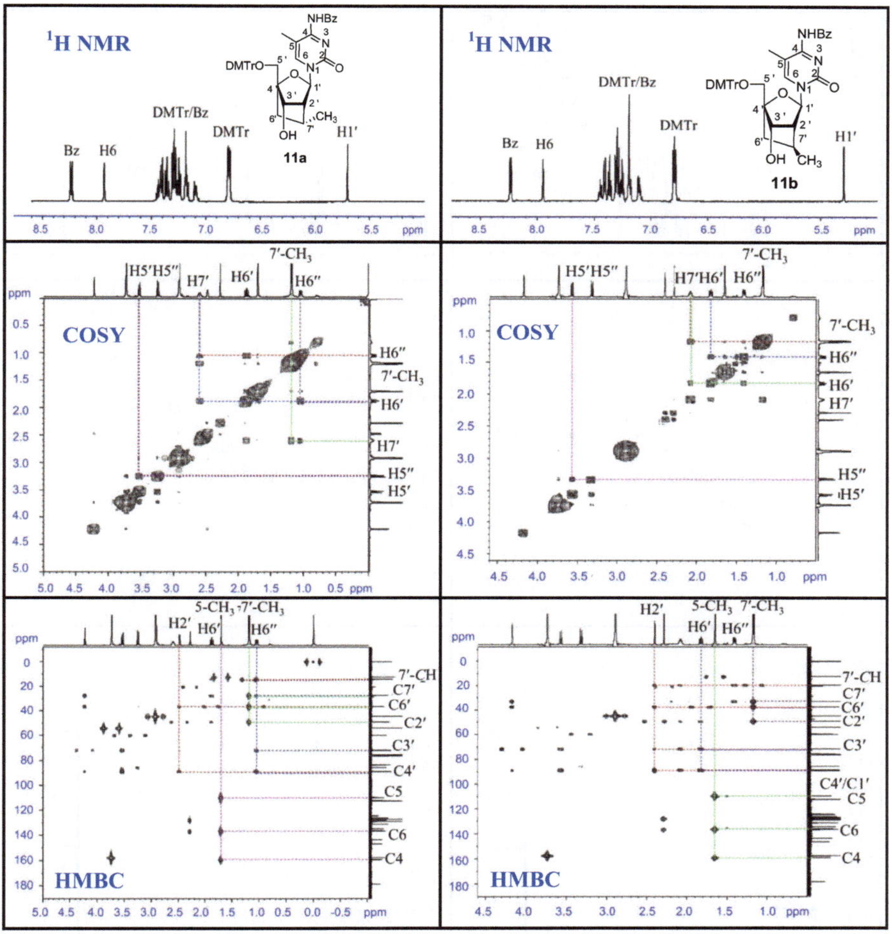

Fig. 3 NMR characterizations (^{1}H, COSY, HMBC) of (1*R*,3*R*,4*R*,5*R*,7*S*)-3-(***N*^{4}**-Benzoyl-5-methylcytosin-1-yl)-1-(4,4′-dimethoxytrityloxymethyl)-7-hydroxyl-5-methyl-2-oxa-bicylco[2.2.1] heptane (**11a** in Fig. 7) and (1*R*,3*R*,4*R*,5*S*,7*S*)-3-(***N*^{4}**-Benzoyl-5-methylcytosin-1-yl)-1-(4,4′-dimethoxytrityloxymethyl)-7-hydroxyl-5-methyl-2-oxa-bicylco[2.2.1] heptane (**11b**). Other 1D and 2D NMR spectra can be found SI Part I (Li et al. 2013)

with *n*-tributyltinhydride (*n*-Bu_3SnH) and azobisisobutyronitrile (AIBN) in toluene under reflux to yield three different isomers, i.e. 5-*exo*-cyclization products, 7′*R*-Me-cLNAs **2–5a** (38–50 %), 7′*S*-Me-cLNAs **2–5b** (4–7 %), and 6-*endo*-cyclization products cENAs **2–5c** (4–5 %) (Fig. 6).

Our results show that the free-radical ring closure of the precursor **1a–d**, with various nucleobase moieties (A, G, ^{Me}C, and T), can create both *exo*- and *endo*-

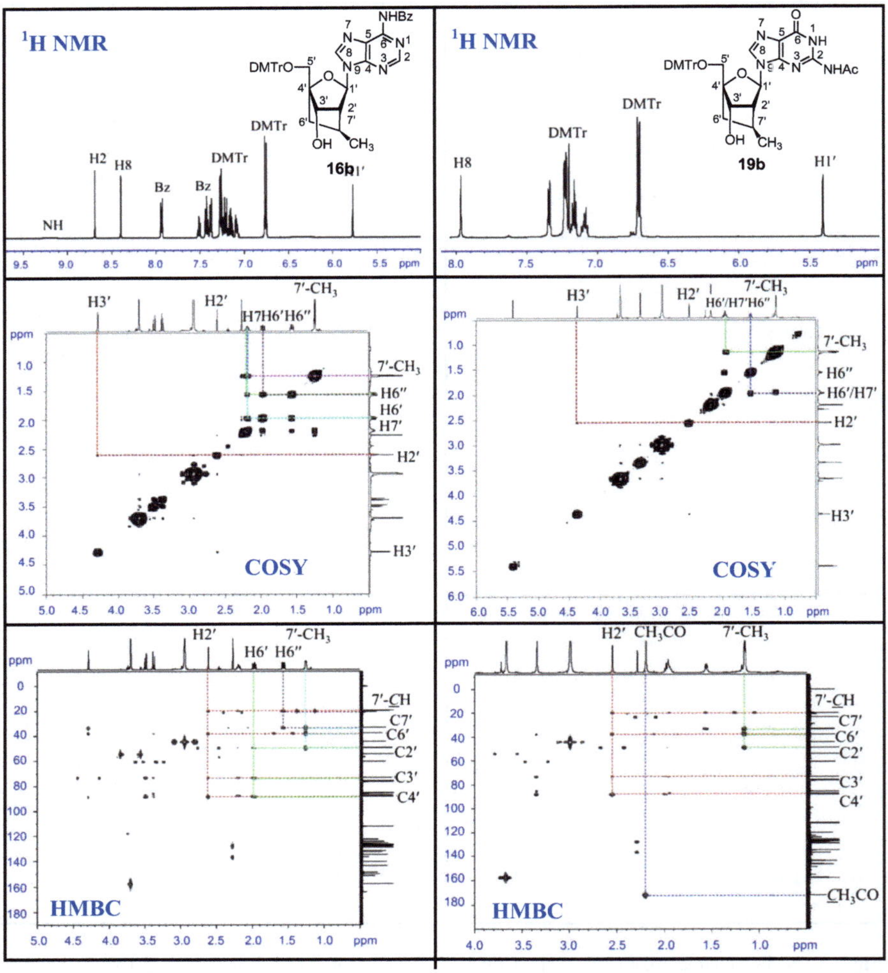

Fig. 4 NMR characterizations (^{1}H, COSY, HMBC) of (1*R*,3*R*,4*R*,5*S*,7*S*)-3-(N^6-benzoyladenin-9-yl)-1-(4,4′-dimethoxytrityloxy)methyl-7-hydroxyl-5-methyl-2-oxa-bicyclo[2.2.1]heptane (**16b** in Fig. 7) and (1*R*,3*R*,4*R*,5*S*,7*S*)-3-(N^2-Acetylguanin-9-yl)-1-(4,4′-dimethoxytrityloxymethyl)-7-hydroxyl-5-methyl-2-oxa-bicylco[2.2.1]heptane (**19b**). Other 1D and 2D NMR spectra can be found in SI Part I (Li et al. 2013)

cyclization products, in an approximate *exo-to-endo* ratio of 10:1 (^{1}H NMR). This is consistent with theoretical study of alkenyl radical ring closure by Beckwith et al., (Beckwith and Schiesser 1985; Beckwith 1981) where it was pointed out that cyclization of 5-hexenyl-1-radical is a kinetically controlled process and generally prefers the *exo*-cyclization products (*exo*/*endo* > 98/2). What is important to note here is that this classical 5-hexenyl type free-radical cyclization reaction has neither been used earlier to understand the ring-closure stereochemistry at the chiral radical

Entry	Modified LNA structures	AON-Sequence* (*Containing modifications in the minor groove*)	$(M+H)^+$	
			Cacl.	Found
Native		5'-d (TCC CGC CTG TGA CAT GCA TT)	-	-
AON1	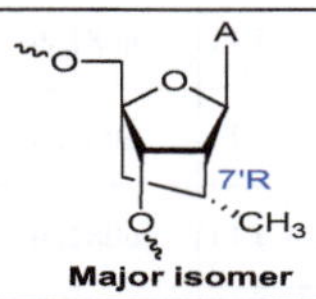	5'-d (TCC CGC CTG TGA CAT GC*A* TT)	**6082.06**	**6082.15**
AON2		5'-d (TCC CGC CTG TGA C*A*T GCA TT)	**6082.06**	**6082.05**
AON3		5'-d (TCC CGC CTG TG*A* CAT GCA TT)	**6082.06**	**6082.10**
AON4	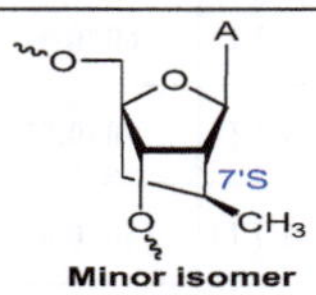	5'-d (TCC CGC CTG TGA CAT GC*A* TT)	**6082.06**	**6082.00**
AON5		5'-d (TCC CGC CTG TGA C*A*T GCA TT)	**6082.06**	**6081.90**
AON6		5'-d (TCC CGC CTG TG*A* CAT GCA TT)	**6082.06**	**6082.01**
AON7	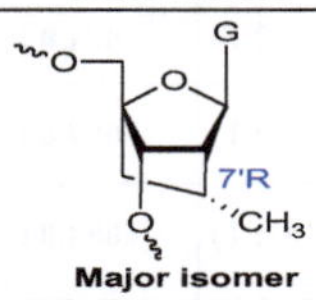	5'-d (TCC CGC CTG TGA CAT *G*CA TT)	**6082.06**	**6082.00**
AON8		5'-d (TCC CGC CTG T*G*A CAT GCA TT)	**6082.06**	**6081.99**
AON9		5'-d (TCC C*G*C CTG TGA CAT GCA TT)	**6082.06**	**6081.89**
AON10	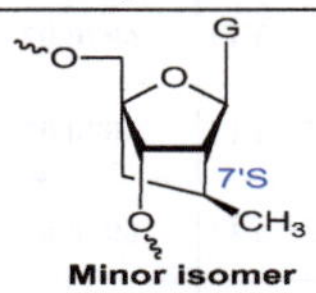	5'-d (TCC CGC CTG TGA CAT *G*CA TT)	**6082.06**	**6081.89**
AON11		5'-d (TCC CGC CTG T*G*A CAT GCA TT)	**6082.06**	**6082.05**
AON12		5'-d (TCC C*G*C CTG TGA CAT GCA TT)	**6082.06**	**6081.99**
AON13	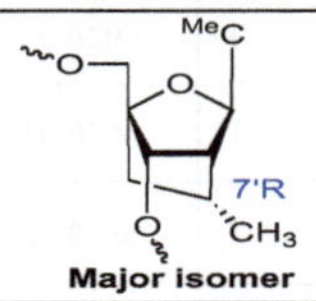	5'-d (TCC CGC CTG TGA CAT G*C*A TT)	**6096.08**	**6096.06**
AON14		5'-d (TCC CGC CTG TGA *C*AT GCA TT)	**6096.08**	**6095.94**
AON15		5'-d (TCC CG*C* CTG TGA CAT GCA TT)	**6096.08**	**6095.95**
AON16	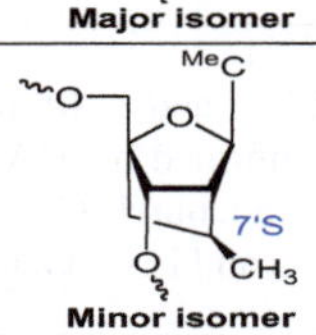	5'-d (TCC CGC CTG TGA CAT G*C*A TT)	**6096.08**	**6095.78**
AON17		5'-d (TCC CGC CTG TGA *C*AT GCA TT)	**6096.08**	**6095.90**
AON18		5'-d (TCC CG*C* CTG TGA CAT GCA TT)	**6096.08**	**6095.89**

Fig. 5 (continued)

center, nor to control the stereochemistry of the fused sugar containing chiral bicyclic systems, which we have developed and utilized (Li et al. 2010; Liu et al. 2010; Seth et al. 2010b; Zhou and Chattopadhyaya 2010; Kumar et al. 2009; Xu et al. 2009; Zhou et al. 2008, 2009a, b; Srivastava et al. 2007; Albæk et al. 2006) as a convenient and reliable method for synthesis of conformationally locked 2′,4′-carbocyclic cLNAs and cENAs building blocks.

AON19	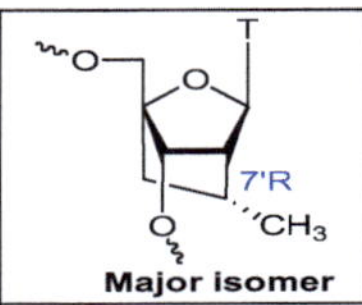	5'-d (TCC CGC CTG TGA CA*T* GCA TT)	**6082.06**	**6081.88**
AON20		5'-d (TCC CGC CTG *T*GA CAT GCA TT)	**6082.06**	**6081.90**
AON21		5'-d (TCC CGC C*T*G TGA CAT GCA TT)	**6082.06**	**6082.15**
AON22	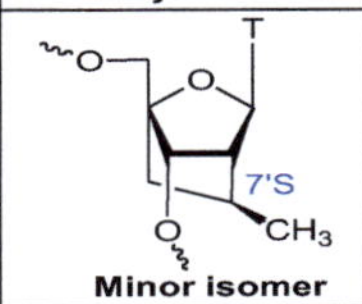	5'-d (TCC CGC CTG TGA CA*T* GCA TT)	**6082.06**	**6081.94**
AON23		5'-d (TCC CGC CTG *T*GA CAT GCA TT)	**6082.06**	**6081.84**
AON24		5'-d (TCC CGC C*T*G TGA CAT GCA TT)	**6082.06**	**6082.02**
AON28	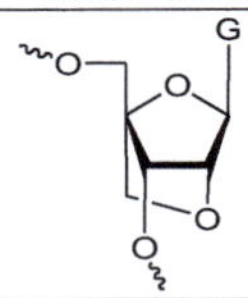	5'-d (TCC CGC CTG TGA CAT *G*CA TT)	**6070.02**	**6069.25**
AON29		5'-d (TCC CGC CTG T*G*A CAT GCA TT)	**6070.02**	**6069.53**
AON30		5'-d (TCC C*G*C CTG TGA CAT GCA TT)	**6070.02**	**6069.57**
AON31	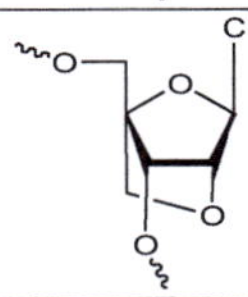	5'-d (TCC CGC CTG TGA CAT G*C*A TT)	**6084.04**	**6083.63**
AON32		5'-d (TCC CGC CTG TGA *C*AT GCA TT)	**6084.04**	**6083.53**
AON33		5'-d (TCC CG*C* CTG TGA CAT GCA TT)	**6084.04**	**6083.67**
AON34	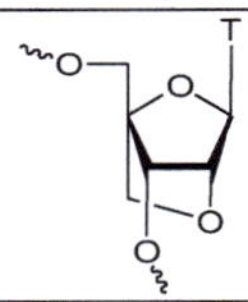	5'-d (TCC CGC CTG TGA CA*T* GCA TT)	**6070.02**	**6069.36**
AON35		5'-d (TCC CGC CTG *T*GA CAT GCA TT)	**6070.02**	**6069.70**
AON36		5'-d (TCC CGC C*T*G TGA CAT GCA TT)	**6070.02**	**6069.70**
AON25	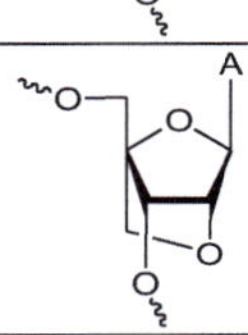	5'-d (TCC CGC CTG TGA CAT GC*A* TT)	**6070.02**	**6069.93**
AON26		5'-d (TCC CGC CTG TGA C*A*T GCA TT)	**6070.02**	**6069.94**
AON27		5'-d (TCC CGC CTG TG*A* CAT GCA TT)	**6070.02**	**6069.77**

Fig. 5 MALDI-TOF mass spectrometry characterization of cLNA and LNA modified AONs (AON**1-36**). Molecular weights of all antisense sequences are confirmed by MALDI-TOF mass spectrum (see Figs. SII.13-48 SI (Li et al. 2013) Part II). *A = adeninyl, G = guaninyl, C = cytosinyl, T = thyminyl. *A* = cLNA-A and LNA-A, *G* = cLNA-G and LNA-G, *C* = cLNA-^{Me}C and LNA-C, T = cLNA-T and LNA-T

2.2 *Synthesis of Phosphoramidites of 7′*S*- and 7′*R*-Me-cLNA-A, -G, -^{Me}C, and -T*

All intermediates have been converted to the corresponding phosphoramidites (Li et al. 2010; Liu et al. 2010; Xu et al. 2009; Zhou et al. 2008, 2009a, b; Srivastava et al. 2007) for solid-supported DNA synthesis, employing modified synthetic strategy described below (Fig. 7). Further experimental details can be found in SI (Li et al. 2013) Part IV (Fig. 8).

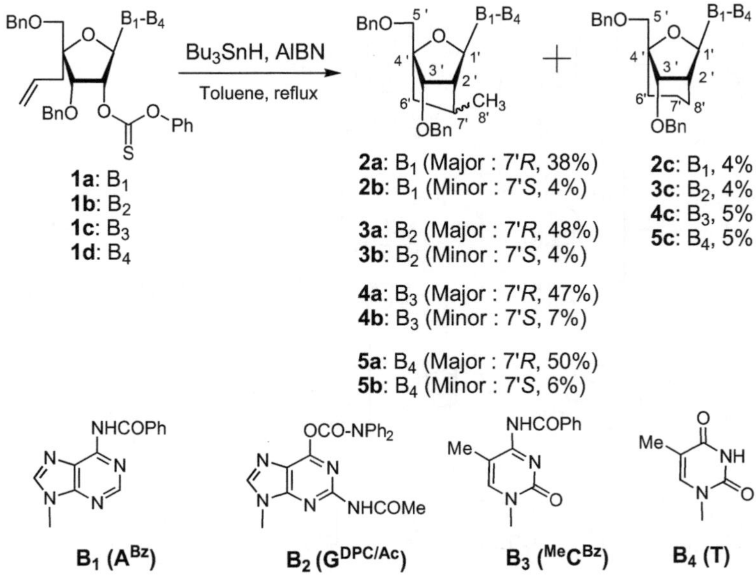

Fig. 6 Key radical cyclization step of the cLNA and cENA synthesis

All isolated intermediates and final phosphoramidites have been characterized by ^{1}H, ^{13}C, ^{31}P, 1D differential NOE, ^{1}H-^{1}H COSY, ^{1}H-^{13}C HMQC, and long-range ^{1}H-^{13}C correlation (HMBC) NMR experiments as well as by mass spectroscopy (see Figs. 2, 3, and 4 as well as SI (Li et al. 2013) Parts I and II).

3 Binding Affinity of cLNA-A, -G, -^{Me}C, -T and LNA-A, -G, -C, -T Single Modified AONs Toward Complementary Native RNA and DNA and Thermal of Stability of the Resulting Duplexes

Phosphoramidites of 7′*S*- and 7′*R*-Me-cLNA-T, 7′*S*- and 7′*R*-Me-cLNA-$^{Me}C^{Bz}$, 7′*S*-Me-cLNA-A^{Bz}, 7′*S*-Me-cLNA-$G^{N2\text{-}Ac}$ (viz. compound **20a/b**, **21a/b**, **22b** and **23b**), 7′*R*-Me-cLNA-A^{Bz} (Upadhayaya et al. 2011), 7′*R*-Me-cLNA-$G^{N2\text{-}Ac}$ (Upadhayaya et al. 2011), as well as LNA-A, -G, -C, -T (commercially available from Link Technologies, UK) were incorporated into a 20 mer AON sequence as a single modification in different sites by the solid-phase synthesis protocol (Beaucage and Caruthers 2001) on an automated DNA/RNA synthesizer. The sequence of AONs and T_m value of duplexes formed by these AONs with complementary RNA and DNA are listed in Fig. 9.

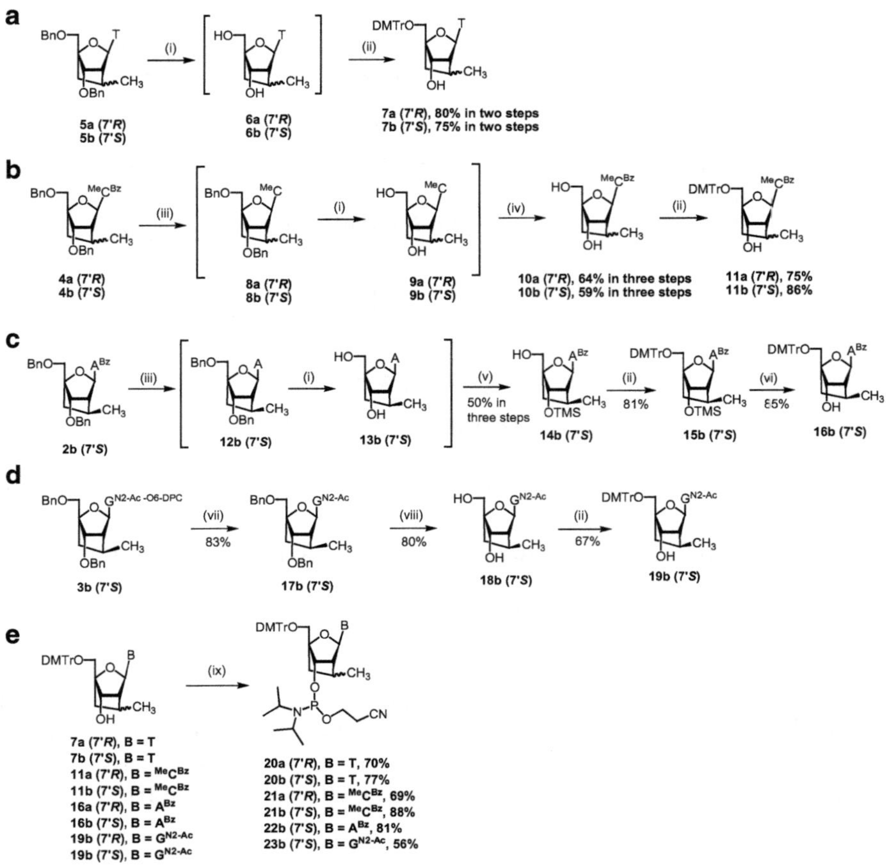

Fig. 7 De-protection steps yielding carbocyclic nucleotides (**7a**) 7′*S*- and (**7b**) 7′*R*-Me-cLNA-T (insert **A**), (**11a**) 7′*S*- and (**11b**) 7′*R*-Me-cLNA-^{Me}C (insert **B**), (**16b**) 7′*R*-Me-cLNA-A (insert **C**), and (**19b**) 7′*R*-Me-cLNA-G (insert **D**) nucleosides as well as their corresponding phosphoramidites (**20a**, **b**), (**21a**, **b**), (**22b**), and (**23b**) (insert **E**). *Reagents and Conditions*: (i) 20 % Pd $(OH)_2/C$, ammonium formate, methanol, reflux, 8 h; (ii) DMTr-Cl, dry pyridine, rt, overnight; (iii) methanolic ammonia, rt, 16 h; (iv) Bz_2O, dry pyridine, rt, overnight; (v) a. TMS-Cl, dry pyridine; b. BzCl, c. aq NH_3; (vi) 1 M TBAF, THF, rt, overnight; (vii) acetic acid, 55 °C; (viii) 20 % Pd $(OH)_2/C$, formic acid, methanol, reflux, 2.5 h; (ix) 2-cyanoethyl-*N*, *N*-diisopropylphosphoramidochloridite, DIPEA, dry DCM, rt, 2 h

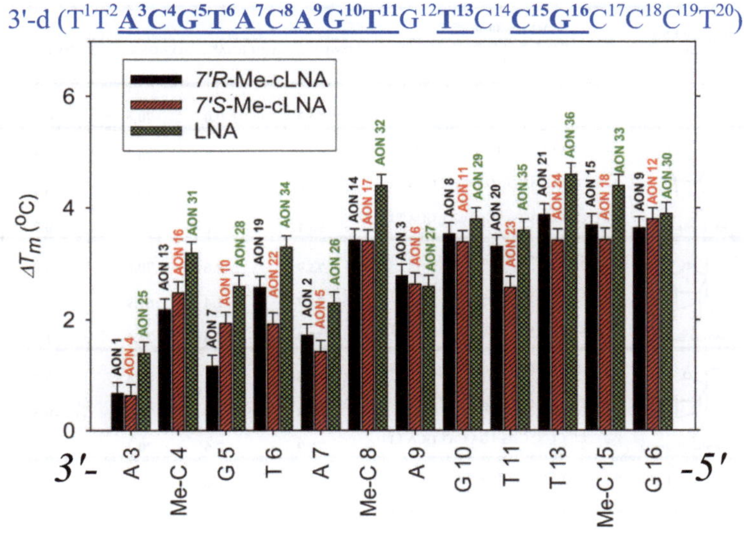

cLNA and LNA incorporation sites in the DNA strand
of the AON/RNA duplexes

Fig. 8 Melting temperature (T_m, °C) variations of 7′R- and 7′S-Me-cLNA-A, -G, -^{Me}C, -T and the corresponding LNA-modified AON/RNA duplexes from those of the native counterparts. Only single-site modified oligos were used, one at a time. These ΔT_ms. have been obtained by subtraction of T_m of the native AON/RNA from T_m of the modified AON/RNA. The melting temperature measurements have been performed with 0.2° precision

3.1 RNA Affinity of AONs Containing cLNA-A, -G, -^{Me}C, -T or LNA-A, -G, -C, -T Single Modification

3.1.1 Affinity Toward Native RNA

From the thermal denaturation study, we have found that AONs containing both 7′R- and 7′S-Me-cLNA-A, -G, -^{Me}C, -T single modifications form AON/RNA duplexes with complementary RNA that exhibit higher T_m values (enhanced by +0.7–3.9 °C/mod, see Figs. 9 and 10) compared to that of the native DNA/RNA. The observed T_m enhancement has been found to be systematically lower than that of the corresponding LNA-A-, -G, -C, and -T-modified AON/RNA duplexes which have ΔT_m values of +1.4–4.6 °C/mod relative to T_m of the native DNA/RNA duplex (Figs. 9 and 10). We have also found that the substitution of native nucleotides by 7′R- and 7′S-Me-cLNA-A at position 3 (counting from 3′-end) leads to moderate ~0.7 °C T_m enhancement of AON/RNA duplex stability, while 7′R- and 7′S-Me-cLNA-A, -G, -^{Me}C, -T modifications at further positions toward 5′-end (positions 4, 5, 6 and 7) lead to higher ~2 °C T_m enhancement compared to T_m of the native counterpart (Figs. 9 and 10). Single cLNA modifications at positions even further in

Entry	Modified LNA structures	AON-Sequence* (*Containing modifications in the minor groove*)	T_m of AON/RNA	ΔT_m of AON/RNA[b]	T_m of AON/DNA	ΔT_m of AON/DNA[c]	ΔT_m RNA selectivity[d]
Native		5'-d (TCC CGC CTG TGA CAT GCA TT)	73.3	0.0	70.6	0.0	2.7
AON1	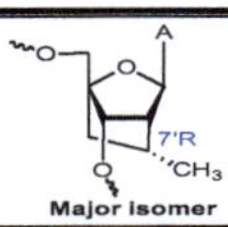	5'-d (TCC CGC CTG TGA CAT GC$\underline{A}^{R}$ TT)	74.0	0.7	70.9	0.2	3.1
AON2		5'-d (TCC CGC CTG TGA C$\underline{A}^{R}$T GCA TT)	75.0	1.7	70.5	-0.1	4.5
AON3		5'-d (TCC CGC CTG TG$\underline{A}^{R}$ CAT GCA TT)	76.1	2.8	71.3	0.7	4.8
AON4	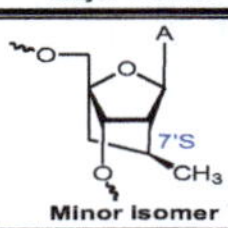	5'-d (TCC CGC CTG TGA CAT GC$\underline{A}^{S}$ TT)	73.9	0.6	70.6	0.0	3.3
AON5		5'-d (TCC CGC CTG TGA C$\underline{A}^{S}$T GCA TT)	74.7	1.4	69.9	-0.7	4.8
AON6		5'-d (TCC CGC CTG TG$\underline{A}^{S}$ CAT GCA TT)	75.9	2.6	71.6	1.0	4.3
AON7	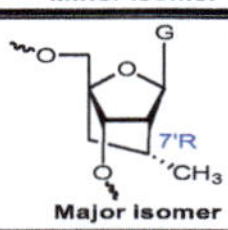	5'-d (TCC CGC CTG TGA CAT $\underline{G}^{R}$CA TT)	74.5	1.2	70.8	0.2	3.7
AON8		5'-d (TCC CGC CTG T$\underline{G}^{R}$A CAT GCA TT)	76.8	3.5	70.1	-0.5	6.7
AON9		5'-d (TCC C$\underline{G}^{R}$C CTG TGA CAT GCA TT)	76.9	3.6	70.4	-0.2	6.5
AON10	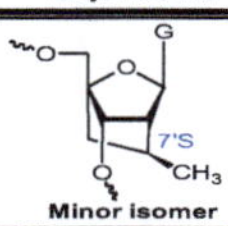	5'-d (TCC CGC CTG TGA CAT $\underline{G}^{S}$CA TT)	75.2	1.9	70.9	0.3	4.3
AON11		5'-d (TCC CGC CTG T$\underline{G}^{S}$A CAT GCA TT)	76.7	3.4	70.4	-0.2	6.3
AON12		5'-d (TCC C$\underline{G}^{S}$C CTG TGA CAT GCA TT)	77.1	3.8	70.6	-0.1	6.5
AON13	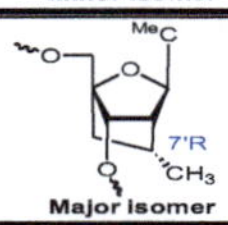	5'-d (TCC CGC CTG TGA CAT G$^{Me}\underline{C}^{R}$A TT)	75.5	2.2	71.4	0.7	4.1
AON14		5'-d (TCC CGC CTG TGA $^{Me}\underline{C}^{R}$AT GCA TT)	76.7	3.4	71.8	1.2	4.9
AON15		5'-d (TCC CG$^{Me}\underline{C}^{R}$ CTG TGA CAT GCA TT)	77.0	3.7	71.8	1.1	5.2
AON16	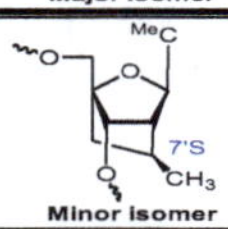	5'-d (TCC CGC CTG TGA CAT G$^{Me}\underline{C}^{S}$A TT)	75.8	2.5	71.7	1.1	4.1
AON17		5'-d (TCC CGC CTG TGA $^{Me}\underline{C}^{S}$AT GCA TT)	76.7	3.4	72.1	1.4	4.6
AON18		5'-d (TCC CG$^{Me}\underline{C}^{S}$ CTG TGA CAT GCA TT)	76.7	3.4	72.1	1.5	4.6
AON19	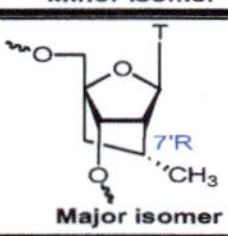	5'-d (TCC CGC CTG TGA CA$\underline{T}^{R}$ GCA TT)	75.9	2.6	70.5	-0.1	5.4
AON20		5'-d (TCC CGC CTG $\underline{T}^{R}$GA CAT GCA TT)	76.6	3.3	70.8	0.1	5.9
AON21		5'-d (TCC CGC C$\underline{T}^{R}$G TGA CAT GCA TT)	77.2	3.9	71.9	1.3	5.2
AON22	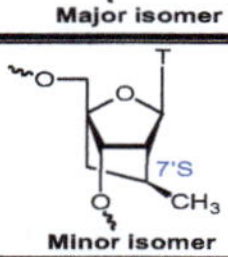	5'-d (TCC CGC CTG TGA CA$\underline{T}^{S}$ GCA TT)	75.2	1.9	70.2	-0.4	5.0
AON23		5'-d (TCC CGC CTG $\underline{T}^{S}$GA CAT GCA TT)	75.9	2.6	70.7	0.0	5.2
AON24		5'-d (TCC CGC C$\underline{T}^{S}$G TGA CAT GCA TT)	76.7	3.4	71.6	1.0	5.1

Fig. 9 (continued)

the direction of 5′-end (positions 8-, 9-, 10-, 11-, 13-, 15-, and 16-positions) have promoted even higher 3–4 °C T_m increase compared to that of the native DNA/RNA (Figs. 9 and 10). Position-dependent T_m variations have also been observed for the corresponding LNA-modified AON/RNA duplexes (AONs **25–36**, Figs. 9 and 10). Thus, LNA modifications at 3-, 4-, 5-, 6-, 7-, and 9-positions led to somewhat lower T_m enhancement toward RNA (+1.4–3.3 °C/mod) than that of the similar modifications at positions toward 5′-end of AONs with LNA modifications at 8-, 10-, 11-, 13-, 15-, and 16-positions (+3.6–4.6 °C/mod) (Figs. 9 and 10). These observations

AON25	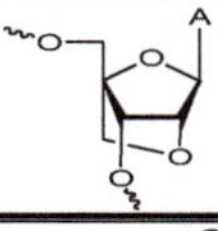	5′-d (TCC CGC CTG TGA CAT GC*A* TT)	74.7	1.4	71.6	1.0	3.1
AON26		5′-d (TCC CGC CTG TGA C*A*T GCA TT)	75.6	2.3	71.5	0.9	4.1
AON27		5′-d (TCC CGC CTG TG*A* CAT GCA TT)	75.9	2.6	72.5	1.9	3.4
AON28	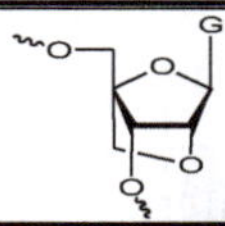	5′-d (TCC CGC CTG TGA CAT *G*CA TT)	75.9	2.6	72.1	1.5	3.8
AON29		5′-d (TCC CGC CTG T*G*A CAT GCA TT)	77.1	3.8	71.4	0.8	5.7
AON30		5′-d (TCC C*G*C CTG TGA CAT GCA TT)	77.2	3.9	71.4	0.8	5.8
AON31	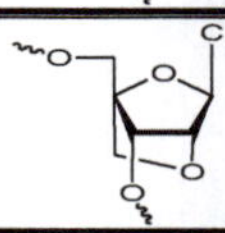	5′-d (TCC CGC CTG TGA CAT G*C*A TT)	76.5	3.2	73.6	3.0	2.9
AON32		5′-d (TCC CGC CTG TGA *C*AT GCA TT)	77.7	4.4	74.1	3.5	3.6
AON33		5′-d (TCC CG*C* CTG TGA CAT GCA TT)	77.7	4.4	73.8	3.2	3.9
AON34	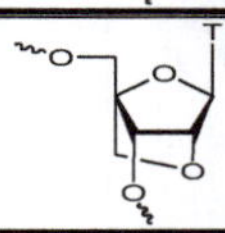	5′-d (TCC CGC CTG TGA CA*T* GCA TT)	76.6	3.3	72.7	2.1	3.9
AON35		5′-d (TCC CGC CTG *T*GA CAT GCA TT)	76.9	3.6	72.6	2.0	4.3
AON36		5′-d (TCC CGC C*T*G TGA CAT GCA TT)	77.9	4.6	73.2	2.6	4.7

Fig. 9 Melting temperatures (T_m) of the duplexes formed by modified AONs with complementary RNA and DNA. [a]Molecular weights of all antisense sequences are confirmed by MALDI-TOF mass spectrum [see Table 1 and Figs. SII.25-60 in SI (Li et al. 2013) Part II]. A = adeninyl, G = guaninyl, C = cytosinyl, T = thyminyl. $\underline{A}^R$ = 7′R-Me-cLNA-A, $\underline{G}^R$ = 7′R-Me-cLNA-G, $^{Me}\underline{C}^R$ = 7′R-Me-cLNA-^{Me}C, $\underline{T}^R$ = 7′R-Me-cLNA-T; $\underline{A}^S$ = 7′S-Me-cLNA-A, $\underline{G}^S$ = 7′S-Me-cLNA-G, $^{Me}\underline{C}^S$ = 7′S-Me-cLNA-^{Me}C, $\underline{T}^S$ = 7′S-Me-cLNA-T; $\underline{A}$ = LNA-A, $\underline{G}$ = LNA-G, $\underline{C}$ = LNA-C, $\underline{T}$ = LNA-T. T_m Values measured at the maximum of the first derivative of the melting curve ($A_{260\ nm}$ vs. temperature) in medium salt buffer (60 mM tris–HCl at pH 7.5, 60 mM KCl, 0.8 mM $MgCl_2$) with temperature 60–90 °C using 1 mM concentrations of two complementary strands. The values of T_m given are averages of three independent measurements (the error of the three consecutive measurements is within + 0.2 °C). [b]T_m difference of RNA was obtained by comparing the T_m of AON/RNA with that of the native AON /RNA. [c]T_m difference of AON/DNA was obtained by comparing the T_m of AON/DNA with that of the native .AON/DNA. [d]T_m based RNA-selectivity: = (T_m of AON/RNA)−(T_m of AON/DNA)

indicate that thermal stabilizations of the AON/RNA duplexes containing either 7′R- or 7′S-Me-cLNA or LNA single modification are significantly dependent on positions of modification while less influenced by specific nucleobase moiety in the modified nucleotide. Since each position of modification in the chosen 20mer AON sequence has unique nearest neighbors, the observed position dependence can be interpreted as the sequence context.

Additionally, we have found that the orientation of C7′-S- and C7′-R-configured methyl group in the cLNAs used in this study has only minor impact on thermal stability of the modified AON/RNA duplexes, comparable with errors of ±0.2 °C in T_m measurements. AON/RNA duplexes containing AONs modified with 7′R-Me-cLNAs (major isomer) at modification positions 3, 6, 7, 9, 10, 11, 13, and 15 have been found to be thermally more stable (by 0.1–0.7 ± 0.3 °C) than those containing AONs modified correspondingly with 7′S-Me-cLNAs (minor isomer), whereas 7′S-Me-cLNAs modifications at position 4, 5, 8, and 16 of the AON strand resulted in up to 0.7 ± 0.3 °C more stabilized modified AON/RNA duplexes compared with those modified by 7′R-Me-cLNAs (Figs. 9 and 10).

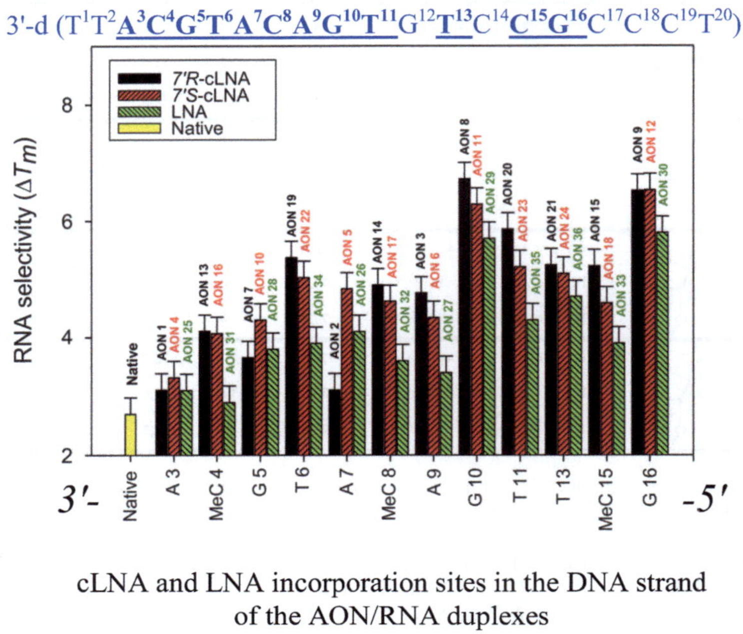

Fig. 10 Relative ΔT_m-based RNA selectivity defined as a difference between melting temperatures of the corresponding cLNA- and LNA-modified AON/RNA and corresponding modified AON/DNA duplexes (AONs **1–36**) as well as between respective native counterparts [RNA selectivity = (T_m of AON/RNA duplex) − T_m of AON/DNA duplex)]. *Yellow-color bar* shows RNA selectivity for the native, *black-color bars* denote RNA selectivity for the duplexes containing AONs modified with 7′*R*-Me-cLNAs (major isomer), *red-color bars* show RNA selectivity of the duplexes containing AONs modified with 7′*S*-Me-cLNAs (minor isomer), and *green-color bars* show RNA selectivity of the duplexes containing AONs modified with LNA, which show that cLNA-modified AONs, irrespective of being 7′*R*- or 7′*S-Me* substituted, are consistently more RNA selective than LNA or native counterparts. The T_m measurements have been performed with 0.2° precision. For details, see Fig. 9

It is, however, important to note that although the difference in T_m modulation by spatial orientation of 7′*S*- and 7′*R*-methyl group in cLNAs has been found to be small (within 0.7 °C), *their respective exonuclease stabilities and RNAse H recruitment properties have been however found in this work to be distinctly different* (*see Sects. 4–6*). *These new findings fingerpoint for further in-depth investigations of diastereomerically pure 7′S- and 7′R-methyl-cLNAs for pharmacological and physico-chemical profiling which is likely to be important in therapeutic or diagnostic context.*

3.1.2 Affinity Toward Native DNA

Influence of the cLNA modifications on T_m values of the cLNA-modified AON/DNA duplexes has been found to be mixed depending on the position of modification and ranging from slight relative thermal destabilization of −0.7 ± 0.3 °C to moderate relative stabilization of +1.5 ± 0.3 °C compared to T_m of the corresponding native DNA/DNAs (Fig. 9), suggesting that those modifications in the AON strand might not significantly change the global or local conformations of duplexes which is reflected in T_m values. On the contrary, the T_m values of LNA-modified AON/DNA duplexes have been found to be systematically higher (ranging from 0.8 to 3.5 °C, Fig. 9) than that of native counterparts, which is in agreement with our previous findings (Srivastava et al. 2007) which for the 15mer AONs modified with LNA-T had shown relative 1–2 °C T_m enhancement compared to that of the corresponding native 15mer DNA/DNA (Srivastava et al. 2007).

3.2 Relative RNA Selectivity of cLNA-A-, -G-, -^{Me}C-, -T-Modified AONs in Comparison with LNA-A-, -G-, -C-, -T-Modified AONs

Differential pair-wise comparison of melting temperatures of AON/DNA and AON/RNA duplexes gives a value of relative selectivity of the AON strand toward the RNA vis-a-vis DNA, called $\Delta T_{m\ (RNA\text{-}selectivity)}$ or RNA selectivity (Li et al. 2010; Nishida et al. 2010) (Fig. 9 and footnotes), and defined as the difference between T_m of AON/RNA duplex and T_m of the corresponding AON/DNA duplex (Figs. 9 and 10).

It is well known (Chatterjee et al. 2005; Acharya et al. 2004) that DNA/RNA duplexes are generally more thermally stable than corresponding DNA/DNA duplexes. Our UV monitored melting experiments have corroborated this conclusion as the 20mer native DNA/RNA hybrid has been found to be 2.7 °C more stable than the native DNA/DNA duplex (Fig. 9). The AONs modified by either 7′*R*- or 7′*S*-Me-cLNA nucleotides have been found to be even more RNA selective than the native, with $\Delta T_{m\ (RNA\text{-}selectivity)}$ values being in the range of 3.1–6.7 °C (thus higher in average by 2.2° ± 0.7°), much higher than that of native type (Fig. 9). The RNA selectivity of the cLNA-modified AONs, irrespective of 7′*R* or 7′*S* configuration, has also been found consistently higher on average by 0.8° ± 0.4° than those of corresponding LNA-modified AONs (Figs. 9 and 10).

4 3′-Exonuclease Stability Studies of cLNA-A-, -G-, -MeC-, -T-Modified AONs

The stability of cLNA-A-, -G-, -^{Me}C-, and -T-modified AONs toward 3′-exonuclease has been investigated using snake venom phosphodiesterase I from *Crotalus adamanteus* venom (SVPDE). The selected native AON and AONs **1**, **4**, **7**, **10**, **13**, **16**, **19**, **22** were labeled at the 5′-end with ^{32}P and then incubated with SVPDE [SVPDE 6.7 ng/μL, 100 mM Tris–HCl (pH 8.0), 15 mM $MgCl_2$, total volume

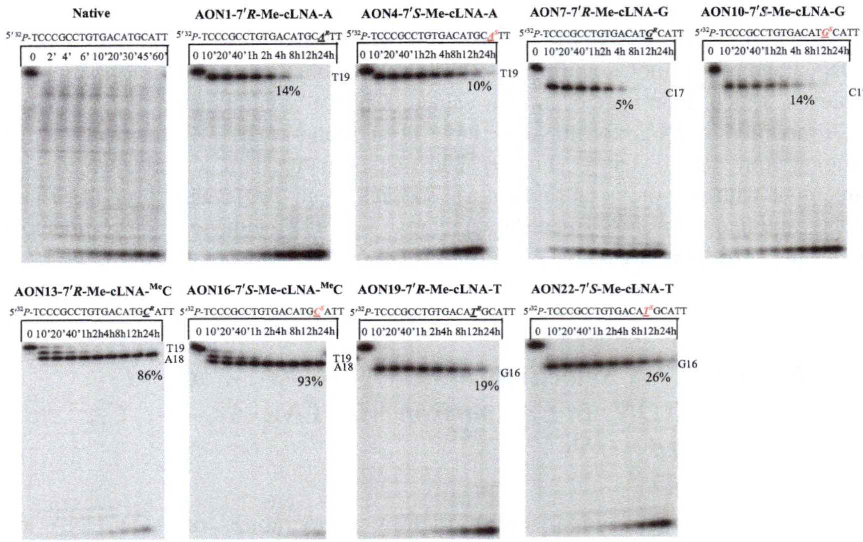

Fig. 11 Denaturing PAGE analysis of the SVPDE degradation of AONs with cLNA modifications. The AON% left after 24 h incubation for each AON is shown below the corresponding band (native—0 %, AON1—14 %, AON4—10 %, AON7—5 %, AON10—14 %, AON13—86 %, AON16—93 %, AON19—19 %, AON22—26 %). Digestion conditions: AON 3 μM (5′-end ^{32}P labeled with specific activity 80,000 cpm) in 100 mM Tris–HCl (pH 8.0) and 15 mM $MgCl_2$, 21 °C, total reaction volume 30 μL, SVPDE concentration (6.7 ng/μL)

30 μL] over a period of 24 h at 21 °C. Aliquots were taken out at appropriate time intervals and analyzed by 20 % denaturing PAGE.

Complete degradation of the native AON within several minutes (Figs. 11 and 12) has demonstrated that the native AON is rather susceptible to SVPDE under the present conditions, whereas cLNA-modified AONs exhibited highly improved 3′-exonuclease resistance to a variable extent. The cleavage bands corresponding to the position of modifications can clearly be seen on PAGE as 19 mer for AON **1** and **4** SVPDE-assisted degradation, as 17mer for AONs **7** and **10**, as 18mer and 19mer for AONs **13** and **16**, and as 16 mer for AONs **19** and **22**.

In order to compare the relative 3′-exonuclease stability of cLNA-A-, -G-, -^{Me}C-, -T-modified AONs in a straightforward manner, the relative band intensities have been quantified after autoradiography to give SVPDE digestion curve for each selected AON (Fig. 12). Pseudo-first-order reaction rates (decaying constants) have been obtained by fitting the curves to single-exponential decay functions (Fig. 12). Additionally, the half-life times ($t_{1/2}$) of the selected modified AONs have also been derived from the obtained pseudo-first-order decaying constants (Fig. 13). Comparison of the SVPDE cleavage rates and lifetimes of the target AONs containing different modification types is presented in Sects. 4.1–4.2 below.

4.1 Relative 3′-Exonucleolytic Stabilities of cLNA-A-, -G-, -^{Me}C-, -T-Modified AONs

Comparison of the pseudo-first-order SVPDE cleavage rates and half-life times ($t_{1/2}$) of the target AONs has revealed that stability of the modified AONs under treatment of SVPDE decreased in the following order: 7′*S*-Me-cLNA-^{Me}C-modified AON **16** ($k = 0.0027 \pm 0.0002$ h^{-1}, $t_{1/2} = 258 \pm 19.0$ h) > 7′*R*-Me-cLNA-^{Me}C-modified AON **13** ($k = 0.0059 \pm 0.0006$ h^{-1}, $t_{1/2} = 119 \pm 13.0$ h) > 7′*S*-Me-cLNA-T-modified AON **22** ($k = 0.0985 \pm 0.0046$ h^{-1}, $t_{1/2} = 7.10 \pm 0.40$ h) > 7′*R*-Me-cLNA-T-modified AON **19** ($k = 0.1107 \pm 0.0084$ h^{-1}, $t_{1/2} = 6.30 \pm 0.50$ h) > 7′*S*-Me-cLNA-A-modified AON **4** ($k = 0.1650 \pm 0.0055$ h^{-1}, $t_{1/2} = 4.20 \pm 0.10$ h) > 7′*R*-Me-cLNA-A-modified AON **1** ($k = 0.3954 \pm 0.0448$ h$_{-1}$, $t_{1/2} = 1.80 \pm 0.20$ h) > 7′*S*-Me-cLNA-G-modified AON **10** ($k = 0.5246 \pm 0.0235$ h^{-1}, $t_{1/2} = 1.35 \pm 0.05$ h) > 7′*R*-Me-cLNA-G-modified AON **7** ($k = 0.7354 \pm 0.0300$ h^{-1}, $t_{1/2} = 0.95 \pm 0.04$ h). These values can be compared to our previously reported (Li et al. 2010) data for the LNA-T-modified 15mer AON ($k = 0.5331 \pm 0.1800$ min^{-1}, $t_{1/2} = 1.47 \pm 0.49$ min) which was found to be very labile toward SVPDE in identical experimental conditions. The remarkable increase of SVPDE exonuclease stability of the cLNA-modified AONs compared to the native as well as to LNA-modified counterparts, which are shown to undergo a complete SVPDE-promoted degradation within several minutes, is an important property for potential application of the modified cLNA in therapeutics.

4.2 The Effect of Sequence Context of a Given AON on 3′-Exonucleolytic Stability

4.2.1 The Effect of the Nucleobase in Carba-LNA on 3′-Exonucleolytic Stability

Experimental SVPDE cleavage reaction rates (Fig. 12) and lifetimes of the selected modified AONs (Fig. 13) clearly show strong dependence of these physico-chemical properties on the type of nucleobase moiety in the modified cLNA. Thus, for example, the most resistant in this series 7′*S*-Me-cLNA-^{Me}C-modified AON **16** ($t_{1/2} = 258 \pm 19.0$ h) has been found to be 41-fold more stable than 7′*S*-Me-cLNA-T-modified AON **22** ($t_{1/2} = 7.10 \pm 0.40$ h) which in turn has been found to be 1.7-fold more stable than corresponding 7′*S*-Me-cLNA-A-modified

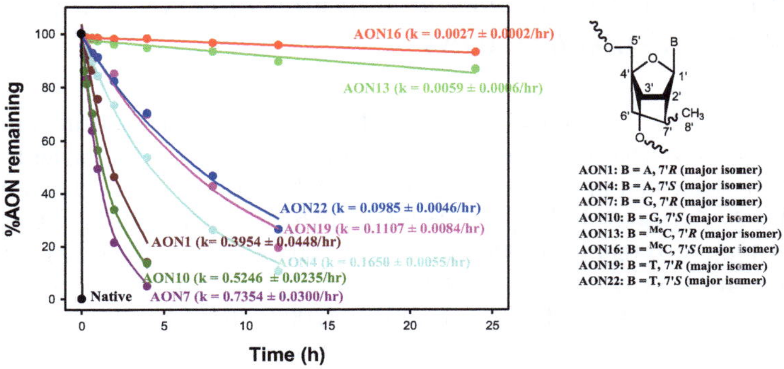

Fig. 12 SVPDE digestion curves of native AON and selected 7′*R*- and 7′*S*-Me-cLNA-A-, -G-, -^{Me}C-, -T-modified AONs. The pseudo-first-order rates shown here were obtained by fitting the curve to the single-exponential decay function. Digestion conditions: AON 3 μM (5′-end ^{32}P labeled with specific activity 80,000 cpm) in 100 mM Tris–HCl (pH 8.0) and 15 mM $MgCl_2$, 21 °C, total reaction volume 30 μL, SVPDE concentration (6.7 ng/μL)

AON **4** ($t_{1/2} = 4.20 \pm 0.10$ h) and 5.3-fold more stable than 7′*S*-Me-cLNA-G-modified AON **10** ($t_{1/2} = 1.35 \pm 0.05$ h). The most stable in this series 7′*S*-Me-cLNA-^{Me}C-modified AON **16** has also been found to be unprecedented 272 times more nucleolytically stable against SVPDE than the least stable in this series 7′*R*-Me-cLNA-G-modified AON **7** (Fig. 13).

4.2.2 The Effect of Configuration of 7′-Me Substitution in Carba-LNA on 3′-Exonucleolytic Stability

Pair-wise comparison of the kinetics of degradation of 7′*R*- and 7′*S*-Me-cLNA-modified AONs by SVRDE has revealed only moderate dependence of the modified AONs' nucleolytical stability on the stereochemical orientation of the 7′-Me group. The results (Figs. 12 and 13) indicate that the 7′-methyl group, which is located toward the vicinal 3′-phosphate (C7′-*S*), has significantly more pronounced effect on the nucleolytic stability of cLNA-modified AONs than its diastereomeric C7′-*R* counterpart. This observation is in agreement with our previously reported (Zhou et al. 2009b) 3′-exonuclease stability comparison between (6′*R*-OH, 7′*R*-Me)-cLNA-T and (6′*R*-OH, 7′*R*-Me)-cLNA-T as well as between (6′*S*-OH, 7′*S*-Me)-cLNA-T- and (6′*S*-OH, 7′*R*-Me)-cLNA-T-modified AONs. It should be noted that although the relative effect of 7′*S*- *versus* 7′*R*-methyl substitution on the lifetime of the cLNA-modified AONs in presence of SVPDE exonuclease has been found to be substantial which is evident from the data showing that 7′*S-Me* diastereomer is lasting 1.1–2.3 times longer than the corresponding 7′*R-Me* counterpart with the same nucleobase, the effect of different nucleobase has however been found to be much more pronounced with the most stable 7′*S*-Me-cLNA-^{Me}C-modified AON **16**

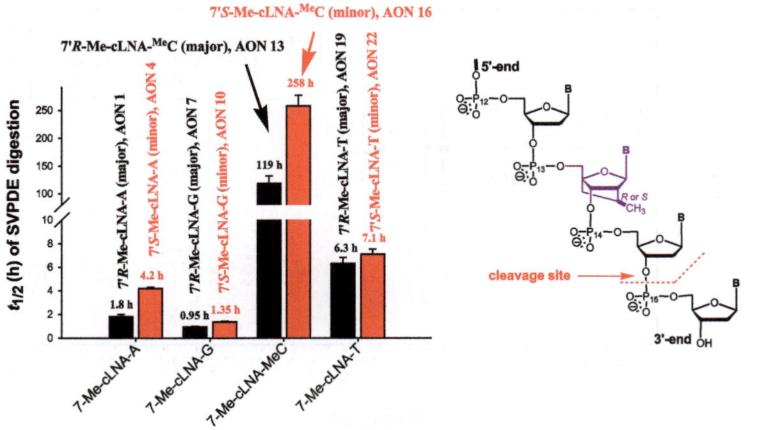

Fig. 13 Half-life ($t_{1/2}$) of SVPDE digestion of selected 7′*R*- and 7′*S*-Me-cLNA-A-, -G-, -^{Me}C-, -T-modified AONs. Note: the native strand and LNA-modified AONs are totally degraded within several minutes

being 272 times more nucleolytically stable against SVPDE than 7′*R*-Me-cLNA-G-modified AON **7** which has been found to be the least stable in this series of cLNAs (Fig. 13). Thus, the nucleolytic resistance of cLNA-modified AONs derived from the 3′-exonucleases degradation kinetics data (Figs. 12 and 13) has been demonstrated to decrease in the following order: 7′*S*- and 7′*R*-Me-cLNA-^{Me}C > 7′*S*- and 7′*R*-Me-cLNA-T > 7′*S*- and 7′*R*-Me-cLNA-A > 7′*S*- and 7′*R*-Me-cLNA-G.

5 Stability of cLNA-A-, -G-, -^{Me}C-, -T-Modified AONs in Human Blood Serum

The selected newly synthesized native AONs, AONs **1**, **4**, **7**, **10**, **13**, **16**, **19**, **22** with cLNA-A, -G, -^{Me}C, and -T modifications at varying positions, have also been investigated for stability in human blood serum. The AONs with 5′-end ^{32}P-labeling were incubated in human blood serum (male, type AB) for up to 48 h at 21 °C, and aliquots were taken out at proper time points and then analyzed by 20 % denaturing PAGE.

The gel pictures obtained by autoradiography are shown in Fig SII.1 in SI (Li et al. 2013) Part II. Due to the presence of alkaline phosphatase in blood serum that gradually removes the 5′-end ^{32}P labeling the determination of the accurate degradation rates for each AON by quantifying the gel band intensities is impossible. By visually comparing the gels, we have found that all cLNA-modified AONs can sustain in human blood serum for more than 48 h, whereas the native AON can only stand up to 12 h under identical condition. Amongst all the

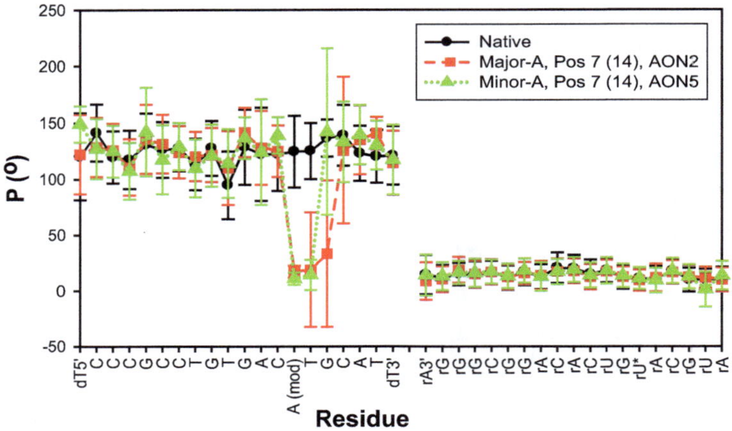

Fig. 14 Pseudorotational phase angle (*P*) of sugar moieties in the native (*black circles*) and 7′(*R*/*S*)-Me-cLNA-modified AON/RNA duplexes [position **7** modification, AON **2** for *R*- (*red squares*) and AON **5** for *S*- (*green triangles*) isomers, respectively], showing the changes of *P* upto 2 nt toward the 3′-end of the AON strand from the modification site [A(mod) in the plot]. The average values and absolute deviations have been obtained from the last 0.5 ns of the 5 ns MD simulations utilizing Amber force (see Computational details in SI (Li et al. 2013) Part IV)

modified AONs, cLNA-^{Me}C-modified AONs **13** and **16** have been found to be the most stable oligonucleotides upon treatment of blood serum, which is consistent with the SVPDE degradation stability. The relative stabilities of all modified AONs in human blood serum are as follows: cLNA-^{Me}C-modified AONs **13** and **16** > cLNA-T-modified AONs **19** and **22** > cLNA-A-modified AONs **1** and **4** > cLNA-G-modified AON **7** and **10** > native AON.

6 Impact of cLNA Modifications on the Duplex Structure and Properties: Theoretical Study

To elucidate structural impact of the introduction of the 7′*R*- and 7′*S*-Me-cLNA modifications into AON/RNA duplexes, a set of Amber force field-based molecular dynamics (MD) simulations of the native and 7′*S*- and 7′*R*-Me-cLNA-A, -G, -^{Me}C,-T-modified AON/RNA duplexes (AON **1**, **2**, **5**, **8**, **11**, **14**, **17**, **20**, **23**) have been performed (see Computational details in SI (Li et al. 2013) Part IV). Here we present a comparison of the structural features of the AON/RNA duplexes modified at position 7 by the 7′*R*-Me-cLNA-A (AON **2**) and 7′*S*-Me-cLNA-A (AON **5**) diastereoisomers with the native DNA/RNA counterpart (Fig. 14).

Total energy and RMSD. All MD simulations of the native and 7′*S*- and 7′*R*-Me-cLNA-A, -G, -^{Me}C,-T-modified AON/RNA duplexes (AONs **1**, **2**, **5**, **8**, **11**, **14**, **17**,

Native AON/RNA (BP 13-15)	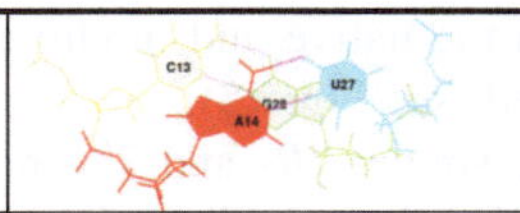	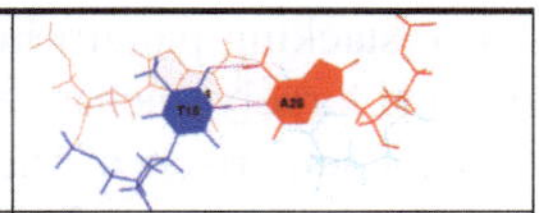
Overlap area (Å^2)	3.39 (0.95)	10.64 (5.51)
AON/RNA (AON2) ***7′R-Me*** **cLNA-A at 7 (14) (Major isomer)**	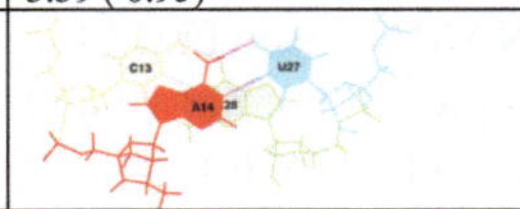	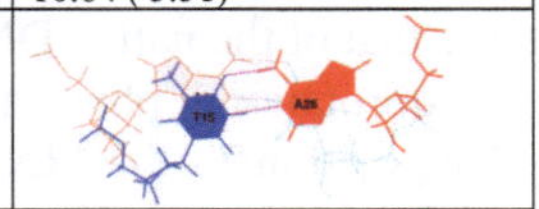
Overlap area (Å^2)	2.99 (1.37)	11.21 (6.48)
AON/RNA (AON5) ***7′S-Me*** **cLNA-A at 7 (14) (Minor isomer)**	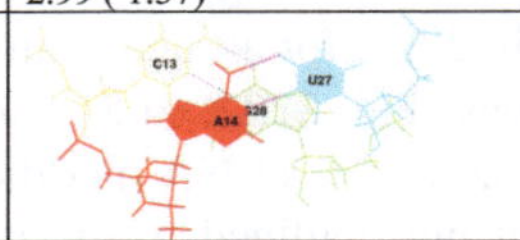	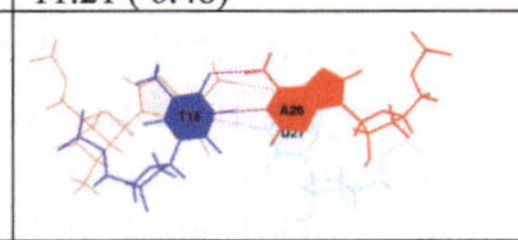
Overlap area (Å^2)	2.70 (0.75)	10.16 (6.36)

Fig. 15 Overlap bp − 1 and bp + 1 base pair steps with the base pair containing modified 7′*R*-Me-cLNA-A and 7′*R*-Me-cLNA-A modifications at position **7** from the 3′-end of the AON strand in the modified AON **2**/RNA and AON **5**/RNA duplexes compared to the same base-pair steps in the native AON/RNA duplex

20, **23**) duplexes have energetically stabilized rapidly within first 0.2 ns after completion of the cooling cycle (see Fig. SIII.2 in SI (Li et al. 2013) Part III) and all simulations remained energetically stable until the end of the 5 ns simulations (RMSd variation 2.35 Å). RMSd comparison with the average structure of the native AON/RNA from the 2–2.5 ns stretch of its MD simulation shows however that the native and the cLNA-modified AON/RNA structures remain fluctuating around this structure with RMSd about 2 Å and amplitude of fluctuation ~0.6 Å with the exception of 7′*S*- and 7′*R*-Me-cLNA-G, containing duplexes (AON **5**, **14**), where it took about 3 ns to stabilize from an average ~3.8 Å to 2.2 Å RMSd. The last 0.5 ns of the corresponding MD simulations (2,500 molecular structures) at two stretches of the simulations (2–2.5 ns and 4.5–5 ns) have been analyzed (Figs. SIII.3–10 in SI (Li et al. 2013) Part III). The average values of the pseudorotational phase angle *versus* corresponding average values of the amplitude of pseudorotation as well as major endocyclic torsions (α, β, γ, δ, ε, ζ, χ) (IUPAC-IUB 1983) and local base-pair (Shear, Stretch, Stagger, Buckle, Propeller, Opening), local base-pair step (Shift, Slide, Rise, Tilt, Roll, Twist), and local base-pair helical parameters (X-disp, Y-disp, h-Rise, Inclination, Tip, h-Twist) of the native and 7′*S*- and 7′*R*-Me-cLNA-A, -G, -^{Me}C,-T-modified AON/RNA duplexes have been collected and analyzed (Figs. SIII.3–10 in SI (Li et al. 2013) Part III).

The hydrogen bonding pattern of all base pairs of the 7′*R*-Me and 7′*S*-Me-cLNA-A modified at pos 7 AON/DNA has been found to be typical for the standard Watson–Crick base pairs. All WC bonds between native as well as with modified residues remained stable within 3.2 Å during 98–100 % of the MD simulation's time, similar to that observed in the MD simulation of the native DNA/RNA duplex. The stacking pattern in the native and modified AON/RNA duplexes has also remained largely unchanged (see Fig. 15).

Pair-wise comparison of the single 7′*R*- and 7′*S*-modified duplexes with their native counterpart has revealed that the overall duplex structure remains essentially the same as that of the native DNA/RNA (see Fig. 14 for representative comparison of the pseudorotational phase angles in the native and 7′*R*- and 7′*S*-Me-cLNA-A-modified at position 7 AON/RNA duplexes, for the other angular and helical duplex parameters for these and other AONs see Figs. SIII.1–8 in SI (Li et al. 2013) Part III) with RMSd remaining within 2.6 Å for all heavy atoms of unmodified residues and <3.5 Å for the modified ones (Fig. SIII.2 in SI (Li et al. 2013) Part III). The structural perturbations introduced by the cLNA modifications have been found to be quite local, originated at but not confined to the modification site and being extending by 2–4 bp in the 3′ direction of the AON strand and 1–2 bp in 5′ direction. Reverse tendency has been observed for the opposite strand where the structural perturbations have been found extending longer distances (2–4 bp *versus* 1–2 bp) in the 5′ direction. Due to similarity of the AON/RNA duplexes containing both 7′*R*-Me and 7′*S*-Me-cLNA diastereomers, it can be concluded that the main significant structural difference between these modification is orientation of the 7′-methyl group relative to minor grooves of the duplexes. The minor 7′*S*-diastereomer has been found to be exposed toward the edge of the duplex backbone, while major 7′*R*-Me-cLNA diastereomer is somewhat buried in the minor groove of the 7′*R*-Me-cLNA-modified AON/RNA duplexes. This orientation has positioned 7′-methyl group in such a way that the key distances from the carbon atom of the 7′-methyl group to O2′ atom of sugar moiety of cLNA found to be within 0.1 Å of 4.1 Å and 5.1 Å for the C7′-*R* and -*S* diastereomers, respectively. This effectively fine-tunes interaction of the 7′-methyl group with the nearest 3′-phosphate positioning its carbon atom C (7′-Me) within 5.1 and 4.3 Å of the phosphorus P (3′-PO_4^-) in C7′-*R*- and -*S*-modified duplexes, respectively, with up to 0.6 Å variations along the trajectories.

7 Conclusions

In this investigation, eight diastereomerically pure 7′*S*- and 7′*R*-Me-cLNA-A, -G, -^{Me}C, and -T nucleosides have been synthesized through a key step of intramolecular 5-*exo* free-radical cyclization, followed by preparative HPLC separation of cyclization mixtures. Physico-chemical and enzymatic properties of AONs containing these modifications have been evaluated and compared with that of the native counterpart. The major conclusions are as follows:

The RNA selectivity values of the cLNA-modified AONs by 7′*S*- or7′*R*-Me-cLNA-A, -G, -^{Me}C, and -T, irrespective of 7′*R* or 7′*S* configuration, are consistently higher by 0.8° ± 0.4° than those of the LNA counterparts and higher by 2.2° ± 0.7°

than those of the native (Figs. 9 and 10). See also conclusions reflecting the stability and RNase H recruitment capability advantages of cLNA modified oligos below.

The variations of T_ms modulated by 7′*S*- and 7′*R*-configured methyl group on cLNAs are strongly influenced by position of the modification and sequence context as well as to some but lesser extent by nucleobase of the cLNA.

Introduction of both 7′*S*- and 7′*R*-configured methyl substituted carba-LNAs leads to similar structural perturbations, except for oppositely oriented 7′*S*- or 7′*R*-methyl group in the respective cLNA-modified AON/RNA duplexes, culminating in the small differences in thermal stabilities of the respective modified duplexes. This relative orientation of the methyl groups toward minor groove and backbone of the duplex has been found to have enzymological implications reflected in experiments studying 3′-exonuclease stability of AONs and RNase H recruitment capability of AON/RNA heteroduplexes (see below).

All cLNA-modified AONs have been found to be much more stable toward the 3′-exonuclease (SVPDE) digestion than the native and LNA-modified counterparts. It has been demonstrated that the 3′-exonucleolytic stability of cLNA-modified AONs varied greatly specially depending on the type of nucleobase moiety in the modified cLNA. Thus, 7′*S*-Me-cLNA-^{Me}C-modified AON **16** has been found to be 272 times more nucleolytically stable against SVPDE than the least stable in this series 7′*R*-Me-cLNA-G-modified AON **7**. The effect of 7′*R*- *versus* 7′*S*-methyl modification has been found to be comparably less pronounced. The 7′*S*-Me-cLNA-modified AONs have however been found lasting 1.1–2.3 times longer than the corresponding 7′*R*-Me counterparts with the same nucleobase. General trend for the nucleolytic resistance of cLNA-modified AONs derived from the 3′-exonucleases degradation kinetics data has been demonstrated to decrease the AON lifetime in SVPDE in the following order: ′*S*-Me-cLNA-^{Me}C > 7′*R*-Me-cLNA-^{Me}C > 7′*S*-Me-cLNA-T > 7′*R*-Me-cLNA-T > 7′*S*-Me-cLNA-A > 7′*R*-Me-cLNA-A > 7′*S*-Me-cLNA-G > 7′*R*-Me-cLNA-G.

Stabilities of cLNA-modified AONs upon blood serum treatment have been found to follow the trend similar to that observed upon treatment of SVPDE 3′-exonuclease.

cLNA-A-, -G-, -^{Me}C-, and -T-modified AON/RNA hybrids as well as LNA-A-, -G-, -C-, and -T-modified AON/RNA hybrids are found to be good substrates for *E. coli* RNase H1 as the native AON/RNA counterpart. A strong dependence of the RNase H digestion rates has been observed with modifications located from the center toward 5′-end of the AON strand leading to 2–8-fold increase of the digestion rates compared to that of modifications from the 3′-end to the center as well as to that of the native AON/RNA counterpart. Thus, by appropriate choice of the modification site in AON sequence of AON/RNA hybrids, the cleavage rates by RNase H1 as well as digestion patterns can be engineered to significant extent.

The ability of 7′*S*- or 7′*R*-Me-cLNA-A, -G, -^{Me}C, and -T nucleotides, irrespective of 7′*R* or 7′*S* configuration, to increase thermal stability of the modified AON/RNA duplexes, to prolong dramatically the lifetime of AONs in human blood serum and in the presence of SVPDE 3′-exonuclease, as well as to recruit RNase H highly efficiently makes our cLNA-modified oligonucleotides potential candidates for RNA targeting therapeutics.

Acknowledgement Generous financial support from the Swedish Natural Science Research Council (Vetenskapsrådet) (to JC) is gratefully acknowledged.

References

Acharya P, Issakson J, Pradeepkumar PI et al (2002) Experimental evidences unequivocally prove the role of stereoelectronics as one of the major forces responsible for the self-assembly of DNA and RNA. Collect Symp Ser 5:99–120

Acharya P, Cheruku P, Chatterjee S et al (2004) Measurement of nucleobase pK_a values in model mononucleotides shows RNA-RNA duplexes to be more stable than DNA-DNA duplexes. J Am Chem Soc 126:2862–2869

Albæk N, Peterson M, Nielsen P (2006) Analogues of a locked nucleic acid with three-carbon 2′,4′-linkages: synthesis by ring-closing metathesis and influence on nucleic acid duplex stability and structure. J Org Chem 71:7731–7740

Altmann KH, Kesselring R, Francotte E et al (1994) 4′,6′-Methano carbocyclic thymidine – a conformationally constrained building-block for oligonucleotides. Tetrahedron Lett 35:2331–2334

Bartel DP (2004) MicroRNAs: genomics, biogenesis, mechanism, and function. Cell 116:281–297

Beaucage SL, Caruthers MH (2001) Synthetic strategies and parameters involved in the synthesis of oligodeoxyribonucleotides according to the phosphoramidite method. In: Beaucage SL (ed) Current protocols in nucleic acid chemistry. Wiley, New Jersey, Chapter 3, Unit 3.3

Beckwith ALJ (1981) Regio-selectivity and stereo-selectivity in radical reactions. Tetrahedron 37:3073–3100

Beckwith ALJ, Schiesser CH (1985) Regio- and stereoselectivity of alkenyl radical ring closure: a theoretical study. Tetrahedron 41:3925–3941

Buchini S, Leumann CJ (2003) Recent improvements in antigene technology. Curr Opin Chem Biol 7:717–726

Chatterjee S, Pathmairi W, Chattopadhyaya J (2005) The 5-Me of thyminyl (T) interaction with the neighboring nucleobases dictate the relative stability of isosequential DNA-RNA hybrid duplexes. Org Biomol Chem 3:3911–3915

Crooke ST (2004) Progress in antisense technology. Annu Rev Med 55:61–95

Enderlin G, Nielsen P (2008) Synthesis of 6′-branched locked nucleic acid by a radical TEMPO-scavanged stereoselective mercury cyclization. J Org Chem 73:6891–6894

Hari Y, Obika S, Ohnishi R et al (2006) Synthesis and properties of 2′-O,4′-C-methyleneoxymethylene bridged nucleic acid. Bioorg Med Chem 14:1029–1038

Hari Y, Kodama T, Imanishi T, Obika S (2011) 2′–O,4′–C-methyleneoxymethylene bridged nucleic acids (2′,4′-BNA^{COC}). In: Goodchild J (ed) Therapeutic oligonucleotides. New York, Humana Press, pp 31–57

Honcharenko D, Varghese OP, Plashkevych O et al (2006) Synthesis and structure of novel conformationally constrained 1′,2′-azetidine-fused bicyclic pyrimidine nucleosides: their incorporation into oligo-DNAs and thermal stability of the heteroduplexes. J Org Chem 71:299–314

Honcharenko D, Barman J, Varghese OP et al (2007) Comparison of the RNase H cleavage kinetics and blood serum stability of the north-conformationally constrained and 2′-alkoxy modified oligonucleotides. Biochemistry 46:5635–5646

Honcharenko D, Zhou C, Chattopadhyaya J (2008) Modulation of pyrene fluorescence in DNA probes depends upon the nature for the conformationally restricted nucleotide. J Org Chem 73:2829–2842

IUPAC-IUB (1983) Joint Commission on Biochemical Nomenclature (JCBN). Abbreviations and symbols for the description of conformations of polynucleotide chains. Recommendations 1982. Eur J Biochem 131:9–15

Koshkin AA, Singh SK, Nielsen P et al (1998) LNA (locked nucleic acids): synthesis of the adenine, cytosine, guanine, 5-methylcytosine, thymine and uracil bicyclonucleoside monomers, oligomerisation, and unprecedented nucleic acid recognition. Tetrahedron 54:3607–3630

Kumar R, Hansen MH, Albæk N et al (2009) Synthesis of functionalized carbocyclic locked nucleic acid analogues by ring-closing diene and enyne metathesis and their influence on nucleic acid stability and structure. J Org Chem 74:6756–6769

Li Q, Yuan F, Zhou C et al (2010) Free-radical ring closure to conformationally locked r-L-carba-LNAs and synthesis of their oligos: nuclease stability, target RNA specificity, and elicitation of RNase H. J Org Chem 75:6122–6140

Li Q, Plashkevych O, Upadhayaya RS et al (2013) Supporting information to "Diastereomer-specific repertoire of 7′R- or 7′S-Me-carba-locked nucleic acids (cLNAs) in antisense oligo/RNA duplexes and engineering of physicochemical and enzymological properties". http://www.boc.uu.se/RS

Liu Y, Xu J, Karimiahmadabadi M et al (2010) Synthesis of 2′,4′-propylene-bridged (carba-ENA) thymidine and its analogues: the engineering of electrostatic and steric effects at the bottom of the minor groove for nuclease and thermodynamic stabilities and elicitation of RNase H. J Org Chem 75:7112–7128

Morita K, Hasegawa C, Kaneko M et al (2002) 2′-O,4′-C-ethylene-bridged nucleic acids (ENA): highly nuclease-resistant and thermodynamically stable oligonucleotides for antisense drug. Bioorg Med Chem Lett 12:73–76

Morita K, Takagi M, Hasegawa C et al (2003) Synthesis and properties of 2′-O,4′-C-ethylene-bridged nucleic acids (ENA) as effective antisense oligonucleotides. Bioorg Med Chem 11:2211–2226

Nishida M, Baba T, Kodama T et al (2010) Synthesis, RNA selective hybridization and high nuclease resistance of an oligonucleotide containing novel bridged nucleic acid with cyclic urea structure. Chem Commun 46:5283–5285

Novina CD, Sharp PA (2004) The RNAi revolution. Nature 430:161–164

Obika S, Nanbu D, Hari Y et al (1997) Synthesis of 2′-O,4′-C-methyleneuridine and -cytidine. Novel bicyclic nucleosides having a fixed C3′-endo sugar puckering. Tetrahedron Lett 38:8735–8738

Obika S, Nanbu D, Hari Y et al (1998) Stability and structural features of the duplexes containing nucleoside analogues with a fixed N-type conformation, 2′-O,4′-C-methyleneribonucleosides. Tetrahedron Lett 39:5401–5404

Peterson M, Bondensgaard K, Wengel J et al (2002) Locked nucleic acid (LNA) recognition of RNA: NMR solution structures of LNA: RNA hybrids. J Am Chem Soc 124:5974–5982

Pradeepkumar PI, Cheruku P, Plashkevych O et al (2004) Synthesis, physicochemical and biochemical studies of 1′,2′-oxetane constrained adenosine and guanosine modified oligonucleotides, and their comparison with those of the corresponding cytidine and thymidine analogs. J Am Chem Soc 126:11484–11499

Prakash TP, Siwkowski A, Allerson CA et al (2010) Antisense oligonucleotides containing conformationally constrained 2′,4′-(N-methoxy)aminomethylene and 2′,4′-aminooxymethylene and 2′-O,4′-C-aminomethylene bridged nucleoside analogues show improved potency in animal models. J Med Chem 53:1636–1650

Rahman SMA, Seki S, Obika S et al (2008) Design, synthesis, and properties of 2′,4′-BNANC: a bridged nucleic acid analogue. J Am Chem Soc 130:4886–4896

Rodriguez JB, Marquez VE, Nicklaus MC et al (1993) Synthesis of cyclopropane-fused dideoxycarbocyclic nucleosides structurally related to neplanocin-C. Tetrahedron Lett 34:6233–6236
Santoro SW, Joyce GF (1997) A general purpose RNA-cleaving DNA enzyme. Proc Natl Acad Sci U S A 94:4262–4266
Schubert S, Kurreck J (2004) Ribozyme- and deoxyribozyme-strategies for medical applications. Curr Drug Targets 5:667–681
Seth PP, Vasquez G, Allerson CA et al (2010a) Synthesis and biophysical evaluation of 2′,4′-constrained 2′O-methoxyethyl and 2′,4′-constrained 2′O-ethyl nucleic acid analogues. J Org Chem 75:1569–1581
Seth PP, Allerso CR, Berdeja A et al (2010b) An exocyclic methylene group acts as a bioisostere of the 2′-oxygen atom in LNA. J Am Chem Soc 132:14942–14950
Singh SK, Nielsen P, Koshkin AA et al (1997) Synthesis and conformation of 3′-O,4′-C-methyleneribonucleosides, novelbicyclic nucleoside analogues for 2′,5′-linked oligonucleotidemodification. Chem Commun 1643–1644
Singh SK, Kumar R, Wengel J (1998) Synthesis of 2′-amino-LNA: a novel conformationally restricted high-affinity oligonucleotide analogue with a handle. J Org Chem 63:10035–10039
Srivastava P, Barman J, Pathmasiri W et al (2007) Five- and six-membered conformationally locked 2′,4′-carbocyclic ribo-thymidines: synthesis, structure, and biochemical studies. J Am Chem Soc 129:8362–8379
Tarkoy ML, Leumann C (1993) Nucleic-acid analogs with constraint conformational flexibility in the sugar-phosphate backbone. 2. Synthesis and pairing properties of decanucleotides from (3′S,5′R)-2′-deoxy-3′,5′-ethano-beta-D-ribofuranosyladenine and thymine. Angew Chem Int Ed Engl 32:1432–1434
Thibaudeau C, Acharya P, Chattopadhyaya J (2005) Stereoelectronic effects in nucleosides and nucleotides and their structural implications. Uppsala University Press, Uppsala. ISBN:91-506-1351-0, available at http://www.boc.uu.se
Upadhayaya RS, Deshpande SG, Li Q et al (2011) Carba-LNA-^{5Me}C/A/G/T modified oligos show nucleobase-specific modulation of 3′-exonuclease activity, thermodynamic stability, RNA selectivity, and RNase H elicitation: synthesis and biochemistry. J Org Chem 76:4408–4431
Varghese OP, Barman J, Pathmairi W et al (2006) Conformationally constrained 2′-N,4′-C-ethylene-bridged thymidine (Aza-ENA-T): synthesis, structure, physical, and biochemical studies of Aza-ENA-T-modified oligonucleotides. J Am Chem Soc 128:15173–15187
Xu J, Liu Y, Dupouy C et al (2009) Synthesis of conformationally locked Carba-LNAs through intramolecular free-radical addition to C=N. Electrostatic and steric implication of the carba-LNA substituents in the modified oligos for nuclease and thermodynamic stabilities. J Org Chem 74:6534–6554
Zhou C, Chattopadhyaya J (2009) The synthesis of therapeutic locked nucleos(t)ides. Curr Opin Drug Discov Dev 12:876–898
Zhou C, Chattopadhyaya J (2010) Why carba-LNA-modified oligonucleotides show considerably improved 3′-exonuclease stability compared to that of the LNA modified or the native counterparts: a Michaelis-Menten kinetic analysis. J Org Chem 75:2341–2349
Zhou C, Plashkevych O, Chattopadhyaya J (2008) Unusual radical 6-endo cyclization to carbocyclic-ENA and elucidation of its solution conformation by 600 MHz NMR and ab initio calculations. Org Biomol Chem 6:4627–4633
Zhou C, Plashkevych O, Chattopadhyaya J (2009a) Double sugar and phosphate backbone-constrained nucleotides: synthesis, structure, stability, and their incorporation into oligodeoxynucleotides. J Org Chem 74:3248–3265
Zhou C, Liu Y, Andaloussi M et al (2009b) Fine tuning of electrostatics around the internucleotidic phosphate through incorporation of modified 2′,4′-carbocyclic-LNAs and -ENAs leads to significant modulation of antisense properties. J Org Chem 74:118–134

Challenges and Opportunities for Oligonucleotide-Based Therapeutics by Antisense and RNA Interference Mechanisms

Ramon Eritja, Montserrat Terrazas, Santiago Grijalvo, Anna Aviñó, Adele Alagia, Sónia Pérez-Rentero, and Juan Carlos Morales

Contents

1 Introduction 228
2 Synthesis of Nuclease-Resistant Oligonucleotides 229
  2.1 A Singular Double Modification at the Sense Strand 230
  2.2 Modified Nucleic Acid Derivatives with North Bicyclo[3.1.0]Hexane Pseudosugars 230
  2.3 N-Coupled Dinucleotide Units at the 3′-Terminal Positions of Oligonucleotides .. 232
3 Synthesis of Oligonucleotide Conjugates 234
  3.1 Synthesis of Oligonucleotide–Peptide Conjugates 234
  3.2 Synthesis of Oligonucleotide–Lipid Conjugates 236
  3.3 Synthesis of Oligonucleotide Conjugates with Intercalating Agents 237
  3.4 Synthesis of Oligonucleotide Conjugates with Carbohydrates 238
References 240

Abstract Oligonucleotide-based therapeutics may be one of the most promising approaches for the treatment of diseases. Although significant progress has been made in developing these agents as drugs, several hurdles remain to be overcome. One of the most promising approaches to overcome these difficulties is the preparation of modified oligonucleotides designed to increase cellular uptake and/or increase stability to nucleases. Herein, we report the developments done by our group in the synthesis of modified oligonucleotides directed to the generation of active compounds for gene inhibition. Specifically we will report the synthesis of novel nuclease-resistant oligonucleotides such as *North* bicyclo[3.1.0]hexane pseudosugars or *N*-coupled dinucleotide units. Also, the design of several siRNA conjugates carrying cell-penetrating peptides, lipids, intercalating agents, and

R. Eritja (✉) • M. Terrazas • S. Grijalvo • A. Aviñó • A. Alagia • S. Pérez-Rentero
IQAC-CSIC, CIBER-BBN, Jordi Girona 18-26, 08034 Barcelona, Spain
e-mail: recgma@cid.csic.es

J.C. Morales
IIQ-CSIC, Americo Vespucio 49, 41092 Sevilla, Spain

V.A. Erdmann et al. (eds.), *Chemical Biology of Nucleic Acids*, RNA Technologies, DOI 10.1007/978-3-642-54452-1_13, © Springer-Verlag Berlin Heidelberg 2014

carbohydrates will be described. Some of these novel derivatives show clear improvements in their biological and inhibitory properties.

Keywords RNA interference • Antisense oligonucleotides • siRNA • Oligonucleotide conjugates • Nuclease-resistant oligonucleotides

1 Introduction

In the past decades new compounds comprising small synthetic nucleic acids have shown promising results as potential drugs (Tiemann and Rossi 2009; Sanghvi 2011). In these cases, nucleic acids are used to inhibit a specific gene by blocking translation or transcription or by stimulating the degradation of a particular messenger RNA. Several strategies can be followed, antisense, short-interference RNA (siRNA), and aptamers being the most important ones. In the antisense strategy, synthetic oligonucleotides (ASOs) complementary to the messenger RNA of a given gene are used to inhibit the translation of messenger RNA to protein (Aboul-Fadl 2005; Chan et al. 2006). In the siRNA strategy, small RNA duplexes (siRNAs) complementary to messenger RNA bind to a protein complex named RNA-induced silencing complex (RISC). The complex formed by the antisense or guide RNA strand and RISC catalyzes the efficient degradation of a specific messenger RNA, thereby lowering the amount of target protein (Brumcot et al. 2006; Tiemann and Rossi 2009). Aptamers are nucleic acid sequences discovered by combinatorial methods that bind with high affinity and specificity to a particular protein. Aptamers can be seen as similar to the antibodies, but they are synthetically produced and made of nucleotides instead of amino acids (Brody and Gold 2000). At the moment, there is one nucleic acid commercialized for the treatment of macular degeneration (Gonzalez 2005). Macugen (Pegaptanib) is an aptamer functionalized with polyethylene glycol that binds to VEGF. An antisense oligonucleotide (Formivirsen, Vitravene) was approved by FDA for the treatment of cytomegalovirus, but it is not produced anymore (Sanghvi 2011). Recently, a new antisense oligonucleotide has been approved by FDA authorities for the treatment of familial hypercholesterolemia (Jiang 2013). Although, at present, there is no commercially available siRNA nearly 30 clinical trials have been started (Burnett and Rossi 2012; Rettig and Behlke 2012) showing an interest of pharmaceutical companies on the siRNA technology as potential strategy for the treatment of a disease caused by the overexpression of a particular protein. Importantly, the first evidence to show that siRNA can be administered intravenously to cancer patients with a significant tumor reduction has been recently published (Tabernero et al. 2013).

The siRNA technology has several advantages compared with the classical drug discovery process. The design of active siRNA is relatively simple. The siRNA is complementary to mRNA sequence and the sequence of the proteins (human genome) is known. There are several bioinformatic tools that can predict with success the optimal sequence for siRNA experiments. Generally the process of designing an active compound is simpler than in the classical drug discovery process. siRNA technology has an universality principle. Once an active compound is developed, other genes can be targeted in a similar way. Finally, there is the possibility of using similar approaches to the ones developed for siRNA to control microRNAs.

However, the siRNA and other nucleic acids technologies have several disadvantages. First the siRNA technology can only be used for silencing upregulated genes. Oligonucleotides are susceptible to degradation by exonucleases under physiological conditions and have low cellular uptake. In addition, they have to be directed to the right cells and some off-target effects have been detected. Finally, nucleic acids have shown to stimulate the innate immunity and they are expensive to produce.

Most of these disadvantages are similar to the ones described for the treatments with proteins or monoclonal antibodies and they may be partially solved by using modified nucleic acids. It has been demonstrated that it is possible to enhance stability to nucleases by introducing small modifications in the structure (Deleavey and Damha 2012; Shukla et al. 2010). In addition, several modifications that improve cellular uptake have been described and may provide specific delivery to target cells. In particular, the conjugation of lipids to oligonucleotides has been shown to generate molecules with improved inhibitory properties (Soutschek et al. 2004; Whitehead et al. 2009). A large development is being made to obtain new formulations that may direct nucleic acids to the specific cells by receptor-mediated uptake (Meares and Yokohama 2012). Finally, chemical modifications of siRNA have been shown to reduce the stimulation of the innate immune response (Sioud 2010; Eberle et al. 2008). However, still there is a need for further development in order to demonstrate the potential use of nucleic acids as drugs in the treatment of human diseases (Davidson and McCray 2011).

Herein, we report the developments done by our group in the synthesis of modified nucleic acids to generate active compounds for gene inhibition. Specifically, we will report the synthesis of novel nuclease-resistant oligonucleotides such as *North* bicyclo[3.1.0]hexane pseudosugars or *N*-coupled dinucleotide units. Also, the design of several siRNA conjugates carrying cell-penetrating peptides, lipids, intercalating agents, and carbohydrates will be described.

2 Synthesis of Nuclease-Resistant Oligonucleotides

The introduction of modifications at the sugar phosphodiester backbone of DNA and RNA has been demonstrated to provide stability to nuclease degradation in oligonucleotides. For example, the phosphodiester backbone of the FDA-approved oligonucleotide aptamer Macugen is modified with phosphorothioate linkages.

In the case of siRNA a large number of modifications are detrimental for the recognition of the duplex RNA to the inhibitory complex RISC and for this reason researchers are using small number of modifications (Deleavey and Damha 2012). Endogenous siRNAs are 21-nt dsRNAs with a 19-nt central duplex and 2-nt 3′ overhangs on each strand. Most of the chemically synthesized siRNAs described in the literature mimicked the natural RISC substrate where the 2-nt 3′-RNA overhangs are replaced with DNA typically via a TT dimer. Some authors have confirmed that the single-stranded 3′-overhangs are highly susceptible to nuclease attack. Protection of this particular site as well as the 5′-end is highly desirable for long-term gene inhibition. In spite of the use of small number of modifications, it has been shown that they have a strong impact not only in the stability to nucleases but also to the reduction of innate immunostimulation.

2.1 A Singular Double Modification at the Sense Strand

In a recent study, we observed an extraordinary anti-inflammatory activity on macrophages induced by one of our chemically modified siRNAs (Ocampo et al. 2012). This compound contained 2′-O-methyl-RNA modifications at the 5′ and a propanediol molecule at the 3′-end, both modifications in the passenger strand. This siRNA was selected for a preclinical study in a mouse model of inflammatory bowel disease (IBD). Local administration of the modified siRNA resulted in extraordinary anti-inflammatory activity. A gene array study on siRNA-treated animals confirmed that anti-inflammatory activity is the result of a reduced inflammatory process caused by the specific action of the siRNA targeting tumor necrosis factor (TNF-α) The exceptional anti-inflammatory activity of this siRNA can be explained by three main reasons: (a) Both of the ends of the passenger strand of the siRNA are protected with exonuclease-resistant modifications, (b) both modifications have very low stimulation of the innate response, and (c) the modification at the 5′-end of the passenger strand prevent the cellular phosphorylation of this end while the unmodified 5′-end of the guide strand is phosphorylated. The unphosphorylated 5′-end of the passenger strand prevents this strand to be bound to RISC avoiding unproductive RNA–RISC complexes and off-target effects.

2.2 Modified Nucleic Acid Derivatives with North Bicyclo [3.1.0]Hexane Pseudosugars

In a different approach, we explored new siRNA analogues with altered ribose rings. It is well known that for efficient gene silencing to take place, the guide siRNA–mRNA duplex must adopt an A-type helical structure (Chiu and Rana 2003). This has prompted several research groups to develop nucleotide analogues with the sugar

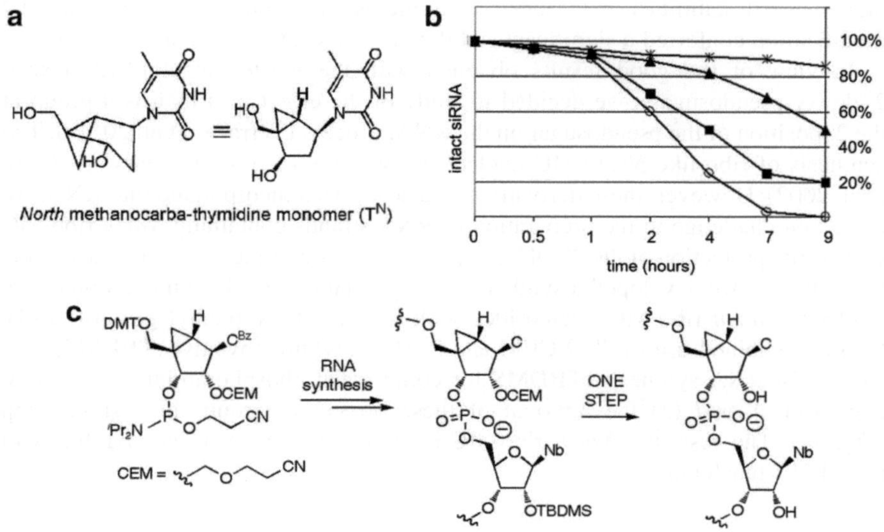

Fig. 1 (**a**) *North* methanocarba 2′-deoxynucleoside used in our first series of studies. (**b**) Serum stabilities of unmodified and T^N-modified siRNAs. siRNAs containing no modifications (*open circles*), six T^N substitutions in the guide strand (*asterisks*), two T^N substitutions in the 3′-dinucleotide overhangs (*filled squares*), and one T^N substitution at position 2 in the sense strand (*filled triangles*) were incubated in human serum (50 %) at 37 °C and withdrawn at the indicated time points. Plots of the remaining intact siRNAs (%) against incubation time. (**c**) Synthetic approach for the preparation of siRNAs containing ribo-like *North* MC nucleoside units

puckering locked in the North conformation and to incorporate them into siRNAs. Examples are Locked Nucleic Acids (LNA) (Singh et al. 1998), which have been found to be accepted by the RNAi machinery and to increase the thermodynamic and serum stability of RNA duplexes to a great extent (Elmén et al. 2005).

With the aim of introducing new features into siRNAs without disrupting the A-type helical structure required for the inhibitory activity, we replaced the sugar ring of natural ribonucleotides of RNA strands by a *North* locked bicyclo[3.1.0] hexane pseudosugar (Marquez et al. 1996). In particular, we incorporated *North* 2′-deoxy-methanocarba (MC)-thymidine derivatives (T^N) (Fig. 1a) into different positions of the guide and the sense strand of an siRNA targeting *Renilla* luciferase. The results of our studies suggested that, in general, incorporation of one to three modifications in the guide strand is well tolerated by the RNAi machinery, with gene silencing activities comparable to those obtained with LNA-modified siRNAs (Terrazas et al. 2011b). Furthermore, two of the most promising siRNA designs that emerged from these studies (one T^N substitution at position 20 in the guide strand and one T^N substitution at position 2 in the sense strand) could be successfully used for targeting the murine tumor necrosis factor (TNF-α) in RAW cells. In this work we also demonstrated that incorporation of T^N units into siRNA duplexes increased their thermal stabilities and decreased immunostimulation. Furthermore, the T^N modification conferred a significant stabilization of siRNAs in serum (Fig. 1b).

In view of the good results obtained with the *North* bicyclo[3.1.0]hexane 2′-deoxy-pseudosugars, we decided to study of the effect of a hydroxyl group at the 2′ position of the pseudosugar on the RNAi process (Terrazas et al. 2011a). The synthesis of ribo-like *North* MC nucleosides was reported a few years ago (Kim et al. 2002). However, these derivatives had never been incorporated into RNA. An important challenge in the preparation of RNA strands containing *North* ribo-MC units is the protection of the 2′-OH group. With the aim of incorporating these units into siRNAs, we developed a synthetic strategy that involved (1) protection of the 2′-OH of *North* ribo-MC nucleosides with a cyanoethoxymethyl group (CEM), (2) the combined use of 2′-*O*-CEM and 2′-*O*-*t*-butyldimethylsilyl (TBDMS) protection for RNA synthesis (TBDMS for commercial phosphoramidites of natural nucleosides), and (3) the removal of these protecting groups in a single step (Fig. 1c). The resulting *North* ribo-MC-modified siRNAs were compatible with the siRNA machinery.

2.3 *N*-Coupled Dinucleotide Units at the 3′-Terminal Positions of Oligonucleotides

On the other hand, our group has also been interested in modifying the 3′-terminal positions of therapeutic oligonucleotides. In particular, in a recent study, we created and analyzed a new class of modification aimed at increasing the stability of oligonucleotides against 3′-exonuclease degradation (the predominant nuclease activity present in serum) without affecting biological action (Terrazas et al. 2013). Rational design showed the possibility of blocking the hydrolytic activity of 3′-exonucleases by creating a new nucleotide scaffold characterized by its lack of phosphodiester bond linking the two 3′-terminal nucleotide building blocks. Our approach was based on the replacement of the two 3′-terminal nucleotides of an oligonucleotide strand (linked through a 3′–5′ phosphodiester bond) by two nucleotide units linked together by an ethyl chain through the exocyclic amino group of the nucleobase. The resulting dimeric nucleoside [N^4-ethyl-N^4 2′-deoxy-5-methylcytidine derivative (BC)] (Fig. 2a) was connected to the oligonucleotide through a normal phosphodiester bond. Molecular dynamics simulations of a 3′-BC-modified DNA: 3′-exonuclease (Klenow Fragment of *E. coli* DNA polymerase I) complex suggested that this kind of modification had negative effects on the correct positioning of the adjacent phosphodiester bond at the active site of the enzyme, due to steric clashes between the alkyl linker and amino acid residues (Leu361) (Fig. 2b).

We verified that functionalization of the 3′-ends of DNA and RNA strands with BC modifications completely blocked the hydrolytic activity of 3′-exonucleases (KF and snake venom phosphodiesterase). Interestingly, the *N*-ethyl-*N* modification

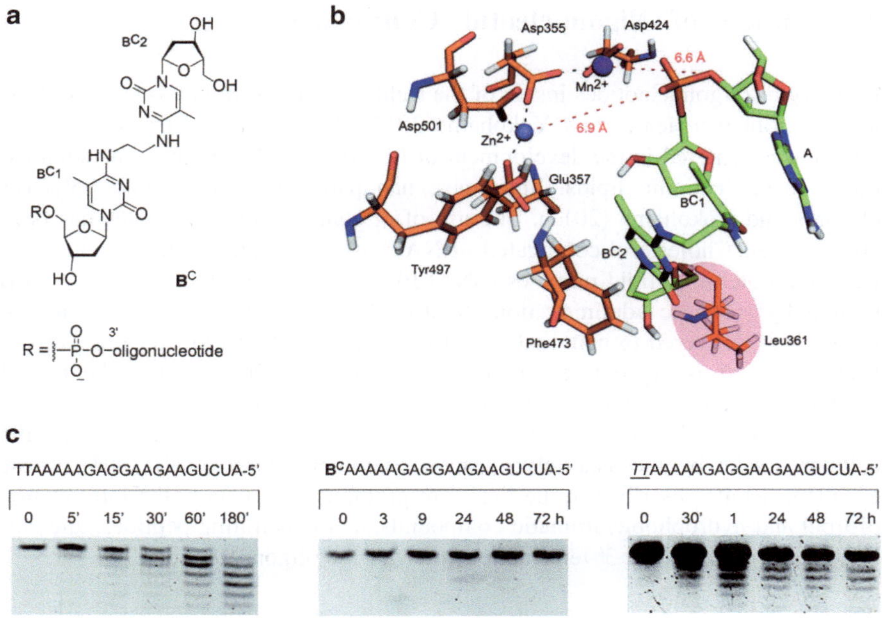

Fig. 2 (**a**) N^4-ethyl-N^4 dimeric pyrimidine nucleoside used in our studies. (**b**) Representative snapshot from the MD trajectory showing the position of relevant KF amino acid residues and the 3′-BC-modified DNA trimer ApBC1-ethyl-BC2. (**c**) 20 % Denaturing polyacrylamide gels depicting the course of the snake venom phosphodiesterase-catalyzed hydrolysis of unmodified, 3′-BC-modified, and 3′-PS-modified single-stranded RNAs

confers higher 3′-exonuclease resistance than phosphorothioate bonds (Fig. 2c). Furthermore, RNA interference experiments with BC-modified siRNAs targeting a luciferase gene and an antiapoptotic gene demonstrated that this modification was accepted by the RNAi machinery.

It has been reported that most of the residues at the 3′–5′-exonuclease active site of DNA polymerase I (among them, Leu361) are conserved among the 3′-exonucleolytic domain of other enzymes that catalyze 3′-exonuclease reactions (Bernad et al. 1989), thus suggesting that the KF can be used as a model to study the effect of oligonucleotide modifications on 3′-exonuclease-catalyzed hydrolysis. Indeed, recent studies have demonstrated that the structures of one of the most abundant mammalian 3′-exonucleases (TREX-1) bound to DNA are closely related to the structures of KF:DNA complexes (Brucet et al. 2007). Thus, the studies performed in this work not only provide a deeper insight into the role of nucleobase–protein interaction on 3′-exonuclease function, but can also help to design new 3′-exonuclease-resistant potential therapeutic agents.

3 Synthesis of Oligonucleotide Conjugates

Delivery of oligonucleotides inside of the right cells in the right tissues is still the largest problem (Meares and Yokohama 2012). Most of the effort done in this direction is centered in the development of formulations for intravenous administration using cationic lipids, liposomes, nanoparticles, and cationic polymers (Meares and Yokohama (2012). The use of oligonucleotide conjugates has been also reported. Cholesterol-conjugated siRNAs have been successfully used together with lipoproteins or lipid particles (SNALP) for the in vivo inhibition of Apo B protein by systemic administration (Soutschek et al. 2004). In another report, siRNA was delivered to neuronal cells by conjugation to a short peptide derived from rabies virus glycoprotein peptide (Kumar et al. 2007). Also, the use of exogenous ligands containing multivalent *N*-acetylgalactosamine clusters allows the intravenous administration of nucleic acids directed to liver cells by specific cellular uptake by the asialoglycoprotein receptors (Maier et al. 2003). These excellent results described in the literature prompted us to study the introduction of lipid and hydrophobic/aromatic compounds, cell-penetrating peptides, and carbohydrates on the 5′- or 3′-terminal positions of the oligonucleotides.

3.1 Synthesis of Oligonucleotide–Peptide Conjugates

Oligonucleotide–peptide conjugates are chimeric molecules composed of a nucleic acid moiety covalently linked to a polypeptide moiety. Oligonucleotide–peptide conjugates can exhibit different properties depending on the peptide sequence and its nature. For instance, if oligonucleotides are covalently linked to neutral peptides, conjugates are often more resistant to nucleases than unmodified oligonucleotides (Robles et al. 1997), whereas in the case of using cationic peptides with oligonucleotides, they can accelerate duplex formation (Corey 1995). From a therapeutic point of view, the use of cationic peptides such as polylysine, arginine, basic, or fusogenic peptides conjugated to oligonucleotides have shown an improved cellular uptake (Said et al. 2010).

To date, several protocols for the chemical synthesis of oligonucleotide–peptide conjugates have been described (Lu et al. 2010; Aviñó et al. 2011a). Conjugation of these large and functionalized molecules is a difficult task and often hindered by several side reactions. There are two main approaches to prepare these conjugates. The first approach is the post-synthetic coupling in solution that is accomplished by a chemoselective ligation mediated by mutually reactive groups that are introduced into the oligonucleotides and peptides. In the second approach, the stepwise solid-phase method, the peptide and oligonucleotide fragments are usually assembled sequentially on the same solid support. Unfortunately, the chemistries of peptide and oligonucleotide synthesis are not compatible; thus modifications of standard protecting groups or activating and deblocking agents are required. In general, the

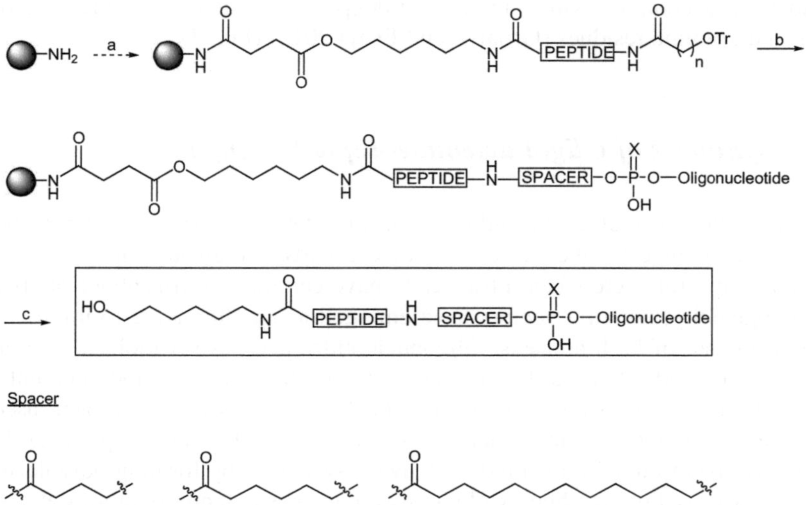

Fig. 3 Synthetic strategy for the synthesis of oligonucleotide–peptide conjugates containing RNA or DNA oligonucleotides through stepwise approach. Reagents and conditions: (a) "Boc-Fmoc" chemistries; (b) DNA synthesis; (c) i. NH_3 (32 %), 55 ºC, overnight, ii. HPLC purification

peptide moiety is synthesized first using tert-butoxycarbonyl (Boc)-amino acids with side chains protected with base-labile groups such as trifluoroacetyl (TFA) or 9-fluorenylmethoxycarbonyl (Fmoc) groups. The oligonucleotide sequence is then assembled using standard DNA synthesis protocols (Fig. 3). The use of stepwise synthesis has allowed us to synthesize oligonucleotide–peptide conjugates which were evaluated in both RNA interference and antisense therapies. In the first case, siRNA duplexes directed against tumor necrosis factor (TNF-α) and carrying a nuclear localization peptide were synthesized (Aviñó et al. 2009). Experiments with lipofectamine confirmed that siRNA duplexes of these conjugates were able to trigger RNAi pathway, thereby silencing gene expression with results similar to those of 3′-cholesterol-modified siRNA duplexes.

Finally, in the case of antisense strategy, we were able to conjugate cell-penetrating peptides [$(Lys)_n$; $(Arg)_n$; $n = 2, 4, 8$; and sweet arrow peptide, SAP $(VXLPPP)_3$, where X = Lys, Orn, homoarginine (HArg), and Arg] with an antisense oligonucleotide which knocked down the expression of *Renilla* luciferase (Grijalvo and Eritja 2012; Grijalvo et al. 2010b). For oligonucleotide–peptide conjugates containing guanidinium groups (X = HArg and Arg, respectively) a post-synthetic approach was carried out by performing a guanidinylation reaction with *O*-methylisourea. The linkage between the antisense oligonucleotide and cell-penetrating peptides was accomplished by introducing alkyl chain spacers with different lengths (4, 6, and 12 atoms of carbons). All antisense conjugates synthesized did not disrupt the antisense mechanism, showing results similar to those of the unmodified oligonucleotide. In order to impart cellular uptake, we observed better transfection efficiencies without using a transfection agent when SAP peptide could form a complex with oligonucleotide–peptide conjugates containing ornithine and arginine residues (Grijalvo and Eritja 2012) (Fig. 3).

3.2 Synthesis of Oligonucleotide–Lipid Conjugates

The development of effective and efficient drug delivery systems continues being the main challenge for the success of nucleic acids in clinical trials. Taking into account that viral vectors can intrinsically have certain concerns about safety and immunogenicity, nonviral vectors have emerged as an alternative to improve the cellular uptake of both antisense oligonucleotides and siRNA molecules in antisense therapy and RNA interference, respectively. Specifically, lipids in combination with nucleic acids are currently one of the strategies that have been used to improve the delivery of nucleic acids (Raouane et al. 2012). Two approaches have been described to obtain this kind of delivery system: i. by forming polyplexes or ii. by covalently linking, thereby obtaining conjugates potentially more stable than those formed electrostatically. Moreover, this kind of conjugation may also prolong the half-life of these conjugates in plasma or even increasing the efficiency of conjugates in some therapies, thereby improving their pharmacological and gene silencing properties in vitro and in vivo. In light of the results, our group has focused attention on introducing series of modified lipids with different length and properties covalently linked with antisense oligonucleotides and siRNA molecules at the 5′- or 3′-termini instead of using lipolexes. Thus, for the synthesis of lipid-3′-conjugates, controlled pore glass (CPG) solid supports were properly functionalized using glycerol molecule as a building block. Lipid modifications are depicted in Fig. 4 which consisted of a saturated hydrocarbonated chain (3′-lipid-C_{14}), a double-unsaturated hydrocarbonated chain (3′-lipid-C_{18}), or several aminoalkyl moieties that contained an amino group alone (3′-lipid-$C_{12}NH_2$) or with a combination of triazolyl and amino groups (3′-lipid-$C_{12}NH_2$-triazolyl). For the synthesis of lipid-5′-conjugates, a double-tailed lipid (5′-lipid-C_{28}) was introduced covalently by using the well-known phosphoramidite approach (Grijalvo et al. 2010a; Grijalvo et al. 2011).

RNA interference experiments of the aforementioned lipid–siRNA conjugates were performed targeting the TNF-α gene. According to transfection experiments in the presence of commercially available transfection agents, our proposed lipid modifications were accepted by the RNAi machinery, thereby obtaining results similar to those of siRNAs without any kind of modification. On the other hand, in order to evaluate the effectiveness of our lipid–siRNA conjugates to impart cellular uptake, transfection experiments were performed in the absence of a transfection agent. Here, we observed promising gene silencing results when 5′-lipid-C_{28} siRNA conjugate was used with silencing activities close to 50 %. Thus, this result allowed us to characterize further the biophysical aspects of the aforementioned conjugate by introducing the 5′-lipid-C_{28} modification into a DNA sequence which served us as a model. The interaction between 5′-lipid-C_{28}-DNA with the cell membrane was

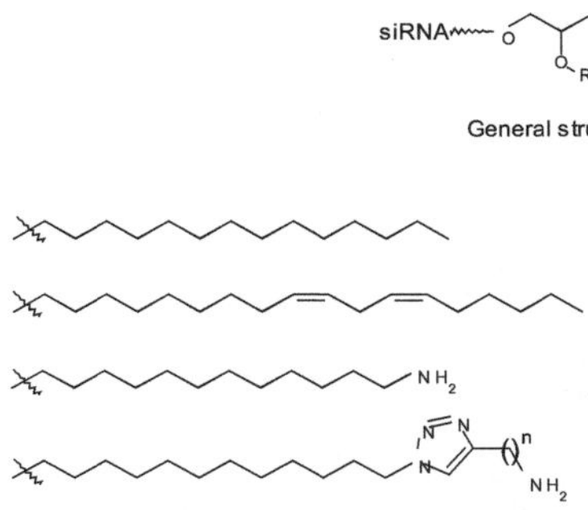

Fig. 4 siRNA conjugates containing a series of lipids at both 3′- and 5′-termini

studied in different cell lines by using three model membrane systems (e.g., monolayers; giant unilamellar vesicles, GUVs; and supported planar bilayers, SPBs). In all cases, a better incorporation into both lipid model membranes and cell systems was showed (Ugarte-Uribe et al. 2013).

3.3 Synthesis of Oligonucleotide Conjugates with Intercalating Agents

Small molecules with affinity to nucleic acids by binding to the major or minor groove or by intercalation between bases have received considerable interest to increase the affinity of inhibitory nucleic acids to their targets (Lönnberg 2009; Manoharan 2002). These conjugates are not directed to enhance cellular uptake or to increase cell-type specific targeting, but they can stabilize nucleic acids interactions with functional π-electron systems since their π–π interactions may provide additional binding energy (Pérez-Rentero et al. 2012; Aviñó et al. 2010). Conjugation at any terminus of oligoribonucleotides could also prevent degradation by nucleases. In addition, these molecules are constituted of chromophores that could be used to study intracellular distribution of nucleic acids.

Duplex siRNA is formed of two complementary strands (sense and antisense) so there are four terminal ends for potential conjugation sites. Previous studies have demonstrated that 3′ and 5′ ends of the sense strand and the 3′ end of the antisense are considered the best sites for conjugation without decreasing RNAi activity. In our group, threoninol was used as a linker to incorporate different modifications at either 3′ or 5′-end of siRNA (Fig. 5).

The effect of aromatic derivatives of different size at the 3′-end of siRNA on RNAi activity was studied. In general, the modifications in the sense strand were well tolerated by the RNAi machinery, but in the antisense strand the inhibition depends on the size of the introduced modification (Somoza et al. 2010).

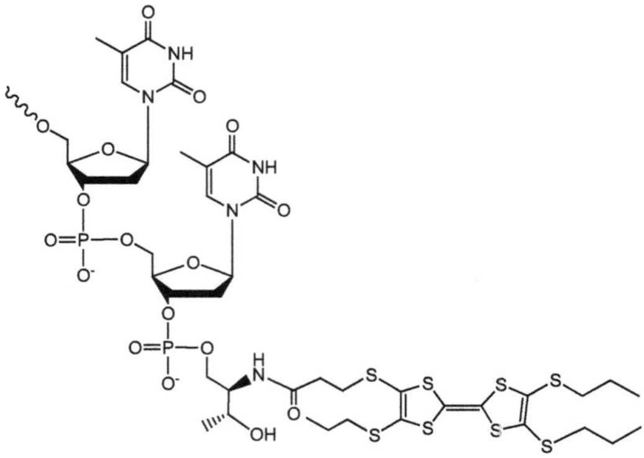

Fig. 5 Attachment of the tetrathiafulvalene (TTF) group at the 3′ overhang of an siRNA using the threoninol linker (Pérez-Rentero et al. 2013)

Oligonucleotide conjugates containing single or multiple acridine or quindoline derivatives have been synthesized and evaluated for siRNA studies against TNF-α protein, which is involved in apoptosis, inflammation, and immunity processes (Aviñó et al. 2012). Multiple acridine or quindoline moieties do not interfere in the RNAi pathway to silence gene expression but have similar efficiency compared to single modifications. No inhibition of siRNA conjugates was observed when cells were transfected without oligofectamine. Fluorescent properties of intercalating agents could be used to study cell delivery. Conjugates of siRNA modified with acridine showed good cellular uptake properties in HeLa cells.

Also the effect of a tetrathiafulvalene (TTF) unit at the 3′-ends of siRNAs was studied by analyzing the inhibition of the luciferase gene in HeLa cells (Pérez-Rentero et al. 2013). The introduction of TTF at the 3′-end can be tolerated by the RNAi machinery only if the passenger strand of an RNA duplex is modified. Although the results were not very promising, the hydrophobicity of these derivatives can provide some extra advantage in the preparation of lipid formulations.

3.4 Synthesis of Oligonucleotide Conjugates with Carbohydrates

The idea of attaching carbohydrates to ODN, RNA, and PNA is to exploit the potential of receptor-mediated endocytosis to improve cellular uptake. The use of sugar units as vectors was first reported under conjugation to oligonucleotide methyl phosphonates by Hangeland et al. (1995). In that case a trident presentation of *N*-acetylgalactosamine (GalNAc) covalently linked to the ODN allowed efficient delivery to human hepatocellular carcinoma cells (Hep G2). Moreover, the uptake was cell-type specific when compared to human fibrosarcoma (HT 1080) and human promyelocytic leukemia (HL-60) cells. The observed specificity is possible due to the presence of asialoglycoprotein (ASGP) receptor in hepatocytes that selectively bind GalNAc moieties. A similar strategy was described by Hamzavi et al. (2003) by attaching a dual presentation of GalNAc to an antisense PNA. Efficient cell uptake and a significantly improved ability of the PNA to inhibit mRNA levels were observed. In the case of siRNA, lactose was the first carbohydrate to be conjugated via an acid-labile spacer (Oishi et al. 2005). Lac-siRNA conjugates were included into polyion complex micelles to improve delivery and stability to serum incubation. They exhibited notable gene silencing for firefly luciferase expression in HuH-7 cells possessing ASGP receptors, which recognize compounds bearing terminal galactose moieties (lactose is β-D-galactopyranosyl-(1→4)-D-glucose disaccharide).

Our group decided to examine glucose or glucose-containing saccharides at the nonreducing end as possible vectors to facilitate oligonucleotide cell entrance. We reasoned that oligonucleotide conjugates carrying glucose moieties could bind to GLUT receptors and facilitate internalization via receptor-mediated endocytosis. We prepared a series of 3′-fluorescently labeled DNA conjugates with carbohydrates attached at the 5′-end via two types of spacers and a dendron scaffold in order to probe a diversity of sugar presentations (Ugarte-Uribe et al. 2010). The synthesis was carried out using carbohydrate phosphoramidite derivatives and standard solid-phase oligonucleotide synthetic procedures. Flow cytometry analysis showed a more efficient uptake in both HeLa and U87.CD4.CXCR4 cells for oligonucleotides containing a terminal glucose unit linked through a long tetraethylene glycol spacer or just one doubler dendrimer than for other glycoconjugates or the unconjugated control. In contrast, a conjugate containing four glucose units showed lower cell uptake in both cell types.

Next, we studied carbohydrate–siRNA conjugates carrying one, two, or four glucose and galactose residues at the 5′-end of the passenger strand of the siRNA duplex (Aviñó et al. 2011b). The tumor necrosis factor (TNF-α) gene was selected as target and was introduced into HeLa and Huh7 cells. When carbohydrate–siRNA conjugates were transfected with oligofectamine similar inhibitory properties to those of unmodified RNA duplexes were observed. When no oligofectamine was used, no inhibition was observed for glucose–siRNA conjugates. However, siRNA carrying galactose residues have slight anti-TNF inhibitory properties (25 % in the best case) when tested on HuH-7 cells without transfecting agent. This inhibition is eliminated if the ASGP receptors are blocked using a competitor. These results indicated a possible mediation of cellular ASGP receptors in the uptake of galactose–siRNA conjugates.

Recently we tried a different approach by investigating apolar carbohydrates linked to siRNA duplexes as new hydrophobic platforms that could improve the oligoribonucleotide cell uptake without disrupting the RNAi machinery during gene inhibition (Vengut-Climent et al. 2013). The design consisted of permethylated glucose covalently linked to the 5′-end of the passenger strand of the siRNA via two different spacers and a dendron scaffold. The synthesis proceeded using the corresponding apolar carbohydrate phosphoramidites and standard solid-phase oligonucleotide automatic synthesis. These modifications did not alter the RNAi machinery on HeLa cells in the dual luciferase assay when conjugates were transfected with oligofectamine. When no transfection agent was used, only some apolar carb–siRNAs showed gene inhibition, up to 26 % in the case of the double-tailed permethylated glucose–siRNA. Most importantly, permethylated glucose C2, permethylated glucose C12, and the corresponding glucose C12 doubler siRNA conjugates presented substantial stability against 5′-exonuclease degradation.

Acknowledgments This research was supported by the European Commission (Grants FP7-FUNMOL 213382 and NMP4-LA-2011-262943, MULTIFUN), by the Spanish Ministry of Education (grant CTQ2010-20541, CTQ2009-13705), by CSIC (intramural PIF06-045), and by the Generalitat de Catalunya (2009/SGR/208). We would like to thank our collaborators for their continuous support on the synthesis and evaluation of the biophysical and biological properties of modified oligonucleotides, especially Dr. S. Ocampo, Dr. J.C. Perales, Dr. E. Fernandez, Dr. C. Romero, Dr. B. Uriarte-Uribe, Dr. I. Alkorta, Dr. F. Goñi, Dr. V.E. Marquez, and Dr. M. Orozco.

References

Aboul-Fadl T (2005) Antisense oligonucleotides: the state of the art. Curr Med Chem 12:2193–2214

Aviñó A, Ocampo SM, Caminal C et al (2009) Stepwise synthesis of RNA conjugates carrying peptide sequences for RNA interference studies. Mol Divers 13:287–293

Aviñó A, Ferreira R, Mazzini S et al (2010) Synthesis and structural properties of oligonucleotides covalently linked to acridine and quindoline derivatives through a threoninol linker. Bioorg Med Chem 18:7348–7356

Aviñó A, Grijalvo S, Pérez-Rentero S et al (2011a) Synthesis of oligonucleotide-peptide conjugates for biomedical and technological applications. Methods Mol Biol 751:223–238

Aviñó A, Ocampo SM, Lucas R et al (2011b) Synthesis and in vitro inhibition properties of siRNA conjugates carrying glucose and galactose with different presentation. Mol Divers 15:751–757

Aviñó A, Ocampo SM, Perales JC et al (2012) Synthesis and in vitro inhibition properties of siRNA conjugates carrying acridine and quindoline moieties. Chem Biodivers 9:557–566

Bernad A, Blanco L, Lázaro JM et al (1989) A conserved 3′-5′ exonuclease active site in prokaryotic and eukaryotic DNA polymerases. Cell 59:219–228

Brody EN, Gold L (2000) Aptamers as therapeutic and diagnostic agents. J Biotechnol 74:5–13

Brucet M, Querol-Audí J, Serra M et al (2007) Structure of the dimeric exonuclease TREX1 in complex with DNA displays a proline-rich binding site for WW domains. J Biol Chem 282:14547–14557

Brumcot D, Manoharan M, Koteliansky V et al (2006) RNAi therapeutics: a potential new class of pharmaceutical drugs. Nat Chem Biol 2:711–719

Burnett JC, Rossi JJ (2012) RNA-based therapeutics: current progress and further prospects. Chem Biol 19:60–71

Chan JH, Lim S, Wong WS (2006) Antisense oligonucleotides: from design to therapeutic application. Clin Exp Pharmacol Physiol 33:533–540

Chiu Y-L, Rana TM (2003) siRNA function in RNAi: a chemical modification analysis. RNA 9:1034–1048

Corey DR (1995) 48000-Fold acceleration of hybridization by chemically-modified oligonucleotides. J Am Chem Soc 117:9373–9374

Davidson BL, McCray PB Jr (2011) Current protocols for RNA interference-based therapies. Nat Rev Genet 12:329–340

Deleavey JF, Damha MJ (2012) Designing chemically modified oligonucleotides for targeted gene silencing. Chem Biol 19:937–954

Eberle F, Giessler K, Deck C et al (2008) Modifications in small interfering RNA that separate immunostimulation from RNA interference. J Immunol 180:3229–3237

Elmén J, Thonberg H, Ljungberg K et al (2005) Locked nucleic acid (LNA) mediated improvements in siRNA stability and functionality. Nucleic Acids Res 33:439–447

Gonzalez CR (2005) Enhanced efficacy associated with early treatment of neovascular age-related macular degeneration with pegaptanib sodium: an exploratory analysis. Retina 25:815–827

Grijalvo S, Eritja R (2012) Synthesis and in vitro inhibition properties of oligonucleotide conjugates carrying amphipathic proline-rich peptide derivatives of the sweet arrow peptide (SAP). Mol Divers 16(2):307–317

Grijalvo S, Ocampo SM, Perales JC et al (2010a) Synthesis of oligonucleotides carrying amino lipid groups at the 3′-end for RNA interference studies. J Org Chem 75:6806–6813

Grijalvo S, Terrazas M, Aviñó A et al (2010b) Stepwise synthesis of oligonucleotide-peptide conjugates containing guanidinium or lipophilic groups in their 3′-termini. Bioorg Med Chem Lett 20(7):2144–2147

Grijalvo S, Ocampo SM, Perales JC et al (2011) Synthesis of lipid-oligonucleotide conjugates for inhibition of gene expression. Chem Biodivers 8:287–299

Hamzavi R, Dolle F, Tavitian B, Dahl O et al (2003) Modulation of the pharmacokinetic properties of PNA: preparation of galactosyl, mannosyl, fucosyl, *N*-acetylgalactosaminyl, and *N*-acetylglucosaminyl derivatives of aminoethylglycine peptide nucleic acid monomers and their incorporation into PNA oligomers. Bioconjug Chem 14:941–954

Hangeland J, Levis JT, Lee YC et al (1995) Cell-type specific and ligand specific enhancement of cellular uptake of oligodeoxynucleoside-methylphosphonates covalently linked with a neoglycopeptide, YEE(ah-Ga1NAc)s. Bioconjug Chem 6:695–701

Jiang K (2013) Biotech comes to its "antisense" after hard-won drug approval. Nat Med 19:252

Kim HS, Ravi RG, Marquez VE et al (2002) Methanocarba modification of uracil and adenine nucleotides: high potency of Northern ring conformation at P2Y1, P2Y2, P2Y4, and P2Y11 but not P2Y6 receptors. J Med Chem 45:208–218

Kumar P, Wu H, McBride JL, Jung KE, Kim MH, Davidson BL, Lee SK, Shankar P, Manjunath N (2007) Transvascular delivery of small interfering RNA to the central nervous system. Nature 448:39–43

Lönnberg H (2009) Solid-phase of oligonucleotide conjugates useful for delivery and targeting of potential nucleic acid therapeutics. Bioconjug Chem 20:1065–1094

Lu K, Duan QP, Ma L et al (2010) Chemical strategies for the synthesis of peptide-oligonucleotide conjugates. Bioconjug Chem 21:187–202

Maier MA, Yannopolus CG, Mohamed N et al (2003) Synthesis of antisense oligonucleotides conjugated to a multivalent carbohydrate cluster for cellular targeting. Bioconjug Chem 14:18–29

Manoharan M (2002) Oligonucleotide conjugates as potential antisense drugs with improved uptake, biodistribution, targeted delivery and mechanism of action. Antisense Nucleic Acid Drug Dev 129:103–128

Marquez VE, Siddiqui MA, Ezzitouni A et al (1996) Nucleosides with a twist. Can fixed forms of sugar ring pucker influence biological activity in nucleosides and oligonucleotides? J Med Chem 39:3739–3747

Meares CF, Yokohama M (2012) Introduction to gene silencing and delivery. Acc Chem Res 45 (7):959–1171, special issue in RNA delivery

Ocampo SM, Romero C, Aviñó A et al (2012) Functionally enhanced siRNA targeting TNFα attenuates DSS-induced colitis and TLR-mediated immunostimulation in mice. Mol Ther 20:382–390

Oishi M, Nagasaki Y, Itaka K et al (2005) Lactosylated poly(ethylene glycol)-siRNA conjugate through acid-labile β-thiopropionate linkage to construct pH-sensitive polyion complex

micelles achieving enhanced gene silencing in hepatoma cells. J Am Chem Soc 127:1624–1625
Pérez-Rentero S, Gállego I, Somoza A et al (2012) Interstrand interactions on DNA duplexes modified by TTF units at the 3′ or 5′-ends. RSC Adv 2:4069–4071
Pérez-Rentero S, Somoza A, Grijalvo S et al (2013) Biophysical and RNA Interference inhibitory properties of oligonucleotides carrying tetrathiafulvalene groups at terminal positions. J Chem 2013:article ID 650610, 11 pages. doi:10.1155/2013/650610
Raouane M, Desmaële D, Urbinati G et al (2012) Lipid conjugated oligonucleotides: a useful strategy for delivery. Bioconjug Chem 23:1091–1110
Rettig GR, Behlke MA (2012) Progress towards in vivo use of siRNAs-II. Mol Ther 20:483–512
Robles J, Mased M, Beltrán M et al (1997) Synthesis and enzymatic stability of phosphodiester-linked peptide-oligonucleotide hybrids. Bioconjug Chem 8:785–788
Said HF, Saleh AF, Abes R et al (2010) Cell penetrating peptides: overview and applications to the delivery of oligonucleotides. Cell Mol Life Sci 67:715–726
Sanghvi Y (2011) A status update of modified oligonucleotides for chemotherapeutics applications. Curr Protoc Nucleic Acid Chem Chapter 4:Unit 4.1.1-22
Shukla S, Sumaria CS, Pradeepkumar PI (2010) Exploring chemical modifications for siRNA therapeutics: a structural and functional outlook. ChemMedChem 5:328–349
Singh SK, Nielsen P, Koshkin AA et al (1998) LNA (locked nucleic acids): synthesis and high affinity nucleic acid recognition. Chem Commun 34:455–456
Sioud M (2010) Advances in RNA sensing by the immune system: separation of siRNA unwanted effects from RNA interference. Methods Mol Biol 629:33–52
Somoza A, Terrazas M, Eritja R (2010) Modified siRNAs for the study of the PAZ domain. Chem Commun 46:4270–4272
Soutschek J, Akinc A, Bramlage B et al (2004) Therapeutic silencing of an endogenous gene by systemic administration of modified siRNAs. Nature 432:173–178
Tabernero J, Shapiro GI, LoRuss PM et al (2013) First-in-man trial of an RNA interference targeting VEGF and KSP in cancer patients with liver involvement. Cancer Discov 3:406–417
Terrazas M, Aviñó A, Siddiqui M et al (2011a) A direct, efficient method for the preparation of siRNAs containing ribo-like *North* bicyclo[3.1.0]hexane pseudosugars. Org Lett 13:2888–2891
Terrazas M, Ocampo SM, Perales JC et al (2011b) Effect of *North* bicyclo[3.1.0]hexane pseudosugars on RNA interference. A novel class of siRNA modification. Chembiochem 12:1056–1065
Terrazas M, Alagia A, Faustino I et al (2013) Functionalization of the 3′-ends of DNA and RNA strands with *N*-ethyl-*N* bridged nucleosides: a promising approach to avoid 3′-exonuclease-catalyzed hydrolysis and to improve the biological properties of therapeutic oligonucleotides. Chembiochem 14:510–520
Tiemann K, Rossi JJ (2009) RNAi-based therapeutics-current status, challenges and prospects. EMBO Mol Med 1:142–151
Ugarte-Uribe B, Pérez-Rentero S, Lucas R et al (2010) Synthesis, cell-surface binding and cellular uptake of fluorescently labeled glucose DNA conjugates with different carbohydrate presentation. Bioconjug Chem 21:1280–1287
Ugarte-Uribe B, Grijalvo S, Busto JV et al (2013) Double-tailed lipid modification as a promising candidate for oligonucleotide delivery in mammalian cells. Biochim Biophys Acta 1830 (10):4872–4884
Vengut-Climent E, Terrazas M, Lucas R et al (2013) Synthesis, RNAi activity and nuclease-resistant properties of apolar carbohydrates siRNA conjugates. Bioorg Med Chem Lett 23:4048–4051
Whitehead KA, Langer R, Anderson DG (2009) Nat Rev Drug Discov 8:129–138

Progress in Chemically Modified Nucleic Acid Aptamers

Masayasu Kuwahara

Contents

1 SELEX Technology 244
 1.1 Background and Basic Principles 244
 1.2 Post-SELEX Modifications 246
 1.3 Simplification of SELEX Process 248
2 Polymerase and Substrate Development 249
 2.1 Problems with SELEX When Using Modified DNA/RNA 249
 2.2 Enzymatic Syntheses of Modified DNA/RNA 251
 2.3 Enzymatic Base-Modified DNA Syntheses 252
 2.4 Enzymatic BNA/LNA Syntheses 253
3 Chemically Modified DNA/RNA Aptamers 254
 3.1 RNA-Based Aptamers 254
 3.2 DNA-Based Aptamers 256
 3.3 XNA Aptamers 257
4 Efficacy of Introducing Foreign Functionalities 258
5 Future Outlook 263
References 264

Abstract Nucleic acid aptamers are DNA/RNA-based ligands that specifically bind to their targets. Unlike antibodies, they can be created using in vitro selection systems and produced on a large-scale level at low cost using solid-phase oligonucleotide syntheses. After the methodology of systematic evolution of ligands by exponential enrichment (SELEX) for DNA/RNA aptamer selection was established in the early 1990s, their applications for diagnostic and therapeutic drugs as well as research reagents and biosensors have attracted considerable attention. In 2004, the first aptamer medicine, MACUGEN® (pegaptanib sodium injection), a chemically

M. Kuwahara (✉)
Graduate School of Science and Technology, Gunma University, 1-5-1 Tenjin-cho, Kiryu, Gunma 376-8515, Japan
e-mail: mkuwa@gunma-u.ac.jp

V.A. Erdmann et al. (eds.), *Chemical Biology of Nucleic Acids*, RNA Technologies,
DOI 10.1007/978-3-642-54452-1_14,

modified RNA aptamer, was approved by United States Food and Drug Administration (U.S. FDA).

However, further improvements in biostability and functional expandability are needed in order to achieve modified nucleic acid aptamers with high binding affinities and specificities for a broad range of targets. Hence base-modified aptamers like SOMAmer™ and sugar-modified aptamers like xeno-nucleic acid (XNA) aptamers have currently come into the spotlight. Furthermore, the remarkable advances that have been made in recent years in sequencing technologies should significantly contribute to the development of nucleic acid aptamers as molecular tags for omics research and for use as clinical biomarkers.

This chapter focuses on the initial attempts, recent advances, and future prospects for modified nucleic acid aptamer development and the related technologies.

Keywords Systematic evolution of ligands by exponential enrichment • Nucleic acid aptamer • Xeno-nucleic acid • 2′,4′-Bridged/locked nucleic acids • Polymerase • Slow off-rate

1 SELEX Technology

1.1 Background and Basic Principles

After the discovery of RNA enzymes (ribozymes) that could catalyze RNA self-cleavage or RNA transesterification during splicing (Zaug and Cech 1986; Guerrier-Takada et al. 1983), considerable interest was focused on the creation and application of functional nucleic acids that could catalyze chemical reactions and specifically bind to target molecules. Around 1990, three research groups, Szostak et al., Joyce et al., and Gold et al. independently developed a selection methodology to recover RNA molecules that could catalyze a specific reaction (ribozyme) or bind to a specific molecule (aptamer) from an RNA library (RNA pool of miscellaneous random sequences) (Ellington and Szostak 1990; Tuerk and Gold 1990; Robertson and Joyce 1990). This method was called in vitro selection; when used especially for selecting aptamers it is often called systematic evolution of ligands by exponential enrichment (SELEX) (Gold et al. 1995).

First, selections were performed for RNA, after which the creation of a DNA enzyme as a catalyst and a DNA aptamer as a specific binder was attempted (Bock et al. 1992). Single-stranded DNA/RNA oligomers with particular sequences that were selected by in vitro selection methods could exert enzyme/aptamer activities. These were similar to those of protein enzymes and antibodies, as they formed specific steric structures with intramolecular hydrogen bonding, stacking interactions, electrostatic interactions, and metal coordination.

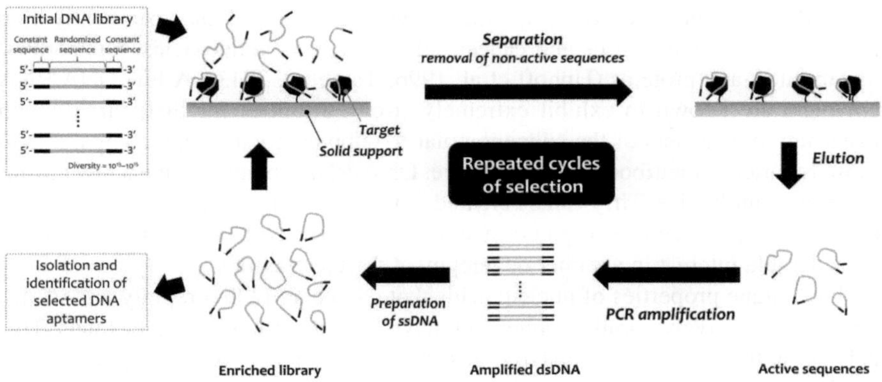

Fig. 1 Schematic illustration of typical SELEX for DNA aptamer selection

The basic principles of in vitro selection are quite simple. They involve multiple repetitions of two main procedures such as separation and amplification to enrich active sequences (Fig. 1). For example, for DNA aptamer selection, synthetic oligodeoxynucleotides (ODNs) that contain a randomized sequence flanked by constant sequences are used as the initial library. A randomized sequence of approximately 20–80 mer in length flanked by constant sequences of approximately 15–25 mer in length on each end is generally used.

An initial ODN library containing approximately 10^{13}–10^{15} different miscellaneous sequences is mixed with a target that is immobilized on a solid support such as a gel or a membrane. Furthermore, the sequences that do not exhibit binding affinity for the immobilized target are washed away and separated from the active sequences that do bind to the target. These active sequences are subsequently eluted by heat denaturation or by the addition of a denaturing agent for the target.

Subsequently, these sequences are amplified by polymerase chain reaction (PCR) to yield multiple copies accompanied by their complementary strands as double-stranded forms. These complementary strands can be selectively removed, by digestion with λ-exonuclease if they are 5′-phosphate-modified-primer-elongated products or by biotin–avidin techniques if they are 5′-biotin-modified-primer-elongated products. The obtained single stranded ODN library is purified and subsequently used for the next selection round. Thus, repeating these procedures can achieve further enrichment of active sequences. The numbers of repetitions depend on the selection system used and typically involve approximately 2–20 rounds.

The active sequences in the enriched library are isolated by cloning technology or by emulsion PCR technology to analyze their nucleotide sequence arrangements and to characterize their binding properties (Shao et al. 2011). For RNA aptamer selection, transcripts of ODNs that contained a randomized sequence region are used as the library, although a promoter sequence for transcription is required in the constant region in addition to primer regions for amplification.

To date, various DNA/RNA aptamers have been reported that specifically bind to targets involving viruses and cancer cells as well as amino acids, nucleotides, metabolites, and proteins (Uphoff et al. 1996; Tan et al. 2013). A few DNA/RNA aptamers are known to exhibit extremely strong affinities for their targets with dissociation constants of the subnanomolar to subpicomolar range, comparable to those reported for antibodies. Furthermore, DNA/RNA aptamers have a few advantages over antibodies. They can be created in vitro without using laboratory animals and can be produced by organic synthesis at a low cost, which has generated considerable interest in aptamer development for therapeutics.

The unique properties of nucleic acids that can be amplified rapidly and identified using massively parallel sequencing have been a particular area of focus. This is because these advantages make it possible to produce DNA/RNA aptamers that can be applied as molecular tags for quantitative and parallel analyses for multiple biomarkers (Ulrich and Wrenger 2009; Brody et al. 2010). However, for applications as medical treatments, enhancing nuclease resistance (i.e., biostability in serum or cells), controlling the pharmacokinetics, and expanding the fitness to a broader spectrum of targets with sufficient affinity are important issues. This final issue is a common concern for in vitro applications as diagnostic reagents and any other indicators. Thus, chemical modifications are required to overcome these problems.

1.2 Post-SELEX Modifications

One of the most difficult problems is applying chemically modified nucleic acids to the SELEX library. This is because modified nucleoside 5′-triphosphates are generally inferior substrates for polymerase reactions compared with the natural nucleoside 5′-triphosphates. In many cases, using chemically modified substrates decreases the amplification and reproduction efficiencies of selected active modified nucleic acids during each round. Post-SELEX modification is an alternative for introducing modified groups into DNA/RNA aptamers (Fig. 2). Some nucleotide residues in SELEX-selected aptamers can be replaced with modified nucleotide residues. However, these replacements can cause a loss of original binding activity, depending on the position and chemical structure of the foreign functionality that is introduced. Thus, in a post-SELEX method, chemical variants of DNA/RNA aptamers should be prepared and analyzed for their binding activity one by one.

Some successful examples of post-SELEX modifications have been reported (Kuwahara and Sugimoto 2010). A post-SELEX method was employed during the development of the fist aptamer drug, Macugen® (pegaptanib sodium injection), which is used for age-related macular degeneration (AMD) therapy (Ng et al. 2006). Pegaptanib is a human vascular endothelial growth factor (VEGF)-binding RNA aptamer that contains 2′-deoxy-2′-fluoro (2′-F) pyrimidine nucleotides (U, C) and 2′-deoxy-2′-methoxy (2′-OMe) purine nucleotides (A, G).

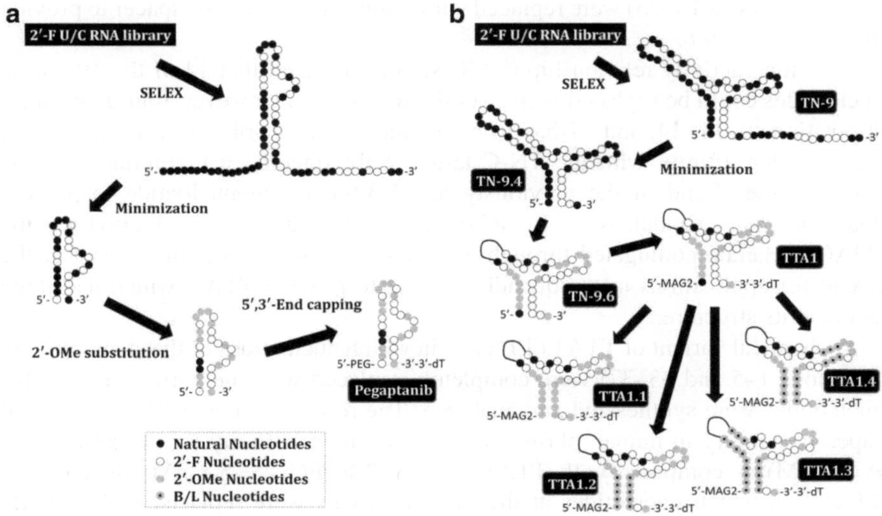

Fig. 2 Examples of post-SELEX modification; (**a**) Pegaptanib and (**b**) human tenascin-C-binding aptamers

In addition, a branched polyethylene glycol strand (40 kDa) and 3′-thymidylic acid are introduced at its 5′-and 3′-ends, respectively.

The prototype of this aptamer was obtained by SELEX using a modified RNA library that included 2′-F pyrimidine nucleotides. Subsequently, natural purine nucleotides (A, G) were replaced with 2′-OMe purine nucleotides to the greatest extent possible to enhance nuclease resistance and to remain intact under physiological conditions. Finally, these modifications were trimmed at both the 5′- and 3′-ends to prolong the circulation time of this aptamer in vivo and/or to enhance its exonuclease resistance.

Many types of sugar-modified nucleotide analogs have been developed that are effective for improving biostability without losing their hybridization properties to DNA/RNA (Barciszewski et al. 2009). Among these sugar modifications are 2′,4′-bridged/locked nucleic acids (2′,4′-BNA/LNA) that contain 2′-*O*,4′-*C*-methylene bridged/locked (B/L) nucleotides, which exhibit particularly excellent performance (Yamamoto et al. 2011; Lundin et al. 2013). Thus, the development of DNA/RNA aptamers that contain B/L nucleotides is significant when considering in vivo uses for nucleic acid aptamers.

Using post-SELEX modifications, Schmidt et al. successfully developed an LNA aptamer, TTA1, which was specific for human tenascin-C (TN-C) (Schmidt et al. 2004). A 71-mer of prototype TN-9 was selected from a modified RNA library that contained 2′-F pyrimidine nucleotides. This was then shortened by removing 16 nucleotides from the 3′-end, which were selected based on the predicted secondary structure (Hicke et al. 2001). In the sequence of the resulting 55-mer, TN-9.4, which could form a three-way junction structure, 17 nucleotides from the

5′-end (positions 10–26) were replaced with a single $(CH_2CH_2O)_6$ spacer to provide the 39-mer TN-9.6.

Structure–activity relationship (SAR) studies revealed that 14 of the 19 purine nucleotides could be replaced with 2′-OMe nucleotides. However, four guanosines at positions 9, 11, 14, and 17 had to be retained as the 2′-ribo form to maintain a high binding affinity with the TN-C target. Subsequently, a thymidine cap was added at the 3′-end of the previously noted 39-mer oligonucleotide to prevent digestion by exonucleases. In addition, a mercapto-acetyl-glycinyl-glycine (MAG2) chelate conjugated through a hexyl-amino linker was introduced at the 5′-end for applications in tumor radioimaging to provide TTA1, which had three stems in its structure.

A chemical variant of TTA1 (TTA1.2) in which nucleotides of the stem I moiety (positions 1–5 and 33–37) were completely replaced with the corresponding B/L nucleotides were synthesized and analyzed. The result was that TTA1.2 exhibited superior stability in human plasma and target binding affinity ($t_{1/2} = 53$ h, $EC_{50} = 2.0$ nM) as compared with TTA1 ($t_{1/2} = 42$ h, $EC_{50} = 5.8$ nM). In contrast, TTA1.1 in which nucleotides at the same positions were replaced with 2′-OMe nucleotides exhibited a >2-fold decrease in binding affinity compared with TTA1, although its stability was substantially improved ($t_{1/2} = 49$ h, $EC_{50} = 13.7$ nM). Incidentally, TTA 1.3 and TTA 1.4 in which the nucleotides of stems II or III in addition to the stem I modification of TTA1.2 were replaced with B/L nucleotides had no binding activity, although their plasma lifetimes were increased ($t_{1/2} = 69$ h and 72 h, respectively). Similarly, B/L nucleotides have been shown to be partially compatible with those in DNA aptamers.

Spiegelmer® technology may be categorized under post-SELEX modification method (Vater and Klussmann 2003). In this method, aptamer is selected from a natural DNA/RNA library by targeting a mirror image of the target, after which SELEX-selected D-DNA/RNA aptamers comprising β-D-ribofuranosyl units are synthetically converted into their mirror images, i.e., L-DNA/RNA aptamers comprising α-L-ribofuranosyl units, to fit to the target. Although preparations for mirror images of targets are difficult in some cases and viruses and cells cannot be treated as targets, L-DNA/RNA is known to have excellent nuclease resistance.

Thus, post-SELEX modifications can provide options for creating modified DNA/RNA aptamers. However, this still requires considerable time and effort because binding affinity can either be markedly decreased or non-existent, as noted above.

1.3 Simplification of SELEX Process

At present, several tens of modified SELEX methods have been devised (McKeague and Derosa 2012). To acquire aptamers more rapidly and conveniently, researchers attempted to reduce routine cycles of the SELEX process. One solution is to improve the precision of the separation process, i.e., how strictly inactive sequences can be excluded from active sequences. The other solution is to identify a method to analyze sequences and binding activities of the selected species such that aptamers can be recovered even though enrichment is immature.

Methods that require only a single selection round have been proposed and substantiated. For example, Berezovski et al. acquired DNA aptamers bound to MutS as a mismatch repair protein by non-SELEX selection using non-equilibrium capillary electrophoresis of equilibrium mixture method (Berezovski et al. 2006). In addition, Nitsche et al. (2007) acquired DNA aptamers bound to vaccinia virus particles by MonoLEX, which functions by segmenting the affinity column with cutting, after loading the library on the target immobilized column and washing with appropriate buffers. These methods may no longer be categorized under SELEX because the exponential enrichment by repeated cycles is not involved in the selection process. Although technical issues are yet to be overcome, these are considered to be favorable, particularly for selection of modified DNA/RNA aptamers. This is because the frequency of enzymatic modified DNA/RNA synthesis, which can often result in an insufficient yield, is minimized in the entire process until aptamer acquisition.

2 Polymerase and Substrate Development

2.1 Problems with SELEX When Using Modified DNA/RNA

DNA/RNA polymerases produce DNA/RNA strands by incorporating substrate triphosphates (NTPs/dNTPs) that correspond to the types of bases on a template strand and successively add these to the 3′-end of an extending strand to form 3′,5′-phosphodiester linkages. Therefore, in principle, it is possible to produce nucleic acid polymers that contain foreign functionalities if chemically modified NTPs/dNTPs are acceptable as polymerase substrates (Fig. 3). Modified substrate triphosphates have been applied to DNA sequencing (Ju et al. 1995; Gharizadeh et al. 2002; Eid et al. 2009; Kajiyama et al. 2011), fluorophore and redox labeling (Zhu et al. 1994; Anne et al. 2001), and expanding the genetic alphabet (Piccirilli et al. 1991; Matsuda et al. 2006; Kimoto et al. 2010), in addition to preparing libraries for SELEX (Ono et al. 1997; Raines and Gottlieb 1998; Sakthivel and Barbas 1998; Padilla and Sousa 1999, 2002; Vaish et al. 2000; Andreola et al. 2000; Vastmans et al. 2000; Noronha et al. 2000; Schoning et al. 2000; Sawai et al. 2001, 2007; Thum et al. 2001; Lee et al. 2001; Held and Benner 2002; Lato et al. 2002; Kuwahara et al. 2003, 2006a, b, 2008; Tasara et al. 2003; Pavey et al. 2004; Ohbayashi et al. 2005; Jager et al. 2005; Kempeneers et al. 2005; Inoue et al. 2007; Tsai et al. 2007; Capek et al. 2007; Peng and Damha 2007; Veedu et al. 2008, 2009; Borsenberger et al. 2009; Yu et al. 2012).

For DNA enzymatic functional labeling, a modified dNTP is often used in the presence of the corresponding natural dNTP to increase product yields. In contrast,

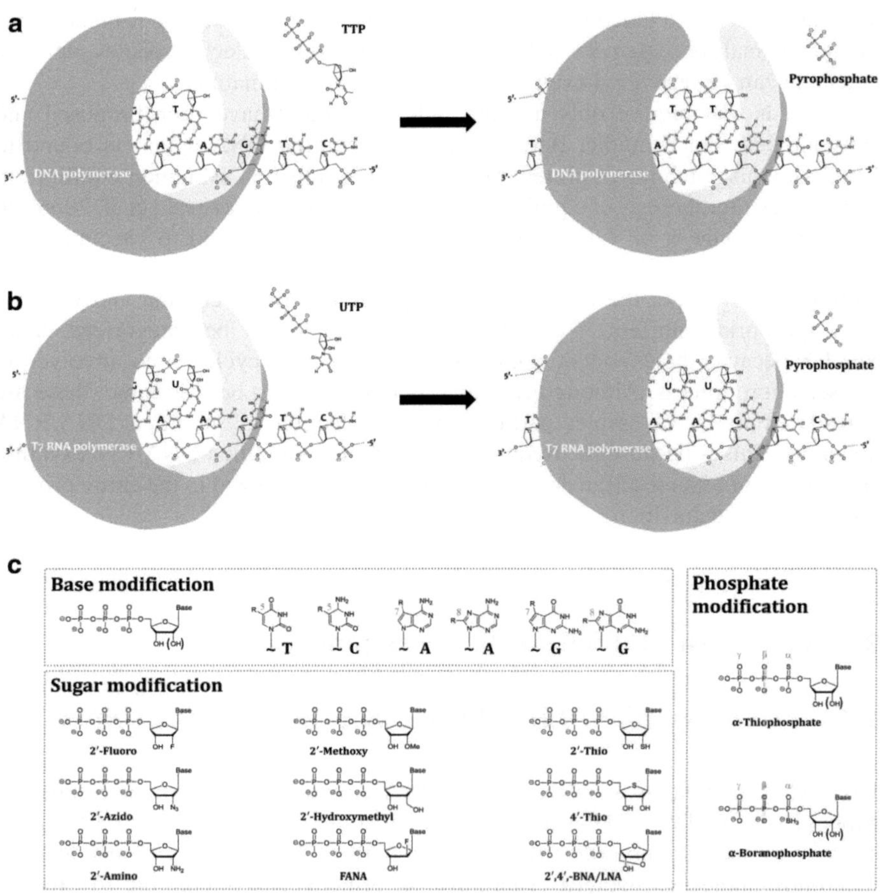

Fig. 3 Enzymatic syntheses of DNA (**a**) and RNA (**b**), and examples of modified nucleoside-5′-triphosphates (**c**)

a modified NTP/dNTP is generally used in the absence of a corresponding natural NTP/dNTP for SELEX to prepare modified nucleic acid libraries, because the natural nucleotide should be completely replaced with the corresponding modified nucleotide at all sites that are incorporated into the extending strand. However, product yield (i.e., extended strands) is generally reduced by total replacement with modified nucleotides. The catalytic efficiencies of polymerases are known to be affected by these modifications not only for substrate triphosphates but also on the extending strand and template (Held and Benner 2002; Kuwahara et al. 2006a).

During a SELEX procedure, the inefficiency of modified DNA/RNA polymerization could naturally bias the selection outcomes. Unintentional exclusion of sequences with the highest binding affinities could occur. Therefore, reactions are often run under extremely high enzyme and/or substrate concentrations to achieve large reaction velocities and reduce this effect. However, there remains a problem in that these conditions are prone to a high frequency of misincorporation. To overcome this, researchers have been developing polymerase variants, triphosphate analogs, and their combinations that improve reaction efficiency.

2.2 Enzymatic Syntheses of Modified DNA/RNA

Thermophilic DNA polymerases are generally used to prepare a modified DNA library for SELEX. Evolutionary family B DNA polymerases such as *Pwo*, *Pfu*, *Vent (exo-)*, *Deep Vent (exo-)*, and *KOD Dash* are particularly preferable to other types of DNA polymerases. During successive incorporations of modified nucleotides, family B DNA polymerases are known to exhibit a much superior performance than family A DNA polymerases such as *Taq*, *Tth*, and *Thermo Sequenase*.

The yields of enzymatic modified DNA syntheses are affected by the sites where the foreign functionality as well as its chemical structure is introduced. For example, during base modifications, dNTP analogs with a pyrimidine substituted at the fifth position and a purine substituted at the seventh position of the base moiety tend to be good substrates for DNA polymerases (Lee et al. 2001; Tasara et al. 2003). Although C8-modified purine nucleoside triphosphates can be used as substrates for polymerase reactions and are applicable in in vitro selection, they are incorporated with lower efficiencies (Kajiyama et al. 2011; Capek et al. 2007). A few DNA polymerases are known to tolerate sugar modifications, such as 2′-fluoro, 2′-fluoro-D-arabino (FANA), and 2′-O,4′-*C*-methylene (2′,4′-BNA/LNA) substitutions in nucleosides (Veedu et al. 2008; Kuwahara et al. 2008). Furthermore, phosphate modifications such as 5′-(α-thio) and 5′-(α-borano) substations (Kunkel et al. 1981; He et al. 1999) in a triphosphate are also acceptable and applicable for SELEX.

In PCR amplification in SELEX, the modified DNA is normally amplified indirectly using PCR in order to prepare a library for the successive steps, even if PCR amplification is available for the modified DNA production. In particular, after affinity separation, the selected modified DNA is reverse transcribed and amplified using PCR to form natural DNA and then transcribed to form modified DNA. This approach is used because indirect synthesis can minimize the influence of sequence dependencies on the yield of the modified DNA as much as possible, compared with direct synthesis.

Unlike DNA polymerases, the repertoire of RNA polymerases available for modified oligonucleotide syntheses is currently limited. This may be because transcription is initiated by a process that is much more complicated than that for replication. For SELEX using a modified RNA library, T7 RNA polymerase and its variants have primarily been used, whereas SP6 RNA polymerase is also known to accept modified NTP analogs. Using these RNA polymerases, NTP analogs of 2′-thio (Raines and Gottlieb 1998), 2′-amino (Padilla and Sousa 1999), 2′-azido (Padilla and Sousa 2002), 2′-hydroxymethyl (Pavey et al. 2004), and 4′-thio (Inoue et al. 2007) in addition to 2′-F and 2′-OMe can act as substrates to provide the corresponding modified RNA. Furthermore, it has also been reported that NTP analogs with base modifications (e.g., C5 modified uridines and cytidines) and with phosphate-modified nucleosides [e.g., 5′-(α-thio) triphosphates and 5′-(α-borano) triphosphates] (Andreola et al. 2000; Lato et al. 2002) are available for modified RNA polymerization.

Two steps are involved in RNA synthesis using T7 RNA polymerases. In the initiation step, T7 RNA polymerases generate purine-rich oligonucleotides with a length of approximately 10 residues. In the elongation step, T7 RNA polymerases successively incorporate a ribonucleotide complementary to the sequence of the DNA template, leading to strand extension. The former step can cause a problem because transcript yields are greatly influenced by the base composition of the leader oligonucleotide sequence, preferably with guanosine residues (Milligan et al. 1987). Furthermore, the polymerases are sensitive to modifications of the 2′-hydroxyl group during the initiation step, whereas recognition of the 2′-position is tolerated during the elongation step (Gudima et al. 1998). Therefore, in the synthesis of RNA containing 2′-modifications, a low ratio of natural GTP in addition to modified NTPs is often necessary to increase the transcript yields. In fact, a T7 RNA polymerase variant (Y639F/H784A) can provide a sufficient yield of highly modified 2′-OMe RNA library in a reaction containing four 2′-OMe NTPs and natural GTP, as well as initiation through the formation of a leader oligonucleotide with a purine-rich sequence GGGAGAGGAGAGAA (Burmeister et al. 2005). Further development of T7 RNA polymerase variants may provide modified RNA free of natural GTP/ATP contamination, which can interfere with the enrichment of modified RNA sequences with aptamer activities in SELEX processes.

2.3 Enzymatic Base-Modified DNA Syntheses

Compared with sugar and phosphate modifications, DNA polymerases are relatively tolerant to base modification. To date, there have been many reports of polymerase reactions such as PCR and primer extension that involve base-modified dNTP analogs, in particular, C5-substituted dUTPs. Among foreign functionalities that may improve aptamer abilities, proteinogenic amino acids and their side chains are the first candidates for base modifications (Kuwahara et al. 2006b; Ohsawa et al. 2008) because proteins perform various important biofunctions. Hence various C5-substituted dUTPs have been synthesized and analyzed for their substrate properties in polymerase reactions. Furthermore, analogs carrying saccharides and nucleobases as foreign functionalities have been reported (Matsui et al. 2007; Kasahara et al. 2013a). Kimoto et al. (2013) recently successfully designed and synthesized triphosphate analogs bearing unnatural bases that can form the third basepair to provide the corresponding modified DNA library. In addition, simultaneous incorporations of modified bases to raise the density of foreign functionalities on a strand have been studied. Four natural nucleotides (A, G, C, and T) were totally replaced with four base-modified nucleotides by the addition of manganese chloride and betaine (Tasara et al. 2003; Jager et al. 2005). Use of these additives results in the improvement of efficiency and yield in the enzymatic production of modified DNA; however, it also simultaneously increases the frequency of misincorporation (Goodman et al. 1983; Beckman et al. 1985).

2.4 Enzymatic BNA/LNA Syntheses

To obtain BNA aptamers by SELEX, BNA libraries should be synthesized by polymerase reactions (Kuwahara and Obika 2013). Enzymatic incorporations of B/L nucleotides as the first analog of BNA were first independently developed by Obika and Imanishi et al. and Wengel et al. in the late 1990s (Obika et al. 1997, 1998; Singh et al. 1998). In 2007, Veedu et al. reported B/L nucleotide incorporation using *Phusion High Fidelity* DNA polymerase, *9°Nm* DNA polymerase, and *Pfu* DNA polymerase. They showed that the first and second of these three polymerases could catalyze primer extension reactions using B/L nucleoside 5′-triphosphates as substrates to yield DNA-based 2′,4′-BNA/LNA strands (Veedu et al. 2007a, b). Furthermore, they assessed *Taq* DNA polymerase, Klenow fragments, *T4* DNA polymerase, *Pfx* DNA polymerase, *Speed STAR HS* DNA polymerase, *T7* RNA polymerase, *E. coli* RNA polymerase, *AMV* reverse transcriptase, and mutant *T7* R&DNA polymerase (Veedu et al. 2007c) and concluded that *Phusion High-Fidelity* DNA polymerase was the only enzyme that could efficiently incorporate B/L nucleotides.

On the other hand, in 2007 and 2008, Kuwahara et al. first reported that *KOD Dash* DNA polymerase, derived from hyperthermophilic archaeon *Pyrococcus kodakaraensis* (*KOD*), was superior to *Phusion High-Fidelity* DNA polymerase (Kuwahara et al. 2008; Nagashima et al. 2007) because of its reduced 3′,5′ exonuclease activity. Furthermore, they demonstrated that *KOD Dash* DNA polymerases are applicable for not only the synthesis of 2′,4′-BNA/LNA but also other types of BNA such as 2′,4′-BNA^{COC} and 2′,4′-BNA^{NC}. In fact, they previously discovered that the usefulness of *KOD* DNA polymerases can be extended for the enzymatic syntheses of base-modified DNAs and studied their application in SELEX methods since 2001 (Sawai et al. 2001; Obayashi et al. 2002; Masud et al. 2003). Thus, they examined *KOD*-related DNA polymerases as the first candidate and successfully demonstrated their usefulness for sugar-modified DNA syntheses. Furthermore, several genetically engineered *KOD* variants were prepared and their suitability for BNA syntheses was examined and promising candidates were identified for applications in BNA aptamer selection. A few variants can be used as parent enzymes to improve the results. At present, *KOD*-related DNA polymerases are widely used for enzymatic syntheses of various modified DNAs (Veedu et al. 2009, 2010; Wheeler et al. 2012; Højland et al. 2012; Johannsen et al. 2012; Goubet et al. 2013; Kuwahara et al. 2009; Vaught et al. 2010; Gold et al. 2010; Kasahara and Kuwahara 2012; Dubois et al. 2012). For example, a DNA polymerase named

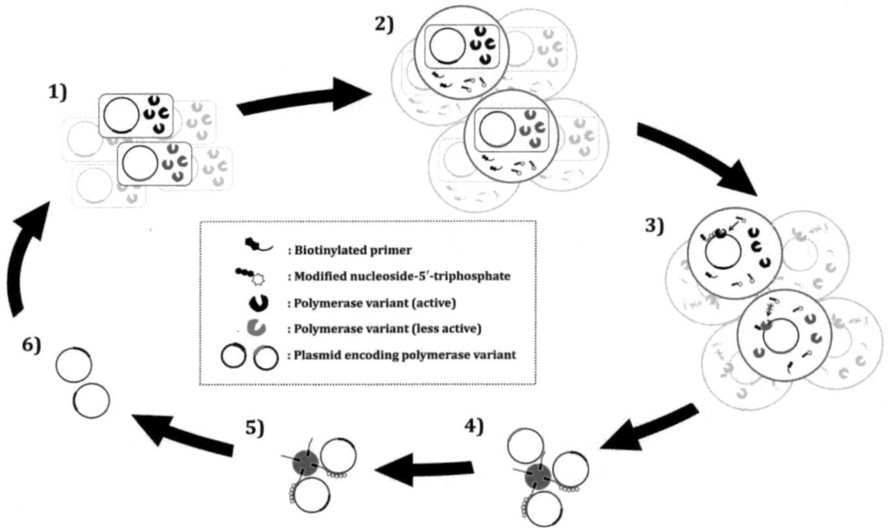

Fig. 4 Schematic illustration of compartmentalized self-tagging (CST) method

KOD XL, which is essentially the same as *KOD Dash*, is employed for selection of SOMAmers as described later.

Recently, *Tgo* DNA polymerase variants, which enable transcription and reverse transcription of six different XNAs (xeno-nucleic acids) including 2′,4′-BNA/LNA, FANA, ANA (arabinonucleic acid), HNA (1,5-anhydrohexitol nucleic acid), CeNA (cyclohexenyl nucleic acid), and TNA (α-L-threofuranosyl nucleic acid) have been developed (Pinheiro et al. 2012). *Tgo* DNA polymerase is derived from *Thermococcus gorgonarius* and belongs to the evolutional family B, which is same as *KOD* DNA polymerase. It is noteworthy that some variants were created not by conventional gene mutation and recombination, but by a unique selection method called compartmentalized self-tagging (CST) (Fig. 4). This method may generate further improvements in polymerases for BNA syntheses in the future.

3 Chemically Modified DNA/RNA Aptamers

3.1 RNA-Based Aptamers

Because T7 RNA polymerase and its variants can accept modified nucleoside-5′-triphospates as substrates, several modified RNA libraries have been examined under SELEX selection since the 1990s. Some trials have successfully acquired modified RNA aptamers (Fig. 5); for example, modified RNA libraries that contained both 2′-amino uridine (U) and 2′-amino cytidine (C) instead of natural U and C provided modified RNA aptamers for human neutrophil elastase (HNE)

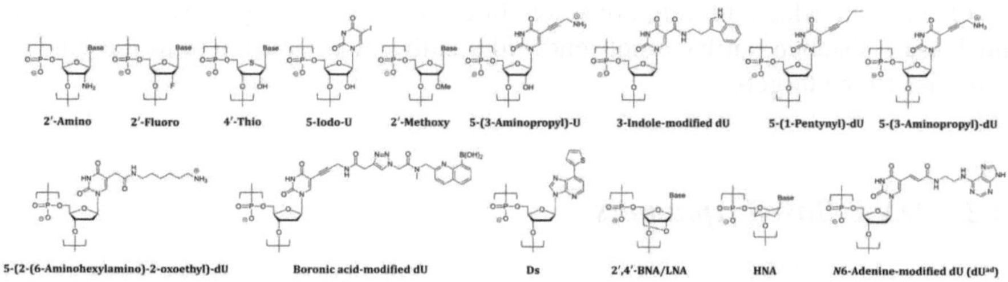

Fig. 5 Examples of modified nucleotides used for aptamer selections

(Lin et al. 1994), immunoglobulin E (IgE) (Wiegand et al. 1996), human thyroid stimulating hormone (TSH) (Lin et al. 1996), keratinocyte growth factor (KGF) (Pagratis et al. 1997), interferon-γ (IFN-γ) (Kubik et al. 1997), and basic fibroblast growth factor (bFGF) (Jellinek et al. 1995). In addition, modified RNA aptamers for KGF (Pagratis et al. 1997), IFN-γ (Kubik et al. 1997), human thrombin (White et al. 2001), and cluster of differentiation 4 (CD4) (Davis et al. 1998) were obtained from libraries that contained both 2′-fluoro U and 2′-fluoro C. Modified RNA aptamers for human thrombin were also obtained from a library that was composed entirely of 4′-thionucleotides (Minakawa et al. 2008). Modified RNA aptamers for bFGF were also obtained from an RNA-based library in which phosphodiester linkages were completely replaced with phosphorothioate linkages (*cf.* thioaptamers) (Jhaveri et al. 1998). Furthermore, libraries that contained 2′-fluoro U provided aptamers for respiratory syncytial virus (RSV) (Pan et al. 1995) and streptavidin (Tahiri-Alaoui et al. 2002).

As examples of base-modified RNA aptamers, photoaptamers that are specific for the human immunodeficiency virus (HIV) Rev protein have been provided (Jensen et al. 1995). These were recovered from a library that contained 5-iodouridine, which can be photocrosslink able with an adjacent residue of the target.

Among proteins, VEGF is one of the most actively pursued targets of aptamers for therapeutic uses. Similar to pegaptanib, VEGF-binding modified RNA aptamers that contained 2′-amino U/C and 2′-OMe A/G were also created using the post-SELEX method[111], before which the precursors had been recovered from a 2′-amino U/C RNA library. Furthermore, using a T7 RNA polymerase variant (Y639F/H784A), VEGF-binding modified RNA aptamers consisting entirely of 2′-*O*-methyl nucleotides were successfully selected from a 2′-OMe U/C/A/G RNA library (Burmeister et al. 2005).

Modified RNA aptamers for small molecules have also been reported, although there are relatively fewer examples as compared with those for proteins. Examples include moenomycin A-binding modified RNA aptamers that contain 2′-amino U/C (Schürer et al. 2001) and adenosine-5′-triphospate (ATP)-binding RNA aptamers that contain boranophosphate G/U (Lato et al. 2002) or 5-(3-aminopropyl) U (Vaish et al. 2003).

Overall, introducing foreign functionalities into RNA have greatly improved the nuclease resistance but have not rendered significant enhancement to the binding affinities to the targets.

3.2 DNA-Based Aptamers

Compared with modified RNA aptamers, modified DNA aptamers were limited until slow off-rate modified aptamers, called SOMAmers, were reported by Gold et al. in 2010 (Vaught et al. 2010; Gold et al. 2010). SOMAmers contain foreign functionalities at the C5 position of 2′-deoxyuridine (dU) and exhibit superior affinity and specificity, which can distinguish the targets from their cognate proteins. In particular, SOMAmer libraries that contained 3-indolyl groups observed in tryptophan as a partial structure were demonstrated to provide high affinity aptamers for a broad range of protein targets (Fig. 5). Hence, SOMAmers have now become the best candidate reagents as alternatives to antibodies in assays used for protein biomarker discoveries and identifications.

However, the idea that introducing foreign functionalities into base moieties of DNA/RNA could expand the adaptability of aptamers to a variety of targets had already existed prior to their publication in 2010. For example, in 1994, Latham et al. reported that 5-(1-pentynyl)-dU contains modified DNA aptamers for human thrombin (Latham et al. 1994). Battersby et al. (1999) and Vaish et al. (2003) reported that ATP-binding aptamers were recovered from libraries of modified DNA and modified RNA containing 5-(3-aminopropyl)-dU/U in 1999 and 2003, respectively. In 2007, Shoji et al. demonstrated that a library containing 5-(2-(6-aminohexylamino)-2-oxoethyl)-dU could provide a modified DNA aptamer that binds to an (*R*)-thalidomide derivative with high enantioselectivity (Shoji et al. 2007). Furthermore, in 2008, Li et al. successfully obtained a modified DNA aptamer that could sensitively identify glycosylation sites of fibrinogen as a glycoprotein from a library that contained boronic acid-modified dU[117]. Thus, base-modified DNA aptamer selection has been utilized since the 1990s.

Although the abovementioned modified DNA libraries were prepared using DNA polymerases, the T7 RNA polymerase variant (Y639F) has also been available for the preparation of 2′-OMe A/G DNA libraries, which eventually provided modified DNA aptamers for interleukin (IL)-23 or human thrombin (Burmeister et al. 2006). It is interesting that only a single mutation in T7 RNA polymerase enabled the production of DNA-based polymers with sufficient yield.

Using expanded genetic alphabets, high-affinity modified DNA aptamers for VEGF-165 and IFN-γ have recently been obtained from libraries containing the four natural and one unnatural nucleotide with the hydrophobic base 7-(2-thienyl) imidazo[4,5-b]pyridine (Ds) (Kimoto et al. 2013). The unnatural Ds base can pair with a diol-modified 2-nitro-4-propynylpyrrole (Px) base to form a third base-pair. Using 5′-triphosphate analogs of Ds and Px, such as dDsTP and dPxTP as substrates and AccuPrimePfx DNA polymerase as a catalyst, SELEX libraries containingthe

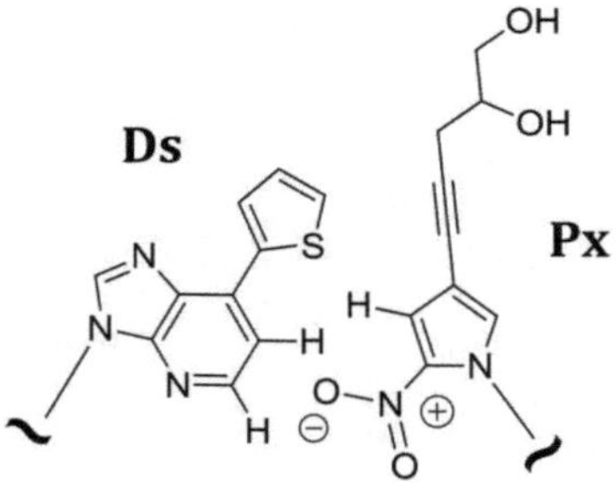

Fig. 6 The third base pair; Ds versus Px

extra genetic alphabet (Ds) have been prepared and selections successfully conducted (Fig. 6).

3.3 XNA Aptamers

Although XNAs may involve modified nucleic acids with conventional sugar modifications like 2′-OMe, 2′-F, and 2′-NH_2, XNAs having unusual sugars like TNA, HNA, CeNA, and 2′,4′-BNA/LNA are attracting attention. Using *Tgo* DNA polymerase variants, Pinheiro et al. have successfully obtained HNA aptamers for HIV trans-activating response RNA (TAR), and hen egg lysozyme (HEL) using the standard SELEX protocol comprising magnetic beads for the capture and isolation of active sequences (Pinheiro et al. 2012). Furthermore, using a 19:1 enzyme mix of *KOD Dash* and *KOD* mutant DNA polymerases (*KOD2*), Kasahara et al. (2013b) have first obtained DNA-based 2′,4′-BNA/LNA aptamers for human thrombin using a capillary electrophoresis (CE)-SELEX method that can separate active sequences form non-active sequences in the liquid phase by non-equilibrium capillary electrophoresis of equilibrium mixtures (NECEEM). In NECEEM, electro-osmotic flow (EOF) occurs from the anode to the cathode and the target–aptamer complexes migrate before unbound free oligonucleotides do (Geiger et al. 2012).

Although various advanced SELEX methods have been developed to date, CE-SELEX may be one of the best methods to enable efficient enrichment of sequences with aptamer activities (Mendonsa and Bowser 2004, 2005; Mosing et al. 2005; Drabovich et al. 2006) because it maximally excludes contamination with non-active sequences (Fig. 7). In contrast, in typical SELEX methods, sequences that are non-specifically bound to the solid support cannot completely be washed out while maintaining the desired specific binding, which inevitably results in non-active sequence contamination in the elution of active sequences.

Thus, improvements in polymerase performance and separation techniques have successfully achieved the first BNA (LNA) aptamers using DNA-based libraries that contained B/L nucleotides as the first model of 2′,4′-linked nucleotide analogs. Further development of selection systems will enable acquisition of BNA aptamers

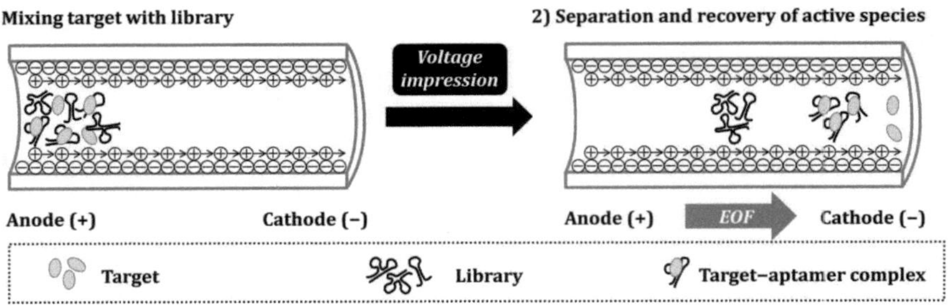

Fig. 7 Separation process in CE-SELEX

entirely composed of B/L nucleotides or those that contain advanced types of BNA that exhibit improved nuclease resistance.

4 Efficacy of Introducing Foreign Functionalities

Natural DNA/RNA strands are rapidly degraded under biophysical conditions. However, introduction of foreign functionalities generally results in improved biostability, which is supported by several research reports. For example, half of a 2′-NH_2 U/C RNA aptamer for HNE remained intact after incubation at 37 °C for 20 h and 9.3 h in 95 % human serum and 95 % human urine, respectively (Lin et al. 1994). In addition, the half-lives of 2′-NH_2 U/C and 2′-F U/C RNA aptamers for KGF in 90 % human serum at 37 °C were reported to be 86 h and 174 h, respectively (Pagratis et al. 1997). In contrast, the effects of these modifications on binding affinities are likely to be limited in many cases. For example, aptamers for human thrombin, VEGF, and ATP generated affinity data for inferring the effects of introduced functionalities.

Compared with the best aptamers for human thrombin (Table 1), the K_d values (dissociation constants) of a natural RNA aptamer (20-mer) (Kubik et al. 1994), a 2′-F U/C RNA aptamer (25-mer) (White et al. 2001), and a 4′-thio A/G/C/U RNA aptamer (73-mer) (Minakawa et al. 2008) were 9.3 nM, 0.54 nM, and 7.2 nM, respectively. In addition, those of a natural DNA aptamer (29-mer) (Tasset et al. 1997), a 2′-OMe U/C DNA aptamer (75-mer) (Burmeister et al. 2006), and a 5-(1-pentynyl)-dU-containing DNA aptamer (60-mer) (Latham et al. 1994) were reported to be 0.5 nM, 26 nM, and 400 nM, respectively. A recently reported DNA-based BNA aptamer (78-mer) that contained B/L nucleotides bearing thymine bases exhibited a K_d value of 18 nM (Kasahara et al. 2013b). In addition, a DNA aptamer that contained B/L nucleotides in the primer region (70-mer), one that contained (*E*)-5-(2-(*N*-(2-(N^6-adeninyl)ethyl))carbamylvinyl)-dU (dU^{ad}) in the non-primer region (70-mer), and one that contained B/L nucleotides in the primer region and dU^{ad} in the non-primer region (70-mer) exhibited K_d values of 1.9 nM, 0.57 nM, and 0.093 nM, respectively (Kasahara et al. 2013a).

Table 1 Natural/modified DNA/RNA aptamers for human α-thrombin

Aptamer	Modification type, inserted position, and length[a]	K_d (nM)	Authors and publication year[b]
RNA	Non, 20-mer 5′-●●●●●●●●●●●●●●●●●●●●	9.3	Kubik et al. (1994)
RNA	2′-F U/C(○), 25-mer 5′-●●●●●○●●●●○○●●●●○●○○○●○○○	0.54	White et al. (2001)
RNA	4′-thio A/G/C/U(○), 73-mer 5′-○○○	7.2	Minakawa et al. (2008)
DNA	Non, 29-mer 5′-●●●●●●●●●●●●●●●●●●●●●●●●●●●●●	0.5	Tasset et al. (1997)
DNA	2′-OMe U/C(○), 75-mer 5′-●●●●●●●●●●●●●●●○●○○○○●○○○○●○○○●●○●●●●●●○●●○●○●○○●●○○○●○○●○○●●○○●●○○●○○○●●○●	26	Burmeister et al. (2006)
DNA	5-(1-pentynyl)-dU(○), 60-mer 5′-●●●●●●●●●●●●●●●●●●●●○●○●○●●○●○●●○●○○●●●●●○○○●●●○●●●●●●●●●○●●	400	Latham et al. (1994)
DNA	B/L-nucleotide A/G/C[me]/T(*), B/L-nucleotide T(○), 78-mer 5′-●●*●●*●●*●●*●●*●●*●●●○●○○●●●●●●●●●○●●○●●●●●●●●●●●●●○○●●○●●●○●●●●●●○●●●●●●●●●●●	18	Kasahara et al. (2013b)
DNA	B/L-nucleotide A/G/C/T(*), 70-mer 5′-●●*●●*●●*●●*●●*●●*●●	1.9	Kasahara et al. (2013a)
DNA	dU[ad](○), 70-mer 5′-●●●●●●●●●●●●●●●●●●●●●●●●●●●○●●●●○●●●●●○○○○○●○○●●○●○●●○●●●○●●●●●○●●●●●●	0.57	Kasahara et al. (2013a)
DNA	B/L-nucleotide A/G/C/T(*), dU[ad](○), 70-mer 5′-●●*●●*●●*●●*●●*●●*●●○○●●○●●●●●●○○●●○●●●●●●○○●○●○●●○●●○●●●○●●●●●○●●●●●●	0.093	Kasahara et al. (2013a)

[a]Chemical structures of modified nucleotides were shown in Fig. 5. Closed circles (•) indicate natural nucleotides

[b]Authors and publication year of the respective aptamers

Table 2 Natural/modified DNA/RNA aptamers for human vascular endothelial growth factor (VEGF)

Aptamer	Modification type, inserted position, and length[a]	K_d (pM)	Authors and publication year[b]
RNA	Non, 80-mer 5′-●●	140	Jellinek et al. (1994)
RNA	2′-F U/C(○), post-SELEX-introduced 2′-OMe U/C(*), 27-mer 5′-○**●●○○**○***○*○○○*○*○*○○○*	49	Ruckman et al. (1998)
RNA	2′-NH_2 U/C(○), post-SELEX-introduced 2′-OMe A/G(*), 3′,5′-capping phosphorothioate T(+), 32-mer 5′−++++*○○○○●●○*●○**●○*○○****○*++++	140	Green et al. (1995)
RNA	2′-OMe U/C/A/G(○), 23-mer 5′-○○○○○○○○○○○○○○○○○○○○○○○	1,400	Burmeister et al. (2005)
DNA	Non, 60-mer 5′-●●	30	Nonaka et al. (2013)
DNA	Ds(○), 47-mer 5′-●●●●●●●●●●●●●●●●●●●●●●●○●●●●●●●●○●●●●●●●●●●●●●●	0.65	Kimoto et al. (2013)

[a]Chemical structures of modified nucleotides were shown in Fig. 5. Closed circles (•) indicate natural nucleotides
[b]Authors and publication year of the respective aptamers

With regard to VEGF-binding aptamers (Table 2), pegaptanib exhibits high binding affinity in a Ca^{2+}-dependent manner with a K_d value of 200 pM at 37 °C (Bell et al. 1999). The K_d values of the natural RNA (80-mer) and natural DNA (60-mer) aptamers were reported to be 140 pM (at 37 °C in phosphate-buffered saline without Ca^{2+}) (Jellinek et al. 1994) and 30 pM (at 20 °C in Tris-buffered saline without Ca^{2+}) (Nonaka et al. 2013), respectively. These values were lower than those of the anti-VEGF antibody, Avastin (bevacizumab), which is used in cancer therapy ($K_d = 1.1$ nM at 25 °C).

A precursor of pegaptanib was a modified RNA aptamer (27-mer) that contained 2′-F U/C and post-SELEX-introduced 2′-OMe A/G and lacked the 5′- and 3′-end capping (Ruckman et al. 1998). This aptamer had a higher affinity with a K_d value of 49 pM at 37 °C in phosphate-buffered saline containing 2 mM Ca^{2+}. In addition, a modified RNA aptamer (32-mer) that contained 2′-NH_2 U/C and post-SELEX-introduced 2′-OMe A/G (Green et al. 1995), and an aptamer (23-mer) that contained 2′-OMe U/C/A/G (Burmeister et al. 2005) had K_d values of 140 pM at 37 °C and 1,400 pM at ambient temperature, respectively. The 32-mer modified RNA aptamer had oligo-thymidine segments containing four contiguous internucleoside phosphorothioate linkages at both the 5′ and 3′ ends to reduce degradation by the exonucleolytic pathway.

These abovementioned modified DNA/RNA aptamers exhibited binding affinities in the picomolar to nanomolar range. However, a modified DNA aptamer (47-mer) containing unnatural Ds bases exhibited a subpicomolar affinity to its target with a K_d value of 0.65 pM at 25 °C in phosphate buffered saline containing 0.05 % (wt/vol) Nonidet P-40 and no divalent cations (Kimoto et al. 2013). Thus, examples that demonstrate the positive effects of foreign functionalities on binding affinity have recently increased, and it is becoming widely recognized that certain modifications can be highly effective for enhancing aptamer avidity.

In addition, such modification efficacies have recently been demonstrated during aptamer selection for small molecular targets as well as protein targets (Davies et al. 2012). Imaizumi et al. recently obtained high affinity modified DNA aptamers (70-mer and 59-mer) that contained dU^{ad}, which specifically bound to a camptothecin (CPT) derivative with K_d values of 39 and 86 nM and very low dissociation rate constants ($k_d = 0.00033$ and $0.00037\ s^{-1}$), respectively (Fig. 8) (Imaizumi et al. 2013). In contrast, using the same target and conditions, DNA aptamers that possess such low k_d values were not obtained from a natural DNA library using SELEX. Even the best of the selected DNA aptamers (36-mer) presented much higher values of K_d (1.1 μM) and k_d ($>10^{-1}\ s^{-1}$), which are generally observed with natural DNA/RNA aptamers for small molecules except for nucleic acid binders.

These aptamers showed extremely rapid rates for both association and dissociation that could be confirmed by surface plasmon resonance (SPR) analyses. That is, as shown in Fig. 8 using an RNA aptamer (35-mer) for riboflavin-5′-phosphate (FNM) (Burgstaller and Famulok 1994) as a typical example, DNA/RNA aptamers for small molecules generate a box-shaped profile. This profile indicates that although the target quickly binds to the aptamer when they are mixed using buffer

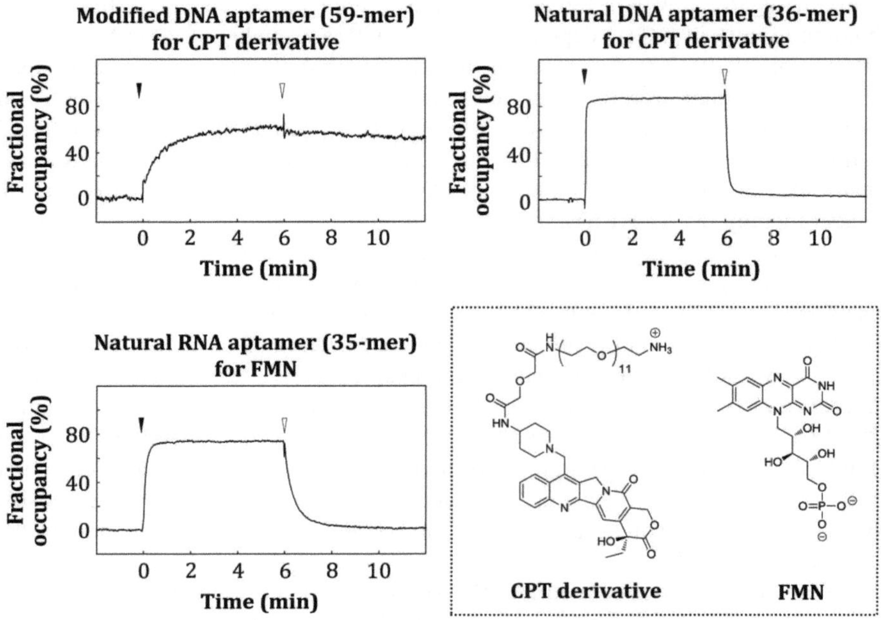

Fig. 8 SPR sensorgrams of aptamers for small molecules. *Closed* and *open inverted triangles* indicate start points of association and dissociation, respectively

solutions, the target is immediately washed out from the aptamer immobilized on a solid-surface. This restricts the applications of aptamers to bioanalyses that involve rinsing as part of the protocol, such as enzyme-linked immunosorbent assay (ELISA). Thus, a high binding affinity and a slow-off-rate property are important for broadening aptamer applications such as small molecule sensing and drug delivery systems (DDS) (Bagalkot et al. 2006; Wochner et al. 2008; Wang et al. 2011; Zhu et al. 2012).

However, until recently the positive effects of modifications have not necessarily been accepted because earlier studies indicated considerably marginal effects of the introduced foreign functionalities for enhancing binding affinities. Using SELEX that targeted ATP, Sassanfar and Szostak (1993) and Sazani et al. (2004) obtained natural RNA aptamers, respectively. These natural RNA aptamers, as well as a 5-(3-aminopropyl)-U-containing RNA aptamer (Vaish et al. 2003), were observed to bind to ATP in a 1:1 stoichiometry and their binding affinities were dependent on Mg^{2+} concentration Although there were certain differences in the selection protocols used, affinity analyses showed that there were no notable differences between natural and modified RNA aptamers, with K_d values for those aptamers ranging approximately from 10^{-7} to 10^{-6}. Furthermore, specificity analyses using other nucleotides, such as adenosine-5′-diphosphate (ADP) and adenosine-5′-monophosphate (AMP), indicated that introducing a cationic functional group could be alternated with that of native functionalities in RNA and divalent metal ions in solution.

Similarly, natural DNA and modified DNA aptamers for ATP have also been reported, respectively. The natural DNA aptamer reported by Huizenga and Szostak (1995) bound to ATP with an apparent K_d value of 6 μM, which was determined assuming a 1:1 binding ratio, and showed similar affinities to AMP and adenosine. However, more precise analyses showed that this natural DNA aptamer cooperatively bound to two ATP molecules with a binding ratio of 1:2 and a K_d of 9 μM^2. In addition, the modified DNA aptamer containing 5-(3-aminopropyl)-dU (Battersby et al. 1999) formed a 1:2 complex with ATP with a K_d value of 6 μM. Furthermore, a DNA aptamer in which thymidine replaced a modified dU was observed to retain its binding affinity, although its affinity for ATP was decreased by approximately 2-fold ($K_d = 13\ \mu M^2$). As with natural/modified RNA aptamers for ATP, these results also indicated that introducing a functionality did not dramatically influence the activities of the selected aptamers.

5 Future Outlook

After an interval of over 10 years since the earlier studies, base modifications have once again gained prominence owing to some successful examples as aforementioned. In addition, base-modified libraries can be applicable for aptamer selections targeting cells and viruses as well as proteins and small molecules. For further development, novel functionalities will be designed and synthesized, and their suitability for the respective targets will then be evaluated to provide optimized combinations between library types and targets. Moreover, various base-modified libraries, in which dual, triple, or quadruple types of functionalities are introduced simultaneously using modified dC/dA/dG in addition to modified dU, will be prepared and applied to SELEX.

Not only base modifications but also sugar and phosphate modifications will be researched further to acquire highly stable aptamers for biological and therapeutic use. Polymerase variants with both a broad tolerance toward modifications and high fidelity, which may possess properties contrary to each other, will be necessary to achieve this goal. In addition, separation techniques to exclude contamination of non-active sequences in the selection processes should be refined further.

Overall, chemical modifications will be the most promising method to exploit the large potential of nucleic acid aptamers. This prospect should be confident, although aptamer development is still in its intermediate stage. Thus, in future, progress in the selection methodologies accompanied with the improvement of polymerase variants (Kuwahara et al. 2010) may generate alternatives to costly antibodies obtained from artificial nucleic acid-based biopolymer libraries involving possible modifications.

References

Andreola ML, Calmels C, Michel J et al (2000) Towards the selection of phosphorothioate aptamers optimizing *in vitro* selection steps with phosphorothioate nucleotides. Eur J Biochem 267:5032–5040

Anne A, Blanc B, Moiroux J (2001) Synthesis of the first ferrocene-labeled dideoxynucleotide and its use for 3′-redox end-labeling of 5′-modified single-stranded oligonucleotides. Bioconjug Chem 12:396–405

Bagalkot V, Farokhzad OC, Langer R et al (2006) An aptamer-doxorubicin physical conjugate as a novel targeted drug-delivery platform. Angew Chem Int Ed Engl 45:8149–8152

Barciszewski J, Medgaard M, Koch T et al (2009) Locked nucleic acid aptamers. Methods Mol Biol 535:165–186

Battersby TR, Ang DN, Burgstaller P et al (1999) Quantitative analysis of receptors for adenosine nucleotides obtained via *in vitro* selection from a library incorporating a cationic nucleotide analog. J Am Chem Soc 121:9781–9789

Beckman RA, Mildvan AS, Loeb LA (1985) On the fidelity of DNA replication: manganese mutagenesis *in vitro*. Biochemistry 24:5810–5817

Bell C, Lynam E, Landfair DJ et al (1999) Oligonucleotide NX1838 inhibits VEGF165-mediated cellular responses *in vitro*. In Vitro Cell Dev Biol Anim 35:533–542

Berezovski M, Musheev M, Drabovich A et al (2006) Non-SELEX selection of aptamers. J Am Chem Soc 128:1410–1411

Bock LC, Griffin LC, Latham JA et al (1992) Selection of single-stranded DNA molecules that bind and inhibit human thrombin. Nature 355:564–566

Borsenberger V, Kukwikila M, Howorka S (2009) Synthesis and enzymatic incorporation of modified deoxyuridine triphosphates. Org Biomol Chem 7:3826–3835

Brody EN, Gold L, Lawn RM et al (2010) High-content affinity-based proteomics: unlocking protein biomarker discovery. Expert Rev Mol Diagn 10:1013–1022

Burgstaller P, Famulok M (1994) Isolation of RNA aptamers for biological cofactors by *in vitro* selection. Angew Chem Int Ed Engl 33:1084–1087

Burmeister PE, Lewis SD, Silva RF et al (2005) Direct *in vitro* selection of a 2′-*O*-methyl aptamer to VEGF. Chem Biol 12:25–33

Burmeister PE, Wang C, Killough JR et al (2006) 2′-Deoxy purine, 2′-*O*-methyl pyrimidine (dRmY) aptamers as candidate therapeutics. Oligonucleotides 16:337–351

Capek P, Cahova H, Pohl R et al (2007) An efficient method for the construction of functionalized DNA bearing amino acid groups through cross-coupling reactions of nucleoside triphosphates followed by primer extension or PCR. Chemistry 13:6196–6120326

Davies DR, Gelinas AD, Zhang C et al (2012) Unique motifs and hydrophobic interactions shape the binding of modified DNA ligands to protein targets. Proc Natl Acad Sci U S A 109:19971–19976

Davis KA, Lin Y, Abrams B et al (1998) Staining of cell surface human CD4 with 2′-F-pyrimidine-containing RNA aptamers for flow cytometry. Nucleic Acids Res 26:3915–3924

Drabovich AP, Berezovski M, Okhonin V et al (2006) Selection of smart aptamers by methods of kinetic capillary electrophoresis. Anal Chem 78:3171–3178

Dubois C, Campbell MA, Edwards SL et al (2012) Stepping towards highly flexible aptamers: enzymatic recognition studies of unlocked nucleic acid nucleotides. Chem Commun 48:5503–5505

Eid J, Fehr A, Gray J et al (2009) Real-time DNA sequencing from single polymerase molecules. Science 323:133–138

Ellington AE, Szostak JW (1990) *In vitro* selection of RNA molecules that bind specific ligands. Nature 346:818–822

Geiger M, Hogerton AL, Bowser MT (2012) Capillary electrophoresis. Anal Chem 84:577–596

Gharizadeh B, Nordstrom T, Ahmadian A et al (2002) Long-read pyrosequencing using pure 2′-deoxyadenosine-5′-*O*′-(1-thiotriphosphate) Sp-isomer. Anal Biochem 301:82–90

Gold L, Polisky B, Uhlenbeck O et al (1995) Diversity of oligonucleotide functions. Annu Rev Biochem 64:763–797

Gold L, Ayers D, Bertino J et al (2010) Aptamer-based multiplexed proteomic technology for biomarker discovery. PLoS One 5:e15004

Goodman MF, Keener S, Guidotti S et al (1983) On the enzymatic basis for mutagenesis by manganese. J Biol Chem 258:3469–3475

Goubet A, Chardon A, Kumar P et al (2013) Synthesis of DNA oligonucleotides containing C5-ethynylbenzenesulfonamide-modified nucleotides (EBNA) by polymerases towards the construction of base functionalized nucleic acids. Bioorg Med Chem Lett 2:761–763

Green LS, Jellinek D, Bell C et al (1995) Nuclease-resistant nucleic acid ligands to vascular permeability factor/vascular endothelial growth factor. Chem Biol 2:683–695

Gudima SO, Kostyuk DA, Grishchenko OI et al (1998) Synthesis of mixed ribo/deoxyribopolynucleotides by mutant T7 RNA polymerase. FEBS Lett 439:302–306

Guerrier-Takada C, Gardiner K, Marsh T et al (1983) The RNA moiety of ribonuclease P is the catalytic subunit of the enzyme. Cell 35:849–857

He K, Porter KW, Hasan A et al (1999) Synthesis of 5-substituted 2′-deoxycytidine 5′-(α-P-borano)triphosphates, their incorporation into DNA and effects on exonuclease. Nucleic Acids Res 27:1788–1794

Held HA, Benner SA (2002) Challenging artificial genetic systems: thymidine analogs with 5-position sulfur functionality. Nucleic Acids Res 30:3857–3869

Hicke BJ, Marion C, Chang YF et al (2001) Tenascin-C aptamers are generated using tumor cells and purified protein. J Biol Chem 276:48644–48654

Højland T, Veedu RN, Vester B et al (2012) Enzymatic synthesis of DNA strands containing α-L-LNA (α-L-configured locked nucleic acid) thymine nucleotides. Artif DNA PNA XNA 3:14–21

Huizenga DE, Szostak JW (1995) A DNA aptamer that binds adenosine and ATP. Biochemistry 34:656–665

Imaizumi Y, Kasahara Y et al (2013) Efficacy of base-modification on target binding of small molecule DNA aptamers. J Am Chem Soc 135:9412–9419

Inoue N, Shionoya A, Minakawa N et al (2007) Amplification of 4′-thioDNA in the presence of 4′-thio-dTTP and 4′-thio-dCTP, and 4′-thioDNA-directed transcription *in vitro* and in mammalian cells. J Am Chem Soc 129:15424–15425

Jager S, Rasched G, Kornreich-Leshem H et al (2005) A versatile toolbox for variable DNA functionalization at high density. J Am Chem Soc 127:15071–15082

Jellinek D, Green LS, Bell C et al (1994) Inhibition of receptor binding by high-affinity RNA ligands to vascular endothelial growth factor. Biochemistry 33:10450–10456

Jellinek D, Green LS, Bell C et al (1995) Potent 2′-amino-2′-deoxypyrimidine RNA inhibitors of basic fibroblast growth factor. Biochemistry 34:11363–11372

Jensen KB, Atkinson BL, Willis MC et al (1995) Using *in vitro* selection to direct the covalent attachment of HIV-1 Rev protein to high affinity RNA ligands. Proc Natl Acad Sci USA 92:12220–12224

Jhaveri S, Olwin B, Ellington AD (1998) *In vitro* selection of phosphorothiolated aptamers. Bioorg Med Chem Lett 8:2285–2290

Johannsen MW, Veedu RN, Madsen AS et al (2012) Enzymatic polymerisation involving 2′-amino-LNA nucleotides. Bioorg Med Chem Lett 22:3522–3526

Ju J, Ruan C, Fuller CW et al (1995) Fluorescence energy transfer dye-labeled primers for DNA sequencing and analysis. Proc Natl Acad Sci USA 92:4347–4351

Kajiyama T, Kuwahara M, Goto M et al (2011) Optimization of pyrosequencing reads by superior successive incorporation efficiency of improved 2′-deoxyadenosine-5′-triphosphate analogs. Anal Biochem 416:8–17

Kasahara Y, Kuwahara M (2012) Artificial specific binders directly recovered from chemically modified nucleic Acid libraries. J Nucleic Acids 2012:156482

Kasahara Y, Irisawa Y, Fujita H et al (2013a) Capillary electrophoresis-systematic evolution of ligands by exponential enrichment selection of base- and sugar-modified DNA aptamers: target binding dominated by 2′-*O*,4′-*C*-methylene-bridged/locked nucleic acid primer. Anal Chem 85:4961–4967

Kasahara Y, Irisawa Y, Ozaki H et al (2013b) 2′,4′-BNA/LNA aptamers: CE-SELEX using a DNA-based library of full-length 2′-*O*,4′-*C*-methylene-bridged/linked bicyclic ribonucleotides. Bioorg Med Chem Lett 23:1288–1292

Kempeneers V, Renders M, Froeyen M et al (2005) Investigation of the DNA-dependent cyclohexenyl nucleic acid polymerization and the cyclohexenyl nucleic acid-dependent DNA polymerization. Nucleic Acids Res 33:3828–3836

Kimoto M, Mitsui T, Yamashige R et al (2010) A new unnatural base pair system between fluorophore and quencher base analogues for nucleic acid-based imaging technology. J Am Chem Soc 132:15418–15426

Kimoto M, Yamashige R, Matsunaga K et al (2013) Generation of high-affinity DNA aptamers using an expanded genetic alphabet. Nat Biotechnol 31:453–457

Kubik MF, Stephens AW, Schneider D et al (1994) High-affinity RNA ligands to human α-thrombin. Nucleic Acids Res 22:2619–2626

Kubik MF, Bell C, Fitzwater T (1997) Isolation and characterization of 2′-fluoro-, 2′-amino-, and 2′-fluoro-/amino-modified RNA ligands to human IFN-γ that inhibit receptor binding. J Immunol 159:259–267

Kunkel TA, Eckstein F, Mildvan AS et al (1981) Deoxynucleoside [1-thio]triphosphates prevent proofreading during *in vitro* DNA synthesis. Proc Natl Acad Sci U S A 78:6734–6738

Kuwahara M, Obika S (2013) *In vitro* selection of BNA (LNA) aptamers. Artif DNA PNA XNA 4:39–48

Kuwahara M, Sugimoto N (2010) Molecular evolution of functional nucleic acids with chemical modifications. Molecules 15:5423–5444

Kuwahara M, Takahata Y, Shoji A et al (2003) Substrate properties of C5-substituted pyrimidine 2′-deoxynucleoside 5′-triphosphates for thermostable DNA polymerases during PCR. Bioorg Med Chem Lett 13:3735–3738

Kuwahara M, Nagashima J, Hasegawa M et al (2006a) Systematic characterization of 2′-deoxynucleoside-5′-triphosphate analogs as substrates for DNA polymerases by polymerase chain reaction and kinetic studies on enzymatic production of modified DNA. Nucleic Acids Res 34:5383–5394

Kuwahara M, Hanawa K, Ohsawa K et al (2006b) Direct PCR amplification of various modified DNAs having amino acids: convenient preparation of DNA libraries with high-potential activities for *in vitro* selection. Bioorg Med Chem 14:2518–2526

Kuwahara M, Obika S, Nagashima J et al (2008) Systematic analysis of enzymatic DNA polymerization using oligo-DNA templates and triphosphate analogs involving 2′,4′-bridged nucleosides. Nucleic Acids Res 36:4257–4265

Kuwahara M, Takeshima H, Nagashima J et al (2009) Transcription and reverse transcription of artificial nucleic acids involving backbone modification by template-directed DNA polymerase reactions. Bioorg Med Chem 17:3782–3788

Kuwahara M, Takano Y, Kasahara Y et al (2010) Study on suitability of KOD DNA polymerase for enzymatic production of artificial nucleic acids using base/sugar modified nucleoside triphosphates. Molecules 15:8229–8240

Latham JA, Johnson R, Toole JJ (1994) The application of a modified nucleotide in aptamer selection: novel thrombin aptamers containing 5-(1-pentynyl)-2′-deoxyuridine. Nucleic Acids Res 22:2817–2822

Lato SM, Ozerova ND, He K et al (2002) Boron-containing aptamers to ATP. Nucleic Acids Res 30:1401–1407

Lee SE, Sidorov A, Gourlain T et al (2001) Enhancing the catalytic repertoire of nucleic acids: a systematic study of linker length and rigidity. Nucleic Acids Res 29:1565–1573

Li M, Lin N, Huang Z et al (2008) Selecting aptamers for a glycoprotein through the incorporation of the boronic acid moiety. J Am Chem Soc 130:12636–12638

Lin Y, Qiu Q, Gill SC et al (1994) Modified RNA sequence pools for *in vitro* selection. Nucleic Acids Res 22:5229–5234

Lin Y, Nieuwlandt D, Magallanez A et al (1996) High-affinity and specific recognition of human thyroid stimulating hormone (hTSH) by *in vitro*-selected 2′-amino-modified RNA. Nucleic Acids Res 24:3407–3414

Lundin KE, Højland T, Hansen BR et al (2013) Biological activity and biotechnological aspects of locked nucleic acids. Adv Genet 82:47–107

Masud MM, Ozaki AN, Kuwahara M et al (2003) Modified DNA bearing 5-(methoxycarbonylmethyl)-2′-deoxyuridine: preparation by PCR with thermophilic DNA polymerase and postsynthetic derivatization. Chembiochem 4:584–588

Matsuda S, Henry AA, Romesberg FE (2006) Optimization of unnatural base pair packing for polymerase recognition. J Am Chem Soc 128:6369–6375

Matsui M, Nishiyama Y, Ueji S et al (2007) Construction of saccharide-modified DNAs by DNA polymerase. Bioorg Med Chem Lett 17:456–460

McKeague M, Derosa MC (2012) Challenges and opportunities for small molecule aptamer development. J Nucleic Acids 2012:748913

Mendonsa SD, Bowser MT (2004) *In vitro* evolution of functional DNA using capillary electrophoresis. J Am Chem Soc 126:20–21

Mendonsa SD, Bowser MT (2005) *In vitro* selection of aptamers with affinity for neuropeptide Y using capillary electrophoresis. J Am Chem Soc 127:9382–9383

Milligan JF, Groebe DR, Witherell GW et al (1987) Oligoribonucleotide synthesis using T7 RNA polymerase and synthetic DNA templates. Nucleic Acids Res 15:8783–8798

Minakawa N, Sanji M, Kato Y et al (2008) Investigations toward the selection of fully-modified 4′-thioRNA aptamers: optimization of *in vitro* transcription steps in the presence of 4′-thioNTPs. Bioorg Med Chem 16:9450–9456

Mosing RK, Mendonsa SD, Bowser MT (2005) Capillary electrophoresis-SELEX selection of aptamers with affinity for HIV-1 reverse transcriptase. Anal Chem 77:6107–6112

Nagashima J, Minezaki S, Obika S et al (2007) Polymerisation of a DNA strand using oligo-DNA template with modified bases, sugars and phosphates. Nucleic Acids Symp Ser 51:55–56

Ng EW, Shima DT, Calias P et al (2006) Pegaptanib, a targeted anti-VEGF aptamer for ocular vascular disease. Nat Rev Drug Discov 5:123–132

Nitsche A, Kurth A, Dunkhorst A et al (2007) One-step selection of Vaccinia virus-binding DNA aptamers by MonoLEX. BMC Biotechnol 7:48

Nonaka Y, Yoshida W, Abe K et al (2013) Affinity improvement of a VEGF aptamer by in silico maturation for a sensitive VEGF-detection system. Anal Chem 85:1132–1137

Noronha AM, Wilds CJ, Lok CN et al (2000) Synthesis and biophysical properties of arabinonucleic acids (ANA): circular dichroic spectra, melting temperatures, and ribonuclease H susceptibility of ANA.RNA hybrid duplexes. Biochemistry 39:7050–7062

Obayashi T, Masud MM, Ozaki AN et al (2002) Enzymatic synthesis of labeled DNA by PCR using new fluorescent thymidine nucleotide analogue and superthermophilic *KOD dash* DNA polymerase. Bioorg Med Chem Lett 12:1167–1170

Obika S, Nanbu D, Hari Y et al (1997) Synthesis of 2′-*O*,4′-*C*-methyleneuridine and -cytidine. Novel bicyclic nucleosides having a fixed *C*3′-endo sugar puckering. Tetrahedron Lett 38:8735–8738

Obika S, Nanbu D, Hari Y et al (1998) Stability and structural features of the duplexes containing nucleoside analogs with a fixed N-type conformation, 2′-*O*,4′-*C*-methyleneribonucleosides. Tetrahedron Lett 39:5401–5404

Ohbayashi T, Kuwahara M, Hasegawa M et al (2005) Expansion of repertoire of modified DNAs prepared by PCR using *KOD Dash* DNA polymerase. Org Biomol Chem 3:2463–2468

Ohsawa K, Kasamatsu T, Nagashima J et al (2008) Arginine-modified DNA aptamers that show enantioselective recognition of the dicarboxylic acid moiety of glutamic acid. Anal Sci 24:167–172

Ono T, Scalf M, Smith LM (1997) 2′-Fluoro modified nucleic acids: polymerase-directed synthesis, properties and stability to analysis by matrix-assisted laser desorption/ionization mass spectrometry. Nucleic Acids Res 25:4581–4588

Padilla R, Sousa R (1999) Efficient synthesis of nucleic acids heavily modified with non-canonical ribose 2′-groups using a mutant T7 RNA polymerase (RNAP). Nucleic Acids Res 27:1561–1563

Padilla R, Sousa R (2002) A Y639F/H784A T7 RNA polymerase double mutant displays superior properties for synthesizing RNAs with non-canonical NTPs. Nucleic Acids Res 30:e138

Pagratis NC, Bell C, Chang YF et al (1997) Potent 2′-amino-, and 2′-fluoro-2′-deoxyribonucleotide RNA inhibitors of keratinocyte growth factor. Nat Biotechnol 15:68–73

Pan W, Craven RC, Qiu Q et al (1995) Isolation of virus-neutralizing RNAs from a large pool of random sequences. Proc Natl Acad Sci U S A 92:11509–11513

Pavey JB, Lawrence AJ, O'Neil IA et al (2004) Synthesis and transcription studies on 5′-triphosphates derived from 2′-C-branched-uridines: 2′-homouridine-5′-triphosphate is a substrate for T7 RNA polymerase. Org Biomol Chem 2:869–875

Peng CG, Damha MJ (2007) Polymerase-directed synthesis of 2′-deoxy-2′-fluoro-β-D-arabinonucleic acids. J Am Chem Soc 129:5310–5311

Piccirilli JA, Moroney SE, Benner SA (1991) A C-nucleotide base pair: methylpseudouridine-directed incorporation of formycin triphosphate into RNA catalyzed by T7 RNA polymerase. Biochemistry 30:10350–10356

Pinheiro VB, Taylor AI, Cozens C et al (2012) Synthetic genetic polymers capable of heredity and evolution. Science 336:341–344

Raines K, Gottlieb PA (1998) Enzymatic incorporation of 2′-thio-CTP into the HDV ribozyme. RNA 4:340–345

Robertson DL, Joyce GF (1990) Selection *in vitro* of an RNA enzyme that specifically cleaves single-stranded DNA. Nature 344:467–468

Ruckman J, Green LS, Beeson J et al (1998) 2′-Fluoropyrimidine RNA-based aptamers to the 165-amino acid form of vascular endothelial growth factor (VEGF165). Inhibition of receptor binding and VEGF-induced vascular permeability through interactions requiring the exon 7-encoded domain. J Biol Chem 273:20556–20567

Sakthivel K, Barbas CF III (1998) Expanding the potential of DNA for binding and catalysis: highly functionalized dUTP derivatives that are substrates for thermostable DNA polymerases. Angew Chem Int Ed Engl 37:2872–2875

Sassanfar M, Szostak JW (1993) An RNA motif that binds ATP. Nature 364:550–553

Sawai H, Ozaki AN, Satoh F et al (2001) Expansion of structural and functional diversities of DNA using new 5-substituted deoxyuridine derivatives by PCR with superthermophilic *KOD* Dash DNA polymerase. Chem Commun 24:2604–2605

Sawai H, Nagashima J, Kuwahara M et al (2007) Differences in substrate specificity of *C*(5)-substituted or *C*(5)-unsubstituted pyrimidine nucleotides by DNA polymerases from thermophilic bacteria, archaea, and phages. Chem Biodivers 4:1979–1995

Sazani PL, Larralde R, Szostak JW (2004) A small aptamer with strong and specific recognition of the triphosphate of ATP. J Am Chem Soc 126:8370–8371

Schmidt KS, Borkowski S, Kurreck J et al (2004) Application of locked nucleic acids to improve aptamer *in vivo* stability and targeting function. Nucleic Acids Res 32:5757–5765

Schoning K, Scholz P, Guntha S et al (2000) Eschenmoser, A. Chemical etiology of nucleic acid structure: the α-threofuranosyl-(3′→2′) oligonucleotide system. Science 290:1347–1351

Schürer H, Stembera K, Knoll D et al (2001) Aptamers that bind to the antibiotic moenomycin A. Bioorg Med Chem 9:2557–2563

Shao K, Ding W, Wang F et al (2011) Emulsion PCR: a high efficient way of PCR amplification of random DNA libraries in aptamer selection. PLoS One 6:e24910

Shoji A, Kuwahara M, Ozaki H et al (2007) Modified DNA aptamer that binds the (*R*)-isomer of a thalidomide derivative with high enantioselectivity. J Am Chem Soc 129:1456–1464

Singh SK, Koshkin AA, Wengel J et al (1998) LNA (locked nucleic acids): synthesis and high-affinity nucleic acid recognition. Chem Commun 4:455–456

Tahiri-Alaoui A, Frigotto L, Manville N et al (2002) High affinity nucleic acid aptamers for streptavidin incorporated into bi-specific capture ligands. Nucleic Acids Res 30:e45

Tan W, Donovan MJ, Jiang J (2013) Aptamers from cell-based selection for bioanalytical applications. Chem Rev 113:2842–2862

Tasara T, Angerer B, Damond M et al (2003) Incorporation of reporter molecule-labeled nucleotides by DNA polymerases. II. High-density labeling of natural DNA. Nucleic Acids Res 31:2636–2646

Tasset DM, Kubik MF, Steiner W (1997) Oligonucleotide inhibitors of human thrombin that bind distinct epitopes. J Mol Biol 272:688–698

Thum O, Jager S, Famulok M (2001) Functionalized DNA: a new replicable biopolymer. Angew Chem Int Ed Engl 40:3990–3993

Tsai CH, Chen J, Szostak JW (2007) Enzymatic synthesis of DNA on glycerol nucleic acid templates without stable duplex formation between product and template. Proc Natl Acad Sci U S A 104:14598–14603

Tuerk C, Gold L (1990) Systematic evolution of ligands by exponential enrichment. Science 249:505–510

Ulrich H, Wrenger C (2009) Disease-specific biomarker discovery by aptamers. Cytometry A 75:727–733

Uphoff KW, Bell SD, Ellington AD (1996) *In vitro* selection of aptamers: the dearth of pure reason. Curr Opin Struct Biol 6:281–288

Vaish NK, Fraley AW, Szostak JW et al (2000) Expanding the structural and functional diversity of RNA: analog uridine triphosphates as candidates for *in vitro* selection of nucleic acids. Nucleic Acids Res 28:3316–33122

Vaish NK, Larralde R, Fraley AW et al (2003) A novel, modification-dependent ATP-binding aptamer selected from an RNA library incorporating a cationic functionality. Biochemistry 42:8842–8851

Vastmans K, Pochet S, Peys A et al (2000) Enzymatic incorporation in DNA of 1,5-anhydrohexitol nucleotides. Biochemistry 39:12757–12765

Vater A, Klussmann S (2003) Toward third-generation aptamers: Spiegelmers and their therapeutic prospects. Curr Opin Drug Discov Devel 6:253–261

Vaught JD, Bock C, Carter J et al (2010) Expanding the chemistry of DNA for *in vitro* selection. J Am Chem Soc 132:4141–4151

Veedu RN, Vester B, Wengel J (2007a) *In vitro* incorporation of LNA nucleotides. Nucleosides Nucleotides Nucleic Acids 26:1207–1210

Veedu RN, Vester B, Wengel J (2007b) Novel applications of locked nucleic acids. Nucleic Acids Symp Ser 51:29–30

Veedu RN, Vester B, Wengel J (2007c) Enzymatic incorporation of LNA nucleotides into DNA strands. Chembiochem 8:490–492

Veedu RN, Vester B, Wengel J (2008) Polymerase chain reaction and transcription using locked nucleic acid nucleotide triphosphates. J Am Chem Soc 130:8124–8125

Veedu RN, Vester B, Wengel J (2009) Efficient enzymatic synthesis of LNA-modified DNA duplexes using *KOD* DNA polymerase. Org Biomol Chem 7:1404–1409

Veedu RN, Burri HV, Kumar P et al (2010) Polymerase-directed synthesis of C5-ethynyl locked nucleic acids. Bioorg Med Chem Lett 20:6565–6568

Wang K, You M, Chen Y et al (2011) Self-assembly of a bifunctional DNA carrier for drug delivery. Angew Chem Int Ed Engl 50:6098–6101

Wheeler M, Chardon A, Goubet A et al (2012) Synthesis of selenomethylene-locked nucleic acid (SeLNA)-modified oligonucleotides by polymerases. Chem Commun 48:11020–11022

White R, Rusconi C, Scardino E et al (2001) Generation of species cross-reactive aptamers using "toggle" SELEX. Mol Ther 4:567–573

Wiegand TW, Williams PB, Dreskin SC et al (1996) High-affinity oligonucleotide ligands to human IgE inhibit binding to Fc epsilon receptor I. J Immunol 157:221–230

Wochner A, Menger M, Orgel D et al (2008) A DNA aptamer with high affinity and specificity for therapeutic anthracyclines. Anal Biochem 373:34–42

Yamamoto T, Nakatani M, Narukawa K et al (2011) Antisense drug discovery and development. Future Med Chem 3:339–365

Yu H, Zhang S, Chaput JC (2012) Darwinian evolution of an alternative genetic system provides support for TNA as an RNA progenitor. Nat Chem 4:183–187

Zaug AJ, Cech TR (1986) The intervening sequence RNA of *Tetrahymena* is an enzyme. Science 231:470–475

Zhu Z, Chao J, Yu H et al (1994) Directly labeled DNA probes using fluorescent nucleotides with different length linkers. Nucleic Acids Res 22:3418–3422

Zhu G, Meng L, Ye M et al (2012) Self-assembled aptamer-based drug carriers for bispecific cytotoxicity to cancer cells. Chem Asian J 7:1630–1636

Aptamers as Molecular Smugglers

Eileen Magbanua and Ulrich Hahn

Contents

1 Aptamers 272
  1.1 Aptamer Stability 273
2 Receptor-Mediated Endocytosis Pathway 273
3 Endosomal Escape 275
  3.1 Naturally Occurring Endosomal Escape Agents 276
  3.2 Chemical Agents 277
  3.3 Photosensitizers 277
4 Targets on Cell Surfaces 278
  4.1 Cell Surface Receptors 278
  4.2 Proteins 279
  4.3 Tumor Antigens 280
5 Delivery Agents 281
  5.1 Fluorophores 281
  5.2 Radionuclides 282
  5.3 Chemotherapeutics 283
  5.4 Proteins 285
  5.5 Nucleic Acid-Based Therapeutics 286
  5.6 Encapsulated Agents in Nanocarriers 286
References 287

Abstract Transporting therapeutics to their final destination within the body is still a challenge. For this purpose aptamers are particularly advantageous because of their ability to discriminate between different tissues or cell types. If a specific cell type or diseased tissue presents a characteristic marker like a protein, lipid, or sugar on its cell surface, aptamers can be selected that target exclusively this type of cells.

E. Magbanua • U. Hahn (✉)
Chemistry Department, Institute for Biochemistry and Molecular Biology, Hamburg University, 20146 Hamburg, Germany
e-mail: uli.hahn@uni-hamburg.de

V.A. Erdmann et al. (eds.), *Chemical Biology of Nucleic Acids*, RNA Technologies,
DOI 10.1007/978-3-642-54452-1_15,

The concentrated localization of an aptamer at defined tissues can also reduce the necessary dose of the therapeutic. Additionally, aptamers not only target cells specifically but are also able to carry a cargo inside the cell via receptor-mediated endocytosis for instance. After the internalization process the aptamer together with its cargo is located in vesicles inside the cell, in so-called endosomes. Receptors are normally recycled back to the cell surface or degraded and the aptamer with its cargo has to escape these vesicles to become operative. We here give a brief overview about the destiny of the internalized receptor with its cargo inside the cell, how the cargo can escape, and what type of cargos can be utilized.

Keywords Aptamer • Receptor recycling • Drug delivery

1 Aptamers

Aptamers are small oligonucleotides, which obtain a specific three-dimensional structure by intramolecular interactions like base pairing or stacking. In doing so not only conventional Watson and Crick base pairing but also Hoogsteen base pairing can occur. As a result, various tertiary structures such as G-quadruplexes, triple-stranded helices or coaxial stacking as in pseudoknots can be formed, structures which are found especially in functional RNAs. In addition, these structures are normally stabilized by complexed cations. Such complex structures can also be formed by aptamers and enable them to interact intermolecularly with specific targets to bind them with high affinity. In the majority of cases the destruction or destabilization of the aptamer structure involves a significant loss of affinity. This can be achieved by simple nucleotide replacement or removal of essential stabilizing cations.

The traditional technique for aptamer selection is the SELEX (systematic evolution of ligands by exponential enrichment) process (Ellington and Szostak 1990; Tuerk and Gold 1990). Basically the selection procedure can be divided into three main parts: (1) incubation of an oligonucleotide pool with the target, (2) separation of bound and unbound molecules, and (3) amplification of bound molecules to generate an enriched pool which is then again incubated with the target. This iterative process is repeated several times, followed by aptamer separation and characterization. Depending on the function of the aptamer various SELEX procedures can be performed. Here we focus on RNA aptamers that bind to specific cell surfaces and are internalized receptor mediated. They can further serve as delivery vehicles to "smuggle goods" as therapeutics, for instances, into the cell. Cell surface receptors can serve as appropriate "smugglers" because they can be internalized specifically due to ligand binding or during recycling processes. Practically, receptor recycling processes can be stopped via cooling the cells down on ice and can be induced by warming up the cells at 37 °C. However, the location of the receptor within the cell might differ between these internalization pathways and is even dependent on the type of molecule that is delivered into the cell via the aptamer.

The selection of receptor-specific aptamers can be challenging if the receptor or the domain of choice cannot be soluble expressed. A proper alternative represents the cell-SELEX where living cells are used. Cell-SELEX is not further described in this chapter, but well described elsewhere (Zhang et al. 2010; Guo et al. 2008; Cerchia and de Franciscis 2010; Meyer et al. 2011).

1.1 Aptamer Stability

The most challenging problem with aptamers, especially in case of RNA aptamers, is the stability. With respect to the usage of aptamers as therapeutic agents, the pharmacokinetic plays a decisive role. To increase aptamer resistance against hydrolysis, the 2′-hydroxyl group can be substituted with 2′-fluoro, 2′-*O*-methyl or 2′-*O*-methoxyethyl. The backbone can be modified with phosphorothioate, and "locked" nucleic acids (2′-*O*,4′-*C*-methylene linked bicyclic ribonucleotides), *L*-nucleic acids ("spiegelmers"), or terminal caps (e.g., inverted nucleotides) can be used as well. The alternative odds are to circulate the aptamer to hinder exonucleases. The fast clearance rate of the aptamer due to its small size can be critical in some cases, but can also be advantageous in case of delivering chemotherapeutics which should have a short medication time, for instance. Otherwise the size of the aptamer can be increased by adding polyethylene glycol or cholesterol.

2 Receptor-Mediated Endocytosis Pathway

The following recycling processes of the receptor and their regulations are not completely understood and further in-depth information can be found in several excellent review publications (Maxfield and McGraw 2004; Grant and Donaldson 2009; Huotari and Helenius 2011).

Basically, receptors can be internalized by clathrin-dependent or clathrin-independent vesiculation. After internalization the receptor is delivered into the early endosome, which can fuse with other endosomes to form sorting endosomes (Fig. 1). The fusion of these new endosomes with the sorting endosomes is regulated by Rab5 and the early endosome antigen 1 (EEA1) (Lawe et al. 2002). Rab proteins are small GTPases that play a critical role in most regulatory transportation processes. Within the endosomal system, Rab4, Rab5, and Rab11 regulate recycling processes. Sorting endosomes are transient (about 10 min) and tubular–vesicular structured with a mild acidic lumen (pH 5.9–6.0), which enables bound ligands to dissociate from the receptor. These endosomes represent the first junction for receptor sorting, whereby three main routes are possible: to the plasma membrane, to the endocytic recycling compartment (ERC) or the endosome maturates to

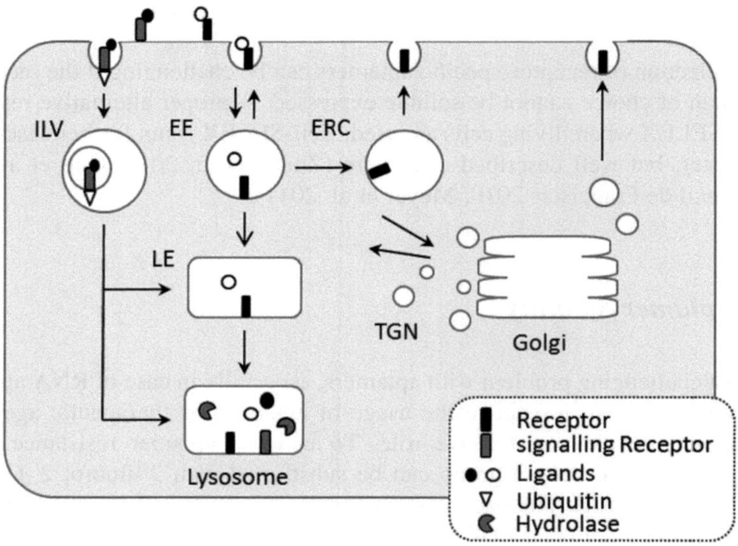

Fig. 1 Endosomal pathways of internalized receptors. Signaling receptors are directly internalized into intraluminal vesicles (ILV) to prevent further stimulation. Such receptors are labeled with ubiquitin. Other internalized receptors are transported into the early endosome (EE), where the ligands can dissociate from the receptor due to pH reduction. Starting from EE three ways back to cell surface are possible: directly from EE, via the endocytic recycling compartment (ERC) or via the *trans*-Golgi network (TGN). Otherwise the receptor reaches the late endosome (LE) and further the lysosome where hydrolases degrade all contents

the late endosome. The fastest receptor recycling route is the direct way back to the plasma membrane (Maxfield and McGraw 2004). About half of the molecules within the sorting endosomes go directly back to the plasma membrane and exit the sorting endosome by fission tubules.

There is a second way back to the plasma membrane via ERC (Fig. 1). If receptors or proteins are not directly transported to the plasma membrane, they are translocated to ERC to prevent them from degradation within the maturating endosome. ERC is another early endosome that consists of tubular compartments and receives membrane proteins from the early endosome. The ERC is more persistent compared to the sorting endosome and also serves as a sorting check point for proteins. Translocation from the ERC to the plasma membrane presents the slower recycling route and transportation as well as participating regulatory proteins differ from clathrin-dependent to clathrin-independent receptors. But one of the primarily responsible proteins that regulate recycling processes at ERC is Rab11, again one representative of the Rab family. The absence of Rab11 can prevent receptors from translocation to the plasma membrane (Wilcke et al. 2000).

If the internalized receptor is not sorted out of the endosome, this compartment maturates to the late endosome, comprising all dissociated ligands in the lumen, as well as remained receptors and proteins. As maturation proceeds, translocation of receptors to the plasma membrane is no longer possible. However, a large proportion of the receptors are recycled in advance. During the maturation process of the endosome Rab proteins are exchanged. Rab5 is replaced by Rab7, as well as Rab11, Rab4, Rab22, and Rab9 are added. Additionally, the lumen acidifies to a pH between 4.9 and 6.0 (Maxfield and Yamashiro 1987) and the concentration of cations like Ca^{2+}, Na^{+}, and K^{+} changes (Jentsch 2008; Huynh and Grinstein 2007). This process goes hand in hand with morphological changes of the endosome, which in its later state becomes bigger and more globular shaped.

Receptors cannot only be recycled but also degraded. Especially signaling receptors are delivered to the late endosomes to stop a continuous stimulation of a cell and to make the cell insensible for a new stimulation (Katzmann et al. 2002). Some of these signaling receptors are ubiquitylated at the cytoplasmatic side and thus labeled for degradation. To strictly hinder internalized signaling receptors to stay in contact with the cytosol, these receptors are delivered into intraluminal vesicles (ILVs). ILV formation already begins in the early endosomes. For this kind of sorting, special machinery is responsible: the endosomal sorting complex required for transport (ESCRT) (Raiborg et al. 2002; Sachse et al. 2002). However, in course of terminating maturation, the late endosome can fuse with other late endosomes or with lysosomes. The lysosome represents the final destination of the receptor plus its components within the lumen since most of them will be degraded by hydrolysis.

The early and the late endosome stay continuously in dialog with the *trans*-Golgi network (TGN). Proteins can be dispatched from the endosome by vesicles and fuse with Golgi compartments. Via the TGN proteins can be sorted to *post*-Golgi compartments like plasma membrane, lysosomes, and endosomes as well. This transportation routes can function bidirectional, with the result that receptors can be recycled via TGN to the plasma membrane but also be degraded in the lysosome. Furthermore, correlated to endosome maturation, the endosome receives hydrolases and other proteins from the TGN as well.

3 Endosomal Escape

Targeting receptors on cell surfaces by aptamers with additional cargos for drug delivery, the final destination is generally clear. But usually the destinations of the receptor and the cargo do not coincide, especially in case of signaling receptors. Additionally, the cargos often dissociate from the receptor due to acidic environment within the endosomal lumen and are further degraded independently from the recycled receptor in the lysosome. If the lysosome does not represent the final destination, the cargo needs to escape the endosome before entering the late endosome or lysosome (Fig. 2).

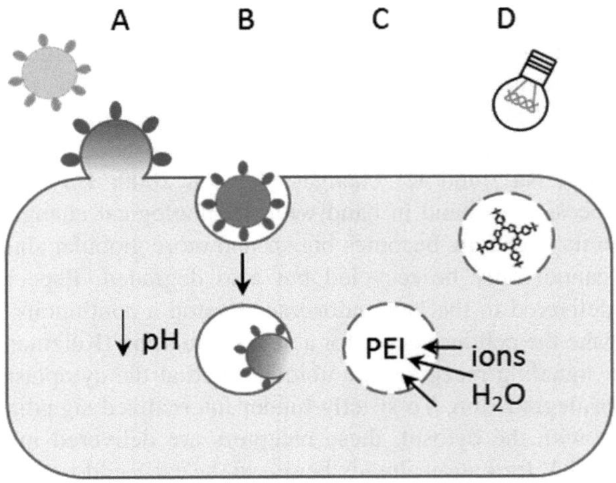

Fig. 2 Endosomal escape. Several strategies exist to enter a host cell or to escape the endosome after internalization process. (A) A bacterium or a virus can fuse directly with a host cell to release its contents into the cytoplasm. (B) Alternatively, the virus or bacterium is internalized and bacterial or viral proteins are able to interfere with endosomal membrane after pH reduction. (C) Introduced chemical agents like polyethylenimine (PEI) can cause disruption of the endosome due to osmotic stress or (D) photosensitizers can be activated by light to disrupt the endosome

3.1 Naturally Occurring Endosomal Escape Agents

Viruses and bacteria have developed different strategies and mechanisms to enter a host cell and to escape the endosome efficiently. Additionally, several toxins of plant origin or of bee venom enter the cell via endocytosis. These naturally occurring endosomal escape agents are predominantly peptides or proteins, which escape the endosome via two main mechanisms. The first strategy is the fusion of the virus or bacterial membrane with the endosomal membrane, resulting in release of viral contents into the cytosol. Cell surface proteins or peptides (e.g., HA2 subunit of influenza haemagglutinin, L2 from Papillomavirus, envelope protein E from West Nile virus, Diphtheria toxin, and glycoprotein H from herpes simplex virus) undergo conformational changes due to the reduction of pH within the endosome and penetrating its membrane. Subsequently, the spatial proximity promotes fusion of both membranes. Several new synthetic peptides or proteins have been designed based on the HA2 subunit of haemagglutinin, which have an enhanced efficiency. The second strategy is the perforation of the endosomal membrane due to pore formation. For this pore formation, different models exist, which are based on interference of endosomal membrane tension by peptides or proteins (e.g., Melittin, a component of bee venom, or bacterial exotoxins like exotoxin A of *P. aeruginosa*, Shiga toxin, Diphtheria toxin, or Cholera toxin). The interference causes a short-time enlargement of pores, which are then occupied and stabilized by the peptides or proteins. After pore formation, the bacterium can fuse with the endosomal membrane or lyse the endosome to enter the cell.

Some of these agents are not only competent endosomal escape agents but also effective toxins. For this reason, lots of these toxins are used as immunotoxins fused to antibodies and are also appropriate tools for aptamer-mediated delivery. Those include bacterial exotoxins as already mentioned as well as plant-derived toxins like Ricin, Gelonin, or Saporin (Varkouhi et al. 2011).

3.2 *Chemical Agents*

There are several other endosomal escape agents that are inconclusively used by bacteria or viruses and operate differently. One class of agents are chemicals like polyethylenimine, poly(aminoamine) polymer, or poly(propylacylic) acid. Those have a high buffer capacity, which leads to a high influx of ions due to protonation. This in turn leads to disruption of endosomal membrane in consequence of osmotic stress (Lin and Engbersen 2008; Varkouhi et al. 2011).

3.3 *Photosensitizers*

Another class of agents comprises photosensitizers, which are used for cancer therapy to release therapeutics into the cell efficiently. Structurally, they are based on tetrapyrrols or rather porphyrins and usually coordinate metal ions, but are more efficient endosomal escape agents in absence of these ions. Photosensitizers have to be activated after internalization via visible light of defined wavelength. As a result the photosensitizers form free radicals or reactive oxygen, which leads to disruption of the endosomal membrane due to oxidation of amino acids, unsaturated fatty acids and cholesterol. Furthermore, due to their substituents photosensitizers can be separated into amphiphilic and hydrophilic items and can be used either as endosomal escape agents or as toxins. Amphiphilic sensitizers (e.g., chlorine e_6, tetraphenylporphine disulfonate like $TPPS_{2a}$, or aluminum phthalocyanine disulfonate like $AlPcS_{2a}$), which have no membrane penetrating ability, localize in the plasma membrane and, after activation, the endosomal membrane is destroyed. In comparison, hydrophilic sensitizers (e.g., $TPPS_4$, mTHPC, $AlPcS_4$) are preferably localized in the lumen of endosomes or lysosomes and can photoinactivate contents which are also present inside the lumen. Depending on the sensitizer, treatment conditions, and cell type, photosensitizers can induce apoptosis (Berg et al. 2010).

4 Targets on Cell Surfaces

4.1 Cell Surface Receptors

Receptors represent the classical target for cell-specific internalization due to their regulated recycling processes. In principle, all receptors are suitable drug delivery vehicles, but regarding specificity, destination, and the main intend to reduce side effects, ubiquitous presented receptors are not always advantageous for this purpose. Therefore, receptors should be used which are associated to a special disease, tissue, or organ. Several aptamers have been selected which channel in its cargo efficiently without any additional endosomal escape agents. Among their targets are only a few receptors.

The transferrin receptor (TfR) has been a favored delivery target for a while, since this receptor is upregulated in malignant cells and needs iron for cell proliferation. Additionally, the amount of presented TfR is directly proportional to tumor stage and progression. As delivery vehicles, the natural ligand transferrin, antibodies, antibody fragments, or peptides have been coupled to chemotherapeutics (e.g., chlorambucil, mitomycin-C, artemisinin, doxorubicin, cisplatin), toxins (e.g., Saporin, Ricin A chain, Diphtheria toxin, Cholera exotoxin, Gelonin, Restrictoxin), enzymes (e.g., RNases), oligonucleotides (e.g., shRNA), as well as radionuclides, reviewed in Daniels et al. (2012). Dependent on application, the delivered agents have additionally been packed in transport carriers like micelles, dendrimers or nanoparticles (Daniels et al. 2012). Furthermore, DNA as well as RNA aptamers have been selected with affinity for the transferrin receptor coupled to a lysosomal enzyme like α-L-iduronidase (Chen et al. 2008).

Another well-known and established receptor for drug delivery is the epidermal growth factor receptor (EGFR), which plays an oncogenic role in several tumors. Besides the wild-type EGFR, malignant cells also produce a truncated form, EGFRvIII, which is tumor specific and does not bind natural EGFR ligands or inhibitors. However, the EGFRvIII-specific antibody mAb80 solely binds and inactivates EGFRvIII as well as the overproduced form of EGFR due to different protein conformations (Gan et al. 2012). Furthermore, a wild-type-specific EGFR aptamer has been selected, which was conjugated to gold nanoparticles and efficiently internalized via endocytosis (Li et al. 2010). Additionally, a polymer consisting of the nucleoside analogue gemcitabine was coupled to an EGFR-specific aptamer and successfully internalized into pancreatic cells (Ray et al. 2012).

The vesicular epidermal growth factor (VEGF) receptor is also upregulated in malignant cells, plays a crucial role in tumor angiogenesis, and is used as a delivery system as well. As delivery agents oligopeptide ligands have been designed and conjugated to HA20 (derived from influenza haemagglutinin (Li et al. 1999)). But primarily anti-VEGF agents are constructed to counteract angiogenesis. One well-known inhibitor, which is mainly applied to defend different types of neovascular age-related macular degeneration, is the RNA aptamer Pegaptanib. Due to its low molecular size it is directly applied into the ocular bulb and also the first aptamer-based therapeutic (Ruckman et al. 1998; Ng et al. 2006).

Furthermore, aptamers have been used for efficient cargo delivery via CD4 receptor (e.g., siRNAs, fluorophores (Guo et al. 2005; Khaled et al. 2005)), HIV-1 gp120 (siRNAs (Zhou et al. 2009)), tyrosine kinase 7 receptor (e.g., chemotherapeutics, enzymes (Shangguan et al. 2006; Huang et al. 2009; Xiao et al. 2008)), and the interleukin-6 receptor (fluorophores (Meyer et al. 2012)) as well.

Although there have only been a few receptor-specific aptamers yet, there are additional promising aptamer targets which are already examined in the context with other delivery strategies, such as the folate receptor which is upregulated in some malignant cells (Leamon and Low 2005). It is also known that especially in ovarian cancer, cell division is dependent on cellular folic acid uptake by the folate receptor. Blocking this upregulated receptor in malignant cells is one anticancer strategy, but another strategy is the delivery of therapeutics via the folate receptor. For this purpose several macromolecules were directly conjugated to folic acid, such as toxins or anti-neoplastic agents (e.g., momordin, *Pseudomonas* exotoxin, doxorubicin, maytansinoids), radioisotopes, antibodies, siRNAs, or antisense DNA (e.g., of c-myb, epidermal growth factor). Healthy cells also internalize the conjugates but not in such a high amount that delivered agents could be effective. One limiting factor seems to be the size of the delivered substance, reviewed in Gruner and Weitman (1998).

4.2 Proteins

A few proteins also serve as efficient delivery targets to enter specific cells. Nucleolin is one of these. It is ubiquitously expressed in proliferating cells, predominantly within the nucleolus. This multifunctional protein can be translocated to the cytosol in case that it is phosphorylated. But nucleolin is also localized on the cell membrane—prevalently on surfaces of cancer cells. The functions of nucleolin are manifold since it binds DNA as well as RNA and is able to unwind both. Additionally, it exhibits several phosphorylation sites and binding sites for proteins (Abdelmohsen and Gorospe 2012). A nucleolin-specific aptamer (Bates et al. 2009) has been selected, which was used for cargo delivery of photosynthesizes (Shieh et al. 2010) or drug-encapsulated liposomes (Cao et al. 2009), for instance.

Tenascin-C is a glycoprotein that is produced in diseased tissues in a high extend, associated with inflammation and infections or cancer. Since tenascin-C exhibits multiple functions, it activates oncogenic signaling and thereby promotes tumor progression, reviewed in Midwood and Orend (2009). Several tensacin-C targeting antibodies are in clinical trials (Akabani et al. 2005) and tenascin-C-specific aptamers also exist, (Daniels et al. 2003; Hicke et al. 2001) which are transported into different tumors, too (Hicke et al. 2006).

4.3 Tumor Antigens

Cell surfaces often differ in the context of diseases. Especially tumor cells change their protein expression levels, modify their protein compositions on the cell surface, or start to present protein variants. These tumor antigens on malignant cells or tissues are optimal targets for drug delivery. Tumor antigens themselves are defined as agents that induce an immune response.

Some genes are solely activated in tumor cells and otherwise completely silent. Such genes are called cancer-germline genes, and representatives of this gene family are *MAGE*, *GAGE*, and *SSX*. All these genes encode proteins or peptides being tumor antigens (Coulie et al. 2001).

One such very promising tumor antigen is the prostate-specific membrane antigen (PSMA), which is faintly present in healthy epithelia cells of prostate, kidney, and in the brain, but in higher levels on prostate cancer cells. The portion of PSMA on cancer cells increases proportionally to disease grade (Ross et al. 2003) and additionally, it is known that PSMA is endocytosed via clathrin-coated pits in LNCaP cells (Liu et al. 1998). For this reason, PSMA was used as delivery vehicle applying an RNA aptamer with high affinity to PSMA (Lupold et al. 2002). To date, a large number of various cargos have successfully been internalized mainly to destroy tumor cells (e.g., via siRNAs, toxins or chemotherapeutics).

Mucin-1 (MUC1) is another tumor antigen that is overproduced in malignant cells—especially in adenocarcinomas. This over-presented MUC1 differs in glycosylation state from the mature MUC1, insofar as it is incompletely sialylated. But during several recycling processes, MUC1 is finally successively glycosylated in extenso (Litvinov and Hilkens 1993). Comparing both MUC1 proteins, the under-glycosylated MUC1 is endocytosed with higher frequency via clathrin-coated pits (Altschuler et al. 2000) and already serves as target for aptamer-mediated delivery (e.g., chlorine e_6) (Ferreira et al. 2006, 2009).

Tumor-specific variants of peptides or proteins derive its origin of aberrant translational start points (e.g., *N*-acetylglucosaminyltransferase V (Guilloux et al. 1996)) or false splicing products (e.g., *TRP2* (Lupetti et al. 1998)). Another reason for tumor-specific protein variants are point mutations, which can lead to changed presentation and interaction with a human leucocyte antigen or to an alteration or progression of the tumor (e.g., CDK4, β-catenin, ras, caspase 8). But these kinds of protein point variants are not suitable aptamer targets due to their uniqueness in special tumors. However, chromosome translocations can produce fusion proteins, which can be tumor antigenic as well as oncogenic. One example is the *BCR-Abl* protein, which is mainly associated with chronic myelogenous leukemia but is only present in the cytosol (Kurzrock et al. 2003). Anyway, specific chromosome translation products mostly correlate with a cancer disease, which makes them appropriate targets if presented on cell surfaces.

Another class of tumor antigens is presented on cells, which are infected with oncogenic viruses. These presented proteins or peptides are virus specific and can serve perfectly as delivery targets in case that they are internalized. After infection by the Epstein–Barr virus for instance, the cells present viral EBNA2 and EBNA3 on their surfaces (Raab-Traub 2012). Another example is the human Papilloma virus (HPV), which infects mucosal and cutaneous epithelia cells and plays a role in cervical cancer. Especially the HPV types HPV16 and 18 present the viral oncoproteins E6 and E7 exclusively and continuously on cell surfaces of cervical tumor cells (Ressing et al. 1995; Seedorf et al. 1987), reviewed in Coulie et al. (2001).

Targeting tumor antigens by antibodies to deliver toxins into malignant cells and solid tumors is a well-established strategy. These agents are named immunotoxins and all immunotoxin targets can perfectly serve as targets for aptamer-mediated delivery as well. Among these are several interleukin receptors (IL-R) like IL-2R, IL-4R, and the IL-13Rα2, which are all up-regulated in malignant cells. Other targets are presented on B- and T-cells, like CD30, a membrane protein of the necrosis factor receptor family, CD19, which is also named B-lymphocyte antigen 19 and CD22, a sugar binding transmembrane antigen, reviewed in Kreitman (2006). A further immunotoxin target is mesothelin a cell surface protein of mesothelioma (Li et al. 2004).

5 Delivery Agents

By now a wide range of delivery agents were investigated, not only in the context of aptamers but also in the context of antibodies or antibody fragments, proteins, polymeric carriers as dendrimers and micelles, liposomes, as well as nanoparticles which all can serve as delivery vehicles. Specific targeted delivery, however, can only be ensured by cell-specific vehicles like aptamers, antibodies, or other specific compounds.

The choice of the delivered substance is dependent on the subsequent application and desired effect. Specific cells, tissues, or tumors can be detected by the usage of fluorophores like dyes, quantum dots, or fluorescent proteins. Further, chemotherapeutics as well as toxins lead to cell death of targeted cells or tissues, genes can be either up- or downregulated, inhibitors can block essential signaling or specific enzymes, otherwise enzymes can be delivered to promote homeostasis in special cells. In the following diverse delivery agents are described which have been used as delivery targets in combination with aptamers or other agents or else would work as well-suited delivery targets.

5.1 Fluorophores

To detect or to localize tissues, tumors, or specific cells the aptamer of choice can be labeled with a fluorescing dye. On a cell the binding and the internalization process of the labeled aptamer per se, as well as the localization of the aptamer inside the cell, can be detected optically. As label a nontoxic, stable fluorophor should be chosen, which exhibits a high quantum yield as well as a harmless excitation wavelength and a dye which does not interfere with the aptamer functionality or binds unspecifically to the cell. Furthermore, dyes can be used that either start to fluoresce after metabolization, change color due to its environment, or work as a toxin.

Fluorescence-based pH sensors can help to identify the localization of the aptamer inside the cell, for instance, due to differing pH values within the cellular compartments, which lead to shifted excitation wavelengths. Besides, pH values change during lots of cellular processes as apoptosis, proliferation, or endocytosis. For this purpose several fluorophores can be used as xanthene dyes, fluoresceins, and pyrene dyes, reviewed in Han and Burgess (2010). To overcome the problem that most of these dyes are monodirectional active, ratiomeric fluorophores can be used which act in both directions (Zhou et al. 2013; Chauhan et al. 2011). There also exist indicators for metal or miscellaneous ions.

In addition, lots of delivery agents exhibit a native fluorescence as chemotherapeutics (e.g., doxorubicin (Bagalkot et al. 2006), triapine (Kowol et al. 2010), topotecan (Gryczynski et al. 1999)), photosensitizers (e.g., chlorin e_6), or toxins (e.g., mycotoxins, like aflatoxin B_1 (Maragos et al. 2008)). Another possibility is to use fluorogenic substrates that obtain their fluorescence character not until they are metabolized by hydrolytic enzymes as phosphatases, peptidases, or peroxidases. Among these agents are fluorogenic peptide substrates or small molecules. Fluorogenic peptides consist of a fluorophor and a proper quencher, which can be separated due to cleavage site in between by a cell-specific active enzyme (Matayoshi et al. 1990; Packard and Komoriya 2008).

5.2 Radionuclides

Radionuclides are commonly used as radioactive tracers for tumor detection and localization in autoradiography and nuclear medicine. To reduce the administered dose the radionuclide is attached to a specific targeting agent as ligands, antibodies, or aptamers. The main benefit of aptamers is the small size which enables its penetration into tissues on the one hand and a fast clearance rate by the kidney on the other hand. For this purpose, a widely used radionuclide is technetium-99m (^{99m}Tc) due to its half-life period of 6 h. Labeling a tenascin-C-specific aptamer with ^{99m}Tc allows detecting several solid tumors (Hicke et al. 2001). Additionally, radionuclides cannot only be used for imaging but also for directed radiotherapy to achieve a high focal dose. Therefore aptamers would be perfectly suited, too.

5.3 Chemotherapeutics

Chemotherapeutics are agents that are commonly used to treat cancer and mainly act as cytostatic agents. They can be active cell cycle-phase-specific within one of the four phases of the cell cycle or nonspecific active during all phases. Comparatively few of these agents have been delivered aptamer-mediated yet. In the following chemotherapeutics are classified regarding their chemical characteristics.

Anti-metabolites are small molecules that replace metabolites due to their analogous chemical structure. With regard to cytostatic agents the anti-metabolites are mostly nucleoside analogs, which interfere with DNA as well as RNA and inhibit their biological processes. The consequence being that both cell division and cell growth is massively affected which can lead to cell death. The cellular and cytotoxic effects of anti-metabolites are often not that simple. Within the cell the anti-metabolites are catabolized or metabolized and can have various effects.

As a pyrimidine analogue for uracil 5-fluorouracil (5-FU) can be used. 5-FU is commonly used against cancer and normally applied directly. It works as an inhibitor of the thymidylate synthase which leads to diminished thymidine synthesis and causes cell death, reviewed in Longley et al. (2003). Alternatively, floxuridine could be used due to its analogous activity or capecitabine, which is a 5-FU prodrug and is converted to 5-FU within three steps. Another pyrimidine analogue or more precisely a nucleoside analogue of deoxycytidine is gemcitabine. The integration of gemcitabine during DNA replication leads to apoptosis. Further, gemcitabine inhibits ribonucleotide reductase which causes cell death as well (Cerqueira et al. 2007). It should be pointed out that gemcitabine was the first chemotherapeutic that was directly linked to an aptamer for targeted delivery (Ray et al. 2012).

Purine analogues are more presented and several of them can be summed up as thiopurines. Among these are azathioprine, 6-mercaptopurine, and 6-thioguanine which are all nonactive prodrugs. All these prodrug thiopurines are further metabolized into cytotoxic and immunosuppressive thiopurine nucleotides. Thiopurines are used to treat inflammatory disorders or as immunosuppressants (Aarbakke et al. 1997; Karran 2006). Other cytostatic purine nucleoside analogues are pentostatin, fludarabine, clofarabine, and cladribine which all have lymphotoxic effects. To reach their active form they have to be phosphorylated by deoxycytidine kinase (dCK). Within the cell pentostatin and cladrine are adenosine deaminase inhibitors (Brogden and Sorkin 1993; Spurgeon et al. 2009). Clofarabine interferes with DNA, inhibits ribonucleotide reductase, and induces apoptosis, reviewed in Hijiya et al. (2012) and Bonate et al. (2006).

Methotrexate (MTX) is another well-known anti-metabolite but not for nucleotides. It is structurally similar to dihydrofolic acid and inhibits dihydrofolate reductase efficiently with a thousand-fold increased affinity compared to its natural ligand. Additionally, MTX inhibits thymidylate synthase and 5-aminoimidazole-4-carboxamide ribonucleotide transformylase, which leads to interference of DNA synthesis and purine synthesis, reviewed in Tian and Cronstein (2007). MTX is a well-efficient anticancer drug and has been adapted to diverse solid tumors and malignancies. To reduce side effects due to high MTX doses, different delivery strategies have been developed and employed. MTX has been conjugated to albumin, to polymers (e.g., dextran, polyethylene glycol, hyaluronic acid, gelatin), to tumor-specific antibodies, to amino acids and peptides (e.g., insulin), to liposomes, and to polymeric microspheres, nanoparticles, or dendrimers, reviewed in Abolmaali et al. (2013).

Another class of chemotherapeutics is that of anthracyclines, which comprises agents that intercalate into DNA and DNA/RNA strands. This leads to inhibition of replication and inhibition of topoisomerase II and causes cell death. A very familiar representative of this class is doxorubicin (Dox). Due to its high cardio-toxicity derivatives or prodrugs have been developed like daunorubicin, epirubicin, and idarubicin (Monsuez et al. 2010). Nevertheless, to reduce side effects efficiently diverse delivery systems have been established. Taking advantage of dox' nature that it emits fluorescence and intercalates into DNA, Dox was physically conjugated to an aptamer which led to reduced fluorescence (Bagalkot et al. 2006; Huang et al. 2009). Another approach used also an aptamer for targeted delivery, but in complex with daunorubicin (Taghdisi et al. 2010). In addition, Dox was also covalently linked to a hormone analogue (Schally and Nagy 2004). In another alternative strategy, Dox has been encapsulated into nanoparticle–aptamer-bioconjugates (Wang et al. 2008).

To achieve a higher efficiency, combinations of different chemotherapeutics could be applied. One example is an aptamer approach, in which aptamer-conjugated nanoparticles were filled with Dox and docetaxel simultaneously (Chu et al. 2006). Docetaxel is a taxane and belongs to the class of anti-microtubule agents. Taxanes promote microtubule assembly and inhibition of microtubule dynamics. As a direct consequence, the cell cycle arrests and several cellular functions are impaired; this causes inhibition of cell proliferation and apoptosis (Herbst and Khuri 2003). Other cytostatic agents that also work as DNA intercalators are actinomycin-D and mitomycin and both agents derive from bacteria.

Cisplatin is another chemotherapeutic agent that contains platinum and damages DNA by crosslinking intra-strand purine bases, thus inducing apoptosis. The prodrug of cisplatin, Pt(IV), has alternatively been used for several delivery purposes, due to diverse side effects caused by cisplatin. Among these approaches are covalently linked cisplatin to folic acid (Elliott et al. 1988) or encapsulated agents within nanoparticles, coupled to different substances as delivery vehicle on the particles' surface, like gold-nanoparticles with folic acids (Patra et al. 2010), carbon-nanotubes with folic acid derivative (Dhar et al. 2008b), or PEG-nanoparticles with PSMA aptamer (Dhar et al. 2008a).

5.4 Proteins

The targeted delivery of proteins, especially of toxins and enzymes, has been widely used for other delivery vehicles (e.g., antibodies, lectins, cytokines, growth factors) before and are used for some aptamer-mediated delivery processes as well (e.g., PSMA-aptamer conjugated with Gelonin (Chu et al. 2006)).

Numerous toxins occur naturally in plants, microorganisms, and fungi; and those, which are used for targeted delivery, are cytotoxic in most cases. Cytotoxic agents interfere with cellular processes, and toxins that exclusively target ribosomes and affect protein synthesis are called ribosome inactivating proteins (RIP). Among these are bacterial exotoxins like Shiga toxins or Shiga-like toxins and fungal RIPs like Restrictocin. The majority of RIPs is of plant origin and is classified based on their structure into three types. Type I consists of a single peptide chain with enzymatic activity (e.g., Gelonin, Saporin-S6 and -R1, Trichosanthin). Type II is composed of an enzymatically-active peptide and fused to a lectin domain (e.g., Ricin). Type III is a nontoxic RIP precursor (ProRIPs, like maize b-32). RIPs effect can also differ in respect of lethal dose and organisms, reviewed in Stirpe and Battelli (2006).

Bacterial toxins are differentiated into exotoxins and endotoxins, whereas exotoxins are more suitable delivery agents. Dependent on their activity exotoxins can be further classified into cell-surface active, membrane damaging, and intracellular toxins.

Interesting for a toxin-based therapy are intracellular exotoxins but also RIPs. Among these, frequently used toxins are Diphtheria toxin, *Pseudomonas* exotoxin A, and Anthrax toxin which all have been used as immunotoxins, protease-activated toxins, and for suicide gene therapy as well. Suicide genes encode toxins which are solely expressed by disease-associated transcription factors (e.g., prostate-specific antigen (Li et al. 2002; Peng et al. 2005) or human carcinoembryonic antigen (Cao et al. 1998)).

For toxins that are extracellular-activated by proteases the toxin of choice has to be composed of a receptor-binding domain, a protease cleavage site, and a toxic domain. To obtain tumor specificity, the targeted receptor as well as the active protease can be chosen tumor specific. On the contrary, intracellular protease-activated toxins are cleaved and activated by viral proteases (e.g., retroviruses, piconaviruses, flaviviruses, herpesviruses, or nidoviruses). One example is to fuse the toxin to a peptide that destabilizes it and harboring a viral cleavage site in between. The other example is to use a ProRIP which is modified regarding viral cleavage site and activation. All these nontoxic proteins achieve their intoxication only in presence of active viral protease, reviewed in Shapira and Benhar (2010). Alternatively, the receptor specificity can be achieved via an aptamer.

Besides, enzymes can be delivered as well. This was shown for TfR-mediated delivery of transferrin coupled to different RNases, reviewed in Daniels et al. (2012), or via a TfR-specific aptamer conjugated to L-iduronidase (Chen et al. 2008).

5.5 Nucleic Acid-Based Therapeutics

Since nucleic acids can be easily modified the application of RNA or DNA as therapeutics increases. Thus, lots of nucleic acid-based therapeutics targeting different diseases are in clinical trials, reviewed in Burnett and Rossi (2012). The aptamer can be either covalently bound to the additional oligonucleotide or hybridized via a complementary linker, considering non-impairment of aptamer and therapeutic oligonucleotide function, respectively.

To manipulate gene regulation in targeted cells the aptamer can be attached to short interfering RNAs (siRNA) or microRNAs (miRNAs). Both target messenger RNAs (mRNAs) and induce RNA interference process which leads to degradation of mRNA or its translational inhibition. Different strategies to connect aptamer and siRNA have been reviewed in Zhou and Rossi (2010) and Dassie et al. (2009).

MiRNAs were identified to be either up- or downregulated in malignant cells, reviewed in Davis et al. (2006), therefore methods to inactivate these miRNAs were established based on antisense oligonucleotides like microRNA sponges (Ebert et al. 2007) and AntagomiRs (Krutzfeldt et al. 2005). The efficiency of miRNA inhibitors strongly correlates with the type of modification; as AntagomiRs harboring 2′-*O*-methyl are less efficient than 2′-*O*-methoxyethyl (Davis et al. 2006). In the end the aptamer–siRNA conjugate/chimera has to be delivered to cytoplasm and has to be made accessible for all components of RNAi machinery.

Ribozymes are catalytic RNAs lacking in protein portion. Interaction with targeted agent can lead to either cis- or trans-active RNA cleavage. Thus, ribozymes can also be used for gene regulation. To obtain application-orientated ribozymes that can serve as therapeutics, in vitro directed evolution, as well as other artificial engineering methods can be utilized. One ribozyme that has been tested in clinical trials is Angiozyme which downregulates VEGF receptor-1 (Sandberg et al. 2000). Aptamer–ribozyme conjugates which are working as therapeutics haven't been published yet, but in the context of biosensors (Breaker 2002).

5.6 Encapsulated Agents in Nanocarriers

For targeted drug delivery, agents or toxins can be packed up in nanostructured systems which are then conjugated to aptamers. The usage of such nanocarriers enables to deliver agents that are highly toxic or unstable. To comply with the requirement for this purpose the nanostructures have to exhibit key properties i.e. being biocompatible, able to incorporate the drug without interfering functionality, being stable until target location, and able to release drug within the target cell. In addition, these nanostructures have to consist of biodegradable polymers to allow controlled release due to environmental changes (e.g., pH, enzymatic activities, temperature) or external impact (e.g., ultrasound, electric, or magnetic fields). Furthermore the nanocarriers can be subdivided into liposomes, polymeric nanoparticles, nanoparticles which consist of lipids and polymers (= lipid–polymer hybrids), and dendrimers.

Liposome-based nanoparticles have been the first applied nanocarriers and consist of amphiphilic phospholipid molecules that form spherical bilayers spontaneously. Hydrophilic components are dissolved inside the lumen of the liposome and hydrophilic agents within the bilayer. Ligand-conjugated liposomes have been constructed to target the TfR, for instance, to deliver chemotherapeutics, reviewed in Daniels et al. (2012). An aptamer-mediated approach has also been realized to deliver cisplatin with nucleolin–aptamer-conjugated liposomes (Cao et al. 2009) or dextran with protein tyrosine kinase 7-aptamer-conjugated liposomes (Kang et al. 2010). Polymeric nanoparticles can be formed by amphiphilic polymers that built spontaneously polymeric micelles which can be either filled with hydrophilic or hydrophobic drugs. The PSMA aptamer was the first aptamer which was used for aptamer-mediated nanoparticle delivery (Farokhzad et al. 2004). However, Polycefin is a multifunctional polymer consisting of β-poly(L-malic acid) and is not only biocompatible and biodegradable but also includes components for receptor-mediated endocytosis (TfR), pH-sensitive drug release and endosomal escape (Lee et al. 2006). Polycefin shows that the usage of nanoparticles allows the construction of individual-designed delivery agents.

In contrast to the further mentioned nanoparticles dendrimers don't encapsulate drugs as such. Dendrimers are defined of a central core with structures that radially disperse and possess functional groups. Coupling different agents to dendrimers multifunctional particles can be formed and chemotherapeutics can be encapsulated within cavities.

References

Aarbakke J, Janka-Schaub G, Elion GB (1997) Thiopurine biology and pharmacology. Trends Pharmacol Sci 18:3–7

Abdelmohsen K, Gorospe M (2012) RNA-binding protein nucleolin in disease. RNA Biol 9:799–808

Abolmaali SS, Tamaddon AM, Dinarvand R (2013) A review of therapeutic challenges and achievements of methotrexate delivery systems for treatment of cancer and rheumatoid arthritis. Cancer Chemother Pharmacol 71:1115–1130

Akabani G, Reardon DA, Coleman RE et al (2005) Dosimetry and radiographic analysis of 131I-labeled anti-tenascin 81C6 murine monoclonal antibody in newly diagnosed patients with malignant gliomas: a phase II study. J Nucl Med 46:1042–1051

Altschuler Y, Kinlough CL, Poland PA et al (2000) Clathrin-mediated endocytosis of MUC1 is modulated by its glycosylation state. Mol Biol Cell 11:819–831

Bagalkot V, Farokhzad OC, Langer R et al (2006) An aptamer-doxorubicin physical conjugate as a novel targeted drug-delivery platform. Angew Chem Int Ed 45:8149–8152

Bates PJ, Laber DA, Miller DM et al (2009) Discovery and development of the G-rich oligonucleotide AS1411 as a novel treatment for cancer. Exp Mol Pathol 86:151–164

Berg K, Berstad M, Prasmickaite L et al (2010) Photochemical internalization: a new tool for gene and oligonucleotide delivery. Top Curr Chem 296:251–281

Bonate PL, Arthaud L, Cantrell WR Jr et al (2006) Discovery and development of clofarabine: a nucleoside analogue for treating cancer. Nat Rev Drug Discov 5:855–863
Breaker RR (2002) Engineered allosteric ribozymes as biosensor components. Curr Opin Biotechnol 13:31–39
Brogden RN, Sorkin EM (1993) Pentostatin. A review of its pharmacodynamic and pharmacokinetic properties, and therapeutic potential in lymphoproliferative disorders. Drugs 46:652–677
Burnett JC, Rossi JJ (2012) RNA-based therapeutics: current progress and future prospects. Chem Biol 19:60–71
Cao GW, Qi ZT, Pan X et al (1998) Gene therapy for human colorectal carcinoma using human CEA promoter controlled bacterial ADP-ribosylating toxin genes: PEA and DTA gene transfer. World J Gastroenterol 4:388–391
Cao Z, Tong R, Mishra A et al (2009) Reversible cell-specific drug delivery with aptamer-functionalized liposomes. Angew Chem Int Ed 48:6494–6498
Cerchia L, de Franciscis V (2010) Targeting cancer cells with nucleic acid aptamers. Trends Biotechnol 28:517–525
Cerqueira NM, Fernandes PA, Ramos MJ (2007) Understanding ribonucleotide reductase inactivation by gemcitabine. Chemistry 13:8507–8515
Chauhan VM, Burnett GR, Aylott JW (2011) Dual-fluorophore ratiometric pH nanosensor with tuneable pK(a) and extended dynamic range. Analyst 136:1799–1801
Chen CH, Dellamaggiore KR, Ouellette CP et al (2008) Aptamer-based endocytosis of a lysosomal enzyme. Proc Natl Acad Sci U S A 105:15908–15913
Chu TC, Marks JW 3rd, Lavery LA et al (2006) Aptamer:toxin conjugates that specifically target prostate tumor cells. Cancer Res 66:5989–5992
Coulie PG, Hanagiri T, Takenoyama M (2001) From tumor antigens to immunotherapy. Int J Clin Oncol 6:163–170
Daniels DA, Chen H, Hicke BJ et al (2003) A tenascin-C aptamer identified by tumor cell SELEX: systematic evolution of ligands by exponential enrichment. Proc Natl Acad Sci U S A 100:15416–15421
Daniels TR, Bernabeu E, Rodriguez JA et al (2012) The transferrin receptor and the targeted delivery of therapeutic agents against cancer. Biochim Biophys Acta 1820:291–317
Dassie JP, Liu XY, Thomas GS et al (2009) Systemic administration of optimized aptamer-siRNA chimeras promotes regression of PSMA-expressing tumors. Nat Biotechnol 27:839–849
Davis S, Lollo B, Freier S et al (2006) Improved targeting of miRNA with antisense oligonucleotides. Nucleic Acids Res 34:2294–2304
Dhar S, Gu FX, Langer R et al (2008a) Targeted delivery of cisplatin to prostate cancer cells by aptamer functionalized Pt(IV) prodrug-PLGA-PEG nanoparticles. Proc Natl Acad Sci U S A 105:17356–17361
Dhar S, Liu Z, Thomale J et al (2008b) Targeted single-wall carbon nanotube-mediated Pt (IV) prodrug delivery using folate as a homing device. J Am Chem Soc 130:11467–11476
Ebert MS, Neilson JR, Sharp PA (2007) MicroRNA sponges: competitive inhibitors of small RNAs in mammalian cells. Nat Methods 4:721–726
Ellington AD, Szostak JW (1990) In vitro selection of RNA molecules that bind specific ligands. Nature 346:818–822
Elliott RL, Stjernholm R, Elliott MC (1988) Preliminary evaluation of platinum transferrin (Mptc-63) as a potential nontoxic treatment for breast-cancer. Cancer Detect Prev 12:469–480
Farokhzad OC, Jon S, Khademhosseini A et al (2004) Nanoparticle-aptamer bioconjugates: a new approach for targeting prostate cancer cells. Cancer Res 64:7668–7672
Ferreira CS, Matthews CS, Missailidis S (2006) DNA aptamers that bind to MUC1 tumour marker: design and characterization of MUC1-binding single-stranded DNA aptamers. Tumour Biol 27:289–301
Ferreira CS, Cheung MC, Missailidis S et al (2009) Phototoxic aptamers selectively enter and kill epithelial cancer cells. Nucleic Acids Res 37:866–876

Gan HK, Burgess AW, Clayton AH et al (2012) Targeting of a conformationally exposed, tumor-specific epitope of EGFR as a strategy for cancer therapy. Cancer Res 72:2924–2930
Grant BD, Donaldson JG (2009) Pathways and mechanisms of endocytic recycling. Nat Rev Mol Cell Biol 10:597–608
Gruner BA, Weitman SD (1998) The folate receptor as a potential therapeutic anticancer target. Invest New Drugs 16:205–219
Gryczynski I, Gryczynski Z, Lakowicz JR et al (1999) Fluorescence spectral properties of the anticancer drug topotecan by steady-state and frequency domain fluorometry with one-photon and multi-photon excitation. Photochem Photobiol 69:421–428
Guilloux Y, Lucas S, Brichard VG et al (1996) A peptide recognized by human cytolytic T lymphocytes on HLA-A2 melanomas is encoded by an intron sequence of the N-acetylglucosaminyltransferase V gene. J Exp Med 183:1173–1183
Guo SC, Tschammer N, Mohammed S et al (2005) Specific delivery of therapeutic RNAs to cancer cells via the dimerization mechanism of phi29 motor pRNA. Hum Gene Ther 16:1097–1109
Guo KT, Paul A, Schichor C et al (2008) Cell-SELEX: novel perspectives of aptamer-based therapeutics. Int J Mol Sci 9:668–678
Han JY, Burgess K (2010) Fluorescent indicators for intracellular pH. Chem Rev 110:2709–2728
Herbst RS, Khuri FR (2003) Mode of action of docetaxel – a basis for combination with novel anticancer agents. Cancer Treat Rev 29:407–415
Hicke BJ, Marion C, Chang YF et al (2001) Tenascin-C aptamers are generated using tumor cells and purified protein. J Biol Chem 276:48644–48654
Hicke BJ, Stephens AW, Gould T et al (2006) Tumor targeting by an aptamer. J Nucl Med 47:668–678
Hijiya N, Barry E, Arceci RJ (2012) Clofarabine in pediatric acute leukemia: current findings and issues. Pediatr Blood Cancer 59:417–422
Huang YF, Shangguan D, Liu H et al (2009) Molecular assembly of an aptamer-drug conjugate for targeted drug delivery to tumor cells. Chembiochem 10:862–868
Huotari J, Helenius A (2011) Endosome maturation. EMBO J 30:3481–3500
Huynh KK, Grinstein S (2007) Regulation of vacuolar pH and its modulation by some microbial species. Microbiol Mol Biol Rev 71:452–462
Jentsch TJ (2008) CLC chloride channels and transporters: from genes to protein structure, pathology and physiology. Crit Rev Biochem Mol Biol 43:3–36
Kang HZ, O'Donoghue MB, Liu HP et al (2010) A liposome-based nanostructure for aptamer directed delivery. Chem Commun 46:249–251
Karran P (2006) Thiopurines, DNA damage, DNA repair and therapy-related cancer. Br Med Bull 79–80:153–170
Katzmann DJ, Odorizzi G, Emr SD (2002) Receptor downregulation and multivesicular-body sorting. Nat Rev Mol Cell Biol 3:893–905
Khaled A, Guo SC, Li F et al (2005) Controllable self-assembly of nanoparticles for specific delivery of multiple therapeutic molecules to cancer cells using RNA nanotechnology. Nano Lett 5:1797–1808
Kowol CR, Trondl R, Arion VB et al (2010) Fluorescence properties and cellular distribution of the investigational anticancer drug Triapine (3-aminopyridine-2-carboxaldehyde thiosemicarbazone) and its zinc(II) complex. Dalton Trans 39:704–706
Kreitman RJ (2006) Immunotoxins for targeted cancer therapy. AAPS J 8:E532–E551
Krutzfeldt J, Rajewsky N, Braich R et al (2005) Silencing of microRNAs in vivo with 'antagomirs'. Nature 438:685–689
Kurzrock R, Kantarjian HM, Druker BJ et al (2003) Philadelphia chromosome-positive leukemias: from basic mechanisms to molecular therapeutics. Ann Intern Med 138:819–830
Lawe DC, Chawla A, Merithew E et al (2002) Sequential roles for phosphatidylinositol 3-phosphate and Rab5 in tethering and fusion of early endosomes via their interaction with EEA1. J Biol Chem 277:8611–8617

Leamon CP, Low PS (2005) Receptor-mediated drug delivery. In: Wang B, Siahaan T, Soltero R (eds) Drug delivery: principles and applications, Wiley series in drug discovery. Wiley, Hoboken, NJ, pp 167–187

Lee BS, Fujita M, Khazenzon NM et al (2006) Polycefin, a new prototype of a multifunctional nanoconjugate based on poly(beta-L-malic acid) for drug delivery. Bioconjug Chem 17:317–326

Li JM, Han JS, Huang Y et al (1999) A novel gene delivery system targeting cells expressing VEGF receptors. Cell Res 9:11–25

Li Y, McCadden J, Ferrer F et al (2002) Prostate-specific expression of the diphtheria toxin A chain (DT-A): studies of inducibility and specificity of expression of prostate-specific antigen promoter-driven DT-A adenoviral-mediated gene transfer. Cancer Res 62:2576–2582

Li Q, Verschraegen CF, Mendoza J et al (2004) Cytotoxic activity of the recombinant anti-mesothelin immunotoxin, SS1(dsFv)PE38, towards tumor cell lines established from ascites of patients with peritoneal mesotheliomas. Anticancer Res 24:1327–1335

Li N, Larson T, Nguyen HH et al (2010) Directed evolution of gold nanoparticle delivery to cells. Chem Commun (Camb) 46:392–394

Lin C, Engbersen JF (2008) Effect of chemical functionalities in poly(amido amine)s for non-viral gene transfection. J Control Release 132:267–272

Litvinov SV, Hilkens J (1993) The epithelial sialomucin, episialin, is sialylated during recycling. J Biol Chem 268:21364–21371

Liu H, Rajasekaran AK, Moy P et al (1998) Constitutive and antibody-induced internalization of prostate-specific membrane antigen. Cancer Res 58:4055–4060

Longley DB, Harkin DP, Johnston PG (2003) 5-Fluorouracil: mechanisms of action and clinical strategies. Nat Rev Cancer 3:330–338

Lupetti R, Pisarra P, Verrecchia A et al (1998) Translation of a retained intron in tyrosinase-related protein (TRP) 2 mRNA generates a new cytotoxic T lymphocyte (CTL)-defined and shared human melanoma antigen not expressed in normal cells of the melanocytic lineage. J Exp Med 188:1005–1016

Lupold SE, Hicke BJ, Lin Y et al (2002) Identification and characterization of nuclease-stabilized RNA molecules that bind human prostate cancer cells via the prostate-specific membrane antigen. Cancer Res 62:4029–4033

Maragos CM, Appell M, Lippolis V et al (2008) Use of cyclodextrins as modifiers of fluorescence in the detection of mycotoxins. Food Addit Contam 25:164–171

Matayoshi ED, Wang GT, Krafft GA et al (1990) Novel fluorogenic substrates for assaying retroviral proteases by resonance energy-transfer. Science 247:954–958

Maxfield FR, McGraw TE (2004) Endocytic recycling. Nat Rev Mol Cell Biol 5:121–132

Maxfield FR, Yamashiro DJ (1987) Endosome acidification and the pathways of receptor-mediated endocytosis. Adv Exp Med Biol 225:189–198

Meyer C, Hahn U, Rentmeister A (2011) Cell-specific aptamers as emerging therapeutics. J Nucleic Acids 2011:904750

Meyer C, Eydeler K, Magbanua E et al (2012) Interleukin-6 receptor specific RNA aptamers for cargo delivery into target cells. RNA Biol 9:67–80

Midwood KS, Orend G (2009) The role of tenascin-C in tissue injury and tumorigenesis. J Cell Commun Signal 3:287–310

Monsuez JJ, Charniot JC, Vignat N et al (2010) Cardiac side-effects of cancer chemotherapy. Int J Cardiol 144:3–15

Ng EW, Shima DT, Calias P et al (2006) Pegaptanib, a targeted anti-VEGF aptamer for ocular vascular disease. Nat Rev Drug Discov 5:123–132

Packard BZ, Komoriya A (2008) Intracellular protease activation in apoptosis and cell-mediated cytotoxicity characterized by cell-permeable fluorogenic protease substrates. Cell Res 18:238–247

Patra CR, Bhattacharya R, Mukherjee P (2010) Fabrication and functional characterization of goldnanoconjugates for potential application in ovarian cancer. J Mater Chem 20:547–554

Peng W, Chen J, Huang YH et al (2005) Tightly-regulated suicide gene expression kills PSA-expressing prostate tumor cells. Gene Ther 12:1573–1580
Raab-Traub N (2012) Novel mechanisms of EBV-induced oncogenesis. Curr Opin Virol 2:453–458
Raiborg C, Bache KG, Gillooly DJ et al (2002) Hrs sorts ubiquitinated proteins into clathrin-coated microdomains of early endosomes. Nat Cell Biol 4:394–398
Ray P, Cheek MA, Sharaf ML et al (2012) Aptamer-mediated delivery of chemotherapy to pancreatic cancer cells. Nucleic Acid Ther 22:295–305
Ressing ME, Sette A, Brandt RMP et al (1995) Human CTL epitopes encoded by human papillomavirus type-16 E6 and E7 identified through in-vivo and in-vitro immunogenicity studies of HLA-A*-0201-binding peptides. J Immunol 154:5934–5943
Ross JS, Sheehan CE, Fisher HAG et al (2003) Correlation of primary tumor prostate-specific membrane antigen expression with disease recurrence in prostate cancer. Clin Cancer Res 9:6357–6362
Ruckman J, Green LS, Beeson J et al (1998) 2′-Fluoropyrimidine RNA-based aptamers to the 165-amino acid form of vascular endothelial growth factor (VEGF165). Inhibition of receptor binding and VEGF-induced vascular permeability through interactions requiring the exon 7-encoded domain. J Biol Chem 273:20556–20567
Sachse M, Urbe S, Oorschot V et al (2002) Bilayered clathrin coats on endosomal vacuoles are involved in protein sorting toward lysosomes. Mol Biol Cell 13:1313–1328
Sandberg JA, Parker VP, Blanchard KS et al (2000) Pharmacokinetics and tolerability of an antiangiogenic ribozyme (ANGIOZYME) in healthy volunteers. J Clin Pharmacol 40:1462–1469
Schally AV, Nagy A (2004) Chemotherapy targeted to cancers through tumoral hormone receptors. Trends Endocrinol Metab 15:300–310
Seedorf K, Oltersdorf T, Krammer G et al (1987) Identification of early proteins of the human papilloma viruses type-16 (Hpv 16) and type-18 (Hpv 18) in cervical-carcinoma cells. EMBO J 6:139–144
Shangguan D, Li Y, Tang Z et al (2006) Aptamers evolved from live cells as effective molecular probes for cancer study. Proc Natl Acad Sci U S A 103:11838–11843
Shapira A, Benhar I (2010) Toxin-based therapeutic approaches. Toxins 2:2519–2583
Shieh YA, Yang SJ, Wei MF et al (2010) Aptamer-based tumor-targeted drug delivery for photodynamic therapy. ACS Nano 4:1433–1442
Spurgeon S, Yu M, Phillips JD et al (2009) Cladribine: not just another purine analogue? Expert Opin Investig Drugs 18:1169–1181
Stirpe F, Battelli MG (2006) Ribosome-inactivating proteins: progress and problems. Cell Mol Life Sci 63:1850–1866
Taghdisi SM, Abnous K, Mosaffa F et al (2010) Targeted delivery of daunorubicin to T-cell acute lymphoblastic leukemia by aptamer. J Drug Target 18:277–281
Tian H, Cronstein BN (2007) Understanding the mechanisms of action of methotrexate: implications for the treatment of rheumatoid arthritis. Bull NYU Hosp Jt Dis 65:168–173
Tuerk C, Gold L (1990) Systematic evolution of ligands by exponential enrichment: RNA ligands to bacteriophage T4 DNA polymerase. Science 249:505–510
Varkouhi AK, Scholte M, Storm G et al (2011) Endosomal escape pathways for delivery of biologicals. J Control Release 151:220–228
Wang AZ, Bagalkot V, Vasilliou CC et al (2008) Superparamagnetic iron oxide nanoparticle-aptamer bioconjugates for combined prostate cancer imaging and therapy. ChemMedChem 3:1311–1315
Wilcke M, Johannes L, Galli T et al (2000) Rab11 regulates the compartmentalization of early endosomes required for efficient transport from early endosomes to the trans-golgi network. J Cell Biol 151:1207–1220
Xiao Z, Shangguan D, Cao Z et al (2008) Cell-specific internalization study of an aptamer from whole cell selection. Chemistry 14:1769–1775

Zhang YF, Chen Y, Han D et al (2010) Aptamers selected by cell-SELEX for application in cancer studies. Bioanalysis 2:907–918

Zhou J, Rossi JJ (2010) Aptamer-targeted cell-specific RNA interference. Silence 1:4

Zhou J, Swiderski P, Li H et al (2009) Selection, characterization and application of new RNA HIV gp 120 aptamers for facile delivery of Dicer substrate siRNAs into HIV infected cells. Nucleic Acids Res 37:3094–3109

Zhou J, Fang CL, Chang TJ et al (2013) A pH sensitive ratiometric fluorophore and its application for monitoring the intracellular and extracellular pHs simultaneously. J Mater Chem B 1:661–667

Biochemical Aspects of Subcellular RNA Transport and Localization

Diana Bauermeister, Maike Claußen, and Tomas Pieler

Contents

1 Three Distinct Transport Mechanisms 294
2 RNA Localization Elements: Structural Features 295
3 Transport Particles: Protein Components 297
4 Motor-Dependent mRNA Localization Involves Multiple Steps 299
  4.1 Initiation of mRNA Transport in the Nucleus 299
  4.2 Cytoplasmic Remodeling of Transport RNPs 300
  4.3 Motor-Proteins and RNA Transport 300
  4.4 RNP Anchoring 301
5 RNA Transport and Translational Regulation 302
6 Formation of "Super-complexes?" 302
7 An Additional Function for Localization RNPs 303
References 304

Abstract Following their synthesis in and export from the nucleus, quite a few mRNA molecules get sorted to specific cellular locations. These events are linked to particular biological activities in which localized protein expression or asymmetrical RNA inheritance after cell division has regulatory relevance. The most prominent examples for regulatory phenomena which are based upon RNA localization are illustrated in Fig. 1. Here, rather than providing a detailed reflection of such biological implications, we will focus on the RNA biochemistry and molecular mechanisms that allow for subcellular RNA localization.

D. Bauermeister • M. Claußen • T. Pieler (✉)
Department of Developmental Biochemistry, University of Göttingen, 37077 Göttingen, Germany

Göttingen Center for Molecular Biosciences (GZMB), University of Göttingen, 37077 Göttingen, Germany
e-mail: diana.obermann@med.uni-goettingen.de; mclauss1@gwdg.de; tpieler@gwdg.de

V.A. Erdmann et al. (eds.), *Chemical Biology of Nucleic Acids*, RNA Technologies,
DOI 10.1007/978-3-642-54452-1_16, © Springer-Verlag Berlin Heidelberg 2014

Keywords RNA localization • RNA transport • Localization element • RNA zipcode • Xenopus

1 Three Distinct Transport Mechanisms

The localization of individual mRNA molecules to specific subcellular locations is mediated by cis-acting, regulatory RNA signal sequences, also referred to as localization elements (LEs). Together with a specific set of proteins, such LEs assemble into complex RNPs. Different sorting pathways localize distinct subsets of RNAs by use of at least three distinct mechanisms.

One mechanism to achieve RNA localization is by selective RNA stabilization at defined intracellular locations and degradation elsewhere. *Hsp83* mRNA, for example, is degraded during early *Drosophila* embryogenesis, but protected from degradation by pole plasm components at the posterior end of the embryo (Ding et al. 1993).

A second mechanism for RNA localization is mediated by "diffusion and entrapment"; here, the underlying concept is that freely diffusing RNA molecules become anchored at specific sites within a cell. *Xenopus nanos1* (formerly *cat2*) has served as an example for RNA localization by this mechanism; during early oogenesis, *nanos1* gets entrapped in the Balbiani body (Chang et al. 2004), which then moves to the vegetal cortex to form the germ plasm. However, there is also evidence that localization to the Balbiani body is an energy-and motor protein-dependent process and thus at least partially mediated by active transport (Heinrich and Deshler 2009). A multitude of other such "early localizing" RNAs becomes equally enriched in the germ plasm (Kloc et al. 1996) and many of these encode RNA-binding-proteins essential for germ cell development, such as *dazl* (Houston and King 2000) or *ddx25* (formerly *deadsouth*) (MacArthur et al. 2000; Yamaguchi et al. 2012).

The most intensely studied, third mechanism of RNA localization is via active transport in a complex with motor-proteins migrating along elements of the cytoskeleton (Fig. 2). In *Xenopus*, such "late localizing" RNAs are initially homogeneously distributed in early oocytes, accumulating in a wedge-like structure in later stages of oogenesis and becoming actively transported in a microtubule and motor-protein-dependent manner to the vegetal cortex where they are anchored (Fig. 1). Many such late localizing mRNAs encode germ layer determinants, as for example VegT, a T-box transcription factor essential for endoderm formation, or Vg1, a TGFβ-like signaling molecule involved in early patterning events (Joseph and Melton 1998; Zhang et al. 1998). Motor molecule-driven RNA transport has also been described for other biological systems, including yeast and *Drosophila* (Fig. 1) (Gagnon and Mowry 2011).

Since the majority of localizing RNAs studied in detail so far travels via active transport, we will mainly focus on this mechanism in the context of this review.

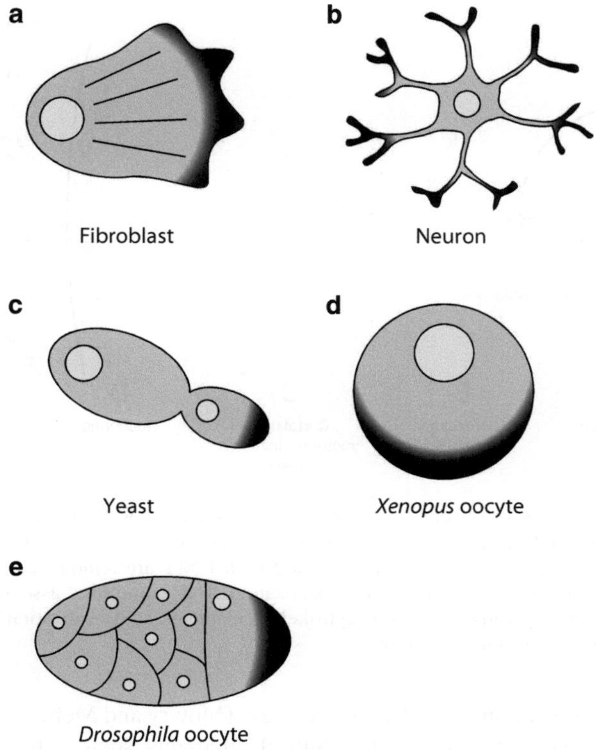

Fig. 1 RNA localization occurs in different biological systems. (**a**) *β-actin* mRNA localizes to the leading edge of chicken fibroblasts and serves as a source for the high local concentrations of actin that are required for cell motility. (**b**) mRNA encoding for the α subunit of calcium/calmodulin kinase II (α-CaMKII) localizes to dendrites of mouse neurons, where its protein acts as a modulator of synaptic plasticity. (**c**) *ASH1* localizes to the tip of the daughter cell in budding yeast and suppresses mating type switching in the daughter cell. (**d**) *Vg1* mRNA localizes to the vegetal cortex of *Xenopus* oocytes and thereby participates in early embryonic patterning (**e**) *Oskar* mRNA is localized to the posterior pole in *Drosophila* oocytes, to act in germ plasm assembly

2 RNA Localization Elements: Structural Features

RNA localization elements are mostly defined by single-copy signal sequences located within the 3′-UTR of the corresponding mRNAs; exceptions are defined by a small subset of localizing RNAs containing either multiple such RNA localization elements, as well as LEs located within the 5′-UTR or coding region of the transcript. *ASH1*, for example, carries as many as four independent localization signals (or "zipcode" elements) within its 3′-UTR and coding region, driving localization to the bud-tip of dividing *Saccharomyces cerevisiae* cells (Chartrand et al. 1999; Gonzalez et al. 1999). Furthermore, RNA localization signal sequences can substantially differ in length, ranging from a few nucleotides (e.g., *fatvg* (Chan

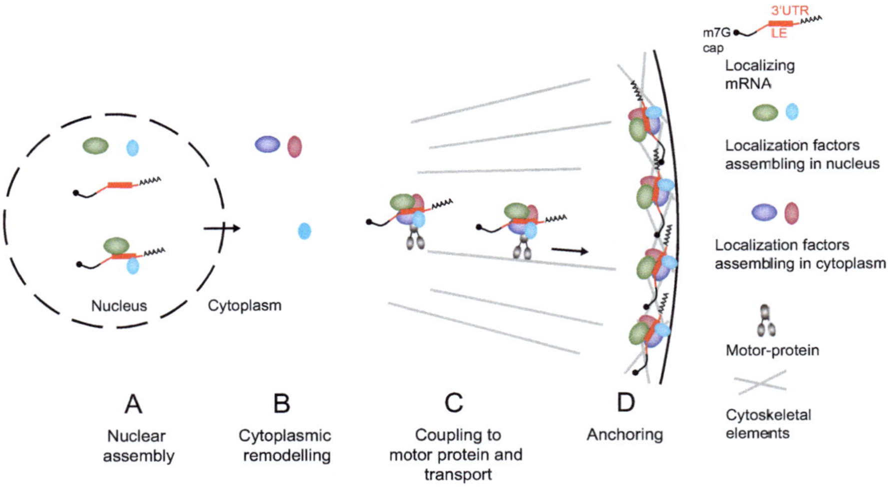

Fig. 2 Schematic view of different steps in motor-dependent mRNA localization. (**a**) mRNA localization is initiated by interaction of a first set of localization factors with the RNA LE in the nucleus. (**b**) After export to the cytoplasm, localization RNPs are remodelled and cytoplasmic localization factors join the complex. (**c**) Localization RNP complexes associate with motor-proteins and are actively transported along cytoskeletal elements. (**d**) Localization RNP complexes become anchored at their target location

et al. 1999)) to several hundred bases (e.g., *vg1*; (Mowry and Melton 1992)). Finally LEs are also found to be composed of multiple elements, such as the bipartite LE of *Drosophila oskar* mRNA (Ghosh et al. 2012).

Characterization of several different LEs that function in RNA localization during *Xenopus* oogenesis has not allowed for the definition of a consensus sequence that would be common to all localization elements from early and/or late pathway RNAs. Analysis of the *vg1*- and *vegT*-LEs identified reiterated copies of relatively short (up to six nucleotide long), diverse sequence elements, which are also present in other localizing RNAs, as being required and sufficient for localization (Betley et al. 2002; Bubunenko et al. 2002; Kwon et al. 2002). Such a requirement of different RNA sequence elements for efficient localization in *Xenopus* oocytes could reflect a combinatorial code of individual protein-binding RNA sequence motifs that identifies a given RNA as a target for the vegetal localization machinery. Such a combinatorial code might also involve additional sequence and/or structural elements that could affect the localization pathway to be chosen, mode of cortical anchoring as well as translational status of the localized RNA. However, other vegetally localizing RNAs, such as *bicc1* and *velo1*, localize independently of such clustered motifs (Betley et al. 2002; Claussen and Pieler 2004).

The involvement of double-stranded RNA-binding proteins, such as Staufen, which recognize RNA duplexes rather than making base-specific contacts in single-stranded RNA regions (Tian et al. 2004), indicates that RNA localization elements could be defined by combinations of double-stranded and single-stranded regions with a single RNA stem–loop being the simplest versions of such motif. Indeed, examples for stem–loop structures driving RNA transport have been defined in *Drosophila* (Bullock and Ish-Horowicz 2001; Serano and Cohen 1995; Van De Bor et al. 2005). The importance of base pairing in the *wingless* LE (*WLE3*) was verified by systematic mutagenesis studies (dos Santos et al. 2008). Interestingly the *WLE3* structure closely resembles the secondary structures predicted for the localization signals in *hairy*, *fushi tarazu*, *fs(1)K10* and *orb*, arguing for the presence of common structural features in localization signals (dos Santos et al. 2008). NMR spectroscopy of one of the stem–loop containing localization sequences, namely the K10TLS of *fs(1)K10*, revealed an unexpected formation of A′ helices with a wider major groove than the classical A form that is usually adopted by RNA helices. The ability to form an A′ conformation was shown to be critical for localization and it will be interesting to see if such an RNA conformation is equally important for other LEs (Bullock et al. 2010).

Information about LE secondary structure in *Xenopus laevis* is quite limited. The localization element of the late localizing *velo1* RNA is predicted to fold into a stem–loop structure and interfering with stem–loop structure formation by point-mutagenesis indeed affects vegetal localization activity. However, compensatory point mutations predicted to restore the stem–loop structure, did not rescue localization activity, indicating that nucleotide identities in the stem–loop could also contribute to localization machinery recruitment (Claussen and Pieler 2004). How the bases in the stem–loop region are made accessible to the localization machinery is not known, however one could imagine, RNA LE recognition to be a stepwise process; initially the *velo1* stem–loop could serve as a primary signal for the recruitment of double-stranded RNA-binding transport proteins such as Staufen. Subsequent remodelling of the complex could involve RNA-helicase activities making the LE accessible for single-stranded RNA-binding proteins and allowing for base specific contacts in theses now single-stranded RNA regions.

3 Transport Particles: Protein Components

One of the best-characterized examples for RNA transport complexes is the one that mediates localization of *ASH1* mRNA through a myosin-motor along the actin cytoskeleton in budding yeast (Fig. 1). In the nucleus, *ASH1* mRNA associates with She2p (Shen et al. 2009, 2010), an unusual RNA-binding protein forming a symmetric homodimer with a basic helical hairpin as RNA recognition site (Niessing et al. 2004). Although She2p binds to all four zip-code elements of *ASH1* (Chartrand et al. 1999; Gonzalez et al. 1999), the in vitro binding affinity and specificity is rather low. Interestingly, a specific and strong interaction with *ASH1* RNA is established only if She2p and She3p act together synergistically (Müller et al. 2011). In addition, She3p mediates the recruitment of the motor-protein Myo4p (Böhl et al. 2000). These studies have provided a framework for how a specific localization RNP is generated. However, mechanistic insights into assembly and cooperativity of localization RNP components in higher organisms are quite limited.

A common feature of many localization factors is the modular arrangement of multiple RNA-binding motifs. Single RNA recognition motifs (RRM) bind to RNA with high affinity, but generally they can only recognize a short RNA sequence, often too short to define a unique binding site (Auweter et al. 2006). In contrast, multiple RNA interaction motifs could ensure a more selective binding of target RNAs (Lunde et al. 2007). For instance, Zipcode-binding protein 1 (ZBP1), which localizes *β-actin* mRNA to the leading edges of fibroblasts and neuron growth-cones in chicken (Fig. 1) (Bassell et al. 1998; Lawrence and Singer 1986), contains two RRMs and four KH (hnRNP K homology) domains. The third and fourth KH domains (KH34) form an antiparallel pseudodimer, which binds the bipartite β-actin zipcode element cooperatively with high specificity and induces an approx. 180° looping of the RNA (Chao et al. 2010; Patel and Mitra 2012). This modulation of RNA structure might be a mechanism to facilitate the subsequent assembly of additional localization factors. Vertebrate homologs of ZBP1 are highly conserved (Yisraeli 2005). *Xenopus* Vg1RBP (or Vera) is required for *vg1* and *vegT* mRNA localization in *Xenopus* oocytes (Bubunenko et al. 2002; Deshler et al. 1997; Havin et al. 1998; Kwon et al. 2002). Similar to ZBP1, the KH domains of Vg1RBP contribute cooperatively to RNA binding (Git and Standart 2002). However, structural details of Vg1RBP function in RNA binding, like shown for ZBP1, are still missing.

The double-stranded (ds) RNA-binding protein Staufen was first identified as being required for *bicoid* and *oskar* mRNA localization in *Drosophila* oocytes (Kim-Ha et al. 1991; St Johnston et al. 1989, 1991). Furthermore, the co-localization of mammalian Staufen with RNP granules in dendrites of hippocampal rat neurons (Kiebler et al. 1999; Mallardo et al. 2003; Tang et al. 2001) and the identification of Staufen as localization factor for *vg1* and *vegT* mRNA in *Xenopus* oocytes revealed the structural and functional conservation of a localization protein for invertebrate and vertebrate systems (Yoon and Mowry 2004). It has been demonstrated that both *Drosophila* and *Xenopus* Staufen are part of localization RNPs in the cytoplasm (Kress et al. 2004; Mhlanga et al. 2009). Ds RNA-binding domains, as those in Staufen, bind their RNA targets generally in a non-sequence-specific manner (St Johnston et al. 1992). This raises the question of how Staufen can specifically interact with its target RNAs in vivo. One mechanism to gain specificity might be the assembly of a core RNP by sequence specific proteins, and subsequent recognition of the complex by Staufen through interactions with the other complex components and dsRNA (Yoon and Mowry 2004).

Mutagenesis studies with the *Xenopus dnd1* localization element revealed a clear correlation between the presence of uridine-rich regions, Elr-type protein binding and localization activity (Arthur et al. 2009). However, while a gradual loss of VM1-elements severely affected localization, interfering with Elr-type protein binding in the context of the *grip2* localization element did not reveal a critical role of Elr-type protein binding during vegetal localization, which could reflect a certain degree of redundancy in respect to the action of different sequence motifs (Claussen et al. 2011).

4 Motor-Dependent mRNA Localization Involves Multiple Steps

4.1 Initiation of mRNA Transport in the Nucleus

Several localization factors have nucleo-cytoplasmic shuttling abilities and bind their target RNAs already in the nucleus. These proteins might act to mark RNAs for their final destination in the cytoplasm and provide an initial step for the assembly of localization RNPs. Yeast She2p, for example, is recruited by the polymerase II machinery prior to binding to *ASH1* mRNA in the nucleus, suggesting that the *ASH1* localization RNP might assemble co-transcriptionally (Shen et al. 2010).

Although ZBP1 is predominantly cytoplasmic, FRAP experiments showed that it binds *β-actin* mRNA shortly after transcription in the nucleus and that its shuttling activity is essential for transport and localization of *β-actin* mRNA (Oleynikov and Singer 2003). Similarly, *Xenopus* Vg1RBP contains potential shuttling domains and binds its target RNAs in the nucleus and the cytoplasm (Havin et al. 1998; Kress et al. 2004; Ross et al. 1997). In addition, Kress et al. could show that *Xenopus* hnRNP I and Vg1RBP directly interact in the oocyte nucleus and bind to *vg1* and *vegT* mRNAs in both, nucleus and cytoplasm (Kress et al. 2004). 40LoVe, a *Xenopus* hnRNP D family protein, is distributed between nucleus and cytoplasm throughout oogenesis and also associates with *vg1* mRNA already in the nucleus (Czaplinski and Mattaj 2006). However, the exact role of these proteins in the complex process of RNA localization remains to be determined.

Interestingly, nuclear splicing can also be tightly coupled to RNA transport; in *Drosophila* oocytes, the exon junction complex is involved in *oskar* mRNA localization. Association of the *Drosophila* homologs for the human EJC components Y14/Tsunagi, Mago nashi, eIF4AIII, and Barentsz with the nascent *oskar* mRNA in the nucleus is required for *oskar* localization to the posterior pole of the oocyte (Hachet and Ephrussi 2001; Mohr et al. 2001; Newmark and Boswell 1994; Palacios et al. 2004). More recently, it could be demonstrated that splicing of the first *oskar* mRNA intron creates a splicing dependent *oskar* localization element, which is required for proper *oskar* motility in the oocyte (Ghosh et al. 2012).

4.2 Cytoplasmic Remodeling of Transport RNPs

After nuclear export, the localization RNP complex undergoes a series of coordinated transport steps; these include the incorporation into larger granules, coupling to motor-proteins, translational repression, and anchoring at the site of its destination. The coordination of this highly dynamic and complex process might involve post-translational, chemical modifications and/or alterations in complex composition.

In *Xenopus* oocytes, Vg1RBP and hnRNP I directly interact in the nucleus, but in the cytoplasm the interaction is sensitive to RNase and therefore thought to be mediated by the RNA scaffold. This observation seems to indicate that a remodeling step occurs during or after export of *vg1* and *vegT* containing RNPs from the nucleus, without removal of either one of these factors (Kress et al. 2004). This transition might be mediated by chemical modification of the RNP complex. Indeed, an un-phosphorylated form of hnRNP I was reported to be restricted to the nucleus, while in the cytoplasm, it is phosphorylated (Xie et al. 2003). Additional remodeling steps for *vg1* and *vegT* transport RNPs are defined by the association of Prrp and Staufen in the cytoplasm. Staufen might be recruited by hnRNP I, as these two proteins were found to interact directly (Kress et al. 2004) and a mutant form of Staufen, showing reduced interaction with hnRNP I, interferes with vegetal RNA localization (Yoon and Mowry 2004).

4.3 Motor-Proteins and RNA Transport

A central step in motor-based RNA localization is the coupling of the RNP complex to a motor-protein and subsequent transport along the cytoskeleton. Whereas the localization of *Drosophila bicoid* and *gurken* mRNAs is predominantly mediated by dynein, localization of *oskar* RNPs can be divided into a dynein-dependent step that leads from the nurse cell to the oocyte, and a kinesin-dependent transport step to the posterior pole of the oocyte (Kugler and Lasko 2009). Bicaudal-D and Egalitarian are two trans-acting factors which might couple *oskar* RNPs to dynein for minus end directed transport (Bullock and Ish-Horowicz 2001; Dienstbier et al. 2009). Kinesin-dependent transport is operating towards the plus end of microtubules. Staufen is a promising candidate factor for mediating the association of localization RNPs with kinesin, as in Staufen mutants *oskar* RNPs exhibit a reduced frequency of fast directional movement (Zimyanin et al. 2008), however, in the same situation, a small proportion of *oskar* RNPs localizes properly at the posterior pole and exhibits fast movement, indicating that additional factors are involved in coupling *oskar* RNPs to kinesin.

Xenopus vg1 mRNA transport was shown to be kinesin-1 and -2 dependent and is mediated along a subpopulation of microtubules, oriented with their plus-ends toward the vegetal cortex (Messitt et al. 2008). Recently it has been suggested that, similar to the transport of *Drosophila oskar*, the motor-based transport of RNAs in

Xenopus is a two-step mechanism (Gagnon et al. 2013). An initial highly directional step requires dynein and localizes the RNA in the upper vegetal cytoplasm. This is followed by a bidirectional kinesin-dependent step, which brings the RNA to the vegetal cortex to be anchored (Gagnon et al. 2013). As observed for *Drosophila* Staufen, *Xenopus* Staufen also associates with kinesin and might thus play a role in motor-protein-recruitment of *vg1* and *vegT* containing RNPs in oocytes (Yoon and Mowry 2004). Therefore, a conserved function for Staufen in coupling localization RNPs to kinesin is possible. However, Staufen does not interact with dynein (Yoon and Mowry 2004) and trans-acting factors mediating the first dynein-dependent step of vegetal RNA localization in *Xenopus* are so far unknown. Interestingly, a recent study in *Xenopus* oocytes revealed that animally and vegetally localizing RNPs contain a core-group of proteins including Vg1RBP, hnRNP I and 40LoVe. Staufen, on the other hand, is specific for vegetally localizing RNPs, indicating that this protein might contribute to the directionality of RNA movement (Snedden et al. 2013).

4.4 RNP Anchoring

Once a localizing RNP reaches its destination, its RNA is anchored preventing diffusion. In vertebrate and invertebrate oocytes, cortical actin-filaments are essential for RNA anchoring (Jankovics et al. 2002; Suzuki et al. 2000; Yisraeli et al. 1990). However, how the transported RNA is transferred from the microtubule to the actin cytoskeleton is unknown. Furthermore, dynein is implicated in functioning as a microtubule-based anchor of apically localizing RNAs in *Drosophila* embryos, suggesting a dual role for this motor-protein in transport and anchoring (Delanoue and Davis 2005).

In *Xenopus* oocytes, the protein Prrp (proline-rich RNA binding protein) was shown to interact with late localizing RNAs such as *vg1* and *vegT* (Zhao et al. 2001). As Prrp also interacts with the microfilament-associated proteins Mena and Profilin, it might provide a link between the localizing RNPs and the actin cytoskeleton (Zhao et al. 2001). Vg1RBP could also play a role in vegetal RNA anchoring, as its phosphorylation correlates with a release of *vg1* mRNA from the vegetal cortex during egg maturation (Git et al. 2009). Furthermore, localizing RNAs themselves can have a role in anchoring mRNAs to the vegetal cortex, as depletion of the noncoding RNA *Xlsirts* and *vegT* mRNA leads to a cortical release of *vg1* (Heasman et al. 2001; Kloc and Etkin 1994). These two RNAs seem to act in the organization of the cytokeratin cytoskeleton at the vegetal cortex as depletion of either RNA leads to the disruption of the cytokeratin network and release of localized RNAs (Kloc et al. 2005). Interestingly, proteins encoded by localizing RNAs could themselves also play a role in vegetal cortex anchoring of other localizing RNAs as recently, Dead-End1 protein was suggested to anchor *trim36* RNA at the vegetal cortex of the *Xenopus* egg (Mei et al. 2013).

5 RNA Transport and Translational Regulation

The restriction of protein expression to a distinct part of a cell is one major aim that appears to be connected to the phenomenon of RNA localization. Thus, a tight link of translational control with RNA localization seems reasonable. Indeed, a global analysis of RNA localization in *Drosophila* embryos revealed a high correlation between sites of the transcript localization and the proteins encoded (Lécuyer et al. 2007).

The first factor identified to play a role in repressing translation of a localizing RNA during transport is Bruno (Kim-Ha et al. 1995). Bruno binds to specific sequences called Bruno response elements (BRE) located in the 3′UTR of *oskar* mRNA (Kim-Ha et al. 1995). Although the detailed mechanism of Bruno-mediated translational repression is not yet known, there are indications that Bruno might interact with the protein Cup to inhibit translational initiation (Nakamura et al. 2004; Wilhelm et al. 2003). Bruno shuttles between the nucleus and the cytoplasm in *Drosophila* oocytes and may bind *oskar* mRNA already in the nucleus (Snee et al. 2008). Further nucleo-cytoplasmic shuttling proteins were shown to regulate *oskar* translation (Besse et al. 2009; Yano et al. 2004) and to potentially bind their target RNA in the nucleus. Thus, RNA binding in the nucleus prior to export might be a mechanism to provide immediate repression upon entering the cytoplasm. Yet another mechanism appears to operate in the case of *β-actin*; its localization factor ZBP1 has a second function in repressing *β-actin* translation (Hüttelmaier et al. 2005). ZBP1 phosphorylation by membrane-associated Src decreases RNA binding and thus, repression of *β-actin* translation is relieved at the cell periphery (Hüttelmaier et al. 2005). Finally, *Xenopus vg1* mRNA is translationally repressed before localization to the vegetal hemisphere of stage IV oocytes (Tannahill and Melton 1989; Wilhelm et al. 2000). Misexpression of vg1 in the animal hemisphere causes severe patterning defects (Dale et al. 1993; Thomsen and Melton 1993), emphasizing the importance of *vg1* mRNA localization and translational repression. The RNA-binding protein ElrB was reported to be responsible for translational repression of *vg1* via binding to a region in the 3′UTR named *vg1* translational element (VTE) (Colegrove-Otero et al. 2005; Otero et al. 2001; Wilhelm et al. 2000), which is distinct from its LE. However, the exact mechanism of how ElrB mediates translational inhibition remains to be defined.

6 Formation of "Super-complexes?"

Localizing RNAs such as *Drosophila oskar* were observed to be part of larger RNP granules, suggesting that multiple localization RNPs might assemble into super-complexes (Besse et al. 2009; Chekulaeva et al. 2006). It is unclear if formation of such granules has a function in RNA transport. Two *Drosophila* proteins, namely Bruno and hnRNPI (PTB), appear to be involved in the aggregation of large RNP granules containing multiple copies of *oskar* mRNA, these granules are inaccessible for ribosomes and thus translationally silent (Besse et al. 2009; Chekulaeva et al. 2006). Similarly, *Xenopus* Vg1RBP and its vertebrate homologs appear to act in the formation of large RNP granules via self-dimerization and interaction with other proteins (Farina et al. 2003; Git and Standart 2002; Nielsen et al. 2004). Furthermore, formation of super-complexes could also be mediated by their RNA components, as shown for *oskar* (Jambor et al. 2011). However, it is still unclear if transport granules contain only one type of transcript in multiple copies or multiple different RNAs. Experimental evidence for co-transport of multiple different RNA molecules in one RNP comes from studies on *ASH1* and *IST2* mRNAs in yeast (Lange et al. 2008).

Although it is apparent that some RNPs form granules, there are also results supporting the idea of independent localization of single RNPs. Use of an in vitro RNP reconstitution assay using *Drosophila* components revealed that most complexes assembled contain only a single RNA molecule (Amrute-Nayak and Bullock 2012). Furthermore, dendritic RNAs were found to localize independently from each other (Mikl et al. 2011). Staufen seems to regulate the number of transcripts in localizing RNPs, suggesting a tight regulation of the RNP composition in neurons (Mikl et al. 2011). Thus, in contrast to a higher transport efficiency and co-regulation of large RNP granules, separate localization of different RNAs might contribute to higher degree of temporal and spatial control of protein production in neurons (Mikl et al. 2011).

7 An Additional Function for Localization RNPs

As mentioned above, vegetal RNA localization in the *Xenopus* oocyte is linked to germ plasm formation and hence to germ cell specification. Quite a few of the germ plasm associated mRNAs encode for RNA-binding proteins which appear to be part of a self sustaining regulatory network; (our own unpublished results). Such germ cell-specific RNA-binding proteins bind to the 3′-UTR of each other's mRNAs, thereby protecting their target RNAs from miRNA-mediated RNA degradation during early phases of embryogenesis, when the maternal to zygotic transition of gene expression occurs (MZT). Similarly, we were able to demonstrate that proteins involved in mRNA localization in the oocyte also function in the germ cell-specific stabilization of maternal mRNAs, where the miRNA target sequence colocalizes with the RNA localization element (Koebernick et al. 2010). Thus, at least in part, it appears to be the same proteins which regulate RNA localization in the oocyte and RNA stability in the embryo.

References

Amrute-Nayak M, Bullock SL (2012) Single-molecule assays reveal that RNA localization signals regulate dynein-dynactin copy number on individual transcript cargoes. Nat Cell Biol 14: 416–423

Arthur PK, Claussen M, Koch S et al (2009) Participation of Xenopus Elr-type proteins in vegetal mRNA localization during oogenesis. J Biol Chem 284:19982–19992

Auweter SD, Oberstrass FC, Allain FH-T (2006) Sequence-specific binding of single-stranded RNA: is there a code for recognition? Nucleic Acids Res 34:4943–4959

Bassell GJ, Zhang H, Byrd AL et al (1998) Sorting of beta-actin mRNA and protein to neurites and growth cones in culture. J Neurosci 18:251–265

Besse F, López de Quinto S, Marchand V et al (2009) Drosophila PTB promotes formation of high-order RNP particles and represses oskar translation. Genes Dev 23:195–207

Betley JN, Frith MC, Graber JH et al (2002) A ubiquitous and conserved signal for RNA localization in chordates. Curr Biol 12:1756–1761

Böhl F, Kruse C, Frank A et al (2000) She2p, a novel RNA-binding protein tethers ASH1 mRNA to the Myo4p myosin motor via She3p. EMBO J 19:5514–5524

Bubunenko M, Kress TL, Vempati UD et al (2002) A consensus RNA signal that directs germ layer determinants to the vegetal cortex of Xenopus oocytes. Dev Biol 248:82–92

Bullock SL, Ish-Horowicz D (2001) Conserved signals and machinery for RNA transport in Drosophila oogenesis and embryogenesis. Nature 414:611–616

Bullock SL, Ringel I, Ish-Horowicz D et al (2010) A′ form RNA helices are required for cytoplasmic mRNA transport in Drosophila. Nature 17:703–709

Chan AP, Kloc M, Etkin LD (1999) fatvg encodes a new localized RNA that uses a 25-nucleotide element (FVLE1) to localize to the vegetal cortex of Xenopus oocytes. Development 126: 4943–4953

Chang P, Torres J, Lewis RA et al (2004) Localization of RNAs to the mitochondrial cloud in Xenopus oocytes through entrapment and association with endoplasmic reticulum. Mol Biol Cell 15:4669–4681

Chao JA, Patskovsky Y, Patel V et al (2010) ZBP1 recognition of beta-actin zipcode induces RNA looping. Genes Dev 24:148–158

Chartrand P, Meng H, Singer RH et al (1999) Structural elements required for the localization of ASH1 mRNA and of a green fluorescent protein reporter particle in vivo. Curr Biol 9:333–336

Chekulaeva M, Hentze MW, Ephrussi A (2006) Bruno acts as a dual repressor of oskar translation, promoting mRNA oligomerization and formation of silencing particles. Cell 124:521–533

Claussen M, Pieler T (2004) Xvelo1 uses a novel 75-nucleotide signal sequence that drives vegetal localization along the late pathway in Xenopus oocytes. Dev Biol 266:270–284

Claussen M, Tarbashevich K, Pieler T (2011) Functional dissection of the RNA signal sequence responsible for vegetal localization of XGrip2.1 mRNA in Xenopus oocytes. RNA Biol 8: 873–882

Colegrove-Otero LJ, Devaux A, Standart N (2005) The Xenopus ELAV protein ElrB represses Vg1 mRNA translation during oogenesis. Mol Cell Biol 25:9028–9039

Czaplinski K, Mattaj IW (2006) 40LoVe interacts with Vg1RBP/Vera and hnRNP I in binding the Vg1-localization element. RNA 12:213–222

Dale L, Matthews G, Colman A (1993) Secretion and mesoderm-inducing activity of the TGF-beta-related domain of Xenopus Vg1. EMBO J 12:4471–4480

Delanoue R, Davis I (2005) Dynein anchors its mRNA cargo after apical transport in the Drosophila blastoderm embryo. Cell 122:97–106

Deshler JO, Highett MI, Schnapp BJ (1997) Localization of Xenopus Vg1 mRNA by Vera protein and the endoplasmic reticulum. Science 276:1128–1131

Dienstbier M, Boehl F, Li X et al (2009) Egalitarian is a selective RNA-binding protein linking mRNA localization signals to the dynein motor. Genes Dev 23:1546–1558

Ding D, Parkhurst SM, Halsell SR et al (1993) Dynamic Hsp83 RNA localization during Drosophila oogenesis and embryogenesis. Mol Cell Biol 13:3773–3781

dos Santos G, Simmonds AJ, Krause HM (2008) A stem-loop structure in the wingless transcript defines a consensus motif for apical RNA transport. Development 135:133–143

Farina KL, Hüttelmaier S, Musunuru K et al (2003) Two ZBP1 KH domains facilitate beta-actin mRNA localization, granule formation, and cytoskeletal attachment. J Cell Biol 160:77–87

Gagnon JA, Mowry KL (2011) Molecular motors: directing traffic during RNA localization. Crit Rev Biochem Mol Biol 46:229–239

Gagnon JA, Kreiling JA, Powrie EA et al (2013) Directional transport is mediated by a Dynein-dependent step in an RNA localization pathway. PLoS Biol 11:e1001551

Ghosh S, Marchand V, Gáspár I et al (2012) Control of RNP motility and localization by a splicing-dependent structure in oskar mRNA. Nat Struct Mol Biol 19:441–449

Git A, Standart N (2002) The KH domains of Xenopus Vg1RBP mediate RNA binding and self-association. RNA 8:1319–1333

Git A, Allison R, Perdiguero E et al (2009) Vg1RBP phosphorylation by Erk2 MAP kinase correlates with the cortical release of Vg1 mRNA during meiotic maturation of Xenopus oocytes. RNA 15:1121–1133

Gonzalez I, Buonomo SB, Nasmyth K et al (1999) ASH1 mRNA localization in yeast involves multiple secondary structural elements and Ash1 protein translation. Curr Biol 9:337–340

Hachet O, Ephrussi A (2001) Drosophila Y14 shuttles to the posterior of the oocyte and is required for oskar mRNA transport. Curr Biol 11:1666–1674

Havin L, Git A, Elisha Z et al (1998) RNA-binding protein conserved in both microtubule- and microfilament-based RNA localization. Genes Dev 12:1593–1598

Heasman J, Wessely O, Langland R et al (2001) Vegetal localization of maternal mRNAs is disrupted by VegT depletion. Dev Biol 240:377–386

Heinrich B, Deshler JO (2009) RNA localization to the Balbiani body in Xenopus oocytes is regulated by the energy state of the cell and is facilitated by kinesin II. RNA 15:524–536

Houston DW, King ML (2000) A critical role for Xdazl, a germ plasm-localized RNA, in the differentiation of primordial germ cells in Xenopus. Development 127:447–456

Hüttelmaier S, Zenklusen D, Lederer M et al (2005) Spatial regulation of β-actin translation by Src-dependent phosphorylation of ZBP1. Nat Cell Biol 438:512–515

Jambor H, Brunel C, Ephrussi A (2011) Dimerization of oskar 3′UTRs promotes hitchhiking for RNA localization in the Drosophila oocyte. RNA 17:2049–2057

Jankovics F, Sinka R, Lukácsovich T et al (2002) MOESIN crosslinks actin and cell membrane in Drosophila oocytes and is required for OSKAR anchoring. Curr Biol 12:2060–2065

Joseph EM, Melton DA (1998) Mutant Vg1 ligands disrupt endoderm and mesoderm formation in Xenopus embryos. Development 125:2677–2685

Kiebler MA, Hemraj I, Verkade P et al (1999) The mammalian staufen protein localizes to the somatodendritic domain of cultured hippocampal neurons: implications for its involvement in mRNA transport. J Neurosci 19:288–297

Kim-Ha J, Smith JL, Macdonald PM (1991) oskar mRNA is localized to the posterior pole of the Drosophila oocyte. Cell 66:23–35

Kim-Ha J, Kerr K, Macdonald PM (1995) Translational regulation of oskar mRNA by bruno, an ovarian RNA-binding protein, is essential. Cell 81:403–412

Kloc M, Etkin LD (1994) Delocalization of Vg1 mRNA from the vegetal cortex in Xenopus oocytes after destruction of Xlsirt RNA. Science 265:1101–1103

Kloc M, Larabell C, Etkin LD (1996) Elaboration of the messenger transport organizer pathway for localization of RNA to the vegetal cortex of Xenopus oocytes. Dev Biol 180:119–130

Kloc M, Wilk K, Vargas D et al (2005) Potential structural role of non-coding and coding RNAs in the organization of the cytoskeleton at the vegetal cortex of Xenopus oocytes. Development 132:3445–3457

Koebernick K, Loeber J, Arthur PK (2010) Elr-type proteins protect Xenopus Dead end mRNA from miR-18-mediated clearance in the soma. Proc Natl Acad Sci U S A 107:16148–16153

Kress TL, Yoon YJ, Mowry KL (2004) Nuclear RNP complex assembly initiates cytoplasmic RNA localization. J Cell Biol 165:203–211

Kugler J-M, Lasko P (2009) Localization, anchoring and translational control of oskar, gurken, bicoid and nanos mRNA during Drosophila oogenesis. Fly 3:15–28

Kwon S, Abramson T, Munro TP et al (2002) UUCAC- and vera-dependent localization of VegT RNA in Xenopus oocytes. Curr Biol 12:558–564

Lange S, Katayama Y, Schmid M et al (2008) Simultaneous transport of different localized mRNA species revealed by live-cell imaging. Traffic 9:1256–1267

Lawrence JB, Singer RH (1986) Intracellular localization of messenger RNAs for cytoskeletal proteins. Cell 45:407–415

Lécuyer E, Yoshida H, Parthasarathy N et al (2007) Global analysis of mRNA localization reveals a prominent role in organizing cellular architecture and function. Cell 131:174–187

Lunde BM, Moore C, Varani G (2007) RNA-binding proteins: modular design for efficient function. Nat Rev Mol Cell Biol 8:479–490

MacArthur H, Houston DW, Bubunenko M et al (2000) DEADSouth is a germ plasm specific DEAD-box RNA helicase in Xenopus related to eIF4A. Mech Dev 95:291–295

Mallardo M, Deitinghoff A, Müller J (2003) Isolation and characterization of Staufen-containing ribonucleoprotein particles from rat brain. Proc Natl Acad Sci U S A 100:2100–2105

Mei W, Jin Z, Lai F et al (2013) Maternal Dead-End1 is required for vegetal cortical microtubule assembly during Xenopus axis specification. Development 140:2334–2344

Messitt TJ, Gagnon JA, Kreiling JA et al (2008) Multiple kinesin motors coordinate cytoplasmic RNA transport on a subpopulation of microtubules in Xenopus oocytes. Dev Cell 15:426–436

Mhlanga MM, Bratu DP, Genovesio A et al (2009) In vivo colocalisation of oskar mRNA and trans-acting proteins revealed by quantitative imaging of the Drosophila oocyte. PLoS One 4: e6241

Mikl M, Vendra G, Kiebler MA (2011) Independent localization of MAP2, CaMKIIα and β-actin RNAs in low copy numbers. EMBO Rep 12:1077–1084

Mohr SE, Dillon ST, Boswell RE (2001) The RNA-binding protein Tsunagi interacts with Mago Nashi to establish polarity and localize oskar mRNA during Drosophila oogenesis. Genes Dev 15:2886–2899

Mowry KL, Melton DA (1992) Vegetal messenger RNA localization directed by a 340-nt RNA sequence element in Xenopus oocytes. Science 255:991–994

Müller M, Heym RG, Mayer A et al (2011) A cytoplasmic complex mediates specific mRNA recognition and localization in yeast. PLoS Biol 9:e1000611

Nakamura A, Sato K, Hanyu-Nakamura K (2004) Drosophila cup is an eIF4E binding protein that associates with Bruno and regulates oskar mRNA translation in oogenesis. Dev Cell 6:69–78

Newmark PA, Boswell RE (1994) The mago nashi locus encodes an essential product required for germ plasm assembly in Drosophila. Development 120:1303–1313

Nielsen J, Kristensen MA, Willemoës M et al (2004) Sequential dimerization of human zipcode-binding protein IMP1 on RNA: a cooperative mechanism providing RNP stability. Nucleic Acids Res 32:4368–4376

Niessing D, Hüttelmaier S, Zenklusen D et al (2004) She2p is a novel RNA binding protein with a basic helical hairpin motif. Cell 119:491–502

Oleynikov Y, Singer RH (2003) Real-time visualization of ZBP1 association with β-actin mRNA during transcription and localization. Curr Biol 13:199–207

Otero LJ, Devaux A, Standart N (2001) A 250-nucleotide UA-rich element in the 3′ untranslated region of Xenopus laevis Vg1 mRNA represses translation both in vivo and in vitro. RNA 7:1753–1767

Palacios IM, Gatfield D, St Johnston D et al (2004) An eIF4AIII- containing complex required for mRNA localization and nonsense-mediated mRNA decay. Nature 427:753–757

Patel VL, Mitra SRH (2012) Spatial arrangement of an RNA zipcode identifies mRNAs under post-transcriptional control. Genes Dev 26:43–53

Ross AF, Oleynikov Y, Kislauskis EH et al (1997) Characterization of a beta-actin mRNA zipcode-binding protein. Mol Cell Biol 17:2158–2165

Serano TL, Cohen RS (1995) A small predicted stem-loop structure mediates oocyte localization of Drosophila K10 mRNA. Development 121:3809–3818

Shen Z, Paquin N, Forget A et al (2009) Nuclear shuttling of She2p couples ASH1 mRNA localization to its translational repression by recruiting Loc1p and Puf6p. Mol Biol Cell 20: 2265–2275

Shen Z, St-Denis A, Chartrand P (2010) Cotranscriptional recruitment of She2p by RNA pol II elongation factor Spt4-Spt5/DSIF promotes mRNA localization to the yeast bud. Genes Dev 24:1914–1926

Snedden DD, Bertke MM, Vernon D et al (2013) RNA localization in Xenopus oocytes uses a core group of trans-acting factors irrespective of destination. RNA 19:889–895

Snee M, Benz D, Jen J et al (2008) Two distinct domains of Bruno bind specifically to the oskar mRNA. RNA Biol 5:1–9

St Johnston D, Driever W, Berleth T et al (1989) Multiple steps in the localization of bicoid RNA to the anterior pole of the Drosophila oocyte. Development 107:13–19

St Johnston D, Beuchle D, Nüsslein-Volhard C (1991) Staufen, a gene required to localize maternal RNAs in the Drosophila egg. Cell 66:51–63

St Johnston D, Brown NH, Gall JG et al (1992) A conserved double-stranded RNA-binding domain. Proc Natl Acad Sci U S A 89:10979–10983

Suzuki H, Maegawa S, Nishibu T et al (2000) Vegetal localization of the maternal mRNA encoding an EDEN-BP/Bruno-like protein in zebrafish. Mech Dev 93:205–209

Tang SJ, Meulemans D, Vazquez L et al (2001) A role for a rat homolog of staufen in the transport of RNA to neuronal dendrites. Neuron 32:463–475

Tannahill D, Melton DA (1989) Localized synthesis of the Vg1 protein during early Xenopus development. Development 106:775–785

Thomsen GH, Melton DA (1993) Processed Vg1 protein is an axial mesoderm inducer in Xenopus. Cell 74:433–441

Tian B, Bevilacqua PC, Diegelman-Parente A et al (2004) The double-stranded-RNA-binding motif: interference and much more. Nat Rev Mol Cell Biol 5:1013–1023

Van De Bor V, Hartswood E, Jones C et al (2005) Gurken and the I factor retrotransposon RNAs share common localization signals and machinery. Dev Cell 9:51–62

Wilhelm JE, Vale RD, Hegde RS (2000) Coordinate control of translation and localization of Vg1 mRNA in Xenopus oocytes. Proc Natl Acad Sci U S A 97:13132–13137

Wilhelm JE, Hilton M, Amos Q et al (2003) Cup is an eIF4E binding protein required for both the translational repression of oskar and the recruitment of Barentsz. J Cell Biol 163:1197–1204

Xie J, Lee J-A, Kress TL et al (2003) Protein kinase A phosphorylation modulates transport of the polypyrimidine tract-binding protein. Proc Natl Acad Sci U S A 100:8776–8781

Yamaguchi T, Taguchi A, Watanabe K et al (2012) DEADSouth protein localizes to germ plasm and is required for the development of primordial germ cells in Xenopus laevis. Biol Open 2: 191–199

Yano T, de Quinto SL, Matsui Y et al (2004) Hrp48, a Drosophila hnRNPA/B homolog, binds and regulates translation of oskar mRNA. Dev Cell 6:637–648

Yisraeli JK (2005) VICKZ proteins: a multi-talented family of regulatory RNA-binding proteins. Biol Cell 97:87–96

Yisraeli JK, Sokol S, Melton DA (1990) A two-step model for the localization of maternal mRNA in Xenopus oocytes: involvement of microtubules and microfilaments in the translocation and anchoring of Vg1 mRNA. Development 108:289–298

Yoon YJ, Mowry KL (2004) Xenopus Staufen is a component of a ribonucleoprotein complex containing Vg1 RNA and kinesin. Development 131:3035–3045

Zhang J, Houston DW, King ML et al (1998) The role of maternal VegT in establishing the primary germ layers in Xenopus embryos. Cell 94:515–524

Zhao WM, Jiang C, Kroll TT et al (2001) A proline-rich protein binds to the localization element of Xenopus Vg1 mRNA and to ligands involved in actin polymerization. EMBO J 20: 2315–2325

Zimyanin VL, Belaya K, Pecreaux J et al (2008) In vivo imaging of oskar mRNA transport reveals the mechanism of posterior localization. Cell 134:843–853

Small Size, Big Impact: Bacterial Functional Nucleic Acids and Their Applications

Wendy W.K. Mok, Simon A. McManus, and Yingfu Li

Contents

1 Introduction 310
2 Riboswitches: RNAs as Receptors and Regulators 311
2.1 Overview of Discovery and Properties of Natural Riboswitches 311
2.2 Riboswitches as Therapeutic Targets 311
2.3 Engineering Riboswitches into Functional Tools 313
3 Small RNAs: Targeted Gene Regulation Through RNA Binding 315
3.1 sRNA Targets 315
3.2 sRNAs for Fast and Tight Regulation 317
3.3 Optimizing sRNAs for Specifically Target Gene Expression 318
4 Concluding Remarks and Future Prospects 320
References 321

W.W.K. Mok
Department of Biochemistry and Biomedical Sciences, McMaster University, Hamilton, ON, Canada L8S 4K1

Department of Chemical and Biological Engineering, Princeton University, Princeton, NJ 08544, USA

S.A. McManus
Department of Biochemistry and Biomedical Sciences, McMaster University, Hamilton, ON, Canada L8S 4K1

Department of Chemistry, Princeton University, Princeton, NJ 08544, USA

Y. Li (✉)
Department of Biochemistry and Biomedical Sciences, McMaster University, Hamilton, ON, Canada L8S 4K1

Department of Chemistry and Chemical Biology, McMaster University, Hamilton, ON, Canada

Michael G. Degroote Institute for Infectious Disease Research, McMaster University, Hamilton, ON, Canada
e-mail: liying@mcmaster.ca

V.A. Erdmann et al. (eds.), *Chemical Biology of Nucleic Acids*, RNA Technologies,
DOI 10.1007/978-3-642-54452-1_17,

Abstract Genome mining efforts carried out across diverse bacterial genomes over the past two decades have paid off handsomely, leading to the discovery of a plethora of regulatory RNAs. These RNAs include riboswitches—a class of ligand responsive gene regulating elements—and small RNAs (sRNA)—short RNA or protein targeting sequences. Together, this ensemble of RNAs orchestrates and fine-tunes metabolic, stress response, and virulence pathways. These RNAs are key players in genetic networks. Searches for novel regulatory RNAs and their subsequent characterization remain an exciting area of research. Due to the ingenuity of their design and important functions they execute, recent research has also focused on engineering synthetic mimics of naturally occurring riboswitches and sRNAs and exploring these elements as potential therapeutics. In this chapter, we will present an overview on the discovery, general properties, and key functions of riboswitches and sRNAs annotated in different bacterial genomes. We will examine these RNAs as possible targets for novel antimicrobials. We will also discuss efforts in creating synthetic riboswitches and sRNAs, as well as the possibility of using them in biotechnology and as ammunition in our continued fight against multidrug-resistant pathogens.

Keywords Riboswitch • sRNA • Metabolite • Gene regulation • Drug discovery • Programmable cells

1 Introduction

As exemplified by other chapters of this book, RNAs are versatile and vital workhorses in the cell. They are indispensable for mediating the transfer of genetic information from DNA to proteins (in the form of messenger RNAs) and are important structural and functional components of ribosomes (in the form of ribosomal RNAs). RNAs have also been found to have enzymatic activity, perhaps as remnants of the hypothetical RNA world. Naturally occurring and synthetic ribozymes can catalyze an interesting repertoire of reactions, ranging from self-cleavage and ligation (Pan and Uhlenbeck 1992; Bartel and Szostak 1993) to more complex chemical reactions, such as carbon–carbon bond formation (Tarasow et al. 1997; Seelig and Jaschke 1999). Over the past two decades, our toolbox of functional RNAs has continued to diversify and expand. Genome mining efforts have contributed to the discovery of numerous regulatory RNAs in eukaryotic and prokaryotic genomes. In bacteria, regulatory RNAs including riboswitches and small RNAs (sRNAs) are key players in governing the expression of key metabolic, stress response, and virulence genes. These two classes of regulatory RNAs will be the main focus of this chapter. The discovery, structure, and mechanism of action of regulatory RNAs have been the subject of many extensive and comprehensive reviews (Storz et al. 2011; Gottesman and Storz 2011; Garst et al. 2011; Liu and

Camilli 2010; Frohlich and Vogel 2009; Dambach and Winkler 2009). Here, we will primarily discuss the applications of these RNA elements in research, bioremediation, and medicine.

2 Riboswitches: RNAs as Receptors and Regulators

2.1 *Overview of Discovery and Properties of Natural Riboswitches*

Proteins have long been considered the main structural and regulatory elements of the cell, serving as the building blocks of enzymes and receptors. The discovery that the mRNAs encoding enzymes for thiamine (vitamin B1) biosynthesis can regulate their own translation shifted this paradigm (Winkler et al. 2002). The 5′-untranslated region (5′-UTR) of the *thiM* mRNA was found to bind thiamine via its ligand-binding "receptor" (or "aptamer") domain and elicit structural changes in its neighboring "expression platform" domain, causing the ribosomal binding site (RBS) of the mRNA to be sequestered in a hairpin loop. Adopting this inhibitory alternative structure in *thiM* subsequently represses its translation and downregulates thiamine biosynthesis when the metabolite is abundant in the cell. This allosteric regulatory RNA element was coined a "riboswitch." Since the discovery of the first riboswitch over a decade ago, more than 20 classes of riboswitches capable of binding a range of small metabolites and metal ions (refer to Serganov and Nudler 2013 for inventory) have been isolated across different species of bacteria, plants, and fungi. These intracellular sensors can regulate gene expression by modulating transcription (Fig. 1a), translation (Fig. 1b), RNA stability (Fig. 1c), and splicing (Fig. 1d). Depending on the alternative structure adopted in the riboswitches, the expression of the downstream gene can be enhanced or repressed following ligand binding.

2.2 *Riboswitches as Therapeutic Targets*

Riboswitches have been predicted to regulate up to 2 % of genes in certain bacterial genomes and many modulate the activities of vital metabolic and virulence genes. Consequently, they have been explored and exploited as potential targets of antimicrobial compounds. Chemical analogs of lysine (Blount et al. 2007), thiamine pyrophosphate (Sudarsan et al. 2005), flavin mononucleotide (Lee et al. 2009), and guanine (Mulhbacher et al. 2010) that can bind their associated riboswitches with high affinity have been found to inhibit gene expression and cell growth. For instance, administrating nanomolar amounts of 2, 5, 6-triaminopyrimidine-4-one, a pyrimidine compound that can bind the guanine riboswitch in the GMP synthetase (*guaA*) mRNA, is bactericidal to a number of clinically relevant nosocomial

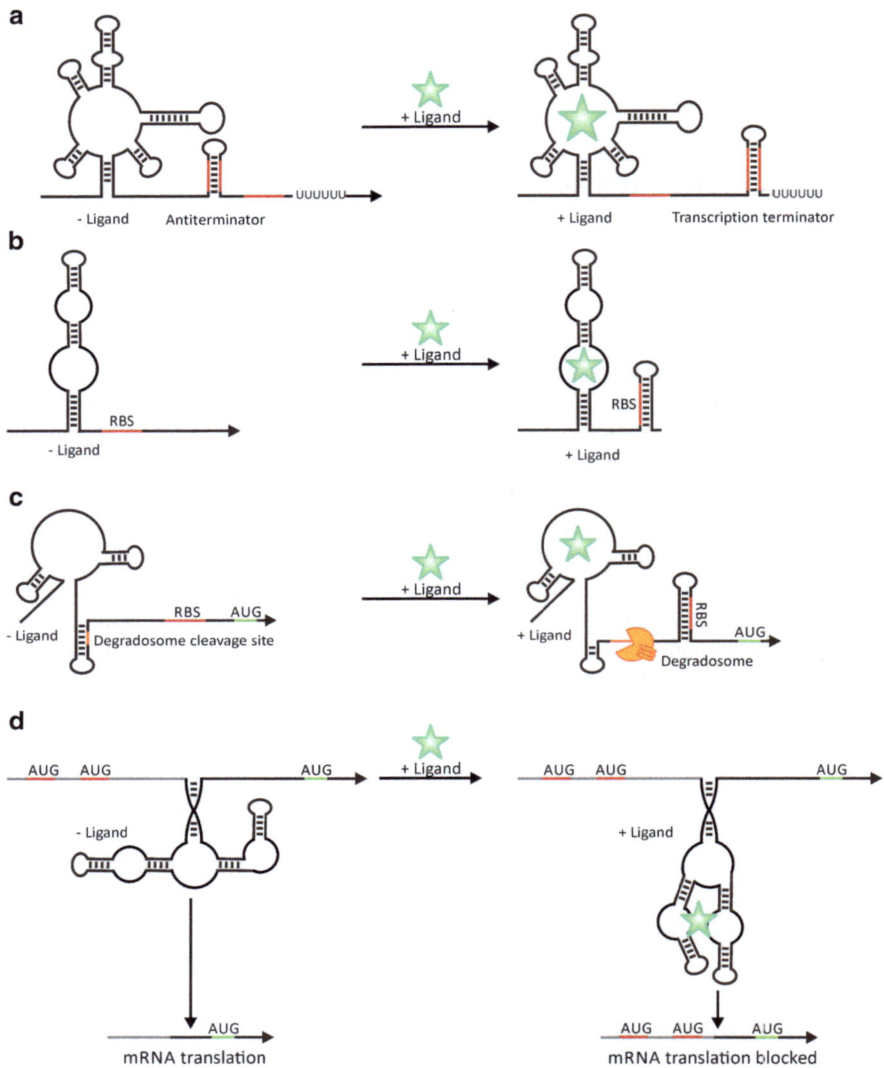

Fig. 1 Mechanisms of riboswitch action. Following ligand binding, structure switching in the aptamers and expression platform domains, resulting in changes in the expression of the target gene. (**a**) Transcription repression. Ligand binding disrupts the antiterminator stem loop and promotes the formation of the transcription terminator, consequently resulting in early transcription termination. (**b**) Translation repression. Ligand binding promotes the formation of a stem that sequesters the ribosome binding site, thus inhibiting translation. (**c**) mRNA stability. Ligand binding results in structural changes in the riboswitch, which exposes a degradosome targeting sequence. (**d**) Alternative splicing. Ligand binding promotes structure switching and alternative slicing of the 5′-UTR, consequently altering the expression of the target transcript

pathogens, including *Staphylococcus aureus* and multidrug-resistant *Clostridium difficile* (Mulhbacher et al. 2010). This agonistic compound represses *guaA* expression, causing cells to be starved for guanine. Riboswitches are attractive drug targets, because these elements often govern genes involved in the transport and biosynthesis of key metabolites. Thus metabolite analogs that can bind the aptamers and repress the expression of riboswitch-associated genes can exert pleiotropic effects that can potentially be lethal to pathogens. As some riboswitches are distributed in diverse bacterial lineages, they can serve as targets for broad-spectrum antibacterial agents. While point mutations that disrupt the binding of riboswitch-targeting antibiotics can arise in the aptamer domains and confer resistance to the drug, these mutations would also impair binding of the natural ligand and cause growth defects (Mulhbacher et al. 2010). Although resistance attributed to riboswitch mutations is expected to be deleterious to bacteria, riboswitch-active compounds should be designed so that they are structurally dissimilar from natural metabolites to avoid being substrates of export and modification proteins. Riboswitches have not been discovered in humans, and eukaryotic riboswitches that have been found function by vastly different mechanisms (Cheah et al. 2007; Li and Breaker 2013); thus these compounds are not expected to act on riboswitch homologs in the human genome. However, as riboswitch-targeting compounds are mimics of natural metabolites, care should be taken in their design to prevent undesirable off-target toxicity arising from competitive inhibition of mammalian proteins.

2.3 Engineering Riboswitches into Functional Tools

Following the architecture of natural riboswitches as a blueprint, researchers have engineered synthetic riboswitches with new ligand specificities and riboswitch circuits targeting novel genes. Even before the discovery of riboswitches, Werstuck and Green demonstrated that RNA aptamers isolated via in vitro selection can be introduced to the 5′-UTR of genes in order to regulate their translation in vitro and in eukaryotic cells (Werstuck and Green 1998). The Gallivan group later placed an aptamer responsive to the small molecule theophylline upstream of the RBS of *lacZ*, *cat*, and *cheZ* (Topp and Gallivan 2007). In *E. coli* transformed with plasmids containing these riboswitches, the administration of theophylline can induce β-galactosidase activity, confer resistance to chloramphenicol, and stimulate cellular migration, respectively. These riboswitch constructs are often plasmid encoded; they can easily be transferred to different bacterial strains, where they can potentially be used to monitor the presence of small molecules in cells. To optimize the activity of synthetic riboswitches and the expression of their associated genes, nucleotides within the riboswitch can be randomized, and these putative riboswitches can be subjected to selection. Depending on the activity of the target gene, different methods can be implemented to isolate the best performing riboswitches. Fowler and colleagues partially randomized nucleotides in the region between the theophylline aptamers and the green fluorescent protein (GFP) gene and subjected *E. coli* harboring the library of riboswitch variants to fluorescence-activated cell sorting (FACS). This powerful technique allows over 10^4 cells to be sorted per second based on their fluorescence intensity. Using FACS, cells that were highly fluorescent in the presence of theophylline were isolated, and riboswitches with high ligand specificity and fluorescence enhancement were produced (Fowler et al. 2008). In fusing the library of theophylline riboswitches upstream of *cheZ*, Topp and Gallivan developed a cost-effective high-throughput motility assay, where they monitored the migration of *E. coli* on semisolid agar in response to theophylline that can be used to isolate functional synthetic riboswitches (Topp and Gallivan 2008).

In recent years, riboswitch research has become an exciting playing field. Current research does not only focus on the identification and characterization of novel riboswitches, but it also encompasses the engineering of riboswitch-based tools and exploring the use of these tools to solve laboratory and real-world problems. Reynoso et al. (2012) used a theophylline riboswitch to regulate a virulence gene in the intracellular murine pathogen *Francisella novicida*, thus enabling its characterization. Through this study, they showed that riboswitches could be used to regulate gene expression in bacterial strains for which genetic tools, such as inducible promoters and cloning vectors, are scarce. Riboswitches can be used in conjunction with native promoters to minimize complications arising from incompatibility of foreign genetic regulation systems. Riboswitches can also be coupled with various reporter genes to monitor intracellular concentrations of small molecules. This strategy greatly simplifies current methods used to quantify cellular metabolites, which involves cell lysis followed by metabolite separation and identification. Fowler et al. (2010) demonstrated that these riboswitch biosensors can be used to study genes and gene products involved in small molecule transport and biosynthesis. Kellenberger et al. (2013) recently linked variants of the GEMM-I riboswitch to the Spinach RNA aptamer and engineered fluorescent biosensors for cyclic dinucleotide molecules. Upon binding cyclic di-GMP or cyclic AMP-GMP, Spinach binds and activates a fluorescent dye. These biosensors can be used in vitro, and they can also be used to image these second messengers in live cells.

Riboswitches can be used to regulate genes and gene circuits, and these elements can also be integrated into bacterial genomes to produce programmable cells. Upon fusing *cheZ* to the theophylline riboswitch, the Gallivan group showed that they could reprogram *E. coli* to migrate up a theophylline gradient and localize to theophylline-rich regions of semisolid media (Topp and Gallivan 2007). Following similar engineering principles, we can envision the possibility of creating synthetic riboswitches against pollutants and disease biomarkers using in vitro selection and coupling these riboswitches to motility genes (Sinha et al. 2010). These constructs can be introduced into bacteria that have been modified to encode genes that can degrade pollutants or synthesize and release therapeutics, thereby producing micro machines for bioremediation and site-specific drug delivery, respectively. Riboswitch-based sensors can also be used to expedite the evolution of metabolite-producing bacteria in order to generate strains that can maximize yields of desired molecules. As a proof of concept, Yang et al. (2013) coupled the lysine riboswitch derived from the 5′-UTR of *lysC* with *tetA*, which encodes the tet/H^+antiporter. The presence of this construct promotes the survival of *E. coli* that produces high levels of lysine in the presence of toxic metal salts, such as $NiCl_2$. After four enrichment cycles, 75 % of the population was enriched with strains that are optimized for L-lysine production. Based on these design principles, natural and synthetic riboswitches can be incorporated into these gene circuits to engineer efficient "microbial factories" for the production of an array of metabolites.

3 Small RNAs: Targeted Gene Regulation Through RNA Binding

While riboswitches are attached to the transcripts they regulate, another class of regulatory RNAs, coined small RNAs (sRNAs), is transcribed separately from their target genes and acts by binding either the mRNA of a targeted gene or the gene product itself. In the past, many sRNAs were overlooked as they are often found in opposite strands of their targets or in what were once believed to be intergenic regions (IGRs), masked by larger flanking genes. Their small size further impeded their detection by mutational screens (Wassarman et al. 2001). Advances in bioinformatics and experimental techniques, including microarrays, tiling arrays, and deep sequencing, contributed to the emergence of higher resolution maps of bacterial genomes and the discovery of a plethora of sRNAs. Despite ranging from only 50 to 500 nucleotides (nt) in size, sRNAs constitute a major method of regulation in bacteria, with over 100 sRNAs identified in *E. coli*.

3.1 sRNA Targets

sRNAs can interact with protein and mRNA targets. Protein-targeting sRNAs often act on global regulatory proteins under stress conditions, thereby affecting multiple metabolic pathways. For example, CsrB in *E. coli* binds CsrA, a small polypeptide that acts as a regulator for carbon storage, during stationary phase growth or under nutrient-poor conditions. Due to the presence of repetitive sequences that mimic the leader regions of the genes bound by CsrA in CsrB, one molecule of CsrB can bind up to 18 copies of CsrA (Liu et al. 1997; Liu and Romeo 1997). This interaction sequesters CsrA away from its target genes, resulting in an increase in glycogen levels in the cell.

mRNA targeting sRNAs can be further subdivided into two classes: *trans*-acting and *cis*-acting. *trans*-acting sRNAs act on distally encoded mRNAs and share limited base-pair complementarity with their targets. Although the region of potential base pairing between *trans*-acting sRNAs and their targets often encompasses 10–25 nt, it has been reported that only a few nucleotides are critical for their interaction (Kawamoto et al. 2006). Due to this limited sequence complementarity, *trans*-acting sRNAs can coordinate the expression of multiple genes in a regulatory circuit, allowing bacteria to mount specific physiological responses and cope with specific stress conditions (Jacques et al. 2006; Gottesman et al. 2006; Altuvia et al. 1997; Johansen et al. 2006). *trans*-acting sRNAs often require mediation by chaperone Hfq, which binds RNAs as a hexamer, leading to RNA remodeling, alleviation of inhibitory secondary structures, and target binding (Beisel et al. 2012; Brennan and Link 2007). Hfq can also bind RNAs at sites that are recognized by RNase E, thereby preventing the degradation of sRNAs that have not yet hybridized with their targets. Alternatively, Hfq binding may recruit RNases and promote the degradation of sRNA::mRNA complexes.

Contrary to *trans*-acting sRNAs, their *cis*-acting counterparts are encoded on the antisense strand of the mRNA they regulate. Consequently, they have the propensity to form extensive base-pairing interactions. Despite their extensive complementarity to their target mRNAs, these sRNAs are typically highly structured containing three to four stem loops. The stems in these integrate structures typically contain bulges to prevent RNA degradation. Particular loops containing "YUNR" motifs are responsible for initial interaction and form kissing loops with their mRNA target (Brantl 2007; Kolb et al. 2000). Following initial interactions involving these few nucleotides, mRNA::sRNA duplex can be further extended and stabilized. As they are transcribed in close proximity, local concentrations of sRNAs can easily be enhanced and tuned to facilitate interactions and ensure efficient target regulation. sRNA binding can alter the fate of their target mRNAs by affecting their stability, translation, and transcription (reviewed in Georg and Hess 2011; Sesto et al. 2013; summarized in Fig. 2).

Similar to eukaryotes, *cis*-acting regulatory RNAs are found to be increasingly abundant across diverse bacterial genomes (Dornenburg et al. 2010; Rasmussen et al. 2009; Georg and Hess 2011). For instance, antisense RNAs constitute up to 46 % of the *H. pylori* transcriptome (Sharma et al. 2010). Antisense RNAs can vary greatly in size, ranging from short transcripts to ones that can overlap and regulate entire operons (recently reviewed in Sesto et al. 2013). *Cis*-acting sRNAs have been found to regulate the replication and copy number of plasmids (as in RNAI and CopA (Stougaard et al. 1981; Tomizawa and Itoh 1981)), transposons (Simons and Kleckner 1983), and phage DNA (Guo et al. 1991). They are also paramount to bacterial physiology as they regulate a number of fundamental processes, including replication, conjugation, ion transport, metabolism, and various stress response pathways (reviewed in Brantl 2007).

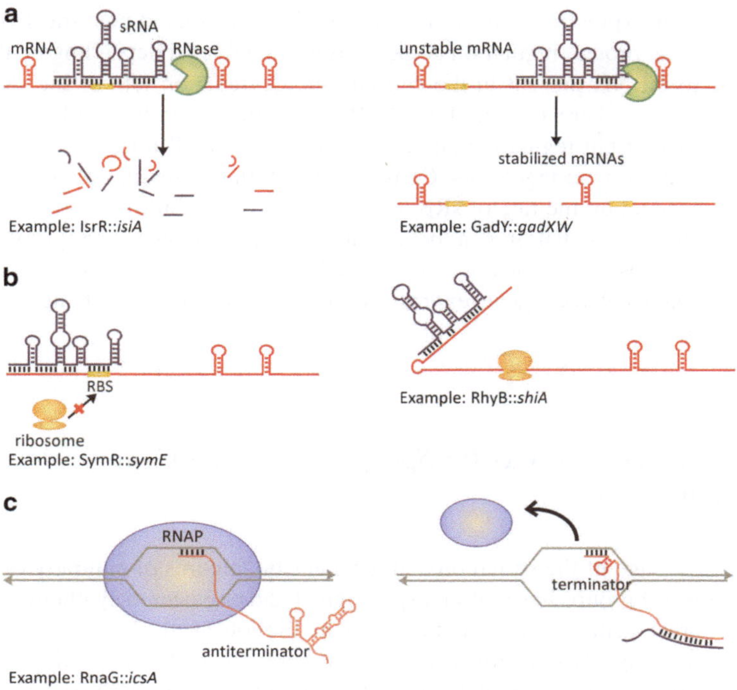

Fig. 2 Mechanisms of cis-acting sRNA. (**a**) Effects of sRNA on mRNA stability. Interactions between sRNA and its target mRNA can target the duplex for RNase degradation (*left panel*). Alternatively, binding of sRNA to an unstable mRNA can target it for processing in order to produce a stable product (*right panel*). (**b**) Effects of sRNA on mRNA translation. sRNA binding can occlude the ribosome binding site (RBS), the start codon, and other codons within the open reading frame (ORF) in order to prevent translation initiation (*left panel*). sRNA binding can also alleviate structures near the translation initiation region (TIR), thereby enabling the binding of ribosomes and translation initiation (*right panel*). (**c**) Effects on transcription. Antisense sRNAs can interact with transcription antiterminators on elongating mRNAs in order to promote the formation of transcription terminators, dissociation of the RNA polymerase (RNAP), and the production of truncated transcripts

3.2 sRNAs for Fast and Tight Regulation

To understand the rationale behind maintaining vast collections of sRNAs in bacterial genomes, we examine gene regulation offered by sRNAs and proteins. Compared with protein-based gene regulation, post-transcriptional regulation of gene expression using sRNAs offers several advantages (reviewed by Mehta et al. 2008; Shimoni et al. 2007). sRNA regulation allows cells to more rapidly control transcription in response to environmental signals compared with protein transcription factors. Under instances where regulators have to be synthesized de novo, sRNAs enable faster action than protein regulators acting via protein–protein interactions. Once the external signal is removed, recovery of the target gene to its initial level of expression is also faster under sRNA-mediated post-transcriptional regulation than protein regulation at the transcriptional level depending on the ratio of sRNA to mRNA present in the cell and the degradation rate of the sRNA. In repressed states where the number of sRNAs outnumbers their mRNA targets, sRNAs provide tight regulation and are less prone to small fluctuations in signals compared with protein regulators. Considering that the repression of target expression is dependent on the rate of sRNA production, sRNAs allow for fine-tuning of gene expression so that it can be maintained at a desired steady-state level. Collectively, these properties of sRNA-based regulation allow cells to rapidly and reliably switch between gene expression states in response to changes to their environment.

3.3 Optimizing sRNAs for Specifically Target Gene Expression

Taking advantage of the structural and mechanistic features of naturally occurring sRNAs, several groups have rationally designed and engineered synthetic sRNA to regulate target gene expression (reviewed in Lioliou et al. 2010). As described previously, many sRNAs affect gene expression through RNA–RNA binding, consequently preventing translation and targeting the RNA–RNA complex for degradation. Research on the mechanism of action of some sRNAs has given insight into how they interact with their target mRNA and identified sequence and structural elements necessary for sRNA–mRNA interaction.

Over the past 15 years, multiple groups have used antisense RNAs to control the expression of metabolic genes (Desai and Papoutsakis 1999; Tummala et al. 2003a, b) and to knockdown and identify essential genes in bacterial genomes, a feat that is difficult to achieve by gene deletions (Ji et al. 2001; Forsyth et al. 2002). Ji and colleagues showed that antagonizing the expression of essential genes in *S. aureus* using their reverse complements produced growth defects in *S. aureus* and reduced bacterial survival in murine models. They propose that this method can be used to identify gene products that can serve as targets for antibiotics discovery. Antisense knockdown of gene expression studies demonstrates that this strategy is more efficient in Gram-positive bacteria than in Gram-negative species. In Gram-negative bacteria, certain structural features may be needed to ensure adequate target binding. Using the kissing loop of CopA, Engdahl and colleagues first designed a synthetic antisense RNA by fusing the recognition stem loop of CopA, which has been modified to recognize the 5′-UTR of *lacI*, to a sequence that is either complementary to the RBS of *lacI* or can act as a ribozyme to cleave the target mRNA (coined the inhibitory unit). They demonstrated that this sequence could reduce the production of LacI–LacZ fusion protein by 50 % (Engdahl et al. 1997). Upon modifying their design to contain a stability element at the 5′ end of their antisense RNA, they were able to achieve 75 % repression (Engdahl et al. 2001). Nakashima et al. (2006) demonstrated that inverted repeats can be introduced to the ends of antisense sRNAs targeting the translation initiation region of desired genes to produce structures with paired termini. Such structures can improve the stability and accumulation of the antisense RNAs, thereby promoting their activity. Using this design, the authors reduced the expression of *ackA*, which is central to carbon flux to and from acetate. Reducing carbon flux to acetate can ultimately improve growth, macromolecular synthesis, and product yields in industrial fermentation.

Naturally occurring sRNAs can also have extensive secondary and tertiary structures that can serve as structural scaffolds for artificial sRNAs. An sRNA that has been well characterized is SibC, which can downregulate the expression of the IbsC toxin peptide through binding to its mRNA transcript (Fozo et al. 2008). Investigation has revealed target recognition sites (TRDs) that are necessary for initial interaction with IbsC mRNA (Han et al. 2010). By changing the target sequence and its surrounding region, the Lee group has made an artificial RNA that can target a LacZ fusion gene and reduce its expression as reported by a reduction in β-galactosidase activity (Park et al. 2013). They found that this regulation was dependent on ensuring that the changes in the TRD did not alter the overall sRNA structural scaffold and also on the site targeted in the lacZ gene. Na et al. (2013) recently generated a library of synthetic sRNAs composed of scaffolds designed based on *E. coli* sRNAs discovered to date and sequences complementary to translation initiation region (TIR) of target genes, and they screened for sRNAs that exhibit optimal binding energy and target repression capability. Based on the sequence of the optimized sRNA, the authors designed sRNAs that can repress four genes involved in tyrosine biosynthesis in 14 strains of *E. coli* and engineered a strain that can produce up to 2.0 g/l tyrosine. They also expressed a library of 122 synthetic sRNAs that repress genes involved in the production of cadaverine, an important nylon precursor. Through this screen, they identified two genes that are not part of the main cadaverine biosynthesis pathways but resulted in 30–40 % increase in cadaverine titers when their expression was repressed. This study shows that sRNA-based screening can be employed to identify genes that can increase the production of a metabolite of interest but are not part of the main metabolic pathways. They further showed that sRNAs can be used to modulate target genes and product yield in lieu of genetic manipulations and bacterial strain construction.

In addition to gene knockdown, bacterial sRNAs have also been integrated into artificial gene regulation systems. The Collins group designed a set of constructs coined "riboregulators" that allow post-transcriptional regulation of gene expression (Isaacs et al. 2004; Callura et al. 2010, 2012). The riboregulators are composed of an RNA element appended upstream of the RBS of the gene to be regulated. This RNA element is complementarity to the RBS, thus it can sequester the RBS by forming a stem loop after its transcription, resulting in inhibition of translation. To relieve translational inhibition, a second trans-acting sRNA placed under the control of an inducible promoter in a plasmid is used. Once its expression is induced, this sRNA can hybridize with the cis-regulatory element through their complementarity. This new base pairing releases the RBS of the gene of interest and allows translation to take place. These elements can be used in conjunction with any promoter of interest to ensure tight regulation of target genes. Recently, the Collins Laboratory used "riboregulators" to regulate the expression of four metabolic genes and constructed a metabolic switchboard to control carbon flow through three *E. coli* glucose-utilization pathways (Callura et al. 2012).

The aforementioned studies demonstrate that the activity of synthetic antisense RNAs can be tuned and optimized through base pairing and structural changes. They further showcase the possibility of applying synthetic sRNAs in constructing synthetic gene circuits, metabolic engineering, as well as drug and drug target discovery.

4 Concluding Remarks and Future Prospects

Since their discovery in bacteria, riboswitches and sRNAs have emerged from being oddities of biomolecular gene regulation to well-conserved critical elements of microbial cells. They act, at times in concert with protein regulators, to orchestrate key physiological processes in bacteria and ensure specific, timely, and tight regulation of their target genes. Riboswitches have been found to control aspects of metabolism, virulence, and antibiotic resistance. As such, they have become popular targets of many potential therapeutic and adjuvant lead molecules. Future studies could lead to a new class of antibiotics that bind and inhibit the function of these RNA sequences leading to bacteriostatic, bactericidal, or virulence reducing effects. The ability to control gene expression through binding of a specific small molecule ligand has led to the use of riboswitches in many characterization and biosensor schemes that detect the concentrations of the small molecule binding partners of riboswitches attached to reporter genes. The modularity of riboswitches into binding and expression domains has also allowed the engineering and selection of many synthetic riboswitches. This further expands the number of small molecules and genes that can be monitored and studied, and it allows the induction and repression of a gene of interest by a small molecule of choice.

sRNAs, the other major class of RNA regulators in bacteria, have the potential to regulate single, specific targets or multiple genes, allowing for pleotropic responses to signals in a fast and efficient manner. Structural studies have facilitated to the engineering of artificial sRNAs, expanding the repertoire of genes that can be regulated by small nucleic acids. The activity and target specificities of synthetic sRNAs can be tuned via nucleotide substitutions. Such sRNAs can be used to regulate genes of interest and explore the function of genes, especially in bacteria for which genetic tools, such as cloning vectors and inducible promoters, are limited. These sRNAs can be used to identify essential and virulence genes in pathogens, thereby facilitating the discovery of possible drug targets. As demonstrated throughout this chapter, the potential to regulate multiple genes using riboswitch and sRNA-based tools allows for the construction of synthetic gene circuits and organisms. They can also be used to improve the use of bacterial strains for metabolite production.

Despite their small size, riboswitches and sRNAs play major regulatory roles in bacteria. While great strides have been made toward their discovery and characterization, additional efforts are needed to find and study these elements, particularly in less well-characterized organisms. Understanding their function and mechanism of action is integral to improving the design of synthetic riboswitch and sRNA-based tools. Such tools have the potential to expand our knowledge of microbes that surround us, augment our chances of overcoming bacterial infections, and enhance the health of our environment.

References

Altuvia S, Weinstein-Fischer D, Zhang A et al (1997) A small, stable RNA induced by oxidative stress: role as a pleiotropic regulator and antimutator. Cell 90:43–53

Bartel DP, Szostak JW (1993) Isolation of new ribozymes from a large pool of random sequences [see comment]. Science 261:1411–1418

Beisel CL, Updegrove TB, Janson BJ et al (2012) Multiple factors dictate target selection by Hfq-binding small RNAs. EMBO J 31:1961–1974

Blount KF, Wang JX, Lim J et al (2007) Antibacterial lysine analogs that target lysine riboswitches. Nat Chem Biol 3:44–49

Brantl S (2007) Regulatory mechanisms employed by cis-encoded antisense RNAs. Curr Opin Microbiol 10:102–109

Brennan RG, Link TM (2007) Hfq structure, function and ligand binding. Curr Opin Microbiol 10:125–133

Callura JM, Dwyer DJ, Isaacs FJ et al (2010) Tracking, tuning, and terminating microbial physiology using synthetic riboregulators. Proc Natl Acad Sci U S A 107:15898–15903

Callura JM, Cantor CR, Collins JJ (2012) Genetic switchboard for synthetic biology applications. Proc Natl Acad Sci U S A 109:5850–5855

Cheah MT, Wachter A, Sudarsan N et al (2007) Control of alternative RNA splicing and gene expression by eukaryotic riboswitches. Nature 447:497–500

Dambach MD, Winkler WC (2009) Expanding roles for metabolite-sensing regulatory RNAs. Curr Opin Microbiol 12:161–169

Desai RP, Papoutsakis ET (1999) Antisense RNA strategies for metabolic engineering of Clostridium acetobutylicum. Appl Environ Microbiol 65:936–945

Dornenburg JE, Devita AM, Palumbo MJ et al (2010) Widespread antisense transcription in Escherichia coli. MBio 1:pii: e00024-10

Engdahl HM, Hjalt TA, Wagner EG (1997) A two unit antisense RNA cassette test system for silencing of target genes. Nucleic Acids Res 25:3218–3227

Engdahl HM, Lindell M, Wagner EG (2001) Introduction of an RNA stability element at the 5′-end of an antisense RNA cassette increases the inhibition of target RNA translation. Antisense Nucleic Acid Drug Dev 11:29–40

Forsyth RA, Haselbeck RJ, Ohlsen KL et al (2002) A genome-wide strategy for the identification of essential genes in Staphylococcus aureus. Mol Microbiol 43:1387–1400

Fowler CC, Brown ED, Li Y (2008) A FACS-based approach to engineering artificial riboswitches. Chembiochem 9:1906–1911

Fowler CC, Brown ED, Li Y (2010) Using a riboswitch sensor to examine coenzyme B(12) metabolism and transport in *E. coli*. Chem Biol 17:756–765

Fozo EM, Kawano M, Fontaine F et al (2008) Repression of small toxic protein synthesis by the Sib and OhsC small RNAs. Mol Microbiol 70:1076–1093
Frohlich KS, Vogel J (2009) Activation of gene expression by small RNA. Curr Opin Microbiol 12:674–682
Garst AD, Edwards AL, Batey RT (2011) Riboswitches: structures and mechanisms. Cold Spring Harb Perspect Biol 3:pii: a003533
Georg J, Hess WR (2011) cis-antisense RNA, another level of gene regulation in bacteria. Microbiol Mol Biol Rev 75:286–300
Gottesman S, Storz G (2011) Bacterial small RNA regulators: versatile roles and rapidly evolving variations. Cold Spring Harb Perspect Biol 3:pii: a003798
Gottesman S, McCullen CA, Guillier M et al (2006) Small RNA regulators and the bacterial response to stress. Cold Spring Harb Symp Quant Biol 71:1–11
Guo PX, Rajagopal BS, Anderson D et al (1991) sRNA of phage phi 29 of Bacillus subtilis mediates DNA packaging of phi 29 proheads assembled in *Escherichia coli*. Virology 185:395–400
Han K, Kim KS, Bak G et al (2010) Recognition and discrimination of target mRNAs by Sib RNAs, a cis-encoded sRNA family. Nucleic Acids Res 38:5851–5866
Isaacs FJ, Dwyer DJ, Ding C et al (2004) Engineered riboregulators enable post-transcriptional control of gene expression. Nat Biotechnol 22:841–847
Jacques JF, Jang S, Prevost K et al (2006) RyhB small RNA modulates the free intracellular iron pool and is essential for normal growth during iron limitation in *Escherichia coli*. Mol Microbiol 62:1181–1190
Ji Y, Zhang B, Van SF et al (2001) Identification of critical staphylococcal genes using conditional phenotypes generated by antisense RNA. Science 293:2266–2269
Johansen J, Rasmussen AA, Overgaard M et al (2006) Conserved small non-coding RNAs that belong to the sigmaE regulon: role in down-regulation of outer membrane proteins. J Mol Biol 364:1–8
Kawamoto H, Koide Y, Morita T et al (2006) Base-pairing requirement for RNA silencing by a bacterial small RNA and acceleration of duplex formation by Hfq. Mol Microbiol 61:1013–1022
Kellenberger CA, Wilson SC, Sales-Lee J et al (2013) RNA-based fluorescent biosensors for live cell imaging of second messengers cyclic di-GMP and cyclic AMP-GMP. J Am Chem Soc 135:4906–4909
Kolb FA, Engdahl HM, Slagter-Jager JG et al (2000) Progression of a loop-loop complex to a four-way junction is crucial for the activity of a regulatory antisense RNA. EMBO J 19:5905–5915
Lee ER, Blount KF, Breaker RR (2009) Roseoflavin is a natural antibacterial compound that binds to FMN riboswitches and regulates gene expression. RNA Biol 6:187–194
Li S, Breaker RR (2013) Eukaryotic TPP riboswitch regulation of alternative splicing involving long-distance base pairing. Nucleic Acids Res 41:3022–3031
Lioliou E, Romilly C, Romby P et al (2010) RNA-mediated regulation in bacteria: from natural to artificial systems. N Biotechnol 27:222–235
Liu JM, Camilli A (2010) A broadening world of bacterial small RNAs. Curr Opin Microbiol 13:18–23
Liu MY, Romeo T (1997) The global regulator CsrA of *Escherichia coli* is a specific mRNA-binding protein. J Bacteriol 179:4639–4642
Liu MY, Gui G, Wei B et al (1997) The RNA molecule CsrB binds to the global regulatory protein CsrA and antagonizes its activity in *Escherichia coli*. J Biol Chem 272:17502–17510
Mehta P, Goyal S, Wingreen NS (2008) A quantitative comparison of sRNA-based and protein-based gene regulation. Mol Syst Biol 4:221
Mulhbacher J, Brouillette E, Allard M et al (2010) Novel riboswitch ligand analogs as selective inhibitors of guanine-related metabolic pathways. PLoS Pathog 6:e1000865
Na D, Yoo SM, Chung H et al (2013) Metabolic engineering of Escherichia coli using synthetic small regulatory RNAs. Nat Biotechnol 31:170–174

Nakashima N, Tamura T, Good L (2006) Paired termini stabilize antisense RNAs and enhance conditional gene silencing in Escherichia coli. Nucleic Acids Res 34:e138
Pan T, Uhlenbeck OC (1992) In vitro selection of RNAs that undergo autolytic cleavage with Pb2+. Biochemistry 31:3887–3895
Park H, Bak G, Kim SC et al (2013) Exploring sRNA-mediated gene silencing mechanisms using artificial small RNAs derived from a natural RNA scaffold in Escherichia coli. Nucleic Acids Res 41:3787–3804
Rasmussen S, Nielsen HB, Jarmer H (2009) The transcriptionally active regions in the genome of Bacillus subtilis. Mol Microbiol 73:1043–1057
Reynoso CM, Miller MA, Bina JE et al (2012) Riboswitches for intracellular study of genes involved in Francisella pathogenesis. MBio 3:pii: e00253-12
Seelig B, Jaschke A (1999) A small catalytic RNA motif with Diels-Alderase activity. Chem Biol 6:167–176
Serganov A, Nudler E (2013) A decade of riboswitches. Cell 152(1–2):17–24
Sesto N, Wurtzel O, Archambaud C et al (2013) The excludon: a new concept in bacterial antisense RNA-mediated gene regulation. Nat Rev Microbiol 11:75–82
Sharma CM, Hoffmann S, Darfeuille F et al (2010) The primary transcriptome of the major human pathogen Helicobacter pylori. Nature 464:250–255
Shimoni Y, Friedlander G, Hetzroni G et al (2007) Regulation of gene expression by small non-coding RNAs: a quantitative view. Mol Syst Biol 3:138
Simons RW, Kleckner N (1983) Translational control of IS10 transposition. Cell 34:683–691
Sinha J, Reyes SJ, Gallivan JP (2010) Reprogramming bacteria to seek and destroy an herbicide. Nat Chem Biol 6:464–470
Storz G, Vogel J, Wassarman KM (2011) Regulation by small RNAs in bacteria: expanding frontiers. Mol Cell 43:880–891
Stougaard P, Molin S, Nordstrom K (1981) RNAs involved in copy-number control and incompatibility of plasmid R1. Proc Natl Acad Sci U S A 78:6008–6012
Sudarsan N, Cohen-Chalamish S, Nakamura S et al (2005) Thiamine pyrophosphate riboswitches are targets for the antimicrobial compound pyrithiamine. Chem Biol 12:1325–1335
Tarasow TM, Tarasow SL, Eaton BE (1997) RNA-catalysed carbon-carbon bond formation. Nature 389:54–57
Tomizawa J, Itoh T (1981) Plasmid ColE1 incompatibility determined by interaction of RNA I with primer transcript. Proc Natl Acad Sci U S A 78:6096–6100
Topp S, Gallivan JP (2007) Guiding bacteria with small molecules and RNA. J Am Chem Soc 129:6807–6811
Topp S, Gallivan JP (2008) Random walks to synthetic riboswitches – a high-throughput selection based on cell motility. Chembiochem 9:210–213
Tummala SB, Junne SG, Papoutsakis ET (2003a) Antisense RNA downregulation of coenzyme A transferase combined with alcohol-aldehyde dehydrogenase overexpression leads to predominantly alcohologenic Clostridium acetobutylicum fermentations. J Bacteriol 185:3644–3653
Tummala SB, Welker NE, Papoutsakis ET (2003b) Design of antisense RNA constructs for downregulation of the acetone formation pathway of Clostridium acetobutylicum. J Bacteriol 185:1923–1934
Wassarman KM, Repoila F, Rosenow C et al (2001) Identification of novel small RNAs using comparative genomics and microarrays. Genes Dev 15:1637–1651
Werstuck G, Green MR (1998) Controlling gene expression in living cells through small molecule-RNA interactions. Science 282:296–298
Winkler W, Nahvi A, Breaker RR (2002) Thiamine derivatives bind messenger RNAs directly to regulate bacterial gene expression. Nature 419:952–956
Yang J, Seo SW, Jang S et al (2013) Synthetic RNA devices to expedite the evolution of metabolite-producing microbes. Nat Commun 4:1413

Towards Defined DNA and RNA Delivery Vehicles Using Nucleic Acid Nanotechnology

Anders Hauge Okholm, David Schaffert, and Jørgen Kjems

Contents

1 Introduction 326
2 Creating Nanostructures from DNA and RNA Building Blocks 327
2.1 DNA Structures 328
2.2 RNA Structures 331
3 Expanding the Functionality of DNA and RNA Structures 334
4 Delivery Vehicles of DNA and RNA Nanostructures 335
4.1 DNA Structures in Drug Delivery 336
4.2 RNA Structures in Drug Delivery 338
5 Future Perspectives 339
6 Conclusion 341
References 341

A.H. Okholm (✉) • J. Kjems
Interdisciplinary Nanoscience Center, Aarhus University, Gustav Wieds Vej 14, DK-8000 Aarhus C, Denmark

Department of Molecular Biology, Aarhus University, Gustav Wieds Vej 14, DK-8000 Aarhus C, Denmark

Centre for DNA Nanotechnology, Aarhus University, Gustav Wieds Vej 14, DK-8000 Aarhus C, Denmark
e-mail: ao07@mb.au.dk

D. Schaffert
Interdisciplinary Nanoscience Center, Aarhus University, Gustav Wieds Vej 14, DK-8000 Aarhus C, Denmark

Department of Molecular Biology, Aarhus University, Gustav Wieds Vej 14, DK-8000 Aarhus C, Denmark

Centre for DNA Nanotechnology, Aarhus University, Gustav Wieds Vej 14, DK-8000 Aarhus C, Denmark

BASF, Carl-Bosch-Str. 38, Ludwigshafen 67056, Germany

V.A. Erdmann et al. (eds.), *Chemical Biology of Nucleic Acids*, RNA Technologies,
DOI 10.1007/978-3-642-54452-1_18,

Abstract Both DNA and RNA nanostructures show exceptional programmability, modularity, and self-assembly ability. Using DNA or RNA molecules it is possible to assemble monodisperse particles that are homogeneous in size and shape and with identical positioning of surface modifications. For therapeutic applications such nanoparticles are of particular interest as they can be tailored to target cells and reduce unwanted side effects due to particle heterogeneity. Recently, DNA and RNA nanostructures have demonstrated this potential by delivery of drugs to specific cells in vitro and in vivo. This has launched an increasing interest to engineer-defined DNA and RNA vehicles for drug delivery. However, before this can be realized, key challenges must be overcome including structure integrity, efficient cell targeting, and drug release. The tunable nature of DNA and RNA assemblies allows for thorough investigations into various structural and functional features, which can address these challenges. To facilitate the synthesis process novel methods enable the construction of sophisticated structures and attachment of relevant functionalities. In this chapter we will discuss the state-of-the-art molecular designs and approaches to harness and use DNA and RNA nanostructures in drug delivery.

Keywords Drug delivery • Nanomedicine • DNA nanotechnology • RNA nanotechnology • Origami • Self-assembly

1 Introduction

Programmed self-assembly of DNA and RNA molecules has proven to be a simple but powerful technique for bottom-up construction of complex two- and three-dimensional nanostructures (Tørring et al. 2011). It has been suggested that DNA and RNA nanostructures have potential applications in a wide range of fields such as biophysics, photonics, therapeutics, and diagnostics. However, nanostructures of nucleic acids have limited functional capacities and may not be able to realize this potential on their own. The past few years have witnessed a growing interest in the functionalization of DNA and RNA nanostructures with various molecules. Recent reports have demonstrated the ability to position a diverse set of functionalities on DNA and RNA nanostructures with atomic precision in defined molecular patterns (Wilner and Willner 2012; Saccà and Niemeyer 2011). This has launched an increasing scientific interest to harness DNA and RNA nanostructures as drug delivery vehicles (Smith et al. 2013; Pinheiro et al. 2011; Krishnan and Bathe 2012). Targeted drug delivery is an attractive therapeutic strategy to reduce drug dosage and the risk of side effects and drug-resistant cells by delivery of the pharmaceutical compound to its site of action. Conventional nanoparticle delivery systems (Peer et al. 2007; Zhang et al. 2008) often suffer from polydispersity, undefined ligand stoichiometry, and a low degree of design modularity. In contrast, nucleic acid nanotechnology provides a unique platform for construction of addressable nanoparticles, in principle identical in chemical composition, size, shape, and surface modifications with tunable stoichiometry and spatial arrangement.

The past 5 years have seen the emergence of nucleic acid nanostructures applied for delivery of molecular payloads. For example, DNA nanostructures have been used for the controlled release of chemotherapeutic agents in cancer cells (Jiang et al. 2012; Zhao et al. 2012) and targeted RNA nanostructures have been reported to deliver siRNA in vitro and in vivo (Shu et al. 2011). These promising reports encourage the further exploration of DNA and RNA nanocarriers; however, several challenges have to be addressed. In-depth investigations of structural integrity, cell targeting, and penetration efficiency together with thorough profiling of biodistribution patterns are required. Understanding the interactions of nucleic acid nanostructures with the complex biological environment found in whole organisms is essential for delivery applications. This chapter will discuss state-of-the-art approaches to design and functionalize DNA and RNA nanostructures for their application as defined drug delivery vehicles.

2 Creating Nanostructures from DNA and RNA Building Blocks

The typical workflow from design to production of DNA and RNA nanostructures includes iterative optimization rounds of sequence design, assembly, functionalization, and subsequent characterization (Fig. 1). The design of nucleic acid self-assembled nanostructures is ultimately based on the inherent structural information imprinted in the sequences. However, even though DNA and RNA are very similar in structural and chemical composition, approaches to design and synthesis of two- and three-dimensional structures are considerably different. Self-assembly of DNA and RNA complexes depends heavily on canonical base pairing to form helical segments. However, noncanonical base pairings, including G:U and G:A interactions, are common in RNA molecules complicating both RNA structure prediction and design. In addition, a variety of long-range interacting RNA tertiary motifs (Jaeger and Chworos 2006), based on nonconventional base pairing, make RNA folding prediction particularly difficult. These structural differences have diversified the principles for constructing nanostructures from DNA and RNA. In general, folding a given shape in DNA is approached by secondary sequence design based on standard Watson–Crick base pairing, whereas for RNA, the approach is to combine and manipulate known tertiary motifs found in nature to generate new structures. Although difficult to predict, it is this structural complexity of RNA that generally outcompetes DNA in terms of antigen binding or catalysis of biochemical reactions as found in RNA aptamers, ribozymes, and riboswitches. Furthermore, RNA can form duplexes with higher thermal stability and undergo important

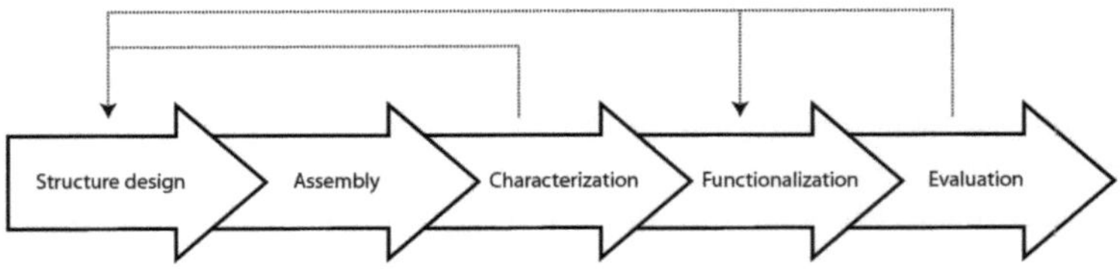

Fig. 1 From conception to application of DNA or RNA nanostructures. The realization of functional DNA and RNA nanostructures starts with conception, structure design, and assembly. Characterization by, e.g., EMSA, AFM, TEM, and FRET is used to evaluate the resulting structures and several rounds of redesign and characterization can be necessary (*Dashed lines* indicate feedback). When a stable structure is assembled, relevant functionalities can be introduced. The resulting functionalized DNA or RNA nanostructure is evaluated on its performance as a drug delivery vehicle

biological processes like translation, RNA interference, and with option to be produced in vivo by the cellular transcription machinery. Despite these differences both DNA and RNA can be used to produce defined, nano-sized structures.

2.1 DNA Structures

In nature DNA is generally encountered as linear double helices that can take more complex shapes when interacting with proteins. This occurs for instance during meiotic recombination, where two symmetric chromosomal DNA double strands cross over to form mobile Holliday junctions. The natural Holliday junction inspired Seeman and coworkers in 1987 to create immobilized DNA junctions by designing asymmetric DNA branches (Seeman 1982). Seeman exploited that immobilized branches can connect via sticky ended cohesion to form DNA crystals in two and three dimensions (Fig. 2a). Immobilized junctions are important motifs for construction of stable DNA nanostructures and were used to create a connected wireframe cube of DNA with edges of 20 base pairs (Chen and Seeman 1991). The significance of this result was captured in the final remark of the article stating that "*The synthesis of this object establishes that it may be feasible to make larger and more complex objects.*"

Since then sticky-ended cohesion and immobilized branches have indeed been essential for creating a variety of wireframe DNA nanostructures of different geometries (Zhang and Seeman 1994; Andersen et al. 2008; Oliveira et al. 2010). One of the most extensively studied is the DNA tetrahedron presented by Turberfield and coworkers in 2005 (Fig. 2b) (Goodman et al. 2005). Single-step assembly by hybridization of four synthetic DNA strands gave rigid tetrahedrons with edge lengths of ~7 nm in near-quantitative yields. Tuning the spatial orientation of sticky ends protruding from one edge of the DNA tetrahedron enabled precise prediction and control of dimer formation. Due to its simple design, assembly, and high structural integrity, the tetrahedron has been the subject of

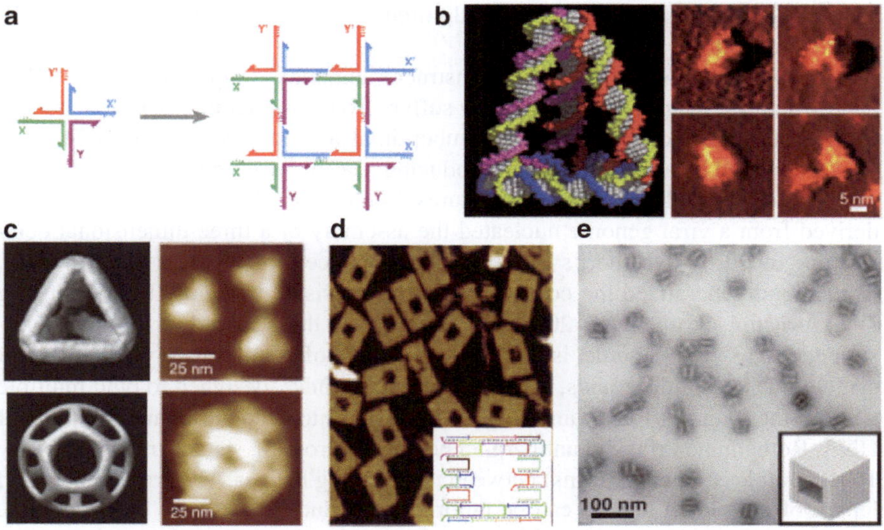

Fig. 2 DNA nanostructures assembled by sticky-ended cohesion. (**a**) Immobilized DNA branches forming 2D crystals via sticky-ended cohesion (Seeman 1982). (**b**) Four synthetic DNA strands assemble into DNA tetrahedrons (Goodman et al. 2005). (**c**) A three-point-star motif folds into wireframe 3D DNA polyhedral (He et al. 2008). (**d**) Single-stranded tiles form a 2D canvas (**e**) or a 3D block to create arbitrary nanoshapes of DNA (Wei et al. 2012; Ke et al. 2012). All figures are reprinted with permission

many later studies (Erben et al. 2006; Kim et al. 2013; Walsh et al. 2011). However, the assembly strategy of the DNA tetrahedron was not believed to be suitable for creation of more complex structures, as this would require several unique synthetic DNA strands. To overcome this, a tile-based hierarchical self-assembly strategy was developed by Mao and coworkers (He et al. 2008). By controlling concentration and motif flexibility, a three-point-star motif consisting of seven DNA strands self-assembled into symmetric polyhedra such as a tetrahedron, dodecahedron, or buckyball (truncated icosahedron) (Fig. 2c). The tile-based assembly provides an elegant strategy to form enclosed objects with a minimum of material, though sacrificing specific addressability to a certain degree. Recently, a single-stranded tile (SST) of DNA was employed to form two-dimensional supramolecular DNA sheets while keeping each tile addressable (Fig. 2d) (Wei et al. 2012). Individual SSTs consist of four domains, which specifically bind their intended neighbors upon self-assembly and correspond to predefined pixels in the canvas. By carefully choosing SSTs for the assembly reaction and introducing so-called "edge protectors" that hybridize to SST domains at the edges of the structure, any continuous two-dimensional shape can be formed with yields ranging from 6 to 40 %. This concept of single DNA "tiles" was subsequently generalized to "bricks" that constitute a cube (Fig. 2e) (Ke et al. 2012). Like carving in a block of wood, solid three-dimensional DNA nanostructures including complex geometrical shapes and enclosed cavities can be sculptured by excluding particular DNA bricks from the assembly.

Although DNA tiles can construct complex supramolecular DNA nanostructures, the procedure generally suffers from strict requirements on stoichiometry between individual tiles and suboptimal assembly yields. In 2004, Shih et al. addressed these problems by introducing scaffold-assisted self-assembly as a new strategy to fold DNA nanostructures (Shih et al. 2004). A 1.7 kb ssDNA derived from a viral genome nucleated the assembly of a three-dimensional octahedron guided by five short synthetic strands in excess. Two years later, in 2006, Rothemund generalized the concept of scaffold-assisted assembly by introducing DNA origami (Rothemund 2006). In this method the scaffold, a 7.3 kb single-stranded M13 viral genome, is folded by an excess of approximately 200 different synthetic DNA single strands, so-called "staple strands", designed to bind multiple sites on the long scaffold strand (Fig. 3a). Taking into account the natural twist of DNA, Rothemund arranged antiparallel helices by crossing staple strands for every 16 nucleotides (about 3/2 turns) between neighboring helices. The arrangement and length of individual helices can be tailored to produce almost any desired shape. DNA staple strands not only fold the scaffold strand into shape, but also serve as uniquely addressable molecular pixels for precise decoration and patterning of DNA origami structures. This powerful technique enables folding of arbitrary-shaped two-dimensional DNA molecular pegboards with an average radius of ~100 nm and more than 200 uniquely addressable sites represented by the staple strands.

The development of DNA containers with the potential to encapsulate or release molecules of interest was realized by extending DNA origami into the third dimension. In 2009, Andersen et al. produced a DNA box $42 \times 36 \times 36\ \mathrm{nm}^3$ in size made up of six interconnected single-layer faces with a controllable lid (Fig. 3b) (Andersen et al. 2009). The lid was locked by hybridization of protruding DNA strands, but opened reliably by strand displacement upon addition of specific DNA "key" strands. The DNA box embodied the idea of creating nanoscopic containers that can detect and respond to external signals. Publications describing other three-dimensional nanostructures rapidly followed (Ke et al. 2009; Kuzuya and Komiyama 2009), in part facilitated by the development of the software package CADnano (Douglas et al. 2009b). Stacking DNA layers of antiparallel helices enabled the assembly of solid three-dimensional objects with remarkable details (Fig. 3c) (Douglas et al. 2009a). Adjacent helices were arranged in a honeycomb lattice by crossing over staple strands between helices for every seven nucleotides (about 2/3 turns). Shortenings or extensions of the binding region of selected staple strands were used to engineer twisted and curved shapes (Fig. 3d) (Dietz et al. 2009). Further explorations into complex shapes of DNA include Möbius strips (Han et al. 2010), tensegrity structures (Liedl et al. 2010), and three-dimensional curved surfaces as demonstrated by the design and assembly of a flask-shaped structure (Fig. 3e) (Han et al. 2011). As DNA origami nanostructures are partially confined by the length, sequence, and production of M13 single-stranded template, alternatives of both shorter (Said et al. 2013), longer (Zhang et al. 2012) and double-stranded

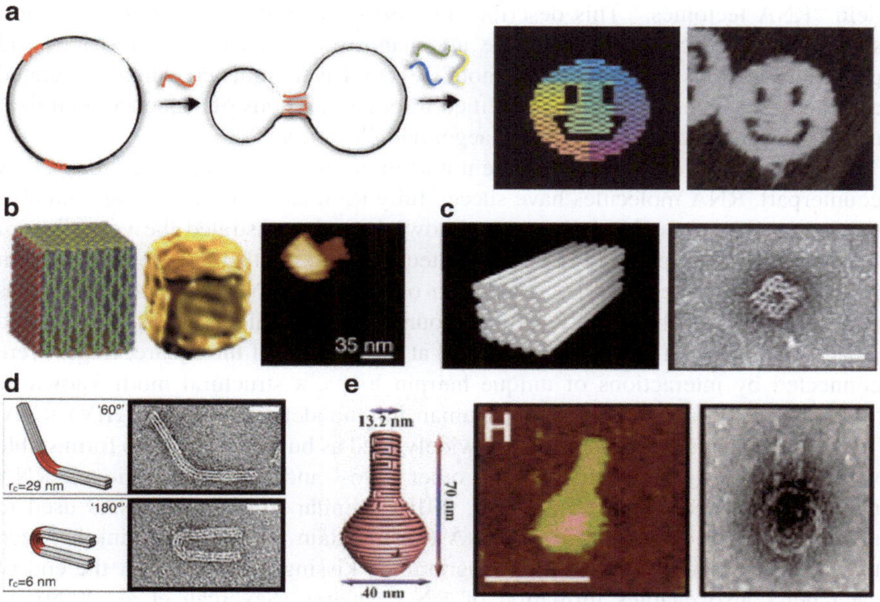

Fig. 3 DNA nanostructures assembled by DNA origami. (**a**) DNA origami uses hundreds of short synthetic staple strands to fold a long single-stranded scaffold strand into various 2D shapes (Smith et al. 2013). (**b**) A 3D DNA origami box made of six interconnected layers with a controllable lid (Andersen et al. 2009). (**c**) *Solid* 3D DNA origami structures (**d**) with twists and curves (Douglas et al. 2009a; Dietz et al. 2009). (**e**) A *curved* 3D nanoflask of a single DNA origami layer (Han et al. 2011). All figures are reprinted with permission

(Högberg et al. 2009; Yang et al. 2012) scaffold strands have been investigated. For example, a small DNA box with a reversible dynamic lid was folded by a 2.0 kb scaffold strand derived from a pUC plasmid (Zadegan et al. 2012) and larger structures have been folded using a combination of M13 and a PhiX174 scaffold to create assemblies of multiple DNA origami structures (Zhao et al. 2011).

2.2 *RNA Structures*

Due to the inferior predictability of RNA structures the prevailing strategy for the design of RNA nanostructures is adoption of sequential/structural motifs from stable RNA folds found in nature. Over the past few decades, high-resolution X-ray crystallography and NMR have provided structural detail with atomic resolution of various natural RNA motifs and interactions. Jaeger and coworkers predicted in 1996 (Westhof et al. 1996) that the growing catalogue of natural tertiary motifs could provide potential building blocks for constructing new synthetic RNA objects (Jaeger and Leontis 2000; Jaeger et al. 2001) and coined the field “RNA tectonics.” This describes the use of natural tertiary RNA modules, so-called “tectoRNAs,” to assemble into nanostructures with novel shapes and potential functions. For a particular motif, several RNA sequences can be generated and the best sequences may be identified through iterations of sequence optimization, folding, and characterization (Jaeger and Chworos 2006).

Although RNA nanostructures are not as thoroughly investigated as their DNA counterpart, RNA molecules have successfully been used to create programmable nanoscale objects. In 2004, Jaeger and coworkers demonstrated the versatility of RNA tectonics by creating a modular “tectosquare,” with the potential to form jigsaw puzzles, assembled by interaction of four tectoRNAs (Fig. 4a) (Chworos et al. 2004). The right angle (RA) motif found in ribosomal RNA was used to bend adjacent helices to a relative angle of 90° at the vertices of the square. Edges were connected by interactions of unique hairpin loops, a structural motif known as “kissing loops”, naturally found in human immunodeficiency virus (HIV) RNA. These motifs have subsequently been widely used as building blocks to form stable vertices and edge connections in other two- and three-dimensional RNA nanostructures (Severcan et al. 2009, 2010). Similar to the tectoRNAs used to assemble the tectosquare, class II tRNAs(Ser) contain a natural 90° kink between two adjacent helices. Insertion of programmed kissing-loop motifs at the end of each helix also enables formation of RNA squares (Severcan et al. 2009). In addition, tRNA(Ser) contains a variable arm oriented out of the plane from the two perpendicular helices. By introducing sticky ends on the variable arm, eight modified tRNAs assembled into antiprisms with the ability to encapsulate proteins (Severcan et al. 2010). In 2007, Shapiro and coworkers predicted a hexagonal RNA nanoring in silico inspired by a 120° loop–loop interaction of transcripts from a natural *E. coli* plasmid (Fig. 4b) (Yingling and Shapiro 2007). Six dumbbell RNA molecules of complementary loops were designed to mediate the formation of a hexagon with no need for sticky-ended cohesion and the realization of the structure was subsequently reported (Grabow et al. 2011). As envisioned in 2007, the hexagon was able to bind up to six siRNAs (Afonin et al. 2011). Another widely used RNA structural motif for engineering RNA nanostructures is the DNA-packaging motor *pRNA* discovered in bacterial virus phi29 in 1987 (Guo et al. 1987) and thoroughly studied by Guo and coworkers. Individual pRNAs can interact via a left- and a right-hand loop extending from a three-way junction. The interactions between six pRNAs form a stable hexamer (Fig. 4c) (Guo et al. 1998), which has the biological function to guide double-stranded genomic DNA into preformed procapsids upon viral packaging and replication. However, by mutating the loop regions, pRNAs can effectively be engineered to form dimers, trimers, tetramers, and hexamers in vitro (Chen et al. 2000; Shu et al. 2003).

While the architectural principles of crossover and sticky-ended cohesion founded the field of DNA nanotechnology, RNA nanotechnology has mainly explored the use of natural structural motifs for engineering RNA nanostructures. To challenge this paradigm, Afonin et al. designed and produced an RNA cube connected entirely by sticky-ended cohesion (Fig. 4d) (Afonin et al. 2010). Six unique RNA strands were designed to effectively form each face in a cube with

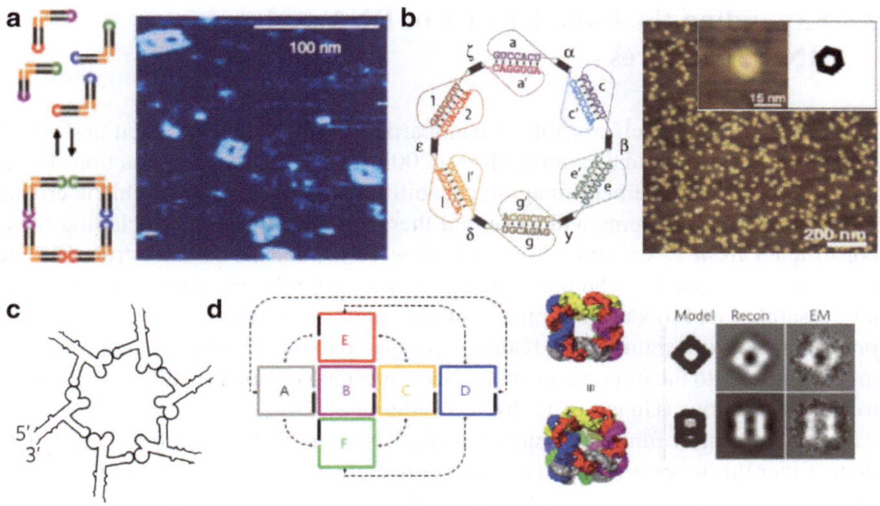

Fig. 4 RNA nanostructures. (**a**) TectoRNAs with right angle (RA) motifs and kissing loops (KL) assemble into RNA tectoscquares (Chworos et al. 2004). (**b**) The RNA nanoring based on RNAI/II inverse kissing complexes (Grabow et al. 2011). (**c**) pRNA hexamer formed by loop–loop interactions (Shu et al. 2011). (**d**) RNA cube assembled by hybridization of complementary RNA strands (Afonin et al. 2010). All figures are reprinted with permission

radii of ~6.5 nm. This demonstrates the potential for future RNA nanostructures assembled this way, although supramolecular RNA structures of similar complexity as DNA origami structures have yet to be accomplished. To approach this DNA/RNA hybrid tiles have been employed to form various polyhedral three-dimensional nanostructures (Ko et al. 2010), and recently, RNA-templated origami structures were folded by annealing of DNA staple strands (Endo et al. 2013). The emergence of supramolecular hybrid assemblies and the increasing complexity of RNA nanostructures in general indicate that artificial supramolecular RNA structures will be realized in the near future. Although RNA nanostructures are still less mature than DNA nanostructures, reports from the past decade have proved that RNA nanostructures can be designed in silico to form predictable structures in vitro. An intriguing feature for RNA nanostructures is the possibility for in vivo transcription and assembly. A realization of this was reported by Pamela Silver and coworkers who demonstrated that one- and two-dimensional RNA structures could fold isothermally in vivo to control the spatial arrangement of proteins in living cells (Delebecque et al. 2011).

3 Expanding the Functionality of DNA and RNA Structures

Functionalization of self-assembled nanoparticle (NP) systems is often performed to, e.g., increase colloidal stability (Rossi 2005), specify cellular interaction (Davis et al. 2010), and implement imaging capabilities (Gao et al. 2004). This is crucial for most delivery systems with potential therapeutic applications, including those constructed from DNA and RNA molecules. DNA and RNA nanostructures are easily functionalized by chemical and enzymatic methods, and due to their inherent addressability and modularity, modification patterns can be assembled with atomic precision. This is a significant advantage compared to other NP systems and enables investigation into the importance of surface patterning of functionalities with regard to cell interactions (Huang et al. 2009; Fishler et al. 2012).

Incorporation of functional modules into DNA and RNA nanostructures can be done either during or after synthesis of the nucleic acid oligomer. For site-specific modification various smaller molecules such as fluorophores, ligands, and chemical handles can be integrated during solid phase synthesis using functionalized phosphoramidites (Lönnberg 2009). Subsequent binding or covalent conjugations are then employed to position larger ligands like proteins, inorganic NP, polymers, or even functional nucleic acid motifs.

Protein patterning on nucleic acid nanostructures can be performed by several specific and usually orthogonal biological or chemical interactions. Most commonly used are biotin–streptavidin (Yan et al. 2003), peptide–antibody (Williams et al. 2007), aptamer–protein (Rinker et al. 2008), alkyne–azide conjugations (Duckworth et al. 2007), thiol–malemide conjugations (Wilner et al. 2009), and ligand binding or coupling to His-tagged (Shen et al. 2009), SNAP-tagged, or HaloTagged proteins (Meyer and Niemeyer 2011). AuNPs are frequently used in nucleic acid nanotechnology in part due to their intriguing photodynamic properties and the reliable and specific linkage between Au surfaces and thiols. Using this strategy, DNA nanostructures have been decorated with AuNP of various sizes (Sharma et al. 2008; Mastroianni et al. 2009).

To preserve functional capability, components conjugated to DNA and RNA molecules prior to self-assembly must be insensitive to the thermal annealing ramp often used for creation of multicomponent nucleic acid nanostructures. However, most proteins are incompatible with these conditions and therefore require alternative strategies for successful incorporation. The most prominent strategy to label DNA or RNA nanostructures employs single-stranded extensions protruding from the structure, so-called handles. They serve as attachment sites for functionalized DNA or RNA *anti*-sense handles that can hybridize to the nanostructure post-assembly. This strategy has been used to arrange proteins and inorganic NPs on DNA origami structures at room temperature and lowers the risk of affecting the self-assembly process by unwanted interference of the biomacromolecules or NPs.

A great advantage for DNA and RNA nanostructures is the intrinsic capacity to anneal nucleic acid-based therapeutic modules like siRNAs (Afonin et al. 2011),

DNAzymes and ribozymes (Hoeprich et al. 2003), riboswitches (Win and Smolke 2007), and DNA and RNA aptamers (Rinker et al. 2008), with general applications in nanomedicine, catalysis, and computing. These modules can be either incorporated directly into the nucleic acid sequences of the nanostructure components or annealed to handles in a sequence-specific fashion.

Understanding the relationship between different functionalities, their spatial arrangement and cellular interactions are essential for the development of therapeutically valuable nanostructures. Intensive screenings will be necessary to identify molecular surface decorations that implement desirable characteristics to these structures. However, synthesis of many individual nucleic acid strands with a complex set of ligands is cumbersome and economically demanding. To overcome this problem, enzymatic labeling has proven to be a robust, fast, and versatile method to label staple strands in parallel with a variety of molecular modifications. The enzyme terminal transferase ligates nucleoside triphosphates to the 3′ end of single-stranded DNA molecules and has been shown to accept small molecules (Jahn et al. 2011) and even larger macromolecules such as dendrimers and proteins (Sørensen et al. 2013) when conjugated to the incoming nucleoside triphosphates. This method enables easy parallelized incorporation of almost any modification into the 3′ end of native DNA strands and represents a rapid and cost-efficient modification method for DNA nanostructures.

4 Delivery Vehicles of DNA and RNA Nanostructures

Efficient cell targeting, uptake, and payload delivery are the main challenges faced by NP systems for drug delivery applications. These challenges are generally approached by modularization: the incorporation of discrete functional modules into the NP to introduce particular relevant traits into the drug delivery system. For cell targeting, modules such as molecular ligands (Low et al. 2008), aptamers (Hicke et al. 2006), or antibodies (Dinauer et al. 2005) have been widely studied in the literature for promoting endocytotic uptake of NP. NPs, internalized this way, usually become entrapped in endosomes and are eventually degraded. Escaping the endosome before degradation is therefore one of the great challenges for most drug delivery systems. DNA nanostructures are found to co-localize with endosomal markers (Schüller et al. 2011). However, several approaches already exist to functionalize NPs with modules that facilitate endosomal escape (Varkouhi et al. 2011), a solution that could be applied for DNA and RNA nanostructures. An alternative strategy for cytoplasmic delivery of DNA and RNA is offered by the large selection of available transfection reagents, typically cationic polymers or liposomes. However, it needs to be investigated how this formulation affects the three-dimensional structure of the complexed nanostructure. In addition, the decomplexation step inside the cell to liberate the RNA or DNA nanostructure must be addressed.

4.1 DNA Structures in Drug Delivery

Even though nucleic acid nanostructures have still not been investigated in much detail at cellular interfaces, a few two- and three-dimensional DNA and RNA nanostructures have already been employed to carry and deliver molecular payloads to cells. The first study of targeted DNA nanostructures was reported in 2008 and concerned DNA nanotubes assembled from a single DNA strand tile modified with folate or the fluorophore Cy3. KB cells overexpressing the folate receptor showed increased internalization of folate-modified DNA nanotubes (Ko et al. 2008). However, the setup provided no stringent control of DNA nanotube structure or surface ligand distribution and no in-depth study of what was taken up was performed. In another study from 2011, a defined DNA icosahedron, encapsulating fluorescent polymers, showed co-localization upon uptake through anionic ligand-binding receptors in *Drosophila* hemocytes and *C. elegans* (Bhatia et al. 2011). This demonstrated intracellular delivery of a molecular cargo using a synthetic DNA delivery vehicle. However, the integrity of the applied DNA nanostructures after cellular internalization was not investigated. To address this, Walsh et al. studied the uptake and structural integrity of the DNA tetrahedron in HEK cells (Goodman et al. 2005). Cy3 and Cy5 molecules were positioned ~3 nm apart on separate strands and in vivo FRET signals indicated that the structural integrity of intracellular tetrahedrons was preserved. More recently, it was shown that the DNA tetrahedron can act as a carrier for the chemotherapeutic agent doxorubicin in the treatment of drug-resistant cancer cells (Kim et al. 2013) and for targeted delivery of folate-conjugated siRNA hybridized to DNA handles protruding from the edges (Fig. 5a) (Lee et al. 2012). Interestingly, both delivery and gene silencing efficiency in KB cells depended on the number of folate ligands and their spatial distribution. The results showed that at least three folate molecules were required for efficient gene knockdown in cells and support the principle of multivalent targeting for these systems. In addition, the assembled tetrahedron offered longer circulation time compared to free siRNAs, presumably due to decreased renal clearance and performed targeting and gene silencing (<50 %) in a tumor xenograft mouse model. Although knockdown efficiencies may have to be improved, these experiments emphasize the importance of ligand orientation at the surface of nucleic acid nanostructures and demonstrate targeted delivery of nucleic acid nanostructures without the need of transfection reagents.

DNA origami structures have also delivered cargo to cells; for example, doxorubicin has been delivered and released in breast cancer cells (Jiang et al. 2012; Zhao et al. 2012). Doxorubicin intercalates with DNA double helices without disrupting the nanostructure and is supposedly released upon intracellular degradation of the DNA origami. DNA origami structures loaded with doxorubicin exhibited cytotoxicity in cells otherwise resistant to free doxorubicin (Jiang et al. 2012). The size and shape dependence of cellular origami uptake was suggested to explain the improved doxorubicin activity when co-delivered with the DNA structures. Notably, loading and release rates of doxorubicin can be tuned

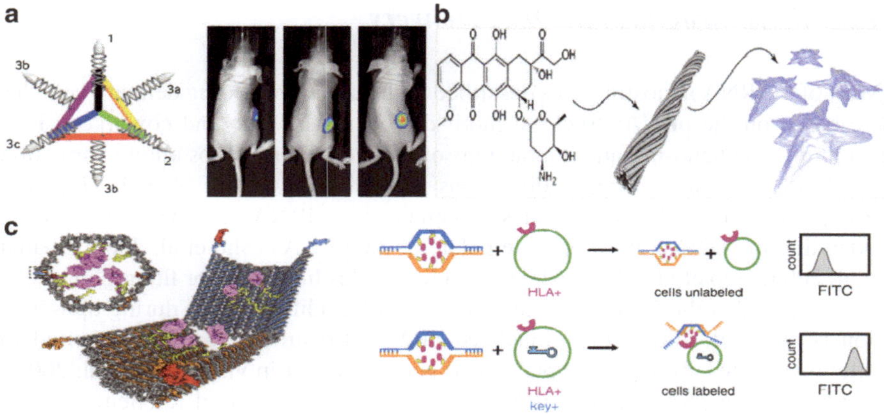

Fig. 5 Delivery vehicles of DNA nanostructures. (**a**) DNA tetrahedron loaded with folate-conjugated siRNA targets tumor grafts in mice (Lee et al. 2012). (**b**) DNA nanostructures deliver the chemotherapeutic agent doxorubicin to otherwise resistant breast cancer cells (Zhao et al. 2012). (**c**) A logic-gated DNA device delivers molecular signals to targeted cells and arrests cell growth (Douglas et al. 2012). All figures are reprinted with permission

by introducing global twist in DNA origami structures optimized for these parameters resulting in an increased cytotoxicity compared to its unstrained counterpart (Fig. 5b) (Zhao et al. 2012).

A vision for DNA nanostructures is to produce therapeutic devices encompassing logic gate-controlled dynamics capable of releasing or exposing molecular payloads in drug delivery. An early example of this approach was given in 2012, when Douglas and Bachelet et al. presented a barrel-shaped DNA origami loaded with cell signaling molecules, exposed only upon recognition of two individual molecular targets (Fig. 5c) (Douglas et al. 2012). In a closed state, the structure resembles an entropic spring, hinged at one end and locked in the other end by hybridization of DNA aptamer sequences. Upon binding of extracelluar ligands, structural rearrangements in the aptamers release the locks and the barrel opens, exposing its interior surface. DNA handles, protruding from the interior of the barrel, were used to carry various fluorescently labeled antibody fragments (Fabs) against human leukocyte antigens (HLA) able to induce cell signaling pathways in lymphoma and leukemia cells lines. Importantly, only cells expressing the right combination of ligand keys matching the aptamer lock complex were exposed to Fabs. This was demonstrated by the ability of the DNA device to discriminate target NKL cells in a complex cellular background of whole-blood leukocytes. Depending on the choice of Fab', the DNA barrel was able to arrest growth in aggressive NKL cells or induce activation of T cells. Although this device operates entirely in the extracellular environment, it does demonstrate the potential of DNA nanostructures for delivery of molecular signals to control biological systems in a programmed manner.

4.2 RNA Structures in Drug Delivery

Most of the RNA nanostructures described in the literature as drug delivery vehicles are based on the pRNA structural motif discovered by Guo and coworkers. Programming the hand-in-hand interactions of its interlocking loops allows formation of pRNA multimers with modifications on extended 5′ and 3′ ends (Fig. 6a) (Hoeprich et al. 2003). Ribozymes, aptamers, and siRNAs can be introduced by extending the DNA transcription template of the pRNAs (Shu et al. 2003; Khaled et al. 2005; Guo et al. 2005), while small molecules like folate or fluorophores can be introduced at the 5′ end by using an excess of modified AMPs during transcription (Guo et al. 2006). Functionalized pRNA structures have been reported to exhibit specific cell binding, uptake, and gene silencing in vitro (Guo et al. 2006). In particular, a pRNA-based siRNA delivery system was used to silence a viral protease in HeLa cells resulting in reduced virus replication and stabilized cell survival (Zhang et al. 2009). An alternative structural motif derived from pRNA emerged in 2011, when it was realized that the core structure of pRNA molecules forms highly stable RNA three-way junctions around the left-hand loop (Fig. 6b). Each of the three arms protruding from the core structure can be modified with combinations of folate, ribozymes, aptamers, and siRNAs and was able to bind and enter targeted cells in vitro and in vivo (Shu et al. 2011). Recently, the core motif was expanded by substituting the right-hand loop with an extra helix enabling the formation of stable tetravalent RNA nanoparticles (Fig. 6b) (Haque et al. 2012). This allowed the combinatorial screening of different designs containing targeting agents and various siRNAs attached to the pRNA core. Due to their structural integrity and modular design the pRNA system appears as the most promising RNA-derived delivery systems for in vivo applications. Only a few alternative RNA assemblies have demonstrated their potential in vitro. By the use of transfection reagents, a "nanoring" comprising multiple siRNAs showed gene silencing activity (<20 %) in HeLa cells (Nakashima et al. 2011). This assembly, described previously (Yingling and Shapiro 2007; Grabow et al. 2011; Afonin et al. 2011), can template the assembly of up to six siRNAs during in vitro transcription in a one-pot reaction and was able to reduce the expression of a target gene in breast cancer cells (Fig. 6c) (Afonin et al. 2012). These reports indicate that siRNA can be delivered by RNA nanostructures and remain functional both with and without the aid of transfection reagents. However, some issues remain to be addressed including potential off-target effects derived from nonspecific triggering of immune surveillance receptors including toll-like receptors.

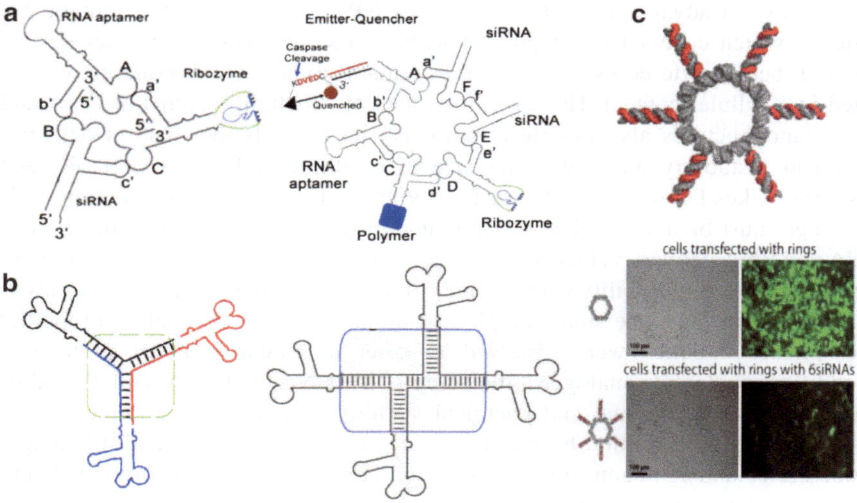

Fig. 6 RNA nanostructures for delivery. (**a**) pRNA trimer and hexamer modified at the 5′ or 3′ end (Shu et al. 2011). (**b**) pRNA-derived core structures produce defined trivalent and tetravalent RNA assemblies (here modified with additional pRNA motifs) (Shu et al. 2011; Haque et al. 2012). (**c**) RNA nanoring modified with siRNAs mediates knockdown of eGFP in human breast cancer cells (Grabow et al. 2011). All figures are reprinted with permission

5 Future Perspectives

DNA and RNA nanostructures have been proposed as new platforms for the development of drug delivery systems. Due to their exceptional structural control, addressability, and modularity, they have the potential for being more readily tailored for specific tasks. But much has to be learnt about how these new structures interact with living organisms. More thorough examinations of stability, cell interactions, and pharmacokinetics of DNA and RNA nanostructures are needed. This requires a new set of methodologies to evaluate the robustness of nucleic acid-based delivery systems. Until now, fluorescence-based assays have been the preferred choice to evaluate cellular interfacing, such as binding, internalization, and intracellular trafficking. However, fluorescent signals do not necessarily remain associated with the DNA or RNA nanostructures complicating the signal interpretation and evaluation. More reliable approaches are needed to assess the fate and structural integrity of DNA and RNA nanostructures for example by detection of scaffold strands (Okholm et al. 2014) or analysis of corporative effects mediated by multiple functionalities on separate strands.

Solutions to this may be facilitated by the highly developed toolbox of specialized enzymes such as polymerases, nucleases, transferases, and ligases that can mediate the synthesis, analysis, modification, detection, and degradation of DNA and RNA nanostructures. Natural and recombinant enzymes acting on nucleic acids provide a great advantage to DNA and RNA NPs over many other synthetic NP systems, which cannot be manipulated with similar specificity and efficiency. A potential benefit offered by these enzymes remains to be more thoroughly investigated in a cellular context. However, the inherent structural features of DNA and RNA nanostructures also impose a major barrier towards their use in delivery. Chemical instability along with an arsenal of different DNases, RNases, and opsonins makes DNA and RNA structures vulnerable to in vivo degradation. This challenge must be addressed as stable structures are crucial for the realization of DNA and RNA delivery vehicles. It was recently shown that a DNA prism structure has increased serum stability when folded compared to its individual parts (Conway et al. 2013). In the same study simple end modifications with small hydrophilic/hydrophobic moieties were observed to offer a similar level of protection. In addition, nucleotide analogues, including LNA, UNA, and PNA, and 2′ modifications can affect thermal and chemical stability of nucleic acid duplexes and reduce their susceptibility to enzymatic degradation (Karkare and Bhatnagar 2006; Freier and Altmann 1997). Such modifications may be incorporated into larger and more complex nucleic acid nanostructures with similar effects.

Another challenge for DNA and RNA nanostructures is the innate immune system of mammalian organisms, which evolved to protect against foreign genetic material and pathogens. Many of the building blocks in DNA and RNA nanostructures harbor motifs from bacteria or viruses and can trigger various recognition mechanisms of the immune system such as toll-like receptors (TLR) (Hemmi et al. 2000; Diebold et al. 2004; Schüller et al. 2011), dsRNA-binding protein kinase (PKR) (Clemens and Elia 1997), and retinoic acid-inducible gene-I (RIG-I) (Hornung et al. 2006). If these sequence motifs cannot be avoided, a strategy to dampen the immune response is to incorporate chemical modifications at nucleotides situated within sequence motifs known to trigger immunogenic surveillance (Bramsen 2011). Alternatively, functionalization with "stealth" moieties can make nucleic acid structures less prone to phagocytosis and elongate circulation times (Moghimi and Szebeni 2003). In this regard, polyethylenglycol (PEG) is of particular interest and has been intensively studied for other NP systems (Bazile et al. 1995; Immordino et al. 2006). Although DNA and RNA nanostructures intended for therapeutic applications should avoid interactions with the innate and adaptive immune systems intentional activation for induction of immunological memory may be attractive in certain cases. For this purpose a DNA tetrahedron has been employed to position antigens and adjuvants to elicit production of specific antibodies in mice (Liu et al. 2012).

We also expect that recent advances in structural manipulation of DNA and RNA will be translated to the field of protein nanostructures in the foreseeable future (Koga et al. 2012; Dill and MacCallum 2012). Already, cage-like structures assembled from protein fragments have demonstrated the potential for proteins in defined nanoscopic structures (King et al. 2012). Protein nanostructures are still far from the complexity of those of nucleic acids; however, their unique chemical diversity and functional properties make them attractive subjects for nanoscale manipulation.

6 Conclusion

During the last two decades nucleic acid nanotechnology has been one of the most prominent techniques to construct defined, bottom-up nanoscale structures. A plethora of DNA and RNA nanostructures of increasing structural complexity have emerged. As the potential for functionalized DNA and RNA nanostructures begins to unravel, it collides with the structure-centric focus of the field. The future will not only lie in the design of better structures but also in broadening their functional repertoire. Recently, first reports have demonstrated the potential of multifunctional DNA and RNA nanostructures as cell targeting vehicles able to carry payload. Further insight into the relationship between cellular interfacing, structural characteristics' and functional properties of DNA and RNA nanostructures is essential. It is likely that DNA and RNA structures with optimized assembly routes and programmable dynamic behavior will be focal points in the search for devices with joint diagnostic and therapeutic capability. It will provide a platform for systematic screening of functional modules in regard to the effects of spatial positioning, stoichiometry, and combinations needed to identify optimal surface decorations for probing their cellular targets. The clinical future for DNA and RNA nanostructures is dependent not only on reduced production costs, but also on the development of new methodologies for establishing structural and functional integrity in the complex biological environment found in and outside the cell. By sophisticated manipulation of DNA and RNA assemblies we can hope to approach the function of nature's own programmable molecular vehicles like viruses and exosomes. Fine-tuned over billions of years these natural systems have created a library of solutions for solving problems associated with delivery of molecules to cells. Inspired by their structure and composition we are likely to take the next big leap forward to realization of defined nucleic acid nanostructures in drug delivery.

References

Afonin KA, Bindewald E, Yaghoubian AJ et al (2010) In vitro assembly of cubic RNA-based scaffolds designed in silico. Nat Nanotechnol 5:676–682

Afonin KA, Grabow WW, Walker FM et al (2011) Design and self-assembly of siRNA-functionalized RNA nanoparticles for use in automated nanomedicine. Nat Protoc 6: 2022–2034

Afonin KA, Kireeva M, Grabow WW et al (2012) Co-transcriptional assembly of chemically modified RNA nanoparticles functionalized with siRNAs. Nano Lett 12:5192–5195

Andersen FF, Knudsen B, Oliveira CLP et al (2008) Assembly and structural analysis of a covalently closed nano-scale DNA cage. Nucleic Acids Res 36:1113–1119

Andersen ES, Dong M, Nielsen MM et al (2009) Self-assembly of a nanoscale DNA box with a controllable lid. Nature 459:73–76

Bazile D, Prud'homme C, Bassoullet MT et al (1995) Stealth Me.PEG-PLA nanoparticles avoid uptake by the mononuclear phagocytes system. J Pharm Sci 84:493–498

Bhatia D, Surana S, Chakraborty S et al (2011) A synthetic icosahedral DNA-based host-cargo complex for functional in vivo imaging. Nat Commun 2:339
Bramsen JB (2011) Chemical modification of small interfering RNA. Methods Mol Biol 721: 77–103
Chen J, Seeman N (1991) Synthesis from DNA of a molecule with the connectivity of a cube. Nature 350:631–633
Chen C, Sheng S, Shao Z et al (2000) A dimer as a building block in assembling RNA. A hexamer that gears bacterial virus phi29 DNA-translocating machinery. J Biol Chem 275:17510–17516
Chworos A, Severcan I, Koyfman AY et al (2004) Building programmable jigsaw puzzles with RNA. Science 306:2068–2072
Clemens MJ, Elia A (1997) The double-stranded RNA-dependent protein kinase PKR: structure and function. J Interferon Cytokine Res 17:503–524
Conway JW, McLaughlin CK, Castor KJ et al (2013) DNA nanostructure serum stability: greater than the sum of its parts. Chem Commun (Camb) 49:1172–1174
Davis ME, Zuckerman JE, Choi CHJ et al (2010) Evidence of RNAi in humans from systemically administered siRNA via targeted nanoparticles. Nature 464:1067–1070
Delebecque CJ, Lindner AB, Silver P et al (2011) Organization of intracellular reactions with rationally designed RNA assemblies. Science 333:470–474
Diebold SS, Kaisho T, Hemmi H et al (2004) Innate antiviral responses by means of TLR7-mediated recognition of single-stranded RNA. Science 303:1529–1531
Dietz H, Douglas SM, Shih WM (2009) Folding DNA into twisted and curved nanoscale shapes. Science 325:725–730
Dill K, MacCallum JL (2012) The protein-folding problem, 50 years on. Science 338:1042–1046
Dinauer N, Balthasar S, Weber C et al (2005) Selective targeting of antibody-conjugated nanoparticles to leukemic cells and primary T-lymphocytes. Biomaterials 26:5898–5906
Douglas SM, Dietz H, Liedl T et al (2009a) Self-assembly of DNA into nanoscale three-dimensional shapes. Nature 459:414–418
Douglas SM, Marblestone AH, Teerapittayanon S et al (2009b) Rapid prototyping of 3D DNA-origami shapes with caDNAno. Nucleic Acids Res 37:5001–5006
Douglas SM, Bachelet I, Church GM (2012) A logic-gated nanorobot for targeted transport of molecular payloads. Science 335:831–834
Duckworth BP, Chen Y, Wollack JW et al (2007) A universal method for the preparation of covalent protein-DNA conjugates for use in creating protein nanostructures. Angew Chem Int Ed Engl 46:8819–8822
Endo M, Yamamoto S, Tatsumi K et al (2013) RNA-templated DNA origami structures. Chem Commun (Camb) 49:2879–2881
Erben CM, Goodman RP, Turberfield AJ (2006) Single-molecule protein encapsulation in a rigid DNA cage. Angew Chem Int Ed Engl 45:7414–7417
Fishler R, Artzy-Schnirman A, Peer E et al (2012) Mixed alkanethiol monolayers on submicrometric gold patterns: a controlled platform for studying cell-ligand interactions. Nano Lett 12:4992–4996
Freier SM, Altmann KH (1997) The ups and downs of nucleic acid duplex stability: structure-stability studies on chemically-modified DNA:RNA duplexes. Nucleic Acids Res 25: 4429–4443
Gao X, Cui Y, Levenson RM et al (2004) In vivo cancer targeting and imaging with semiconductor quantum dots. Nat Biotechnol 22:969–976
Goodman RP, Schaap IAT, Tardin CF et al (2005) Rapid chiral assembly of rigid DNA building blocks for molecular nanofabrication. Science 310:1661–1665
Grabow WW, Zakrevsky P, Afonin K et al (2011) Self-assembling RNA nanorings based on RNAI/II inverse kissing complexes. Nano Lett 1:878–887
Guo P, Erickson S, Anderson D (1987) A small viral RNA is required for in vitro packaging of bacteriophage phi 29 DNA. Science 236:690–694
Guo P, Zhang C, Chen C et al (1998) Inter-RNA interaction of phage phi29 pRNA to form a hexameric complex for viral DNA transportation. Mol Cell 2:149–155

Guo S, Tschammer N, Mohammed S et al (2005) Specific delivery of therapeutic RNAs to cancer cells via the dimerization mechanism of phi29 motor pRNA. Hum Gene Ther 16:1097–1109

Guo S, Huang F, Guo P (2006) Construction of folate-conjugated pRNA of bacteriophage phi29 DNA packaging motor for delivery of chimeric siRNA to nasopharyngeal carcinoma cells. Gene Ther 13:814–820

Han D, Pal S, Liu Y et al (2010) Folding and cutting DNA into reconfigurable topological nanostructures. Nat Nanotechnol 5:712–717

Han D, Pal S, Nangreave J et al (2011) DNA origami with complex curvatures in three-dimensional space. Science 332:342–346

Haque F, Shu D, Shu Y et al (2012) Ultrastable synergistic tetravalent RNA nanoparticles for targeting to cancers. Nano Today 7:245–257

He Y, Ye T, Su M et al (2008) Hierarchical self-assembly of DNA into symmetric supramolecular polyhedra. Nature 452:198–201

Hemmi H, Takeuchi O, Kawai T et al (2000) A Toll-like receptor recognizes bacterial DNA. Nature 408:740–745

Hicke BJ, Stephens AW, Gould T et al (2006) Tumor targeting by an aptamer. J Nucl Med 47: 668–678

Hoeprich S, Zhou Q, Guo S et al (2003) Bacterial virus phi29 pRNA as a hammerhead ribozyme escort to destroy hepatitis B virus. Gene Ther 10:1258–1267

Högberg B, Liedl T, Shih W (2009) Folding DNA origami from a double-stranded source of scaffold. J Am Chem Soc 131:9154–9155

Hornung V, Ellegast J, Kim S et al (2006) 5′-Triphosphate RNA is the ligand for RIG-I. Science 314:994–997

Huang J, Grater SV, Corbellini F et al (2009) Impact of order and disorder in RGD nanopatterns on cell adhesion. Nano Lett 9:1111–1116

Immordino ML, Dosio F, Cattel L (2006) Stealth liposomes: review of the basic science, rationale, and clinical applications, existing and potential. Int J Nanomedicine 1:297–315

Jaeger L, Chworos A (2006) The architectonics of programmable RNA and DNA nanostructures. Curr Opin Struct Biol 16:531–543

Jaeger L, Leontis NB (2000) Tecto-RNA: one-dimensional self-assembly through tertiary interactions. Angew Chem Int Ed Engl 39:2521–2524

Jaeger L, Westhof E, Leontis NB (2001) TectoRNA: modular assembly units for the construction of RNA nano-objects. Nucleic Acids Res 29:455–463

Jahn K, Tørring T, Voigt NV et al (2011) Functional patterning of DNA origami by parallel enzymatic modification. Bioconjug Chem 22:819–823

Jiang Q, Song C, Nangreave J et al (2012) DNA origami as a carrier for circumvention of drug resistance. J Am Chem Soc 134:13396–13403

Karkare S, Bhatnagar D (2006) Promising nucleic acid analogs and mimics: characteristic features and applications of PNA, LNA, and morpholino. Appl Microbiol Biotechnol 71:575–586

Ke Y, Sharma J, Liu M et al (2009) Scaffolded DNA origami of a DNA tetrahedron molecular container. Nano Lett 9:2445–2447

Ke Y, Ong LL, Shih WM et al (2012) Three-dimensional structures self-assembled from DNA bricks. Science 338:1177–1183

Khaled A, Guo S, Li F et al (2005) Controllable self-assembly of nanoparticles for specific delivery of multiple therapeutic molecules to cancer cells using RNA nanotechnology. Nano Lett 5:1797–1808

Kim K-R, Kim D-R, Lee T et al (2013) Drug delivery by a self-assembled DNA tetrahedron for overcoming drug resistance in breast cancer cells. Chem Commun (Camb) 5:3–5

King NP, Sheffler W, Sawaya MR et al (2012) Computational design of self-assembling protein nanomaterials with atomic level accuracy. Science 336:1171–1174

Ko S, Liu H, Chen Y et al (2008) DNA nanotubes as combinatorial vehicles for cellular delivery. Biomacromolecules 9:3039–3043

Ko SH, Su M, Zhang C et al (2010) Synergistic self-assembly of RNA and DNA molecules. Nat Chem 2:1050–1055
Koga N, Tatsumi-Koga R, Liu G et al (2012) Principles for designing ideal protein structures. Nature 491:222–227
Krishnan Y, Bathe M (2012) Designer nucleic acids to probe and program the cell. Trends Cell Biol 22:624–633
Kuzuya A, Komiyama M (2009) Design and construction of a box-shaped 3D-DNA origami. Chem Commun (Camb) 28:4182–4184
Lee H, Lytton-Jean AKR, Chen Y et al (2012) Molecularly self-assembled nucleic acid nanoparticles for targeted in vivo siRNA delivery. Nat Nanotechnol 7:389–393
Liedl T, Högberg B, Tytell J et al (2010) Self-assembly of three-dimensional prestressed tensegrity structures from DNA. Nat Nanotechnol 5:520–524
Liu X, Xu Y, Yu T et al (2012) A DNA nanostructure platform for directed assembly of synthetic vaccines. Nano Lett 12:4254–4259
Lönnberg H (2009) Solid-phase synthesis of oligonucleotide conjugates useful for delivery and targeting of potential nucleic acid therapeutics. Bioconjug Chem 20:1065–1094
Low PS, Henne W, Doorneweerd DD (2008) Discovery and development of folic-acid-based receptor targeting for imaging and therapy of cancer and inflammatory diseases. Acc Chem Res 41:120–129
Mastroianni AJ, Claridge S, Alivisatos P (2009) Pyramidal and chiral groupings of gold nanocrystals assembled using DNA scaffolds. J Am Chem Soc 131:8455–8459
Meyer R, Niemeyer CM (2011) Orthogonal protein decoration of DNA nanostructures. Small 7:3211–3218
Moghimi SM, Szebeni J (2003) Stealth liposomes and long circulating nanoparticles: critical issues in pharmacokinetics, opsonization and protein-binding properties. Prog Lipid Res 42: 463–478
Nakashima Y, Abe H, Abe N et al (2011) Branched RNA nanostructures for RNA interference. Chem Commun (Camb) 47:8367–8369
Okholm AH, Nielsen JS, Vinther M et al (2014) Quantification of cellular uptake of DNA nanostructures by qPCR. Methods. http://dx.doi.org/10.1016/j.ymeth.2014.01.013
Oliveira CLP, Juul S, Jørgensen HL et al (2010) Structure of nanoscale truncated octahedral DNA cages: variation of single-stranded linker regions and influence on assembly yields. ACS Nano 4:1367–1376
Peer D, Karp JM, Hong S et al (2007) Nanocarriers as an emerging platform for cancer therapy. Nat Nanotechnol 2:751–760
Pinheiro AV, Han D, Shih WM et al (2011) Challenges and opportunities for structural DNA nanotechnology. Nat Nanotechnol 6:763–772
Rinker S, Ke Y, Liu Y et al (2008) Self-assembled DNA nanostructures for distance-dependent multivalent ligand-protein binding. Nat Nanotechnol 3:418–422
Rossi JJ (2005) RNAi therapeutics: SNALPing siRNAs in vivo. Gene Ther 13:583–584
Rothemund PWK (2006) Folding DNA to create nanoscale shapes and patterns. Nature 440: 297–302
Saccà B, Niemeyer CM (2011) Functionalization of DNA nanostructures with proteins. Chem Soc Rev 40:5910–5921
Said H, Schüller VJ, Eber FJ et al (2013) M1.3–a small scaffold for DNA origami. Nanoscale 5:284–290
Schüller VJ, Heidegger S, Sandholzer N et al (2011) Cellular immunostimulation by CpG-sequence-coated DNA origami structures. ACS Nano 5:9696–9702
Seeman NC (1982) Nucleic acid junctions and lattices. J Theor Biol 99:237–247
Severcan I, Geary C, Verzemnieks E et al (2009) Square-shaped RNA particles from different RNA folds. Nano Lett 9:1270–1277
Severcan I, Geary C, Chworos A et al (2010) A polyhedron made of tRNAs. Nat Chem 2:772–779
Sharma J, Chhabra R, Andersen CS et al (2008) Toward reliable gold nanoparticle patterning on self-assembled DNA nanoscaffold. J Am Chem Soc 130:7820–7821

Shen W, Zhong H, Neff D et al (2009) NTA directed protein nanopatterning on DNA origami nanoconstructs. J Am Chem Soc 131:6660–6661
Shih WM, Quispe JD, Joyce GF (2004) A 1.7-kilobase single-stranded DNA that folds into a nanoscale octahedron. Nature 427:618–621
Shu D, Huang LP, Hoeprich S et al (2003) Construction of phi29 DNA-packaging RNA monomers, dimers, and trimers with variable sizes and shapes as potential parts for nanodevices. J Nanosci Nanotechnol 3:295–302
Shu D, Shu Y, Haque F et al (2011) Thermodynamically stable RNA three-way junction for constructing multifunctional nanoparticles for delivery of therapeutics. Nat Nanotechnol 6:658–667
Smith D, Schüller V, Engst C et al (2013) Nucleic acid nanostructures for biomedical applications. Nanomedicine 8:105–121
Sørensen RS, Okholm AH, Schaffert D et al (2013) Enzymatic ligation of large biomolecules to DNA. ACS Nano 7:1–16
Tørring T, Voigt NV, Nangreave J et al (2011) DNA origami: a quantum leap for self-assembly of complex structures. Chem Soc Rev 40:5636–5646
Varkouhi AK, Scholte M, Storm G et al (2011) Endosomal escape pathways for delivery of biologicals. J Control Release 151:220–228
Walsh AS, Yin H, Erben CM et al (2011) DNA cage delivery to mammalian cells. ACS Nano 5:5427–5432
Wei B, Dai M, Yin P (2012) Complex shapes self-assembled from single-stranded DNA tiles. Nature 485:623–626
Westhof E, Masquida B, Jaeger L (1996) RNA tectonics: towards RNA design. Fold Des 1:R78–R88
Williams BR, Lund K, Liu Y et al (2007) Self-assembled peptide nanoarrays: an approach to studying protein-protein interactions. Angew Chem Int Ed Engl 46:3051–3054
Wilner OI, Willner I (2012) Functionalized DNA nanostructures. Chem Rev 112:2528–2556
Wilner O, Weizmann Y, Gill R (2009) Enzyme cascades activated on topologically programmed DNA scaffolds. Nat Nanotechnol 4:249–254
Win MN, Smolke CD (2007) A modular and extensible RNA-based gene-regulatory platform for engineering cellular function. Proc Natl Acad Sci U S A 104:14283–14288
Yan H, Park SH, Finkelstein G et al (2003) DNA-templated self-assembly of protein arrays and highly conductive nanowires. Science 301:1882–1884
Yang Y, Han D, Nangreave J et al (2012) DNA origami with double-stranded DNA as a unified scaffold. ACS Nano 6:8209–8215
Yingling YG, Shapiro B (2007) Computational design of an RNA hexagonal nanoring and an RNA nanotube. Nano Lett 7:2328–2334
Zadegan RM, Jepsen MDE, Thomsen KE et al (2012) Construction of a 4 zeptoliters switchable 3D DNA box origami. ACS Nano 6:10050–10053
Zhang Y, Seeman NC (1994) Construction of a DNA-truncated octahedron. J Am Chem Soc 116:1661–1669
Zhang L, Gu FX, Chan JM et al (2008) Nanoparticles in medicine: therapeutic applications and developments. Clin Pharmacol Ther 83:761–769
Zhang HM, Su Y, Guo S et al (2009) Targeted delivery of anti-coxsackievirus siRNAs using ligand-conjugated packaging RNAs. Antiviral Res 83:307–316
Zhang H, Chao J, Pan D et al (2012) Folding super-sized DNA origami with scaffold strands from long-range PCR. Chem Commun (Camb) 48:6405–6407
Zhao Z, Liu Y, Yan H (2011) Organizing DNA origami tiles into larger structures using preformed scaffold frames. Nano Lett 11:2997–3002
Zhao Y-X, Shaw A, Zeng X et al (2012) DNA origami delivery system for cancer therapy with tunable release properties. ACS Nano 6:8684–8691

Targeted Editing of Therapeutic Genes Using DNA-Based Transcriptional Activators: Scope and Challenges

Ganesh N. Pandian and Hiroshi Sugiyama

Contents

1 Introduction 348
2 Transcriptional Activation in Nature 350
3 Disorders Associated with Faulty Transcriptional Machinery 351
3.1 Tumorigenesis 351
3.2 Neurological Disorders 352
3.3 Congenital Heart Disease 353
3.4 Diabetes 354
4 Artificial Transcription Factors for Gene Regulation 354
4.1 Synthetic Transcription Modulators: Design and Overview 354
4.2 Genome Editing with Targeted Nucleases 355
4.3 DNA Binding Small Molecules as Transcriptional Modulators 357
5 Targeting Small Molecules as Transcriptional Activators 357
6 Future Perspectives 360
References 361

Abstract In recent years, there has been an increasing interest among the clinicians to devise genetic knowledge-based therapeutic strategies to treat complex diseases. Since targeted treatments ensure better drug efficacy and fewer long-term side effects, development of tailor-made drugs is of rising demand. Through modern biological and analytical techniques, we could now predict the transcription factors associated with disorders and design chemical and/or biological tools to reset the transcriptional machinery of the diseased cell and restore them back to healthy state. However, therapeutic targeting of the defected genes in the body is not straightforward owing to the critical influence of the dynamic epigenome. In this chapter, we give a detailed overview of the customizable artificial transcription activators and therapeutically significant transcription factors and suggest the need

G.N. Pandian • H. Sugiyama (✉)
Department of Chemistry, Graduate School of Science, Institute for Integrated Cell Material Sciences (WPI-iCeMS), Kyoto University, Sakyo, Kyoto 606 8502, Japan
e-mail: hs@kuchem.kyoto-u.ac.jp

V.A. Erdmann et al. (eds.), *Chemical Biology of Nucleic Acids*, RNA Technologies, DOI 10.1007/978-3-642-54452-1_19,

to gain inspiration from the coordinated chromatin modifications observed in natural cellular environment and design targeting transcriptional activators with epigenetic activity.

Keywords Genome editing • Transcription activators • TALENs • Multifunctional molecules • Transcription factor-based disorders

1 Introduction

Living systems develop themselves regularly and gain different functional characteristics to meet the demands needed to maintain themselves (Lamb and Jablonka 2005). Gaining insights into this natural well-organized pattern of dynamic, living systems could encourage us to acquire novel approaches for solving some of the unresolved clinical issues that confront humanity (Frances et al. 2012). In living cells, the chromosomal DNA stores and encodes the heritable information through the functional unit called genes over a long period using only four nucleobases: adenine (A), thymine (T), guanine (G), and cytosine (C). Each gene contains a set of information that gets interpreted through a series of processes including transcription, RNA splicing, translation, and posttranslational modification to synthesize a functional gene product usually proteins. DNA is amenable to the accidental changes in its code called mutations. Germ-line mutations that occur in the heritable germ cells lead to a constitutional mutation where every single cell of the offspring carries a mutation. Such causal mutations are difficult to repair. The rise of gene therapy, a technique of replacing the defected genes with new genes in the body, offered opportunities to therapeutically modulate the mutated gene (Pfeifer and Verma 2001). However, most of the diseases could be linked with the malfunctioning transcriptional machinery that gets controlled by multiple genes and not a single gene. Aberrant transcription patterns get correlated with most of the disorders that are still incurable (Jiang et al. 2004). Hence, development of artificial transcription activator with functional capability of their natural counterparts is of increasing demand. However, transcription is governed not only by the genetic code but also by numerous transcription factors, cofactors, and chromatin regulators. This phenomenon got substantiated with the successful completion of Human Genome Project, which revealed that the number of protein-coding genes in humans is only twice than that could be found in the fruit fly (Cloud 2010). Consequently, the critical role of epigenetics, an additional layer that causes the transmissible changes in gene expression beyond the fundamental structure of DNA, gained higher recognition. In contrast to irreversible mutations in DNA, pathological epigenetic modifications could be edited and restored (Pandian and Sugiyama 2013). However, therapeutic targeting of the dynamic epigenetic modifications has complications as one gene could undergo several types of epigenetic

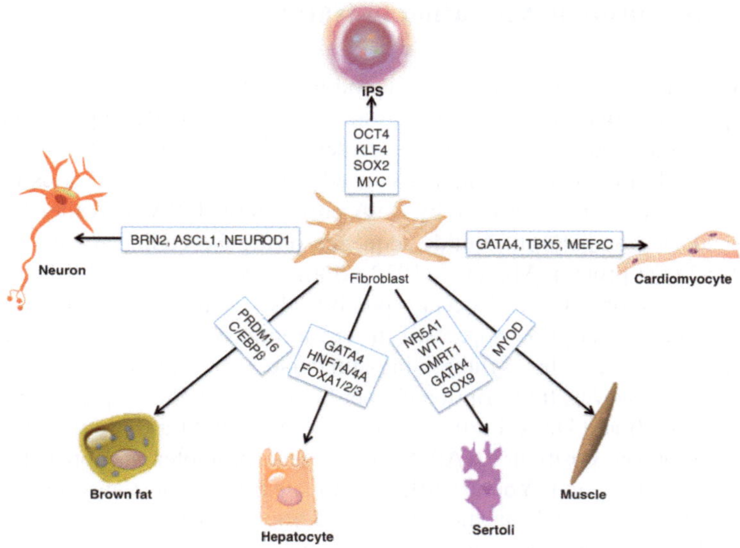

Fig. 1 Master transcriptional regulators that can shift the transcriptional machinery of somatic fibroblast to various distinctive cell states [adapted from Lee and Young (2013)]

modifications. Also, each cell type could have their own-programmed version of epigenome that is very different from the other cell type (Hamm and Costa 2011). Thus, in nature, transcriptional activation gets controlled at various distinctive levels through coordinated changes occurring in both genome and epigenome. Hence, artificial transcriptional activators should not only contain a component for recognizing DNA sequences but also an unit capable of triggering protein–protein interactions that recruit transcriptional machinery for gene activation. Identification of such key transcription factors/activators conferring to cell fate has therapeutic significance. Because cellular reprogramming of somatic cell to any desired cell type could now be achieved through enforced transcriptional activation of defined factors (Fig. 1).

This chapter aims to give a detailed overview of the customizable, DNA-based transcriptional activators by referring certain major disorders associated with mutations in the transcription factors. Briefly, we explain the background studies that inspired the strategy of artificial transcriptional activation and summarize the scope and the challenges of the developing strategies that aim to achieve targeted editing of therapeutic genes. Among the major class of transcriptional activators, small molecules get proclaimed to have the potential to enter the clinic owing to the ease of use. Small molecules also may have relatively lesser side effects than the naturally occurring proteins as they eliminate the risk of genetic modifications (Schreiber 2005). Hence, we provide an introductory note on the novel innovative strategy of developing multifunctional small molecule to achieve concomitant gene expression for cell fate specification.

2 Transcriptional Activation in Nature

In 1961, François Jacob and Jacques Monod first demonstrated the key concept of transcriptional regulation in bacterial systems that get operated by the gene of a short-lived intermediate to synthesize proteins (Jacob and Monod 1961). This landmark study facilitated the studies carried out in several laboratories for demonstrating RNA synthesis in vitro by RNA polymerase (RNAP). Walter Fiers first established the complete nucleotide sequence of a gene coding for the bacteriophage MS2 coat protein (Min et al. 1972). Subsequent studies established that the transcription factors bind and occupy precise DNA sequences for recruiting and regulating the transcription apparatus. In eukaryotic cells, RNAP II controls the transcription process by synthesizing precursors of mRNA, snRNAs, and microRNAs. However, transcription cannot be initiated with simple incubation of purified RNAP II and DNA. Further factors called general transcription factors and protein complexes position RNAP II at the gene promoter and open the DNA double helix (Lee and Young 2013). Activated level of transcription under in vitro conditions can be achieved with additional factors, which are usually DNA-binding proteins targeting the DNA sequences typically located upstream of the promoter region. Such basal transcription enhancing factors are the so-called "transcriptional activators." Eukaryotic transcriptional activators have two modules, one for sequence-specific DNA binding and the other for activating components of the RNAP II transcriptional apparatus that bind promoters. Concomitant operation of both *cis*-acting elements (DNA sequences like promoter) and *trans*-acting elements (proteins that bind specific *cis*-acting elements) activates gene(s) at the right place and time (Ptashne and Gann 2002).

Nature has empowered transcriptional activators with versatile mechanisms not only for regulating other genes but also for regulating themselves. One such regulatory mechanism involves the modulation of the activator NFkB entry into the nucleus by the specific signals from its inhibitor IkB. Likewise, activation domain of the protein like Gal4 gets concealed by Gal80 (Ptashne and Gann 2002). Individual activators also shut themselves down like that observed in the yeast transcriptional activator Gcn4, which activates the kinase Srb10 and sends signals for destruction by ubiquitin-mediated proteolysis upon contact with transcriptional machinery (Chi et al. 2001). Thus, transcriptional machinery can precisely destruct the activators and keep them under control. Inside the eukaryotic cells, specific DNA sequences are not always accessible to transcriptional activators and other complexes as DNA gets wrapped around the histone octamer complexes to form nucleosomes. Conditional ablation of histone proteins in yeast led to transcriptional activation of several genes to demonstrate that specific transcriptional activators are always not required (Durrin et al. 1992). Robinson and Kadonaga demonstrated the modulation of the basal transcription level under in vitro conditions by either reconstituting nucleosomes with a DNA template or by the addition of transcriptional activators targeting DNA sequences located upstream of the promoter (Robinson and Kadonaga 1998). Taken together these results indicated that multiple parameters like state of the DNA template (Linker or nucleosomal), the type of transcription factors and transcriptional activators that stimulate RNAP, and the number of their binding sites on DNA govern the level of transcription of any given gene.

Along with this layer of control, mediator complexes also play an essential role in elevating the transcription level by acquiring the information from transcriptional activators, repressors, signaling pathways, and other regulators across initiation and elongation processes (Conaway and Conaway 2011). Transcriptional activation by recruitment leads to the formation of a protein–DNA complex at the promoter, which in turn delivers the signal for the RNAP II to initiate the transcription (Ptashne and Gann 2002). Keaveney and Struhl supported the recruitment model in vivo by demonstrating transcriptional inactivation after disconnecting the activation domain of the transcriptional activator away from the enhancer-bound protein and then transferring them to the components of RNAPII machinery (Keaveney and Struhl 1998). Recent studies suggest that DNA loop formation and stabilizing factor cohesin can modulate gene expression (Seitan and Merkenschlager 2012).

Epigenetic activation of genes could be achieved by the recruitment of ATP-dependent SW1/SNF complexes, which could facilitate access of the transcription apparatus and its regulators to DNA by remodeling the chromatin architecture. Chromatin remodeling also occurs through the histone-modifying enzymes that could trigger acetylation, methylation, and ubiquitinylation, which in turn facilitates the interaction surfaces for protein complexes conferring to transcriptional control (Rando 2012). Protein–protein interactions can also lead to the formation of Polycomb complex that silences the gene until later stage of development and differentiation (Lee and Young 2013). Certain regions that contain retrotransposons and other repetitive elements can lead to transcriptionally repressive chromatin architecture. Numerous studies on the transcription factors, their cofactors, and chromatin regulators of the eukaryotic systems provided the platform to the currently accepted model that operate at distinctive levels to achieve programmed gene control (Rando 2012; Conaway and Conaway 2011; Lee and Young 2013; Zhou et al. 2012).

3 Disorders Associated with Faulty Transcriptional Machinery

3.1 *Tumorigenesis*

Transcription factors play a fundamental role in cell cycle and determine the switching of cells to either proliferated or differentiated state. Accordingly, mutations in the tumor suppressing transcription factors stimulate tumorigenesis. The acute lymphoblastic leukemia-associated transcription factor TAL1 could trigger oncogenesis by forming an intricate gene network along with other transcription factors (Sanda et al. 2012). Some transcription factors like c-Myc cause tumorigenesis by not regulating the expression of oncogenic genes but by recruiting the transcription elongation factor to generate elevated levels of transcripts within the cellular transcriptional machinery (Lin et al. 2012). Thus, transcriptional amplification is also a tumor cell proliferation determining rate-limiting factor. Tumor development could also occur with mutations in the gene-encoding mediators like MED12 as this coactivator is either deleted or mutated in prostate cancer and benign tumors (Barbieri et al. 2012). Expression level of cohesin gets affected in some cancer cells; however, their direct role in tumorigenesis is under debate (Mannini and Musio 2011). Most of the factors belonging to STAT family and HOX get associated with a variety of cancers including breast cancer (Lee and Young 2013).

Feinberg and Vogelstein first demonstrated that alteration in the hypomethylation process could distinguish a cancer cell from its natural counterparts (Feinberg and Vogelstein 1983). Subsequently, mutations in many epigenetic modulators like ARID1A, SMARCA4, and SMARCB1 direct the cells to develop cancer phenotype (Dawson and Kouzarides 2012). Aberrant gene silencing caused by mutations in the Polycomb factors like EZH2 and SUZ12 and the overexpression of histone-modifying enzymes like SETDB1 get associated with tumorigenesis (Ceol et al. 2011). Activation and repression of histone deacetylases (HDAC) could modulate cell fate decision. This critical enzyme capable of triggering global level changes in gene expression gets implicated in acute promyelocytic leukemia, a disorder triggered by the fusion proteins of retinoic acid receptor-α. The transcription elongation modulating factors like the chromatin regulator MLL and their translocation components get associated with leukemia (Luo et al. 2012). Mutations in the tumor suppressor gene p53 get associated with a rare autosomal dominant hereditary disorder, which increases susceptibility to cancer (Iwakuma et al. 2005). Modern analytical methods facilitate us to identify the novel role of several such transcription factors. These transcription factors are of clinical significance as they could serve as potential therapeutic targets.

3.2 Neurological Disorders

Neurological disorder refers to structural and biochemical abnormalities of the body nervous system including those occurring in the brain and spinal cord. Rett Syndrome, an X-linked neurodevelopmental disorder of the gray matter of the brain, could be associated with mutations in the gene MECP2 (Amir et al. 1999). The transcription factor, FOXP2, gets implicated in developmental speech dyspraxia, a severe speech, and language disorder as the gene encoding this factor got mutated in the patients manifesting this disease (Lai et al. 2001). Mediators integrate information from activators, repressors, and signaling pathways for neural development. Hence, mutations in the mediator subunit MED23 could distort the transcription factors-mediator complexes to trigger transcriptional dysregulation of mitogen-responsive immediate-early genes associated with brain development and plasticity (Goh and Grants 2012). MED12 mutations get associated with the intellectual disability called Opitz–Kaveggia syndrome (Rump et al. 2011). Epigenetic chromatin remodeling plays an integral role in the transcription-dependent long-term potentiation (LTP), a phenomenon that converts short-term memories into stable long-term memories. Transcriptional activation of BDNF, a trκB-acting neurotrophin, provides support for long-term synaptic plasticity in the adult hippocampus (Minichiello 2009). Elevated levels of Reelin, the memory-stimulating gene, and repressed level of the memory-suppressing PPI1 gene improve memory formation (Beffert et al. 2005). LTP-associated factors such as GAP43, Arc, Integrin-associated protein, Synapsin I, and EGR1 can alter cognitive processes in murine models (Ryan et al. 2012). Essential role of Neurotensin in regulating the working memory got demonstrated in humans (Li et al. 2011). Mutations in the genes encoding SWI/SNF components have been associated with neurological syndromes associated with intellectual disability (Santen et al. 2012).

3.3 Congenital Heart Disease

Congenital heart disease (CHD) is a general term referring to the defect in the structure of the heart, which is present from birth and later develops to cause morbidity and mortality. A number of transcription factors and interactions between them have been associated with CHD (McCulley and Black 2012). In particular, the homeodomain protein Nkx2-5; GATA family zinc finger proteins GATA4, 5, and 6; MEF2 factors and SRF (MADS box proteins); T-box factors, including Tbx1, Tbx2, Tbx3, Tbx5, Tbx18, and Tbx20; and the Lim-homeodomain protein Isl1 play an integral role for heart development (He et al. 2011; McCulley and Black 2012). Many of these above-mentioned transcription factors are also important to control cardiac chamber maturation, conduction system development, and endocardial cushion remodeling (McCulley and Black 2012; Olson 2006). Each of these core cardiac transcription factors coordinates with themselves and others to establish a tightly regulated transcriptional gene network (He et al. 2011). Consequently, mutations in some of these core cardiac transcription factors get associated with CHD (Liu et al. 2011; Ouyang et al. 2011; Swaby et al. 2011). Among the growing number of novel transcription factors that are implicated in patients with CHD, Nkx2-5, GATA4, and Tbx5 are the well-characterized factors critical for normal heart development (McCulley and Black 2012). A deeper understanding of this intricate network of transcriptional regulators through several different model systems and genetic studies could aid us to realize therapeutic targets. Along with the transcription factors, microRNAs have also been linked to various cardiovascular diseases (Grueter et al. 2012).

3.4 Diabetes

Diabetes is a chronic hormone-related disorder in which a person has elevated blood sugar, which results in the damage of body parts. Loss-of-function mutation studies revealed a number of transcription factors and their binding sequences, which could be associated with diabetes (Thandavarayan et al. 2011). Some of the major factors conferring to transcriptional machinery pertaining to pancreatic cell fate include HNF1a, HNF1b, HNF4a, PDX1, and NEUROD1 (Odom et al. 2004). Maturity-onset diabetes of the young (MODY) could be developed due to mutations in these key factors, which also affect their target genes to perturb this transcriptional circuitry (Maestro et al. 2007). Since single nucleotide polymorphisms frequently occur in the binding sites of diverse transcription factors, mechanistic studies may uncover the unknown pathways that affect the regulatory circuitry of pancreatic cells to cause diabetes (Maurano et al. 2012).

4 Artificial Transcription Factors for Gene Regulation

4.1 Synthetic Transcription Modulators: Design and Overview

Recent advances in experimental and computational methods reveal therapeutically relevant transcription factors and their mode of operation, which in turn facilitate us to develop innovative strategies for artificially engineering gene circuits (Majmudar et al. 2005). To construct a transcription factor mimic, the presence of a DNA-binding domain (DBD), linker domain, and activation domain (AD) is essential. The constructed artificial transcriptional activators should also be capable of localizing inside the nucleus. Precise identifying of the target DNA sequences (usually the transcription factor binding sites) plays a vital role in achieving successful transcriptional regulation. Comprehensive mapping studies of the binding-site locations available in yeast model suggested binding sites to be shorter and fuzzier in the promoter regions that bind multiple transcription factors (Bilu and Barkai 2005). Transcriptional modeling assists us to map the binding of all known transcription factors in model organisms. The linker domain needs to be flexible, and its length should be variable to achieve high local AD concentration without squelching to inhibit the expression of other genes (Mapp et al. 2000). High level of expression can be achieved with different types of AD that could effectively interact with other proteins and recruit transcriptional machinery. Natural transcription factors can now be replaced with naturally occurring DNA-binding proteins or with synthetic alternatives, such as peptide nucleic acids, triplex-forming oligonucleotides, and hairpin polyamides (Wu et al. 2003). In the subsequent sections, we will briefly discuss the recent progress in transcription activator like effector nucleases and hairpin pyrrole imidazole polyamides, which are the efficient tools available to achieve targeted genome editing.

4.2 Genome Editing with Targeted Nucleases

Genome editing is a strategy available to fix the defected genes in the body by adding, removing, or exchanging DNA sequences. Genome editing using engineered netted the 2011 "Method of the Year" prize owing to its broad research and therapeutic applications (Baker 2012). Engineered nucleases are the artificial proteins comprised of a programmable, sequence-specific DBD and a nuclease capable of cleaving DNA. Targeted genome editing was first carried out with zinc finger nucleases (ZFNs), which are the initial version of engineered nucleases (Joung and Sander 2013). Engineered ZFNs artificially trigger point mutations, deletions, insertions, inversions, duplications, and translocations, which are then repaired by the endogenous DNA repair machinery. A ZFN called SB-728-T can introduce mutations in the HIV host coreceptor CCR5 gene (Perez et al. 2008) and now is in phase II clinical trials for the treatment of HIV/AIDS. However, immunogenicity and off-target effects are a major concern.

Recently, the advanced version of nucleases called transcription activator-like effector nucleases (TALENs) has emerged as an alternative to ZFNs to achieve targeted genome editing. TALENs have a nonspecific FokI nuclease domain and a customizable TAL effector protein that could recognize DNA sequences through a central repeat domain made up of repeat variable diresidues (Bogdanove et al. 2010). Nonhomologous end-joining (NHEJ) repair of a nuclease-induced double strand breaks (DSB) permits the introduction of an insertion or deletion mutation of variable lengths (Fig. 2). In just 2 years after its inception, TALENs have already been utilized in modulating endogenous expression of genes in various mammalian and plant models (Joung and Sander 2013; Reyon et al. 2012; Miller et al. 2011; Mussolino et al. 2011). In humans, endogenous expression of several biologically significant genes including OCT4, BCL6, EZH2, FGFR1, HDAC1, 2, and 6, KRAS, MYC, SUZ12, SIRT6, SMAD2, and HPRT could now be altered with TALENs (Joung and Sander 2013; Reyon et al. 2012).

While most of the therapeutic approaches aim to treat the symptoms of genetic diseases, targeted nucleases operate to alter the gene products and/or sequences that cause conferring to a certain disorder. ZFN-induced homology directed repair (HR) could correct mutations in the gene associated with sickle cell anemia (Sebastiano et al. 2011). Targeted gene correction of α1-antitrypsin deficiency in induced pluripotent stem cells (Yusa et al. 2011) and mutation in the Parkinson's disease-associated alpha-synuclein (SNCA) gene in patient-specific induced pluripotent stem (iPS) cells suggested the possibility of autologous transplant strategies (Soldner et al. 2011). TALENs have now been shown to trigger HR in human pluripotent stem cells and somatic cells (Hockemeyer et al. 2011; Miller et al. 2011). Hence, it is possible now to use engineered nucleases to perturb the

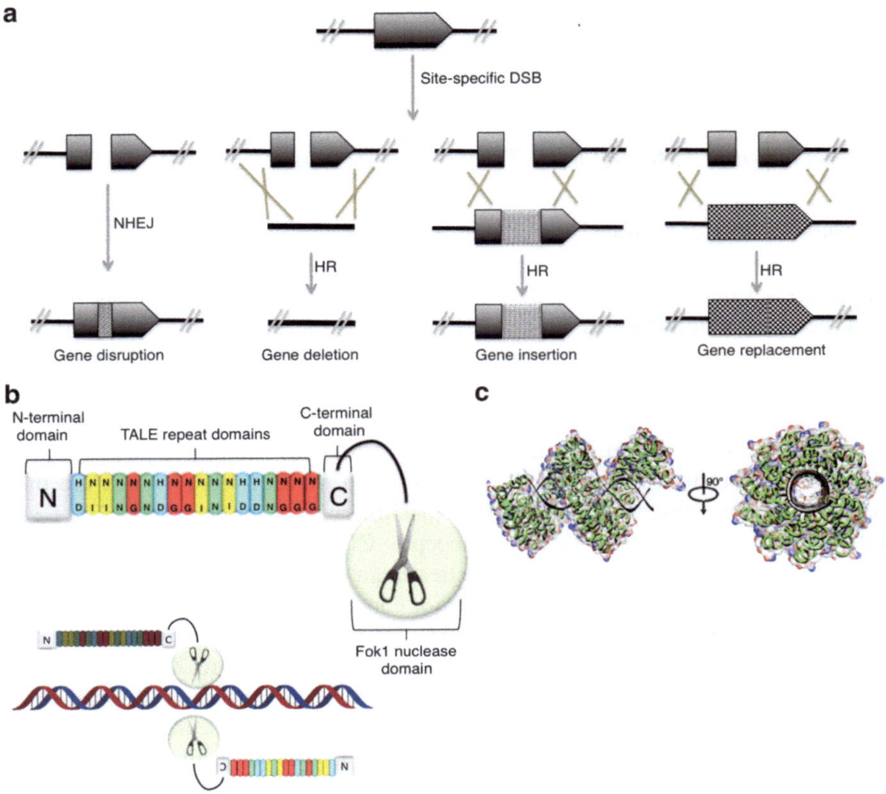

Fig. 2 TALEN-mediated genome editing [adapted from Sun and Zhao (2013)]. (**a**) Engineered nucleases trigger the site-specific chromosomal double strand breaks (DSB), which get repaired by either NHEJ or HR, resulting in gene disruption, gene deletion, gene insertion, or gene replacement. (**b**) Schematic of TALEN architecture. A TALEN is composed of a N-terminal and a C-terminal segment, a FokI catalytic domain (scissors in a *circle*). The central TALE repeat domain comprises a series of repeat units responsible for specific recognition. The two TALENs form a heterodimer in a tail-to-tail orientation at the target site for executing a site-specific DNA DSB. (**c**) Overall structure of the TALE PthXo1 central repeat domain in complex with its target site [adapted from Mak et al. (2012), protein data bank (PDB) identification code: 3UGM, Sun and Zhao (2013)]

activity of therapeutically significant gene like HIV-associated CCR5 (Perez et al. 2008). Targeted transcriptional disruption of any DNA sequence with TALENs could be harnessed to correct the faulty transcriptional machinery associated with gene of interest. However, care should be taken while designing the TALENs as lack of specificity causes off-target cleavage resulting in the chromosomal rearrangements and/or apoptosis (Mussolino et al. 2011).

4.3 DNA Binding Small Molecules as Transcriptional Modulators

DNA stores heritable information over a long period using only four nucleobases. Hence, small molecules that interact with these nucleobases can significantly alter the mutations that confer a specific neurological disorder. Pyrrole–imidazole polyamides (PIPs) are cell-permeable small molecules that have the ability to bind to the minor groove of DNA with an affinity that is comparable to that of transcription factors (Mapp et al. 2000; Kashiwazaki et al. 2012). PIPs bind to DNA following a binding rule, where an antiparallel pairing of I opposite P (I/P) recognizes a G-C base pair while a P/P pair recognizes A-T or T-A base pairs. Synthetic PIPs possessing an alkylating moiety could cause selective gene silencing (Pandian and Sugiyama 2012). PIPs could also interfere with transcriptional factors and modulate both basal and activated transcription. PIP targeting a GAA TTC repeats can potentially be used to reverse the effects and alter the FXN gene associated with the neurological disorder Friedrich ataxia (FRDA) (Burnett et al. 2006). Accordingly, PIP administration increased the FXN transcription by about two- to threefold in the cell lines of a FRDA patient who had a very low level (6–13 %) of Frataxin protein. However, this PIP does not fit into single-stranded RNA or duplex regions of RNA to suggest a lack of effect on the translation of Frataxin mRNA (Chenoweth et al. 2013). Pharmacokinetic profiles of PIPs in rat models validated its therapeutic potential (Synold et al. 2012).

5 Targeting Small Molecules as Transcriptional Activators

Targeting PIPs could also activate transcription by hampering the DBD of a repressor protein. For example, PIPs targeting the *cis*-regulatory sequence of human cytomegalovirus early protein-2 derepressed the intermediate early protein-86 (IE86)-mediated transcription (Dickinson et al. 1999). PIP-based artificial transcriptional activator tethered to a 20-residue activating peptide stimulates promoter-specific transcription (Mapp et al. 2000). A detailed overview on some of the chemical approaches to gene regulation gets reviewed well in (Lee and Mapp 2010). The programmable DNA binding small molecules and/or natural proteins often overlook the need to remodel chromatin architecture, an essential module required to achieve targeted transcriptional activation. Recently, novel strategies have emerged to achieve DNA-based transcriptional activators that could match the versatility of their natural counterparts. One such novel small molecule termed SAHA-PIP comprises (1) the cell-permeable PIPs as DNA recognition module and (2) SAHA, an epigenetically active histone deacetylases inhibitor as functional module (Ohtsuki et al. 2009). SAHA-PIP targeting the promoter region of the tumor suppressor gene p16 in HeLa cells resulted in selective hyperacetylation (Fig. 3a). Mismatched SAHA-PIP and SAHA alone did not induce such activity, suggesting

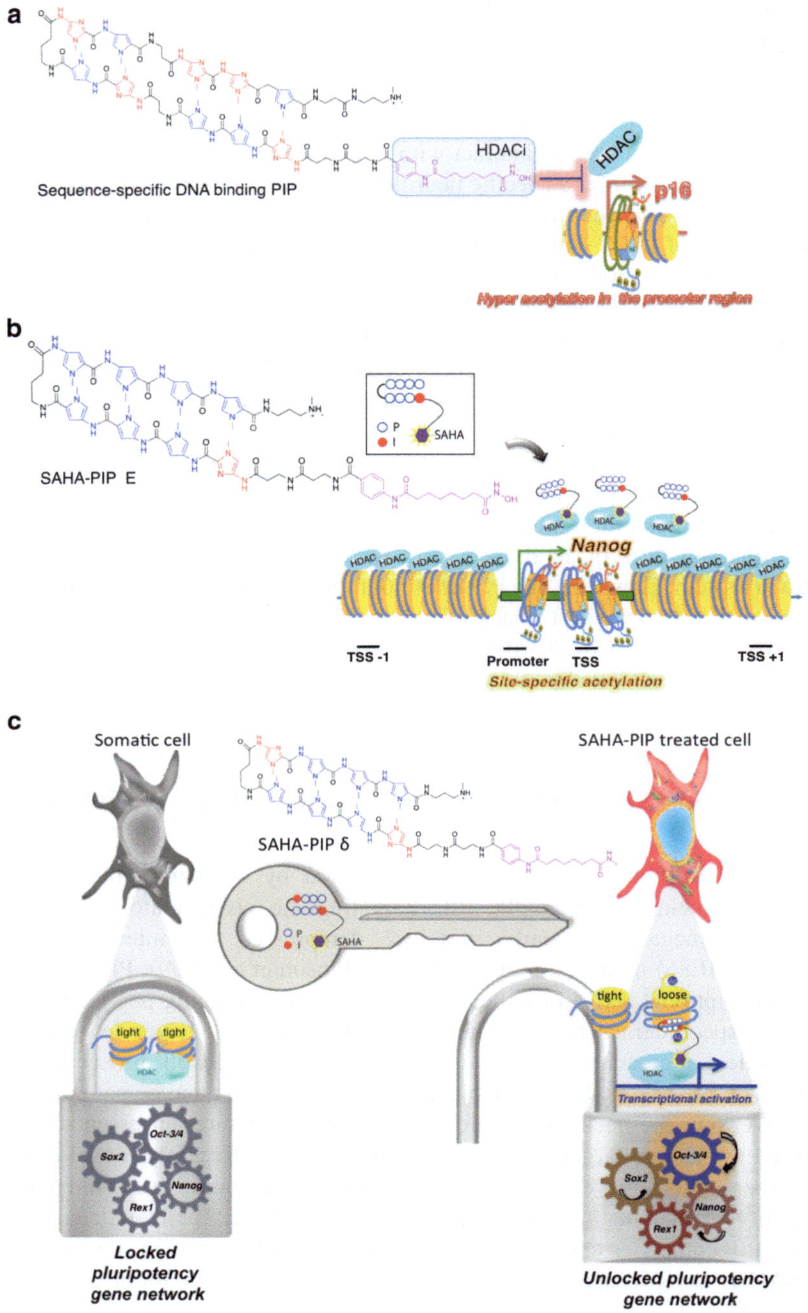

Fig. 3 SAHA-PIPs and their biological activities. (**a**) SAHA-PIP encompassing sequence-specific PIP conjugated with SAHA (*Blue box*) was shown to induce hyper acetylation (loosened chromatin in *color*) in the promoter region of p16. (**b**) SAHA-PIP called **E** (simplified illustration of structure in the *inset*) could cause site-specific acetylation by hindering the HDACs only in the

the sequence-specific capability of SAHA-PIP. Evaluation of the effect of SAHA-PIP library on the epigenetic-dependent and complicated pluripotency gene network in mouse embryonic fibroblasts (MEFs) identified certain SAHA-PIPs capable of differentially inducing pluripotency genes. SAHA-PIPs initiated epigenetic marks conferring to transcriptionally permissive chromatin including histone H3 Lys9, Lys14 acetylation, and Lys4 trimethylation (Fig. 3b) (Pandian et al. 2011).

Tunability of SAHA-PIPs got substantiated with the discovery of a potent SAHA-PIP, termed "δ" among a library of SAHA-PIPs with improved recognition of GC-rich sequences (Pandian et al 2012a, b). Microarray analysis suggested that δ-treated MEFs triggered transcriptional activation but not repression to overcome the mesenchymal epithelial transition (MET) stage, an important rate-limiting step during dedifferentiation of the somatic genome. However, interpretation of a definite mechanism is challenging as SAHA-PIPs have both HDAC inhibitory activity and sequence-specific binding ability. Nevertheless, based on the match site analysis, it is reasonable to suggest site-specific epigenetic activation to unlock the typical conserved transcriptional machinery as the probable mechanism (Fig. 3c). However, a thorough understanding of the SAHA-PIP-mediated epigenetic activation requires further studies. Synthetic PIPs gain an advantage over other natural DNA-binding proteins as effective transcriptional activators because they possess flexible covalent sites and can bind to the methylated DNA sequences and disrupt the packed chromatin structure (Han et al. 2013). Also, subtle modification in the structure of SAHA altered the specificity of "δ" towards a different HDAC enzyme (Saha et al. 2013). Thus, it is possible to attach enzyme inhibitors to PIPs for different functional purposes. Although it is difficult to predict the off-rate of PIPs inside the cells, the high binding affinity of SAHA-PIP to target DNA sequences and their ability to induce pluripotency genes even at 72 h and 100 nM concentration suggest it to be slow (Vaijayanthi et al. 2012, 2013; Pandian and Sugiyama 2013). A 15–16 base pair recognizing may specify a single site within the three billion base pair human genome (Trauger et al. 1998). Cell permeability is both an advantage and bottleneck of the PIPs as both the molecular size and pyrrole/imidazole content of the PIPs could hamper its ability to permeate cells (Pandian and Sugiyama 2012). Recent studies reveal the possibility to increase the permeability of PIPs by engineering their chemical architecture (Meier et al. 2012).

Fig. 3 (continued) promoter and transcription region and not in the region that is 1 Kbp before or after the transcription start site (TSS). (**c**) A lock and key model to demonstrate the activation of pluripotency gene network. The key (SAHA-PIP **δ**) unlocks pluripotency gene network from tight chromatin by site-specific chromatin remodeling of Oct-3/4 to trigger the core pluripotency gene network

6 Future Perspectives

Through intellectual integration of fragmented scientific disciplines and development of cutting-edge technologies, it is now possible to gain a deeper understanding about gene regulation in a wide variety of cells. Subsequently, functional roles of various transcription regulators including the noncoding RNA, microRNA, and piwi-interacting RNA have now been realized (Lee and Young 2013). The clinical significance of enforced transcriptional activation got substantiated with the Nobel Prize-winning discovery of iPS cell technology that allows us to achieve desired cell type by artificially inducing a set of defined factors (Wu et al. 2013). Consequently, there have been a growing number of reports on the development of new strategies to target and edit the transcriptional machinery pertaining to cell fate specifications (Pandian et al. 2014). Cellular reprogramming with chemicals alone got achieved recently in mouse cells and could soon be accomplished in human cells (Masuda et al. 2013). However, generation of iPS cells still requires several small molecules and concomitant expression of multiple factors. In this regard, development of SAHA-PIPs that could induce multiple pluripotency genes may enhance the generation of clinical-grade iPS cells (Pandian et al. 2012a, b). SAHA-PIPs could also be employed for differentiating iPS cells and/or somatic cells into a desired cell type. Despite the exciting progress made with TALENs, many barriers like undesirable mutations and genome-wide specificity still hinder their routine use for therapeutic applications. Moreover, permeability of the purified native or modified TALENs needs further validation. Targeting recombinant events can now be achieved by fusing a recombinase domain with TALE repeat arrays (Mercer et al. 2012). Such strategies to complement the customizable TALE repeat arrays with functional domains with epigenetic activity may achieve heritable alterations in gene expression. In this regard, programmable, epigenetically active SAHA-PIPs gain advantage as it could interact with both genetic and epigenetic environment inside the cell. Defect in the autoregulatory circuitry that maintains the cellular homeostasis may lead to various disorders. Genome engineering tools like TALENs and SAHA-PIP capable of targeting therapeutic gene(s) may potentially target the key gene(s) associated with certain disorders (Fig. 4).

To achieve such complex feat(s), factors such as cell permeability, accessibility, and stochastic variations in epigenome of a cell should be considered during their design and development. Nevertheless, the remarkable success with these innovative gene editing tools may encourage researchers to invent such versatile transcriptional activators capable of therapeutic gene modulation.

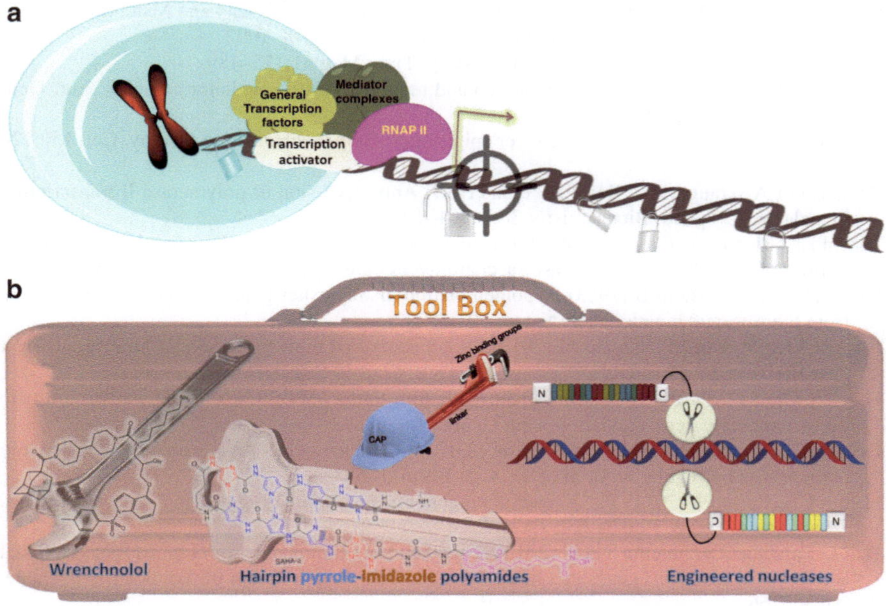

Fig. 4 Genome engineering. (**a**) Targeted transcriptional activation within the cellular genome (epigenetic modifications illustrated as locks). (**b**) Customizable DNA-binding tools available to edit genome DNA, Wrenchnolol (Shimogawa et al. 2004) (illustrated as wrench), SAHA-PIP (Key), Engineered nucleases as scissors, Transcriptional regulation by HAT-HDAC [Zinc monkey wrench Kalin et al. (2009)]

References

Amir RE, Van den Veyver IB, Wan M et al (1999) Rett syndrome is caused by mutations in X-linked MECP2, encoding methyl-CpG-binding protein 2. Nat Genet 23:185–188

Baker M (2012) Gene-editing nucleases. Nat Methods 9:23–26

Barbieri CE, Baca SC, Lawrence MS et al (2012) Exome sequencing identifies recurrent SPOP, FOXA1 and MED12 mutations in prostate cancer. Nat Genet 44:685–689

Beffert U, Weeber EJ, Durudas A et al (2005) Modulation of synaptic plasticity and memory by Reelin involves differential splicing of the lipoprotein receptor Apoer2. Neuron 47:567–579

Bilu Y, Barkai N (2005) The design of transcription-factor binding sites is affected by combinatorial regulation. Genome Biol 6:R103

Bogdanove AJ, Schornack S, Lahaye T (2010) TAL effectors: finding plant genes for disease and defense. Curr Opin Plant Biol 13:394–401

Burnett R, Melander C, Puckett JW et al (2006) DNA sequence-specific polyamides alleviate transcription inhibition associated with long GAA · TTC repeats in Friedreich's ataxia. Proc Natl Acad Sci U S A 103:11497–11502

Ceol CJ, Houvras Y, Jane-Valbuena J et al (2011) The histone methyltransferase SETDB1 is recurrently amplified in melanoma and accelerates its onset. Nature 471:513–517

Chenoweth DM, Meier JL, Dervan PB (2013) Pyrrole-Imidazole polyamides distinguish between double-helical DNA and RNA. Angew Chem Int Ed 52:415–418

Chi Y, Huddleston MJ, Zhang X et al (2001) Negative regulation of Gcn4 and Msn2 transcription factors by Srb10 cyclin-dependent kinase. Genes Dev 15:1078–1092
Cloud J (2010) Why your DNA isn't your destiny. Time Mag 61:175–189
Conaway RC, Conaway JW (2011) Function and regulation of the mediator complex. Curr Opin Genet Dev 21:225–230
Dawson MA, Kouzarides T (2012) Cancer epigenetics: from mechanism to therapy. Cell 150:12–27
Dickinson LA, Trauger JW, Baird EE et al (1999) Anti-repression of polymerase II transcription by a designed ligand. Biochemistry 38:10801–10807
Durrin LK, Mann RK, Grunstein M (1992) Nucleosome loss activates CUP1 and HIS3 promoters to fully induced levels in the yeast Saccharomyces cerevisiae. Mol Cell Biol 12:1621–1629
Feinberg AP, Vogelstein B (1983) Hypomethylation distinguishes genes of some human cancers from their normal counterparts. Nature 301:89–92
Frances MB, Mukund T, Satyajit M (2012) Evolutionary cell biology: lessons from diversity. Nat Cell Biol 14:651
Goh YS, Grants JM (2012) Mutations in the mediator subunit MED23 link intellectual disability to immediate early gene regulation. Clin Genet 81:430–432
Grueter CE, van Rooij E, Johnson BA et al (2012) A cardiac microRNA governs systemic energy homeostasis by regulation of MED13. Cell 149:671–683
Hamm CA, Costa FF (2011) The impact of epigenomics on future drug design and new therapies. Drug Discov Today 16:626–635
Han L, Pandian GN, Junetha S et al (2013) A synthetic small molecule for targeted transcriptional activation of germ cell genes in a human somatic cell. Angew Chem Int Ed. doi:10.1002/anie.201306766
He A, Kong SW, Ma Q et al (2011) Co-occupancy by multiple cardiac transcription factors identifies transcriptional enhancers active in heart. Proc Natl Acad Sci U S A 108:5632–5637
Hockemeyer D, Wang H, Kiani S et al (2011) Genetic engineering of human pluripotent cells using TALE nucleases. Nat Biotechnol 29:731–734
Iwakuma T, Lozano G, Flores ER (2005) Li-Fraumeni syndrome: a p53 family affair. Cell Cycle 4:865–867
Jacob F, Monod J (1961) Genetic regulatory mechanisms in the synthesis of proteins. J Mol Biol 3:318–356
Jiang YH, Bressler J, Beaudet AL (2004) Epigenetics and human disease. Annu Rev Genomics Hum Genet 5:479–510
Joung JK, Sander JD (2013) TALENs: a widely applicable technology for targeted genome editing. Nat Rev Mol Cell Biol 14:49–55
Kalin JH, Butler KV, Kozikowski AP (2009) Creating zinc monkey wrenches in the treatment of epigenetic disorders. Curr Opin Chem Biol 13:263–271
Kashiwazaki G, Bando T, Yoshidome T et al (2012) Synthesis and biological properties of highly sequence specific-alkylating N-methylpyrrole–N-methylimidazole polyamide conjugates. J Med Chem 55:2057–2066
Keaveney M, Struhl K (1998) Activator-mediated recruitment of the RNA polymerase II machinery is the predominant mechanism for transcriptional activation in yeast. Mol Cell 1:917–924
Lai CSL, Fisher SE, Hurst JA et al (2001) A forkhead-domain gene is mutated in a severe speech and language disorder. Nature 413:519–523
Lamb MJ, Jablonka E (2005) Evolution in four dimensions: genetic, epigenetic, behavioral, and symbolic variation in the history of life. MIT Press, Cambridge, MA, pp 5–34
Lee LW, Mapp AK (2010) Transcriptional switches: chemical approaches to gene regulation. J Biol Chem 285:11033–11038
Lee TI, Young RA (2013) Transcriptional regulation and its misregulation in disease. Cell 152:1237–1251
Li J, Chen C, Chen C et al (2011) Neurotensin receptor 1 gene (NTSR1) polymorphism is associated with working memory. PLoS ONE 6(3):e17365

Lin CY, Lovén J, Rahl PB et al (2012) Transcriptional amplification in tumor cells with elevated c-Myc. Cell 151:56–67
Liu XY, Wang J, Zheng JH et al (2011) Involvement of a novel GATA4 mutation in atrial septal defects. Int J Mol Med 28:17–23
Luo Z, Lin C, Guest E et al (2012) The super elongation complex family of RNA polymerase II elongation factors: gene target specificity and transcriptional output. Mol Cell Biol 32:2608–2617
Maestro MA, Cardalda C, Boj SF et al (2007) Distinct roles of HNF1beta, HNF1alpha, and HNF4alpha in regulating pancreas development, beta-cell function and growth. Endocr Dev 12:33–45
Majmudar CY, Lum JK, Prasov L et al (2005) Functional specificity of artificial transcriptional activators. Chem Biol 12:313–321
Mak AN, Bradley P, Cernadas RA et al (2012) The crystal structure of TAL effector PthXo1 bound to its DNA target. Science 335:716–719
Mannini L, Musio A (2011) The dark side of cohesin: the carcinogenic point of view. Mutat Res 728:81–87
Mapp AK, Ansari AZ, Ptashne M et al (2000) Activation of gene expression by small molecule transcription factors. Proc Natl Acad Sci U S A 97:3930–3935
Masuda S, Wu J, Hishida T et al (2013) Chemically induced pluripotent stem cells (CiPSCs): a transgene-free approach. J Mol Cell Biol 5:354–355
Maurano MT, Humbert R, Rynes E et al (2012) Systematic localization of common disease-associated variation in regulatory DNA. Science 337:1190–1195
McCulley DJ, Black BL (2012) Transcription factor pathways and congenital heart disease. Curr Top Dev Biol 100:253–277
Meier JL, Montgomery DC, Dervan PB (2012) Enhancing the cellular uptake of Py–Im polyamides through next-generation aryl turns. Nucleic Acids Res 40:2345–2356
Mercer AC, Gaj T, Fuller RP (2012) Chimeric TALE recombinases with programmable DNA sequence specificity. Nucleic Acids Res 40:11163–11172
Miller JC, Tan S, Qiao G et al (2011) A TALE nuclease architecture for efficient genome editing. Nat Biotechnol 29:143–148
Min JW, Haegeman G, Ysebaert M et al (1972) Nucleotide sequence of the gene coding for the bacteriophage MS2 coat protein. Nature 237:82–88
Minichiello L (2009) TrkB signalling pathways in LTP and learning. Nat Rev Neurosci 10:850–860
Mussolino C, Morbitzer R, Lütge F et al (2011) A novel TALE nuclease scaffold enables high genome editing activity in combination with low toxicity. Nucleic Acids Res 39:9283–9293
Odom DT, Zizlsperger N, Gordon DB et al (2004) Control of pancreas and liver gene expression by HNF transcription factors. Science 303:1378–1381
Ohtsuki A, Kimura MT, Minoshima M et al (2009) Synthesis and properties of PI polyamide-SAHA conjugate. Tetrahedron Lett 50:7288–7292
Olson EN (2006) Gene regulatory networks in the evolution and development of the heart. Science 313:1922–1927
Ouyang P, Saarel E, Bai Y et al (2011) A de novo mutation in NKX2.5 associated with atrial septal defects, ventricular noncompaction, syncope and sudden death. Clin Chim Acta 412:170–175
Pandian GN, Sugiyama H (2012) Programmable genetic switches to control transcriptional machinery of pluripotency. Biotechnol J 7:798–809
Pandian GN, Sugiyama H (2013) Strategies to modulate heritable epigenetic defects in cellular machinery: lessons from nature. Pharmaceuticals 6:1–24
Pandian GN, Shinohara K, Ohtsuki A et al (2011) Synthetic small molecules for epigenetic activation of pluripotency genes in mouse embryonic fibroblasts. ChemBioChem 12:2822–2828

Pandian GN, Ohtsuki A, Bando T et al (2012a) Development of programmable small DNA-binding molecules with epigenetic activity for induction of core pluripotency genes. Bioorg Med Chem 20(8):2656–2660
Pandian GN, Nakano Y, Sato S et al (2012b) A synthetic small molecule for rapid induction of multiple pluripotency genes in mouse embryonic fibroblast. Sci Rep 2:e544
Pandian GN, Taniguchi J, Junetha S et al (2014) Distinct DNA-based epigenetic switches trigger transcriptional activation of silent genes in human dermal fibroblasts. Sci Rep 4:e3843
Perez EE, Wang J, Miller JC et al (2008) Establishment of HIV-1 resistance in CD4+ T cells by genome editing using zinc-finger nucleases. Nat Biotechnol 26:808–816
Pfeifer A, Verma IM (2001) Gene therapy: promises and problems. Annu Rev Genomics Hum Genet 2:177–211
Ptashne M, Gann A (2002) Genes and signals. Cold Spring Harbor Laboratory, Cold Spring Harbor, NY
Rando OJ (2012) Combinatorial complexity in chromatin structure and function: revisiting the histone code. Curr Opin Genet Dev 22:148–155
Reyon D, Tsai SQ, Khayter C et al (2012) FLASH assembly of TALENs for high-throughput genome editing. Nat Biotechnol 30:460–465
Robinson KM, Kadonaga JT (1998) The use of chromatin templates to recreate transcriptional regulatory phenomena in vitro. Biochim Biophys Acta 1378:1–6
Rump P, Niessen RC, Verbruggen KT et al (2011) A novel mutation in MED12 causes FG syndrome (OpitzKaveggia syndrome). Clin Genet 79:183–188
Ryan MM, Ryan B, Smith MK et al (2012) Temporal profiling of gene networks associated with the late phase of long-term potentiation in vivo. PLoS ONE 7:e40538
Saha A, Pandian GN, Sato S et al (2013) Synthesis and biological evaluation of a targeted DNA-binding transcriptional activator with HDAC8 inhibitory activity. Bioorg Med Chem 21:4201–4209
Sanda T, Lawton LN, Barrasa MI et al (2012) Core transcriptional regulatory circuit controlled by the TAL1 complex in human T cell acute lymphoblastic leukemia. Cancer Cell 22:209–221
Santen GW, Kriek M, van Attikum H (2012) SWI/SNF complex in disorder: SWItching from malignancies to intellectual disability. Epigenetics 7:1219–1224
Schreiber SL (2005) Nat Chem Biol 1:64–66
Sebastiano V, Maeder ML, Angstman JF et al (2011) In situ genetic correction of the sickle cell anemia mutation in human induced pluripotent stem cells using engineered zinc finger nucleases. Stem Cells 29:1717–1726
Seitan VC, Merkenschlager M (2012) Cohesin and chromatin organisation. Curr Opin Genet Dev 22:93–100
Shimogawa H, Kwon Y, Mao Q et al (2004) A wrench-shaped synthetic molecule that modulates a transcription factor–coactivator interaction. J Am Chem Soc 126:3461–3471
Soldner F, Laganière J, Cheng AW et al (2011) Generation of isogenic pluripotent stem cells differing exclusively at two early onset Parkinson point mutations. Cell 146:318–331
Sun N, Zhao H (2013) Transcription activator-like effector nucleases (TALENs): a highly efficient and versatile tool for genome editing. Biotechnol Bioeng 110:1811–1821
Swaby JA, Silversides CK, Bekeschus SC et al (2011) Complex congenital heart disease in unaffected relatives of adults with 22q11.2 deletion syndrome. Am J Cardiol 107:466–471
Synold TW, Xi B, Wu J et al (2012) Single-dose pharmacokinetic and toxicity analysis of pyrrole–imidazole polyamides in mice. Cancer Chemother Pharmacol 70:617–625
Thandavarayan RA, Giridharan VV, Sari FR et al (2011) Depletion of 14-3-3 protein exacerbates cardiac oxidative stress, inflammation and remodeling process via modulation of MAPK/NF-κB signaling pathways after streptozotocin-induced diabetes mellitus. Cell Physiol Biochem 28:911–922
Trauger JW, Baird EE, Dervan PB (1998) Recognition of 16 base Pairs in the minor groove of DNA by a pyrrole-imidazole polyamide dimer. J Am Chem Soc 120:3534–3535

Vaijayanthi T, Bando T, Pandian GN et al (2012) Progress and prospects of pyrrole-imidazole polyamide–fluorophore conjugates as sequence-selective DNA probes. ChemBioChem 13:2170–2185

Vaijayanthi T, Bando T, Hashiya K et al (2013) Design of a new fluorescent probe: pyrrole/imidazole hairpin polyamides with pyrene conjugation at their γ-turn. Bioorg Med Chem 21:852–855

Wu ZQ, Belanger G, Brennan BB et al (2003) Targeting the transcriptional machinery with unique artificial transcriptional activators. J Am Chem Soc 125:12390–12391

Wu YL, Pandian GN, Ding YP et al (2013) Clinical grade iPS cells: need for versatile small molecules and optimal cell sources. Chem Biol 20:1311–1322

Yusa K, Rashid ST, Marchand HS et al (2011) Targeted gene correction of α1-antitrypsin deficiency in induced pluripotent stem cells. Nature 478:391–394

Zhou Q, Li T, Price DH (2012) RNA polymerase II elongation control. Annu Rev Biochem 81:119–143

Interaction of DNA Intramolecular Structures with Their Complementary Strands: A Thermodynamic Approach for the Control of Gene Expression

Irine Khutsishvili, Sarah E. Johnson, Calliste Reiling, Iztok Prislan, Hui-Ting Lee, and Luis A. Marky

Contents

1 Introduction 368
2 Materials and Methods 369
  2.1 Materials 369
  2.2 Isothermal Titration Calorimetry 370
  2.3 Differential Scanning Calorimetry 371
  2.4 UV Melting Curves 371
3 Results and Discussion 372
  3.1 Overall Experimental Approach 372
  3.2 Targeting a Simple Hairpin Loop 372
  3.3 Targeting Hairpin Loops Containing Bulges and Internal Loops in Their Stem ... 376
  3.4 Targeting of a Three-Way Junction 378
  3.5 Targeting a DNA Pseudoknot 379
  3.6 Targeting a Combined Triplex–Duplex Complex 379
  3.7 Thermodynamic Profiles for the Targeting Reactions 380
4 Conclusions 381
References 382

Abstract The folding of mRNA sequences into secondary/tertiary structures plays an important role in RNA and DNA function and expression. Disruption of these structures can potentially be used in the control of gene expression. However, a detailed understanding of the physicochemical properties of nucleic acid structures is needed before this targeting approach can be used. In this chapter, we have examined six intramolecular DNA structures and have investigated their reaction thermodynamics with single strands partially complementary to their stems and loops. We measured the heat of each reaction directly using isothermal titration calorimetry. These are compared with the heat measured indirectly using Hess

I. Khutsishvili • S.E. Johnson • C. Reiling • I. Prislan • H.-T. Lee • L.A. Marky (✉)
Department of Pharmaceutical Sciences, College of Pharmacy, University of Nebraska Medical Center, Omaha, NE 68198-6025, USA
e-mail: lmarky@unmc.edu

V.A. Erdmann et al. (eds.), *Chemical Biology of Nucleic Acids*, RNA Technologies,
DOI 10.1007/978-3-642-54452-1_20,

cycles obtained from differential scanning calorimetric unfolding thermodynamic profiles. Each reaction yielded favorable free energy terms that were enthalpy driven, indicating each complementary strand was able to disrupt the intramolecular complex. In short, we have developed a thermodynamic approach that can be used in the control of gene expression that targets the loops of secondary structures formed by mRNA.

Keywords Targeting DNA secondary structures • Antisense • Heat • Thermodynamics

1 Introduction

The formation of DNA secondary structures, such as hairpin loops, triplexes, G-quadruplexes, and i-motifs, is well documented (Bochman et al. 2012; Darlow and Leach 1998; Gehring et al. 1993; Rich 1993). These non-canonical DNA secondary structures have been postulated to be involved in a variety of biological functions (Juliano et al. 2001; Crooke 1999; Helene 1991, 1994; Firulli et al. 1994; Fox 1990; Han and Hurley 2000; Mills et al. 2002; Bock et al. 1992; Wang et al. 1993; Rando et al. 1995) and may also be important causal factors in human diseases such as cancer and the aging of the cell (Mills et al. 2002; Simonsson et al. 1998; Huard and Autexier 2002; Brown et al. 1998; Beal and Dervan 1991; Zahler et al. 1991). The targeting of these structures with nucleic acid oligonucleotides may stop their biological function (Folini et al. 2002; Koeppel et al. 2001). In general, nucleic acid oligonucleotides, as drugs, present an exquisite selectivity and are able to discriminate targets that differ by a single nucleotide (Crooke 1999). Therefore, oligonucleotides may be used for the control of gene expression (Helene 1991, 1994). There are several approaches for the use of oligonucleotides as modulators of gene expression: the antigene, antisense, and siRNA, all target-specific nucleic acid structures (Mahato et al. 2005). For instance, in the antisense strategy, a single strand binds to messenger RNA, forming a DNA/RNA hybrid duplex that inhibits translation, by sterically blocking the correct assembly of the translational machinery or by inducing an RNase H-mediated cleavage of the DNA/RNA hybrid duplex (Helene 1991).

There are advantages and disadvantages in these strategies. In the direct targeting of a gene, the antigene strategy offers some advantages over the other strategies; there are only two copies of a particular gene, whereas there is a large continuous supply of the mRNA gene transcript. Therefore, blocking the transcription of the gene itself prevents repopulation of the mRNA pool, allowing a more efficient and lasting inhibition of gene expression (Fox 2000; Vasquez et al. 2001). The main disadvantage is that the oligonucleotide needs to cross the nuclear membrane and access its DNA target within the densely packed chromatin structure

(Brown et al. 1998); while in the targeting of mRNA, there is a need to know the potential secondary/tertiary structures that the mRNA target sequence is forming. Common disadvantages of the use of oligonucleotides for targeting purposes are (a) oligonucleotides need to cross the hydrophobic cellular membranes; for instance, nucleic acid duplexes do not cross these membranes because of their hydrophilicity (Cantor and Schimmel 1980) and (b) the fast degradative action of nucleases. These disadvantages can be circumvented by using single strands that are somewhat hydrophobic, allowing them to interact better with polycationic micelles and/or enabling them to cross cellular membranes. These polycations can be used as delivery vectors, protecting the oligonucleotide from the action of nucleases. Furthermore, modification of the oligonucleotide phosphate or sugar groups also confers protection against nuclease degradation.

From a thermodynamic point of view, successful control of gene expression depends on the effective binding of a DNA sequence to its target with tight affinity and specificity. This can be provided by using a long sequence of 15–20 bases in length (Crooke 1999); strong specificity is conferred by hydrogen bonding in the formation of Watson–Crick and/or Hoogsteen base pairs, while high affinity is provided by the large negative free energy upon formation of a duplex or triplex; thereby, competing efficiently with the proteins involved in transcription or translation.

In this work, we have investigated reactions involving the interaction of a variety of intramolecular DNA structures (hairpin loops, hairpins containing bulges and internal loops, three-arm junctions, pseudoknots, and joined triplex–duplex complexes) with their respective complementary strands. We are actually mimicking the targeting of the potential secondary structures formed by mRNA using complementary strands. We would like to answer the following questions: Is the single strand able to invade the intramolecular secondary structure, disrupt it, and form a stable duplex product? Consequently, what are the thermodynamic contributions favoring the formation of these product duplexes? To this end, we have used a combination of isothermal titration (ITC) and differential scanning (DSC) calorimetric techniques to determine reaction enthalpies and their associated free-energy terms. The results show that all reactions yielded favorable free-energy contributions, resulting from exothermic enthalpies, which are due to the formation of additional base-pair stacks in the formation of duplex products involving the loop unpaired nucleobases of the putative secondary structures.

2 Materials and Methods

2.1 Materials

Oligonucleotides were synthesized by the Core Synthetic Facility of the Eppley Research Institute at UNMC or Integrated DNA Technologies, HPLC purified, and

Table 1 Sequences, designations, and molar extinction coefficients of ODNs[a]

5′ to 3′ Oligonucleotide sequence	Designation	$\varepsilon^{260} \times 10^{-5}$ (M^{-1} cm^{-1})
d(GCGCT_5GCGC)	*Hp*	1.08
d(C$\mathrm{A_5}$GCGC)	*HpCS*	1.10
d(GCGCT_5GTAACT_5GTTACGCGC)	*Blg_5*	2.49
d(AAGTTAC$\mathrm{A_5}$GCGC)	*BlgIICS*	1.77
d(GCGCT_5GTAACT_5GTTACT_5GCGC)	*IL_{10}*	2.88
d(G$\mathrm{A_3T_2}$GCGCT_5GCGCGTGCT_5GCAC$\mathrm{A_2T_3}$C)	*Hammer*	3.47
d(G$\mathrm{A_3T_2}$GAGC$\mathrm{A_5}$GC)	*HammerCS*	1.94
d(TCTCT$T_5$$\mathrm{A_7}$AGAGA$T_5$$\mathrm{T_7}$)	*PsKn*	3.31
d(C$\mathrm{T_7A_5}$AGAGA)	*PsKnCS*	1.81
d($\mathrm{A_7}$GTAAC*GCAA*GTTAC$\mathrm{T_7}$*C*T_3*C*$\mathrm{T_7}$)	*Complex*	3.86
d($\mathrm{A_7}$G$\mathrm{A_3}$G$\mathrm{A_7}$GTAACTTGCG)	*ComplexCS*	3.54

[a]Italics indicate loop sequences

desalted by column chromatography using G-10 Sephadex exclusion chromatography. The sequences of oligonucleotides (ODNs) used in this work and their designations are shown in Table 1. ODN solutions were prepared by dissolving the dry and desalted ODNs in buffer. Before each experiment, these solutions were heated to 90 °C and cooled to room temperature. Their concentrations were determined at 260 nm and 90 °C using an AVIV 14DS spectrophotometer. Molar extinction coefficients, shown in the last column of Table 1, were obtained by extrapolation of the tabulated values for dimers and monomeric bases (Cantor et al. 1970) at 25 °C to 90 °C, using procedures reported previously (Lee et al. 2008, 2011; Marky and Kupke 2000; Marky et al. 1983, 2007). The extinction coefficients of the duplexes (not shown) are simply calculated by averaging the molar extinction coefficients of its component complementary strands. Measurements were made in appropriate buffer solutions: 10 mM sodium phosphate with 0.1 M NaCl at pH 7.0, adjusted to the desired salt concentration with NaCl, and pH with either NaOH or HCl. Reagent grade salts from Sigma were used without further purification.

2.2 Isothermal Titration Calorimetry

The heat of reaction for a particular DNA secondary structure with its complementary strand was measured directly by isothermal titration calorimetry (ITC) using the iTC_{200} (GE-Healthcare). A 40 μL syringe was used for four to five injections of 1.00 μL (*Blg_5* and *IL_{10}*) or 2.00 μL (*Hp*, *Hammer*, *PsKn* and *Complex*) of titrant with a similar concentration to the oligomer in the 0.2-mL reaction cell, and mixing was affected by stirring this syringe at 1,000 rpm. Complementary strands were placed in the reaction cell to decrease their endotermic heat contribution from disrupting base–base stacking interactions, if any. The reaction heat of each injection was measured by integration of the area of the injection curve, corrected for the dilution heat of the titrant, and normalized by the moles of titrant added to yield the reaction enthalpy, ΔH_{ITC} (Wiseman et al. 1989). All titrations were designed to obtain mainly the ΔH_{ITC} for each targeting reaction, by averaging the heat of the last four to six injections, which corresponds to formation of duplex products.

2.3 Differential Scanning Calorimetry

The total heat required for the unfolding of each oligonucleotide, either secondary structure or duplex product, of these targeting reactions was measured with a VP-DSC differential scanning calorimeter (DSC) from Microcal (Northampton, MA). Standard thermodynamic profiles and transition temperatures, T_{M}s, are obtained from a DSC experiment using the following relationships (Marky and Breslauer 1987): $\Delta H_{\mathrm{DSC}} = \int \Delta C_{\mathrm{P}}(T)\ \mathrm{d}T$; $\Delta S_{\mathrm{DSC}} = \int (\Delta C_{\mathrm{P}}(T)/T)\ \mathrm{d}T$, and the Gibbs equation, $\Delta G^{\circ}_{(\mathrm{T})} = \Delta H_{\mathrm{DSC}} - T\ \Delta S$; where $\Delta C_{\mathrm{P}}(T)$ is the change in the heat capacity of the ODN solution during the unfolding process, ΔH_{DSC} is the unfolding enthalpy, ΔS_{DSC} is the entropy of unfolding, and $\Delta G^{\circ}_{(\mathrm{T})}$ is the unfolding free energy at temperature T, and with the assumption of negligible heat capacities.

2.4 UV Melting Curves

Absorbance versus temperature profiles were measured at 260 nm and/or 275 nm, with a thermoelectrically controlled Aviv Spectrophotometer Model 14DS UV–VIS (Lakewood, NJ). The temperature was scanned at a heating rate of ~0.6 °C/min, and shape analysis of the melting curves yielded T_{M}s (Marky and Breslauer 1987). The transition molecularity for the unfolding of a particular complex was obtained by monitoring the T_{M} as a function of the strand concentration. Intramolecular complexes have T_{M}s that are independent of strand concentration, while intermolecular complexes have T_{M}s dependent on strand concentration (Marky and Breslauer 1987). Furthermore, these T_{M} dependencies were used to estimate the T_{M}s of the concentration of duplex products formed in the ITC titrations, allowing corrections for the experimental reaction heats, if any. This data is not shown for sake of space.

3 Results and Discussion

3.1 Overall Experimental Approach

A cartoon of all targeting reactions is shown in Fig. 1a. Each reaction was designed to yield a favorable free energy contribution in enthalpy-driven reactions. We have concentrated on measuring the T_Ms and reaction heats, because $\Delta G°$ depends on these two parameters, as will be illustrated later. However, we also determined standard thermodynamic profiles for each targeting reaction. Initially, we used ITC titrations to measure directly the heat (ΔH_{ITC}) for each reaction, by averaging the heats of the last four to five injections, under unsaturated conditions, of each titration. We then use DSC unfolding to determine the T_Ms and unfolding heats for the reactants and products of a given reaction. The unfolding thermodynamic profiles were then used to set up Hess cycles (Fig. 1b) to yield the enthalpy (ΔH_{HC}) and other parameters for each targeting reaction, allowing us to compare the results obtained from DSC experiments with the ITC data. UV melting curves were carried out for the oligonucleotide reactants and products as a function of strand concentration to determine their T_M dependencies on strand concentration. The resulting data of these latter experiments were then used to correct the ΔH_{HC}s for both the ODN concentration used in the ITC experiments, by obtaining the T_Ms for the unfolding of the duplex products at the concentration of the ITC experiments, if needed.

To determine the free energy, $\Delta G°_{ITC}$, for each targeting reaction, we use the following relationship, $\Delta G°_{ITC} = \Delta G°_{HC}\ (\Delta H_{ITC}/\Delta H_{HC})$ (Lee et al. 2008, 2011; Marky et al. 1996; Rentzeperis et al. 1994), where $\Delta G°_{HC}$ is calculated from the DSC data in a similar way as with the ΔH_{HC} terms. Finally, the Gibbs equation is used to determine the $T\Delta S_{ITC}$ parameter, where T is the temperature of the ITC experiments.

3.2 Targeting a Simple Hairpin Loop

As shown in the first reaction of the cartoon of Fig. 1a (Fig. 1a-1), a simple hairpin loop (*Hp*) is targeted with a partial complementary strand (*HpCS*) to form a duplex with a dangling end (*Hp-Dup*). In the ITC titrations, *Hp* was placed in the reaction cell with a similar concentration of *HpCS* in the syringe, the time between injections was set at 4–12 min to allow for complete reaction at 10 °C (Fig. 2a). After correcting each injection for the dilution heat of the titrant (*HpCS*), the average heat of the four injections yielded an exothermic heat of 7.0 μcal, which after normalization by the concentration of the limiting reagent yielded an average ΔH_{ITC} of −27.7 kcal/mol for the formation of the duplex product (*Hp-Dup*). The favorable heat of this reaction is due to a net compensation of exothermic heat from the

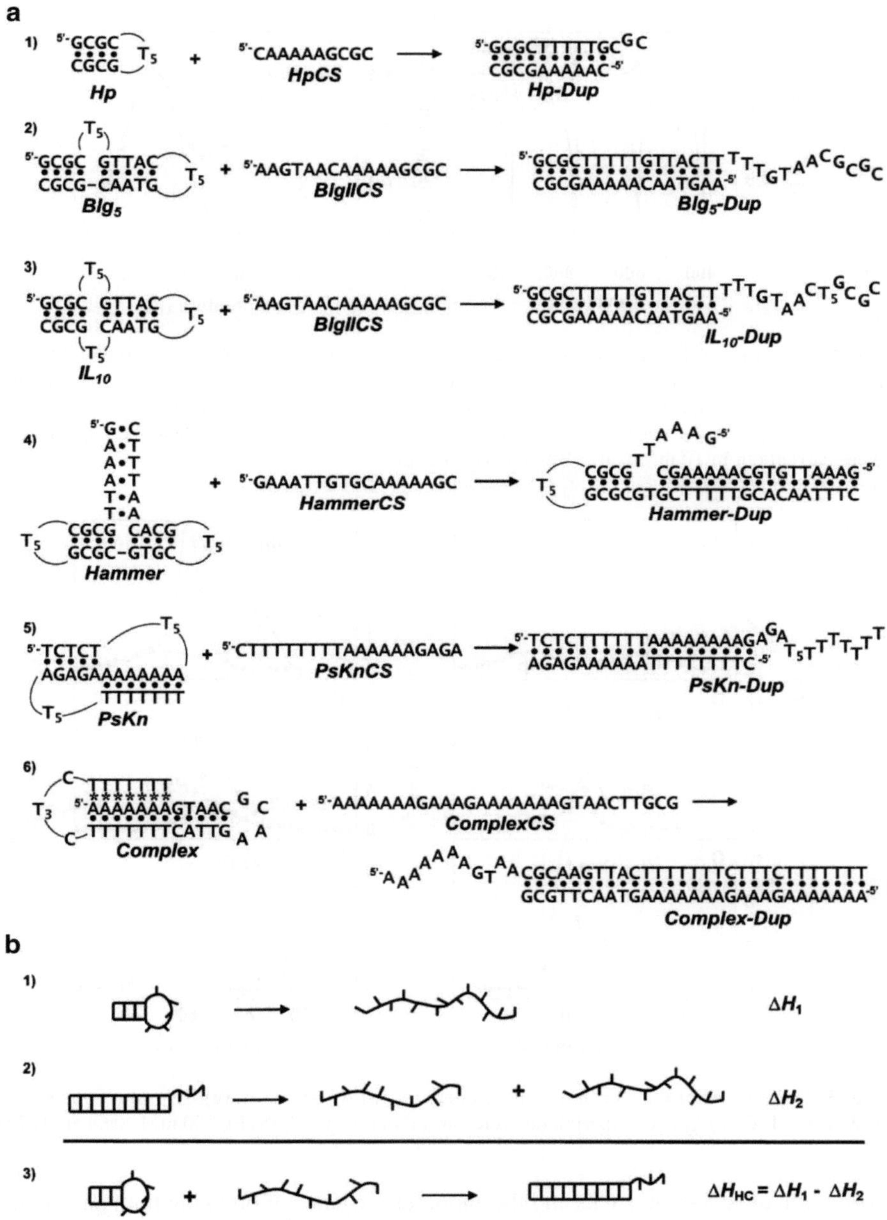

Fig. 1 Cartoon of targeting reactions. Complementary strands were used to target the following DNA secondary structures: (**a**)—(1) *Hp*, (2) Blg_5, (3) IL_{10}, (4) *Hammer*, (5) *PsKn*, and (6) *Complex*. (**b**) Example of a Hess cycle, the coupling of the DSC unfolding reactions of reactants and products yields the ITC targeting reaction

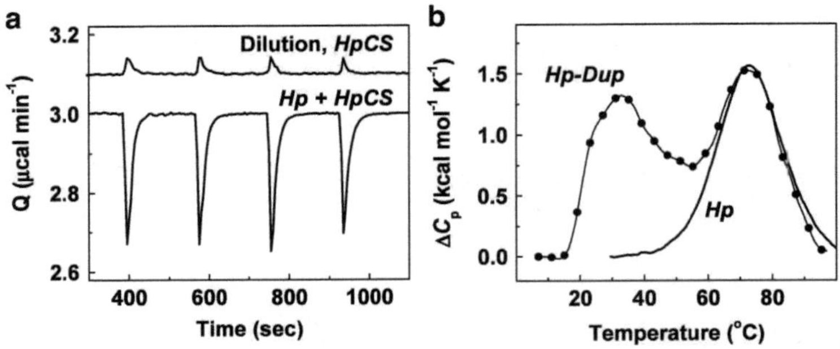

Fig. 2 Targeting of the hairpin loop *Hp*. (**a**) ITC titration curve of *HpCS* (syringe, 0.05 mM) into *Hp* (reaction cell, 0.06 mM) and the corresponding *HpCS* dilution curve, both at 10 °C. *Curves* are offset for clarity. The ITC titrations for the other secondary structures have similarly shaped curves and for the sake of space are not shown. (**b**) DSC curves of reactant and product. All experiments were carried out in 10 mM NaPi, 100 mM NaCl at pH 7.0

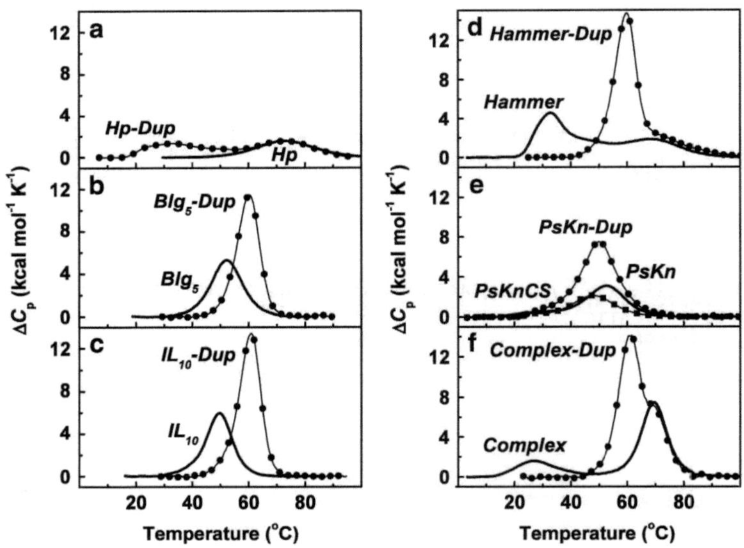

Fig. 3 DSC curves of the reactant(s) and product for (**a**) *Hp*, (**b**) Blg_5, (**c**) IL_{10}, (**d**) *Hammer*, (**e**) *PsKn*, and (**f**) *Complex*. All experiments were carried out in 10 mM NaPi, 100 mM NaCl at pH 7.0

formation of base-pair stacks and the endothermic heats of breaking both base-pairs stacks in the hairpin stem and base–base stacking in the loop.

The DSC unfolding curves of the reactant and product are shown in Figs. 2b and 3a, and the curves show transitions with negligible heat capacity effects between the initial and final states. *Hp* unfolded in a reversible monophasic transition with a T_M of 73.3 °C and ΔH_{DSC} of 38.1 kcal/mol, indicating 100 % formation of *Hp* at temperatures below 35 °C. *Hp-Dup* unfolds in a reversible biphasic transition with T_Ms of 34.9 °C and 72.4 °C and ΔH_{DSC}s of 37.5 kcal/mol and 39.8 kcal/mol, respectively; *Hp-Dup* is formed 100 % at temperatures below 18 °C. The total enthalpy value is in agreement with the enthalpy estimated from nearest-neighbor parameters of 77.9 kcal/mol (Breslauer et al. 1986; SantaLucia et al. 1996). Furthermore, the similarity in T_M and ΔH_{DSC} values of the second transition of *Hp-Dup* with the *Hp* transition suggests that the first transition of *Hp-Dup* corresponds to the unfolding of its duplex state into *Hp* and *HpCS* and followed by the unfolding of *Hp* (Fig. 2a).

UV melting experiments for the hairpin reactant and the duplex product were carried out as a function of strand concentration (data not shown). The optical T_Ms for the hairpin remains constant, while the T_M of each product depends on strand concentration, indicating that the reactant forms an intramolecular hairpin loop while the product forms a bimolecular duplex at low temperatures. We could use these T_M dependencies on strand concentration to determine the actual heat contribution for the % formation of the duplex products of future reactions at temperatures of the ITC experiments. However, this particular duplex forms 100 % at 15 °C. The duplex products of the reactions in the following sections are longer; therefore, all form 100 % duplexes at the ITC experimental temperatures.

We use the DSC unfolding thermodynamic profiles of Table 2 to create a Hess cycle for this targeting reaction. This Hess cycle corresponds to temperatures equal to the T_Ms, which are extrapolated to the temperature of the ITC targeting reaction assuming a negligible heat capacity effect. This procedure yielded a ΔH_{HC} value of −37.5 kcal/mol for the formation of *Hp-Dup*, which is ~10 kcal/mol more exothermic than the ΔH_{ITC} value. To understand the measured exothermic heats obtained isothermally and from the DSC unfolding reactions, we need to consider additional hydration contributions. In the ITC reactions, there are endothermic contributions from release of electrostricted water molecules from the base-pair stacks of *Hp* (Gasan et al. 1990), which are partially overridden by exothermic contributions from the release of structural water around the complementary strand and from the immobilization of electrostricted water by the duplex product (*Hp-Dup*) (Zieba et al. 1991; Marky and Kupke 2000). In the DSC Hess cycle, these hydration contributions, release of electrostricted water molecule by *Hp*, and the uptake of electrostricted water molecules by *Hp-Dup* have similar magnitudes compared to those in the ITC reactions. By contrast, the removal of structural water from the complementary strand and the extent of its base–base stacking contributions contribute to a lesser extent due to their formation in DSC experiments at temperatures above their T_Ms.

Table 2 Thermodynamic profiles for the unfolding of reactants and products[a]

Molecule	Transition	T_M (°C)	ΔH_{DSC} (kcal/mol)	$T\Delta S_{DSC}$ (kcal/mol)	$\Delta G°_{(T)}$ (kcal/mol)
Hp		73.3	38.1	31.1	7.0
Hp-Dup	1st	34.9	37.5	34.5	3.0
	2nd	72.4	39.8	32.6	7.2
	Total		75.6	67.1	10.2
Blg$_5$		52.3	77.8	67.7	10.1
Blg$_5$-Dup		60.9	121.7	103.2	18.5
IL$_{10}$		49.8	76.4	69.4	7.0
IL$_{10}$-Dup		60.9	139.6	122.5	17.1
Hammer	1st	32.3	49.1	45.5	3.6
	2nd	45.6	34.0	30.2	3.8
	3rd	69.2	41.6	34.4	7.2
	Total		124.7	110.1	14.6
Hammer-Dup	1st	59.4	150.0	127.7	22.3
	2nd	75.0	36.2	29.4	6.8
	Total		186.2	157.1	29.1
PsKn	1st	36.9	8.4	7.5	0.9
	2nd	52.8	51.7	44.1	7.6
	Total		60.1	51.6	8.5
PsKnCS	1st	27.0	2.1	1.9	0.2
	2nd	47.9	38.4	33.3	5.1
	Total		40.5	40.5	5.3
PsKn-Dup	1st	34.0	11.0	10.0	1.0
	2nd	50.2	124.0	106.6	17.4
	Total		135.0	116.6	18.4
Complex	1st	27.4	25.9	24.0	1.9
	2nd	69.5	86.9	70.5	16.4
	Total		112.8	94.5	18.3
Complex-Dup	1st	60.9	152.0	126.6	25.4
	2nd	71.1	49.5	40.0	9.5
	Total		201.5	166.6	34.9

[a]The $\Delta G°_T$ and $T\Delta S_{DSC}$ terms correspond to temperatures of 10 °C (*Hp* and *Blg$_5$*) and 20 °C (*IL$_{10}$*) and 5 °C (all others). All experiments were carried out in 10 mM phosphate buffer at pH 7.0 and 100 mM NaCl. Experimental errors as follows: T_M (±0.5 °C), ΔH_{DSC} (±5 %), $T\Delta S_{DSC}$ (±5 %), $\Delta G°_T$ (±7 %)

3.3 *Targeting Hairpin Loops Containing Bulges and Internal Loops in Their Stem*

In these reactions, a complementary strand, *BlgIlCS*, is used to target both a hairpin loop with a bulge of five thymine residues (*Blg$_5$*) and a hairpin loop with an internal loop of ten thymine residues (*IL$_{10}$*). Each reaction yielded duplex products of 15 base-pair duplexes with dangling ends of 12 (*Blg$_5$-Dup*) and 17 (*IL$_{10}$-Dup*) bases, respectively. The ITC titrations for these reactions were carried out at

10 °C and 20 °C, respectively (data not shown), *BlgIlCS* was placed in the reaction cell to diminish single-stranded stacking contributions, if any. After correcting each injection for dilution heats of the titrant, the average heat of five to six injections yielded exothermic heats of 8.6 μcal and 7.6 μcal, which after normalization by the concentration of the limiting reagent, yielded ΔH_{ITC}s of −42.6 kcal/mol (*Blg₅-Dup*) and −49.9 kcal/mol (*IL₁₀-Dup*) for the formation of the duplex products. The favorable reaction heats are due to a net compensation of exothermic heat measured from the formation of ~15 base-pair stacks and endothermic heats from breaking 9 base-pairs stacks of the hairpins, base–base stacking of the thymine loops, and potential higher removal of water from the hairpins, relative to the water uptake of the duplexes.

The DSC unfolding curves of the reactants and products for each reaction are shown in Fig. 3b, c, and the resulting unfolding thermodynamic profiles are shown in Table 2. The DSC curves of the reactants and of each duplex product show monophasic transitions with negligible heat capacity effects. The reactants unfolded with T_Ms and ΔH_{DSC}s of 52.3 °C and 77.8 kcal/mol (*Blg₅*) and 49.8 °C and 76.4 kcal/mol (*IL₁₀*), respectively, indicative of stable hairpin loops that form completely at temperatures below 35 °C. The product duplexes, *Blg₅-Dup* and *IL₁₀-Dup*, unfolded with T_Ms of 60.9 °C and 60.9 °C, and ΔH_{DSC}s of 121.7 kcal/mol and 139.6 kcal/mol, respectively. Furthermore, the higher enthalpy (17.9 kcal/mol) of *IL₁₀-Dup* may be due to its longer dangling end, yielding additional stacking contributions and/or higher uptake of water molecules.

UV melting experiments for each reactant and duplex product, as a function of strand concentration (data not shown), yielded similar T_Ms for each hairpin molecule, but the T_M of each product depended on strand concentration. This indicates that the reactants formed intramolecularly while each product formed bimolecularly.

We used the DSC unfolding enthalpies of the reactant and duplex product to create Hess cycles that correspond to the targeting reaction at temperatures equal to the T_Ms, which are extrapolated to the temperature of the ITC targeting reaction without correction for heat capacity effects because these values were negligible. This exercise yielded ΔH_{HC} values of −43.9 kcal/mol (*IL₁₀-Dup*) and −63.2 kcal/mol (*IL₁₀-Dup*). The net exothermic enthalpies for these targeting reaction correspond to a complete override of endothermic heat contributions by the exothermic heat contributions, which is due to a net gain of at least seven base-pair stacks and to associated hydration changes in each reaction (Zieba et al. 1991; Marky and Kupke 2000). In spite of similar targeting strand, reaction 3 has a higher exothermic enthalpy (−19.2 kcal), which may be due to higher stacking and hydration contributions of the extra loop of *IL₁₀-Dup* at low temperatures.

3.4 Targeting of a Three-Way Junction

In the following reaction, we target an intramolecular three-way junction, *Hammer* (Fig. 1a-4). *Hammer* has two helical arms with four base pairs and a third arm with six base pairs, and is targeted with a single strand, (*HammerCS*) complementary to its right arm and right loop. The molecularity of each reactant and product is assessed initially by carrying out UV melting curves as a function of strand concentration; *Hammer* forms intramolecularly, while *Hammer-Dup* forms bimolecularly.

The ITC titrations for this reaction were carried out at 10 °C (data not shown). After correcting each injection for the titrant dilution heat and normalizing by the concentration of the limiting reagent, the resulting ΔH_{ITC} was −39.8 kcal/mol for the formation of *Hammer-Dup*.

The DSC unfolding curves of the reactant (*Hammer*) and product (*Hammer-Dup*) are shown in Fig. 3d and the resulting thermodynamic profiles are shown in Table 2. The DSC curve of *Hammer* (Fig. 3d) shows three transitions with T_{M}s of 32.3 °C, 45.6 °C, and 69.2 °C, negligible heat capacity effects, and ΔH_{DSC}s of 49.1 kcal/mol, 34.0 kcal/mol, and 41.6 kcal/mol, respectively. This indicates a sequential melting of its helical domains: $^{5'}$-GAAATT/$^{5'}$-AATTTC, $^{5-'}$-GTGCT_5GCAC, and $^{5'}$-GCGCT_5GCGC. The DSC curves of *Hammer-Dup* (Fig. 3d) shows a biphasic transition with T_{M}s and ΔH_{DSC}s, of 59.4 °C and 150 kcal/mol (first transition), and 75.0 °C and 36.2 kcal/mol (second transition), corresponding to the unfolding of a heptadecamer duplex, followed by melting of a hairpin loop with non-complementary dangling ends.

The DSC Hess cycle yielded a ΔH_{HC} of −61.5 kcal/mol for the targeting reaction of *Hammer*. This net exothermic enthalpy correspond to formation of 16 base-pair stacks (*Hammer-Dup*) that completely overrides the endothermic contributions, which includes disruption of both eight base-pair stacks of *Hammer* and base–base stacking contributions of *HammerCS*, the release of structural water from both reactants, and the immobilization of electrostricted water by the product duplex (Zieba et al. 1991; Marky and Kupke 2000).

Furthermore, close inspection of the structure of the duplex products (Fig. 1a-4) indicates that we should get similar ΔH_{HC} and ΔH_{ITC} values for these two reactions; instead we obtained a less favorable ΔH_{ITC} value, by 21.7 kcal/mol. The difference may be explain in terms of a higher hydration level of *Hammer* at the junction site (Seeman and Kallenbach 1994; Zhong et al. 1997); therefore, additional energy is needed to remove this water at the lower temperatures of the ITC experiments.

3.5 Targeting a DNA Pseudoknot

In this targeting reaction, the strand at the top of *PsKn* (Fig. 1a-5) is targeted with its 19-nucleotide complementary strand, *PsKnCS*, to form a duplex (*PsKn-Dup*). The ITC titrations of five injections were carried out at 5 °C (data not shown), after averaging heat of the last four injections, corrected for the dilution heat, and normalized by the concentration of the limiting reagent, yielded a ΔH_{ITC} of −28.2 kcal/mol. This exothermic term is due to the net compensation of an exothermic heat, formation of 18 base-pair stacks, and endothermic heat, disrupting 10 base-pair stacks and base–base stacking of the thymine loops.

UV melting experiments as a function of strand concentration showed the T_Ms remain constant (*PsKn* and *PsKnCS*) and increased (*PsKn-Dup*) with the increase in strand concentration, indicating their intramolecular and bimolecular formation.

The DSC unfolding curves of *PsKn* (reactant) and *PsKn-Dup* (product) are shown in Fig. 3e and the corresponding thermodynamic profiles are shown in Table 2. The unfolding of *PsKn*, shows a shoulder and a well-defined peak with T_Ms of 36.9 °C and 52.8 °C and heats of 8.4 kcal/mol and 51.7 kcal/mol, respectively. The targeting strand, *PsKnCS*, formed a hairpin with biphasic unfolding: a shoulder (T_M = 27.0 °C, ΔH_{DSC} = 2.1 kcal/mol) and a peak (T_M = 47.9 °C, ΔH_{DSC} of 38.4 kcal/mol). The duplex product shows also similar biphasic transition; *PsKn-Dup* unfolds with T_Ms of 34.0 °C and 50.2 °C and endothermic heats of 11.0 kcal/mol and 124.0 kcal/mol, respectively.

The Hess cycle for this targeting reaction yielded a ΔH_{HC} of −34.4 kcal/mol, which is similar to the heat obtained by ITC. Therefore, this net exothermic enthalpy corresponds to a complete override of the endothermic heat contributions, disruption of both 10 base-pair stacks of *PsKn* and base-pairs formed by the *PsKnCS* hairpin, by the exothermic heat contributions of the formation of 18 base-pair stacks in *PsKn-Dup*.

3.6 Targeting a Combined Triplex–Duplex Complex

In this targeting reaction, an intramolecular triplex–duplex complex (*Complex*) is targeted with its partial complementary strand (*ComplexCS*) to form a duplex with a 11 base dangling end (*Complex-Dup*), see Fig. 1a-6. The sequence of the left loop (CTTTC) is for the proper alignment of the thymine third strand of the triplex portion.

The ITC titration for the reaction of *Complex* with *ComplexCS* was carried out at 5 °C. The average heat of five injections was normalized by the concentration of the limiting reagent, *Complex* in the syringe, yielded a ΔH_{ITC} of −82.8 kcal/mol. This heat corresponds to the net compensation of an exothermic heat (formation of 28 base-pair stacks) and endothermic heat of breaking six base–triplet stacks and five base–base stacking of the constrained loops of *Complex*.

Table 3 Thermodynamic profiles of targeting reactions

	ITC		DSC (Hess Cycle)		
Targeted structure	ΔH_{ITC} (kcal/mol)	$\Delta G°_{ITC}$ (kcal/mol)	ΔH_{HC} (kcal/mol)	$T\Delta S_{HC}$ (kcal/mol)	$\Delta G°_{HC}$ (kcal/mol)
Hp	−27.7	−1.1	−37.5	−35.9	−1.6
Blg₅	−42.6	−8.2	−43.9	−35.5	−8.4
IL₁₀	−49.9	−7.1	−63.2	−53.2	−10.0
Hammer	−39.8	−9.4	−61.5	−47.0	−14.5
PsKn	−28.2	−3.8	−34.4	−29.8	−4.6
Complex	−82.8	−15.5	−88.7	−72.1	−16.7

All parameters measured in 10 mM NaPi, 0.1 M NaCl at pH 7.0. Experimental errors as follows: ΔH_{ITC} (±5 %), $\Delta G°_{ITC}$ (±7 %), ΔH_{HC} (±10 %), $T\Delta S_{HC}$ (±10 %), $\Delta G°_{HC}$ (±14 %)

The DSC-unfolding curves of the reactant and product of this reaction are shown in Fig. 3f, and standard unfolding thermodynamic profiles of each reactant and product at 5 °C are shown in Table 2. The unfolding of both *Complex* and *Complex-Dup* shows biphasic transitions, while no transition was observed with the reactant single strand (data not shown). *Complex* unfolds with T_Ms of 27.4 °C and 69.5 °C and endothermic ΔH_{DSC}s of 25.9 and 86.9 kcal/mol, respectively. The total heat of 112.8 kcal/mol is in excellent agreement with the sum of the unfolding heats of the separate triplex and hairpin domains (Lee et al. 2011). *Complex-Dup* yielded T_Ms and ΔH_{DSC}s of 60.9 °C and 152.0 kcal/mol and 71.1 °C and 49.5 kcal/mol, respectively; the sum of these two enthalpies is in fair agreement with a nearest-neighbor estimate of 224.5 kcal/mol (Breslauer et al. 1986; SantaLucia et al. 1996). UV melting curves showed that the T_Ms of *Complex* do not depend on strand concentration, while the T_Ms of *Complex-Dup* do depend on strand concentration, indicating intramolecular and bimolecular unfolding.

The DSC Hess cycle yielded a ΔH_{HC} of −88.7 kcal/mol for the formation of *Complex-Dup*, in good agreement with its ΔH_{ITC} value (Table 3). The net exothermic enthalpy of this targeting reaction corresponds to a complete override of the endothermic heat contributions (disruption of six base–triplet stacks of *Complex*, disruption of base–base stacking contributions of *ComplexCS*, release of electrostricted water molecules from *Complex*) by exothermic heat contributions [formation of 18 base-pair stacks in *Complex-Dup*, release of structural water from *Complex* and *ComplexCS*, and immobilization of electrostricted water by *Complex-Dup* (Lee et al. 2008; Marky and Kupke 2000; Zieba et al. 1991)].

3.7 Thermodynamic Profiles for the Targeting Reactions

Table 3 shows thermodynamic profiles for each reaction. The first column of this table shows the ΔH_{ITC} values determined directly from the ITC titrations, while the third column lists the ΔH_{HC} values determined indirectly from Hess cycles using the unfolding thermodynamic profiles from DSC experiments. These enthalpy values have been discussed in previous sections for each targeting reaction. The $T\Delta S_{HC}$ values of the fourth column are determined using Hess cycles at the temperatures of the ITC experiments. The $\Delta G°_{HC}$ terms (last column) are calculated from the Gibbs equation: $\Delta G°_{HC} = \Delta H_{HC} - T\Delta S_{HC}$, while the $\Delta G°_{ITC}$ parameters (second column of Table 3) are determined from the relationship, $\Delta G°_{ITC} = \Delta G°_{HC}\ (\Delta H_{ITC}/\Delta H_{HC})$. We obtained favorable $\Delta G°_{ITC}$ values, ranging from −1.1 kcal/mol (targeting of *Hp*) to −15.5 kcal/mol (targeting of *Complex*). Overall, the favorable free-energy term of each targeting reaction resulted from a large compensation of a favorable enthalpy and unfavorable entropy contributions (Table 2). The enthalpy contributions have been discussed extensively. Unfavorable entropy contributions include the bimolecular association of two strands and the putative immobilization of electrostricted water molecules by the duplex product. The release of water molecules from each reactant will also contribute with favorable entropy contributions.

In summary, the major driving force of each targeting reaction is the exothermic enthalpy contribution of product formation, which corresponds to the formation of base-pair stacks that involved the unpaired bases of the loops of each intramolecular DNA secondary structure investigated. In other words, the targeting complementary strand of each reaction is able to invade the particular DNA secondary structure, forming stable duplex products at low temperatures.

4 Conclusions

We have investigated several reactions using DNA single strands as reagents to target intramolecular DNA complexes. Specifically, we have used a combination of ITC, DSC, and spectroscopy techniques to determine standard thermodynamic profiles for the reaction of a variety of DNA secondary structures (hairpin loops, hairpins containing bulges and internal loops, three-arm junctions, pseudoknots, and joined triplex–duplex complexes) with their complementary strands. The enthalpies are measured directly in ITC titrations and are compared with those obtained indirectly from Hess cycles using the DSC unfolding data. All reactions investigated yielded favorable free energy terms, i.e., each single strand is able to invade and disrupt the corresponding intramolecular DNA structure. Resulting from the typical compensation of favorable enthalpy-unfavorable entropy contributions, these exothermic heat contributions are due primarily to the formation of additional base-pair stacks in the duplex products.

In summary, the results show the melting behavior of a variety of secondary DNA structures that can be used, together with nearest-neighbor parameters, in predicting the unfolding thermodynamics of a DNA molecule from knowledge of its sequence. Furthermore, the thermodynamic approach developed here can be used for the control of gene expression, by targeting the loops of the secondary structures formed by mRNA. In general, the larger the number of unpaired bases in a loop, the larger the number of base-pairs and base-pair stacks that are formed in the duplex product, yielding larger free-energy terms. Similarly, the simultaneous targeting of several loops of the secondary/tertiary structure of mRNA will also yield favorable free-energy terms that most likely will end up stopping transcription.

Furthermore, this investigation of nucleic acid targeting reactions has enabled us to develop a method, based on physicochemical principles, to examine the molecular forces that stabilized specific motifs in a variety of DNA secondary structures.

Acknowledgments This work was supported by Grant MCB-1122029 from the National Science Foundation and a GAANN graduate fellowship (C.R.) of Grant P200A120231 from the US Department of Education.

References

Beal PA, Dervan PB (1991) Second structural motif for recognition of DNA by oligonucleotide-directed triple-helix formation. Science 251:1360–1363

Bochman ML, Paeschke K, Zakian VA (2012) DNA secondary structures: stability and function of G-quadruplex structures. Nat Rev Genet 13:770–780

Bock LC, Griffin LC, Latham JA et al (1992) Selection of single-stranded DNA molecules that bind and inhibit human thrombin. Nature 355:564–566

Breslauer KJ, Frank R, Blocker H et al (1986) Predicting DNA duplex stability from the base sequence. Proc Natl Acad Sci USA 83:3746–3750

Brown PM, Madden CA, Fox KR (1998) Triple-helix formation at different positions on nucleosomal DNA. Biochemistry 37:16139–16151

Cantor CR, Schimmel PR (1980) Biophysical chemistry. Part III: The behavior of biological macromolecules. W. H. Freeman, New York

Cantor CR, Warshaw MM, Shapiro H (1970) Oligonucleotide interactions. 3. Circular dichroism studies of the conformation of deoxyoligonucleotides. Biopolymers 9:1059–1077

Crooke ST (1999) Molecular mechanisms of action of antisense drugs. Biochim Biophys Acta Gene Struct Expr 1489:31–43

Darlow JM, Leach DRF (1998) Evidence for two preferred hairpin folding patterns in d(CGG)·d (CCG) repeat tracts in vivo. J Mol Biol 275:17–23

Firulli AB, Maibenco DC, Kinniburgh AJ (1994) Triplex forming ability of a c-myc promoter element predicts promoter strength. Arch Biochem Biophys 310:236–242

Folini M, Pennati M, Zaffaroni N (2002) Targeting human telomerase by antisense oligonucleotides and ribozymes. Curr Med Chem Anti-Cancer Agents 5:605–612

Fox KR (1990) Long (dA)n × (dT)n tracts can form intramolecular triplexes under superhelical stress. Nucleic Acid Res 18:5387–5391

Fox KR (2000) Targeting DNA with triplexes. Curr Med Chem 7:17–37

Gasan AI, Maleev VY, Semenov MA (1990) Role of water in stabilizing the helical biomacromolecules DNA and collagen. Stud Biophys 136:171–178

Gehring K, Leroy J-L, Gueron A (1993) A tetrameric, DNA structure with protonated cytosine · cytosine base pairs. Nature 363:561–565

Han H, Hurley LH (2000) A potencial target for anti-cancer drug design. Trends Pharmacol Sci 21:136–142

Helene C (1991) Rational design of sequence-specific oncogene inhibitors based on antisense and antigene oligonucleotides. Eur J Cancer 27:1466–1471

Helene C (1994) Control of oncogene expression by antisense nucleic acids. Eur J Cancer 30A:1721–1726

Huard S, Autexier C (2002) Targeting human telomerase in cancer therapy. Curr Med Chem Anti-Cancer Agents 2:577–587
Juliano RL, Astriab-Fisher A, Falke D (2001) Macromolecular therapeutics: emerging strategies for drug discovery in the postgenome era. Mol Interv 1:40–53
Koeppel F, Riou J-F, Laoui A et al (2001) Ethidium derivatives bind to G-quartets, inhibit telomerase and act as fluorescent probes for quadruplexes. Nucleic Acid Res 29:1087–1096
Lee HT, Olsen CM, Waters L et al (2008) Thermodynamic contributions of the reactions of DNA intramolecular structures with their complementary strands. Biochimie 90:1052–1063
Lee HT, Carr C, Siebler H et al (2011) A thermodynamic approach for the targeting of nucleic acid structures using their complementary single strands. Methods Enzymol 492:1–26
Mahato RI, Cheng K, Guntaka RV (2005) Modulation of gene expression by antisense and antigene oligodeoxynucleotides and small interfering RNA. Expert Opin Drug Deliv 2:3–28
Marky LA, Breslauer KJ (1987) Calculating thermodynamic data for transitions of any molecularity from equilibrium melting curves. Biopolymers 26:1601–1620
Marky LA, Kupke DW (2000) Enthalpy-entropy compensations in nucleic acids: contribution of electrostriction and structural hydration. Methods Enzymol 323:419–441
Marky LA, Blumenfeld KS, Kozlowski S et al (1983) Salt-dependent conformational transitions in the self-complementary deoxydodecanucleotide d(CGCGAATTCGCG): evidence for hairpin formation. Biopolymers 22:1247–1257
Marky LA, Rentzeperis D, Luneva NP et al (1996) Differential hydration thermodynamics of stereoisomeric DNA-benzo[a]pyrene adducts derived from diol epoxide enantiomers with different tumorigenic potentials. J Am Chem Soc 118:3804–3810
Marky LA, Maiti S, Olsen CM et al (2007) Building blocks of nucleic acid nanostructures: unfolding thermodynamics of intramolecular DNA complexes. In: Labhasetwar V, Leslie-Pelecky D (eds) Biomedical applications of nanotechnology. Wiley, New York, pp 191–225
Mills M, Lacroix L, Arimondo PB et al (2002) Unusual DNA conformations: implications for telomeres. Curr Med Chem Anti-Cancer Agents 2:627–644
Rando RF, Ojwang J, Elbaggari A et al (1995) Suppression of human immunodeficiency virus type 1 activity in vitro by oligonucleotides which form intramolecular tetrads. J Biol Chem 270:1754–1760
Rentzeperis D, Kupke DW, Marky LA (1994) Differential hydration of dA.dT base pairs in parallel-stranded DNA relative to antiparallel DNA. Biochemistry 33:9588–9591
Rich A (1993) DNA comes in many forms. Gene 135:99–109
SantaLucia J Jr, Allawi HT, Seneviratne PA (1996) Improved nearest-neighbor parameters for predicting DNA duplex stability. Biochemistry 35:3555–3562
Seeman NC, Kallenbach NR (1994) DNA branched junctions. Annu Rev Biophys Biomol Struct 23:53–86
Simonsson T, Pecinka P, Kubista M (1998) DNA tetraplex formation in the control region of c-myc. Nucleic Acid Res 26:1167–1172
Vasquez KM, Dagle JM, Weeks DL et al (2001) Chromosome targeting at short polypurine sites by cationic triplex-forming oligonucleotides. J Biol Chem 276:38536–38541
Wang KY, Krawczyk SH, Bischofberger N et al (1993) The tertiary structure of a DNA aptamer which binds to and inhibits thrombin determines activity. Biochemistry 32:11285–11292
Wiseman T, Williston S, Brandts JF et al (1989) Rapid measurement of binding constants and heats of binding using a new titration calorimeter. Anal Biochem 179:131–137
Zahler AM, Williamson JR, Cech TR et al (1991) Inhibition of telomerase by G-quartet DNA structures. Nature 350:718–720
Zhong M, Marky LA, Kallenbach NR et al (1997) Thermodynamics of dT–dT base pair mismatching in linear DNA duplexes and three-arm DNA junctions. Biochemistry 36:2485–2491
Zieba K, Chu TM, Kupke DW et al (1991) Differential hydration of dA $\times$ dT base pairing and dA and dT bulges in deoxyoligonucleotides. Biochemistry 30:8018–8026

Site-Directed Spin Labeling of RNA for Distance Measurements by EPR

Joachim W. Engels, Christian Grünewald, and Lena Wicke

Contents

1 Introduction 386
2 EPR 387
2.1 Methodology 387
2.2 Comparison of FRET and PELDOR-EPR 388
2.3 Comparison of EPR and NMR 389
2.4 Impact of EPR on Structure Determinations 389
3 Spin Labels 390
4 Synthesis of Labeled RNA/DNA 392
5 Applications and Examples 396
5.1 Preface 396
5.2 Ruler for DNA and RNA 397
5.3 Structure of DNA/RNA Hybrid 398
5.4 Structure of cUUCGg Tetraloop 400
5.5 Structure of Neomycin Riboswitch 401
5.6 Structure of Nep1 Protein/RNA Complex 402
5.7 In Cell EPR 404
6 Conclusion 404
References 405

Abstract Electron paramagnetic resonance (EPR) is a structural method and can in addition to X-ray, nuclear magnetic resonance (NMR), and FRET elucidate the structure of different macromolecular systems and determine local surroundings of paramagnetic centers in DNA and RNA. This technique permits structural characterization as well as dynamic structural changes of the macromolecular systems. In order to do so free radicals with good stability have to be introduced, most prominently nitroxide radicals. Here, we describe the site-directed spin labeling

J.W. Engels (✉) • C. Grünewald • L. Wicke
Institute of Organic Chemistry and Chemical Biology, Goethe-University Frankfurt, Max-von-Laue-Str. 7, 60438 Frankfurt am Main, Germany
e-mail: joachim.engels@chemie.uni-frankfurt.de

V.A. Erdmann et al. (eds.), *Chemical Biology of Nucleic Acids*, RNA Technologies,
DOI 10.1007/978-3-642-54452-1_21,

(SDSL) of DNA and RNA based on the Sonogashira cross-coupling reaction and compare to other methods available. In our method the appropriate building blocks, either 5-iodo-substituted pyrimidine or 2-iodopurine deoxy- or ribo-nucleoside phosphoramidites, were prepared and incorporated by solid-phase synthesis. Following this the protected oligonucleotides were "on column" reacted with the acetylenic nitroxide spin labels and subsequently purified. As a result the so-called "RNA ruler" for distance measurements of double-stranded RNA was developed. We present applications of this technique for DNA duplexes, DNA/RNA hybrids, RNA hairpins, riboswitches, and aptamers in vitro and "*in cell.*"

Keywords Spin label • RNA synthesis • PELDOR-EPR • SDSL • Sonogashira cross-coupling

1 Introduction

RNA in its structural and functional diversity is more and more a central theme for the development of new analytical tools (Gesteland et al. 2006). Traditionally X-ray spectroscopy has started to first elucidate the three-dimensional structure of tRNAs. More recent results solved the ribosome structure and some of its functional activities, a highlight crowned with the Nobel Prize in chemistry 2009 (Ehrenberg 2009).

Nuclear magnetic resonance (NMR) spectroscopy being able to measure in solution added dynamic views on RNA (Zhang et al. 2007). Now, one is also able to solve more rigid structures at high resolution, with an RMS of 0.3 Å (Nozinovic et al. 2010). Electron paramagnetic resonance (EPR) and NMR are very sensitive to the molecular environment of the electrons/nuclei under observation and can provide a wealth of detailed information about the three-dimensional structure, dynamics, and interactions with itself and other partners (Qin and Dieckmann 2004).

For EPR measurements, biomacromolecules such as RNA/DNA usually need the addition of spin labels in order to be analyzed by EPR. Site-directed spin labeling (SDSL) was originally introduced and popularized in the field of protein structure and dynamics, particularly membrane proteins (Hubbell and Altenbach 1994). The concept was to replace a nitroxide-containing amino acid for the native residue at selected sites. In most cases a cysteine was introduced and selectively modified with (1-oxyl-2,2,5,5-tetramethylpyrroline-3-methyl) methanethiosulfonate resulting in a disulfide-bonded spin label. This procedure has successfully been used and several reviews appeared, the latest recently (Hubbell et al. 2013) where the authors discuss the advantages of spin-labeling EPR (SDSL-EPR) to complement NMR. EPR is unconstrained by the size or complexity of the system and theoretically allows a dynamic window from picoseconds to seconds.

In this chapter, we concentrate on chemical synthesis that is key to derive at properly modified RNAs with spin labels either during oligonucleotides solid-phase synthesis or afterwards (Strube et al. 2001; Schiemann et al. 2003, 2004). In particular, we focus on RNA examples where we introduced these methods.

2 EPR

2.1 Methodology

EPR or electron spin resonance spectroscopy (ESR) is a powerful technique to determine local surroundings of paramagnetic centers in proteins (Hubbell et al. 2000), in DNA (Schiemann et al. 2004), in RNA (Piton et al. 2005), and also in nonbiological polymers (Godt et al. 2000; Wautelet et al. 2003; Jeschke et al. 2006). This technique permits structural characterization as well as dynamic structural changes of the macromolecular systems. In order to do so free radicals with good stability have to be incorporated by syntheses.

Pulsed electron–electron double resonance (PELDOR) or, as it is also called, double electron–electron resonance (DEER) spectroscopy, a double resonance technique, was originally introduced in the 1980s (Milov et al. 1981). During the past few years this technique has been further developed (Jeschke et al. 2000) and applied to elucidate the structure and function of biomacromolecules (Schiemann and Prisner 2007). The technique is based on the detection of the modulation of the echo amplitude of a spin population caused by the dipolar interaction with another spin population, which has been excited by a microwave pulse of different microwave frequency. Here, in a simplified description, the frequency υ of this modulation is inversely proportional to the cube of the distance r between the two spin labels (Böhme et al. 2010):

$$v(r) = \frac{52.04\,\text{MHz nm}^{-3}}{r^3}$$

The PELDOR experiments are usually performed at low temperatures (~50 K), where a frozen ensemble of molecular conformers contributes to the total signal (Krstić et al. 2012).

At this temperature the electron spin relaxation time is long enough to measure dipolar couplings in a corresponding distance range of up to 8 nm. This frequency after background correction for interstrand resonances and Tikhonov regularization gives the distance distribution (Fig. 1).

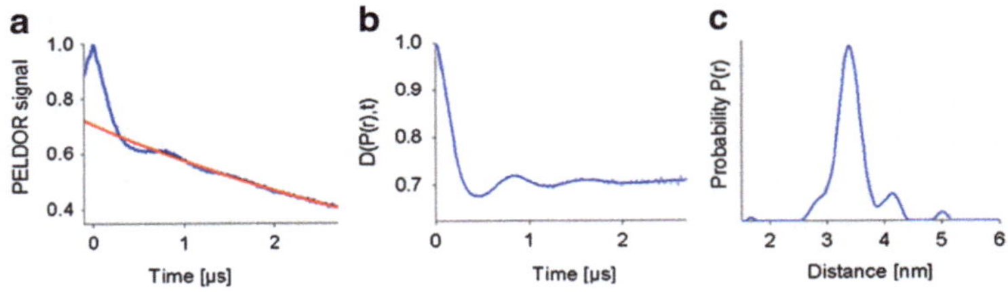

Fig. 1 PELDOR signal analysis: (**a**) time domain PELDOR, (**b**) time domain PELDOR signal after baseline correction (**c**) Tikhonov regularization, distances of a spin pair can be deduced

The distance distribution is encoded in the damping of the modulation. The faster the modulation is damped the broader is the distance distribution and vice versa.

2.2 *Comparison of FRET and PELDOR-EPR*

An alternative method to measure nanometer distances in biomacromolecules is FRET (fluorescence resonance energy transfer or Foerster resonance energy transfer). In the following we compare FRET with EPR-PELDOR.

1. FRET gives distances of biomacromolecules in solution, whereas PELDOR experiments are measured in frozen solution.
2. FRET measurements can be performed on single molecules; for PELDOR one needs micromolar concentrations (e.g., 10 µl, 100 µM).
3. FRET experiments allow time-resolved distance measurements; PELDOR results show frozen distance distributions only.
4. PELDOR experiments are based on an unambiguous coupling mechanism. For FRET experiments the mechanism of acceptor photobleaching is not always unambiguous but needs further reference measurements.
5. Distance determinations in PELDOR are parameter free, whereas in FRET the orientation parameter κ needs additional assumptions.
6. The size of the label used is usually in favor of PELDOR where small molecules (radicals, see Sect. 3) can be used compared to larger fluorophores, additionally fixed via flexible linkers. In any case the influence of the label and linker on the actual biomolecule needs careful considerations whether it changes the biomacromolecules structure (Schiemann 2012).

In conclusion, one can argue the two methods are rather more complementary than exclusive. In general, FRET is easier to perform and mostly commercial equipment can be used, but the results are very often qualitative. PELDOR is still more a specialist method that can furnish atomistic detailed studies.

2.3 Comparison of EPR and NMR

The two complementary magnetic resonance methods EPR und NMR will also be compared for their pros and cons (Schiemann 2012).

1. NMR spectroscopy detects the spin of magnetic nuclei in mostly diamagnetic probes and due to the omnipresent nuclei of H, C, N is able to solve three-dimensional (global) structures. EPR spectroscopy on the other hand relies on the spin of unpaired electrons and thus only detects the environment around it (the radical).
2. This restricted observation of EPR though allows studying biomolecules without size restriction unlike NMR (at present limited around 50 kDa).
3. The sensitivity of EPR spectroscopy is higher (nanomolar) compared to NMR spectroscopy (millimolar). The ratio of magnetic moment of the electron spin to (a) proton spin is ($\mu_e/\mu_H = 658$) and (b) carbon spin 2052. The polarization for a 9.4 T (400 MHz) magnetic field and T = 300 K for (a) electron spin is 2.1 %, (b) proton spin: 0.0032 %, and (c) carbon spin: 0.0010 %.
4. Larger magnetic moments also mean faster relaxation and thus faster timescales for pulsed EPR experiments (nanosecond regime) compared to NMR (millisecond to second regime). This advantage has to be paid for by higher technical requirements, broader line width (MHz vs. Hz).
5. The electron–nuclear spin couplings in EPR due to the larger magnetic moment are larger than the nuclear spin–spin couplings (MHz vs. Hz). This allows EPR spectroscopy to measure larger distances between electron and nucleus (up to 10 Å) or electron and electron (up to 80 Å) than NMR.
6. The theoretical demand for the simulation of EPR spectra is significantly higher than for liquid state NMR spectra due to anisotropic components and very fast relaxations.
7. The transformation of EPR or also NMR parameter into structural data based on quantum chemical methods (DFT) is still rather complicated, especially for transition elements but adequate for nitroxide radicals.
8. NMR as well as EPR measurements can be performed in liquid buffer solutions or frozen/powder; one does not need single crystals. Both methods provide dynamic data though in different time regimes (Schiemann 2012).

2.4 Impact of EPR on Structure Determinations

Furthermore, Hubbell (Hubbell et al. 2013) discusses the development of structure elucidation from the golden age of X-ray crystallography for finding out how the weak non-covalent interaction cooperated to stabilize secondary and tertiary structures of proteins. But the information gained by X-ray is incomplete to understand molecular mechanism, since reactions often rely on conformational flexibility. The same is true for nucleic acids and especially for RNA but with a time delay of roughly 15 years of research. Following X-ray solution NMR added residue-specific information on protein dynamics in solution though on short range internuclear distances were measured. For NMR to obtain dynamics on a μs scale is a challenge, but not so for EPR (Schweiger and Jeschke 2001). One of the most popular applications in the field of SDLS-EPR proved to be inter-spin distance measurements by pulsed dipolar spectroscopy, notably double electron–electron resonance DEER and pulsed electron double resonance PELDOR. Two nitroxide labels are introduced at arbitrary sites and the inter-nitroxide distances are determined. The spatial distribution and the local environment have to be considered, mostly by molecular dynamics (MD) calculations. PELDOR experiments as stated are conducted in water at cryogenic temperatures and the distances typically quoted are between 17 and 60 Å for commercial X-band (9.5 GHz and Q-band 35 GHz). The issue of freezing the solution is especially delicate resulting potentially in unequal distribution of conformational states or substates. Relatively, rigid or better stiff labels with unrestricted rotation about the triple bond are the best solution so far. In general it is advisable to resolve structural ambiguity with an increased number of pair-wise distance data sets (Klose et al. 2012). Here on a model protein FRET and DEER/PELDOR experimental distances are compared with MD, MD with explicit water, and MC. The stochastic Monte Carlo (MC) method gave for the EPR distance determinations the best results. The difference in MD with or without water was a broader distribution of distances for MD results with explicit water.

3 Spin Labels

As introduced above, site-specific attachment of either fluorophores or spin labels (SDSL) is key to the success of the method applied.

First of all, three major questions arise for spin labeling of DNA or RNA:

1. What chemical structure should be used as spin label?
2. Which attachment site for derivatization is optimal?
3. What point of time in synthesis scheme is advisable for introduction?

The reason for most problems occurs from questions of chemical stability of spin labels and chemical reactivity for the application in RNA synthesis (see Sect. 4).

There are two major groups of spin probes used for EPR of biochemically interesting molecules. On the one hand paramagnetic metal ions (e.g., Cu^{2+} or Mn^{2+}) or metal clusters (e.g., FeS) are used for analysis using EPR. These are especially valuable as reporter groups when they are naturally occurring, such as FeS-Clusters in the respiratory chain proteins, or when they are mimicking naturally occurring ions such as Mg^{2+} that can be replaced by Mn^{2+}. The drawback of these paramagnetic centers is their sometimes-complicated fine structure due to complex spin multiplicity and the fact that metal ions need complexing structures for site-directed attachment.

On the other hand, small organic molecules with stable radicals are used for spin labeling. The most prominent organic spin labels are nitroxide radicals (Fig. 2).

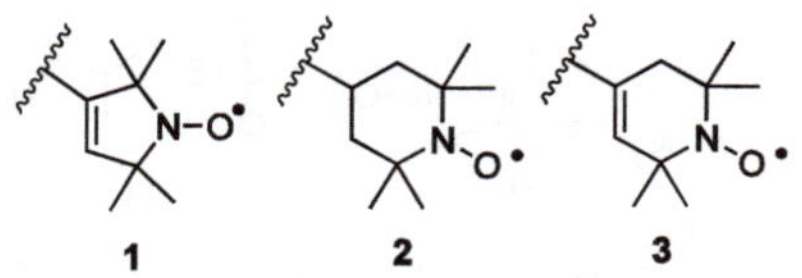

Fig. 2 Basic structures of most commonly used organic spin labels

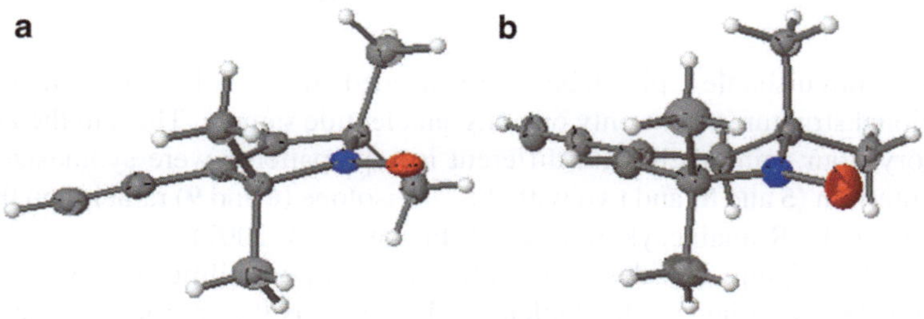

Fig. 3 X-ray structures of (**a**) TPA(**4**) and (**b**) TEMPA(**7**) (Frolow et al. 2009; Bats et al. 2009)

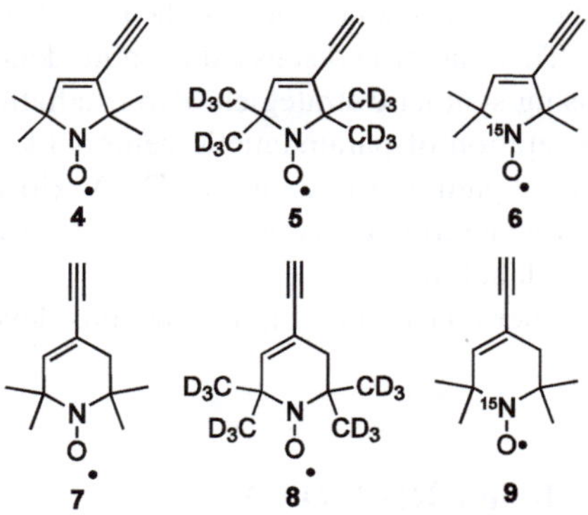

Fig. 4 Isotope spin-label derivatives of TPA and TEMPA synthesized in our lab

These stable nitroxide radicals are commonly referred to, and most spin labels are derived from just two chemical structures: the 5-membered ring 2,2,5,5-tetramethyl-pyrrolin-1-oxyl (TP, **1**) and the 6-membered ring 2,2,6,6-tetramethyl-piperidine-1-oxyl (TEMPO, **2**), as well as its chemical derivative, **3** (Fig. 2).

These spin labels are stable radicals but can react with redox-active chemicals what can cause problems in synthesis and handling. The cyclic nitroxides themselves show limited structural flexibility. The five-membered TP ring 2,2,5,5-tetramethyl-pyrrolin-1-oxyl-3-acetylene(TPA **4**) is almost planar and shows less structural flexibility. The six-membered TEMPO derivative 2,2,6,6-tetramethyl-3,4-dehydro-piperidin-1-oxyl-4-acetylene (TEMPA, **7**) on the other hand exhibits more flexibility as can be seen in the X-ray structure solved in our group (Fig. 3).

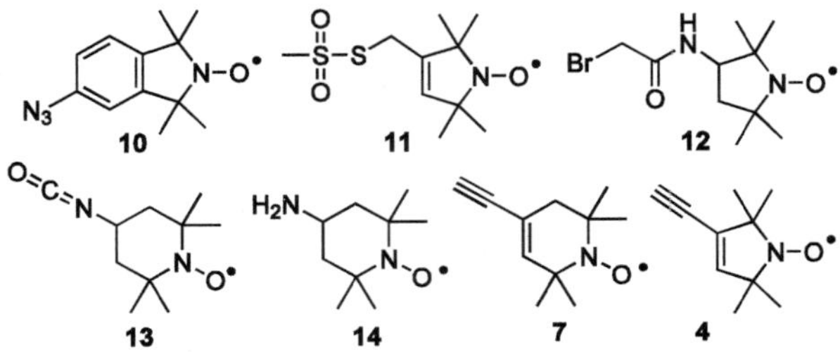

Fig. 5 Spin labels exhibiting functionalities for attachment on RNA/DNA

EPR-distinguishable spin labels are needed to be able to detect three-dimensional structures from only one oligonucleotide sample. Thus, in the author's laboratory, four spin labels with different isotope patterns were synthesized, two with deuterium (**5** and **8**) and two with the ^{15}N-isotope (**6** and **9**) rather than the ^{14}N-isotope (Fig. 4) (Romainczyk et al. 2012; Frolow et al. 2007).

Generally, oligonucleotides do not offer a lot of possibilities for derivatization by nature. In consequence, the nucleic acid as well as the spin labels often needs specific functionalities enabling chemical reactions for attachment to RNA or DNA. Figure 5 shows typical functionalities being used, which result for example in acetylene, amide, carbamate, triazole, or disulfide linkage between modification and oligonucleotide. The choice of attachment at the RNA is mainly governed by its chemical accessibility. This matter is discussed in more detail in Sect. 4.

At last, instead of using synthetic strategies, also irradiation or oxidizing agents can be used for the generation of paramagnetic centers like for DNA 3′-sugar or thymyl radicals created by gamma irradiation of DNA. However, both irradiation and oxidizing agents for generation of paramagnetic centers are rather statistic and not used for site-specific labeling.

In this chapter, we rather concentrate on the methods developed in our group.

4 Synthesis of Labeled RNA/DNA

Since the introduction of the so-called phosphoramidite methodology first DNA and later RNA synthesis has matured.

Oligonucleotide synthesis in most cases relies on the solid-phase phosphoramidite method (Beaucage and Caruthers 1981). Several modifications over the years have streamlined the protocol to a high yielding, simple automated procedure. This is particularly true for the DNA synthesis. More recently, also RNA synthesis has undergone an impressive renaissance. It is a widely accepted fact that RNA synthesis is more challenging compared to DNA synthesis: Coupling efficiency in the synthesis cycle is lower, and it is necessary to permanently block the 2′-hydroxyl function complicating the setup for RNA (Beaucage and Iyer 1992; Davis 1995; Somoza 2008). The most commonly used protecting group strategy follows the "classic" TBDMS approach (Fig. 6, **15**) (Usman et al. 1985, 1987),

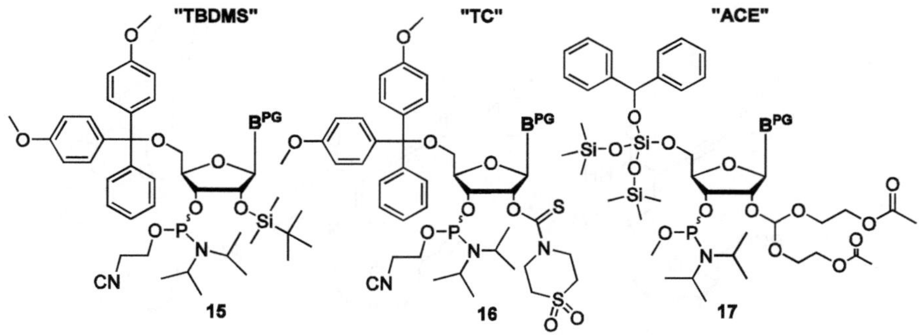

Fig. 6 Phosphoramidite building blocks for commercialized RNA synthesis: TBDMS: *tert*-butyldimethylsilyl, TC: 1,1-dioxo-1λ^6-thiomorpholine-4-carbothioate, ACE: bis (2-acetoxyethoxy)methyl orthoester

which can also involve alternative utilization of 2′-*O*-TOM groups (Wu and Pitsch 1998; Pitsch et al. 2001) resulting in reduced steric hindrance and, thus, in increased coupling rates. The emergence of ACE chemistry (Fig. 6, **17**) (Scaringe et al. 1998, 2004) in the first instance and that of recently developed TC strategy (Dellinger et al. 2011) (compare to Fig. 6, **16**), too, are described to improve RNA synthesis. Actually, these advances are reported to make the outcome of the procedure almost as reliable as DNA synthesis (Scaringe et al. 1998; Dellinger et al. 2011).

Generally, there is a growing demand in the synthesis of oligonucleotides and efficient introduction of specific labels such as spin labels to obtain desired modified nucleic acid compounds, and thus, several approaches have been reported so far (Shelke and Sigurdsson 2012).

In addition to apply enzymatic methods (not being discussed here), chemical derivatization of synthetic oligonucleotides mainly consists of the synthesis of modified phosphoramidites. They can be either directly incorporated to oligonucleotides during solid-phase synthesis, or by postsynthetic approaches, in which stable precursor nucleotides are inserted and afterwards functionalized via site-specific labeling reaction (Höbartner et al. 2010).

The last mentioned concept is nonlinear, profits from greater flexibility in regard to the use of different labels eventually, and requires only very small amounts of possibly expensive (and/or complex) labels. Furthermore, some modifications, and particularly labels such as nitroxide radicals, are potentially sensitive to the conditions of automated oligonucleotide synthesis. This limits the feasibility of successful incorporation of already spin-labeled phosphoramidites to oligonucleotides during synthesis (first mentioned strategy) since decomposition of nitroxide radicals by reduction or disproportionation had been observed (Piton et al. 2007).

Also, this might be the reason why only few groups report the introduction of spin labels to oligonucleotides via this method, and we will only present examples being most important within frame of this book chapter. Nevertheless, in 1988, Hopkins and coworkers (Spaltenstein et al. 1988) were first describing the nontrivial synthetic route to a spin-labeled phosphoramidite of TPA (**18**) that was

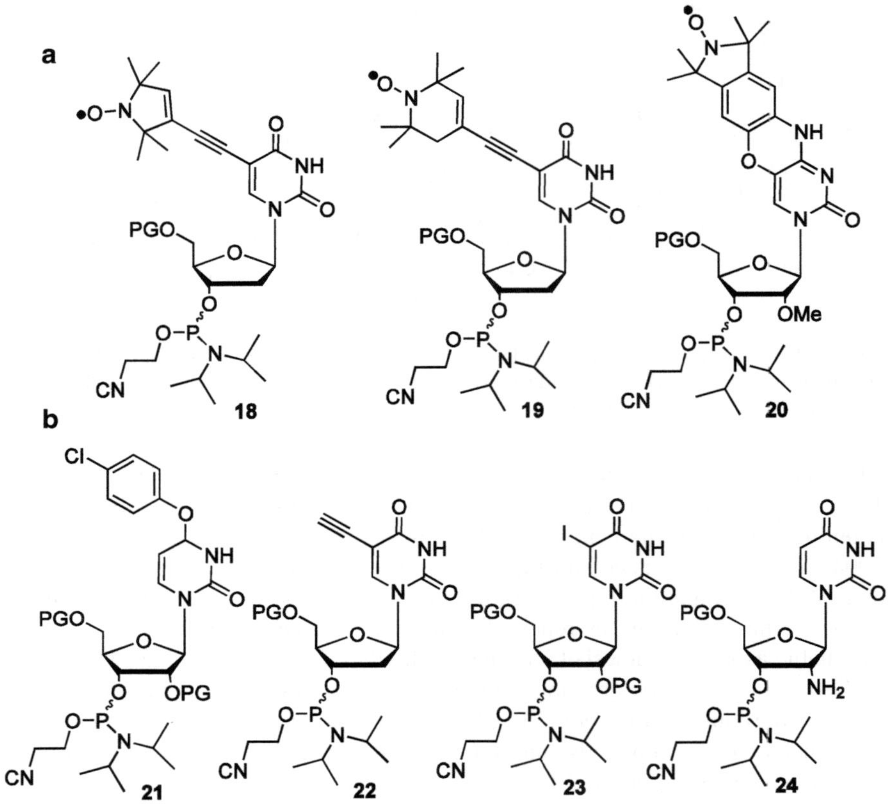

Fig. 7 Commonly used phosphoramidite building blocks: (**a**) Prefabricated spin labeled phosphoramidites for incorporation, (**b**) Precursor phosphoramidites enabling postsynthetic labeling

incorporated to DNA via phosphoramidite approach in automated oligonucleotide synthesis using SDSL. In a similar manner, acetylene-linked TEMPA phosphoramidites (**19**) were applied (Gannett et al. 2001). Alternatively, spin labels such as TEMPO were located at exocyclic amine groups, and modified phosphoramidites afterwards inserted to DNA (Bannwarth and Schmidt 1994; Giordano et al. 2001).

The only example for RNA is presented by researchers around Sigurdsson and Höbartner who recently extended their concept of the C-spin nucleoside (Ç) with a ring-fused, rigid nitroxide label being fluorescent after reduction for DNA (Barhate et al. 2007). In case of RNA they reported synthesis and incorporation of the analog Çm phosphoramidite building block **20** (Höbartner et al. 2012), compare to Fig. 7a.

However, more prominently, a range of postsynthetic approaches has been described in recent years. We would like to focus on this topic and discuss a selection of postsynthetic methods—particularly for RNA and with center of attention to base modifications.

Already in the 1970s, tRNA containing thiopyrimidines was S-alkylated by bromo or iodoacetamide spin label derivatives (Sprinzl et al. 1974), compare to Sect. 3, Fig. 5, **12**. Alternatively, disulfide bridges were built up to spin label RNA in analogy to the approach for proteins with cysteine residues (Qin et al. 2003), Fig. 5, **11**. Another method is the convertible approach consisting of nucleophilic substitution by an amino modified nitroxide label **14** attacking the RNA that exhibits a leaving group such as 4-chlorophenyl (Fig. 7b, **21**) or fluor (Sicoli et al. 2010; Buttner et al. 2013). A prominent example for sugar labeling is provided by coupling of an amine function (**24**) with an isocyanato-modified spin label (Sect. 3, Fig. 5, **13**) (Schiemann et al. 2003) while attachment at phosphorthioate backbone was performed by, e.g., the group around Qin (Cai et al. 2007). So far, the so-called click reaction ([3 + 2] Huisgen–Sharpless–Meldal cycloaddition) had been only applied for DNA at the base moiety (Jakobsen et al. 2010; Ding et al. 2010), compare to Fig. 7b, **22**, and Sect. 3, Fig. 5, **10**.

Anyway, all mentioned postsynthetic methods are based on precursor oligonucleotides containing a reactive functional group, which was mainly introduced as modified phosphoramidite before. However, most of these modified building blocks are either commercially available or simple synthetic access is given. A selection of common precursor phosphoramidites is shown in Fig. 7b.

Our research group aimed to select a labeling reaction compatible with postsynthetic derivatization on-column, which benefits from advantages in solid-phase synthesis such as high yields and ease in separation and purification. This approach depends on RNA stability, solid support conditions, and chosen protecting group chemistry. Generally, the kind of the chemical labeling reaction also defines the linkage between oligonucleotide and modification, as well as it requires a coupling partner with special functionality (e.g., an amino spin label, compare to Sect. 3, Fig. 5, **14**, is used for nucleophilic substitution in the above-mentioned convertible approach, Fig. 7, **21**). Sonogashira cross-coupling meets the requirements of compatibility with RNA and solid support conditions, reacts selectively on well accessible halogenated building blocks (such as **23**) in the precursor oligonucleotide, and additionally, is characterized by linear and rigid acetylene as linkage pattern in the product. Thus, in structural analogy to the formerly mentioned investigations on TPA- and TEMPA-modified phosphoramidites (**18** and **19**) we intended to introduce respective nitroxide spin labels via postsynthetic reaction.

While on-column Sonogashira cross-couplings on DNA have been first reported by Grinstaff and Khan (1999) to attach different functionalities, followed by Wagenknecht and coworkers (Rist et al. 2003) for fluorescent modifications, and us (Strube et al. 2001; Schiemann et al. 2004, 2007), we pioneered the application also on RNA for spin labeling (Schiemann et al. 2007; Piton et al. 2005, 2007; Romainczyk et al. 2012). For RNA synthesis we employed TBDMS as well as ACE protection strategies. Doing so, we achieved—particularly for ACE chemistry affording high coupling rates and using rather mild synthesizer reagents such as *tert*-butylhydroperoxide and dichloroacetic acid—excellent conversion rates to pure products and very good yields. It must be noted that initial setups consisted of Sonogashira cross-couplings during oligonucleotide synthesis in a semi-

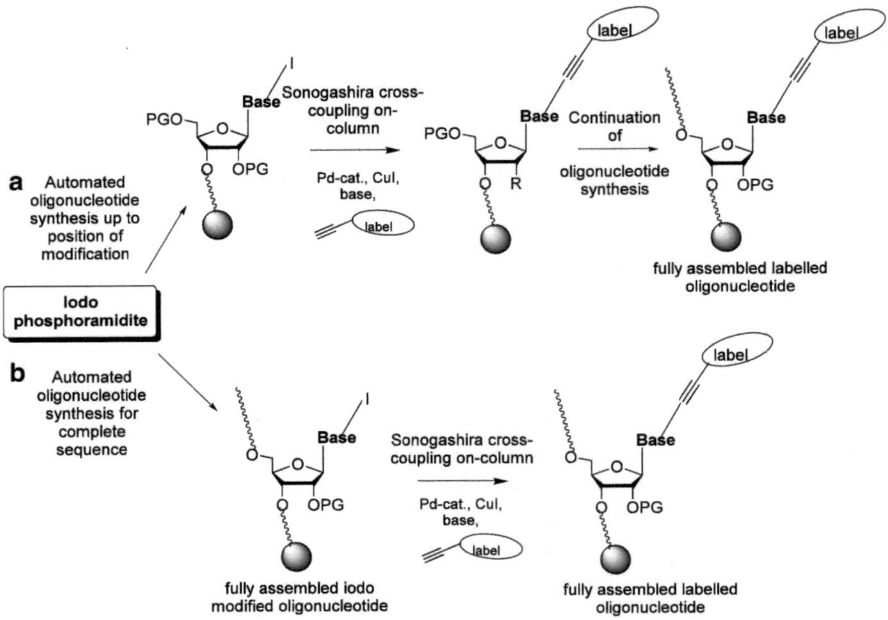

Fig. 8 Schematic representation of postsynthetic Sonogashira cross-coupling on RNA: (**a**) Reaction on 5′-end during solid-phase synthesis (semi-postsynthetic); (**b**) Reaction on fully assembled oligonucleotide after synthesis (totally postsynthetic)

automated manner at the 5′-end and subsequent continuation of synthesis to fully assembled oligonucleotides; while more recently, Sonogashira couplings could be employed after the complete synthesis without interruption (Romainczyk et al. 2011, 2012) (see Fig. 8). This "total postsynthetic" strategy avoids damage of the sensitive spin label and lowers coupling yields after functionalization during automated synthesis such as discussed above for introduction of spin-labeled phosphoramidites. It should be mentioned though that the modification during oligonucleotide synthesis allows the addition of different labels providing functional diversity.

5 Applications and Examples

5.1 Preface

In this section we will present and discuss our examples of spin-labeled nucleic acids to prove the applicability of our approach.

Initially we started with the synthesis of smaller DNA and RNA oligomers in length between 12 and 18 mers to optimize our synthetic procedures and establish

Table 1 Synthesized oligonucleotides and the comparison of the measured and calculated distances

Name	Sequence	r(PELDOR)[Å]	r(MD) [Å]
DNA1	3′-GCG-AT*A-CAT-GCG-5′ 5′-CGC-TAT-G*TA-CGC-3′	19.2 ± 0.5	19.6 ± 1.0
DNA2	3′-CGA-CT*A-TAG-TCG-5′ 5′-GCT-GATAT*C-AGC-3′	23.3 ± 0.6	21.4 ± 1.6
DNA3	3′-CT*G-ACT-AGT-CAG-5′ 5′-GAC-TGA-TCA-GT*C-3′	34.7 ± 1.4	33.0 ± 2.7
DNA4	3′-GCT*-GAC-TAT-AGT-CAG-C-5′ 5′-C-GAC-TGA-TAT-CAG-T*CG-3′	44.8 ± 5.0	43.3 ± 2.5
DNA5	3′-GCT*-GAC-TAT-ATA-GTC-AGC-5′ 5′-CGA-CTG-ATA-TAT-CAG-T*CG-3′	52.5 ± 8.0	52.5 ± 3.0
RNA1	3′-CGA-CU*A-UAG-UCG-5′ 5′-GCU-GAU-AU*C-AGC-3′	19.3 ± 1.2	18.0 ± 2.4
RNA2	3′-CU*G-ACU-AGU-CAG-5′ 5′-GAC-UGA-UCA-GU*C-3′	33.7 ± 3.9	30.5 ± 2.4
RNA3	3′-GCU*-GAC-UAU-AGU-CAG-C-5′ 5′-C-GAC-UGA-UAU-CAG-U*CG-3′	38.7 ± 1.3	36.2 ± 3.1
RNA4	3′-CGA-CA*U-AUG-UCG-5′ 5′-GCU-GUA-UA*C-AGC-3′	21.9 ± 0.8	24.7 ± 0.8
RNA5	3′-GCA-C*AU-ACG-UAU-GUG-C-5′ 5′-C-GUG-UAU-GCA-UAC*-ACG-3′	33.6 ± 2.6	34.3 ± 1.8
RNA6	3′-CGA-GU*G-AUA-CAU-CGC 5′ 5′-GCU-CAC-UAU-GUA*-GCG-3′	26.9 ± 1.3	24.6 ± 2.4

Asterisk marked bases are labeled with TPA, data from Schiemann et al. (2004) and Piton et al. (2007)

the analytical methods. In these projects, proof and validation of our concept were achieved. Based on these results more complex constructs were synthesized and analyzed by different methods. To prove the accuracy of our method for different helical conformations, the project of DNA/RNA hybrids was undertaken. Further challenges were overcome in the application of a tetraloop sequence, the neomycin riboswitch (aptamer) construct. In case of protein/RNA complex, the combination of EPR with NMR is shown and ultimately the "in cell" experiments pave the way for real live measurements.

5.2 *Ruler for DNA and RNA*

Derived from published data for linear model systems and known applications of the EPR technique for detection of distances in the range of 17–60 Å, we synthesized different DNA/RNA oligomers. For the evaluation of the method, not only distances were varied but also helical forms B (DNA1-5) (Schiemann et al. 2004) and A (RNA1-6) (Piton et al. 2007). The different oligonucleotides are shown in Table 1. Self-prepared iodinated phosphoramidites except commercial 5-iodo

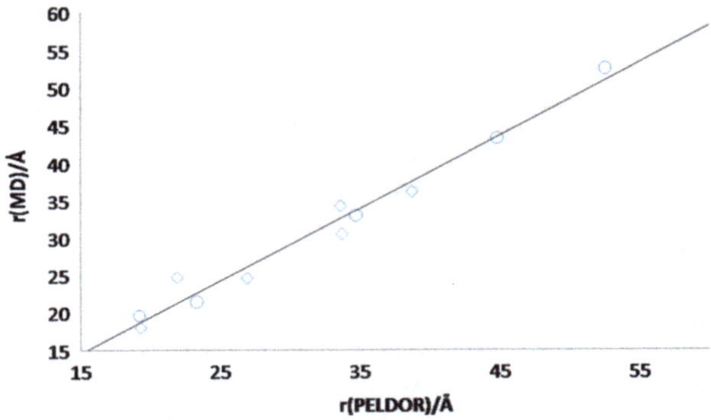

Fig. 9 Correlation of PELDOR and MD distances for RNAs1–6 (*open squares*) and DNAs 1–5 (*open circles*): $y = 0.99x + 0$

uridine were incorporated site specifically and exchanged by Sonogashira cross-coupling for spin labels.

The comparison of measured and calculated data is in very good agreement and allowed validation of the method. This is demonstrated in the following diagram (Fig. 9).

Since we used identical bases for some of the sequences either as DNA or RNA, one can compare the distances. For one turn of B-helix the pitch is ca. 33.2A and for A-helix 28.2A, condensed for about 15–20 %. If we compare DNA2 (23.3 ± 0.6) with RNA1 (19.3 ± 1.2) we observe a perfect match (17 % difference), less pronounced for DNA3 with RNA2 and good again for DNA4 with RNA3. Thus, in general we can discriminate RNA from sequence identical DNA by PELDOR.

In order to unambiguously determine the A- or B-helix geometry of the duplexes CD spectra were recorded, giving atypical B-type or A-type CD spectrum.

When we were trying to incorporate the six-membered TEMPA **7** into different RNA constructs they showed a destabilization of the helix depending on the labeling position. Labeling in the minor groove gives better results than in the major groove (see Sect. 5.3). This destabilization could not be detected with TEMPA-labeled DNA.

5.3 *Structure of DNA/RNA Hybrid*

Considering the differences between the major and minor grooves in A-and-B-helices it is obvious that a bulky spin label (as seen in the X-ray structures of TPA **4** and TEMPA **7**, Sect. 3, Fig. 3), (Frolow et al. 2009; Bats et al. 2009), is less well accommodated in the narrow major groove of an A-helix than in a B-helical form with a wide open major groove.

We prepared a set of deoxy and ribo-oligonucleotides where spin labels were either on the 2 position of adenine or the 5 position of uracil. Whereas for A the position 2 substituents direct towards the minor groove, the substituents on the 5 position of U point towards the major groove. In Fig. 10 the circular dichroism

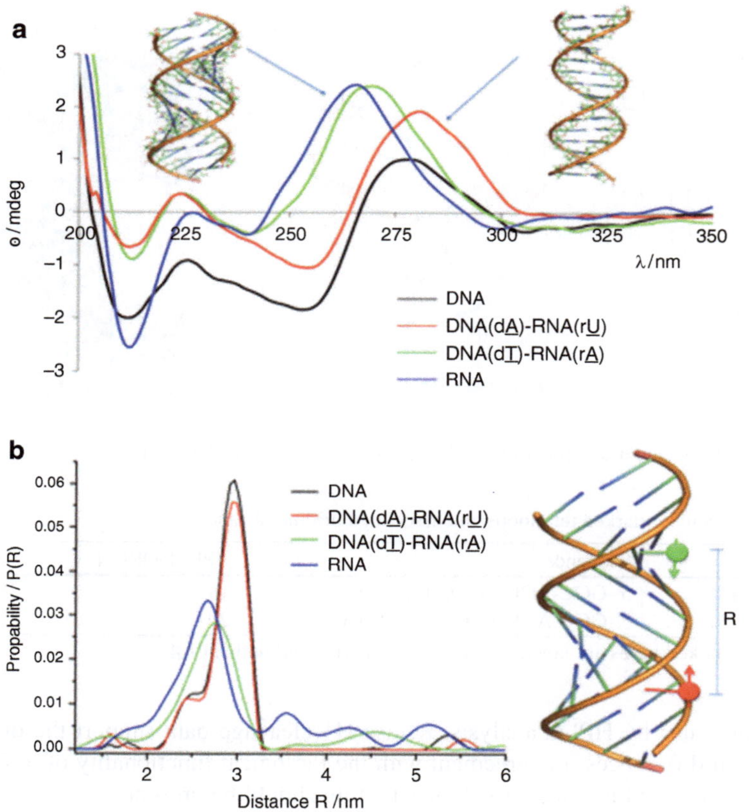

Fig. 10 (**a**) CD Spectra of spin-labeled duplexes, *left* A-, *right* B-helix and (**b**) PELDOR, (Romainczyk et al. 2011), reproduced by permission of The Royal Society of Chemistry

(CD) spectra and the Tikhonov regularizations are shown. Obviously one observes two different cases: one for the B-helix (DNA like) and one for the A-helix (RNA-like). Considering the above discussion, spin-labeled uridine (rU, Fig. 10) in RNA points towards the major groove and does not fit well into the narrow major groove of RNA. Thus, the RNA strand adopts more a B-helix or DNA like conformation. Vice versa, spin labeled adenosine (rA, Fig. 10) the spin label in the wide minor groove of RNA is well adapted and the RNA strand stays in the A-helix, RNA-like.

As a further proof for our assignment we checked how RNase H1 recognizes these hybrids. RNase H1 lacks a consensus sequence for cleavage and preferentially cleaves 7–10 nucleotides from the 5′-RNA/3′-DNA terminus.

It is known that preferred cleavage by the enzyme is a product of fitting sequential/helical geometry, predominantly A-like. We incubated the helices, the DNA and RNA double strands, as well as the four mixed hybrids and measured the

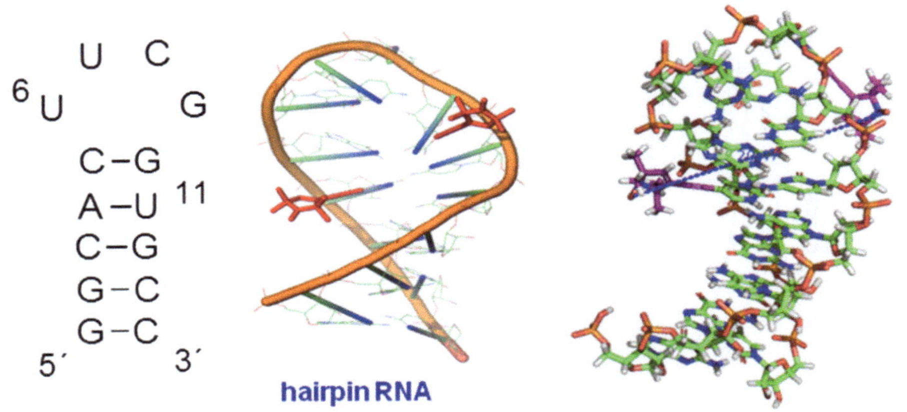

Fig. 11 cUUCGg tetraloop hairpin with spin labels and distance U6-U11

Table 2 Doubly marked tetraloops and associated uridine distances

Name	Sequence	r(PELDOR[Å]	r(MD)[Å]
U(6)-U(11)	5′-GGC-ACU*-UCG-GU*G-CC-3′	18.2	18
U(7)-U(11)	5′-GGC-ACU-U*CG-GU*G-CC-3′	20	14.6

Asterisk marked bases are labeled; labeling was performed with TPA**4**

cleavage rates by HPLC analysis. RNase H1 cleavage data support the observed structural differences. In agreement with the biological functionality of RNase H1 for faster cleaved hybrids an A-type structure should be present.

In summary, we have combined PELDOR with CD for the characterization of RNA/DNA hybrid structures. We were able to show excellent agreement of PELDOR data with CD spectra.

5.4 Structure of cUUCGg Tetraloop

Hairpins or stem loops are a common RNA secondary structure motif found originally as represented by Woese et al. (1990). Tetraloops are an often-found type of four-base hairpin loop motifs that cap many double helices (Cheong et al. 1990; Muth and Engels 1995). The loop structure influences their biological function and for our study the most stable tetraloop sequence (cUUCGg) was selected (Fig. 11).

We incorporated the nitroxyl spin label TPA **4** on uracil bases in the cUUCGg-tetraloop construct (sequence shown in Table 2) in order to be able to detect the differences between the fixed uridine bases U(6) and U(11) and the flexible base U(7).

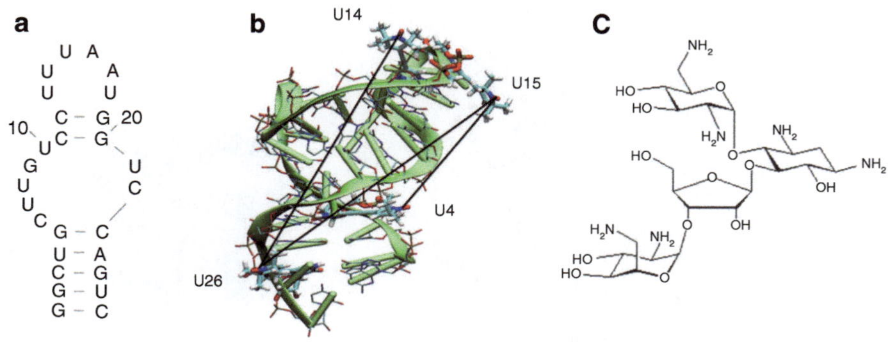

Fig. 12 (**a**) secondary structure of neomycin riboswitch; (**b**) MD modeled structure based on NMR structure with spin labels; (**c**) Neomycin-B

The measurements were done by PELDOR-EPR spectroscopy on double-labeled constructs. In order to explain the structural results MD simulations were consulted.

Our data did not confirm the flexibility of U7 base but is in full agreement with the NMR structure published by Nozinovic et al. (2010).

5.5 Structure of Neomycin Riboswitch

Aptamers are synthetic oligonucleotides that bind with high affinity and specificity small molecules, proteins, and virus particles. The investigated aptamer construct was found by Suess and coworkers (Weigand et al. 2008) using an in vitro selection and in vivo screening and has a high binding affinity for neomycin B. The binding of the aminoglycoside causes a conformational change and proved to be a riboswitch. Riboswitches are structured noncoding RNA domains that selectively bind metabolites and control gene expression (Mandal and Breaker 2004).

For detection of a conformational change introduced by addition of neomycin, we synthesized three doubly labeled mutants of the wild-type riboswitch (Krstić et al. 2010). The positions (Fig. 12) were selected based on the NMR data (Duchardt-Ferner et al. 2010). One of the labels was placed in the loop region of the construct and the second one in the bottom region of the stem. The published structure changes include the reorganization of the base pairing triggered through the conformational change of the whole RNA structure. Addition of neomycin causes the formation of a binding pocket by the oligomer.

Before detection via EPR the conformation was checked by UV and CD spectroscopy for being properly folded.

We were not able to see the conformational change, but our results favor the assumption that the conformation is already preformed. The effect could be based on the high concentration or low temperature (Krstić et al. 2010).

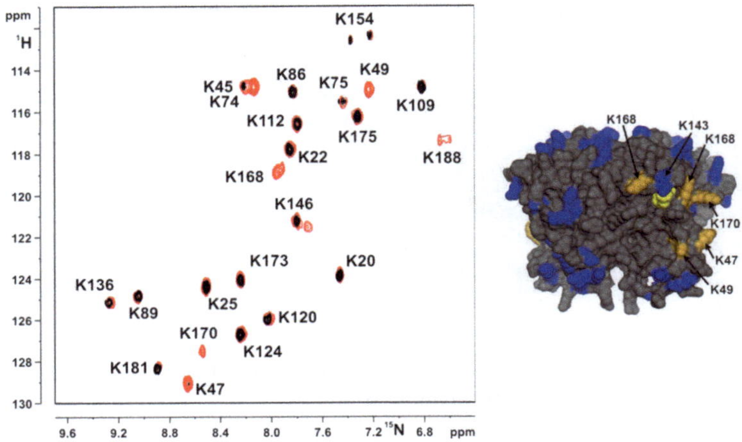

Fig. 13 2D-NMR data before and after reduction of spin label. To prove the similarity of the wild-type and labeled complex by NMR, the reduction of the spin label is necessary. The reduction was performed directly in the NMR tube with ascorbate under normal condition; 1 h at room temperature was sufficient. Reprinted with permission of Current Protocols in Nucleic Acid Chemistry (Romainczyk et al. 2012), with permission of John Wiley & Sons, Inc

5.6 Structure of Nepl Protein/RNA Complex

Besides the application of spin-labeled compounds for EPR, the application for NMR spectra of protein nucleic acid complexes is known. The advantage of spin-labeled complexes is the reduction of signals in the immediate neighborhood of the label in each NMR detection method due to its paramagnetic character. This allows an accurate viewing of a particular region and helps to overcome the size limitation of NMR techniques.

In cooperation with the group of Jens Wöhnert we analyzed the structural interaction of a protein/RNA complex (Wurm et al. 2010). The *Nep*1 complex is a highly conserved nucleoprotein playing an essential role in the ribosome biogenesis. A mutation in the human Nep1is known as the Bowen-Conradi syndrome (Bowen-Conradi syndrome is a very rare genetic disorder characterized by growth delays before birth, failure to thrive during infancy).

We synthesized RNA sequence: 5′-U*UCAACGCC-3′ (*U = TPA **4** labeled U), and by combination of labeling techniques, ^{15}N labeling of the amino acid lysine in *Nep* and spin labeling of a nucleobase (uridine at the 5′ end of the model RNA), the orientation of RNA in the protein-binding pocket was solved (Fig. 13).

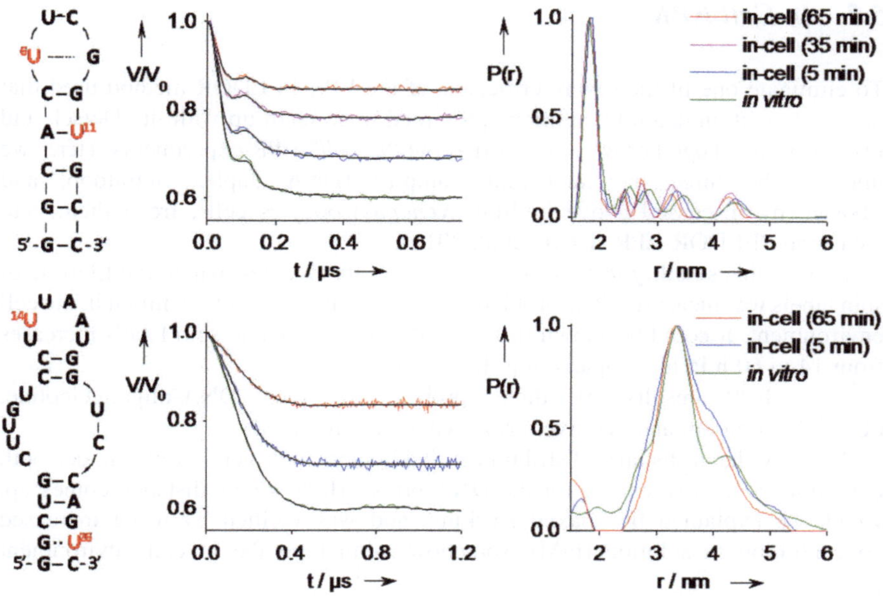

Fig. 14 Secondary structures (with spin-labeled nucleotides in *red*), baseline corrected PELDOR time traces and distance distributions for double-labeled 14-mer cUUCGgtetraloop hairpin RNA (*upper panel*) and the 27-mer neomycin-sensing riboswitch (*lower panel*). In-cell data after different incubation times are compared with in vitro data (*green*). Krstić et al. 2011; reproduced by permission of Angewandte Chemie

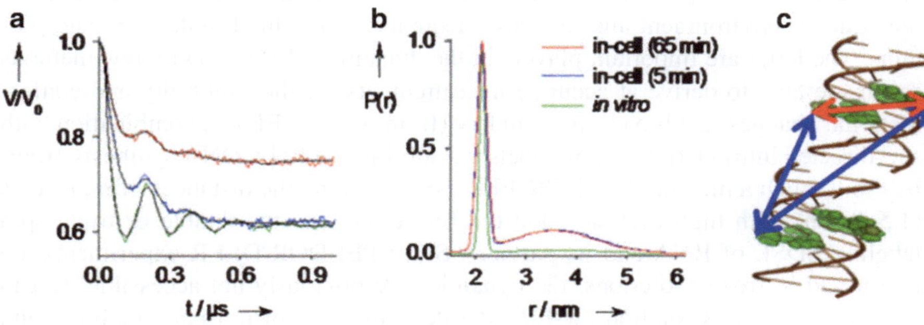

Fig. 15 Baseline corrected PELDOR time traces (**a**) and distance distribution (**b**) for 12-bp double-labeled DNA 5′-GCTGATAT*CAGC-3′ (T* = TPA**4** labeled T) in vitro (*green*) and *in-cell* (*red*); (**c**) Model of coaxially stacked DNA with *blue arrows* indicating additional distances. Krstić et al. 2011, reproduced by permission of Angewandte Chemie

5.7 *In Cell EPR*

To eliminate one of the severe criticisms of the EPR-PELDOR method used that one deals with unnatural conditions prompted us to team up with the Dötsch and Prisner group. Together we have carried out "*in cell*" EPR experiments. Here, we injected the three oligonucleotide samples (DNA duplex, tetraloop, and riboswitch), discussed above in frog (*Xenopus*) oocytes cells, froze them, and measured PELDOR-EPR (Krstić et al. 2011).

Prior to this stability tests were performed in the oocytes, where the lifetime of spin labels was measured. The problem lies indeed in the reductive milieu in the cell environment. It could be shown that half-life of RNA-bound spin labels increases from 14 to 60 min in comparison to free ones.

Here EPR results of three probes, 12 mer DNA-oligonucleotide, cUUCGgtetraloop, and neomycin riboswitch are presented.

The calculated distance distributions (Tikhonov) look very similar under both conditions (Fig. 14), except for the DNA probe. Here, a new distance comes up, which we explained by coaxial stacking, and was verified by using increased concentration in solution (mM), somehow mimicking the in cell environment (Fig. 15).

6 Conclusion

Nucleic acids and especially RNA are highly flexible molecules relying strongly on the aqueous environment and its constituents like ions, small molecules, and proteins. The latter are important players in the dynamics of RNA. Only few methods so far are able to derive at accurate measurements for their mobility and conformational changes in a broad time window (from ps to s). EPR in combination with site-directed introduction of spin labels via solid-phase RNA/DNA synthesis promises to be such a method. PELDOR-EPR can determine the distances of molecules (1.5–8 nm) with high accuracy and the future will see the double or triple spin labeling SDSL of RNA and its partners. Since PELDOR/DEER experiments are performed in frozen solutions, the dynamics are obviously not accessible. But in other techniques as solution cw-EPR the determination of nitroxide mobility and orientation is a good measure for RNA motions. Thus, the proper combination of nitroxide spin label with site-directed spin labeling (SDSL) opens a whole new playground for EPR technologies.

References

Bannwarth W, Schmidt D (1994) Oligonucleotides containing spin-labeled 2′-deoxycytidine and 5-methyl-2′-deoxycytidine as probes for structural motifs of DNA. Bioorg Med Chem Lett 4:977–980

Barhate N, Cekan P, Massey AP et al (2007) A nucleoside that contains a rigid nitroxide spin label: a fluorophore in disguise. Angew Chem Int Ed 46:2655–2658

Bats JW, Frolow O, Engels JW (2009) 4-Ethynyl-2,2,6,6-tetramethyl-1,2,5,6-tetrahydropyridine N-oxide. Acta Crystallogr E 65:o529

Beaucage SL, Caruthers MH (1981) Deoxynucleoside phosphoramidites - a new class of key intermediates for deoxypolynucleotide synthesis. Tetrahedron Lett 22:1859–1862

Beaucage SL, Iyer RP (1992) Advances in the synthesis of oligonucleotides by the phosphoramidite approach. Tetrahedron 48:2223–2311

Böhme S, Steinhoff HJ, Klare JP (2010) Accessing the distance range of interest in biomolecules: site-directed spin labeling and DEER spectroscopy. Spectrosc Int J 24:283–288

Buttner L, Seikowski J, Wawrzyniak K et al (2013) Synthesis of spin-labeled riboswitch RNAs using convertible nucleosides and DNA-catalyzed RNA ligation. Bioorg Med Chem 21:6171–6180

Cai Q, Kusnetzow AK, Hideg K et al (2007) Nanometer distance measurements in RNA using site-directed spin Labeling. Biophys J 93:2110–2117

Cheong C, Varani G, Tinoco IJ (1990) Solution structure of an unusually stable RNAhairpin, 5′GGAC(UUCG)GUCC. Nature 346:680–682

Davis RH (1995) Large-scale oligoribonucleotide production. Curr Opin Biotechnol 6:213–217

Dellinger DJ, Timar Z, Myerson J et al (2011) Streamlined process for the chemical synthesis of RNA using 2′-O-thionocarbamate-protected nucleoside phosphoramidites in the solid phase. J Am Chem Soc 133:11540–11556

Ding P, Wunnicke D, Steinhoff HJ et al (2010) Site-directed spin-labeling of DNA by the azide-alkyne 'click' reaction: nanometer distance measurements on 7-deaza-2′-deoxyadenosine and 2′-deoxyuridine nitroxide conjugates spatially separated or linked to a 'dA-dT' base pair. Chem Eur J 16:14385–14396

Duchardt-Ferner E, Weigand JE, Ohlenschläger O et al (2010) Highly modular structure and ligand binding by conformational capture in a minimalistic riboswitch. Angew Chem Int Ed 49:6216–6219

Ehrenberg M (2009) The nobel prize in chemistry 2009 - advanced information. http://www.nobelprize.org/nobel_prizes/chemistry/laureates/2009/advanced.html

Frolow O, Bode BE, Engels JW (2007) The synthesis of EPR differentiable spinlabels and their coupling to uridine. Nucleosides Nucleotides Nucleic Acids 26:655–659

Frolow O, Bats JW, Engels JW (2009) 3-ethynyl-2,2,5,5-tetramethyl-1-oxyl-3-pyrroline. Acta Crystallogr E 65:o1848

Gannett PM, Darian E, Powell JH et al (2001) A short procedure for synthesis of 4-ethynyl-2,2,6,6-tetramethyl-3,4-dehydro-piperidine-1-oxyl nitroxide. Syn Commun 31:2137–2141

Gesteland RF, Cech TR, Atkins JF (eds) (2006) The RNA world, 3rd edn. Cold Spring Harbor Laboratory, New York

Giordano C, Fratini F, Attanasio D et al (2001) Preparation of spin-labeled 2-amino-dA, dA, dC and 5-methyl-dC phosphoramidites for the automatic synthesis of EPR active oligonucleotides. Synth Stuttg 2001:565–572

Godt A, Franzen C, Veit S et al (2000) EPR probes with well-defined, long distances between two or three unpaired electrons. J Org Chem 65:7575–7582

Grinstaff MW, Khan SI (1999) Palladium(0)-catalyzed modification of oligonucleotides during automated solid-phase synthesis. J Am Chem Soc 121:4704–4705

Höbartner C, Wachowius F (2010) Chemical synthesis of modified RNA. In: Mayer G (ed) The chemical biology of nucleic acids, vol 1. Wiley, Chichester, UK, pp 1–37

Höbartner C, Sicoli G, Wachowius F et al (2012) Synthesis and characterization of RNA containing a rigid and nonperturbing cytidine-derived spin label. J Org Chem 77:7749–7754
Hubbell WL, Altenbach C (1994) Investigation of structure and dynamics in membrane-proteins using site-directed spin-labeling. Curr Opin Struct Biol 4:566–573
Hubbell WL, Cafiso DS, Altenbach C (2000) Identifying conformational changes with site-directed spin labeling. Nat Struct Biol 7:735–739
Hubbell WL, Lopez CJ, Altenbach C et al (2013) Technological advances in site-directed spin labelling of proteins. Curr Opin Struct Biol 23:1–9
Jakobsen U, Shelke SA, Vogel S et al (2010) Site-directed spin-labeling of nucleic acids by click chemistry: detection of abasic sites in duplex DNA by EPR spectroscopy. J Am Chem Soc 132:10424–10428
Jeschke G, Pannier M, Godt A et al (2000) Dipolar spectroscopy and spin alignment in electron paramagnetic resonance. Chem Phys Lett 331:243–252
Jeschke G, Zimmermann H, Godt A (2006) Isotope selection in distance measurements between nitroxides. J Magn Reson 180:137–146
Klose D, Klare JP, Grohmann D et al (2012) Simulation vs. reality: a comparison of in silico distance predictions with DEER and FRET measurements. PLoS ONE 7:e39492
Krstić I, Frolow O, Sezer D et al (2010) PELDOR spectroscopy reveals preorganization of the neomycin-responsive riboswitch tertiary structure. J Am Chem Soc 132:1454–1455
Krstić I, Hänsel R, Romainczyk O et al (2011) Long-range distance measurements on nucleic acids in cells by pulsed EPR spectroscopy. Angew Chem Int Ed 50:5070–5074
Krstić I, Endeward B, Margraf D et al (2012) Structure and dynamics of nucleic acids. Top Curr Chem 321:159–198
Mandal M, Breaker RR (2004) Gene regulation by riboswitches. Nat Rev Mol Cell Biol 5:451–463
Milov AD, Salikohov KM, Shirov MD (1981) Application of endor in electron-spin echo for paramagnetic center space distribution in solids. Sov Phys Solid State 23:975–982
Muth A, Engels JW (1995) Force field calculations of RNA-tetraloops. J Mol Model 1:97–98
Nozinovic S, Fürtig B, Jonker HRA et al (2010) High-resolution NMR structure of an RNA model system: the 14-mer cUUCGg tetraloop hairpin RNA. Nucleic Acids Res 38:683–694
Piton N, Schiemann O, Mu YG et al (2005) Synthesis of spin-labeled RNAs for long range distance measurements by PELDOR. Nucleosides Nucleotides Nucleic Acids 24:771–775
Piton N, Mu YG, Stock G et al (2007) Base-specific spin-labeling of RNA for structure determination. Nucleic Acids Res 35:3128–3143
Pitsch S, Weiss PA, Jenny L et al (2001) Reliable chemical synthesis of oligoribonucleotides (RNA) with 2′-O-[(triisopropylsilyl)oxy]methyl(2′-O-tom)-protected phosphoramidites. Helv Chim Acta 84:3773–3795
Qin PZ, Dieckmann T (2004) Application of NMR and EPR methods to the study of RNA. Curr Opin Struct Biol 14:350–359
Qin PZ, Hideg K, Feigon J et al (2003) Monitoring RNA base structure and dynamics using site-directed spin labeling. Biochemistry 42:6772–6783
Rist M, Amann N, Wagenknecht HA (2003) Preparation of 1-ethynylpyrene-modified DNA via Sonogashira-type solid-phase couplings and characterization of the fluorescence properties for electron-transfer studies. Eur J Org Chem 2003:2498–2504
Romainczyk O, Endeward B, Prisner TF et al (2011) The RNA-DNA hybrid structure determined by EPR, CD and RNase H1. Mol Biosyst 7:1050–1052
Romainczyk O, Elduque X, Engels JW (2012) Attachment of nitroxide spin labels to nucleic acids for EPR. Curr Protoc Nucleic Acid Chem 49:7.17.1–7.17.40
Scaringe SA, Wincott FE, Caruthers MH (1998) Novel RNA synthesis method using 5′-O-silyl-2′-O-orthoester protecting groups. J Am Chem Soc 120:11820–11821
Scaringe SA, Kitchen D, Kaiser R et al (2004) Preparation of 5′-silyl-2′-orthoester ribonucleosides for use in oligoribonucleotide synthesis. Curr Protoc Nucleic Acid Chem 16:2.10.1–2.10.16
Schiemann O (2012) EPR-spektroskopie an biologischen systemen. In: Lottspeich F, Engels JW (eds) Bioanalytik, 3rd edn. Springer, Berlin, pp 509–526

Schiemann O, Prisner TF (2007) Long-range distance determinations in biomacromolecules by EPR spectroscopy. Q Rev Biophys 40:1–53

Schiemann O, Weber A, Edwards TE et al (2003) Nanometer distance measurements on RNA using PELDOR. J Am Chem Soc 125:3434–3435

Schiemann O, Piton N, Mu YG et al (2004) A PELDOR-based nanometer distance ruler for oligonucleotides. J Am Chem Soc 126:5722–5729

Schiemann O, Piton N, Plackmeyer J et al (2007) Spin labeling of oligonucleotides with the nitroxide TPA and use of PELDOR, a pulse EPR method, to measure intramolecular distances. Nat Protoc 2:904–923

Schweiger A, Jeschke G (2001) Principles of pulse electron paramagnetic resonance spectroscopy. Oxford University Press, Oxford, NY

Shelke SA, Sigurdsson ST (2012) Site-directed spin labelling of nucleic acids. Eur J Org Chem 2012:2291–2301

Sicoli G, Wachowius F, Bennati M et al (2010) Probing secondary structures of spin-labeled RNA by pulsed EPR spectroscopy. Angew Chem Int Ed Engl 49:6443–6447

Somoza A (2008) Protecting groups for RNA synthesis: an increasing need for selective preparative methods. Chem Soc Rev 37:2668–2675

Spaltenstein A, Robinson BH, Hopkins PB (1988) A rigid and nonperturbing probe for duplex DNA motion. J Am Chem Soc 110:1299–1301

Sprinzl M, Kramer E, Stehlik D (1974) Structure of phenylalanine transfer-RNA from yeast - spin-label studies. Eur J Biochem 49:595–605

Strube T, Schiemann O, MacMillan F et al (2001) A new facile method for spin-labeling of oligonucleotides. Nucleosides Nucleotides Nucleic Acids 20:1271–1274

Usman N, Pon RT, Ogilvie KK (1985) Preparation of ribonucleoside 3′-O-phosphoramidites and their application to the automated solid-phase synthesis of oligonucleotides. Tetrahedron Lett 26:4567–4570

Usman N, Ogilvie KK, Jiang MY et al (1987) Automated chemical synthesis of long oligoribonucleotides using 2′-O-silylated ribonucleoside 3′-O-phosphoramidites on a controlled-pore glass support - synthesis of a 43-nucleotide sequence similar to the 3′-half molecule of an *Escherichia coli* formylmethionine transfer-RNA. J Am Chem Soc 109:7845–7854

Wautelet P, Le Moigne J, Videva V et al (2003) Spin exchange interaction through phenylene-ethynylene bridge in diradicals based on iminonitroxide and nitronylnitroxide radical derivatives. 1. Experimental investigation of the through-bond spin exchange coupling. J Org Chem 68:8025–8036

Weigand JE, Sanchez M, Gunnesch E-B et al (2008) Screening for engineered neomycin riboswitches that control translation initiation. RNA 14:89–97

Woese CR, Winker S, Gutell RR (1990) Architecture of ribosomal RNA: constraints on the sequence of “tetra-loops”. Proc Natl Acad Sci USA 87:8467–8471

Wu XL, Pitsch S (1998) Synthesis and pairing properties of oligoribonucleotide analogues containing a metal-binding site attached to beta-D-allofuranosyl cytosine. Nucleic Acids Res 26:4315–4323

Wurm JP, Meyer B, Bahr U et al (2010) The ribosome assembly factor Nep1 responsible for Bowen–Conradi syndrome is a pseudouridine-N1-specific methyltransferase. Nucleic Acids Res 38:2387–2398

Zhang Q, Stelzer AC, Fisher CK et al (2007) Visualizing spatially correlated dynamics that directs RNA conformational transitions. Nature 450:1263–1267

Chemo-enzymatic Strategies to Modify RNA in vitro or in Living Cells

Daniela Schulz and Andrea Rentmeister

Contents

1 Introduction 410
2 Click Chemistry, an Unequaled Set of Reactions 410
3 Direct Incorporation of Non-natural Nucleotides and Postsynthetic Chemical Modification 412
3.1 Incorporation of Non-natural Nucleotides by Chemical Synthesis 414
3.2 Incorporation of Non-natural Nucleotides by Enzymes 415
4 Chemo-enzymatic Modification of Native RNAs 417
4.1 Targeting Specific Subsets of RNA by Chemo-enzymatic Modification 417
4.2 Sequence-Specific Labeling of RNAs by a Chemo-Enzymatic Approach 419
5 Conclusions 419
References 420

Abstract The discovery of novel RNA species and functions has generated the need for methods to selectively label different types of RNA in order to study their localization, dynamics, and structure both in vitro and in cells. Progress in the field of bioorthogonal chemistry has led to the development of a toolbox of reactions compatible with cellular components—in the best case even with living cells—that are also suitable for RNA modification. The first step, however, is the introduction of a group suitable for bioorthogonal chemistry. Besides chemical synthesis, this can be achieved by various enzymes that (1) either accept non-natural nucleotide-building blocks or (2) modify RNA with non-natural residues.

This chapter will first briefly introduce the relevant click reactions and then focus on the different state of the art approaches for modification of RNAs by

D. Schulz • A. Rentmeister (✉)
Department of Chemistry, Institute of Biochemistry, University of Muenster, Wilhelm-Klemm-Straße 2, 48149 Muenster, Germany
e-mail: a.rentmeister@uni-muenster.de

V.A. Erdmann et al. (eds.), *Chemical Biology of Nucleic Acids*, RNA Technologies,
DOI 10.1007/978-3-642-54452-1_22, © Springer-Verlag Berlin Heidelberg 2014

chemo-enzymatic labeling. We will try to highlight the potential and limitations for broader applications, including future use in cells.

Keywords Bioorthogonal chemistry • Chemo-enzymatic modification • Click chemistry • Postsynthetic modification • RNA labeling

1 Introduction

Chemo-enzymatic approaches provide a novel strategy to target and analyze biomolecules in complex surroundings or even living cells. In these two-step labeling approaches the biomolecule of interest is first equipped with a reactive functionality. This step ensures the specificity of labeling and can be performed enzymatically. The functionalized biomolecule is then amenable to the second step—a chemical modification, which covalently links the biomolecule to a reporter group. The modularity of this approach in combination with today's cell-compatible click reactions provides a treasure trove of techniques to label biomolecules, including RNA, in vitro and in vivo. These approaches could open up new possibilities in the field of isolation as well as visualization of RNA and thus help to gain insight into RNA pathways.

2 Click Chemistry, an Unequaled Set of Reactions

We will first introduce the set of click reactions suitable for **RNA modification**. This will allow us to refer to these repeatedly used reactions in the following subchapters.

Click reactions—as defined by Sharpless—proceed fast, preferentially in aqueous buffer, and with high yields, and are therefore ideal to label biomolecules (Kolb et al. 2001). A subset of these click reactions have been used successfully to label RNA and will be introduced in the following (Fig. 1).

The most widely used click reactions are the RNA modification (CuAAC) and the strain-promoted azide-alkyne-cycloaddition (SPAAC). The CuAAC has been repeatedly used to label RNAs (Paredes and Das 2012; Jao and Salic 2008; Motorin et al. 2011). The reaction was originally described by Sharpless as well as Meldal who both observed that Cu(I)-ions have an accelerating effect on the triazol formation between an azide and an alkyne (Tornøe et al. 2002; Rostovtsev et al. 2002). It is possible not only to use Cu(I)-salts as catalysts but also to generate the reactive species in situ by reducing Cu(II)-salts. The effect of Cu(I) on the cycloaddition has been proposed to result from lowering the activation barrier of azide-addition to the acetylide intermediate (Himo et al. 2005).

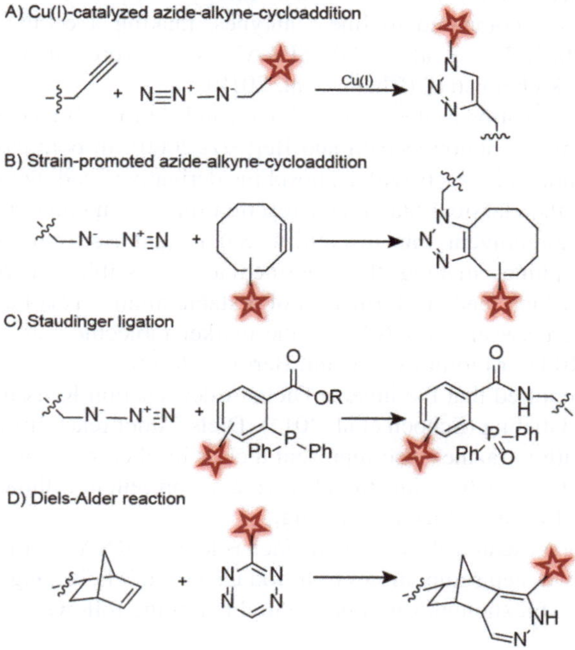

Fig. 1 Reaction schemes of selected click reactions. (**a**) In CuAAC, an alkyne-functionality reacts with an azido-containing molecule in the presence of Cu(I) to form a triazol (Rostovtsev et al. 2002), providing a covalent linkage between target molecule and reporter, e.g., a fluorophore (shown as *star*). Alternatively, a label can be introduced via the alkyne to an azido-modified molecule of interest (Winz et al. 2012). (**b**) In SPAAC, an azide reacts with a strained cyclooctyne in the absence of catalyst to form a triazole (Agard et al. 2004). (**c**) In the Staudinger Ligation an amide bond forms the linkage between target molecule and label. It is generated by reaction of an azide with an ester-containing phosphine. The resulting aza-ylide intermediate (not shown) is rearranged and the amide containing phosphine is formed as product (Saxon and Bertozzi 2000). (**d**) The Diels–Alder reaction describes a [4 + 2]-cycloaddition of diene with dienophile. Using tetrazine as diene, nitrogen is released from the intermediate (Sauer et al. 1965; Devaraj et al. 2008), making the reaction irreversible and leading to formation of a dihydropyridazine as linkage between the target molecule and the label

The Cu(I)-catalyzed cycloaddition was found to be compatible with RNAs if Cu (I)-stabilizing ligands, such as acetonitrile, TBTA (Tris[(1-benzyl-1H-1,2,3-triazol-4-yl)methyl]amine), or THPTA (Tris(3-hydroxy-propyltriazolyl-methyl)-amine) are used (Paredes and Das 2012; Motorin et al. 2011).

Some click reactions, such as the SPAAC (Agard et al. 2004), the Staudinger ligation (Saxon and Bertozzi 2000), or the inverse Diels–Alder reaction (Schoch et al. 2011) can even be performed without catalyst. In these cases, the biomolecule has to be labeled with an azido-moiety first. In RNA modification, the azido-derivatives react with strained cyclooctynes instead of linear alkynes commonly used in CuAAC. Spring-loaded cyclooctynes accelerate the reaction by lowering the activation barrier, because less energy is required for distortions into the transition state compared to linear alkynes, making a catalyst dispensable (Agard et al. 2004; Ess et al. 2008). SPAAC is therefore often referred to as "copper-free click chemistry" (Chang et al. 2010).

The RNA modification as designed by Saxon and Bertozzi is an extension of the classical Staudinger reaction (Saxon and Bertozzi 2000). In both cases, an azido-functionalized molecule reacts with a phosphine derivative, leading to formation of an aza-ylide. In the classical Staudinger reaction this intermediate is degraded by hydrolysis. To circumvent this instability, Saxon and Bertozzi used an ester-containing phosphine, making the intermediate accessible for intramolecular rearrangement, which leads to formation of a stable amide. This bond provides a covalent linkage between the RNA and the marker molecule, e.g., a fluorophore (Aigner et al. 2011) or biotin (Saxon and Bertozzi 2000).

It has been reported that the inverse Diels–Alder reaction RNA modification is useful for RNA labeling (Schoch et al. 2011). Diels–Alder reactions of norbornene as dienophile with tetrazines also represent a class of click reactions that does not require a catalyst. Since these functionalities are not present in cellular systems they may be suitable for bioorthogonal labeling.

The reactions presented above provide means to label RNAs bearing "clickable" moieties. This two-step setup allows to introduce tags of choice, e.g., fluorophores or biotin. Successful strategies will be highlighted in the following sections.

3 Direct Incorporation of Non-natural Nucleotides and Postsynthetic Chemical Modification

A prerequisite for specific chemical modification is to first introduce a unique reactive handle into the molecule (Fig. 2). For RNAs, non-native nucleotides bearing functional groups for subsequent click reactions have been used to achieve labeling in vitro (Paredes and Das 2011; Dojahn et al. 2013) and in vivo (Jao and Salic 2008). Such nucleotide derivatives can be incorporated into RNA either by chemical synthesis (Schoch et al. 2011) or by enzymatic catalysis, e.g., by T7-polymerase (Dojahn et al. 2013). Thus, the RNA of interest is first labeled with reactive moieties that do not fulfill a purpose (e.g., fluorescence) per se, but which can be converted specifically by click reactions [e.g., norbornene-modified nucleotides are accessible to Diels–Alder reaction (Schoch et al. 2011) or alkyne-modified nucleotides are applicable to CuAAC (Dojahn et al. 2013)].

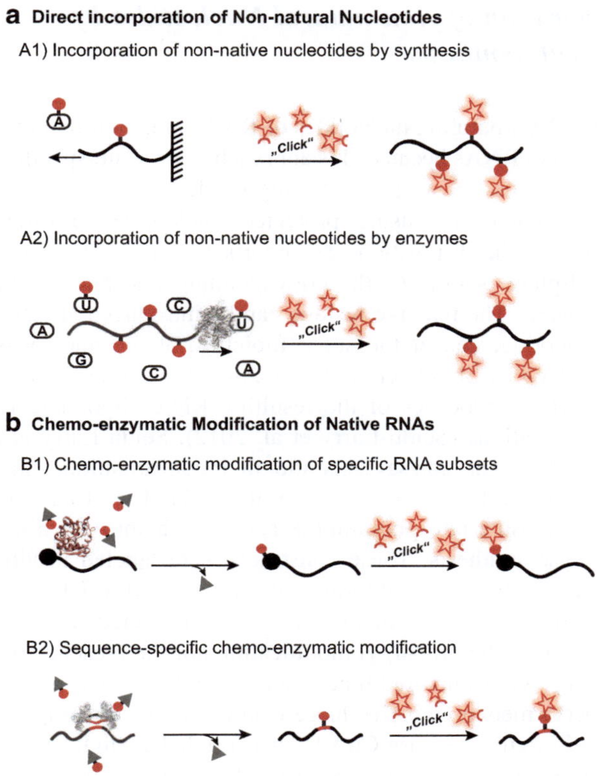

Fig. 2 Schematic illustrations of different RNA-labeling strategies. (**a**) Approaches based on incorporation of non-native nucleotides during synthesis. (**a1**) Incorporation of non-native nucleotides during chemical synthesis. After releasing the oligonucleotide from the solid-phase it can be further modified by a suitable click reaction. This second step leads to a covalent linkage between RNA and a reporter molecule, e.g., a fluorophore. (**a2**) Introduction of non-native nucleotides using enzymes. Incorporation of modified uridine into RNA during transcription using RNA polymerase II is shown (PDB of shown enzyme: 2E2I). For other examples using this strategy, see Sect. 3.2. The resulting RNA is a substrate for further chemical modification. (**b**) Approaches based on chemo-enzymatic modification of native RNAs by methyltransferases (for details see Sect. 4). (**b1**) Methyltransferases recognizing and modifying specific subsets of RNAs can be used to transfer a functional moiety (*red dot*) from AdoMet-analogs (*gray* triangle coupled to *red dot*) to the corresponding target. Labeling by a cap-specific enzyme and subsequent chemical modification, an approach specific for eukaryotic mRNAs, is shown (PDB of shown enzyme: 3GDH). (**b2**) Sequence-specific labeling using a box C/D methyltransferase (PDB of shown enzyme: 3ID5), which is guided to its target RNA by a complementary sequence (shown in *red*), and subsequent chemical modification. Legend: NTPs are indicated as A, C, G, or U. Modifications are indicated by *red dots*. Labels are indicated by *red stars*

3.1 Incorporation of Non-natural Nucleotides by Chemical Synthesis

Incorporation of Non-natural Nucleotides of RNA using non-natural nucleotides is widely used to label RNAs because this approach allows site-specific incorporation of the modifications. Synthesis of oligonucleotides is mainly based on phosphoramidite chemistry, using protected nucleotides bearing the reactive phosphoramidite at the 3′-position. To incorporate non-natural nucleotides into RNAs by solid-phase synthesis, the corresponding phosphoramidite derivatives have to be prepared. The reactive residue can be introduced at different positions of an oligonucleotide, i.e., at the sugar moieties, at the nucleobases, or in the backbone (reviewed by Shukla et al. 2010). Alterations at each of these positions, however, can affect properties of the resulting RNA oligonucleotide, e.g., base pairing or conformations (Seidu-Larry et al. 2012). Seidu-Larry et al. stated that modifications at the sugars can severely alter their conformation, which could affect hybridization and stability (Seidu-Larry et al. 2012). Therefore, they synthesized N^2-propargyl guanosine phosphoramidite and built it into the 5′-end of siRNAs during solid-phase synthesis. These constructs were labeled in vitro by CuAAC with different azido-containing fluorophores and applied to HEK cells. The intracellular distribution of the modified siRNAs was unaffected compared to a control siRNA. Thus, neither the propargyl modification nor the presence of a 5′-modified nucleotide in the siRNA impaired functionality (Seidu-Larry et al. 2012). Although analysis was performed in vivo, the fluorescent labeling of the oligonucleotides had to be performed in vitro because CuAAC was used. The high concentration of Cu (I) needed for the reaction would be toxic for cells, as upon oxidation of Cu (I) damage of biomolecules occurs (Brewer 2010). To protect biomolecules from defects during labeling, Cu(I)-stabilizing ligands were introduced, which minimize oxidation of Cu(I). Acetonitrile was the first stabilizing ligand described by Paredes and Das (2011), but today many alternatives are available and extensively used (Motorin et al. 2011; Winz et al. 2012; Tomkuviene et al. 2012) (see Sect. 2. for more details). Nonetheless, although Cu(I) can be stabilized and biomolecule labeling is possible in vitro, the complexes themselves still cause toxicity (Kennedy et al. 2011).

Regarding azido-modified nucleotides, Micura and coworkers proposed that this reactive group could be suitable for modification at the 2′-position of the ribose, as it would favor the correct conformation of the sugar (Aigner et al. 2011). An additional advantage of azido-labeling is its compatibility with copper-free chemistries, such as SPAAC and Staudinger ligation. However, due to the side-reactivity of azides with phosphor(III)-species (a reactivity also used for the Staudinger reaction), azido-modified nucleotides can only be used at low concentrations in phosphoramidite chemistry (Aigner et al. 2011). This instability was also observed for an alkyne-modified adenosine introduced during synthesis instead of the azido-analog (Jawalekar et al. 2008). Nonetheless, 2′-azido-modified adenosine and uridine could be incorporated into RNAs during solid-phase synthesis (Aigner et al. 2011). In a subsequent ligation step using phophoramidite chemistry, sequence-specifically labeled products were obtained. The azido-modified oligonucleotide showed only a slight decrease in duplex stability and no differences in conformation compared to unmodified counterparts. Furthermore, the obtained constructs could be successfully labeled with fluorescent dyes and were fully functional in vivo (Aigner et al. 2011). Based on these results, Micura expanded their approach and showed that also guanosine and cytidine can be incorporated as azido-derivates into RNA during synthesis (Fauster et al. 2012).

Despite these promising results, labeling RNAs by chemical synthesis, suffers from some limitations, namely the often time- and cost-consuming synthesis of the activated nucleotides, the need for special equipment (e.g., synthesizer) as well as the restricted length of obtained products. Thus, it can be advantageous to use enzymes as catalysts for the incorporation of the required functionalities.

3.2 Incorporation of Non-natural Nucleotides by Enzymes

Bioorthogonal groups can also be introduced into RNAs during incorporation of Non-natural Nucleotides, i.e., transcription. These enzyme-based approaches can be performed easily, fast, and without the need for special equipment (e.g., synthesizer, HPLC for purification) (Dojahn et al. 2013). Furthermore, longer oligonucleotides are accessible, whereas chemical synthesis of RNA longer than 50 nucleotides is still hard (El-Sagheer and Brown 2010).

Various nucleotide-analogs, enzymes, and conditions have been used to obtain modified RNA compatible with CuAAC. One of the first reports on successful RNA labeling by CuAAC was published in 2011 (Paredes and Das 2011). The reactive azido-moiety was introduced during transcription as 5′-azido-5′-deoxyguanosine (applied in a fourfold molar excess relative to GTP as initiator nucleotide for incorporation of Non-natural Nucleotides) (Paredes and Das 2011). The problem of Cu(I)-induced RNA degradation in the chemical labeling step was solved by addition of acetonitrile as cosolvent. This allowed to successfully label RNAs bearing an azido-functionality at the 5′-end using CuAAC.

Apart from 5′-end labeling, internal site-specific modification of RNAs was achieved by in vitro transcription (Ishizuka et al. 2012). In this approach, an unnatural base in the DNA allowed to site-specifically incorporate an ethynyl-modified pyrrole-2-carbaldehyde as base during transcription (opposite of 7-(2-thienyl)-imidazo[4,5-b]pyridine in the template DNA strand). The resulting site-specifically alkyne-modified tRNA was fluorescently labeled by CuAAC using THPTA as stabilizing ligand (Ishizuka et al. 2012).

The combination of enzymatic approaches with CuAAC has been used to target newly transcribed RNAs in cells (Jao and Salic 2008; Grammel et al. 2012). For example, using NIH 3T3 cells, Jao and Salic showed the potential of RNA polymerases I-III to incorporate the alkyne-bearing nucleotide-analog ethynyluridine into newly transcribed RNA. Modified RNAs could be labeled after fixation of cells by CuAAC and snapshots of RNA synthesis and turnover could be obtained (Jao and Salic 2008). Beside ethynyluridine also ethynylcytidine has been used as nucleoside analog for in vivo labeling of RNAs during transcription (Qu et al. 2013).

Grammel et al. also used incorporation of Non-natural Nucleotides for in vivo labeling and introduced an alkyne-modified adenosine-analog that could be chemically modified after fixation of cells (Grammel et al. 2012). This approach did not only allow visualization of RNA synthesis and turnover but also of poly(A)-tail-dynamics. The approaches described by Jao and Salic as well as Grammel et al. enable the in vivo labeling of newly transcribed RNAs by RNA polymerases. This provides a means to examine endogenous RNAs and avoids the need to introduce chemically synthesized or in vitro transcribed RNAs into cells (Jao and Salic 2008; Grammel et al. 2012). Nonetheless, these approaches suffer from some drawbacks for application in vivo, namely (1) the toxicity of the copper catalyst limiting application in living systems and (2) the lack of specificity for one particular target RNA, since only global transcription patterns can be visualized.

To overcome the limitations of the CuAAC, the spectrum of enzymatically incorporated nucleotides has been constantly expanded. Hence, modified RNAs have become accessible to modification by other click reactions—so far in vitro.

Diels–Alder reactions of norbornene with tetrazines proceed without a catalyst (Schoch et al. 2011). Since neither of these functionalities is present in cells, they could be suitable for bioorthogonal labeling. Schoch et al. produced modified RNAs by in vitro transcription with a 5′-norbornene-modified guanosine that was incorporated into the transcripts by incorporation of Non-natural Nucleotides as initiator nucleotide. Efficient labeling was observed if the modified nucleotide was used in tenfold excess compared to guanosine, yielding up to 70 % of labeled transcripts. These acted as substrates for chemical modification by a dansyl-bearing tetrazine (Schoch et al. 2011).

Other click reactions that can be performed without a catalyst are SPAAC as well as the Staudinger ligation. An enzymatic approach was developed to gain azido-modified oligonucleotides because introducing azido-containing nucleotides by solid-phase chemistry turned out to be difficult (Jawalekar et al. 2008; Aigner et al. 2011) (see also chapter 3.1). Incorporation of Non-natural Nucleotides were successfully used to transfer different azido-modified nucleotides to the 3′-end of in vitro-transcribed RNA (Winz et al. 2012). Careful optimization of reaction conditions led to transfer of a single 2′-N_3-dATP to the 3′-end by *E. coli* poly(A)-polymerase. RNAs containing a terminal azido-modified nucleotide could still function as substrate for poly(A)-polymerases and ligases which allowed conversion of terminal into internal modifications. The azido-modified RNAs could be labeled by CuAAC as well as by SPAAC (Winz et al. 2012).

As mentioned above, an azido-functionality is a versatile reactive moiety as it does not only provide the possibility to use CuAAC or SPAAC but also to take advantage of the Staudinger reaction (Rao et al. 2012) or Staudinger ligation (Aigner et al. 2011). Rao et al. synthesized an azido-containing uridine triphosphate that was efficiently incorporated during in vitro transcription. The azido-containing

RNAs could be fluorescently labeled by CuAAC or be reduced to give an amine-functionalized RNA by Staudinger reaction (Rao et al. 2012).

In summary, incorporation of non-native nucleotides containing reactive handles bears great potential for labeling the complex biomolecule RNA. The incorporation can be achieved chemically as well as enzymatically and equips the resulting molecule with a click compatible functionality. Successful modification of these reactive groups mainly relies on improvements in the field of click chemistry, which include the development of RNA-compatible Cu(I)-protocols and of novel click reactions, like the Staudinger ligation first described by Saxon and Bertozzi (2000).

4 Chemo-enzymatic Modification of Native RNAs

As highlighted above, the incorporation of non-natural nucleotides into RNAs during chemical or enzymatic synthesis provides a means to label this biomolecule in vitro as well as in vivo. Apart from these methods, it is also possible to modify native full-length RNAs by chemo-enzymatic approaches. Herein, a bioorthogonal functionality is enzymatically transferred to the native transcript and subsequently used for modification by a compatible click reaction. Enzymes of choice are methyltransferases. Variants as well as wild-type enzymes have been used successfully to transfer moieties containing functional groups, such as alkene, alkyne, or azide, to their target molecule using an appropriate AdoMet-analog as cosubstrate (Motorin et al. 2011; Tomkuviene et al. 2012; Schulz et al. 2013). The enzymatically modified RNAs could then be labeled by suitable click reactions. This strategy yielded promising results for different kinds of RNA in vitro. We will review the state of the art approaches and try to highlight the potential and limitations for broader applications, including future use in cells.

4.1 Targeting Specific Subsets of RNA by Chemo-enzymatic Modification

Chemo-enzymatic approaches targeting native RNAs could prove invaluable to label specific RNAs in the run for in vivo visualization of these biomolecules. Due to their potential to label endogenous RNAs they could render introduction of in vitro manipulated RNAs into the cell unnecessary. These approaches harness RNA-modifying enzymes capable of transferring reactive residues to a defined class of RNAs—in the best case even to a specific RNA. To date, methyltransferases have been most widely used to fulfill this task—not only for RNA but also for other biomolecules, such as DNA or proteins (Wang et al. 2013; Tomkuviene et al. 2012; Schulz et al. 2013; Dalhoff et al. 2006). Some methyltransferases show promiscuous activity on chemo-enzymatic modification, which contain reactive residues like alkene or alkyne instead of the naturally occurring methyl group (Dalhoff et al. 2006; Peters et al. 2010). The promiscuous activity can be further increased using protein engineering (Islam et al. 2011; Schulz et al. 2013). This strategy allows introduction of click-compatible moieties to native RNAs that can subsequently be labeled by suitable click reactions.

Using a chemo-enzymatic approach, Motorin et al. achieved fluorescence labeling of chemo-enzymatic modification in vitro. They synthesized the AdoMet-analog 5′-[(*S*)(3*S*)-3-amino-3-carboxypropyl]pent-2-en-4-ynylsulfonio]-5′-deoxyadenosine (AdoEnYn) containing a pentenynyl residue instead of the methyl group. The tRNA-specific methyltransferase Trm1 was able to transfer this bulky side chain to the target atom, the N^2 of guanosine 26 in $tRNA^{Phe}$ and thus specifically modify the targeted tRNAs, making them accessible to labeling by CuAAC (Motorin et al. 2011).

Besides tRNAs, chemo-enzymatic modification represent interesting targets for modification, because they can have an important role in gene regulation by restricting protein biosynthesis spatially as well as temporally (Bertrand et al. 1998; Rook et al. 2000; Willis et al. 2007; Lécuyer et al. 2007). Modification based on a chemo-enzymatic approach could improve analysis of the fundamental mechanisms, as it could be suitable for labeling endogenous mRNAs in living cells. It has been proved that modification of mRNAs by such an approach is possible. Since in chemo-enzymatic approaches the specificity is determined by the enzymatic step, an enzyme was utilized, which specifically recognizes and modifies the 5′-cap—a hallmark of eukaryotic mRNAs. The trimethylguanosinesynthase from *Giardia lamblia* (GlaTgs), which is known to hypermethylate the cap m^7G at N^2 proved suitable. Schulz et al. generated variants of this enzyme that accepted several AdoMet-analogs as cosubstrates and were able to transfer alkene as well as alkyne residues specifically to correctly capped mRNAs. These reactive handles, covalently bound to the N^2 of the cap, made eukaryotic mRNAs accessible for chemical modification by thiol-ene click reaction as well as CuAAC (Schulz et al. 2013).

This approach could set the basis for visualization of targeted RNAs in living cells. However, to achieve this, not only the enzymes but also the AdoMet-analogs would have to be present intracellularly. To this end, Luo et al. recently succeeded in establishing synthesis of an AdoMet-analog in vivo using a variant of the *S*-adenosyl-L-methionine-synthethase. The potential of their approach was demonstrated by labeling histones in vivo, which acted as substrates for CuAAC in vitro (Wang et al. 2013).

4.2 Sequence-Specific Labeling of RNAs by a Chemo-Enzymatic Approach

The chemo-enzymatic approaches described in Sect. 4.1 make use of enzymes that recognize specific structures only present in the target-RNAs. Besides, also chemo-enzymatic modification of RNA by a methyltransferase and subsequent click reaction was successful. Tomkuvienė et al. made use of the sequence-guided 2′-O-methyltransferase present in box C/D ribonucleoprotein particles (box C/D RNPs). In this case the site of modification is determined by binding of the RNP's guide-RNA to the complementary target. Using this sequence-specific methyltransferase, efficient labeling of different RNAs with an alkyne-moiety by simply altering the guide-sequence was achieved. The modified RNAs could be fluorescently labeled by CuAAC (Tomkuviene et al. 2012). This approach makes a wide spectrum of RNAs accessible to labeling by click reactions in vitro, because in principle every RNA sequence of interest can be modified. However, due to the need of the complementary guide-sequence as marker for the box C/D RNA methyltransferase the application of this method for in vivo-approaches remains to be investigated.

In summary, methyltransferases and AdoMet-analogs bear great potential for labeling endogenous RNAs making these accessible for visualization or isolation.

5 Conclusions

In this chapter we have tried to elaborate chemo-enzymatic approaches as versatile strategies to selectively label different subsets of RNAs. The versatility can be attributed to a combination of recent developments and improvements in the field of click chemistry with strategies to introduce reactive residues into an RNA molecule. The compilation of all of these achievements has resulted in today's toolbox containing approaches to label in principle any RNA of interest with a reporter molecule of choice.

Non-native nucleotides can be incorporated site specifically either by chemical or enzymatic synthesis. Different nucleotide derivatives have been shown to be compatible with this step, which allows using different types of click reactions for the subsequent chemical modification. Harnessing these two-step approaches, different reporter groups can be introduced, such as fluorophores for imaging or biotin for isolation. The plethora of methods provides simple and reliable means to prepare oligonucleotides, e.g., as probes for in vitro but also in vivo applications.

Furthermore, labeling native RNAs using subset-specific or sequence-specific methyltransferases provides a successful alternative. These approaches seize promiscuous activity of these enzymes or their variants on AdoMet-analogs. In addition to in vitro applications these chemo-enzymatic strategies bear potential for application in vivo—as already demonstrated for proteins (Wang et al. 2013). This development is a promising starting point to make endogenous RNAs accessible to live cell imaging. Subcellularly localized mRNAs are key players in emerging mechanisms for the regulation of protein biosynthesis in certain cell types (Rook et al. 2000; Lécuyer et al. 2007). Non-perturbing labeling methods would help to study these processes in vivo and are therefore highly required. Chemo-enzymatic labeling represents a milestone in the development of versatile methods suitable to examine dynamics and transport of any given RNA in vitro and in living cells.

References

Agard NJ, Prescher JA, Bertozzi CR (2004) A strain-promoted [3 + 2] azide-alkyne cycloaddition for covalent modification of biomolecules in living systems. J Am Chem Soc 126: 15046–15047

Aigner M, Hartl M, Fauster K et al (2011) Chemical synthesis of site-specifically 2′-azido-modified RNA and potential applications for bioconjugation and RNA interference. Chembiochem 12:47–51

Bertrand E, Chartrand P, Schaefer M et al (1998) Localization of ASH1 mRNA particles in living yeast. Mol Cell 2:437–445

Brewer GJ (2010) Copper toxicity in the general population. Clin Neurophysiol 121:459–460

Chang PV, Prescher JA, Sletten EM et al (2010) Copper-free click chemistry in living animals. Proc Natl Acad Sci USA 107:1821–1826

Dalhoff C, Lukinavicius G, Klimasǎuskas S et al (2006) Direct transfer of extended groups from synthetic cofactors by DNA methyltransferases. Nat Chem Biol 2:31–32

Devaraj NK, Weissleder R, Hilderbrand S (2008) Tetrazine-based cycloadditions: application to pretargeted live cell imaging. Bioconjug Chem 19:2297–2299

Dojahn CM, Hesse M, Arenz C (2013) A chemo-enzymatic approach to specifically click-modified RNA. Chem Commun 9:3128–3130

El-Sagheer AH, Brown T (2010) New strategy for the synthesis of chemically modified RNA constructs exemplified by hairpin and hammerhead ribozymes. Proc Natl Acad Sci USA 107: 15329–15334

Ess DH, Jones GO, Houk KN (2008) Transition states of strain-promoted metal-free click chemistry: 1,3 dipolar cycloadditions of phenyl azide and cyclooctynes. Org Lett 10: 1633–1636

Fauster K, Hartl M, Santner T et al (2012) 2′-Azido RNA, a versatile tool for chemical biology: synthesis, X-ray structure, siRNA applications, click labeling. ACS Chem Biol 7:581–589

Grammel M, Hang H, Conrad NK (2012) Chemical reporters for monitoring RNA synthesis and poly(A) tail dynamics. Chembiochem 13:1112–1115

Himo F, Lovell T, Hilgraf R et al (2005) Copper (I) -catalyzed synthesis of azoles. DFT study predicts unprecedented reactivity and intermediates. J Am Chem Soc 127:210–216

Ishizuka T, Kimoto M, Sato A et al (2012) Site specific functionalization of RNA molecules by an unnatural base pair transcription system via click chemistry. Chem Commun (Camb) 48: 10835–10837

Islam K, Zheng W, Yu H et al (2011) Expanding cofactor repertoire of protein lysine methyltransferase. ACS Chem Biol 6:679–684

Jao CY, Salic A (2008) Exploring RNA transcription and turnover in vivo by using click chemistry. Proc Natl Acad Sci USA 105:15779–15784

Jawalekar AM, Meeuwenoord N, Cremers JSGO et al (2008) Conjugation of nucleosides and oligonucleotides by [3 + 2] cycloaddition. J Org Chem 73:287–290

Kennedy DC, McKay CS, Legault MCB et al (2011) Cellular consequences of copper complexes used to catalyze bioorthogonal click reactions. J Am Chem Soc 133:17993–18001

Kolb HC, Finn MG, Sharpless KB (2001) Click chemistry: diverse chemical function from a few good reactions. Angew Chem Int Ed Engl 40:2004–2021

Lécuyer E, Yoshida H, Parthasarathy N et al (2007) Global analysis of mRNA localization reveals a prominent role in organizing cellular architecture and function. Cell 131:174–187
Motorin Y, Burhenne J, Teimer R et al (2011) Expanding the chemical scope of RNA: methyltransferases to site-specific alkynylation of RNA for click labeling. Nucleic Acids Res 39:1943–1952
Paredes E, Das SR (2011) Click chemistry for rapid labeling and ligation of RNA. Chembiochem 12:125–131
Paredes E, Das SR (2012) Optimization of acetonitrile co-solvent and copper stoichiometry for pseudo-ligandless click chemistry with nucleic acids. Bioorg Med Chem Lett 22:5313–5316
Peters W, Willnow S, Duisken M et al (2010) Enzymatic site-specific functionalization of protein methyltransferase substrates with alkynes for click labeling. Angew Chem Int Ed Engl 49: 5170–5173
Qu D, Zhou L, Wang W et al (2013) 5-Ethynylcytidine as a new agent for detecting RNA synthesis in live cells by "click" chemistry. Anal Biochem 434:128–135
Rao H, Sawant A, Tanpure A, Srivatsan SG (2012) Posttranscriptional chemical functionalization of azide-modified oligoribonucleotides by bioorthogonal click and Staudinger reactions. Chem Commun (Camb) 48:498–500
Rook MS, Lu M, Kosik KS (2000) CaMKII alpha 3′ untranslated region-directed mRNA translocation in living neurons: visualization by GFP linkage. J Neurosci 20:6385–6393
Rostovtsev VV, Green LG, Fokin VV et al (2002) A stepwise huisgen cycloaddition process: copper(I)-catalyzed regioselective "ligation" of azides and terminal alkynes. Angew Chem Int Ed Engl 41:2596–2599
Sauer J, Mielert A, Lang D et al (1965) Umsetzung von 1.2.4.5-tetrazinen mit olefinen. Zur Struktur von dihydropyridazinen. Chem Ber 98:1435–1445
Saxon E, Bertozzi CR (2000) Cell surface engineering by a modified Staudinger reaction. Science 287:2007–2010
Schoch J, Ameta S, Jäschke A (2011) Inverse electron-demand Diels-Alder reactions for the selective and efficient labeling of RNA. Chem Commun (Camb) 47:12536–12537
Schulz D, Holstein JM, Rentmeister A (2013) A chemo-enzymatic approach for site-specific modification of the RNA cap. Angew Chem Int Ed Engl 52:7874–7878
Seidu-Larry S, Krieg B, Hirsch M et al (2012) A modified guanosine phosphoramidite for click functionalization of RNA on the sugar edge. Chem Commun (Camb) 48:11014–11016
Shukla S, Sumaria CS, Pradeepkumar PI (2010) Exploring chemical modifications for siRNA therapeutics: a structural and functional outlook. ChemMedChem 5:328–349
Tomkuviene M, Clouet-d'Orval B, Cerniauskas I et al (2012) Programmable sequence-specific click-labeling of RNA using archaeal box C/D RNP methyltransferases. Nucleic Acids Res 40:6765–6773
Tornøe CW, Christensen C, Meldal M (2002) Peptidotriazoles on solid phase: [1,2,3]-triazoles by regiospecific copper(i)-catalyzed 1,3-dipolar cycloadditions of terminal alkynes to azides. J Org Chem 67:3057–3064
Wang R, Islam K, Liu Y et al (2013) Profiling genome-wide chromatin methylation with engineered posttranslation apparatus within living cells. J Am Chem Soc 135:1048–1056
Willis DE, van Niekerk EA, Sasaki Y et al (2007) Extracellular stimuli specifically regulate localized levels of individual neuronal mRNAs. J Cell Biol 178:965–980
Winz M-L, Samanta A, Benzinger D et al (2012) Site-specific terminal and internal labeling of RNA by poly(A) polymerase tailing and copper-catalyzed or copper-free strain-promoted click chemistry. Nucleic Acids Res 40:e78

Metal Dependence of Ligand Binding and Heavy-Atom Derivatization of Evolutionarily Distinct $PreQ_1$ Riboswitches

Joseph E. Wedekind, Joseph A. Liberman, Jermaine L. Jenkins, and Mohammad Salim

Contents

1 Introduction 424
  1.1 Learning Objectives of This Chapter 426
2 Isothermal Titration Calorimetry Methods for Riboswitches 426
  2.1 Quantitative Procedures to Test Metal Ion Requirements for Ligand Binding 426
  2.2 Checking the Feasibility of Isothermal Titration Calorimetry Measurements 427
  2.3 General Notes About ITC Instrument Settings 428
  2.4 Preparation of Samples for ITC 428
  2.5 Outcomes for Metal Dependence of Ligand Binding Based on ITC 430
3 General Approaches to Introduce Heavy Atoms into Riboswitches 431
  3.1 Mother Liquor Manipulations to Promote Heavy Atom Binding 431
  3.2 Eliminating SO_4^{2-} to Promote $Os(NH_3)_5^{3+}$ Binding 431
  3.3 The Mode of $Os(NH_3)_5^{3+}$ Binding to the $PreQ_1$-I Riboswitch 432
  3.4 Co-crystallization with Cs^+ for Single-Wavelength Anomalous Diffraction Phasing 433
  3.5 The Binding Mode of Cs^+ to a G•U Wobble Pair and Other Sites 435
References 438

J.E. Wedekind (✉) • J.L. Jenkins
Department of Biochemistry & Biophysics, University of Rochester School of Medicine and Dentistry, Rochester, NY 14642, USA

Center for RNA Biology, University of Rochester School of Medicine and Dentistry, Rochester, NY 14642, USA

Structural Biology & Biophysics Facility, University of Rochester School of Medicine and Dentistry, Rochester, NY 14642, USA
e-mail: joseph.wedekind@rochester.edu

J.A. Liberman • M. Salim
Department of Biochemistry & Biophysics, University of Rochester School of Medicine and Dentistry, Rochester, NY 14642, USA

Center for RNA Biology, University of Rochester School of Medicine and Dentistry, Rochester, NY 14642, USA

V.A. Erdmann et al. (eds.), *Chemical Biology of Nucleic Acids*, RNA Technologies, DOI 10.1007/978-3-642-54452-1_23, © Springer-Verlag Berlin Heidelberg 2014

Abstract Riboswitches are functional RNA elements located most frequently within the 5′-leader sequences of bacterial mRNAs. By directly binding to small molecules via an aptamer domain, a riboswitch can adapt quickly to changes in the concentration of a specific intracellular ligand, thereby establishing a feedback loop that controls gene expression. Here we discuss methods utilized in the structure determination of evolutionarily distinct classes of $preQ_1$ riboswitches known as class I and II, respectively. These riboswitches "sense" the pyrrolopyrimidine metabolite $preQ_1$—an intermediate on the biosynthetic pathway that produces the hypermodified tRNA base queuosine, which imparts translational fidelity. Herein, we describe (1) the use of isothermal titration calorimetry (ITC) to explore metal ion requirements for ligand binding and (2) modifications to crystallization media containing SO_4^{2-} or Na^+ that were necessary for phase determination using site-bound $Os(NH_3)_5^{3+}$ or Cs^+ ions, respectively. Our experience has shown that simple manipulations to the mother liquor can lead to favorable binding of the latter ions without the need to engineer metal-binding sites, thus making our methods a first-choice approach that is broadly applicable to functional RNAs.

Keywords Riboswitch • RNA crystallography • Isothermal titration calorimetry • Cesium • Osmium • Mother liquor • Heavy atom • $PreQ_1$

1 Introduction

RNA is central to life and has proven to be an exceptionally diverse molecule in terms of its functional repertoire (Serganov and Patel 2007; Clancy 2008). One activity of recent note is that certain structural elements, called riboswitches, can regulate gene expression by controlling mRNA transcription and translation (Winkler and Breaker 2003; Bastet et al. 2011) and less frequently by influencing intron excision (Kubodera et al. 2003; Lee et al. 2010; Chen et al. 2011), and mRNA stability via backbone self scission (Winkler et al. 2004); see Serganov and Nudler (2013) for a review. In the most common scenario, a riboswitch detects a small molecule that binds to its 'aptamer domain.' Such binding causes the riboswitch to undergo a conformational change that sequesters or unfurls mRNA sequences in a downstream expression platform, leading to gene "on" or "off" states (Mironov et al. 2002; Winkler et al. 2002; Nahvi et al. 2002; Sudarsan et al. 2003).

In this chapter, we review techniques used to characterize two phylogenetically unrelated riboswitches that bind the same secondary metabolite, $preQ_1$—an intermediate in the queuosine (Q) biosynthetic pathway (Fig. 1a). These riboswitches are designated class 1 ($preQ_1$-I) and class 2 ($preQ_1$-II) and were identified in the genomes of numerous bacteria within the Firmicutes phylum. $PreQ_1$-I riboswitches,

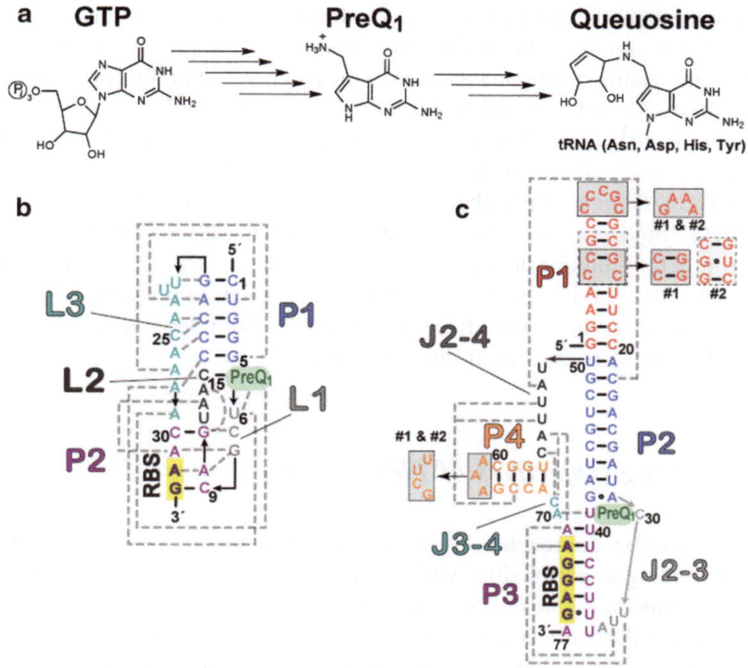

Fig. 1 Queuosine biosynthesis in bacteria and schematic structure of preQ$_1$-I and preQ$_1$-II riboswitches reviewed herein. (**a**) The bacterial biosynthetic pathway from GTP to queuosine (Q) involves several enzymes (*arrows*) that lead to the soluble intermediate preQ$_1$, which is incorporated into tRNA at that wobble position to ultimately give Q as a means to confer translational fidelity (reviewed in McCarty and Bandarian 2012). (**b**) Representative class 1 preQ$_1$ riboswitch from *Thermoanaerobacter tengcongensis* (Tte) depicting the H-type pseudoknot with tertiary hydrogen-bond interactions (*gray*) from crystal structures (Jenkins et al. 2011; Spitale et al. 2009). Watson–Crick pairs are depicted as short, *solid lines*; *arrows* indicate connectivity. The ribosome-binding site (RBS) is highlighted in *yellow*. (**c**) Representative class 2 preQ$_1$ riboswitch from *Lactobacillus rhamnosus* (Lra) depicting the HL$_{out}$ pseudoknot. Interactions and sequence features are as described in (**b**) based on the structure (Liberman et al. 2013). Boxed regions (*gray*) show wild-type sequences that were modified to produce constructs #1 and #2, respectively

such as that from *B. subtilis*, reside in the 5′-leader sequence of the *queC* gene located in the *queCDEF* operon, which encodes enzymes essential for synthesis of preQ$_1$. PreQ$_1$-I functions as a “switch” because high cellular levels of preQ$_1$ favor metabolite binding to its aptamer domain, thus promoting formation of a rho-independent transcription terminator within its downstream expression platform. When preQ$_1$ levels are low, the expression platform forms an anti-terminator stemloop that favors gene transcription (Roth et al. 2007). Other preQ$_1$-I riboswitches appear to regulate translation, such as that from *T. tengcongensis* (Tte) (Fig. 1b). The Tte preQ$_1$-I riboswitch is located within the 5′-leader sequence of COG4708 (DUF988) mRNAs that encode a putative preQ$_1$ transporter. In the

$preQ_1$-bound state, the aptamer adopts a conformation that buries the first two bases of the ribosome-binding site (RBS), thus blocking ribosome access to the expression platform leading to attenuated gene expression. When $preQ_1$ levels are low, the block is relieved and translation ensues (Roth et al. 2007; Spitale et al. 2009; Jenkins et al. 2011). Similar feedback regulation has been ascribed to $preQ_1$-II riboswitches, such as that from *L. rhamnosus* (Fig. 1c). Such riboswitches also govern COG4708 genes (Weinberg et al. 2007) although they bury the entire RBS when $preQ_1$ binds to the aptamer (Liberman et al. 2013; Meyer et al. 2008). Because $preQ_1$ is unique to the bacterial metabolome (McCarty and Bandarian 2012), and class 1 and 2 $preQ_1$ riboswitches are present in a variety of pathogenic bacteria (Meyer et al. 2008; Roth et al. 2007; Weinberg et al. 2007), these riboswitches are of interest as antibacterial targets.

1.1 Learning Objectives of This Chapter

Here we describe approaches used to define the metal dependence of small-molecule binding to $preQ_1$ riboswitches by isothermal titration calorimetry (ITC), as well as methods employed to introduce osmium(III) pentaammine and cesium (I) heavy-atom derivatives into the class 1 and 2 $preQ_1$ riboswitches for multi-wavelength anomalous diffraction (MAD) or single-wavelength anomalous diffraction (SAD) phasing, respectively (Spitale et al. 2009; Jenkins et al. 2011; Liberman et al. 2013). The methods are written for graduate students or post-doctoral trainees with preliminary to intermediate knowledge of these topics; the information is meant as a practical guide rather than a pedagogical one. A key aspect of our methods is that they are broadly applicable to other functional RNAs and should stimulate the development of derivative approaches in the reader's laboratory.

2 Isothermal Titration Calorimetry Methods for Riboswitches

2.1 Quantitative Procedures to Test Metal Ion Requirements for Ligand Binding

Ions play a significant role in RNA folding and function. In this section, we describe methods to probe riboswitch ion requirements by measuring ligand affinity in the absence or presence of various ions. K^+ and Mg^{2+} are likely to be of greatest interest since they are prominent in the bacterial cell. K^+ concentrations range from 0.1 to 0.6 M (McLaggan et al. 1994), whereas free Mg^{2+} is present at 1–2 mM (Alatossava et al. 1985). Mg^{2+} promotes RNA folding and tertiary interactions (Pyle 2002; Draper et al. 2005; Woodson 2005) but can also impart function by participating in the chemical reactions of ribozymes (Nakano et al. 2001; Chen et al. 2010; Toor et al. 2009; Marcia and Pyle 2012; Stahley and Strobel 2006). In the case of riboswitches, Mg^{2+} shifts the folding equilibrium of most aptamer domains into a "binding-competent" conformation that interacts with ligand (Montange and Batey 2008). Some riboswitches recognize their ligand in the absence of Mg^{2+}, such as the guanine (xptG) riboswitch (Serganov et al. 2004), the SAH riboswitch (Edwards et al. 2010), as well as the $preQ_1$-I and $preQ_1$-II translational riboswitches (Sect. 2.5 and Liberman et al. 2013); however, Mg^{2+} still drives these RNAs into a binding-competent state. In addition to imparting structural stability, some riboswitches bind synergistically to Mg^{2+} and ligands, such as the glmS, TPP, FMN, and glycine riboswitches (Klein and Ferre-D'Amare 2006; Serganov et al. 2006, 2009; Lipfert et al. 2010); see Ferre-D'Amare and Winkler (2011) for a review. Such synergy has been noted for the FMN riboswitch interaction with K^+ (Serganov et al. 2009). In addition to binding Mg^{2+}, the *Oceanobacillus iheyensis* group IIc intron selectively binds K^+ based on size, which forms part of its catalytic metal cluster (Marcia and Pyle 2012). Collectively these observations illustrate that complete characterization of a functional RNA requires analysis of the metal dependence of its activity, which should be the preface to any structural investigation.

2.2 Checking the Feasibility of Isothermal Titration Calorimetry Measurements

ITC is a quantitative method to characterize the binding thermodynamics for any two interacting molecules. This label-free approach offers a full characterization of the enthalpy ($\Delta H°$), entropy ($\Delta S°$), and free-energy ($\Delta G°$) changes, as well as the equilibrium association constant, K_a; note K_a equals $1/K_d$ (the equilibrium dissociation constant). Although the approach can require large amounts of material, it is operative over a wide range of affinities between 1 mM and 1 nM. This range, however, is dictated by the respective ligand and receptor solubilities (i.e. low-affinity interactions), or what is detectable in terms of the measured heats of injection (i.e. very high-affinity interactions). Several key experimental design aspects should be considered as a preface to conducting ITC. The most important of these is the shape of the binding isotherm, which is determined by the unitless parameter c, where $c = (([M] \times n) \div K_d)$. The c value is determined by K_d, the concentration of the molecule $[M]$ in the sample cell, and the stoichiometry (n) of binding. In practice, one should strive for $20 < c < 100$ (Myszka et al. 2003) to generate measurements at the bottom, middle (inflection), and top of the titration curve; the overall shape for single-site binding should be sigmoidal. Populating the curve with data points is especially important near the inflection region since these observations are needed to accurately fit the K_a. Because we typically have an estimate of the K_d for the interaction in question, we employ the equation: $[M] = (c \times K_d) \div n$, which allows determination of the necessary molecular concentration in the sample cell. A good starting point for the titrant concentration in the syringe is 7–10 times the value of [*M*]. Upon collecting preliminary data, this relationship should be adjusted to reach saturation. Achieving a regular baseline after each injection is important. The heat of ligand dilution following saturation may be accounted for by subtracting the average of the last three points of the saturated thermogram from the total heat of each prior injection. This will provide greater accuracy in the integrated heats that are used in nonlinear least squares curve fitting to obtain the values of $\Delta H°$, n and K_a. Accurate knowledge of the concentrations of the molecules under investigation is imperative.

2.3 General Notes About ITC Instrument Settings

The following settings apply to a MicroCal VP-ITC system (GE Healthcare). If it has not been filled recently, or if prior experiments were run at elevated temperatures, the reference cell should be emptied and refilled with degassed, distilled H_2O. We begin investigating most binding interactions at a cell temperature of 30 °C; the temperature can be adjusted on a case-by-case basis to alter ΔH (and raw heats) to optimize the binding isotherm. A typical experiment begins with ~30 injections from a 250 μl syringe with an initial injection of 3 μL (discarded during fitting). The remaining injections are 10 μL with 20 s duration per injection. The reference power is 15 μcal s^{-1}, stirring speed 307 rpm, and the feedback mode/gain is 'high'. These settings are suitable to interrogate most macromolecular interactions (Suddala et al. 2013; Liberman et al. 2013; Salter et al. 2012). The time between injections should be sufficient for the interaction heat to return to baseline before the next injection begins; 240 s is a reasonable starting point but can be adjusted as needed. One should also strive for an average integrated heat of 3–5 μcal evolved (or absorbed) per injection for the first two-thirds of the injections. The sensitivity of the VP-ITC is about 0.1 μcal.

2.4 Preparation of Samples for ITC

The Tte $preQ_1$-I riboswitch 33-mer (Fig. 1b) was produced by chemical synthesis and HPLC purified by C18 reverse-phase separation (Wedekind and McKay 2000; Spitale et al. 2009; Spitale and Wedekind 2009). The Lra $preQ_1$-II riboswitch 77-mer (Fig. 1c) was generated by in vitro transcription and purified by denaturing PAGE (Lippa et al. 2012; Liberman et al. 2013). The $preQ_1$ ligand used for ITC analysis was prepared by chemical synthesis (LeadGen Labs, LLC). Lyophilized RNA samples were stored at −20 °C until needed. Samples were dissolved in 0.050 M Na-HEPES pH 7.0 containing 0.10 M NaCl. For $MgCl_2$-free conditions 0.5 mM EDTA at pH 7.0 was included in lieu of multivalent ions. The RNA solution was heated to 65 °C for 5 min, then $MgCl_2$ was added slowly to 6.0 mM

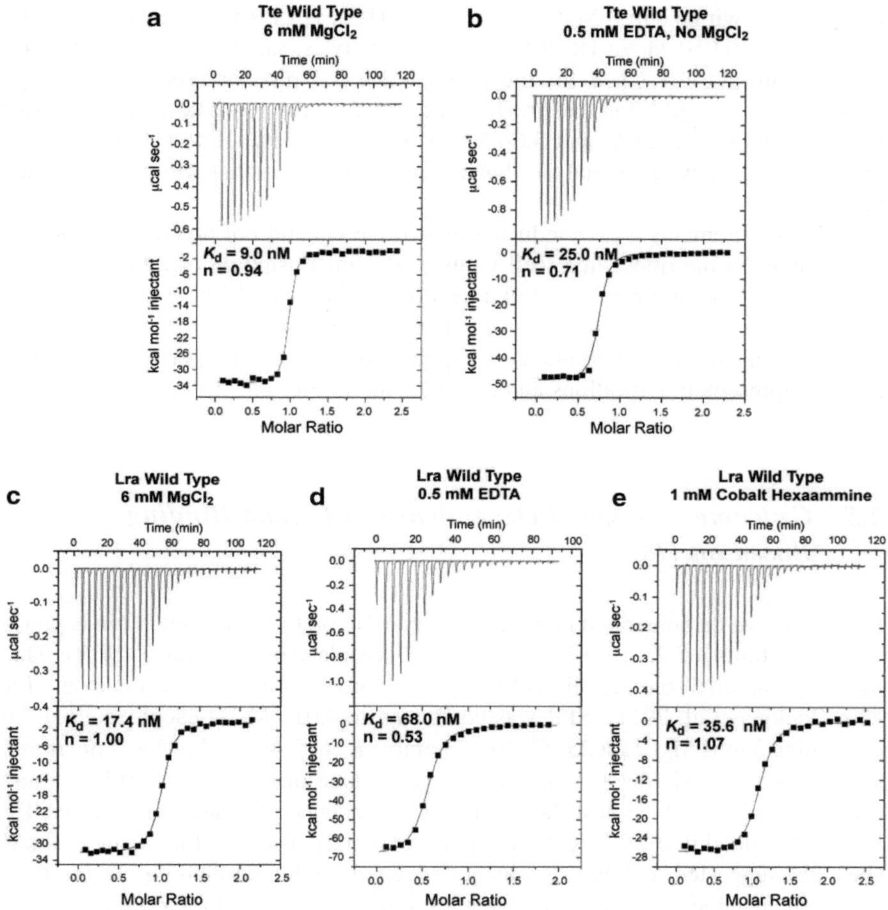

Fig. 2 Representative ITC titrations and curve fits for ligand binding to preQ$_1$-I and preQ$_1$-II riboswitches in the presence of EDTA or various Mg^{2+} or $Co(NH_3)_6^{3+}$ levels. (**a**) PreQ$_1$ interaction with the Tte 33-mer (class 1) wild-type sequence in 0.10 M NaCl, 0.006 M $MgCl_2$, and 0.050 M HEPES pH 7.0 at 25 °C; the *c* value was 522. (**b**) PreQ$_1$ binding to the Tte 33-mer under conditions in (**a**) except that $MgCl_2$ was replaced with 0.5 mM EDTA; the *c* value was 228. (**c**) PreQ$_1$ binding to the wild-type Lra 77-mer under conditions in (**a**); the *c* value was 184. (**d**) PreQ$_1$ interaction with the wild-type Lra 77-mer under conditions described in (**b**); the *c* value was 63. (**e**) PreQ$_1$ binding to the wild-type Lra riboswitch under conditions in (**a**) except that $MgCl_2$ was replaced with 0.001 M $Co(NH_3)_6Cl_3$; the *c* value was 101. In cases where $c > 100$, such values are a compromise because reduced RNA concentrations gave low heats of injection. The resulting K_d values should be treated circumspectly. The average values for two or more measurements are reported in the main text

or $Co(NH_3)_6Cl_3$ was added to 1.0 mM, followed by slow cooling to 24 °C. $MgCl_2$, and $Co(NH_3)_6Cl_3$ were omitted completely for folding conditions containing EDTA. Each riboswitch sample was dialyzed overnight at 4 °C against 4 L of

0.10 M NaCl with 6.0 mM $MgCl_2$, 1 mM $Co(NH_3)_6Cl_3$, or 0.5 mM EDTA, buffered at pH 7.0 by 0.050 M Na-HEPES. Following dialysis, samples were diluted with dialysis buffer to: 4.7 μM for the $preQ_1$-I and 3.3 μM for $preQ_1$-II wild-type riboswitches; 5.7 μM for the $preQ_1$-I and 1.5–5.7 μM for wild-type sequences folded in 0.5 mM EDTA; and 3.6 μM for wild-type $preQ_1$-II with $Co(NH_3)_6Cl_3$. $PreQ_1$ was dissolved in dialysis buffer to a concentration tenfold higher than the RNA.

ITC measurements were conducted by syringe injection of $preQ_1$ into a sample cell containing the riboswitch (cell volume ~1.7 mL) using 28 or 29 injections of 10 μL each—except for the first 3 μL injection. An interval of 120 or 260 s was used between injections. Thermograms were analyzed with Origin 7.0 (MicroCal) using a 1:1 binding model. Experiments for publication should be performed at least twice. Representative titrations and curve fits are in Fig. 2.

2.5 *Outcomes for Metal Dependence of Ligand Binding Based on ITC*

Previously we demonstrated that the wild-type Tte $preQ_1$-I riboswitch binds $preQ_1$ with an affinity of 2.1 ± 0.3 nM using surface plasmon resonance (SPR). This method required 5′-biotinylation of the RNA aptamer for immobilization to an SPR chip (Jenkins et al. 2011). SPR was conducted in 0.010 M Na-cacodylate pH 7.0 containing 3 mM $MgCl_2$ at 25 °C. A comparable K_d of 7.4 ± 2.3 nM was measured at pH 7.0 in 6 mM $MgCl_2$ with an average n value of 0.98 ± 0.07 (Fig. 2a and Suddala et al. 2013). To test the metal-ion dependence of ligand binding, the Tte RNA was folded in 0.5 mM EDTA in place of metals. The resulting K_d of 25.0 nM (Fig. 2b) suggested that $MgCl_2$ was not required for ligand binding but the relatively low n value of 0.7 suggested that a significant portion of the riboswitch does not adopt a binding-competent conformation. A similar trend was observed for the Lra $preQ_1$-II riboswitch in which average K_d values for $preQ_1$ binding were 17.9 ± 0.6 nM and 72.5 ± 6.3 nM in the presence and absence of Mg^{2+}, respectively (Fig. 2c, d and Liberman et al. 2013). Despite a modest factor-of-four decrease in affinity, the average n value in the presence of EDTA was 0.57 ± 0.05, suggesting that slightly more than half of the riboswitch adopts a $preQ_1$ binding-competent state. In this manner, the ion dependence of ligand binding can be explored, as well as the need for inner-sphere or outer-sphere interactions (reviewed in Wedekind 2011). For example, the crystal structure showed Mg^{2+} binding by the $preQ_1$-II riboswitch via inner- and outer-sphere contacts (Liberman et al. 2013). However, ITC measurements indicated $Co(NH_3)_6^{3+}$ can substitute for Mg^{2+} in ligand binding (Fig. 2e), based on a minimal increase in K_d. This result suggests inner-sphere ion coordination is not a strict requirement for $preQ_1$ binding (Liberman et al. 2013).

3 General Approaches to Introduce Heavy Atoms into Riboswitches

3.1 Mother Liquor Manipulations to Promote Heavy Atom Binding

For novel RNA folds, it is often necessary to determine phases for the measured structure factor amplitudes (F_{obs}) by use of experimental approaches. This requires an empirical search for conditions that lead to the identification of one or more "heavy atom" binding sites that can be used to establish phases for the non-hydrogen RNA atoms (Wedekind and McKay 2000). For the preQ_1-I and preQ_1-II riboswitches, determination of the respective phases presented unique challenges that required manipulation of the mother liquor to promote heavy atom binding (Spitale and Wedekind 2009; Liberman et al. 2013). PreQ_1-I riboswitch crystals grew from high-salt mother liquors containing the negatively charged SO_4^{2-} ion, which competes with the RNA for binding $Os(NH_3)_5^{3+}$ and variants thereof, thus requiring sulfate removal from crystals. By contrast, the poor reproducibility of preQ_1-II crystals—approximately one useful crystal for every seven screened for X-ray diffraction—required co-crystallization of Cs^+ with the riboswitch in order to determine the structure by use of single-crystal approaches. Here we review techniques that assisted heavy-atom binding and ultimately led to high-resolution structure determinations of these, heretofore unknown, RNA folds.

3.2 Eliminating SO_4^{2-} to Promote $Os(NH_3)_5^{3+}$ Binding

High-salt mother liquors are useful to grow RNA crystals but can predispose subsequent heavy atom-derivative searches to grave failure. Although strategic placement of halogenated pyrimidines can assist with heavy-atom phasing (Golden 2000; Pley et al. 1994; Correll et al. 1997b), there are examples in which high-salt mother liquors have been supplanted by low ionic strength conditions that favor site-specific ion binding due to reduced electrostatic shielding. Such changes can be made at the level of RNA crystal growth or by in situ methods that entail slow replacement of salt within the crystal. In the case of the minimal hairpin ribozyme, crystals could be grown from solutions of 1.8 M $(NH_4)_2SO_4$ as well as lower ionic strength 0.25 M Li_2SO_4 supplemented with 22–24 % (w/v) poly(ethylene) glycol 2000 monomethyl ether (Alam et al. 2005).

By contrast, useful preQ_1-I riboswitch crystals grew only from solutions of 1.8 M Li_2SO_4 (Spitale et al. 2009). Although a well diffracting, low-ionic-strength crystal form grew from 2-methyl-2,4-pentanediol (Lippa et al. 2012), merohedral twinning thwarted phasing attempts (Torelli 2008). Efforts to prepare heavy-atom derivatives of the Li_2SO_4 crystal form using halogenated pyrimidines and heavy-atom soaks proved unsuccessful. To overcome the limited solubility and poor reactivity of heavy-atom compounds in the high-salt mother liquor, crystals were first placed into a "stabilization" solution comprising 2.0 M Li_2SO_4, 0.10 M Na-cacodylate pH 6.0, 0.02 M $MgSO_4$, 5 % (v/v) 1,3-propanediol, 2 mM spermine, and saturating amounts of ligand. Crystals were then serially transferred into a "derivative" solution of 4.0 M LiOAc, 0.02 M $Mg(OAc)_2$, 0.10 M Na-cacodylate pH 6.5, 5 % (v/v) 1,3-propanediol, 2 mM spermine, saturating ligand, and 0.10 M pentaammine-(trifluoro-methane-sulfonato) Os(III) triflate (Sigma-Aldrich). Our rationale was that we previously grew hammerhead ribozyme crystals from 1.9 to 2.2 M NH_4SO_4 as well as 3.2 M KOAc (Wedekind and McKay 2000), suggesting that both anions were compatible with crystals. Additionally, sulfate adopts multiple modes of metal coordination via uni- or bidentate interactions (Cotton and Wilkinson 1988) and is known to produce insoluble group II and transition-metal complexes. By contrast, acetate coordinates unidentately making it a relatively weak metal binder.

During the exchange of SO_4^{2-} with OAc^-, we maintained constant ionic strength by assuming 4.0 M LiOAc was equivalent to 2.0 M $Li_2(SO_4)$. The exchange solutions were mixed in the following derivative-to-stabilizing solution ratios: 1:3, 1:1, and 3:1; each transfer lasted 15 min and crystals showed no signs of osmotic shock. After the final crystal transfer into the derivative solution, crystals were incubated for 3 h. High acetate concentrations also serve as a cryoprotectant (Wedekind and McKay 2000), which allowed samples to be flash frozen directly by plunging into $N_2(l)$. MAD phasing of preQ$_1$-I crystals has been described elsewhere (Spitale et al. 2009).

3.3 The Mode of $Os(NH_3)_5^{3+}$ Binding to the PreQ$_1$-I Riboswitch

Inspection of the preQ$_1$-I riboswitch structure revealed the mode of $Os(NH_3)_5^{3+}$ binding to RNA (Spitale et al. 2009). Anomalous difference Fourier maps revealed two binding sites: a minor site at the sugar edge of Uri21 and a major site on the twofold axis between preQ$_1$-I riboswitch molecules (Fig. 3). The latter doubly weighted peak had a 30 σ height and the ion received coordination from six oxygens. Due to the isotropy of pentaammine binding, it was not possible to discern the identity of the sixth, inner-sphere ligand in the osmium coordination sphere; such spatial averaging is not always the case (Cate and Doudna 1996). Notably, our observations differ from osmium(III) hexammine binding to the P4–P6 domain of the group I intron, which involved binding at tandem G•U wobble base pairs (Cate and Doudna 1996). Similar G•U binding modes were reported for $Ir(NH_3)_6^{3+}$ in a comprehensive analysis of major-groove ion binding motifs (Keel et al. 2007); at present, it is unclear whether $Os(NH_3)_5^{3+}$ can bind G•U pairs in the manner

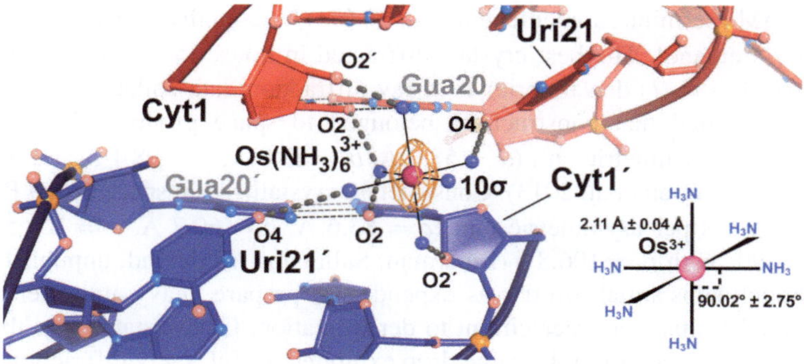

Fig. 3 Site-bound Os(III) at the dyad interface of the Tte $preQ_1$-I riboswitch. The outer-sphere Os $(NH_3)_5^{3+}$ amines bind to the riboswitch via interactions with the O2-keto and O2′ sugar of Cyt1, as well as O4 keto contacts from Uri21 (*red model*). Because the $Os(NH_3)_5^{3+}$ resides on a proper twofold axis, these interaction are duplicated by a symmetry-related molecule (*blue model*) denoted by a prime (′). The anomalous difference electron density map (*orange mesh*) is contoured at 10σ. (*Inset*) The average octahedral geometry for $Os(NH_3)_6^{3+}$ (reviewed in Wedekind 2011)

documented for hexammine ions. Nonetheless, we favor pentaammine-(trifluoromethane-sulfonato) Os(III) triflate as a first choice RNA derivative because it is site selective, commercially available, does not require G•U pairs in the sequence, and circumvents the need for in-house synthesis of $Os(NH_3)_6^{3+}$ or $Ir(NH_3)_6^{3+}$ (reviewed in Wedekind 2011).

3.4 Co-crystallization with Cs^+ for Single-Wavelength Anomalous Diffraction Phasing

To identify crystals of the $preQ_1$-II riboswitch, we screened sequences from several species including: *Streptococcus gordonii*, *S. pneumoniae*, *Lactobacillus casei*, and *L. rhamnosus*. The best X-ray-diffraction-quality crystals grew from a modified variant of the *L. rhamnosus* sequence (Lippa et al. 2012; Liberman et al. 2013) in which nonconserved stem loops in P1 and P4 were replaced with stable GNRA and UNCG tetraloops, respectively (Fig. 1c, construct #1) (Liberman et al. 2013). In another variation, P1 was modified to include a G•U wobble (Fig. 1c, construct #2) that was shown to function as a phasing module due to its avidity for $Os(NH_3)_6^{3+}$, Ir $(NH_3)_6^{3+}$ or Cs^+ (Keel et al. 2007). Constructs #1 and #2 yielded crystals by the vapor-diffusion method from solutions of: 14.4–14.8 % (*w*/*v*) poly(ethylene) glycol 6000 (PEG6K), 0.14–0.16 M $MgOAc_2$, 0.05 M Na-cacodylate pH 6.0, 1–2 mM spermine, and 0.20 M NaCl. Crystals appeared at 20 °C and grew as rectangular plates to 0.20 mm × 0.05 mm × 0.02 mm within 2–3 weeks. Cryo-protection was achieved by a 30 s transfer into synthetic mother liquor comprising: 17.2–17.8 % (*w*/*v*) PEG6K, 168-192 mM $MgOAc_2$, 0.24 M NaCl, 0.06 M Na-cacodylate pH 6.0, and 1.2 mM spermine supplemented with 20 % (*v*/*v*) 2-methyl-2,4-pentanediol and 20 % (*v*/*v*) ethanol. The best crystals diffracted in house to 2.8 Å resolution and were stored in $N_2(l)$ dewars prior to X-ray diffraction data collection. Diffraction analysis revealed that construct #1 belonged to space group $C222_1$ with one molecule per asymmetric unit ($a = 58.0$ Å, $b = 86.0$ Å, $c = 98.1$ Å and $\alpha = \beta = \gamma = 90°$) (Liberman et al. 2013). Construct #2 crystallized in space group $P2_1$ with four molecules per asymmetric unit ($a = 83.6$ Å, $b = 60.7$ Å, $c = 101.5$ Å and $\alpha = \gamma = 90°$ with $\beta = 106.8°$) (Liberman, Salim and Wedekind, unpublished).

Although substantial effort was expended to prepare heavy-atom derivatives, both crystal forms were recalcitrant to derivatization. One obstacle was that only one in seven crystals from the same drop exhibited useful X-ray diffraction, which hindered efforts to generate a starting supply of crystals that could be subjected to heavy-atom screening. Another shortcoming was a paucity of $Ir(NH_3)_6^{3+}$ binding based on the absence of a measurable anomalous signal. The latter phenomenon was most likely the result of elevated Mg^{2+} concentrations in mother liquors that were necessary to obtain high-quality X-ray diffraction. Even though construct #2 possessed a G•U wobble, it had no appreciable $Ir(NH_3)_6^{3+}$ binding.

To surmount difficulties, we undertook co-crystallization in the presence of CsCl in lieu of NaCl. Substitution of Na^+ with Cs^+ caused no apparent changes in space group despite notable differences between these ions. Cs^+ is electron rich (54 e^-) relative to Na^+ (10 e^-) and forms as many as 11 inner-sphere contacts. A search of the Cambridge Structural Database (Allen 2002) version 5.34 using ConQuest v1.15 (Bruno et al. 2002) revealed an average Cs^+ to O–C distance of 3.23 ± 0.16 Å, an average Cs^+ to N–C distance of 3.31 ± 0.15 Å, (where the O–C and N–C bond was of any type), and an average Cs^+ to water distance of 3.27 ± 0.16 Å. Such distances are significantly longer than those for Na^+ to O–C, as well as K^+ to O–C, which exhibited values of 2.53 ± 0.17 and 2.83 ± 0.08 Å, respectively.

The presence of Cs^+ improved the resolution of X-ray diffraction for $preQ_1$-II crystals, as well as the reproducibility of well-diffracting crystals. Of particular significance is the fact that Cs^+ exhibits a +9 electron anomalous-diffraction signal (f'') at 1.70 Å resolution, which enabled use of the site-bound Cs^+ ions of construct #1 as a means to generate high-quality SAD phases for a complete structure determination (Liberman et al. 2013). By comparison, the low-symmetry space group of construct #2 crystals hampered collection of the anomalous signal, preventing an independent structure determination. Nonetheless, the latter structure was solved by molecular replacement, which yielded the locations of a number of Cs^+ ions whose identities were corroborated by anomalous difference Fourier maps (Liberman, Salim and Wedekind, unpublished). We also conducted co-crystallization of the Lra $preQ_1$-II riboswitch from Br^- and I^- salts in lieu of Cl^-. The latter ions produced well diffracting crystals using Na^+ as a counter ion, but no significant anomalous diffraction signal was detected. Our results suggest that halide salts are not useful for RNA phasing, in contrast to proteins (Dauter and Dauter 2001).

3.5 The Binding Mode of Cs^+ to a G•U Wobble Pair and Other Sites

We included a single-site G•U wobble pair in stem P1 of the preQ$_1$-II riboswitch with the goal of using it as a phasing module (Fig. 1c, construct #2). This approach was described in a thorough analysis on the propensity of single and tandem G•U wobble sites to bind $Os(NH_3)_6^{3+}$ or $Ir(NH_3)_6^{3+}$ for heavy-atom phasing (Keel et al. 2007). We chose the sequence 5′-GGC-3′/3′-CUG-5′, which features a single-wobble pair, dubbed phasing module 14 (PM14) (Keel et al. 2007). PM14 was stated to have advantages over other G•U motifs including (1) metal binding with one of the highest anomalous signals and lowest B-factors relative to a reference hexammine site and (2) inclusion of a single wobble requires only three nucleotides to be altered in the target RNA sequence rather than four (Keel et al. 2007). It was also noted that the G•U wobble of PM14 binds Cs^+, which proved useful in phasing the SAM-II riboswitch wherein NH_4^+ in the mother liquor was replaced with Cs^+, introduced by crystal soaking (Keel et al. 2007; Gilbert et al. 2008). As noted above, the preQ$_1$-II riboswitch was co-crystallized with 0.15 M Cs^+ yielding 15 site-bound Cs^+ ions in construct #1 (one molecule per asymmetric unit) (Liberman et al. 2013) and 35 Cs^+ ions in construct #2 (four molecules per asymmetric unit) (Liberman, Salim and Wedekind, unpublished).

The strongest anomalous difference Fourier signal in electron density maps was observed for Cs^+ bound at the G•U wobble of construct #2; the average peak height was $21 \pm 2\ \sigma\ \delta$ for all four riboswitch copies in the asymmetric unit. Because construct #1 did not possess such a phasing module (Fig. 1c), Cs^+ was not detected at a spatially equivalent location. Cs^+ coordination at the G•U wobble site of construct #2 entailed contacts in the P1 major groove (Fig. 4a), which concur with the mode of Cs^+ coordination at PM14 in the SAM-II riboswitch (Gilbert et al. 2008). A variety of other single G•U wobble pairs have been reported to bind Cs^+ including: 5′-GGA-3′/3′-CUG-5′ in the lysine riboswitch (Serganov et al. 2008); 5′-GGA-3′/3′-CUU-5′ in the FMN riboswitch (Serganov et al. 2009); as well as 5′-GGU-3′/3′-CUA-5′ and 5′-CGU-3′/3′-GUA-5′ from the *O. iheyensis* group IIc intron (Marcia and Pyle 2012). Each of the latter, site-bound ions was introduced by soaking crystals in solutions of Cs^+. (Wobble sites engaged in tertiary or crystal contacts were not considered in the sequences above).

The low-symmetry $P2_1$ space group of our preQ$_1$-II riboswitch (construct #2) contains four molecules per asymmetric unit, which impeded high-resolution SAD phasing. As such, we focused on the higher symmetry $C222_1$ space group of construct #1, which proved successful despite the absence of a G•U pair for ion binding. Serendipitous Cs^+ binding prompted us to ask (1) *what sequences common to both preQ$_1$-II constructs exhibited appreciable levels of Cs^+ binding based on the measurable anomalous signal?* and (2) *which of the latter Cs^+ sites are present in secondary structure that does not strictly depend upon tertiary or crystal contacts for ion binding?* Of the 15 Cs^+ sites identified and modeled in the structure of construct #1, the average anomalous peak height was 8.9 σ and the average

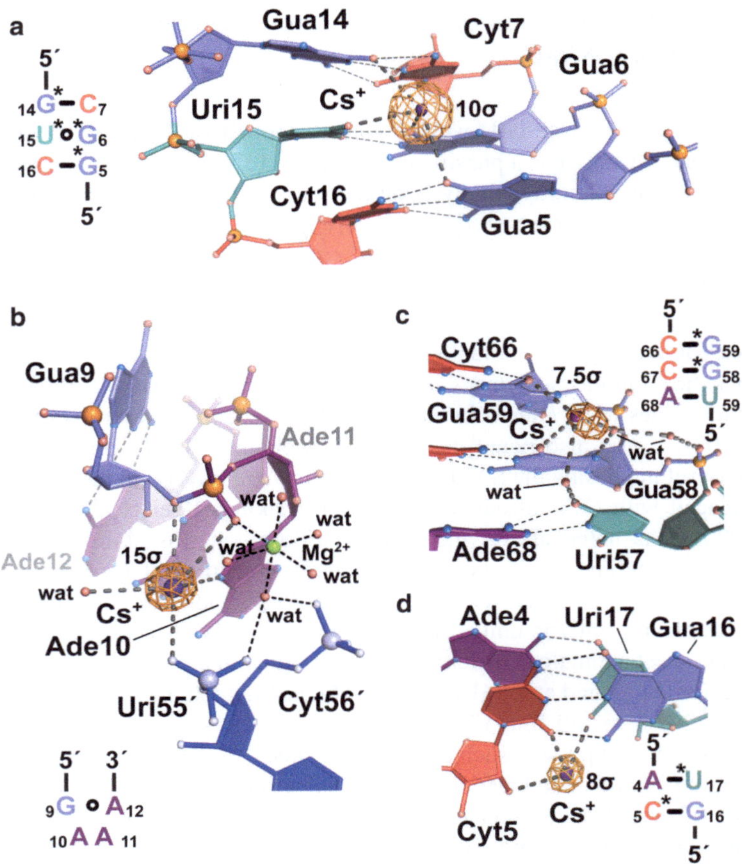

Fig. 4 Representative site-bound Cs^+ ions in the Lra preQ$_1$-II riboswitch G•U wobble, GNRA tetraloop and secondary structures with accompanying anomalous difference Fourier maps. (**a**) The mode of G•U wobble coordination to Cs^+ in construct #2 entails inner-sphere contacts with the keto oxygens of Gua5, 6 and 14, and Uri15 in the P1 major groove. Here and elsewhere the secondary structure is shown; *asterisks* (*) indicate base keto-oxygen coordination to Cs^+. (**b**) GNRA tetraloop coordination to Cs^+ utilizes N7 of Ade10, O3′ of Gua9, and nonbridging phosphate oxygens from Ade10 and Uri55′; the latter interaction arises from a crystal contact (blue ball-and-stick-model). A nearby Mg^{2+} shows inner-sphere coordination to the nonbridging phosphate oxygen from Ade10. (**c**) Major groove Cs^+ binding by tandem Gua bases. The O6 groups of Gua58 and Gua59 make inner-sphere Cs^+ contacts whereas N7 of Gua58 and the O4 keto of Uri57 serve to position waters for ion coordination. (**d**) Minor groove Cs^+ binding at the sugar-edge of Cyt5. Coordination entails the O2 keto and 2′-O groups of Cyt5 and the cross-strand O2 keto group of Uri17

refined *B*-factor was 43 Å^2. Our results revealed the highest occupancy and largest anomalous signal for Cs^+ at Ade10, which resides in apposition to the phosphodiester bond between Gua9 and Ade10 within the GNRA tetraloop that caps P1 (Fig. 4b). This site has an average peak height of 17.4 ± 3.5 σ based on all five molecules of constructs #1 and #2. The strong occupancy is likely derived from four inner-sphere contacts originating from three GNRA ligands, as well as a nonbridging phosphate oxygen donated by a crystal contact (Fig. 4b). A nearby Mg^{2+} coordinates to the backbone of Ade10, which simultaneously makes an inner-sphere contact to the GNRA-bound Cs^+. Mg^{2+} aside, a similar constellation of Cs^+ ligands was observed in the GNRA tetraloop of the *O. iheyensis* group IIc intron, including N7 of base Ade33 in the "N" position of the tetraloop (Marcia and Pyle 2012). Like the $preQ_1$-II riboswitch, the coordination sphere of this ion is completed by inner-sphere coordination to a crystal contact, which draws into question the generality of the GAAA motif as a "sure-fire" Cs^+-binding site.

Two other Cs^+ sites were prominent in $preQ_1$-II structures of constructs #1 and #2. Both sites are located in helical regions devoid of tertiary or crystal contacts. The first site was bound in the major groove of P4 at the sequence 5′-UGG-3′/3′-ACC-5′. The average height of the anomalous difference Fourier peak was $4.9 \pm 2.8\ \sigma$. Here, tandem Gua bases appear important in Cs^+ binding (Fig. 4c). A comparable mode of coordination is observed at the base of a GNRA tetraloop in the *O. iheyensis* group IIc intron in the sequence 5′-UGG-3′/3′-ACA-5′, wherein the G•A pair corresponds to the first and last residues of a GNRA tetraloop. Specifically, Gua275 at position *i* of the tetraloop forms a sheared pair with Ade278 at position $i + 3$. The keto oxygens of the sequence 5′-UGG-3′ each make inner-sphere contacts to Cs^+ (Marcia and Pyle 2012).

The involvement of O6 groups from tandem guanines to coordinate Cs^+ has also been observed at Gua29/30 of the *O. iheyensis* group IIc intron (Marcia and Pyle 2012). Other tandem Gua sequences are present but these make prominent use of tertiary contacts for ion binding. Tandem Gua nucleobases—specifically tandem G•U wobbles—have long been known as hotspots for hexammine binding (Cate and Doudna 1996; Keel et al. 2007). Moreover, tandem Gua bases in RNA duplexes—outside the context of wobbles—are known to be sites of $Mg(H_2O)_6^{2+}$ coordination (Correll et al. 1997a; Jovine et al. 2000; Shi and Moore 2000; Wedekind and McKay 2003). Our work and that of others suggests that such sequences bind Cs^+, although binding is not necessarily mutually exclusive with respect to $Mg(H_2O)_6^{2+}$ based on respective *O. iheyensis* group IIc intron structures soaked in Li^+/Mg^{2+} and Cs^+/Mg^{2+} (Marcia and Pyle 2012).

A second sequence-specific Cs^+ binding site was located in the minor groove of $preQ_1$-II constructs #1 and #2 (Fig. 4d). This site entailed sugar-edge binding at Cyt5 and Cyt16, respectively. In construct #1, Cs^+ bound to Cyt5 with an anomalous difference Fourier peak height of $9.9\ \sigma$. In construct #2, Cs^+ bound to Cyt16 in all four copies in the asymmetric unit with an average peak height of $4.5 \pm 1.8\ \sigma$. Differences in Cs^+ localization were the result of sequence differences resulting from use of the PM14 G•U wobble (Fig. 1c). The site of Cs^+ binding in construct #1 is 5′-AC-3′/3′-UG-5′, whereas that of construct #2 is 5′-AG-3′/3′-UC-5′. At present, it is unclear what features surrounding the respective Cyt nucleobases are important for Cs^+ localization since not all accessible 5′-AC-3′ and 5′-CU-3′ sequences in

$preQ_1$-II riboswitch structures coordinate Cs^+. Higher resolution diffraction is needed to clarify this issue.

Acknowledgments We thank C. Kielkopf and J. Bogue for helpful discussions. J.A.L. was funded in part by NIH training grant T32 GM068411 and a Hooker graduate fellowship. This research was funded by NIH grants RR026501 and GM063162 to J.E.W. Portions of this research were carried out at the Stanford Synchrotron Radiation Lightsource (SSRL) and the Macromolecular Cornell High-Energy Synchrotron Source (MacCHESS). SSRL is a Directorate of SLAC National Accelerator Laboratory and an Office of Science User Facility operated for the U.S. DOE by Stanford University. The SSRL Structural Molecular Biology Program is supported by the DOE Office of Biological and Environmental Research and by NIH grants GM103393 and RR001209. MacCHESS is supported by NSF grant DMR-0936384 and NIH grant GM103485.

References

Alam S, Grum-Tokars V, Krucinska J et al (2005) Conformational heterogeneity at position U37 of an all-RNA hairpin ribozyme with implications for metal binding and the catalytic structure of the S-turn. Biochemistry 44:14396–14408

Alatossava T, Jutte H, Kuhn A et al (1985) Manipulation of intracellular magnesium content in polymyxin B nonapeptide-sensitized *Escherichia coli* by ionophore A23187. J Bacteriol 162:413–419

Allen FH (2002) The Cambridge Structural Database: a quarter of a million crystal structures and rising. Acta Crystallogr B 58:380–388

Bastet L, Dube A, Masse E et al (2011) New insights into riboswitch regulation mechanisms. Mol Microbiol 80:1148–1154

Bruno IJ, Cole JC, Edgington PR et al (2002) New software for searching the Cambridge Structural Database and visualizing crystal structures. Acta Crystallogr B 58:389–397

Cate JH, Doudna JA (1996) Metal-binding sites in the major groove of a large ribozyme domain. Structure 4:1221–1229

Chen JH, Yajima R, Chadalavada DM et al (2010) A 1.9 Å crystal structure of the HDV ribozyme precleavage suggests both Lewis acid and general acid mechanisms contribute to phosphodiester cleavage. Biochemistry 49:6508–6518

Chen AG, Sudarsan N, Breaker RR (2011) Mechanism for gene control by a natural allosteric group I ribozyme. RNA 17:1967–1972

Clancy S (2008) RNA functions. Nat Educ 1(1):102

Correll CC, Freeborn B, Moore PB et al (1997a) Metals, motifs, and recognition in the crystal structure of a 5S rRNA domain. Cell 91:705–712

Correll CC, Freeborn B, Moore PB et al (1997b) Use of chemically modified nucleotides to determine a 62-nucleotide RNA crystal structure: a survey of phosphorothioates, Br, Pt and Hg. J Biomol Struct Dyn 15:165–172

Cotton FA, Wilkinson G (1988) Advanced inorganic chemistry, 5th edn. Wiley, New York

Dauter Z, Dauter M (2001) Entering a new phase: using solvent halide ions in protein structure determination. Structure 9:R21–R26

Draper DE, Grilley D, Soto AM (2005) Ions and RNA folding. Annu Rev Biophys Biomol Struct 34:221–243

Edwards AL, Reyes FE, Heroux A et al (2010) Structural basis for recognition of S-adenosylhomocysteine by riboswitches. RNA 16:2144–2155

Ferre-D'Amare AR, Winkler WC (2011) The roles of metal ions in regulation by riboswitches. Met Ions Life Sci 9:141–173

Gilbert SD, Rambo RP, Van Tyne D et al (2008) Structure of the SAM-II riboswitch bound to S-adenosylmethionine. Nat Struct Mol Biol 15:177–182

Golden BL (2000) Heavy atom derivatives of RNA. Methods Enzymol 317:124–132

Jenkins JL, Krucinska J, McCarty RM et al (2011) Comparison of a $preQ_1$ riboswitch aptamer in metabolite-bound and free states with implications for gene regulation. J Biol Chem 286:24626–24637

Jovine L, Djordjevic S, Rhodes D (2000) The crystal structure of yeast phenylalanine tRNA at 2.0 Å resolution: cleavage by Mg^{2+} in 15-year old crystals. J Mol Biol 301:401–414

Keel AY, Rambo RP, Batey RT et al (2007) A general strategy to solve the phase problem in RNA crystallography. Structure 15:761–772

Klein DJ, Ferre-D'Amare AR (2006) Structural basis of glmS ribozyme activation by glucosamine-6-phosphate. Science 313:1752–1756

Kubodera T, Watanabe M, Yoshiuchi K et al (2003) Thiamine-regulated gene expression of Aspergillus oryzae thiA requires splicing of the intron containing a riboswitch-like domain in the 5′-UTR. FEBS Lett 555:516–520

Lee ER, Baker JL, Weinberg Z et al (2010) An allosteric self-splicing ribozyme triggered by a bacterial second messenger. Science 329:845–848

Liberman JA, Salim M, Krucinska J et al (2013) Structure of a class II $preQ_1$ riboswitch reveals ligand recognition by a new fold. Nat Chem Biol 9:353–355

Lipfert J, Sim AY, Herschlag D et al (2010) Dissecting electrostatic screening, specific ion binding, and ligand binding in an energetic model for glycine riboswitch folding. RNA 16:708–719

Lippa GM, Liberman JA, Jenkins JL et al (2012) Crystallographic analysis of small ribozymes and riboswitches. Methods Mol Biol 848:159–184

Marcia M, Pyle AM (2012) Visualizing group II intron catalysis through the stages of splicing. Cell 151:497–507

McCarty RM, Bandarian V (2012) Biosynthesis of pyrrolopyrimidines. Bioorg Chem 43:15–25

McLaggan D, Naprstek J, Buurman ET et al (1994) Interdependence of K^+ and glutamate accumulation during osmotic adaptation of *Escherichia coli*. J Biol Chem 269:1911–1917

Meyer MM, Roth A, Chervin SM et al (2008) Confirmation of a second natural $preQ_1$ aptamer class in *Streptococcaceae bacteria*. RNA 14:685–695

Mironov AS, Gusarov I, Rafikov R et al (2002) Sensing small molecules by nascent RNA: a mechanism to control transcription in bacteria. Cell 111:747–756

Montange RK, Batey RT (2008) Riboswitches: emerging themes in RNA structure and function. Annual Rev Biophysics 37:117–133

Myszka DG, Abdiche YN, Arisaka F et al (2003) The ABRF-MIRG'02 study: assembly state, thermodynamic, and kinetic analysis of an enzyme/inhibitor interaction. J Biomol Tech 4:247–269

Nahvi A, Sudarsan N, Ebert MS et al (2002) Genetic control by a metabolite binding mRNA. Chem Biol 9:1043–1049

Nakano S, Proctor DJ, Bevilacqua PC (2001) Mechanistic characterization of the HDV genomic ribozyme: assessing the catalytic and structural contributions of divalent metal ions within a multichannel reaction mechanism. Biochemistry 40:12022–12038

Pley HW, Flaherty KM, McKay DB (1994) Three-dimensional structure of a hammerhead ribozyme. Nature 372:68–74

Pyle AM (2002) Metal ions in the structure and function of RNA. J Biol Inorg Chem 7:679–690

Roth A, Winkler WC, Regulski EE et al (2007) A riboswitch selective for the queuosine precursor $preQ_1$ contains an unusually small aptamer domain. Nat Struct Mol Biol 14:308–317

Salter JD, Lippa GM, Belashov IA et al (2012) Core-binding factor beta increases the affinity between human Cullin 5 and HIV-1 Vif within an E3 ligase complex. Biochemistry 51:8702–8704

Serganov A, Nudler E (2013) A decade of riboswitches. Cell 152:17–24

Serganov A, Patel DJ (2007) Ribozymes, riboswitches and beyond: regulation of gene expression without proteins. Nat Rev Genet 8:776–790

Serganov A, Yuan YR, Pikovskaya O et al (2004) Structural basis for discriminative regulation of gene expression by adenine- and guanine-sensing mRNAs. Chem Biol 11:1729–1741

Serganov A, Polonskaia A, Phan AT et al (2006) Structural basis for gene regulation by a thiamine pyrophosphate-sensing riboswitch. Nature 441:1167–1171

Serganov A, Huang L, Patel DJ (2008) Structural insights into amino acid binding and gene control by a lysine riboswitch. Nature 455:1263–1267

Serganov A, Huang L, Patel DJ (2009) Coenzyme recognition and gene regulation by a flavin mononucleotide riboswitch. Nature 458:233–237

Shi H, Moore PB (2000) The crystal structure of yeast phenylalanine tRNA at 1.93 Å resolution: a classic structure revisited. RNA 6:1091–1105

Spitale RC, Wedekind JE (2009) Exploring ribozyme conformational changes with X-ray crystallography. Methods 49:87–100

Spitale RC, Torelli AT, Krucinska J et al (2009) The structural basis for recognition of the $PreQ_0$ metabolite by an unusually small riboswitch aptamer domain. J Biol Chem 284:11012–11016

Stahley MR, Strobel SA (2006) RNA splicing: group I intron crystal structures reveal the basis of splice site selection and metal ion catalysis. Curr Opin Struct Biol 16:319–326

Sudarsan N, Barrick JE, Breaker RR (2003) Metabolite-binding RNA domains are present in the genes of eukaryotes. RNA 9:644–647

Suddala KC, Rinaldi AJ, Feng J et al (2013) Single transcriptional and translational $preQ_1$ riboswitches adopt similar pre-folded ensembles that follow distinct folding pathways into the same ligand-bound structure. Nucleic Acid Res 41:10462–10475

Toor N, Keating KS, Pyle AM (2009) Structural insights into RNA splicing. Curr Opin Struct Biol 19:260–266

Torelli AT (2008) Structural studies of transition-state stabilization by a small catalytic RNA, metabolite-binding by a regulatory RNA sequence and the mechanism of action of activation induced deaminase. Ph.D. Dissertation, University of Rochester, School of Medicine and Dentistry, Rochester, NY

Wedekind JE (2011) Metal ion binding and function in natural and artificial small RNA enzymes from a structural perspective. Met Ions Life Sci 9:299–345

Wedekind JE, McKay DB (2000) Purification, crystallization, and X-ray diffraction analysis of small ribozymes. Methods Enzymol 317:149–168

Wedekind JE, McKay DB (2003) Crystal structure of the leadzyme at 1.8 A resolution: metal ion binding and the implications for catalytic mechanism and allo site ion regulation. Biochemistry 42:9554–9563

Weinberg Z, Barrick JE, Yao Z et al (2007) Identification of 22 candidate structured RNAs in bacteria using the CMfinder comparative genomics pipeline. Nucleic Acid Res 35:4809–4819

Winkler WC, Breaker RR (2003) Genetic control by metabolite-binding riboswitches. Chembiochem 4:1024–1032

Winkler W, Nahvi A, Breaker RR (2002) Thiamine derivatives bind messenger RNAs directly to regulate bacterial gene expression. Nature 419:952–956

Winkler WC, Nahvi A, Roth A et al (2004) Control of gene expression by a natural metabolite-responsive ribozyme. Nature 428:281–286

Woodson SA (2005) Metal ions and RNA folding: a highly charged topic with a dynamic future. Curr Opin Chem Biol 9:104–109

DNA G-Quadruplexes and I-Motifs in Therapeutics and Diagnostics

Yogini P. Bhavsar-Jog, Samantha M. Reilly, and Randy M. Wadkins

Contents

1 Introduction 442
2 Factors Affecting the Formation and Stability of G4 444
2.1 Cations 444
2.2 Loop Length 444
2.3 Backbone Modifications 445
2.4 Molecular Crowding 445
3 Factors Affecting the Formation and Stability of iMs 446
3.1 pH Dependence and Molecular Crowding 446
3.2 Backbone Modification 446
4 G4s as Drug Targets in Telomeres and Promoters 446
4.1 Telomeric G4 as Drug Targets 447
4.2 Promoter G4s as Drug Targets 449
5 G4s/iMs as Biosensors 449
5.1 G4-Based Biosensors 450
5.2 iM-Based Biosensors 451
6 Therapeutic Agents and Drug-Delivery Vehicles 452
6.1 G4-Forming Aptamers as Therapeutic Agents 452
6.2 G4s as Drug-Delivery Vehicles 453
6.3 iMs as Drug-Delivery Vehicles 454
7 Summary 455
References 455

Abstract G-quadruplex- and i-motif-based DNA oligomers are being investigated for their integration into therapeutic and diagnostic micro-assemblies. Examples include quadruplex-forming aptamers as potential anti-HIV agents, and serum-stable quadruplexes as carriers for delivering porphyrins into cancer cells for photodynamic therapy. The i-motifs from C-rich DNA find application in

Y.P. Bhavsar-Jog • S.M. Reilly • R.M. Wadkins (✉)
Department of Chemistry and Biochemistry, The University of Mississippi, University, MS 38677, USA
e-mail: rwadkins@olemiss.edu

V.A. Erdmann et al. (eds.), *Chemical Biology of Nucleic Acids*, RNA Technologies,
DOI 10.1007/978-3-642-54452-1_24,

pH-triggered hydrogels that can carry agents like drugs, proteins, and polymers to their targets. The pH-dependent conformational dynamics of i-motifs also make them useful as biosensors for detecting pH changes in cellular microenvironments. Due to these and many other applications, and in an effort to present a compendia of recent uses of quadruplexes and i-motifs, this chapter will concern itself with formation of G-quadruplexes and i-motifs; their physical and chemical properties; the effects of molecular crowding and hydration on their structure and stability; and their application in therapeutics as drug targets, drug-delivery vehicles, and diagnostic tools.

Keywords DNA tetraplexes • i-DNA • Thermal stability • Telomeres • Promoters

1 Introduction

In addition to canonical Watson–Crick duplex DNA structure, DNA can exist in various noncanonical conformations such as G-quadruplexes (G4s) and i-motifs (iMs). The G4s and iMs are formed from guanine-rich and cytosine-rich sequences, respectively. Gellert et al. (1962) showed that four guanine bases can hydrogen bond with each other to form G-tetrads (Fig. 1a). Further, Arnott et al. (1974) showed that the array of G-tetrads stack to form G4s (Fig. 1b). The foundation of discovery of iMs was laid in 1963 when Langridge and Rich (1963) proposed that two cytosines could hydrogen bond to form hemiprotonated structure (Fig. 1c). However, it was only in 1985 that Brown et al. (1985) were able to establish the structure of iMs (Fig. 1d) using circular dichroism spectroscopy. Ever since these discoveries, the existence of G4s and iMs in living cells has remained a topic of great controversy, and the formation of guanine quadruplexes in human cells has been only recently reported by Biffi et al. (2013) by developing a structure-specific antibody for quantitatively visualizing G4s. This confirmation of existence of G4s in vivo has significantly increased the importance of G4s and iMs in biological roles, especially in transcriptional regulation and drug targeting.

The G4s and iMs can occur in multiple conformations. They can be intramolecular (formed from single strand DNA; ssDNA) or intermolecular (formed from multiple DNA strands, generally bimolecular or tetramolecular). As discussed in the review article by Burge et al. (2006), G4s can be parallel or antiparallel. Parallel G4s are formed when all guanine-rich strands are oriented in the same direction. Antiparallel G4s are those that have at least one of the guanine-rich strands oriented in the opposite direction. In addition, Burge et al. (2006) describe how G4s can vary based on the classification of the loops: propeller, lateral, and diagonal (Fig. 2). Unlike G4s, iMs almost always occur in antiparallel conformations.

The sequences forming G4s and iMs occur abundantly near the transcription start sites (TSS) of several oncogenes (Zhao et al. 2007). Due to their topological

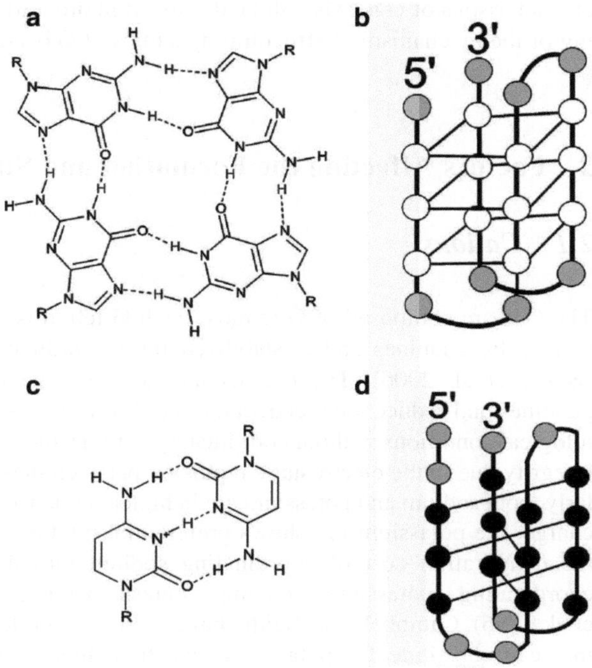

Fig. 1 (**a**) G-tetrad formation that form the basis of the G4 structure; (**b**) a cartoon of a G4 structure; (**c**) C–C^{+} hydrogen bond that forms the basis of the i-motif stability; (**d**) a cartoon of an iM structure

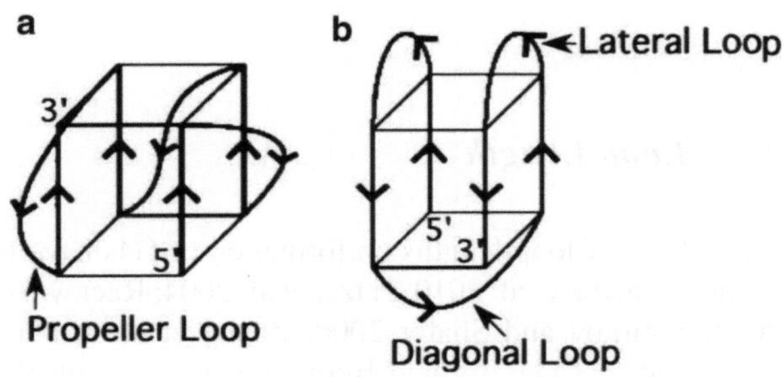

Fig. 2 (**a**) Parallel G4 with propeller loops; (**b**) antiparallel G4 with both lateral and diagonal loops [Modified from a figure reused with permission from Burge et al. (2006)]

variance from duplex DNA and their function as gene regulatory elements, the G4s and iMs are perceived as unique targets for binding of drugs and transcription factors to regulate gene expression [summarized by Düchler (2012)]. Owing to their nontoxic nature and their biodegradability, G4- and iM-based biomaterials are gaining rapid popularity in therapeutics. Although the applicability of G4s and iMs is widely being explored in several nonphysiological applications, including, nanocircuitry and nanomechanical motors, this chapter will highlight the roles of G4s and iMs in physiological applications, such as drug targeting, biosensing, and drug delivery (Liu et al. 2006; Miyoshi et al. 2006). Along with the applications of G4s/iMs, several factors affecting the conformational and biophysical characteristics of G4s/iMs will be discussed in order to provide a clear understanding of the mechanism of structural dynamics of G4s and iMs.

2 Factors Affecting the Formation and Stability of G4

2.1 Cations

The G4s are composed of G-tetrads. Each G-tetrad is involved in Hoogsteen base pairing by guanines and is stabilized by a coordination complex with a cation (Burge et al. 2006). The cations interact with the oxygens on tetrad-forming guanines and reduce the electrostatic repulsion between the oxygens. Under physiological conditions without coordinating cations, the G-tetrads lose their structural integrity due to the electrostatic repulsion between these oxygen atoms. Intracellularly, both sodium and potassium are in high abundance, but due to its ideal size and charge, the potassium ion shows preferential binding to stabilize G-tetrads. Also, the dehydration cost of coordinating sodium into the G-tetrad is greater than coordinating potassium, making potassium even more preferable (Hud et al. 1996). Campbell and Neidle have written an in-depth review on other cations that can substitute for potassium, which include ammonium, barium, calcium, cesium, lead, lithium, rubidium, strontium, and thallium (Campbell and Neidle 2012; Risitano and Fox 2004).

2.2 Loop Length

The effects of loop lengths on formation of G4s have been investigated by several groups (Guedin et al. 2010; Hazel et al. 2004; Rachwal et al. 2007; Risitano and Fox 2004; Smirnov and Shafer 2000; Zhang et al. 2011). Collectively, these studies concluded that G4s formed from short loops (with three or less bases) are more stable relative to those having more than three bases in their loops. The G4s with shorter loops showed a higher melting temperature and had lower entropic cost for folding (Guedin et al. 2010; Hazel et al. 2004). Also, the G4s with longer loops require higher concentrations of cations in order to be thermodynamically as stable as G4s with shorter loops (Rachwal et al. 2007). Although G4s with long loops are less favorable, they can be formed under physiological conditions (Smirnov and Shafer 2000). The loop length also dictates the orientation of G4s; generally, short looped G4s adopt parallel orientation, while G4s with longer loops are more likely to have an antiparallel conformation.

2.3 Backbone Modifications

Studies have shown that altering the DNA backbone composition to locked nucleic acid (LNA), L-DNA, phosphorothioate, and 2′-O-methyl can alter the structure and thermal stability of G4s. When the backbone of entire G4-forming sequences is modified with LNA, the formation of the G4 was inhibited due to steric hindrance by the backbone (Randazzo et al. 2004). The modification of the backbone to L-deoxyribose, a mirror image of naturally occurring D-deoxyribose, has been shown to decrease slightly the G4's thermal stability when compared to the regular DNA backbone, suggesting that the chirality of the backbone has little effect on G4 formation (Li et al. 2012b). However, chemical modification of phosphorothioate, which is a sulfur substituting for 2′ oxygen atom of the phosphate in the DNA backbone, shows a drop in thermal stability caused by the much larger size of the sulfur atoms compared to the oxygen atoms they are replacing (Li et al. 2012b; Neidle and Balasubramanian 2006). The 2′-O-methyl backbone modifications have shown an increase in thermal stability as well as a preference toward adopting compact parallel G4 conformation. The modification of the 2′ position to O-methyl causes the ribose sugar to preferentially adopt the *anti*-conformation over the *syn*-conformation, leading to thermally stable, parallel G4s when compared to unmodified 2′-position (Li et al. 2012a, b; Neidle and Balasubramanian 2006).

2.4 Molecular Crowding

Macromolecular crowding agents like proteins and nucleic acids occupy 20–40 % of intracellular volume (Miyoshi and Sugimoto 2008). Hence, it is important to understand the effects of crowding condition on G4s. In vitro, molecular crowding conditions can be mimicked by using large polymers, such as polyethylene glycol (PEG). The molecular crowding with PEG induces a conformational change of G4s from antiparallel to parallel and lowers the concentration of cations required to stabilize the G4 structures (Kan et al. 2006; Miyoshi et al. 2002). In general, lowering the water activity and dielectric constant around DNA causes G4s to be more stable (Lane et al. 2008), since a decrease in the dielctric constant of the solvent increases the stability of cation–guanine complex (Smirnov and Shafer 2007). However, studies indicate that crowding alone is not sufficient to enhance the stability of G4s, since using bovine serum albumin does not affect the thermal stability or conformation of the G4 structure. Contrarily, the addition of smaller co-solvents, such as acetonitrile and dimethyl sulfoxide, leads to increase in thermal stability as well as change in conformation (antiparallel to parallel) of G4 structures. Together, these experiments suggest that dehydration of the structure is primary requirement for enhancing the thermal stability (Dhakal et al. 2013).

3 Factors Affecting the Formation and Stability of iMs

3.1 pH Dependence and Molecular Crowding

For iM stabilization, the cation necessary is a proton; iMs are composed of intercalated hydrogen bonds formed between hemiprotonated cytosines. In dilute buffer conditions, the typical pK_a of iMs is around pH 6.0 (Chen et al. 2012; Kypr et al. 2009). Most of the studies in the literature on iMs have been conducted at pH 5.5 in order to ensure that all DNA is in the intercalated conformation. These conditions do not mimic realistic biological systems, where the pH is approximately 7.4. However, in 2009, Zhou et al. (2010) found that if the temperature was lowered to 4 °C, the pK_a of the iM structure shifts to neutral pH, but this low temperature is far from physiological temperature for most organisms. Around the same time Rajendran et al. (2010) found that molecular crowding with both PEG-200 and PEG-8000 at 20 % (w/v) could shift the pK_a of iMs to physiologically relevant pHs (7.1–7.4) at room temperature. This experiment is more in agreement with the hypothesis that iMs can be formed in vivo in the presence of macromolecular crowding at physiological pH.

3.2 Backbone Modification

Studies on iMs with backbone modifications to the DNA have shown that different modifications can have varied effects on iM stability. The iMs formed in RNA tend to be far less thermally stable than those in DNA (Mergny and Lacroix 1998). In addition, two different studies have found that having a peptide nucleic acid backbone (PNA) on iMs causes them to have lower thermal stability (Modi et al. 2006; Sharma and Ganesh 2005). However, it was found that thermal melting temperature increased by approximately 7 °C when an intermolecular iM was formed between a strand of DNA and a strand of PNA versus between two strands of DNA, suggesting DNA/PNA iM hybrids are more stable than regular DNA iMs (Modi et al. 2006).

4 G4s as Drug Targets in Telomeres and Promoters

Due to their topological variance from the duplex DNA, G4-tetraplexes offer unique recognition sites for binding of drugs and transcription factors. G4s occurring at the ends of the telomeres have been targeted in order to arrest the cell division during cancer metastasis. The G4s within the promoters of several proto-oncogenes, such as *MYC* and *KIT*, are regarded as prime targets for drugs in cancer therapy because of their ability to silence or activate the expression of these genes.

Over the last few years, several excellent review articles have appeared on this topic (Balasubramanian et al. 2011; Brooks et al. 2010; Düchler 2012; Han and Hurley 2000; Hurley et al. 2000; Neidle and Read 2000), and here we give brief highlights.

4.1 Telomeric G4 as Drug Targets

The length of human telomeric DNA varies depending on the cell type. In somatic cells, the telomeres span from 6 to 8 kb while in germline cells the telomeric length ranges between 10 and 20 kb (Neidle and Parkinson 2002; Steinert et al. 2004). Normally, the length of the telomeres progressively decreases due to the end replication effect, where DNA polymerase is not able to completely replicate the lagging strand, resulting in a loss of 50–200 bases with each cell division (Gowan et al. 2001). Cells with critically shortened telomeres reach the Hayflick limit and cease to divide and die (Hayflick and Moorhead 1961). However, in tumorous cells, the telomeric length remains unchanged (Counter et al. 1998; Neidle and Parkinson 2002). The maintained length of telomeres in tumor cells is attributed to an enzyme called telomerase, which is a RNA-dependent DNA polymerase responsible for elongation of telomeres by adding nucleotides (De Lange 1994; Raymond et al. 1996). Telomerase is highly expressed in 80–85 % of cancer cells (Kim et al. 1994). Due to overexpression of telomerase, the cells do not attain the Hayflick limit, resulting in unrestricted number of cell divisions; thus, regulating telomerase activity is considered as one of the solutions toward controlling tumorigenesis (Counter et al. 1998; Neidle and Parkinson 2002). In addition to the elongation of telomeres, telomerase also caps the ends of telomeres (Blackburn 2000; Saretzki 2003). Due to this capping, the ends of telomeres are protected from being recognized as ssDNA breaks, which would otherwise activate the DNA damage response apparatus of cells (d.r.c) leading to apoptosis (Chan and Blackburn 2002; Longhese 2008). This protective mechanism is shown in Fig. 3.

Both the catalytic and capping functions (Neidle and Parkinson 2002) of telomerase are directly inhibited by the presence of G4s. The ends of human telomeres are composed of tandem repeats of the ssDNA sequence 5′-TTAGGG-3′, which is capable of forming G4s (Moyzis et al. 1988). Several drugs are being developed to stabilize the telomeric quadruplexes. For example, BRACO19 and BRACO20 are trisubstituted acridine derivatives, which have low toxicity and high G4-stabilizing and telomere-cap disturbing ability (Burger et al. 2005; Moorhouse et al. 2006; Neidle and Read 2000; Neidle and Parkinson 2002). The effects of BRACO19 were tested on human uterus carcinoma cells and showed a significant anticancer activity within 2 days of treatment. After 39 days of exposure to BRACO19, telomere length decreased by 17 %. This activity is attributed to stabilization of G4s and inhibition of telomerase (Burger et al. 2005). The BRACO20's activity was tested in ovarian cancer cells. It was found that BRACO20 can bind to G4s with 30 % higher affinity than for duplexes and could inhibit telomerase (Read et al. 2001). Other examples of G4-binding agents are amidoanthracine-9,10-diones (e.g.,

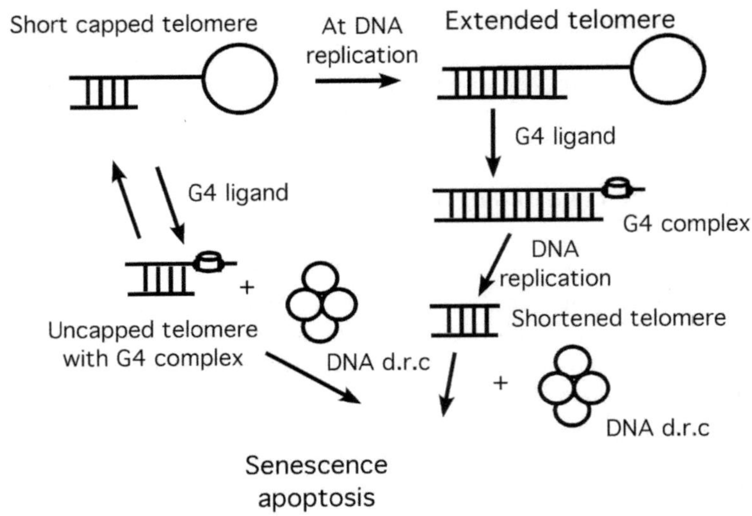

Fig. 3 Capping and elongation by telomerase (*large sphere*) and the formation of G4s (*cylindrical structure*) in uncapped telomeres. When G4s are formed at the telomeric ends, both capping and elongation by telomerase are inhibited and the DNA damage response apparatus of cells (d.r.c.; four *small spheres*) is activated leading to senescence [Reused with permission from Neidle and Read. (2000)]

BSU1051) and pentacyclic acridinium (RHPS4) (Gowan et al. 2001; Han et al. 1999). The BSU1051 showed higher binding affinity toward G4s relative to cruciform, duplex, and ssDNA. This compound inhibits telomerase activity by binding to G4s without affecting telomerase itself. The telomerase activity is blocked in concentration-dependent manner (Han et al. 1999). The RHPS4 showed anticancer activity in breast and vulval carcinoma; however, it had no effect on ovarian carcinoma cells, which possess longer telomeres and are inherently more resistant to the cytotoxic effects of RHPS4 (Gowan et al. 2001).

Natural products such as telomestatin and meridine are known to stabilize G4s and inhibit telomerase activity (Arola and Vilar 2008; Chen et al. 2011; Zhou et al. 2005). Kim et al. (2003) showed that telomestatin can convert linear DNA to G4s even in the absence of cations. Telomestatin exhibits 70-fold higher selectivity for intramolecular G4s than duplex DNA. The G4s stabilized by binding of telomestatin inhibit telomerase activity, leading to accelerated telomere shortening. Guittat et al. (2005) found that, like telomestatin, meridine also shows selectivity toward binding to intramolecular G4s. Although meridine is regarded as inferior to telomestatin in binding to G4s and stabilizing them, it is easier (due to their compact structure relative to telomestatin) to develop meridine analogs with better binding affinities and selectivity for G4s than it is to create telomestatin analogs (Guittat et al. 2005).

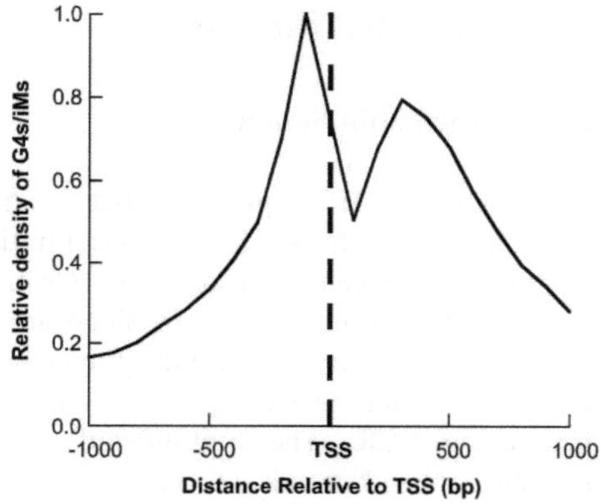

Fig. 4 Relative densities of G4s/iMs around transcription start sites. The density of G4s/iMs is higher in upstream regions relative to intragenic regions

4.2 *Promoter G4s as Drug Targets*

The G4-forming sequences occur abundantly near the transcription start sites of several genes (Fig. 4) (Balasubramanian et al. 2011). G4s located in the promoter regions regulate gene expression and hence are attractive therapeutic targets (Balasubramanian et al. 2011). The *MYC, KIT, KRAS, MYB, BCL-2, VEGF, PDGFA, RB1,* and TERT genes are some of the widely studied oncogenes having G4-forming sequences in their promoter regions (Balasubramanian et al. 2011; Sun and Hurley 2009). The G4s occurring in these genes adopt various conformations; however, the common features like G-tetrads and planarity are being used to develop small molecules that can bind to these planar platforms and silence genes (Balasubramanian et al. 2011). Some of the most common molecules used in regulating the gene expression by binding to G4s are $TMPyP_4$ and $TMPyP_2$ (Siddiqui-Jain et al. 2002), trisubstituted isoalloxazines (Bejugam et al. 2007), Se2SAP (Seenisamy et al. 2005), and Quarfloxin (Drygin et al. 2009).

5 G4s/iMs as Biosensors

Above, we discussed G4 as potential therapeutic targets. We now describe how these structures have been incorporated into diagnostic and stimuli-responsive schemes. DNA-based biosensors are useful in monitoring the changes in the environment around them via structural differences between folded and unfolded secondary structures and are capable of detecting nM to pM concentrations of analyte. Further, the micrometer to sub-micrometer size of these sensors makes them convenient to probe relatively tighter spaces like intracellular micro-environments. Here, we describe a few developing applications of G4/iM-based biosensors.

5.1 G4-Based Biosensors

5.1.1 Thrombin Sensor

Thrombin is a serine protease that is an important marker in the diagnosis of pulmonary metastasis. Higher concentrations of thrombin in the blood are also responsible for venous thrombosis (Grütter et al. 1990). Hence, monitoring and maintaining thrombin levels in blood are important. In 1992, Bock et al. (1992) found multiple thrombin binding aptamers. They started with a pool of $>10^{13}$ different 96-mers DNA sequences. These sequences were biotinylated and amplified using PCR. The biotinylated ss-oligos were then separated from the non-biotinylated strands on an agarose-avidin column and screened for their ability to bind thrombin. One of the aptamers isolated by this process was 5′-GGTTGGTGTGGTTGG-3′, which is now the most prevalent thrombin binding aptamer (TBA). Using this aptamer, Sun et al. have recently reported a quantitative thrombin sensor (Sun et al. 2013). In the presence of thrombin, the TBA on the outermost layer of an electrode binds thrombin and creates a barrier for electrons and inhibits electron flow, resulting in decreased differential pulse voltammetry (DPV) signals. Thrombin concentrations in the range of 1 pM to 160 nM can be measured using this G4-based sensor.

5.1.2 Ion Sensors

Potassium ions (K^+) are involved in many biological functions, including nerve transmission, regulation of blood pressure, and the formation of collagen or elastin. Abnormal levels of K^+ ions could lead to muscle cramps or weakness, nausea, diarrhea, frequent urination, and dehydration (Hu et al. 2012). This makes it necessary to create devices that monitor the concentrations of K^+ ions. Huang et al. (Huang and Chang 2008) have developed a G4-based probe in order to facilitate the rapid detection of K^+ ions in urine. For this probe, Huang et al. used an ATP-binding G4 aptamer (5′-ACCTGGGGGAGTATTGCGGAGGAAGGT-3′) and the cyanine dye oligreen (OG) (Huang and Chang 2008). When K^+ ions bind to the OG–aptamer complex, the aptamer adopts G-quadruplex structure. This leads to dissociation of OG from the aptamer, leading to a decrease in fluorescence. This probe is highly selective for potassium ions; the selectivity is measured to be more than 10,000-fold for K^+ over Na^+. The lower limit of detection of K^+ is approximately 75 nM. Highly specific and selective G4-forming aptamers can also be used for detecting ions like Pb^{2+} and Hg^{2+} (Liu et al. 2009).

5.2 *iM-Based Biosensors*

5.2.1 Glucose Sensors

Glucose is a ubiquitous fuel necessary for performing metabolic functions. High glucose levels in blood/urine serve as a marker for diagnosing diabetes (Sacks 2011). Several different types of probes to monitor glucose levels based on iMs have been developed. Li et al. (2012a) have designed a probe sensitive and selective for glucose detection and the working principle of this probe is based on two phenomena—conformational switching of iMs and aggregation of gold nanoparticles (AuNPs). In the absence of glucose, the C-rich nucleotide (5′-CCCTAACCCTAACCCTAACCCT-3′) has an extended single-stranded conformation at neutral pH. In this ssDNA conformation, the DNA protects the AuNPs from salt-induced aggregation (pink-colored solution). However, in the presence of glucose and glucose oxidase, gluconic acid is formed, which lowers the pH of the medium. At acidic pH, the C-rich strand adopts an iM conformation, deprotecting the AuNPs and leading to their aggregation. The aggregation of the AuNPs in high ionic strength media causes the AuNP solution to change color from pink to blue. The change in the color of AuNPs is used to estimate the concentration of glucose. This probe can detect the glucose levels ranging from 0.5 μM to 5 mM. The high sensitivity of this probe is attributed to extremely high extinction coefficients of AuNPs and efficient conformational switching of the C-rich nucleotide.

5.2.2 Intracellular pH Sensor

The iM-based pH sensors can be used to monitor intracellular pH. The fluctuation in intracellular pH can be used to detect conditions like dysplasia as well as in monitoring other biological processes. Modi et al. have developed a nanodevice known as an I-switch that can map spatial and temporal pH changes in living cells (Modi et al. 2009). This I-switch is built from three different oligos (O1, O2, and O3; Fig. 5). The O1 and O2 are hybridized to O3 to form duplex DNA. The oligos O1 and O2 have overhangs, which form one-half of the bimolecular iM. The overhangs are tagged with the fluorophores Alexa-647 and Alexa-488. At pH above 6.8, the iM-forming overhangs remain as ssDNA, and at pH lower than 5.5, the overhangs from O1 and O2 form intermolecular iM. While going from pH 5.5 to 6.8, the iM unfolds, reducing the fluorescence resonance energy transfer (FRET) intensities, and thereby reporting on pH changes. Thus, the I-switch works as an efficient reporter of pH ranging from 5.5 to 6.8. This pH sensor was successfully employed intracellularly to detect pH changes associated with endosomal maturation.

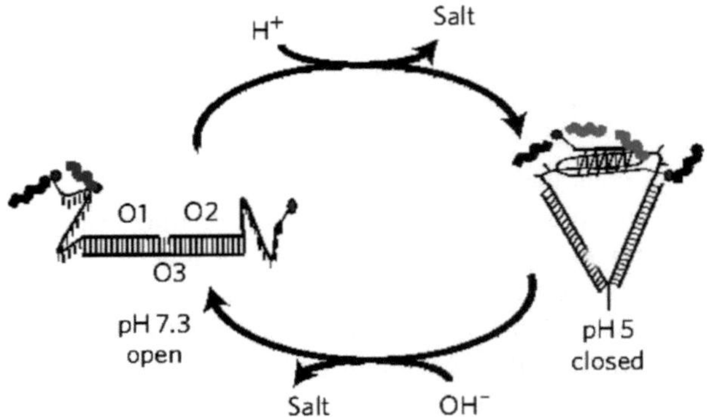

Fig. 5 Schematic of an I-switch, which is composed of oligos (O1, O2, and O3). Oligos O1 and O2 can hybridize to O3 and consist of overhangs tagged with fluorophores (represented by *spheres*). These overhangs each form one-half of a bimolecular iM. At pH 7.3, the overhangs are open (not in iM conformation), while at pH 5.4, the overhangs are closed (interlocked to form iMs), leading to FRET [Reused with permission from Modi et al. (2009)]

6 Therapeutic Agents and Drug-Delivery Vehicles

The G4s/iMs-based nanodevices are rapidly gaining popularity as drug-delivery agents over synthetic polymers. This popularity is attributed to their ability to recognize specific targets, their ability to access intracellular spaces, and their biocompatibility, biodegradability, and nontoxicity. Along with targeted drug-delivery vehicles, some G4-forming aptamers have exhibited therapeutic significance in cancer and HIV treatments.

6.1 G4-Forming Aptamers as Therapeutic Agents

The AS1411 is a 26-mer G4 aptamer that has passed Phase II clinical trials as one of the first anticancer nucleic acid aptamers to treat lymphoma (Shum et al. 2013). Initially, the aptamers were developed based on the fact that guanine-rich oligos (GROs) possess anti-proliferative properties in cancer cells (Bates et al. 2009). However, the anticancer activity of AS1411 was discovered via serendipity and not via systematic evolution of ligand by exponential enrichment (SELEX), which is the technique responsible for most aptamers (Bates et al. 2009). Paula Bates and coworkers proposed that the 26-mer G4 aptamer binds to nucleolin, which is a protein overexpressed in several cancer types (Teng et al. 2007). The binding of G-rich oligonucleotide to nucleolin inhibits one or more of its normal functions (ribosome production, nuclear transport, and cell entry), which suppresses tumor proliferation in cell lines. Recently, Magbanua et al. reported another 16-mer DNA aptamer AID-1 (5′-GGGTGGGTGGGTGGGT-3′) that can bind to interleukin-6

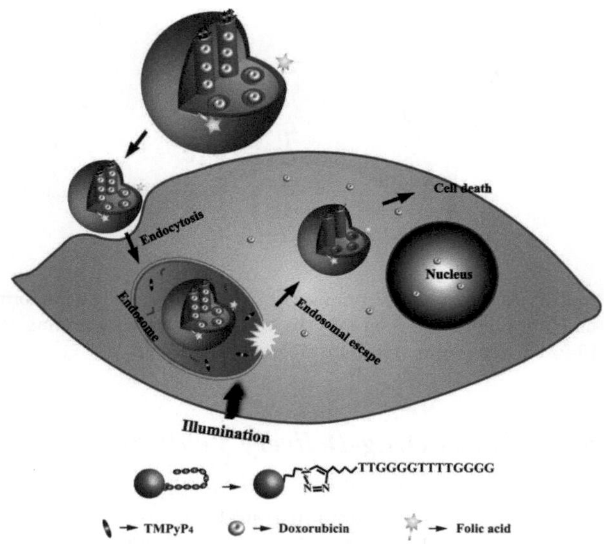

Fig. 6 The G-rich oligonucleotide shown can form a G4. $TMPyP_4$ is conjugated to this G4, which is attached to the MSP. The MSP-DNA-$TMPyP_4$ complex is tagged with folic acid, allowing the complex to enter cells via folate-mediated endocytosis. Upon irradiating with visible or near infrared light, the $TMPyP_4$ generated ROS, which uncaps MSP from DNA, releasing the cargo (shown by doxorubicin in this figure) [Reused with permission from Chen et al. (2013)]

receptor (IL-6R) (Magbanua et al. 2013). In order to locate this aptamer, Magbauna et al. performed 13 rounds of SELEX. The resulting AID-1 binds to IL-6R with K_d values in the nM range and inhibits both HIV infection and HIV-1 integrase.

6.2 G4s as Drug-Delivery Vehicles

Several groups including Shieh et al. (2010) and Chen et al. (2013) have used G4-forming aptamers as carriers for delivering anticancer drugs. Chen et al. have developed light-operated vehicles comprising of photosensitizer-incorporated, G4-capped, mesoporous silica nanoparticles (MSPs), as shown in Fig. 6 (Chen et al. 2013). The oligonucleotide 5′-GGGGTTTTGGGG-3′ forms G4s, which are conjugated to MSP and $TMPyP_4$. This complex of MSP-DNA-$TMPyP_4$ was tagged with folic acid and internalized into cells via folate-mediated endocytosis. Upon irradiating with visible or near-infrared wavelengths of light, *Meso*-$TMPyP_4$ acts as a photosensitizer to generate a reactive oxygen species (ROS). The production of ROS cleaves the DNA capping of the MSPs, releasing the cargo. The G4s in this system act both as a capping agent and as a photodynamic therapy drug carrier, whereas $TMPyP_4$ acts as the ROS generator. Further, $MPyP_4$ stabilizes the G4 DNA, protecting it from degradation. Importantly, Chen et al. successfully loaded MSPs with the anticancer drug doxorubicin (DOX) and demonstrated precise, spatially and temporally controlled drug release for both chemotherapy and photodynamic therapy in human cancer cells.

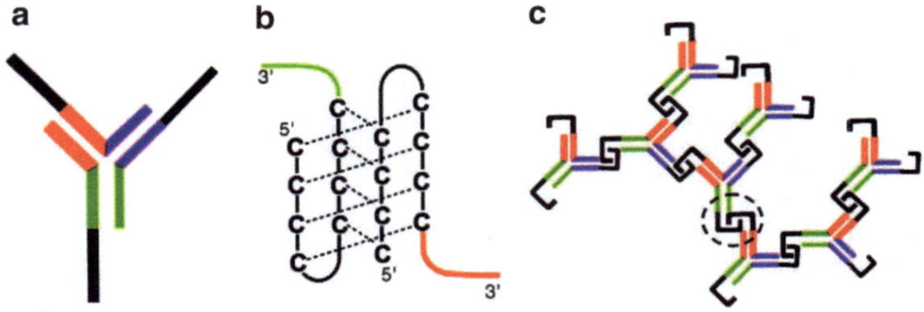

Fig. 7 iM-based oligos forming Y-shaped structure. The iM-forming ends of Y-shaped DNA (**a**) units form intermolecular iMs (**b**) in acidic conditions leading to the formation of hydrogels (**c**) [Reused with permission from Cheng et al. (2009)]

6.3 *iMs as Drug-Delivery Vehicles*

Cheng et al. have demonstrated an assembly of pH-responsive, fast-acting hydrogels based on iMs (Cheng et al. 2009). They have constructed a Y-shaped DNA structure formed from 37-mer ssDNA. These ssDNA contain two functional domains: an 11-mer interlocking iM domain containing two cytosine-rich stretches, and a 26-mer, which forms the double-stranded structure in the Y-shape (Fig. 7). At basic pH, each individual Y-shaped unit is isolated from the other. However, at acidic pH, the iM-forming strands assemble to form intermolecular iMs. This interlocked system results in the formation of three-dimensional structures in solution and in the formation of hydrogels. Cheng et al. have suggested that these structures can be used in pH-sensitive drug-delivery systems.

Keum et al. have developed an iM-based DNA pyramid that can deliver cargo to its targets (Fig. 8) (Keum and Bermudez 2012). The vertices of the pyramids are connected by iM-forming DNA sequences (Fig. 8: strands 2, 3, and 4). At basic pHs, edges of the pyramids are sealed by duplexes formed by C-rich DNA strands and a partially complementary strand (Fig. 8: strand 1). The pyramid's stability was modulated by introducing the mismatched bases in strand 1. Due to these mismatching bases, the melting temperature of the duplex structure forming the edges of the pyramid was reduced. The sequence of the strand 1 was optimized so that under basic pHs the strand formed duplex structure, but under acidic conditions, the strand 1 dissociated from pyramid, allowing the C-rich strands to fold into iMs and rendering the edges of the pyramid open. In order to demonstrate the function of iM-forming DNA pyramids, Keum et al. used an enhanced green fluorescent protein (EGFP) as cargo to be delivered by the pyramid. EGFP was modified with a hexa-histidine tag, while strand 1 of the pyramid was modified with Nickel-nitrilotriacetic acid (Ni^{2+}-NTA). The EGFP was attached to strand 1 via a nickel-mediated interaction between nitrilotriacetic acid (NTA) and histidine tag. At acidic pH, strands 2, 3, and 4 form iMs, opening the pyramid. At the same time, the NTA-Ni^{2+}-His linkage, which is unstable in acidic conditions, is cleaved, releasing the EGFP from the open pyramid.

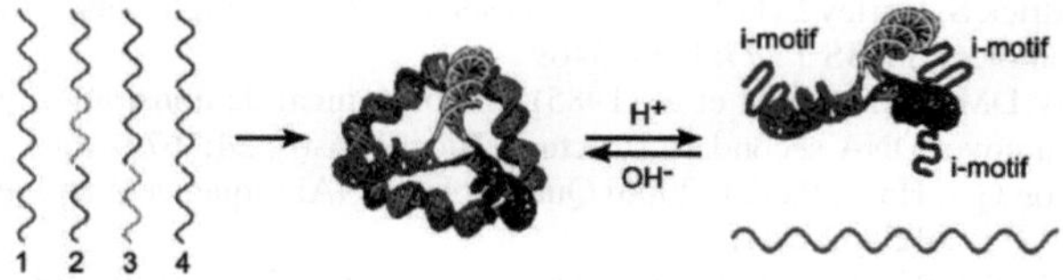

Fig. 8 DNA-pyramid protein delivery system. The edges are composed of iM-forming sequences. Under basic conditions, the edges are sealed, forming duplex DNA. Under acidic conditions, the edges form iMs, leading to the partially open conformation of the pyramid, facilitating the cargo delivery [Reused with permission from Keum and Bermudez (2012)]

7 Summary

In this chapter, we have attempted to provide a brief introduction to G4/iM discovery, the topological sensitivity of G4/iM to solution conditions, and applications of G4s/iMs in therapeutics and clinical research. The occurrence of G4s/iMs in the regions upstream to the TSS has made them very attractive targets for binding of drugs in order to regulate gene expression. Owing to their small sizes, G4/iM-forming oligos can reach within the intracellular micro-environments and hence can be employed as biosensors to monitor intracellular changes. Due to their biocompatibility, G4s/iMs-based aptamers are being developed as drug-delivery vehicles. With the recent confirmation of existence of G4s in living cells and with the increasing demands for nontoxic, biocompatible drugs and drug-delivery vehicles, G4s/iMs will remain significantly important to therapeutics and clinical research for years to come.

References

Arnott S, Chandrasekaran R, Marttila CM (1974) Structures for polyinosinic acid and polyguanylic acid. Biochem J 141:537–543

Arola A, Vilar R (2008) Stabilisation of G-quadruplex DNA by small molecules. Curr Top Med Chem 8:1405–1415

Balasubramanian S, Hurley LH, Neidle S (2011) Targeting G-quadruplexes in gene promoters: a novel anticancer strategy? Nat Rev Drug Discov 10:261–275

Bates PJ, Laber DA, Miller DM et al (2009) Discovery and development of the G-rich oligonucleotide AS1411 as a novel treatment for cancer. Exp Mol Pathol 86:151–164

Bejugam M, Sewitz S, Shirude PS et al (2007) Trisubstituted isoalloxazines as a new class of G-quadruplex binding ligands: small molecule regulation of c-kit oncogene expression. J Am Chem Soc 129:12926–12927

Biffi G, Tannahill D, McCafferty J et al (2013) Quantitative visualization of DNA G-quadruplex structures in human cells. Nat Chem 5:182–186

Blackburn EH (2000) Telomere states and cell fates. Nature 408:53–56

Bock LC, Griffin LC, Latham JA et al (1992) Selection of single-stranded DNA molecules that bind and inhibit human thrombin. Nature 355:564–566

Brooks TA, Kendrick S, Hurley LH (2010) Making sense of G-quadruplex and i-motif functions in oncogene promoters. FEBS J 277:3459–3469
Brown DM, Gray DM, Patrick MH et al (1985) Photochemical demonstration of stacked C · C base pairs in a novel DNA secondary structure. Biochemistry 24:1676–1683
Burge S, Parkinson GN, Hazel P et al (2006) Quadruplex DNA: sequence, topology and structure. Nucleic Acids Res 34:5402–5415
Burger AM, Dai F, Schultes CM et al (2005) The G-quadruplex-interactive molecule BRACO-19 inhibits tumor growth, consistent with telomere targeting and interference with telomerase function. Cancer Res 65:1489–1496
Campbell NH, Neidle S (2012) G-quadruplexes and metal ions. In: Sigel A, Sigel H, Sigel RK (eds) Interplay between metal ions and nucleic acids. Springer, Netherlands, pp 119–134
Chan SW, Blackburn EH (2002) New ways not to make ends meet: telomerase, DNA damage proteins and heterochromatin. Oncogene 21:553–563
Chen JL, Sperry J, Ip NY et al (2011) Natural products targeting telomere maintenance. Med Chem Commun 2:229–245
Chen C, Li M, Xing Y et al (2012) Study of pH-induced folding and unfolding kinetics of the DNA i-motif by stopped-flow circular dichroism. Langmuir 28:17743–17748
Chen C, Zhou L, Geng J et al (2013) Photosensitizer-incorporated quadruplex DNA-gated nanovechicles for light-triggered. Targeted dual drug delivery to cancer cells. Small. doi:10.1002/smll.201201916:
Cheng E, Xing Y, Chen P et al (2009) A pH-triggered, fast-responding DNA hydrogel. Angew Chem 121:7796–7799
Counter CM, Hahn WC, Wei W et al (1998) Dissociation among in vitro telomerase activity, telomere maintenance, and cellular immortalization. Proc Natl Acad Sci USA 95:14723–14728
De Lange T (1994) Activation of telomerase in a human tumor. Proc Natl Acad Sci USA 91:2882
Dhakal S, Cui Y, Koirala D et al (2013) Structural and mechanical properties of individual human telomeric G-quadruplexes in molecularly crowded solutions. Nucleic Acids Res 41:3915–3923
Drygin D, Siddiqui-Jain A, O'Brien S et al (2009) Anticancer activity of CX-3543: a direct inhibitor of rRNA biogenesis. Cancer Res 69:7653–7661
Düchler M (2012) G-quadruplexes: targets and tools in anticancer drug design. J Drug Target 20:389–400
Gellert M, Lipsett MN, Davies DR (1962) Helix formation by guanylic acid. Proc Natl Acad Sci USA 48:2013–2018
Gowan SM, Heald R, Stevens MF et al (2001) Potent inhibition of telomerase by small-molecule pentacyclic acridines capable of interacting with G-quadruplexes. Mol Pharmacol 60:981–988
Grütter MG, Priestle JP, Rahuel J et al (1990) Crystal structure of the thrombin-hirudin complex: a novel mode of serine protease inhibition. EMBO J 9:2361–2365
Guedin A, Gros J, Alberti P et al (2010) How long is too long? Effect of loop size on G-quadruplex stability. Nucleic Acids Res 38:7858–7868
Guittat L, De Cian A, Rosu F et al (2005) Ascididemin and meridine stabilise G-quadruplexes and inhibit telomerase in vitro. Biochim Biophys Acta 1724:375–384
Han H, Hurley LH (2000) G-quadruplex DNA: a potential target for anti-cancer drug design. Trends Pharmacol Sci 21:136–141
Han H, Hurley LH, Salazar M (1999) A DNA polymerase stop assay for G-quadruplex-interactive compounds. Nucleic Acids Res 27:537–542
Hayflick L, Moorhead PS (1961) The serial cultivation of human diploid cell strains. Exp Cell Res 25:585–621
Hazel P, Huppert J, Balasubramanian S et al (2004) Loop-length-dependent folding of G-quadruplexes. J Am Chem Soc 126:16405–16415
Hu K, Huang Y, Zhao S et al (2012) Ultrasensitive detection of potassium ions based on target induced DNA conformational switch enhanced fluorescence polarization. Analyst 137:2770–2773

Huang C, Chang H (2008) Aptamer-based fluorescence sensor for rapid detection of potassium ions in urine. Chem Commun 12:1461–1463

Hud NV, Smith FW, Anet FA et al (1996) The selectivity for K versus Na in DNA quadruplexes is dominated by relative free energies of hydration: a thermodynamic analysis by 1H NMR. Biochemistry 35:15383–15390

Hurley LH, Wheelhouse RT, Sun D et al (2000) G-quadruplexes as targets for drug design. Pharmacol Ther 85:141–158

Kan Z, Yao Y, Wang P et al (2006) Molecular crowding induces telomere G-quadruplex formation under salt-deficient conditions and enhances its competition with duplex formation. Angew Chem Int Ed 45:1629–1632

Keum J, Bermudez H (2012) DNA-based delivery vehicles: pH-controlled disassembly and cargo release. Chem Commun 48:12118–12120

Kim NW, Piatyszek MA, Prowse KR et al (1994) Specific association of human telomerase activity with immortal cells and cancer. Science 266:2011–2015

Kim M, Gleason-Guzman M, Izbicka E et al (2003) The different biological effects of telomestatin and TMPyP4 can be attributed to their selectivity for interaction with intramolecular or intermolecular G-quadruplex structures. Cancer Res 63:3247–3256

Kypr J, Kejnovská I, Renčiuk D et al (2009) Circular dichroism and conformational polymorphism of DNA. Nucleic Acids Res 37:1713–1725

Lane AN, Chaires JB, Gray RD et al (2008) Stability and kinetics of G-quadruplex structures. Nucleic Acids Res 36:5482–5515

Langridge R, Rich A (1963) Molecular structure of helical polycytidylic acid. Nature 198:725–728

Li W, Feng L, Ren J et al (2012a) Visual detection of glucose using conformational switch of i-motif DNA and non-crosslinking gold nanoparticles. Chem Eur J 18:12637–12642

Li C, Zhu L, Zhu Z et al (2012b) Backbone modification promotes peroxidase activity of G-quadruplex-based DNAzyme. Chem Commun 48:8347–8349

Liu D, Bruckbauer A, Abell C et al (2006) A reversible pH-driven DNA nanoswitch array. J Am Chem Soc 128:2067–2071

Liu C, Huang C, Chang H (2009) Highly selective DNA-based sensor for lead (II) and mercury (II) ions. Anal Chem 81:2383–2387

Longhese MP (2008) DNA damage response at functional and dysfunctional telomeres. Genes Dev 22:125–140

Magbanua E, Zivkovic T, Hansen B et al (2013) d (GGGT) 4 and r (GGGU) 4 are both HIV-1 inhibitors and interleukin-6 receptor aptamers. RNA Biol 10:216–227

Mergny J, Lacroix L (1998) Kinetics and thermodynamics of i-DNA formation: phosphodiester versus modified oligodeoxynucleotides. Nucleic Acids Res 26:4797–4803

Miyoshi D, Sugimoto N (2008) Molecular crowding effects on structure and stability of DNA. Biochimie 90:1040–1051

Miyoshi D, Nakao A, Sugimoto N (2002) Molecular crowding regulates the structural switch of the DNA G-quadruplex. Biochemistry 41:15017–15024

Miyoshi D, Inoue M, Sugimoto N (2006) DNA logic gates based on structural polymorphism of telomere DNA molecules responding to chemical input signals. Angew Chem Int Ed 45:7716–7719

Modi S, Wani AH, Krishnan Y (2006) The PNA–DNA hybrid I-motif: implications for sugar–sugar contacts in i-motif tetramerization. Nucleic Acids Res 34:4354–4363

Modi S, Swetha M, Goswami D et al (2009) A DNA nanomachine that maps spatial and temporal pH changes inside living cells. Nat Nanotechnol 4:325–330

Moorhouse AD, Santos AM, Gunaratnam M et al (2006) Stabilization of G-quadruplex DNA by highly selective ligands via click chemistry. J Am Chem Soc 128:15972–15973

Moyzis RK, Buckingham JM, Cram LS et al (1988) A highly conserved repetitive DNA sequence, (TTAGGG) n, present at the telomeres of human chromosomes. Proc Natl Acad Sci USA 85:6622–6626

Neidle S, Balasubramanian S (2006) Quadruplex nucleic acids. RSC Publishing, London

Neidle S, Parkinson G (2002) Telomere maintenance as a target for anticancer drug discovery. Nat Rev Drug Discov 1:383–393
Neidle S, Read MA (2000) G-quadruplexes as therapeutic targets. Biopolymers 56:195–208
Rachwal PA, Findlow IS, Werner JM et al (2007) Intramolecular DNA quadruplexes with different arrangements of short and long loops. Nucleic Acids Res 35:4214–4222
Rajendran A, Nakano S, Sugimoto N (2010) Molecular crowding of the cosolutes induces an intramolecular i-motif structure of triplet repeat DNA oligomers at neutral pH. Chem Commun 46:1299–1301
Randazzo A, Esposito V, Ohlenschläger O et al (2004) NMR solution structure of a parallel LNA quadruplex. Nucleic Acids Res 32:3083–3092
Raymond E, Sun D, Chen S et al (1996) Agents that target telomerase and telomeres. Curr Opin Biotechnol 7:583
Read M, Harrison RJ, Romagnoli B et al (2001) Structure-based design of selective and potent G quadruplex-mediated telomerase inhibitors. Proc Natl Acad Sci USA 98:4844–4849
Risitano A, Fox KR (2004) Influence of loop size on the stability of intramolecular DNA quadruplexes. Nucleic Acids Res 32:2598–2606
Sacks DB (2011) A1C versus glucose testing: a comparison. Diabetes Care 34:518–523
Saretzki G (2003) Telomerase inhibition as cancer therapy. Cancer Lett 194:209–219
Seenisamy J, Bashyam S, Gokhale V et al (2005) Design and synthesis of an expanded porphyrin that has selectivity for the c-MYC G-quadruplex structure. J Am Chem Soc 127:2944–2959
Sharma NK, Ganesh KN (2005) PNA C–C i-motif: superior stability of PNA TC8 tetraplexes compared to DNA TC8 tetraplexes at low pH. Chem Commun 34:4330–4332
Shieh Y, Yang S, Wei M et al (2010) Aptamer-based tumor-targeted drug delivery for photodynamic therapy. ACS Nano 4:1433–1442
Shum K, Zhou J, Rossi JJ (2013) Nucleic acid aptamers as potential therapeutic and diagnostic agents for lymphoma. J Cancer Ther 4:872–890
Siddiqui-Jain A, Grand CL, Bearss DJ et al (2002) Direct evidence for a G-quadruplex in a promoter region and its targeting with a small molecule to repress c-MYC transcription. Proc Natl Acad Sci USA 99:11593–11598
Smirnov I, Shafer RH (2000) Effect of loop sequence and size on DNA aptamer stability. Biochemistry 39:1462–1468
Smirnov IV, Shafer RH (2007) Electrostatics dominate quadruplex stability. Biopolymers 85:91–101
Steinert S, Shay JW, Wright WE (2004) Modification of subtelomeric DNA. Mol Cell Biol 24:4571–4580
Sun D, Hurley LH (2009) The importance of negative superhelicity in inducing the formation of G-quadruplex and i-motif structures in the c-Myc promoter: implications for drug targeting and control of gene expression. J Med Chem 52:2863–2874
Sun C, Wang X, Yang X et al (2013) A label-free electrochemical aptasensor for sensitive thrombin detection in whole blood. Electrochim Acta 106:327–332
Teng Y, Girvan AC, Casson LK et al (2007) AS1411 alters the localization of a complex containing protein arginine methyltransferase 5 and nucleolin. Cancer Res 67:10491–10500
Zhang AYQ, Bugaut A, Balasubramanian S (2011) A Sequence-Independent analysis of the loop length dependence of intramolecular RNA G-quadruplex stability and topology. Biochemistry 50:7251–7258
Zhao Y, Du Z, Li N (2007) Extensive selection for the enrichment of G4 DNA motifs in transcriptional regulatory regions of warm blooded animals. FEBS Lett 581:1951–1956
Zhou J, Zhu X, Lu Y et al (2005) Senescence and telomere shortening induced by novel potent G-quadruplex interactive agents, quindoline derivatives, in human cancer cell lines. Oncogene 25:503–511
Zhou J, Wei C, Jia G et al (2010) Formation of i-motif structure at neutral and slightly alkaline pH. Mol Biosyst 6:580–586

Peptides Targeting G-Quadruplex Structures

Kenji Usui and Arisa Okada

Contents

1 Introduction 460
 1.1 G-Quadruplex Structure 460
 1.2 G-Quadruplex Ligands 461
 1.3 Peptides as G-Quadruplex Ligands 462
2 Classification of Peptide Ligands 463
3 Peptide Ligands 464
 3.1 Peptides with Small Molecules 464
 3.2 Peptides Consisting of Unnatural/Nonstandard Amino Acid/Peptide Monomers . . 466
 3.3 Peptides Composed Entirely of Standard Amino Acids 469
4 Conclusions and Future Prospects 471
References 471

Abstract Research in recent decades has revealed that some DNA and RNA secondary structures modulate a variety of cellular events. One secondary structure, the Guanine(G)-quadruplex, also regulates various cellular events that are mostly related to serious diseases. Systems capable of controlling DNA and RNA G-quadruplex structures would therefore be useful for the modulation of various cellular events to produce biological effects. Because of their biological importance, many G-quadruplex-targeting compounds have been described. However, the next generation of targeting molecules should exhibit increased G-quadruplex

K. Usui (✉)
Faculty of Frontiers of Innovative Research in Science and Technology (FIRST), Konan University, 7-1-20 Minatojima-Minamimachi, Chuo-ku, Kobe 650-0047, Japan

Frontier Institute for Biomolecular Engineering Research (FIBER), Konan University, 7-1-20 Minatojima-Minamimachi, Chuo-ku, Kobe 650-0047, Japan
e-mail: kusui@center.konan-u.ac.jp

A. Okada
Faculty of Frontiers of Innovative Research in Science and Technology (FIRST), Konan University, 7-1-20 Minatojima-Minamimachi, Chuo-ku, Kobe 650-0047, Japan

V.A. Erdmann et al. (eds.), *Chemical Biology of Nucleic Acids*, RNA Technologies, DOI 10.1007/978-3-642-54452-1_25, © Springer-Verlag Berlin Heidelberg 2014

sequence specificity, a higher structure-inducing or -collapsing ability, and a greater degree of functionality, including on–off switches of binding ability and cellular penetration. Peptides might be good candidates for these next-generation G-quadruplex-targeting molecules due to the following advantages: (1) their easy design and synthesis, (2) their ability to mimic protein–G-quadruplex interactions, (3) the possibility of employing artificial amino acids in addition to naturally occurring amino acids, and (4) the ability to combine G-quadruplex-binding sequences with other functional sequences. Accordingly, several peptide-based compounds, such as furan-based cyclic peptides, PNA-conjugated peptides, and small molecule-peptide conjugates, have been developed. In this chapter, we introduce all these peptide ligands and describe most of the approaches for targeting G-quadruplex structures. We then conclude that peptides are among the most promising functional ligands for G-quadruplexes to control various biological events in next-generation approaches.

Keywords G-Quadruplex • Peptide • Ligand • Cancer therapy • Telomere • Promoter

1 Introduction

1.1 G-Quadruplex Structure

Research in the last few decades has revealed that DNA and RNA can form duplexes and other secondary structures, such as triplex, hairpin, and i-motif structures containing noncanonical base pairs (Sühnel 2001). These studies have shown that these secondary structures confer certain functions and modulate a variety of cellular events.

Guanine-rich (G-rich) DNA or RNA sequences can also fold into another secondary structure, a quadruplex, which is defined as a structure containing two or more stacked G-quartets (G-tetrads) in which four guanines are assembled in a planar arrangement via Hoogsteen hydrogen bonding (Majumdar and Patel 2002; Neidle and Balasubramanian 2006) (Fig. 1). The G-quartet is further stabilized by electrostatic interactions between the oxygen atoms of the carbonyl and a potassium cation located at the center of the quartet. There are several potential topologies for guanine-quadruplexes (G-quadruplexes), with the sequences oriented in a parallel or antiparallel arrangement. For example, the d[AG3(T2AG3)3] sequence forms the parallel G-quadruplex in the presence of K^+ (pdb code:1KF1), while in Na^+ solution this sequence forms antiparallel G-quadruplex (pdb code: 143D). The d[G3 (T2AG3)3T] sequence forms the antiparallel G-quadruplex in K^+ solution (pdb code: 2KF8).

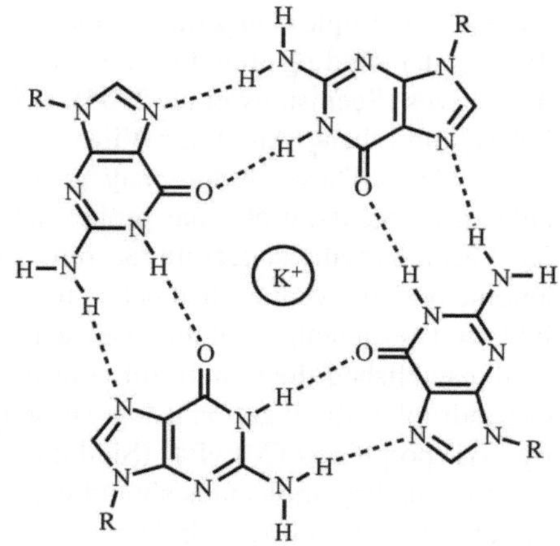

Fig. 1 A guanine quartet highlighting the hydrogen-bonding interactions and the metal cation located at the center of the quartet

Recently, approximately 370,000 G-rich sequences capable of forming G-quadruplex structures have been found in the human genome (Huppert and Balasubramanian 2005; Todd et al. 2005). The distribution of G-rich sequences in the genome is not random, and higher proportions are found in the telomere, promoter, introns, and 5′- and 3′-untranslated regions (UTR) of genes. Consequently, it has been suggested that the G-quadruplex structures generated from these sequences are related to cellular events, including transcription, translation, pre-RNA splicing, and telomerase elongation, all of which play roles in various serious diseases such as cancer and cellular aging (Gomez et al. 2004; Henderson et al. 1987; Kumari et al. 2007; Rankin et al. 2005; Siddiqui-Jain et al. 2002; Yaku et al. 2010; Yu et al. 2006). For example, G-quadruplex formation of the G-rich promoter regions in some proto-oncogenes, such as *c-myc*, *bcl-2*, *c-kit*, and *N-ras*, has been proposed as a regulating switch for gene expression. Siddiqui-Jain et al. showed that a specific G-quadruplex structure formed in the *c-myc* promoter region functioned as a transcriptional repressor element (Siddiqui-Jain et al. 2002). Kumari et al. showed that RNA G-quadruplexes in the 5′ UTR of the gene transcript of the human *N-ras* proto-oncogene, modulated gene expression at the translational level (Kumari et al. 2007). For other examples, the formation of the quadruplexes of G-rich telomeric region shows to decrease the activity of telomerase, which is responsible for telomere elongation (Yaku et al. 2010).

1.2 G-Quadruplex Ligands

In the context as described above, systems capable of controlling DNA and RNA G-quadruplex structures would be useful for the modulation of various cellular events for producing biological effects. Because of their biological importance, many G-quadruplex-targeting ligands (De Cian et al. 2008; Hurley 2002) have been described, including phthalocyanine derivatives (Yaku et al. 2010, 2012), porphyrin derivatives (Seenisamy et al. 2004), LNA (locked nucleic acid) (Kumar and Maiti 2007), and others (Arola and Vilar 2008; Luedtke 2009; Sissi et al. 2011; Tucker et al. 2012). These studies may provide a new avenue for the development of anticancer agents, molecular probes, and biofunctional regulators, and these trials have been partially successful at some applications in research fields. For example, anionic phthalocyanines had been proposed as a telomeric G-quadruplex ligand and inhibited efficiently telomerase activity (Yaku et al. 2010). Siddiqui-Jain et al. established the system for control of *c-myc* transcription by ligand-mediated G-quadruplex stabilization using cationic porphyrin, 5,10,15,20-tetra-(*N*-methyl-4-pyridyl) porphine (TMPyP4) (Siddiqui-Jain et al. 2002). However, the next generation of binding molecules should exhibit a greater G-quadruplex sequence specificity, a higher ability to induce or collapse the structure, and a greater degree of functionality, including on–off switches of binding ability, cellular penetration, and the ability to target organelles in order to apply these molecules to industrial and medical fields.

1.3 Peptides as G-Quadruplex Ligands

Peptides (very small proteins composed of less than 20–30 standard/nonstandard amino acids) are promising next-generation G-quadruplex-binding candidates because of the following advantages: (1) peptides are easier to design and synthesize than antibodies, recombinant proteins, or even highly modified/functionalized small molecules (Chan and White 2000); (2) peptides can mimic protein–G-quadruplex interactions (Tomizaki et al. 2005; Usui et al. 2007); (3) analyses based on peptide libraries can be used to elucidate the binding properties of DNA molecules (Kobayashi et al. 2011; Usui et al. 2009); (4) in addition to naturally occurring amino acids (standard amino acids), various functional moieties (e.g., artificial amino acids or nonstandard amino acids) can be employed as building blocks in designed peptides (Usui et al. 2009); and (5) because certain peptide sequences may exhibit transmembrane (Madani et al. 2011) or hormonal properties, combining G-quadruplex-binding peptides with these functional sequences can produce multifunctional molecular systems that are useful in cell and tissue engineering. Accordingly, many peptide ligands have been described to date. Here, we review a selection of peptides that have been successfully used as G-quadruplex ligands. We then attempt to demonstrate that peptides are among the most promising functional ligands for G-quadruplexes for the control of various biological events in the near future.

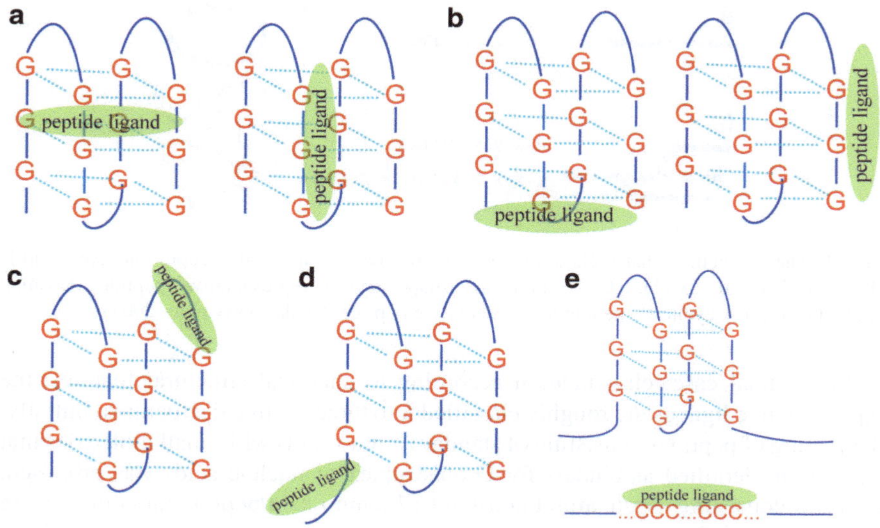

Fig. 2 Binding mode classification. (**a**) Ligands can invade a G-quadruplex. (**b**) Ligands can bind to the surface of a G-quadruplex. (**c**) Ligands can bind to the single-strand DNA/RNA in the exposed loops of a G-quadruplex. (**d**) Ligands can bind to the duplex–quadruplex junction. (**e**) Ligands can bind to the complementary cytosine strand

2 Classification of Peptide Ligands

Almost all peptide ligands developed thus far can be sorted by classification. We propose three different classification criteria, according to the binding modes, shapes of molecules, and chemical structural features.

For the binding mode classification, representative targeting is classified into five different modes (Fig. 2). First, ligands can invade a G-quadruplex, thus facilitating or destroying a G-quadruplex structure (Fig. 2a). Second, ligands can bind to the surface of a G-quadruplex (Fig. 2b). These two modes are predominant, although the following three modes have also been proposed by several authors. Third, ligands can be designed to bind to the single-strand DNA/RNA in the exposed loops of a G-quadruplex (Fig. 2c). Fourth, ligands can bind to the duplex–quadruplex junction (Fig. 2d). Fifth, although ligands do not bind directly to G-quadruplex structures, ligands can bind to the complementary cytosine strand, thus facilitating a G-quadruplex structure with the extrusion of the G-rich strand (Fig. 2e).

In the case of classification according to the shape of the molecule, the representative shapes are classified into the three categories (Fig. 3). The simplest and easiest-to-synthesize shape is linear (Fig. 3a); linear ligands can generally be synthesized using standard peptide synthesis methods (Chan and White 2000). Branch-shaped and cyclic-shaped ligands (Fig. 3b, c) may display a higher affinity and specificity than linear ligands, but the synthesis methods are often difficult (Nagai et al. 2005; Okamoto et al. 2005).

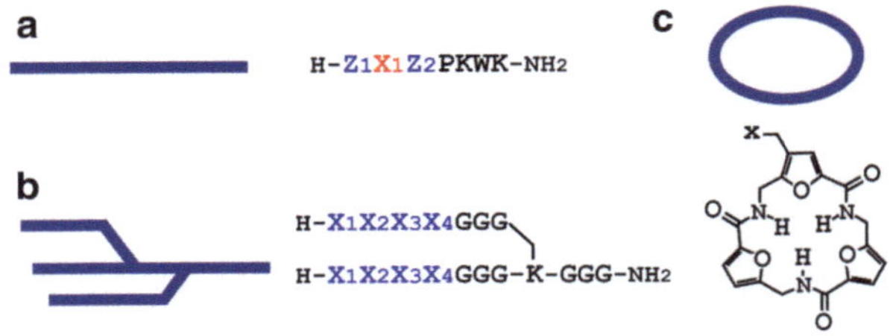

Fig. 3 The molecular shape classification. (**a**) A linear shape and a representative peptide (Kobayashi et al. 2011). (**b**) A branched shape and a representative peptide (Redman et al. 2006). (**c**) A cyclic shape and a representative peptide (Chakraborty et al. 2007)

In the final case, classification according to chemical structural features, the representative ligands are roughly classified into three chemical structures. Initially, the joining of peptides consisting of standard amino acids with small molecules that have been identified as binders for G-quadruplex or nucleic acids was proposed. Peptides consisting of unnatural/nonstandard amino acid/peptide monomers were then developed. Recently, peptides composed entirely of standard amino acids have been reexamined. Although these peptides cannot be expected to obtain a strong affinity for G-quadruplexes, the appeal of utilizing such ligands is enduring because they can be easily conjugated to other functional peptide sequences and can be expressed in *E. coli* or other cells.

The following sections present a selection of peptides that have been successfully used as G-quadruplex ligands. The presented peptides have been classified and arranged according to the chemical structural features (the final classification) in the next section [i.e., peptides consisting of standard amino acids with small molecules (Sect. 3.1), peptides consisting of unnatural/nonstandard amino acid/peptide monomers (Sect. 3.2), and peptides composed entirely of standard amino acids (Sect. 3.3)].

3 Peptide Ligands

3.1 Peptides with Small Molecules

In the early study of ligands using peptides, the conjugation of peptides to small molecules that have been identified as binders for G-quadruplex or nucleic acids [e.g., some carbocyanine dyes (Chen et al. 1996) and derivatives based on an acridine or acridone scaffold (Harrison et al. 1999; Read et al. 2001)] was a promising strategy to generate ligands with superior recognition properties to the peptide or small molecule alone. In the late 1990s, Neidle's group was the first to demonstrate that substituted acridines with amide bonds were effective G-quadruplex stabilizers and telomerase inhibitors (Harrison et al. 1999; Read et al. 2001). For instance, the

3,6,9-trisubstituted acridines showed potent inhibitory activity against human telomerase in vitro, with EC50 values of up to 60 nM (Read et al. 2001). Then, Balasubramanian's group developed peptide-hemicyanine ligands (Green et al. 2006; Schouten et al. 2003); candidate peptide fragments were selected from a native tetrapeptide library (X-X-X-X, where X is Arg, Lys, His, Glu, Asp, Asn, Gln, Pro, Phe, Thr, Ser, Val, and Gly), and the library was screened using the radiolabeled folded human telomeric quadruplex in the presence of competitor DNA [poly(dI-dC)]. The selected peptide was then conjugated to a hemicyanine core, and it was found that the G-quadruplex bound in a 1:1 stoichiometry with ca. 50-fold discrimination in favor of G-quadruplex recognition compared to duplex recognition. At nearly the same time, this group also developed 3,6-bispeptide-acridone and 3,6-bispeptide-acridine conjugates using the selected peptide sequences (Ladame et al. 2002, 2004; Whitney et al. 2004). Although the molecule structure was more complicated, the authors found almost tenfold higher binding (K_d) than with the peptide-hemicyanine ligand. They further discovered linear and branched peptide ligands containing artificial/nonstandard amino acids, including acridone, methylpyrrole, and methylimidazole derivatives for G-quadruplexes, using a direct ELISA (enzyme-linked immunosorbent assay) screen of a one-bead-one-compound library (Redman et al. 2006). Linear peptides showed similar rankings in their affinity for G-quadruplex as their branched counterparts. While the branching peptides showed enhancement of G-quadruplex binding, they did so to a lesser extent than might be predicted and also displayed enhancement of discrimination against duplex. A series of trisubstituted acridine-peptide conjugates was then explored for the recognition and discrimination ability between DNA G-quadruplexes derived from human telomeres and the *c-kit* and *N-ras* proto-oncogenes (Redman et al. 2009). The best ligands had binding affinities in the 1–5 nM range and ca. tenfold discrimination between the G-quadruplexes studied.

Other groups have developed peptide ligands with other small molecules. Huang et al. investigated telomerase inhibitory, cytotoxicity, and the human telomerase reverse transcriptase (hTERT)-repressing effects of a number of synthesized 2,6-diamidoanthraquinones and 1,5-diamidoanthraquinones by comparison of their homologs disubstituted with amino acids. Several of the 1,5-diamidoanthraquinones and 2,6-diamidoanthraquinones inhibited telomerase activity, with IC50 at the sub-micro- to micromolar range. Then these showed acute cytotoxicity to cancer cells with EC50 similar or better than that of mitoxantrone (Huang et al. 2007). Zagotto et al. also constructed a small library of 2,6- and 2,7-amino-acyl/peptidyl anthraquinones with diverse connecting linkers, charge, lipophilicity, and bulk to inhibit enzyme activity by inducing G-quadruplexes in human telomeres. Because cell senescence determinations confirmed delayed toxicity induced by Lys- and Arg-containing anthraquinone derivatives, these molecules were expected to be efficient G-quadruplex binders and telomerase inhibitors (Zagotto et al. 2008). Moreover, they extended the length of side chains by introduction of an amino acid spacer such as Gly, Ala, or Phe, characterized by different flexibility, lipophilicity, and bulkiness. Additionally, 2,6,

2,7, 1,8, and 1,5 regioisomers were examined. Then they found that the Lys-Phe-substituted anthraquinones were poorly cytotoxic, but still able to promote a senescence mechanism in cancer cells (Zagotto et al. 2011).

Furthermore, two cyclic peptides conjugated to a small molecule, which are ligands in more complex structures than those with an almost linear structure, as described above, were discovered. Actinomycin D is known as an anticancer agent and a natural antibiotic that inhibits the transcription of genes by interacting with a GC-rich duplex, single-stranded, or hairpin form of DNA (Wadkins et al. 1996). This natural compound was shown to bind to and induce changes in both the structure and stability of both the Na^+ and K^+ forms of G-quadruplex DNA. Nearly simultaneously in 2009, two different groups demonstrated actinomycin D interactions with human telomeric G-quadruplex DNA (Hudson et al. 2009) and with oncogenic promoter G-quadruplex DNA to repress gene expression (Kang and Park 2009). Both groups proposed that the binding of actinomycin D to G-quadruplex DNA is facilitated by the stacking of the phenoxazone chromophore to the terminal G-quartet of the quadruplex DNA.

3.2 Peptides Consisting of Unnatural/Nonstandard Amino Acid/Peptide Monomers

3.2.1 Distamycin A and Its Derivatives

In the early study of peptides consisting of unnatural/nonstandard amino acid/peptide monomers, the approach was similar to the conjunction of peptides to small molecules, i.e., beginning with the use of compounds identified as G-quadruplex binders. Distamycin A is one example of these peptides. Distamycin A is a basic oligopeptide antibiotic isolated from the fermentation medium of *Streptomyces distallicus* and contains three 1-methylpyrrole-2-carboxamide residues linked to formamido and amidinopropyl groups. Using ^{1}H-NMR, Randazzo et al. first investigated the interaction of distamycin A, known as a minor groove-binding molecule (Kopka et al. 1985), with a synthetic tetraplex DNA structure (Randazzo et al. 2001). More detailed studies of its structure and its applications followed (Cocco et al. 2003; David et al. 2002; Martino et al. 2007) and improvements in affinity have recently been examined via the synthesis of distamycin A analogs. Moore et al. synthesized numerous amide-linked oligopyrroles based on distamycin A and measured their interactions with a human intramolecular G-quadruplex using a melting procedure. These authors found that the binding affinity for G-quadruplex increased with the number of pyrrole groups and that this was consistent with a mixed groove/G-quartet stacking binding mode (Moore et al. 2006). According to their results, they proposed a molecular model of a six-repeat pyrrole polyamide molecule docked onto the crystal structure of the human 22-mer quadruplex structure; two repeats of the polyamide were located in a quadruplex groove, and the others were stacked on a terminal G-quartet surface. In another approach, a functional moiety in distamycin A was replaced with another moiety [e.g., the uncharged N-methyl amide analog (Cosconati et al. 2010) and the carbamoyl analog (Pagano et al. 2007)] and the number of pyrrole units were increased in the carbamoyl analog (Pagano et al. 2008). The carbamoyl analog gave approximately a tenfold increase in the quadruplex affinity (the analog K_b value was $2.3 \times 10^6 M^{-1}$). The presence of one additional pyrrole unit affected the binding of the G-quadruplex. Distamycin A and the carbamoyl analog with one additional pyrrole unit bound quadruplexes in both two different buffered solutions containing either K^+ or Na^+; conversely, the carbamoyl analog with two additional pyrrole unit appeared to have a poor affinity in any case.

3.2.2 Cyclic Peptides Consisting of Nonstandard Amino Acids

Inspired by phthalocyanine derivatives and porphyrin derivatives for G-quadruplex ligands, planar-shaped molecules as building blocks that have extended π-delocalized systems, which show favorable stacking of the molecules on the face of a G-quartet, are promising scaffolds for developing new ligands. Consequently, planar peptide macrocycles are one of the most suitable molecules for scaffolds, although cyclic peptide synthesis is generally more difficult than linear peptide synthesis. Jantos et al. reported the G-quadruplex-binding properties of oxazole-based peptide macrocycles (Jantos et al. 2006). The developed macrocycles were chiral and functionalized with alkylamines to provide water solubility and ability for favorable interactions with the DNA scaffold. Then they found their preferable interaction with the proto-oncogene *c-kit* G-quadruplex (a dissociation constant in the range 4–30 μM) in comparison to the human telomeric quadruplex (a dissociation constant in the range 10–50 μM). Maiti's group also developed a novel furan amino acid, 5-(aminomethyl)-2-furancarboxylic acid, and synthesized its trimer, an 18-member cyclic oligopeptide, which displayed a near-planar geometry with an s-*cis* orientation of all the amide carbonyls (Chakraborty et al. 2002, 2004). This group then synthesized and evaluated the binding ability of two cationic analogs of 18- and 24-membered cyclic oligopeptides of furan amino acids. A comparative analysis of the interaction data of these cyclic peptides with G-quadruplexes and double-strand DNA indicated that 24-membered cyclic peptides are highly selective for telomeric G-quadruplex structures with a dissociation constant in the range 3–30 μM (for duplex with dissociation constants in the range more than 700–950 μM) (Chakraborty et al. 2007). Then, they developed more cationic cyclic peptides based on the furan amino acids by the functionalization with amine groups and demonstrated their repressive effect on *c-myc* transcription and expression (Agarwal et al. 2010a, b). Using real-time PCR (polymerase chain reaction), they observed up to 70 % downregulation of *c-myc* transcripts upon the 18-membered cyclic peptide treatment in HeLa cells (Agarwal et al. 2010a). They also observed up to 90 % downregulation of *c-myc* transcripts upon the 24-membered cyclic peptide treatment in HeLa cells (Agarwal et al. 2010b). Additionally, they showed the downregulation of the protein levels with the 24-membered cyclic peptides using a luciferase assay and the prominent arrest of cells in the sub-G1 stage upon addition of the cyclic peptides that leads toward apoptosis by cell cycle analysis. These results implied that the cyclic peptides provided a suitable scaffold for the development of a therapeutic agent targeting G-quadruplex structures.

3.2.3 PNA Peptides

As described above, many approaches have been reported for developing gene-specific ligands, but a system that takes advantage of nucleobase hydrogen bond recognition would be of great use for G-quadruplex targeting. Peptides conjugated to nucleic acids are attractive ligand candidates over natural nucleic acids, and peptide nucleic acid (PNA) (Nielsen et al. 1991) is a representative of such artificial polymers. PNA consists of nucleotide bases as side chains and *N*-(2-aminoethyl)-glycine units linked by peptide bonds as the backbone rather than the natural sugar phosphate backbone (Nielsen and Egholm 1999). Because the PNA oligomers are uncharged, they exhibit tighter binding to natural nucleic acids than complementary binding of natural nucleic acids due to a lack of poly-phosphate interactions (anionic charge–charge interactions) (Nielsen et al. 1993). Additionally, PNA oligomers tolerate enzyme digestion because PNAs are not easily recognized by either nucleases or proteases. Consequently, these molecules are stable in vivo for use as antigene, antisense, or nucleic acid probes (Demidov et al. 1994). PNA oligomers also have the ability to form various secondary structures, either with itself or in complex with nucleic acids; for instance, they can form very stable triplex structures in addition to DNA duplexes (Egholm et al. 1995; Demidov and Frank-Kamenetskii 2002; Panyutin et al. 2007). Furthermore, these oligomers can form stable quadruplex structures in the presence of appropriate cations, including Na^+ and K^+ (Datta et al. 2005). Randazzo's group has investigated the structure of PNA/DNA chimeras (5′TGGG3′-t, where lower and upper case letters indicate PNA and DNA residues, respectively) forming quadruplex structures (Esposito et al. 2003).

Accordingly, PNA ligands have been designed to target all possible regions of G-quadruplex molecules, i.e., the G-quadruplex loop, body, and DNA/RNA complementary sequences of G-quadruplexes. Almost all were linear ligands because PNA ligands were designed based on their binding to complementary sequences.

In the early study of PNA ligands, PNA oligomers containing cytosines were designed to disrupt a folded quadruplex structure in target DNA sequences and to form DNA/PNA hybrids that are stabilized by overhanging nucleotides on the target strand (Datta and Armitage 2001). Other cytosine(C)-rich PNA ligands have since emerged (Amato et al. 2008, 2009). Amato et al. showed that C-rich PNA ligands interact with quadruplexes and that the binding mode depends on the stability of the quadruplex, binding to the quadruplex structure as "quadruplex-binding agents" in K^+ buffer and disrupting the quadruplex structure by forming PNA/DNA hybrid complexes as "quadruplex openers" in NH_4^+ solution (Amato et al. 2011).

Onyshchenko et al. proposed a different approach based on targeting only the C-rich strand complementary to the quadruplex-forming DNA strand (Onyshchenko et al. 2009, 2011). In this case, a PNA ligand invades a double helix and binds to the C-rich strand then inducing the G-rich strand to form a native DNA G-quadruplex. These authors also assessed the effect of the total PNA charge on the efficiency of the invasion: a PNA that is positively charged or overall neutrally charged invades naturally super-coiled plasmids, whereas a negatively charged PNA does not.

It was then observed that G-rich PNAs formed hetero-quadruplexes with homologous DNA oligomers (Datta et al. 2003; Lusvarghi et al. 2009; Paul et al. 2008; Roy et al. 2007) and RNA oligomers (Marin and Armitage 2005, 2006). A high binding affinity of G-rich PNAs to DNA/RNA G-quadruplexes was observed, with the PNA ligands invading DNA/RNA quadruplexes to form very stable PNA–DNA/RNA hybrid quadruplex structures. The binding stoichiometry depended on the PNA sequences and DNA/RNA sequences. In addition to G-quadruplex invasion and hetero-formation, these PNA ligands can also interact with the C-rich DNA, the complementary strand to G-rich sequence, thus contributing to additional stabilization of the complex of G-quadruplex and PNA ligands. These studies indicated that the formation of these hybrid structures expands the scope of molecular recognition by PNA oligomers. Consequently, we also attempted to develop novel G-rich PNA oligomers combined with a binding-switch module moiety to construct a system to regulate G-quadruplex structures to control the various biological events related to G-quadruplexes.

3.3 Peptides Composed Entirely of Standard Amino Acids

In the study of the G-quadruplex binding of peptides consisting entirely of standard amino acid monomers, appropriate strategies to date include deriving core sequences from natural proteins that have been identified as G-quadruplex binders or screening binding sequences from peptide libraries (Tomizaki et al. 2005; Usui et al. 2007).

3.3.1 Derivation of Core Sequences from Natural Proteins

With regard to the derivation of sequences from natural proteins, it has been found that an arginine-glycine-rich RGG peptide from fragile X mental retardation protein (FMRP) (Darnell et al. 2001, 2009; Schaeffer et al. 2001) could bind to G-quadruplexes and solved the solution structure of the complex between the peptide and an in vitro-selected G-rich *scI* RNA (Phan et al. 2011). The RGG peptide bound to the major groove of the RNA duplex, with the G-quadruplex forcing a sharp turn of this peptide at the junction of duplex and G-quadruplex. These findings may have implications for the recognition of other genomic G-rich

RNAs. Additionally, insulin and the highly homologous insulin-like growth factor 2 (IGF-2), which are relatively long peptides (small proteins), have been shown to bind to a G-quadruplex formed by the repeat G-rich sequence in the insulin-linked polymorphic region (ILPR) (Connor et al. 2006; Schonhoft et al. 2010; Xiao et al. 2009). Insulin shows clear preference for binding to the antiparallel G-quadruplexes in the ILPR and identifies a loop nucleotide critically for binding (Schonhoft et al. 2010). Xiao et al. reported on interactions of insulin and IGF-2 with ILPR variants and showed their Kds were at least as strong as those reported for combinatorially selected aptamers (generally in the nM–μM range)(Xiao et al. 2009). Furthermore, proteins that could be captured by DNA/RNA aptamers containing G-quadruplex structures at the binding site are also attractive. For example, thrombin can be captured specifically at nanomolar concentrations by the thrombin binding aptamer (TBA), which is a well-characterized chair-like, antiparallel G-quadruplex structure (Avino et al. 2012; Nagatoishi et al. 2011; Nagatoishi and Sugimoto 2012). It is expected that core sequences will be derived from these proteins in the near future.

3.3.2 Screening Binding Sequences from Peptide Libraries

Regarding the screening strategy, the separation of mixtures of homodipeptides and of alanyl dipeptides using G-quartet stationary phases in capillary electrochromatography to determine the extent to which selectivity toward individual amino acids may participate in the interactions between G-quartet structures and proteins was carried out (Vo and McGown 2004). These authors demonstrated that certain dipeptides, such as Trp-Trp, showed a particular affinity for the G-quartet phases, suggesting that they might play a role in the binding affinity of larger peptides and proteins for G-quartet structures, even though the dipeptides themselves did not appear to have a strong binding affinity. Balasubramanian's group also selected peptide fragments from a native tetrapeptide library, as described above (Schouten et al. 2003). Although they successfully screened some tetrapeptide sequences for G-quadruplex binding, these peptides still required conjunction to small molecules known to bind G-quadruplexes to obtain several-hundred-fold higher binding affinity. Recently, our group attempted to screen for binders to a particular G-quadruplex sequence using easily designed short peptides consisting of naturally occurring amino acids (Kobayashi et al. 2011). Despite the small size of the library used in this study, candidates as specific binders were identified. In addition, a selected peptide stabilized the G-quadruplex structure of a DNA oligonucleotide derived from the promoter region of the proto-oncogene *c-myc*. Although the peptide also showed micromolar binding affinity, by improving the design of the library peptides and the screening methods using statistical analysis (Usui et al. 2009), our system could be used to screen for peptides that bind to a particular G-quadruplex and alter its thermodynamic properties.

Thus, peptides composed entirely of standard amino acids that strongly bind to a G-quadruplex have rarely been found. Nonetheless, these molecules remain attractive for next-generation applications because they can be expressed in *E. coli* or other cells and it should be possible to identify binders with strong specificity and to add specific functional attributes of previously reported peptides, such as the ability to penetrate cells (Madani et al. 2011), or to target organelles (Sakhrani and Padh 2013).

4 Conclusions and Future Prospects

We have reviewed a selection of peptides that have been successfully used as G-quadruplex ligands. The presented peptides have been classified and arranged according to the chemical structural features.

Targeting G-quadruplex structures using peptides may provide an exciting new avenue for the development of anticancer agents, molecular probes, and biofunctional regulators. However, thus far, nearly all studies have only been demonstrated in vitro, and single-stranded G-rich DNA/RNA sequences have been mainly used in these studies. A major impediment to the validation of G-quadruplex DNA/RNA is the lack of peptides that bind to G-quadruplex structures with high affinity ($K_d < 1$ nM) and high specificity (10,000-fold or lower affinity to all other sequences). However, recent robust advances in peptide-related studies, as reviewed here, generate confidence that these problems will be solved in the near future. Consequently, the next step will be to demonstrate the specificity for a particular G-quadruplex structure and the efficiency of regulation of targeted cellular events using these peptides in cells and in vivo. In addition, the conjunction of these peptides to other functional moieties, such as a probe, cell-penetrating sequence, or binding-switch module, should be applied to attain next-generation systems in a timely fashion. These systems will provide researchers with new tools for controlling various biological events in cell and tissue engineering and medical technology fields.

Acknowledgments We apologize to those authors whose work was not cited directly owing to space limitations. We thank Ms. E. Takekawa (Konan University, Japan) for generous supports. K. U. is also grateful to the Grants-in-Aid for Scientific Research and the "Core research" project (2009–2014) from MEXT.

References

Agarwal T, Roy S, Chakraborty TK et al (2010a) Furan based cyclic homo-oligopeptides bind G-quadruplex selectively and repress c-MYC transcription. Bioorg Med Chem Lett 20:4346–4349

Agarwal T, Roy S, Chakraborty TK et al (2010b) Selective targeting of G-quadruplex using furan-based cyclic homooligopeptides: effect on c-MYC expression. Biochemistry 49:8388–8397

Amato J, Gabelica V, Borbone N et al (2008) A short C-rich PNA fragment capable to form novel G-quadruplex-PNA complexes. Nucleic Acids Symp Ser 52:167–168

Amato J, Oliviero G, De Pauw E et al (2009) Hybridization of short complementary PNAs to G-quadruplex forming oligonucleotides: an electrospray mass spectrometry study. Biopolymers 91:244–255

Amato J, Pagano B, Borbone N et al (2011) Targeting G-quadruplex structure in the human c-Kit promoter with short PNA sequences. Bioconjug Chem 22:654–663

Arola A, Vilar R (2008) Stabilisation of G-quadruplex DNA by small molecules. Curr Top Med Chem 8:1405–1415

Avino A, Fabrega C, Tintore M et al (2012) Thrombin binding aptamer, more than a simple aptamer: chemically modified derivatives and biomedical applications. Curr Pharm Des 18:2036–2047

Chakraborty TK, Tapadar S, Kumar SK (2002) Cyclic trimer of 5-(aminomethyl)-2-furancarboxylic acid as a novel synthetic receptor for carboxylate recognition. Tetrahedron Lett 43:1317–1320

Chakraborty TK, Tapadar S, Raju TV et al (2004) Cyclic trimers of chiral furan amino acids. Synlett 14:2484–2488

Chakraborty TK, Arora A, Roy S et al (2007) Furan based cyclic oligopeptides selectively target G-quadruplex. J Med Chem 50:5539–5542

Chan WC, White PD (2000) Fmoc solid phase peptide synthesis: a practical approach. Oxford University Press, New York, NY

Chen Q, Kuntz ID, Shafer RH (1996) Spectroscopic recognition of guanine dimeric hairpin quadruplexes by a carbocyanine dye. Proc Natl Acad Sci U S A 93:2635–2639

Cocco MJ, Hanakahi LA, Huber MD et al (2003) Specific interactions of distamycin with G-quadruplex DNA. Nucleic Acids Res 31:2944–2951

Connor AC, Frederick KA, Morgan EJ et al (2006) Insulin capture by an insulin-linked polymorphic region G-quadruplex DNA oligonucleotide. J Am Chem Soc 128:4986–4991

Cosconati S, Marinelli L, Trotta R et al (2010) Structural and conformational requisites in DNA quadruplex groove binding: another piece to the puzzle. J Am Chem Soc 132:6425–6433

Darnell JC, Jensen KB, Jin P et al (2001) Fragile X mental retardation protein targets G quartet mRNAs important for neuronal function. Cell 107:489–499

Darnell JC, Fraser CE, Mostovetsky O et al (2009) Discrimination of common and unique RNA binding activities among Fragile X mental retardation protein paralogs. Hum Mol Genet 18:3164–3177

Datta B, Armitage BA (2001) Hybridization of PNA to structured DNA targets: quadruplex invasion and the overhang effect. J Am Chem Soc 123:9612–9619

Datta B, Schmitt C, Armitage BA (2003) Formation of a PNA_2-DNA_2 hybrid quadruplex. J Am Chem Soc 125:4111–4118

Datta B, Bier ME, Roy S et al (2005) Quadruplex formation by a guanine-rich PNA oligomer. J Am Chem Soc 127:4199–4207

David WM, Brodbelt J, Kerwin SM et al (2002) Investigation of quadruplex oligonucleotide-drug interactions by electrospray ionization mass spectrometry. Anal Chem 74:2029–2033

De Cian A, Lacroix L, Douarre C et al (2008) Targeting telomeres and telomerase. Biochimie 90:131–155

Demidov VV, Frank-Kamenetskii MD (2002) PNA openers and their applications. Methods Mol Biol 208:119–130

Demidov VV, Potaman VN, Frank-Kamenetskii MD et al (1994) Stability of peptide nucleic acids in human serum and cellular extracts. Biochem Pharmacol 48:1310–1313

Egholm M, Christensen L, Dueholm KL et al (1995) Efficient pH-independent sequence-specific DNA binding by pseudoisocytosine-containing bis-PNA. Nucleic Acids Res 23:217–222

Esposito V, Galeone A, Mayol L, Messere A et al (2003) PNA-DNA chimeras forming quadruplex structures. Nucleosides Nucleotides Nucleic Acids 22:1681–1684

Gomez D, Lemarteleur T, Lacroix L et al (2004) Telomerase downregulation induced by the G-quadruplex ligand 12459 in A549 cells is mediated by hTERT RNA alternative splicing. Nucleic Acids Res 32:371–379

Green JJ, Ladame S, Ying L et al (2006) Investigating a quadruplex-ligand interaction by unfolding kinetics. J Am Chem Soc 128:9809–9812

Harrison RJ, Gowan SM, Kelland LR et al (1999) Human telomerase inhibition by substituted acridine derivatives. Bioorg Med Chem Lett 9:2463–2468

Henderson E, Hardin CC, Walk SK et al (1987) Telomeric DNA oligonucleotides form novel intramolecular structures containing guanine–guanine basepairs. Cell 51:899–908

Huang HS, Chen IB, Huang KF et al (2007) Synthesis and human telomerase inhibition of a series of regioisomeric disubstituted amidoanthraquinones. Chem Pharm Bull (Tokyo) 55:284–292

Hudson JS, Brooks SC, Graves DE (2009) Interactions of actinomycin D with human telomeric G-quadruplex DNA. Biochemistry 48:4440–4447

Huppert JL, Balasubramanian S (2005) Prevalence of quadruplexes in the human genome. Nucleic Acids Res 33:2908–2916

Hurley LH (2002) DNA and its associated processes as targets for cancer therapy. Nat Rev Cancer 2:188–200

Jantos K, Rodriguez R, Ladame S et al (2006) Oxazole-based peptide macrocycles: a new class of G-quadruplex binding ligands. J Am Chem Soc 128:13662–13663

Kang HJ, Park HJ (2009) Novel molecular mechanism for actinomycin D activity as an oncogenic promoter G-quadruplex binder. Biochemistry 48:7392–7398

Kobayashi K, Matsui N, Usui K (2011) Use of a designed peptide library to screen for binders to a particular DNA g-quadruplex sequence. J Nucleic Acids 572873. doi:10.4061/2011/572873

Kopka ML, Yoon C, Goodsell D et al (1985) The molecular origin of DNA-drug specificity in netropsin and distamycin. Proc Natl Acad Sci U S A 82:1376–1380

Kumar N, Maiti S (2007) Role of locked nucleic acid modified complementary strand in quadruplex/Watson-Crick duplex equilibrium. J Phys Chem B 111:12328–12337

Kumari S, Bugaut A, Huppert JL et al (2007) An RNA G-quadruplex in the 5′ UTR of the NRAS proto-oncogene modulates translation. Nat Chem Biol 3:218–221

Ladame S, Harrison RJ, Neidle S et al (2002) Solid-phase synthesis of symmetrical 3,6-bispeptide-acridone conjugates. Org Lett 4:2509–2512

Ladame S, Schouten JA, Stuart J et al (2004) Tetrapeptides induce selective recognition for G-quadruplexes when conjugated to a DNA-binding platform. Org Biomol Chem 2:2925–2931

Luedtke NW (2009) Targeting G-quadruplex DNA with small molecules. CHIMIA Int J Chem 63:134–139

Lusvarghi S, Murphy CT, Roy S et al (2009) Loop and backbone modifications of peptide nucleic acid improve g-quadruplex binding selectivity. J Am Chem Soc 131:18415–18424

Madani F, Lindberg S, Langel U et al (2011) Mechanisms of cellular uptake of cell-penetrating peptides. J Biophys 414729. doi: 10.1155/2011/414729

Majumdar A, Patel DJ (2002) Identifying hydrogen bond alignments in multistranded DNA architectures by NMR. Acc Chem Res 35:1–11

Marin VL, Armitage BA (2005) RNA guanine quadruplex invasion by complementary and homologous PNA probes. J Am Chem Soc 127:8032–8033

Marin VL, Armitage BA (2006) Hybridization of complementary and homologous peptide nucleic acid oligomers to a guanine quadruplex-forming RNA. Biochemistry 45:1745–1754

Martino L, Virno A, Pagano B et al (2007) Structural and thermodynamic studies of the interaction of distamycin A with the parallel quadruplex structure $[d(TGGGGT)]_4$. J Am Chem Soc 129:16048–16056

Moore MJ, Cuenca F, Searcey M et al (2006) Synthesis of distamycin A polyamides targeting G-quadruplex DNA. Org Biomol Chem 4:3479–3488

Nagai Y, Nakanishi T, Okamoto H et al (2005) IR study on stacking manner of peptide nanorings in peptide nanotubes. Jpn J Appl Phys 44:7654–7661

Nagatoishi S, Sugimoto N (2012) Interaction of water with the G-quadruplex loop contributes to the binding energy of G-quadruplex to protein. Mol Biosyst 8:2766–2770

Nagatoishi S, Isono N, Tsumoto K et al (2011) Hydration is required in DNA G-quadruplex-protein binding. Chembiochem 12:1822–1826

Neidle S, Balasubramanian S (2006) Quadruplex nucleic acids. Royal Society of Chemistry, Cambridge, UK

Nielsen PE, Egholm M (1999) An introduction to peptide nucleic acid. Curr Issues Mol Biol 1:89–104

Nielsen PE, Egholm M, Berg RH et al (1991) Sequence-selective recognition of DNA by strand displacement with a thymine-substituted polyamide. Science 254:1497–1500

Nielsen PE, Egholm M, Berg RH et al (1993) Sequence specific inhibition of DNA restriction enzyme cleavage by PNA. Nucleic Acids Res 21:197–200

Okamoto H, Yamada T, Miyazaki H et al (2005) Difference in self-assembling morphology of peptide nanorings. Jpn J Appl Phys 44:8240–8248

Onyshchenko MI, Gaynutdinov TI, Englund EA et al (2009) Stabilization of G-quadruplex in the BCL2 promoter region in double-stranded DNA by invading short PNAs. Nucleic Acids Res 37:7570–7580

Onyshchenko MI, Gaynutdinov TI, Englund EA et al (2011) Quadruplex formation is necessary for stable PNA invasion into duplex DNA of BCL2 promoter region. Nucleic Acids Res 39:7114–7123

Pagano B, Mattia CA, Virno A et al (2007) Thermodynamic analysis of quadruplex DNA-drug interaction. Nucleosides Nucleotides Nucleic Acids 26:761–765

Pagano B, Virno A, Mattia CA et al (2008) Targeting DNA quadruplexes with distamycin A and its derivatives: an ITC and NMR study. Biochimie 90:1224–1232

Panyutin IG, Panyutin IV, Demidov VV (2007) Targeting linear duplex DNA with mixed-base peptide nucleic acid oligomers facilitated by bisPNA openers. Anal Biochem 362:145–147

Paul A, Sengupta P, Krishnan Y et al (2008) Combining G-quadruplex targeting motifs on a single peptide nucleic acid scaffold: a hybrid (3 + 1) PNA-DNA bimolecular quadruplex. Chem Eur J 14:8682–8689

Phan AT, Kuryavyi V, Darnell JC et al (2011) Structure-function studies of FMRP RGG peptide recognition of an RNA duplex-quadruplex junction. Nat Struct Mol Biol 18:796–804

Randazzo A, Galeone A, Mayol L (2001) ^{1}H-NMR study of the interaction of distamycin A and netropsin with the parallel stranded tetraplex $[d(TGGGGT)]_4$. Chem Commun 11:1030–1031

Rankin S, Reszka AP, Huppert J et al (2005) Putative DNA quadruplex formation within the human *c-kit* oncogene. J Am Chem Soc 127:10584–10589

Read M, Harrison RJ, Romagnoli B et al (2001) Structure-based design of selective and potent G quadruplex-mediated telomerase inhibitors. Proc Natl Acad Sci U S A 98:4844–4849

Redman JE, Ladame S, Reszka AP et al (2006) Discovery of G-quadruplex stabilizing ligands through direct ELISA of a one-bead-one-compound library. Org Biomol Chem 4:4364–4369

Redman JE, Granadino-Roldán JM, Schouten JA et al (2009) Recognition and discrimination of DNA quadruplexes by acridine-peptide conjugates. Org Biomol Chem 7:76–84

Roy S, Tanious FA, Wilson WD et al (2007) High-affinity homologous peptide nucleic acid probes for targeting a quadruplex-forming sequence from a *MYC* promoter element. Biochemistry 46:10433–10443

Sakhrani NM, Padh H (2013) Organelle targeting: third level of drug targeting. Drug Des Devel Ther 7:585–599

Schaeffer C, Bardoni B, Mandel JL et al (2001) The fragile X mental retardation protein binds specifically to its mRNA via a purine quartet motif. EMBO J 20:4803–4813

Schonhoft JD, Das A, Achamyeleh F et al (2010) ILPR repeats adopt diverse G-quadruplex conformations that determine insulin binding. Biopolymers 93:21–31

Schouten JA, Ladame S, Mason SJ et al (2003) G-quadruplex-specific peptide-hemicyanine ligands by partial combinatorial selection. J Am Chem Soc 125:5594–5595

Seenisamy J, Rezler EM, Powell TJ et al (2004) The dynamic character of the G-quadruplex element in the c-MYC promoter and modification by TMPyP4. J Am Chem Soc 126:8702–8709

Siddiqui-Jain A, Grand CL, Bearss DJ et al (2002) Direct evidence for a G-quadruplex in a promoter region and its targeting with a small molecule to repress c-MYC transcription. Proc Natl Acad Sci U S A 99:11593–11598

Sissi C, Gatto B, Palumbo M (2011) The evolving world of protein-G-quadruplex recognition: a medicinal chemist's perspective. Biochimie 93:1219–1230

Sühnel J (2001) Beyond nucleic acid base pairs: from triads to heptads. Biopolymers 61:32–51

Todd AK, Johnston M, Neidle S (2005) Highly prevalent putative quadruplex sequence motifs in human DNA. Nucleic Acids Res 33:2901–2907

Tomizaki KY, Usui K, Mihara H (2005) Protein-detecting microarrays: current accomplishments and requirements. Chembiochem 6:782–799

Tucker WO, Shum KT, Tanner JA (2012) G-quadruplex DNA aptamers and their ligands: structure, function and application. Curr Pharm Des 18:2014–2026

Usui K, Tomizaki K, Mihara H (2007) Screening of alpha-helical peptide ligands controlling a calcineurin-phosphatase activity. Bioorg Med Chem Lett 17:167–171

Usui K, Tomizaki K, Mihara H (2009) A designed peptide chip: protein fingerprinting technology with a dry peptide array and statistical data mining. Methods Mol Biol 570:273–284

Vo TU, McGown LB (2004) Selectivity of quadruplex DNA stationary phases toward amino acids in homodipeptides and alanyl dipeptides. Electrophoresis 25:1230–1236

Wadkins RM, Jares-Erijman EA, Klement R et al (1996) Actinomycin D binding to single-stranded DNA: sequence specificity and hemi-intercalation model from fluorescence and 1H NMR spectroscopy. J Mol Biol 262:53–68

Whitney AM, Ladame S, Balasubramanian S (2004) Templated ligand assembly by using G-quadruplex DNA and dynamic covalent chemistry. Angew Chem Int Ed Engl 43:1143–1146

Xiao J, Carter JA, Frederick KA et al (2009) A genome-inspired DNA ligand for the affinity capture of insulin and insulin-like growth factor-2. J Sep Sci 32:1654–1664

Yaku H, Murashima T, Miyoshi D et al (2010) Anionic phthalocyanines targeting G-quadruplexes and inhibiting telomerase activity in the presence of excessive DNA duplexes. Chem Commun 46:5740–5742

Yaku H, Fujimoto T, Murashima T et al (2012) Phthalocyanines: a new class of G-quadruplex-ligands with many potential applications. Chem Commun 48:6203–6216

Yu HQ, Miyoshi D, Sugimoto N (2006) Characterization of structure and stability of long telomeric DNA G-quadruplexes. J Am Chem Soc 128:15461–15468

Zagotto G, Sissi C, Lucatello L et al (2008) Aminoacyl-anthraquinone conjugates as telomerase inhibitors: synthesis, biophysical and biological evaluation. J Med Chem 51:5566–5574

Zagotto G, Ricci A, Vasquez E et al (2011) Tuning G-quadruplex vs double-stranded DNA recognition in regioisomeric lysyl-peptidyl-anthraquinone conjugates. Bioconjug Chem 22:2126–2135

Synthesis of Site-Specifically Modified Long-mer RNAs

Darko Balke, Jennifer Frommer, Nico Rublack, Danilo Springstubbe, Bettina Appel, and Sabine Müller

Contents

1 Introduction 478
2 Chemical Synthesis of Natural and Modified Oligoribonucleotides 479
3 Monomer-Building Blocks for Post-Synthetic Modification of RNA 480
4 5′- and 3′-Terminal Modification of RNA Molecules 481
5 Enzymatic Ligation of RNA Fragments 484
6 Chemical Ligation of RNA Fragments 488
7 Twin Ribozymes for Internal Modification of Long RNAs 491
8 Summary 493
References 493

Abstract Site-specific functionalization of RNA and DNA molecules has become a major task in nucleic acid chemistry. Synthetic RNAs are required for a large number of applications. Beyond synthetic RNA for antisense, aptamer, ribozyme, and siRNA technologies, oligoribonucleotides carrying site-specific modifications for structure and function studies are needed. A wide variety of monomer-building blocks is commercially available to be used for site-specific incorporation by solid-phase RNA synthesis. However, the efficient chemical preparation of RNA is limited to oligomers of about 60–70 nucleotides. Therefore, efficient strategies for fragment ligation or other alternative novel protocols are required. We here provide an overview on our work on the synthesis of site-specifically modified long-mer RNAs, focusing on enzymatic and chemical ligation of synthetic RNA fragments and on ribozyme-mediated sequence exchanges.

Keywords Chemical ligation • Enzymatic ligation • RNA modification • RNA synthesis • Ribozyme

D. Balke • J. Frommer • N. Rublack • D. Springstubbe • B. Appel • S. Müller (✉)
Institut für Biochemie, Ernst-Moritz-Arndt Universität Greifswald, Felix-Hausdorff Str. 4, 17487 Greifswald, Germany
e-mail: smueller@uni-greifswald.de

V.A. Erdmann et al. (eds.), *Chemical Biology of Nucleic Acids*, RNA Technologies, DOI 10.1007/978-3-642-54452-1_26, © Springer-Verlag Berlin Heidelberg 2014

1 Introduction

Over the past two decades RNA has become a major focus of research into structure and biological function as well as diagnostic and therapeutic application. The cellular functions of RNA are manifold, and they are fulfilled by a large variety of small and long noncoding RNAs (Aalto and Pasquinelli 2012; Serganov and Patel 2012; Moran et al. 2012). Biomolecular function is tightly bound to structure, and therefore, structural studies of functional biomolecules are of utmost importance. In that, crystallization and x-ray analysis gain a key position. However, techniques like NMR-, EPR- or fluorescence spectroscopy that allow looking at the structure and dynamics of biomolecules in solution, have become highly competitive, and nowadays are more and more applied to the elucidation of biomolecular structure. Those methods however, require the site-specific introduction of particular reporter units such as isotops, spin probes, or dyes, in order to make the molecule "visible" for the chosen spectroscopic technique. What holds for structural studies, often also applies to investigations into biomolecular function. Here, modified nucleobases, sugar, and backbone modifications or any kind of functional entities attached to the RNA of interest are required to learn about its function and the mechanism of action.

The method of choice for site-specific incorporation of modified nucleosides into RNA is chemical synthesis using modified nucleoside phosphoramidites. This allows for specific localization of a desired functionality at a predefined position of the RNA strand, something that enzymatic synthesis of RNA cannot achieve. For chemical synthesis, the desired functionality either is available as phosphoramidite and thus is directly incorporated during RNA chain assembly, or alternatively, functional entities such as for example aliphatic amines, thiols, alkynes, or azides are introduced via suitable phosphoramidite-building blocks and used afterwards for post-synthetic attachment of a desired modification. However, the limitation of an adequate product yield of long chemically synthesized RNAs makes this strategy only suitable for functionalization of rather short, synthetically available oligomers. Hence, modification of long transcripts or natural RNA requires alternative chemical or enzymatic techniques allowing for internal or end labeling. The most obvious strategy would be chemical synthesis of short RNA fragments followed by joining these fragments via covalent linkages. Accordingly, a number of protocols for enzymatic or chemical ligation of synthetic oligonucleotides have been developed. Furthermore, natural or in vitro transcribed RNAs are also prone to modification by using either strategies for 5′-or 3′-terminal modification or engineered tools such as twin ribozymes for site-specific introduction of a certain functionalization. In the following subchapters we will discuss methods applied in our laboratory for the preparation of specifically modified long RNAs, starting from chemical synthesis of oligoribonucleotides and suitably functionalized building blocks, over post-synthetic modification, enzymatic and chemical ligation of RNA-fragments, up to ribozyme-mediated RNA functionalization.

2 Chemical Synthesis of Natural and Modified Oligoribonucleotides

For the chemical synthesis of natural and modified oligoribonucleotides, we make use of the phosphoramidite approach in conjunction with 2′-*O*-TBDMS chemistry. Controlled pore glass (CPG) is used as solid support, to which the first ribonucleoside derivative is attached via its 2′- or 3′-OH group as a succinate linkage. The procedure is based on 2′-*O*-*tert*-butyldimethylsilyl (TBDMS) nucleoside protection and involves sequential couplings of β-cyanoethyl-(*N*,*N*-diisopropylamino)-phosphoramidites of 5′-*O*-dimethoxytrityl-2′-*O*-TBDMS nucleosides. The exocyclic amino groups of the nucleobases are protected either with phenoxyacetyl (PAC) for adenosine, isopropyl-phenoxyacetyl (iPr-PAC) for guanosine, or acetyl for cytidine. The standard synthesis is on the micromole scale. First, the 5′-*O*-protecting group of the starting nucleoside on the support is removed under acidic conditions. Subsequently, the incoming phosphoramidite is coupled to the growing chain in the presence of an activator. We use 1-(benzylmercapto)-1*H*-tetrazole activation, allowing for coupling yields in the 98–99 % range (Welz and Müller 2002). After the coupling reaction, the 5′-hydroxyl groups that have not been coupled are capped with acetate groups, and the newly formed phosphite triester is oxidized to the requested phosphotriester. The cycle is repeated until an oligonucleotide of desired length has been synthesized.

Following chain assembly, removal of base-protecting groups is accomplished using aqueous ammonia (30 %) and ethanolic methylamine in a 1:1 mixture for 2 h at room temperature. Further, fluoride ion treatment is necessary to remove the 2′-*O*-TBDMS groups. While 1 M tetrabutylammonium fluoride was conventionally used for this purpose, we prefer the use of triethylamine trihydrofluoride first suggested by Gasparutto et al. (1992). In combination with polar aprotic solvents and at elevated temperatures, this reagent offers advantages in speed and reliability (Wincott et al. 1995). Therefore, we use triethylamine trihydrofluoride/DMF (3:1) at 55 °C for 1.5 h for efficient removal of the 2′-*O*-TBDMS groups. Following deprotection, the oligonucleotide is desalted by precipitation with *n*-butanol and purified by HPLC or polyacrylamide gel electrophoresis (Müller et al. 2004; Rublack et al. 2011).

As mentioned above, site-specific labeling of RNA molecules requires the incorporation of an appropriate modified nucleotide during the chemical solid-phase synthesis of oligomers. Apart from the four natural ribonucleosides, a large number of phosphoramidite-building blocks are available, allowing for either direct introduction of the desired modification, or introduction of a suitable precursor that can be used for post-synthetic attachment of a chosen functional unit. However, as the yield of RNA is directly proportional to its length, this technique is only efficient for small and mid-size RNAs.

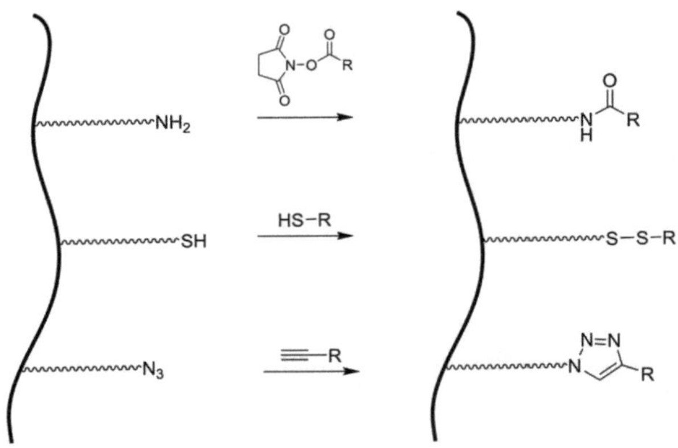

Fig. 1 Strategies for post-synthetic functionalization of oligonucleotides. From *top* to *bottom*: Amide bond formation (NHS ester strategy), disulfide bond formation, triazole formation by "click chemistry," *R* any label of interest, e.g., dye

3 Monomer-Building Blocks for Post-Synthetic Modification of RNA

Among the methods for the synthesis of modified RNA, post-synthetic functionalization has become a major technique. It is of particular importance to labeling of RNAs with spectroscopic reporters such as dyes and spin probes, since those often are not stable enough under the conditions of chemical RNA synthesis. Therefore, suitable functionalities provided as phosphoramidite-building blocks are coupled during RNA chain assembly and upon synthesis and purification of the oligoribonucleotide are selectively reacted with the desired molecular entity (Fig. 1).

Among the chemistries depicted in Fig. 1, the incorporation of aliphatic amines followed by coupling of a desired functionality activated as NHS ester is probably the most common technique (Davies et al. 2000). However, also thiol groups, in most cases provided as 4-thiouridine at a desired position of the synthetic RNA, have been used for conjugation chemistry (Sontheimer 1994; Yu 1999; Wunnicke et al. 2011). More recently, azide alkyne coupling has become a powerful method for post-synthetic RNA functionalization, requiring either an azide to be incorporated in the RNA and the alkyne provided by the functional entity or vice versa (Paredes 2011). All these techniques require modified monomer-building blocks to be incorporated into oligonucleotides and a suitable protocol for post-synthetic conjugation.

A number of monomer-building blocks providing the functionalities shown in Fig. 1 are commercially available. However, to extend this repertoire, we have synthesized a variety of 5′-*O*-dimethoxytrityl-2′-*O*-*tert*-butyldimethylsilyl-protected 3′-*O*-phosphoramidites of nucleosides carrying amino linkers of different length and flexibility at the heterocyclic base. In particular, we have focused on derivatives with the linker attached to C5 of the pyrimidine bases and to C8 and C2 of the purine bases (Rublack et al. 2011; Sindbert et al. 2011) (Fig. 2). The amino linkers were introduced by either Heck or Sonogashira cross-coupling reactions prior to or after introduction of the required protecting groups at exocyclic amines and at the 5′- and 2′-OH groups. Finally the protected linker-modified nucleosides were converted to phosphoramidite-building blocks and used for RNA synthesis.

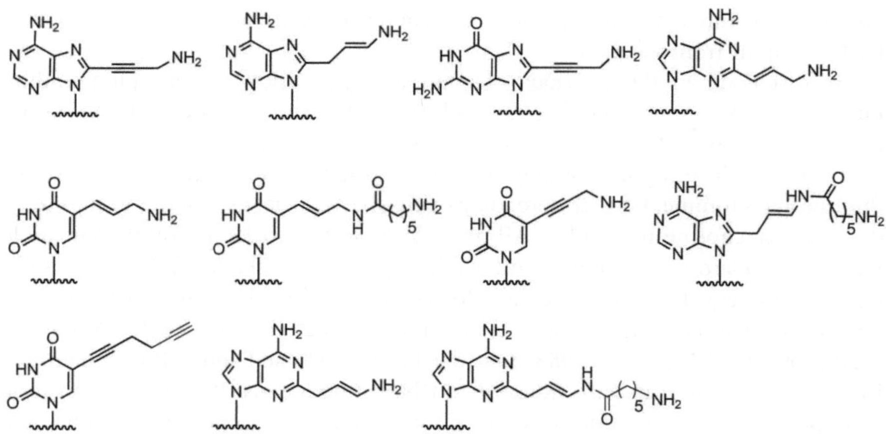

Fig. 2 Monomer RNA-building blocks carrying amino linkers of different length and flexibility

The amino-modified RNAs were labeled with suitable donor and acceptor dyes for fluorescence spectroscopy, in particular FRET measurements (Sindbert et al. 2011) or conjugated to flavine mononucleotide (Rublack et al. 2011) for the investigation of ribozyme action. Furthermore, 2′-aminouridine and 4-thiouridine, commercially available as monomer-building blocks for RNA synthesis, were used to attach spin labels to the tetracycline aptamer, to be investigated by EPR spectroscopy (Wunnicke et al. 2011).

Other commercially available monomers that we have used in a number of studies for post-synthetic functionalization are 5′-terminal amino (Drude et al. 2007, 2011) or thiol linkers (Piekielska et al. 2011), as well as specifically functionalized solid supports for the synthesis of oligonucleotides with terminal 3′-phosphate, 3′-amino or 3′-sulfhydryl groups to conjugate fluorescent probes and affinity labels, as well as to immobilize oligonucleotides to gold surfaces (Piekielska et al. 2011).

4 5′- and 3′-Terminal Modification of RNA Molecules

The insertion of modified nucleotides at internal positions of an RNA bears the risk of altering the RNA topology and thus also its properties. In order to minimize the potential negative impact on the RNA structure and to allow for proper folding of the oligonucleotides, attaching labels onto either the 3′- or the 5′-end of the oligonucleotide may be less dangerous. Furthermore, 3′- or 5′-terminal end-labeling is less dependent on the size of the RNA molecule and thus is also suitable for functionalization of long RNAs. Although, it has to be taken into account that longer RNA molecules have a higher tendency for folding into complex structures, which eventually, even at denaturing conditions, might sequester the termini from reaction.

As discussed in the previous chapter, a large variety of phosphoramidite-building blocks for 3′- and 5′-end labeling of synthetically prepared RNAs are commercially available. These building blocks contain suitable functionalities for post-synthetic labeling such as amino and thiol groups as well as halogen atoms like fluorine and bromine. Furthermore, they themselves can act as reporter groups (e.g., dyes) and anchoring molecules. All of these building blocks are fully compatible with the standard phosphoramidite coupling chemistries like the TBDMS-, ACE-, and TOM-chemistry (Müller et al. 2004). In addition, several strategies were developed that take advantage of the unique reactivity of the RNA termini (Sampson and Uhlenbeck 1988; Oh and Pace 1994) and thus allow functionalization of natural or in vitro transcribed RNA at both ends. It involves activation of either terminus, by addition of EDC and imidazole for the 5′-terminus, and periodate for the 3′-terminus. Subsequently, the RNA can be linked to an amino- or hydrazide-modified tag (Hermanson 2008; Willkomm and Hartmann 2009). Enzymatic modification of the 5′-terminus of RNAs can be performed for example by introduction of phosphate isotopes using T4 PNK (T4 polynucleotide kinase), or by transcription priming, which even allows the incorporation of modified nucleosides. The addition of modified nucleosides to the 3′-terminus is mediated by a range of suitable enzymes, e.g., PUP (poly(U) polymerase), PAP (poly(A) polymerase), or TdT (terminal deoxynucleotidyl transferase) (England and Uhlenbeck 1978a, b; Igloi 1996; Martin and Keller 1998; Rosemeyer et al. 1995; Winz et al. 2012). However, the product yield, when using these methods, still remains rather low. Nevertheless, as a commonly used method, the post-synthetic modification of long RNAs at their 3′-end is achieved by oxidation with metaperiodate ions. In 1928, Malaprade has shown that periodate ions can selectively oxidize adjacent hydroxyl functional groups in acidic, neutral, or weakly alkaline solutions at room temperature (Malaprade 1928). In that, *cis*-diols undergo cleavage of the carbon–carbon bond leading to aldehydes. Accordingly, the *cis*-diol moiety of the ribose sugar of the 3′-terminus of RNA can be subjected to periodate oxidation yielding a dialdehyde (Marinetti and Rouser 1955), which can be further derivatized with a nucleophilic tag (Winston et al. 2001). Most common tags used for this technique like dyes, biotine derivatives, or linkers can be purchased as hydrazides, semicarbazides, thiosemicarbazides, and primary amines (Fig. 3). The reaction between the 3′-oxidized RNA and primary amines forms only slightly stable products. However, the following reduction by borohydride compounds leads to stable morpholino derivatives. Hydrazide-modified labels react with aldehydes to form hydrazone linkages, making the reaction much more efficient. The addition of sodium cyanoborohydride will further increase the reaction efficiency and enhance the bond stability over time and pH changes.

The periodate-mediated oxidation step is very mild and doesn't affect other RNA-specific functionalities. This labeling technique has some advantages over alternative methods. It is a rather cheap way of attaching reporter groups onto any RNAs, independent of the specific sequence and length. Moreover, via this route, a

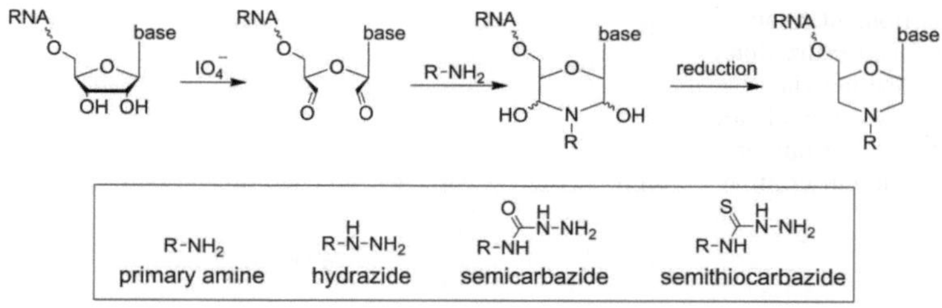

Fig. 3 Periodate-mediated oxidation of the 3′-end of RNA and reductive coupling to amino group bearing derivatives

large variety of molecules can be very efficiently coupled to the 3′-end of the RNA. In addition to attaching reporter groups onto RNA, it is also possible to achieve general RNA functionalization. For example, we have synthesized a cytidine derivative containing a hexaethylene glycol tether bearing a primary amino group (Fig. 4a). At the C4-position of the heterocyclic base, we have incorporated a short linker connected to a biotin moiety. After 3′-end oxidation of a suitable RNA library, the generated 3′-terminal dialdehydes react with the aliphatic primary amino group of the cytidine derivative (Fig. 4a), allowing attaching the resulting modified RNAs onto a solid phase through biotin-streptavidine interaction and to carry out in vitro selection experiments.

With the same purpose, we have synthesized a cytidine derivative coupled to guanosine (Fig. 4b) in order to modify the 5′-end of the members of an RNA library via transcription priming. In 1988, Sampson and Uhlenbeck carried out a run-off in vitro transcription experiment, where GMP was present along with the other NTPs (Sampson and Uhlenbeck 1988). They observed that for the enzymatic preparation of RNA, the T7 RNA polymerase not only accepts GTP as substrate but also modified guanosine monophosphates (Milligan and Uhlenbeck 1989). Prerequisites for successful incorporation of guanosine derivatives into an RNA strand are the guanosine-5′-monophosphate moiety as well as a free 3′-hydroxyl group. These modified guanosines are also referred to as initiator molecules because nucleotides that don't possess a triphosphate at their 5′-end can only be attached at the 5′-end of the nascent RNA chain. The incorporation of the modified guanosine occurs statistically for which reason the initiator molecule should be used in excess over GTP. There are several examples for successful transcription priming reactions described in the literature (Seelig and Jäschke 1999; Eisenführ et al. 2003; Wolf et al. 2008). By using a chemically synthesized biotinylated adenosine coupled to guanosine, we were able to determine the yield of incorporation of the initiator molecule into a 39-mer RNA. In addition, transcription priming seems to be a promising technique for the evolution of ribozymes catalyzing a variety of chemical reactions such as Diels–Alder and Michael reaction (Seelig and Jäschke 1999; Eisenführr et al. 2003).

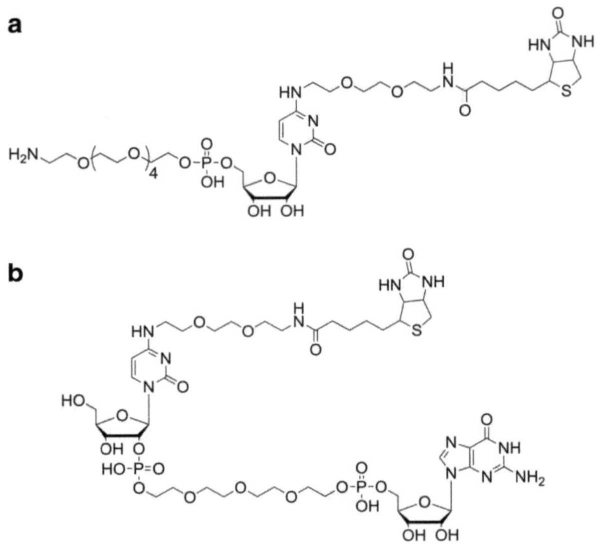

Fig. 4 Cytidine derivatives for RNA functionalization by 3′-end labeling via periodate-induced oxidation (**a**) and 5′-end labeling of RNA through transcription priming (**b**)

5 Enzymatic Ligation of RNA Fragments

Due to the limitations of chemical RNA synthesis mentioned above, ligation of synthetic RNA fragments has become an important option for the preparation of long modified RNAs. Thus, the desired modification can be incorporated in oligoribonucleotides of appropriate lengths by chemical synthesis, followed by joining the individual fragments to the full-length RNA. Basically, also in vitro transcription can efficiently produce long RNAs. However, apart from 5′- and 3′-end modification, this procedure is limited to unmodified or, in the best case, statistically modified molecules. Therefore, fragment synthesis followed by enzymatic or chemical ligation most often is the method of choice. Several protocols exist for the enzymatic ligation of RNA fragments (Turunen et al. 2013). Most commonly used are three different polynucleotide ligases, capable of ligating nicks in single- and/or double-stranded DNA or RNA constructs: T4 DNA ligase (T4 Dnl), T4 RNA ligase 1 (T4 Rnl 1), and T4 RNA ligase 2 (T4 Rnl 2). These three ligases are encoded by the genome of bacteriophage T4 during infection of host bacteria (Remaut et al. 1983; Miller et al. 2003). They catalyze the formation of a phosphodiester bond between 5′-phosphate and 3′-hydroxyl end groups in DNA or RNA. The oligonucleotide carrying the 5′-terminal phosphate acts as a donor strand, whereas the oligomer carrying the terminal 3′-hydroxyl group to be linked to the donor strand is called the acceptor strand. All three ligases require ATP as cofactor.

When using T4 DNA ligase, the ligation site needs to be located within a double-stranded region of the nucleic acid. The duplex may exist of only DNA or RNA or can be a DNA/RNA hybrid (Pheiffer and Zimmerman 1983; Rossi et al. 1997).

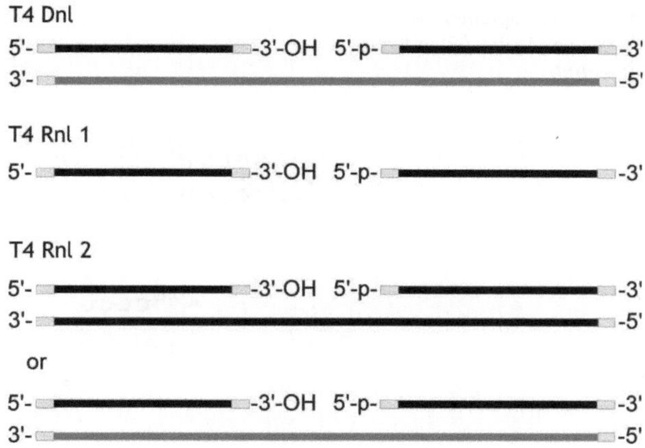

Fig. 5 Ligation strategies involving ligases T4 Dnl, T4 Rnl 1, T4 Rnl 2 with the most common application. *Black lines*: RNA, *gray lines*: DNA

Single-stranded regions are not accepted as substrate. For ligation of two RNA fragments with T4 DNA ligase, it is a good choice to use a DNA complement also known as splint, to create the required double stranded region around the ligation site (Fig. 5). After ligation, the splint can be digested with DNase. Therefore, purification of the ligated RNA product is easily performed by PAA gel electrophoresis. It has been shown that the length of the splint affects the efficiency of the ligation (Kurschat et al. 2005). The temperature optimum for the enzyme is 16 °C. It is also possible, ligating three or more RNA fragments in a one-pot reaction by using several specific or just one full-length DNA splint (Turunen et al. 2013). For each individual system, reaction conditions need to be defined in order to achieve efficient ligation. Varying the equivalents of donor, acceptor, and splint is another important parameter with strong impact on the ligation yield (Rader et al. 2006).

T4 RNA ligase 1 (T4 Rnl) supports the ligation of single-stranded nucleic acids and accepts DNA and RNA fragments as substrates (Middleton et al. 1985; Edwards et al. 1991). T4 RNA ligase 2 (T4 Rnl2) ligates RNA acceptor strands (3′-hydroxyl group) with RNA- or DNA-donor strands (5′-phosphate group) and is much more efficient for joining nicks in ds RNA substrates than connecting the ends of ssRNA (Nandakuman and Shuman 2004; Bullard and Bowater 2006). It is possible to use both types of T4 RNA ligase to cyclize linear ssRNA substrates (Ho and Shuman 2002), as for example the intramolecular circularization of a linear RNA strand to an authentic infectious circular RNA of the citrus exocortis viroid strain A (CEV-A) (Rigden and Rezaian 1992). For joining ssRNA fragments by T4 RNA ligases, it is advantageous if the reactive groups are properly oriented to each other. Good success has been achieved in ligating the anticodon and the D loop of different tRNA molecules (Ohtsuki et al. 1998; Persson et al. 2003). Alternatively, proper orientation of the reactive ends can be achieved by using a splint that by hybridization with both the donor and the acceptor brings the substrates close together, however, leaves the terminal 2–3 nucleotides of both strands single stranded at the ligation junction (Lang and Micura 2008).

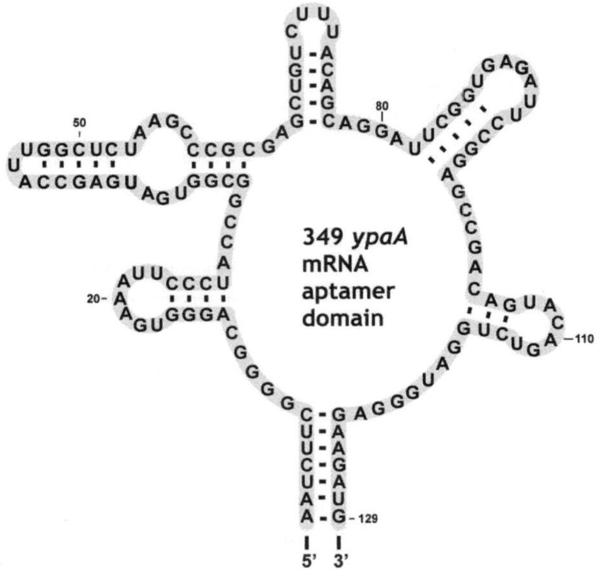

Fig. 6 The aptamer domain of the *ypaA* riboswitch from *B. subtilis*

In addition to the "wild-type" versions of T4 ligases, nowadays, some special truncated variants of T4 RNA ligase as well as thermostable variants of DNA and RNA ligases are commercially available. Those have found some special application such as improved linker ligations in cloning experiments (Aravin and Tusch 2005; Pfeffer et al. 2005) or the application in a primer-free SELEX process for ssDNA aptamers (Lai and DeStefano 2011). However, these engineered ligases have not yet found entry into ligation of chemically synthesized modified RNA fragments at preparative scale.

The design of a ligation experiment should take into account that the ligase has different preferences for the terminal nucleotides of the donor- and acceptor strand carrying the reactive groups. For T4 Rnl1, the preference is $A > G \geq C > U$ for the $3'$-terminal nucleotide of the acceptor strand and $pC > pU > pA > pG$ for the $5'$-terminal nucleotide of the donor strand (Bain and Switzer 1992; England and Uhlenbeck 1978a, b; Romaniuk et al. 1982). It should also be noticed that the ligase catalyzes side reactions like unwanted cyclization or synthesis of oligomers by connecting the $5'$-end of the donor to the $3'$-hydroxyl end of another donor strand (concatemerization). Such side reactions decrease the yield of the ligation reaction. However, the addition of some additives, e.g., PEG, BSA, or cobalt hexamine chloride was shown to support the desired reaction of donor and acceptor and thus increase the ligation yield (Tessier et al. 1986).

In our laboratory, we invented different ligation strategies for the synthesis of the aptamer domain of the flavin mononucleotide (FMN) responsive riboswitch *ypaA* from *B. subtilis* (Fig. 6), which regulates the expression of a gene encoding a

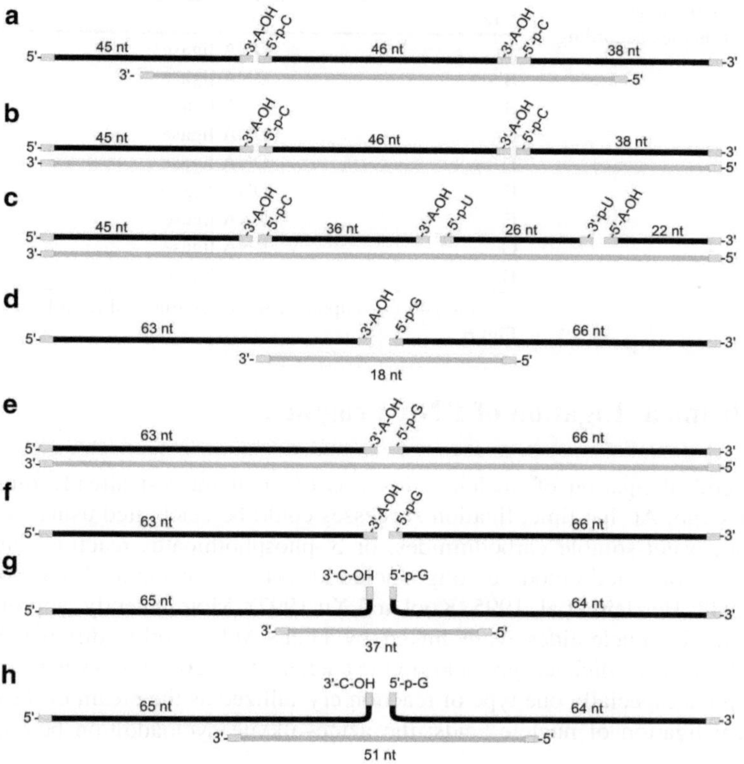

Fig. 7 Different enzymatic ligation strategies for the generation of the *ypaA* aptamer domain. *Black lines*: RNA, *gray lines*: DNA. For ligation positions compare with Fig. 5

riboflavin transport protein by effecting translation initiation (Winkler et al. 2002). To study binding and folding properties of the *ypaA* aptamer domain, natural and 2-aminopurine-modified variants of the 129nts RNA were prepared in preparative scale using a combination of chemical and enzymatic synthesis.

Several ligation experiments were designed, dividing the full-lengths aptamer in three or two strands to be ligated at splints of different lengths and nature, and using either T4 DNA ligase or T4 RNA ligase 1 (Fig. 7, Table 1). Best results were obtained by ligation, with T4 RNA ligase 1, of 2 strands, 64 and 65 nt in length, and a 51 nt DNA splint, forming 23 base pairs with the acceptor oligonucleotide and 24 base pairs with the donor nucleotide, such that two nucleotides of both donor and acceptor strand remained single stranded at the ligation junction. This setup allowed obtaining the full-lengths aptamer with 42 % yield (Turunen et al. 2013).

Table 1 Screening of ligation strategies according to Fig. 7

Strategy	Ligase	Yield
A	DNA ligase	6 %
B	DNA ligase	10 %
B	RNA ligase	–
C[a]	DNA ligase	<5 %
D	DNA ligase	<5 %
E	DNA Ligase	24 %
F	RNA ligase	<5 %
G	RNA ligase	15 %
H	RNA ligase	42 %

[a]Position 46 in wt aptamer sequence mutated from U to C, comp. Fig. 6

6 Chemical Ligation of RNA Fragments

The chemical ligation of nucleic acids was of great interest already more than 20 years ago. At that time, ligation successes could be celebrated using cyanogen bromide, water-soluble carbodiimides, or 3′-phosphothioates reacting with both 5′-tosylate- or 5′-iodo-modified oligonucleotides (Dolinnaya et al. 1988; Sokolova et al. 1988; Herrlein et al. 1995; Kool and Xu 1997). More recently, appropriately modified oligonucleotides were linked by Diels–Alder cycloaddition (Marchán et al. 2006) or a Michael-type addition (Sanchez et al. 2013). However, over the past 5 years especially one type of reaction crystallized as the cream of the crop in chemical ligation of nucleic acids: the azide–alkyne cycloaddition belonging to Click reactions.

Since Click chemistry was introduced (Kolb et al. 2001), it developed a large field of applications. Click chemistry is not only one category of chemical reaction, but it is defined by reaction requirements. A Click reaction has to run under mild conditions, the educts are easy to obtain and to handle, and so are the resulting products (Kolb et al. 2001). The most popular Click reaction is the cycloaddition between azide- and alkyne-functionalized educts, termed 1,3-dipolar Huisgen cycloaddition (Rostovtsev et al. 2002; Tornøe et al. 2002). This reaction runs in buffered aqueous solution as well as in organic solvents, like acetonitrile or DMF (Hein and Fokin 2010). Usually, high temperatures are necessary to form the triazole from the azide and alkyne, thereby forming the 1,4- and the 1,5-triazole regioisomers (Hein and Fokin 2010). However, when the reaction is catalyzed by Cu(I) at room temperature, only the 1,4-disubstituted 1,2,3-triazolo compounds are formed (Rostovtsev et al. 2002; Tornøe et al. 2002) (Fig. 8).

The copper-catalyzed cycloaddition is called CuAAC and offers a great scope of practice. Due to the fact that the CuAAC works in aqueous solutions, biomolecules like proteins and nucleic acids have moved in the field of interest. Thus, with the aid of Click chemistry, an alternative strategy for conjugation and labeling reactions could be established (Seela and Sirivolu 2008). In the area of nucleic acids, the problem of degradation as a result of working with the reactive Cu(I) arose

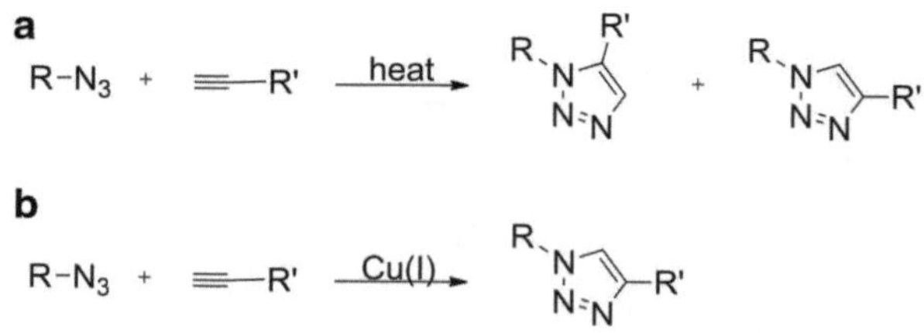

Fig. 8 (**a**) The Huisgen cycloaddition of an azide and an alkyne at ahigh temperature (the 1,4- and the 1,5-triazole regioisomers are formed) and (**b**) catalyzed bei Cu(I) (only the 1,4-triazole regioisomer is formed)

(Burrows and Muller 1998). To solve this problem, a variety of ligands have been developed, forming stable complexes with Cu(I) to shield the DNA/RNA from reaction (Chan et al. 2004). Among the many variants, the water-soluble ligand tris(3-hydroxylpropyltriazol-methyl)amin (THPTA) is the most attractive for working with nucleic acids (Kumar et al. 2007). Furthermore, it has been shown that acetonitrile is not only a favored solvent but it is also known for its ability to coordinate Cu(I) in a complex and thereby provides a stabilizing effect (Paredes and Das 2011). Consequently, additional ligands such as THPTA are not necessary.

In addition to RNA labeling as mentioned above, nucleic acid ligation via Click chemistry possesses great potential to become the chosen method for making large RNAs (Fig. 9). Due to the fact that enzymatic ligation with ligases does not always go hand in hand with good yields, there is a need for an alternative. The essential azide- and alkyne-modification at the 3′- and 5′-end of the RNA strands can be generated in several ways, enzymatically or chemically. The enzymatic options are on the one hand 5′-transcription priming (see chapter "Effects of Ionic Liquid and Liposomes on the Structure, Stability, and Function of Nucleic Acids") with 5′-N_3-G, and on the other hand 3′-tagging with 3′-N_3-dATP by the Poly(A) polymerase (Paredes and Das 2011). The 5′-N_3-G can be synthesized in two steps, 3′-N_3-dATP is commercially available.

For chemical synthesis, a number of azide- and alkyne-modified phosphoramidites are available. Thus, modifications can be introduced easily at the 3′- or 5′-end of RNAs or at internal positions using commercially available building blocks. An often discussed problem is the non-sufficient stability of the azide at the conditions of RNA synthesis by the phosphoramidite approach, in particular the applied P(III)-chemistry (El-Sagheer and Brown 2009). Therefore, in the majority of cases, the azide modification is enzymatically introduced post-synthesis or by transformation of the terminal 5′-OH group into the azide, while the protected RNA is still bound to the solid support (El-Sagheer and Brown 2010).

After initial experiments on synthesis of dinucleotides by Click chemistry (Nuzzi et al. 2007) and follow-up experiments replacing the phosphodiester linkages of a 10 mer by 1,2,3-triazoles (Isobe et al. 2008), DNA-oligonucleotides were effectively "clicked" for the first time (El-Sagheer and Brown 2009). These clicked DNA strands could be used very well as a PCR template. The same group

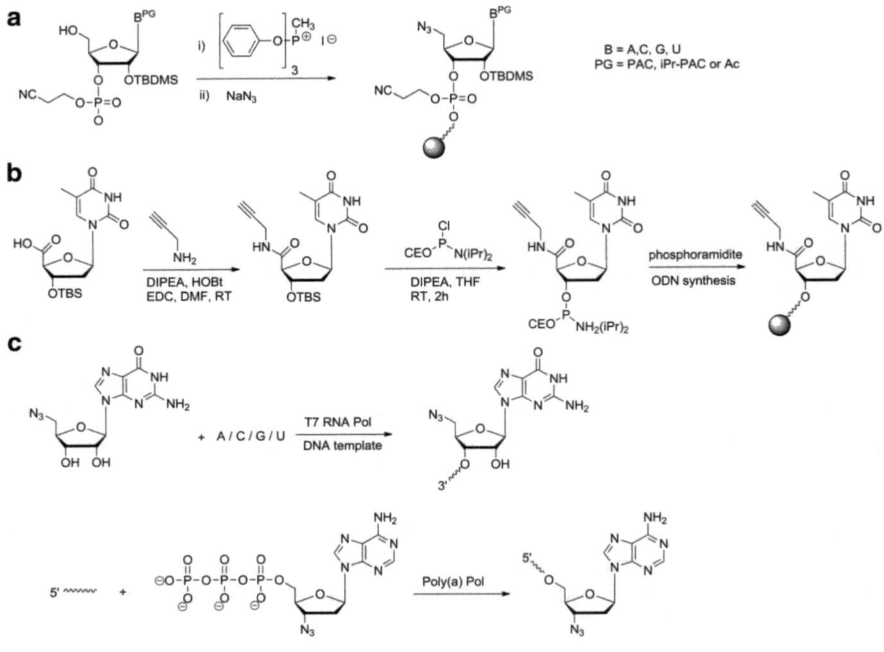

Fig. 9 (**a**) Post-synthetic formation of an 5′-azide-modified oligonucleotide with a full-protected RNA bound to the solid support (El-Sagheer and Brown 2010); (**b**) generation of a 5′-alkyne-modified oligonucleotide, using a synthetically available phosphoramidite (El-Sagheer and Brown 2009). Similarly, support bound nucleosides carrying a 3′-terminal alkyne moiety can be used for 3′-alkyne functionalization; (**c**) generation of a 5′-azide-modified oligonucleotide by transcription priming (*top*) and of a 3′-azide-modified oligonucleotide by 3′-end tagging with Poly(A) polymerase (*bottom*) (Paredes and Das 2011)

introduced Click chemistry as a RNA ligation-method, generating a functional hammerhead ribozyme by ligating a 5′-N_3-U-oligonucleotide to a 3′-alkyne-2′-deoxy-5-methyl-C-oligonucleotide (El-Sagheer and Brown 2010). Miraculously, the unnatural triazol linkage had no negative effect on the ribozyme, although being located in the catalytic pocket; the cleavage reaction proceeded with excellent yield. Furthermore, also the HDV ribozyme was successfully ligated by Click chemistry, and no negative effects in stability and function were observed (Paredes et al. 2011). Both, the hammerhead ribozyme and the HDV ribozyme Click ligation were performed between substrates having a deoxyribonucleoside at the 3′-side and a ribonucleoside at the 5′-side.

We have successfully performed Click-ligation between full RNA fragments, hence having ribonucleotides at both, the 3′- and the 5′-ligation side. The RNA of interest was the *ypaA* aptamer, corresponding to the aptamer region of the *ypaA* riboswitch from *B. subtilis,* already introduced in the previous chapter on enzymatic RNA ligation (Fig. 6). As discussed above, with a full length of 129 nucleotides, chemical synthesis is rather challenging. Therefore, the introduction of

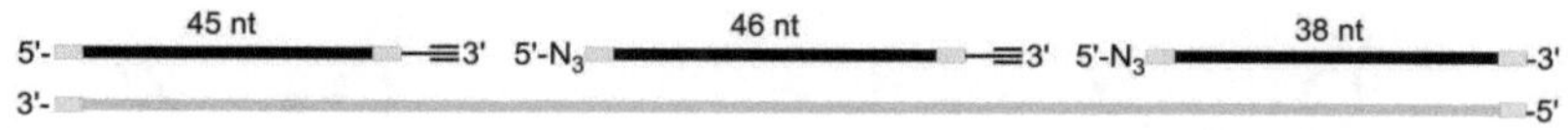

Fig. 10 The *ypaA* aptamer fragmented in three strands with 5′-azide- and 3′-alkyne-modifications for Click ligation at a DNA template. *Black lines*: RNA, *gray lines*: DNA

modifications, as required for structure and function studies, was achieved by fragment synthesis followed by enzymatic or chemical ligation. For the Click ligation experiments, three RNA strands were synthesized, thereby one strand was decorated with a 3′-azide-modification, the second strand with a 3′-azide and a 5′-alkyne-modification, and the third strand with only a 5′-alkyne-modification (Fig. 9a, b). Like most of the other ligation reactions, the Click ligation was performed with the aid of a splint as recommended (El-Sagheer and Brown 2009; Paredes 2011). We succeeded in generating a full-length ligated *ypaA* aptamer by azide–alkyne coupling (Fig. 10), and as in the case of the Click-ligated ribozymes mentioned above, the aptamer proved functional in folding and FMN binding.

7 Twin Ribozymes for Internal Modification of Long RNAs

In addition to chemical or enzymatic methods described above, another sophisticated strategy for RNA modification has been developed in our laboratory, allowing the introduction of basically every desired modification into RNA. This strategy is based on the application of a small-engineered twin ribozyme that mediates the exchange of a short RNA patch against a separately added synthetic RNA fragment, within a suitable RNA substrate (Welz et al. 2003; Ivanov et al. 2005; Vauléon et al. 2005; Drude et al. 2007).

Twin ribozymes are derived from tandem duplication of the hairpin ribozyme and thus consist of two catalytically active motifs. Based on the cleavage/ligation properties of the hairpin ribozyme, twin ribozymes cleave and ligate a substrate RNA at two defined positions in a strictly controlled fashion. Hereby a short RNA patch is excised and replaced by an externally added oligonucleotide bearing the desired modification (Vauléon et al. 2005). The sequence exchange proceeds in several steps. First, the substrate RNA is bound by the twin ribozyme by Watson–Crick base pairing and is cleaved at two defined positions. In the next step, a short RNA patch dissociates and the remaining gap is filled with an externally added oligonucleotide containing the desired modification. However, binding of this oligonucleotide competes with re-association of the internal cleavage fragment. The equilibrium can be shifted towards binding of the external oligonucleotide by promoting the dissociation of the internal cleavage fragment. This is achieved by adapting the twin ribozyme structure to this purpose. In particular, binding of the substrate RNA is designed to lead to formation of a destabilizing structure within the sequence patch that is supposed to be cut out (e.g., mismatches or bulges). In this case, dissociation of the formed cleavage fragment is promoted. In contrast, the

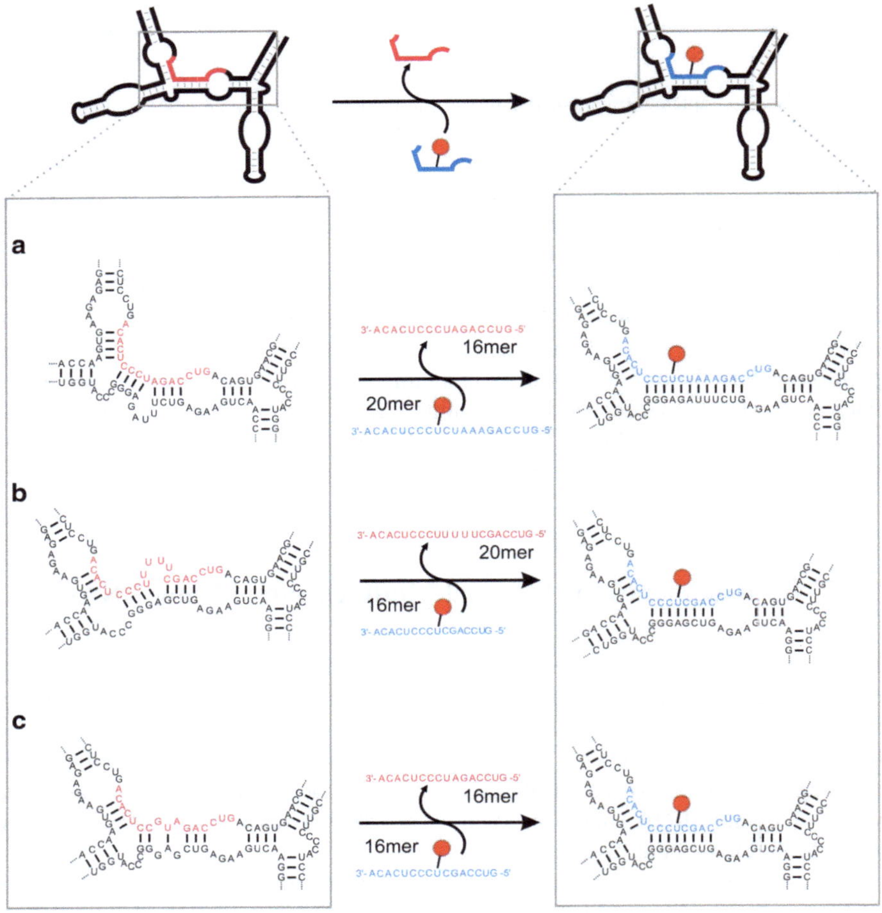

Fig. 11 Twin ribozymes for RNA functionalization

externally added oligonucleotide forms a contiguous duplex with the ribozyme and therefore is preferentially bound because of its favorable stable structure. After twin ribozyme mediated ligation, the desired product is formed. Depending on the design of the twin ribozyme and substrate RNA complex, fragments of the same length can be exchanged (Vauléon et al. 2005), as well as short fragments against longer versions (Welz et al. 2003) and vice versa (Drude et al. 2007) (Fig. 11). This qualifies twin ribozymes as an eminent tool for incorporation of modifications into long RNA transcripts. We have shown that RNA fragments conjugated with fluorescent dyes or biotin are well accepted for strand exchange. Up to 53 % of a dye-labeled oligoribonucleotide could be inserted into an enzymatically prepared 145-mer RNA (Vauléon et al. 2005).

8 Summary

The route to synthetic RNA involves chemical or enzymatic synthesis techniques. While chemical synthesis usually is the method of choice for preparation of shorter (up to about 50 nucleotides) RNA fragments, longer RNAs are preferentially made by enzymatic run-off transcription in vitro or by chemical synthesis of fragments that post-synthesis are joined by chemical or enzymatic ligation. In contrast to enzymatic methods, the chemical synthesis of oligoribonucleotides stands out as the strategy allowing for site-specific incorporation of modifications such as dyes, spin or affinity labels, and a number of other functionalities. As discussed in the chapters above, we have succeeded in the synthesis of long RNAs carrying site-specific modifications by a combination of strategies: chemical synthesis of modified monomer-building blocks and oligoribonucleotides, post-synthetic conjugation with specific functionalities, enzymatic, and chemical fragment ligation, and twin ribozyme-mediated sequence exchanges.

The progress in the chemical synthesis of specifically modified long-mer RNAs will enhance prospects in RNA structure and function studies, as well as in RNA engineering and construction of sophisticated RNA devices. In combination with the development of synthetic routes to appropriately activated modifiers, long RNAs nowadays can be conjugated with a large variety of tags and functionalities, something that for many years has been reserved to short RNAs.

References

Aalto AP, Pasquinelli AE (2012) Small non-coding RNAs mount a silent revolution in gene expression. Curr Opin Cell Biol 24:333–340

Aravin A, Tusch T (2005) Identification and characterization of small RNAs involved in RNA silencing. FEBS Lett 579:5830–5840

Bain JD, Switzer C (1992) Regioselective ligation of oligoribonucleotides using DNA splints. Nucleic Acids Res 20:4372

Bullard DR, Bowater RP (2006) Direct comparison of nick-joining activity of the nucleic acid ligases from bacteriophage T4. Biochem J 398:135–144

Burrows CJ, Muller JG (1998) Oxidative nucleobase modifications leading to strand scission. Chem Rev 98:1109–1151

Chan TR, Hilgraf R, Sharpless KB et al (2004) Polytriazoles as copper(I)-stabilizing ligands in catalysis. Org Lett 6:2853–2855

Davies MJ, Shah A, Bruce IJ (2000) Synthesis of fluorescently labeled oligonucleotides and nucleic acids. Chem Soc Rev 29:97–107

Dolinnaya NG, Sokolova NI, Grayaznova OI et al (1988) Site-directed modification of DNA duplexes by chemical ligation. Nucleic Acids Res 16:3721–3738

Drude I, Vauléon S, Müller S (2007) Twin ribozyme mediated removal of nucleotides from an internal RNA site. Biochem Biophys Res Commun 363:24–29

Drude I, Strahl A, Galla D et al (2011) Design of hairpin ribozyme variants with improved activity for poorly processed substrates. FEBS J 278:622–633

Edwards JB, Delort J, Mallet J (1991) Oligodeoxyribonucleotide ligation to single-stranded cDNAs: a new tool for cloning 5′-ends of mRNAs and for constructing cDNA libraries by in vitro amplification. Nucleic Acids Res 19:5227–5232

Eisenführ A, Arora PS, Sengle G et al (2003) A ribozyme with michaelase activity: synthesis of the substrate precursors. Bioorg Med Chem 11:235–249

El-Sagheer AH, Brown T (2009) Synthesis and polymerase chain reaction amplification of DNA strands containing an unnatural triazol linkage. J Am Chem Soc 131:3958–3964

El-Sagheer AH, Brown T (2010) New strategy for the synthesis of chemically modified RNA constructs exemplified by hairpin and hammerhead ribozymes. Proc Natl Acad Sci USA 35:15329–15334

England TE, Uhlenbeck OC (1978a) Enzymatic oligoribonucleotide synthesis with T4 RNA ligase. Biochemistry 17:2069–2076

England TE, Uhlenbeck OC (1978b) 3′-terminal labelling of RNA with T4 RNA ligase. Nature 275:560–561

Gasparutto D, Livache T, Bazin H et al (1992) Chemical synthesis of a biologically active natural tRNA with its minor bases. Nucleic Acids Res 20:5159–5166

Hein JE, Fokin VV (2010) Copper-catalyzed azide-alkyne cycloaddition (CuAAC) and beyond: new reactivity of copper(I) acetylides. Chem Soc Rev 39:1302–1315

Hermanson GT (2008) Nucleic acid and oligonucleotide modification and conjugation. In: Hermanson GT (ed) Bioconjugate techniques, 2nd edn. Academic, San Diego, pp 969–1002

Herrlein MK, Nelson JS, Letsinger RL (1995) A covalent lock for self-assembled oligonucleotide conjugates. J Am Chem Soc 117:10151–10152

Ho CK, Shuman S (2002) Bacteriophage T4 RNA ligase 2 (gp24.1) exemplifies a family of RNA ligases found in all phylogenetic domains. Proc Natl Acad Sci USA 99:12709–12714

Igloi GL (1996) Nonradioactive labeling of RNA. Anal Biochem 233:124–129

Isobe H, Fujino T, Yamazaki N et al (2008) Triazole-linked analogue of deoxyribonucleic acid (^{TL}DNA): design, synthesis, and double-strand formation with natural DNA. Org Lett 17:3729–3732

Ivanov SA, Vauléon S, Müller S (2005) Efficient RNA ligation by reverse-joined hairpin ribozymes and engineering of twin ribozymes consisting of conventional and reverse-joined hairpin ribozyme units. FEBS J 272:4464–4474

Kolb HC, Finn MG, Sharpless KB (2001) Click chemistry: diverse chemical function from a few good reactions. Angew Chem Int Ed Engl 40:2004–2021

Kool ET, Xu Y (1997) A novel 5′-iodonucleoside allows efficient nonenzymatic of single-stranded and duplex DNAs. Tetrahedron Lett 38:5595–5598

Kumar R, El-Sagheer AH, Tumpane J et al (2007) Template-directed oligonucleotide strand ligation, covalent intramolecular DNA circularization and catenation using click chemistry. J Am Chem Soc 129:6859–6864

Kurschat WC, Müller J, Wombacher R et al (2005) Optimizing splinted ligation of highly structured small RNAs. RNA 11:1909–1914

Lai Y, DeStefano JJ (2011) A primer-free method that selects high-affinity single-stranded DNA aptamers using thermostable RNA ligase. Anal Biochem 414:246–253

Lang K, Micura R (2008) Preparation of site-specifically modified riboswitch domains as an example for enzymatic ligation of chemically synthesized RNA fragments. Nat Protoc 8:1457–1466

Malaprade L (1928) Action of polyalcohols on periodic acid. Analytical application. Bull Soc Chim (France) 43:683–696

Marchán V, Ortega S, Pulido D et al (2006) Diels-Alder cycloaddition in water for the straightforward preparation of peptide-oligonucleotide conjugates. Nucleic Acids Res 34:e24/1–e24/9

Marinetti G, Rouser G (1955) The periodate oxidation of ribose-5-phosphate in acid and alkaline solution. J Am Chem Soc 77:5345–5349

Martin G, Keller W (1998) Tailing and 3′-end labeling of RNA with yeast poly(A) polymerase and various nucleotides. RNA 4:226–230

Middleton T, Herlihy WC, Schimmel PR et al (1985) Synthesis and purification of oligoribonucleotides using T4 RNA ligase and reverse-phase chromatography. Anal Biochem 144:110–117

Miller ES, Kutter E, Mosig G et al (2003) Bacteriophage T4 genome. Microbiol Mol Biol Rev 67:86–156

Milligan JF, Uhlenbeck OC (1989) Synthesis of small RNAs using T7 polymerase. Methods Enzymol 180:51–63

Moran VA, Perera RJ, Khalil AM (2012) Emerging functional and mechanistic paradigms of mammalian long non-coding RNAs. Nucleic Acids Res 40:6391–6400

Müller S, Wolf J, Ivanov SA (2004) Current strategies for the synthesis of RNA. Curr Org Synth 1:293–307

Nandakuman J, Shuman S (2004) How an RNA ligase discriminates RNA versus DNA damage. Mol Cell 16:211–221

Nuzzi A, Massi A, Dondoni A (2007) Model studies toward the synthesis of thymidine oligonucleotides with triazole internucleosidic linkages via iterative Cu(I)-promoted azide-alkyne ligation chemistry. QSAR Comb Sci 26:1191–1199

Oh BK, Pace NR (1994) Interaction of the 3′-end of tRNA with ribonuclease P RNA. Nucleic Acids Res 22:4087–4094

Ohtsuki T, Kawai G, Watanabe K (1998) Stable isotope-edited NMR analysis of *Ascaris suum* mitochondrial tRNAMet having a TV-replacement loop. J Biochem 124:28–34

Paredes E, Das SR (2011) Click chemistry for rapid labeling and ligation of RNA. Chembiochem 12:125–131

Paredes E, Evans M, Das SR (2011) RNA labeling, conjugation and ligation. Methods 54:251–259

Persson T, Cuzic S, Hartmann RK (2003) Catalysis by RNase P RNA: unique features and unprecedented active site plasticity. J Biol Chem 278:43394–43401

Pfeffer S, Sewer A, Lagos-Quintana M et al (2005) Identification of microRNAs of the herpesvirus family. Nat Methods 2:269–276

Pheiffer BH, Zimmerman SB (1983) Polymer-stimulated ligation: enhanced blunt- or cohesive-end ligation of DNA or deoxyribo-oligonucleotides by T4 DNA ligase in polymer solutions. Nucleic Acids Res 11:7853–7871

Piekielska K, Gębala M, Gwiazda S et al (2011) Impedimetric detection of hairpin ribozyme activity. Electroanalysis 23:37–42

Rader SD, Stark MR, Pleiss JA et al (2006) An RNA ligase-mediated method for the efficient creation of large, synthetic RNAs. RNA 12:2014–2019

Remaut E, Tsao H, Fiers W (1983) Improved plasmid vectors with a thermoinducible expression and temperature-regulated runaway replication. Gene 22:103–113

Rigden JE, Rezaian MA (1992) In vitro synthesis of an infectious viroid: analysis of the infectivity of monomeric linear CEV. Virology 186:201–206

Romaniuk E, McLaughlin LW, Neilson T et al (1982) The effect of acceptor oligoribonucleotide sequence on the T4 RNA ligase reaction. Eur J Biochem 125:639–643

Rosemeyer V, Laubrock A, Seibl R (1995) Nonradioactive 3′-end-labeling of RNA molecules of different lengths by terminal deoxynucleotidyltransferase. Anal Biochem 224:446–449

Rossi R, Montecucco A, Ciarrocchi G et al (1997) Functional characterization of the T4 DNA ligase: a new insight into the mechanism of action. Nucleic Acids Res 25:2106–2113

Rostovtsev VV, Green LG, Fokin VV et al (2002) A stepwise Huisgen cycloaddition process: copper(I)-catalyzed regioselective ligation of azides and terminal alkynes. Angew Chem Int Ed Engl 41:2596–2299

Rublack N, Nguyen H, Appel B et al (2011) Synthesis of specifically modified oligonucleotides for application in structural and functional analysis of RNA. J Nucleic Acids 2011:1–19

Sampson JR, Uhlenbeck OC (1988) Biochemical and physical characterization of an unmodified yeast phenylalanine transfer RNA transcribed in vitro. Proc Natl Acad Sci USA 85:1033–1037

Sanchez A, Pedroso E, Grandas A (2013) Oligonucleotide cyclization: the thiol-maleimide reaction revisited. Chem Commun 49:309–311

Seela F, Sirivolu VR (2008) Pyrrolo-dC oligonucleotides bearing alkynyl side chains with terminal triple bonds: synthesis, base pairing and fluorescent dye conjugates prepared by the azide-alkyne "click"reaction. Org Biomol Chem 6:1674–1687

Seelig B, Jäschke A (1999) Ternary conjugates of guanosine monophosphate as Initiator nucleotides for the enzymatic synthesis of 5′-modified RNAs. Bioconjug Chem 10:371–378

Serganov A, Patel DJ (2012) Molecular recognition and function of riboswitches. Curr Opin Struct Biol 22:279–286

Sindbert S, Kalinin S, Nguyen H et al (2011) Accurate distance determination of nucleic acids via Förster resonance energy transfer: implications of dye linker length and rigidity. J Am Chem Soc 133:2463–2480

Sokolova NI, Ashirbekova DT, Dolinnaya NG et al (1988) Chemical reactions within DNA duplexes: cyanogen bromide as an effective oligodeoxyribonucleotide coupling agent. FEBS Lett 232:153–155

Sontheimer EJ (1994) Site-specific RNA crosslinking with 4-thiouridine. Mol Biol Rep 20:35–44

Tessier DC, Brousseau R, Vernet T (1986) Ligation of single-stranded oligodeoxyribonucleotides by T4 RNA Ligase. Anal Biochem 158:171–178

Tornøe CW, Christensen C, Meldal M (2002) Peptidotriazoles on solid phase: [1,2,3]-triazoles by regiospecific copper(I)-catalyzed 1,3-dipolar cycloadditions of terminal alkynes to azides. J Org Chem 67:3057–3064

Turunen JJ, Pavlova LV, Hengesbach M et al (2013) RNA ligation. In: Hartmann R, Bindereif A, Schön A, Westhof E (eds) Handbook of RNA biochemistry, 2nd edn. Wiley, Weinheim

Vauléon S, Ivanov SA, Gwiazda S et al (2005) Site-specific fluorescent and affinity labelling of RNA by using a small engineered twin ribozyme. ChemBioChem 6:2158–2162

Welz R, Müller S (2002) 5-(Benzylmercapto)-1*H*-tetrazole as activator for 2′-*O*-TBDMS phosphoramidite building blocks in RNA synthesis. Tetrahedron Lett 43:795–797

Welz R, Bossmann K, Klug C et al (2003) Site-directed alteration of RNA sequence mediated by an engineered twin ribozyme. Angew Chem Int Ed Engl 42:2424–2427

Willkomm DK, Hartmann RK (2009) 3′-terminal attachment of fluorescent dyes and biotin. In: Hartmann RK, Bindereif A, Schön A, Westhof E (eds) Handbook of RNA biochemistry. Wiley-VCH, Weinheim, pp 86–94

Wincott F, DiRenzo A, Shaffer C et al (1995) Synthesis, deprotection, analysis and purification of RNA and ribozymes. Nucleic Acids Res 23:2677–2684

Winkler WC, Cohen-Chalamish S, Breaker RR (2002) An mRNA structure that controls gene expression by binding FMN. Proc Natl Acad Sci USA 99:15908–15913

Winston SE, Fuller SA, Evelegh MJ et al (2001) Conjugation of enzymes to antibodies. Curr Protoc Mol Biol 11(11.1)

Winz ML, Samanta A, Benzinger D et al (2012) Site-specific terminal and internal labeling of RNA by poly(A) polymerase tailing and copper-catalyzed or copper-free strain-promoted click chemistry. Nucleic Acids Res 40:e78

Wolf J, Dombos V, Appel B et al (2008) Synthesis of guanosine 5′-conjugates and their use as initiator molecules for transcription priming. Org Biomol Chem 6:899–907

Wunnicke D, Strohbach D, Weigand JE et al (2011) Ligand-induced conformational capture of a synthetic tetracycline riboswitch revealed by pulse EPR. RNA 17:182–188

Yu YT (1999) Construction of 4-thiouridine site-specifically substituted RNAs for cross-linking studies. Methods 18:13–21

Synthesis and Exon-Skipping Activity of Chemically Modified RNAs

Yoshiaki Masaki, Takeshi Yamada, Hisao Saneyoshi, Akihiro Ohkubo, Kohji Seio, and Mitsuo Sekine

Contents

1 Introduction 498
2 Overview of Previously Reported Modified Nucleic Acids for DMD Exon-Skipping ... 499
2.1 2′-OMe-PS-RNA 499
2.2 PMO 500
2.3 PNA 501
2.4 LNA/BNA 501
2.5 ENA 501
2.6 TcDNA 502
2.7 2′-OMOE-PS-RNA 502
3 Design and Synthesis of 2′-*O*-Modified Nucleosides 502
4 Synthesis of 2′-*O*-Cyanoethyl and 2′-*O*-[2-(*N*-Methylcarbamoyl)ethyl] Modified Nucleosides 503
5 Effect of Chemically Modified AONs on Exon-51 Skipping in *mdx52* Mice 506
6 Conclusion 507
References 507

Abstract In this chapter, an overview of chemically modified antisense oligonucleotides (AON) for Duchenne muscular dystrophy (DMD) treatment is described. In particular, the promising exon-skipping properties of 2′-*O*-modified phosphorothioate oligoribonucleotides incorporating 2′-*O*-[2-(*N*-methycarbamoyl)ethyl] ribonucleosides that could be readily synthesized by new regioselective oxa-Michael reaction are reviewed in detail.

Keywords Antisense • Exon-skipping • DMD • Chemically modified RNA

Y. Masaki • T. Yamada • H. Saneyoshi • A. Ohkubo • K. Seio • M. Sekine (✉)
Department of Life Science, Tokyo Institute of Technology, 4259 Nagatsuta, Midori-ku, Yokohama 226-8501, Japan
e-mail: msekine@bio.titech.ac.jp

V.A. Erdmann et al. (eds.), *Chemical Biology of Nucleic Acids*, RNA Technologies,
DOI 10.1007/978-3-642-54452-1_27,

1 Introduction

Exon-skipping is a method used to modulate pre-mRNA splicing. Splicing modulation was first demonstrated in 1993 by Dominski and Kole (1993) who designed 2′-*O*-methyl antisense oligonucleotide (AON) to redirect cryptic splicing of the β-globin gene and successfully restored correct mature β-globin mRNA. AON-mediated splicing modulation has since been studied as a useful tool not only for restoring normal splicing to cryptic splicing mutations but also for inducing reading frame shifts to correct or disrupt protein translation, induce exon inclusion, or switch alternative splicing isoforms (van Roon-Mom and Aartsma-Rus 2012; Vacek et al. 2003). The reading frame correction for Duchenne muscular dystrophy (DMD) by exon-skipping is the most appropriate example of a clinical application of this methodology.

DMD is a lethal X-linked disease involving progressive muscular deterioration. This disease affects 1 in 3,500 newborn boys (Muntoni and Wood 2011; Bushby et al. 2010a, b; Davies and Nowak 2006). Affected boys gradually lose their ability to walk in their early teens and die in their twenties because of cardiac or respiratory failure. DMD is caused by dystrophin gene mutations that lead to one or more exon deletions (Hoffman et al. 1987). If the sum of the nucleotides in the deleted exons is not divisible by 3, the reading frame of the mature mRNA is shifted, resulting in nonfunctional dystrophin. AON-mediated exon-skipping works by targeting splicing elements' binding sites within pre-mRNA, inducing an additional exon skip to make the total number of deleted bases divisible by 3. As a consequence, mRNA with the additional exon skip can be translated into shortened dystrophin.

Dystrophin comprises four functional domains: the N-terminus actin-binding domain (N-ABD), central rod domain, cysteine-rich (CR) domain, and C-terminus domain (CT). The N-ABD binds cytoskeletal proteins and actin, and the CR domain binds β-dystroglycan, which connects it to the extracellular matrix. These dystrophin functional domains are linked through the central rod domain, which is a spectrin-like 24-repeat domain (Davies and Nowak 2006). Because the popular site for DMD dystrophin gene mutations is within the spectrin-like repeated domain (e.g., exon 51 in 13 % of all patients), exon-skipped mRNA can be translated into shortened but functional dystrophin (Fig. 1) (Koenig et al. 1989).

The first AON-mediated DMD splicing modulation was performed by Matsuo et al. in 1995 (Takeshima et al. 1995). Since then, a number of chemically modified AONs for the treatment of DMD, including 2′-*O*-methyl-phosphorothioate oligoribonucleotides (2′-OMe-PS-RNA), phosphorodiamidate morpholino oligomers (PMO), peptide nucleic acids (PNA), locked nucleic acids/2′,4′-bridged nucleic acids (LNA/BNA), 2′-*O*, 4′-*C*-ethylene bridged nucleic acids (ENA), tricyclo-DNA (TcDNA), and 2′-*O*-methoxyethyl-phosphorothioate oligoribonucleotides (2′-OMOE-PS-RNA), have been synthesized and extensively evaluated in vitro and/or in vivo (Fig. 2). Among the modified AONs reported, 2′-OMe-PS-RNA and PMO are closest to clinical application. Although these modified AONs are able to restore functional dystrophin, clinical studies have reported that a large

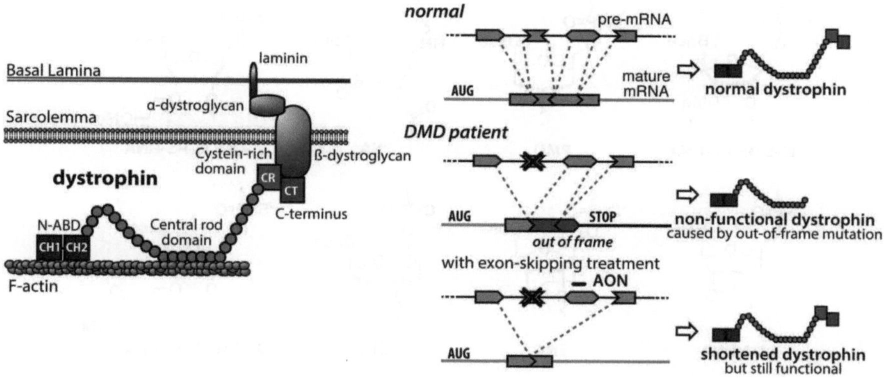

Fig. 1 Principles of exon-skipping for DMD treatment

weekly dose of 2′-OMe-PS-RNA is required (180 mg/30 kg, phase III efficacy and safety study NCT01254019) and possibly larger weekly dosage for PMO treatment (900 mg/30 kg or 1.5 g/30 kg, phase II efficacy study NCT01396239). Although no serious adverse effects were reported, injection of nucleic acid drugs in such a large quantity is apparently unfavorable in terms of cost and potential adverse effects that accompany long-term treatment because AON-mediated exon-skipping treatment is necessary for the patient's entire life. Thus, it is of great importance to develop a novel modified AON that not only requires a smaller dose but also can be synthesized easily. We have reported 2′-*O*-[2-(*N*-methylcarbamoyl)ethyl] (2′-*O*-MCE) AONs as promising candidates for exon-skipping treatment of DMD. In this chapter, we describe the details of our strategy for synthesizing 2′-*O*-MCE AONs and demonstrate their excellent exon-skipping activity.

2 Overview of Previously Reported Modified Nucleic Acids for DMD Exon-Skipping

2.1 2′-OMe-PS-RNA

2′-*O*-Methyl oligoribonucleotides were first synthesized by Inoue et al. (1987). Since then, they have been used for various applications. Drisapersen, a 20-nt long 2′-OMe-PS-RNA, is one of the most advanced antisense drug candidates for DMD treatment using exon-51 skipping therapy. The preclinical pharmacokinetic/pharmacodynamics study of 2′-OMe-PS-RNA revealed that the half-life of this antisense RNA was approximately 10 days in the gastrocnemius and quadriceps muscles, approximately 33 days in the triceps muscle, and 46 days in heart muscle (Heemskerk et al. 2010). This remarkably long half-life may be because of the nonspecific high affinity of the phosphorothioate structure of 2′-OMe-PS-RNA for

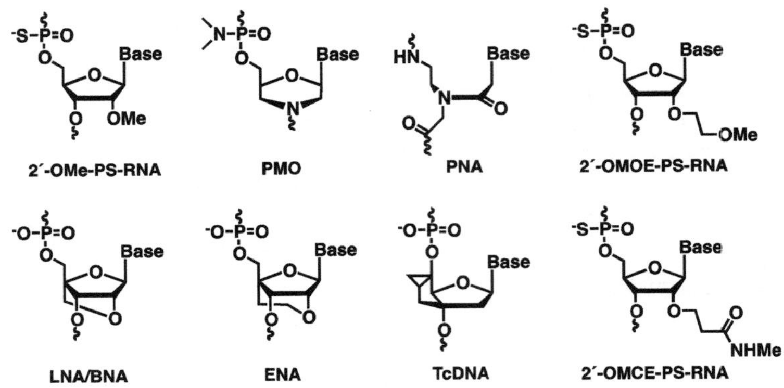

Fig. 2 Chemically modified AONs for exon-skipping treatment of DMD

human serum albumin, which is the most important drug carrier in the plasma (Prakash et al. 2002; Greig et al. 1995; Crooke et al. 1996). In a phase I/IIa study, Drisapersen was administrated for over 5 weeks at a dose of 0.5–6 mg/kg; subsequently, a 12-week open-label extension was performed at a dose of 6 mg/kg/week (Goemans et al. 2011). From this study, dose-dependent molecular efficacy with a modest improvement in the 6-min walk test was confirmed. Although there were no serious adverse events, those administered with the 6 mg/kg dose in the extension study revealed proteinuria that may have been caused by drug-associated kidney damage.

2.2 *PMO*

PMO were developed by Summerton in 1993 (Summerton and Weller 1993). A comparison study of 2′-OMe-PS-RNA and PMO for exon-23 skipping in *mdx* mice revealed that PMO were more effective than 2′-OMe-PS-RNA. However, the efficacy of PMO and 2′-OMe-PS-RNA was reported to be more comparable for human exons (Heemskerk et al. 2009). Eteplirsen (AVI-4658), a 30-nt long PMO, is another promising candidate for DMD treatment through exon-51 skipping. The half-life of PMO is short, as PMO do not bind to plasma proteins. In a phase II study, eteplirsen was administrated at a dose of 0.5–20 mg/kg/week over 12 weeks. Dose-dependent molecular efficacy with no serious adverse events was confirmed (Cirak et al. 2011). A study on the efficacy, safety, and tolerability is ongoing at a dose of 30–50 mg/kg/week.

2.3 *PNA*

PNA were developed as neutral nucleic acid analogs by P. E. Nielsen in 1991 (Nielsen et al. 1991). PNA has been used in biological studies as well as biotechnology-related applications because it has a stronger binding affinity for DNA and RNA and higher biological stability than unmodified oligonucleotides. Wood et al. reported that a single intramuscular injection of PNA for exon-23 skipping in an *mdx* mouse model restored about 3.5-fold higher dystrophin expression than that of the corresponding 2′-OMe-PS-RNA (Yin et al. 2008). However, the exon-skipping activity of PMO was more effective than PNA (approximately 1.5-fold). Gait et al. reported the use of cell-penetrating peptide (CPP)-conjugated PNA for exon-23 skipping resulted in threefold higher dystrophin expression than for naked PNA in *mdx* mice (Ivanova et al. 2008). $(R\text{-}Ahx\text{-}R)_3$IdKILFQNdRRMKWHKBC (Pip1a) and $(R\text{-}Ahx\text{-}R)_3$IHILFQNdRRMKWHKBC (Pip2b), originally designed from R6-Penetratin, were used as CPPs.

2.4 *LNA/BNA*

LNA or BNA were independently synthesized by T. Imanishi et al. in 1997 (Obika et al. 1997) and J. Wengel et al. in 1998 (Koshkin et al. 1998). LNA/BNA has been widely used as a rigid nucleic acid derivative of antisense, siRNA, antagomiR, and aptamer molecules for genetic diagnosis and regulation. LNA/BNA contains the backbone structure of O2′-C4′ methylene "bridged" nucleosides that reveal the "locked" C3′-*endo* conformation. This locked conformation reportedly increases the duplex stability through pre-organization effects and enhances nuclease resistance. An LNA/BNA AON was used for exon-46 skipping in 2004 by Aartsma-Rus et al. (2004). They investigated the exon-skipping activities of an LNA 14-mer, PNA 14-mer, 2′-OMePS 20-mer, and PMO 22-mer. Among these, LNA/BNA exhibited the highest exon-skipping effect in DMD patient-derived (DL279.1) myotubes. However, the sequence specificity of the LNA/BNA AON was limited.

2.5 *ENA*

ENA™ (Sankyo LifeTech Co., Ltd., Tokyo, Japan) is the LNA/BNA analogue that exhibited increased duplex stability and nuclease resistance (Morita et al. 2002). An RNA-ENA chimera effectively induced exon-19 skipping of dystrophin in myocytes from a DMD patient muscle biopsy sample (Yagi et al. 2004). Over 90 % exon-skipping efficacy was observed at $\geq$200 nM of the 31-mer RNA/ENA chimera (a gap-mer with a 5-mer ENA at both termini) in culture medium. The effects appeared at 3 h after transfection, reaching a peak at 12 h and remaining at the plateau until day 7. The effect was observed even after 14 days.

2.6 TcDNA

TcDNA is class of conformationally constrained DNA analogues that was originally developed by C. Leumann et al. (Murray et al. 2012). Tc-DNA oligonucleotide treatment targeting exon 23 in *mdx* mice revealed widespread restoration of dystrophin even in cardiac muscle; exon-skipping was also detected in the brain (Goyenvalle et al. 2012). In addition, Tc-DNA improved the muscle function and dystrophic pathology in utrophin/distrophin double-knockout mice, a more severe mouse model.

2.7 2′-OMOE-PS-RNA

2′-OMOE-PS-RNAs have been developed by Isis pharmaceutical (Altmann et al. 1996; Martin 1995). MOE-PS has been successfully used to downregulate various targeted mRNAs. Kynamuro, which was approved in 2013 by the FDA as an RNase H-dependent antisense drug, has a chimeric structure consisting of 2′-OMOE-PS-RNA and DNA chains (Raal et al. 2010). Lu Yang et al. reported the exon-23 skipping efficiency induced by 2′-OMOE-PS-RNA induced in *mdx* mice (Yang et al. 2013). They compared the exon-skipping efficiency of 2′-OMOE-PS-RNA 25-mer, 2′-OMOE-PS-RNA 20-mer, and 2′-*O*-methyl phosphorothioate RNA 20-mer (2′-OMe-PS-RNA) in vitro and in vivo. Both in vitro and in vivo studies revealed that the 2′-OMOE-PS-RNA 25-mer was more effective than the others. There was no apparent difference in the activity between the 2′-OMOE-PS-RNA 20-mer and 2′-OMe-PS-RNA 20-mer. The exon-skipping efficacy of 2′-OMOE-PS-RNA 25-mer was approximately 95 % at 500 nM in H_2K *mdx* cells. The peak time for measuring exon-23 skipping was 48 h post-transfection. The authors suggested that the origin of the different efficacies may be partially attributed to ameliorate transfection efficiency of the 2′-OMOE-PS-RNA 25-mer than the other AONs.

3 Design and Synthesis of 2′-*O*-Modified Nucleosides

Although 2′-OMe-PS-RNA and PMO have promising efficacy for DMD treatment, it is necessary to develop more effective antisense molecules that can be easily synthesized on a large scale and have similar effects at a lower dose or longer administration time for reducing the potential adverse events and the medication cost. To meet these criteria, we focused on 2′-*O*-modified phosphorothioate AONs, which are more accessible than PMOs from the synthetic point of view. The synthesis of the latter is limited to shorter sequences because of the inherent insufficient coupling efficiency. Synthesis of AONs longer than 20-mers is rather difficult on a Kilogram scale. In addition, 2′-*O*-modification could improve the binding affinity (Masaki et al. 2010; Freier and Altmann 1997) of antisense molecules for target mRNAs and resistance to nuclease digestion. These unresolved problems led us to develop a new strategy for 2′-*O*-modification.

The typical procedure for synthesizing 2′-*O*-modified ribonucleosides involves the use of alkylating reagents in the presence of sodium hydride to introduce alkyl groups into the 2′-alcoholic function. This procedure is practically useful but often forms complex products, which requires tedious separation processes. In addition, protection of the amino groups in the nucleobases and the hydroxyl groups (except the 2′-hydroxyl) is necessary. In some cases, alkylation requires the use of more stable protecting groups, which are not used for oligonucleotide synthesis. Thus, the resulting multistep procedure for protection and deprotection makes the synthesis of modified ribonucleosides laborious. We found an effective 2′-*O*-modification through the Michael reaction that allows direct alkylation of the 2′-hydroxyl group without protection of the amino groups except the imino proton of uridine (Saneyoshi et al. 2005; Yamada et al. 2011). This new regioselective reaction is a more direct approach for synthesizing 2′-*O*-modified ribonucleoside 3′-phosphoroamidite building blocks. In addition, oligoribonucleotides incorporating 2′-*O*-cyanoethyl (CNE) and 2′-*O*-[2-(*N*-methylcarbamoyl)ethyl] (MCE) ribonucleosides, which can be readily synthesized through the Michael reaction, exhibit enhanced binding affinity for the complementary RNA strands and an increased nuclease resistance. These oligoribonucleotides have proven to be superior to the 2′-*O*-methylated oligoribonucleotides used in the current exon-skipping therapy for DMD. In the following sections, we discuss the details of the synthesis of 2′-*O*-modified ribonucleosides and in vivo 2′-*O*-MCE-modified AONs exon-skipping activity assay.

4 Synthesis of 2′-*O*-Cyanoethyl and 2′-*O*-[2-(*N*-Methylcarbamoyl)ethyl] Modified Nucleosides

We initially synthesized 2′-*O*-cyanoethyl ribonucleosides because they have a simple structure with less steric hindrance and a polarized cyano group that may stabilize the hydration structure in the minor groove. To synthesize 2′-*O*-cyanoethyl ribonucleosides, we have extensively studied the conditions used for the Michael reaction to ensure appropriate protection of uridine derivatives (Markiewicz et al. 1984) with acrylonitrile for the introduction of the cyanoethyl group into the 2′-OH group (Saneyoshi et al. 2005). The previous conditions used for the Michael reaction of alcohols with acrylonitrile involved the use of strong bases,

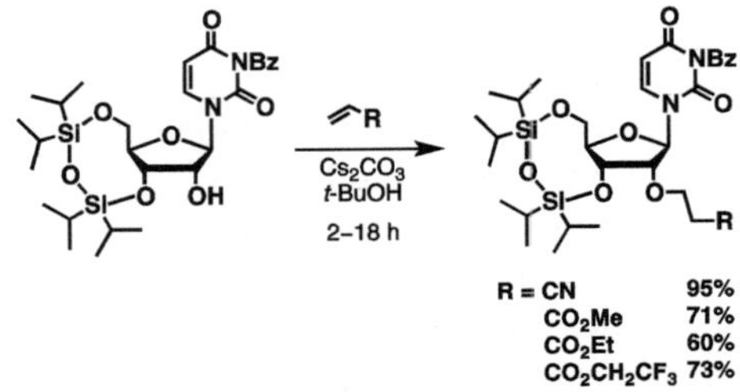

Fig. 3 Oxa-Michael reaction to uridine derivative

such as sodium hydroxide, alkoxides, and benzyltrimethylammonium hydroxides (Triton B) (Krishna and Jayaraman 2003; Chung et al. 2004; Pearson et al. 2002; Simonot and Rousseau 1993). In fact, Triton B gave a poor yield of the cyanoethylated product (24 %). As a result of extensive screening, we determined that cesium carbonate (Cs_2CO_3) was the best base catalyst, yielding moderate levels of the cyanoethylated product (54 %). This reaction caused considerably undesired polymerization of acrylonitrile. It is possible that the polymerization occurred by consecutive Michael addition of the carbanion on the α-carbon atom next to the cyano group, which was generated by Michael reaction of the 2′-hydroxyl group with acrylonitrile. To avoid this additional alkylation, the addition of suitable proton sources was studied, and *tert*-butyl alcohol was selected. Because the acidity of *tert*-butyl alcohol (pK_a 18) was lower than that of the 2′-hydroxyl group (pK_a 12–14.9) (Velikyan et al. 2001; Lyne and Karplus 2000), we presumed that Cs_2CO_3 would preferentially activate the 2′-hydroxyl group over the α-proton on the cyanoethyl group. The carbanion of the cyanoethyl group generated during the normal Michael addition would be promptly protonated by *tert*-butanol. The selection of Cs_2CO_3 in *tert*-butyl alcohol resulted in excellent yields of the cyanoethylated product (95 %; Fig. 3). This method could be expanded to the use of 2′-*O*-methyl, ethyl, and 2,2,2-trifluoroethyl esters of acrylic acids as Michael acceptors to yield 2′-*O*-alkoxycarbonylethylribonucleoside derivatives (Yamada et al. 2011). These resulting ester functions would be used as new scaffolds to introduce modified groups at the 2′-site.

Under the conditions using Cs_2CO_3 in *tert*-butyl alcohol, it turned out that the imido group of 3′,5′-*O*-(1,1,3,3-tetraisopropyldisiloxan-1,3-diyl)uridine (Markiewicz et al. 1984) was more rapidly alkylated than the 2′-hydroxyl group. Because the N3-imino proton of uridine (pK_a 9.25) (Sanger 1984) is apparently more acidic than the 2′-OH proton (pK_a 12–14.9), we presumed that the imino proton of uridine would be easily dissociated by Cs_2CO_3, resulting in the dissociated anion species to be more reactive toward the Michael acceptor. Thus, the imino protons on the uracil and guanine bases would require protection beforehand. The *N*-acylated amino groups of nucleobases were rapidly alkylated with acrylonitrile; therefore, the usual N-protecting groups could not be used for our Michael reaction. In contrast, the unprotected amino group on the adenine and cytosine bases remained intact under the standard conditions since their pK_a values, probably similar to that of aniline (pK_a 30.6) (Bordwell et al. 1977), are much higher than that of the hydroxyl group. This

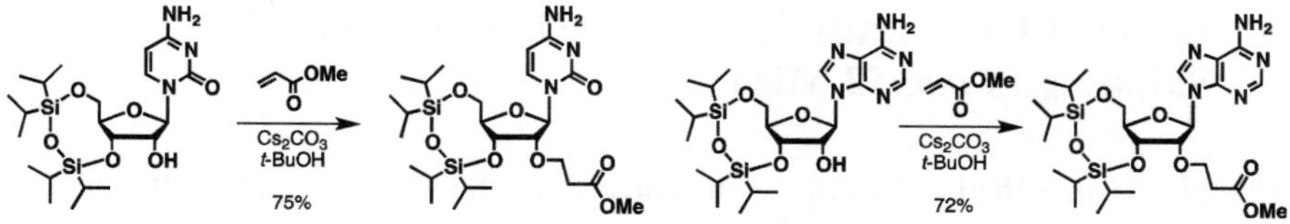

Fig. 4 *O*-selective Michael reaction to cytosine and adenosine derivatives

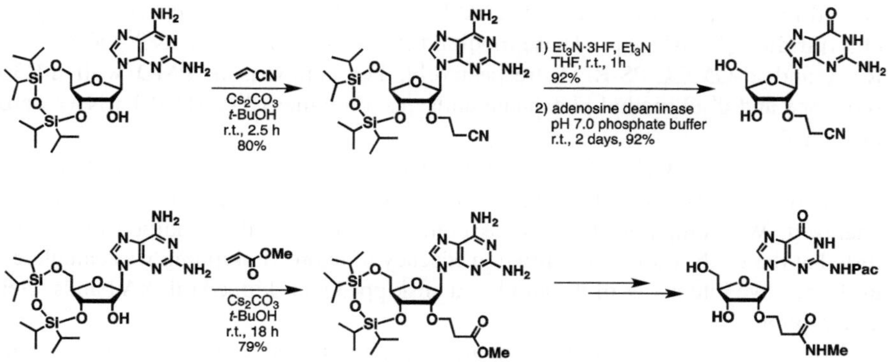

Fig. 5 *O*-selective Michael reaction for the preparation of guanosine derivatives

difference permitted us to alkylate the hydroxyl group of cytosine and adenosine derivatives without any protecting groups on their nucleobases (Fig. 4).

However, the desired 2′-*O*-cyanoethylguanosine derivative could not be synthesized from the usual Michael reaction initiating material 3′,5′-*O*-(1,1,3,3-tetraisopropyldisiloxan-1,3-diyl)guanosine because of its poor solubility. Instead, we selected 3′,5′-*O*-(1,1,3,3-tetraisopropyldisiloxan-1,3-diyl)-2-aminoadenosine because this compound is soluble in *tert*-butanol, and the 6-amino group can be easily deaminated using sodium nitrate or adenosine deaminase, yielding the corresponding guanosine derivatives. In addition, use of this compound in the Michael reaction with methyl acrylate or acrylonitrile was expected to proceed in a functionally selective manner (Fig. 5).

These 2′-*O*-modified ribonucleosides were successfully converted to the corresponding phosphoramidite units and incorporated into oligonucleotides according to the standard procedure for automated RNA synthesis. The synthesized sequence is 5′-CUCCAACAGCAAAGAAGAUGGCAUUUCUAG-3′ [murine B30 (mB30) sequence] (Arechavala-Gomeza et al. 2007; Aoki et al. 2012). This sequence is essentially the same as that of the clinically used PMO with slight modifications because of its use in mice.

5 Effect of Chemically Modified AONs on Exon-51 Skipping in *mdx52* Mice

Chemically modified AONs were synthesized, and their exon-skipping efficiency in *mdx52* mice was evaluated. We selected 2′-*O*-methyl phosphorothioate AON (2′-OMe-PS-RNA), which contains the targeted mB30 sequence, as a reference compound. All six residues of 2′-*O*-methyluridine in this antisense oligomer were altered to 2′-*O*-cyanoethyluridine (2′-*O*-U_{CNE}) and 2′-*O*-[(2-*N*-methyl-carbamoyl) ethyl]uridine (2′-*O*-U_{MCE}). These modified AONs are referred to as 2′-OCNE-PS-RNA and 2′-OMCE-PS-RNA, respectively. For comparison, PMO with mB30 sequence and 2′-*O*-methyl phosphate antisense oligomer (2′-OMe-PO-RNA) were evaluated.

Each synthetic AON (10 μg) was injected into each tibialis anterior muscle of 8-week-old *mdx52* mice. Two weeks after injection, the mice were sacrificed and analyzed. We compared the exon-skipping efficacies of the modified oligomers tested at the mRNA level. The relative efficacy of exon-51 skipping was calculated as the relative intensities of the amplified unskipped and skipped mRNA bands after electrophoresis.

The boxplots of the observed results are summarized in Fig. 6a. The median exon-51 skipping efficiency for 2′-OMe-PO-RNA, 2′-OMe-PS-RNA, 2′-OCNE-PS-RNA, 2′-OMCE-PS-RNA, and PMO was 5.2, 22.8, 43.3, 56.6, and 39.9 %, respectively. As observed, 2′-OMe-PS-RNA was more effective than 2′-OMe-PO-RNA. The phosphorothioate-modified species is clearly more effective than the phosphate-unmodified species. The exon-skipping efficacy of PMO (39.9 %) was higher than that of 2′-OMe-PS-RNA (22.8 %), which is consistent with the results obtained by Heemskerk et al. (2009). Remarkably, 2′-OCNE-PS-RNA and 2′-OMCE-PS-RNA revealed results comparable to those for PMO. It should be noted that the significant enhancement of exon-skipping by 2′-OMe-PS-RNA was caused by modification of only the uridines. Although there is no statistically significant difference between the results seen for 2′-OCNE-PS-RNA and 2′-OMCE-PS-RNA, we focused on MCE derivatives because the median exon-skipping efficacy of 2′-OMCE-PS-RNA was higher than that of 2′-OCNE-PS-RNA.

The results of immunohistochemical staining of TA muscle treated with 2′-OMCE-PS-RNA or PMO is presented in Fig. 6b. There was no expression of dystrophin around the rim of any muscle fibers from untreated *mdx52* mice. In contrast, restoration of dystrophin around the rim of muscle fibers was observed in both muscles from 2′-OMCE-PS-RNA- and PMO-treated *mdx52* mice. This observation indicates that the MCE-modified product has a comparable efficacy to PMO for dystrophin restoration.

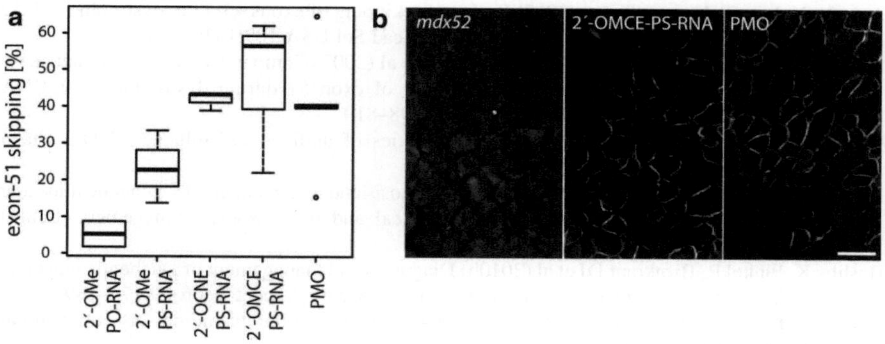

Fig. 6 Effect of chemically modified AONs. (**a**) Boxplot of exon-51 skipping. (**b**) Immunohistochemical staining of dystrophin in TA muscle. Dystrophin was detected using dystrophin C-terminal antibody (Dys-2). Bar = 100 μm (Reprinted with permission from Yamada et al. (2011). Copyright (2011) American Chemical Society)

6 Conclusion

During the past decade, several studies have been conducted on the use of antisense oligonucleotides for exon-skipping treatment of DMD. Only a limited number of studies have reported on the use of chemically modified antisense molecules. Thus, improvements in the efficacy of antisense activity will be achieved using chemically modified antisense oligonucleotides. In addition, there are various unsolved problems with regard to exon-skipping treatment, including drug delivery to cardiac cells. DMD patients often die of heart failure, yet restoration of dystrophin in cardiac cells by exon-skipping using the usual antisense drugs was not effective. This failure is due to the lack of naked PMO delivery and low delivery efficiency of 2′-OMe-PS-RNA. Although we did not address the delivery of such antisense drugs to cardiac cells, others have reported on methods of delivering new compounds such as PMO using cell-penetrating peptides. These methods have revealed promising results but often revealed significant toxicity of the drug materials used. The combination of these delivery methods with chemically modified antisense molecules should be extensively studied in the future.

References

Aartsma-Rus A, Kaman WE, Bremmer-Bout M et al (2004) Comparative analysis of antisense oligonucleotide analogs for targeted DMD exon 46 skipping in muscle cells. Gene Ther 11:1391–1398

Altmann KH, Dean NM, Fabbro D et al (1996) Second generation of antisense oligonucleotides: From nuclease resistance to biological efficacy in animals. Chimia 50:168–176

Aoki Y, Yokota T, Nagata T et al (2012) Bodywide skipping of exons 45-55 in dystrophic mdx52 mice by systemic antisense delivery. Proc Natl Acad Sci USA 109:13763–13768

Arechavala-Gomeza V, Graham IR, Popplewell LJ et al (2007) Comparative analysis of antisense oligonucleotide sequences for targeted skipping of exon 51 during dystrophin pre-mRNA splicing in human muscle. Hum Gene Ther 18:798–810

Bordwell FG, Algrim D, Vanier NR (1977) Acidities of anilines and toluenes. J Org Chem 42:1817–1819

Bushby K, Finkel R, Birnkrant DJ et al (2010a) Diagnosis and management of Duchenne muscular dystrophy, part 1: diagnosis, and pharmacological and psychosocial management. Lancet Neurol 9:77–93

Bushby K, Finkel R, Birnkrant DJ et al (2010b) Diagnosis and management of Duchenne muscular dystrophy, part 2: implementation of multidisciplinary care. Lancet Neurol 9:177–189

Chung HH, Harms G, Seong CM et al (2004) Dendritic oligoguanidines as intracellular translocators. Biopolymers 76:83–96

Cirak S, Arechavala-Gomeza V, Guglieri M et al (2011) Exon skipping and dystrophin restoration in patients with Duchenne muscular dystrophy after systemic phosphorodiamidate morpholino oligomer treatment: an open-label, phase 2, dose-escalation study. Lancet 378:595–605

Crooke ST, Graham MJ, Zuckerman JE et al (1996) Pharmacokinetic properties of several novel oligonucleotide analogs in mice. J Pharmacol Exp Ther 277:923–937

Davies KE, Nowak KJ (2006) Molecular mechanisms of muscular dystrophies: old and new players. Nat Rev Mol Cell Biol 7:762–773

Dominski Z, Kole R (1993) Restoration of correct splicing in thalassemic pre-mRNA by antisense oligonucleotides. Proc Natl Acad Sci USA 90:8673–8677

Freier SM, Altmann KH (1997) The ups and downs of nucleic acid duplex stability: structure-stability studies on chemically-modified DNA:RNA duplexes. Nucleic Acids Res 25:4429–4443

Goemans NM, Tulinius M, van den Akker JT et al (2011) Systemic administration of PRO051 in Duchenne's muscular dystrophy. N Engl J Med 364:1513–1522

Goyenvalle A, Babbs A, Avril A et al (2012) T.O.3 Tricyclo-DNA: a promising chemistry for the synthesis of antisense molecules for splice-switching approaches in DMD. Neuromus Disord 22:907

Greig MJ, Gaus H, Cummins LL et al (1995) Measurement of macromolecular binding using electrospray mass spectrometry. Determination of dissociation constants for oligonucleotide: serum albumin complexes. J Am Chem Soc 117:10765–10766

Heemskerk HA, de Winter CL, de Kimpe SJ et al (2009) In vivo comparison of 2′-*O*-methyl phosphorothioate and morpholino antisense oligonucleotides for Duchenne muscular dystrophy exon skipping. J Gene Med 11:257–266

Heemskerk H, de Winter C, van Kuik P et al (2010) Preclinical PK and PD studies on 2′-*O*-methyl-phosphorothioate RNA antisense oligonucleotides in the mdx mouse model. Mol Ther 18:1210–1217

Hoffman EP, Brown RH Jr, Kunkel LM (1987) Dystrophin: the protein product of the Duchenne muscular dystrophy locus. Cell 51:919–928

Inoue H, Hayase Y, Imura A et al (1987) Synthesis and hybridization studies on two complementary nona(2′-*O*-methyl)ribonucleotides. Nucleic Acids Res 15:6131–6148

Ivanova GD, Arzumanov A, Abes R et al (2008) Improved cell-penetrating peptide-PNA conjugates for splicing redirection in HeLa cells and exon skipping in mdx mouse muscle. Nucleic Acids Res 36:6418–6428

Koenig M, Beggs AH, Moyer M et al (1989) The molecular basis for Duchenne versus Becker muscular dystrophy: correlation of severity with type of deletion. Am J Hum Genet 45:498–506

Koshkin AA, Singh SK, Nielsen P et al (1998) LNA (locked nucleic acids): synthesis of the adenine, cytosine, guanine, 5-methylcytosine, thymine and uracil bicyclonucleoside monomers, oligomerisation, and unprecedented nucleic acid recognition. Tetrahedron 54:3607–3630

Krishna TR, Jayaraman N (2003) Synthesis of poly(propyl ether imine) dendrimers and evaluation of their cytotoxic properties. J Org Chem 68:9694–9704

Lyne PD, Karplus M (2000) Determination of the pKa of the 2′-hydroxyl group of a phosphorylated ribose: implications for the mechanism of hammerhead ribozyme catalysis. J Am Chem Soc 122:166–167

Markiewicz WT, Biala E, Kierzek R (1984) Application of the tetraisopropyldisiloxane-1,3-diyl group in the chemical synthesis of oligoribonucleotides. Bull Pol Acad Sci Chem 32:433–451

Martin P (1995) Ein neuer zugang zu 2′-*O*-alkylribonucleosiden und eigenschaften deren oligonucleotide. Helv Chim Acta 78:486–504

Masaki Y, Miyasaka R, Ohkubo A et al (2010) Linear relationship between deformability and thermal stability of 2′-*O*-modified RNA hetero duplexes. J Phys Chem B 114:2517–2524

Morita K, Hasegawa C, Kaneko M et al (2002) 2′-*O*,4′-*C*-ethylene-bridged nucleic acids (ENA): highly nuclease-resistant and thermodynamically stable oligonucleotides for antisense drug. Bioorg Med Chem Lett 12:73–76

Muntoni F, Wood MJ (2011) Targeting RNA to treat neuromuscular disease. Nat Rev Drug Discov 10:621–637

Murray S, Ittig D, Koller E et al (2012) TricycloDNA-modified oligo-2′-deoxyribonucleotides reduce scavenger receptor B1 mRNA in hepatic and extra-hepatic tissues–a comparative study of oligonucleotide length, design and chemistry. Nucleic Acids Res 40:6135–6143

Nielsen PE, Egholm M, Berg RH et al (1991) Sequence-selective recognition of DNA by strand displacement with a thymine-substituted polyamide. Science 254:1497–1500

Obika S, Nanbu D, Hari Y et al (1997) Synthesis of 2′-O,4′-C-methyleneuridine and -cytidine. Novel bicyclic nucleosides having a fixed C3′-endo sugar puckering. Tetrahedron Lett 38:8735–8738

Pearson WI, Guo LY, Jewell TM (2002) Preparation of immobilized swainsonine analogs on solid support. Tetrahedron Lett 43:2175–2178

Prakash TP, Manoharan M, Kawasaki AM et al (2002) 2′-*O*-[2-(methylthio)ethyl]-modified oligonucleotide: an analogue of 2′-*O*-[2-(methoxy)-ethyl]-modified oligonucleotide with improved protein binding properties and high binding affinity to target RNA. Biochemistry 41:11642–11648

Raal FJ, Santos RD, Blom DJ et al (2010) Mipomersen, an apolipoprotein B synthesis inhibitor, for lowering of LDL cholesterol concentrations in patients with homozygous familial hypercholesterolaemia: a randomised, double-blind, placebo-controlled trial. Lancet 375:998–1006

Saneyoshi H, Seio K, Sekine M (2005) A general method for the synthesis of 2′-*O*-cyanoethylated oligoribonucleotides having promising hybridization affinity for DNA and RNA and enhanced nuclease resistance. J Org Chem 70:10453–10460

Sanger W (1984) Principles of nucleic acid structure. Springer, New York

Simonot B, Rousseau G (1993) Preparation of 3-(*N*-alkenoxy)propanoic acids. Synth Commun 23:549–560

Summerton JE, Weller DD (1993) Uncharged morpolino-based polymers having phosphorous containing chiral intersubunit linkages US 5185444 A

Takeshima Y, Nishio H, Sakamoto H et al (1995) Modulation of in vitro splicing of the upstream intron by modifying an intra-exon sequence which is deleted from the dystrophin gene in dystrophin Kobe. J Clin Invest 95:515–520

Vacek M, Sazani P, Kole R (2003) Antisense-mediated redirection of mRNA splicing. Cell Mol Life Sci 60:825–833

van Roon-Mom WM, Aartsma-Rus A (2012) Overview on applications of antisense-mediated exon skipping. Methods Mol Biol 867:79–96

Velikyan I, Acharya S, Trifonova A et al (2001) The pKa's of 2′-hydroxyl group in nucleosides and nucleotides. J Am Chem Soc 123:2893–2894

Yagi M, Takeshima Y, Surono A et al (2004) Chimeric RNA and 2′-*O*,4′-*C*-ethylene-bridged nucleic acids have stronger activity than phosphorothioate oligodeoxynucleotides in induction of exon 19 skipping in dystrophin mRNA. Oligonucleotides 14:33–40

Yamada T, Okaniwa N, Saneyoshi H et al (2011) Synthesis of 2′-*O*-[2-(*N*-methylcarbamoyl)ethyl] ribonucleosides using oxa-Michael reaction and chemical and biological properties of oligonucleotide derivatives incorporating these modified ribonucleosides. J Org Chem 76:3042–3053

Yang L, Niu H, Gao X et al (2013) Effective exon skipping and dystrophin restoration by 2′-*O*-methoxyethyl antisense oligonucleotide in dystrophin-deficient mice. PLoS ONE 8:e61584

Yin H, Lu Q, Wood M (2008) Effective exon skipping and restoration of dystrophin expression by peptide nucleic acid antisense oligonucleotides in mdx mice. Mol Ther 16:38–45

mRNA and snRNA Cap Analogs: Synthesis and Applications

Janusz Stepinski and Edward Darzynkiewicz

This article is dedicated to the memory of Aaron Shatkin (1934–2012), the discoverer of the cap structure, my mentor and close friend (E. Darzynkiewicz)

Contents

1 Introduction: Cap Structures, Their Occurrence, and Functions 512
2 Chemical Synthesis of 5′ mRNA Cap Analogs 515
2.1 Mononucleotide Cap Analogs 516
2.2 Dinucleotide Cap Analogs 519
2.3 Capped Oligonucleotides 525
2.4 Capping on Solid Support 529
2.5 Mixed Chemical–Biochemical Capping 529
2.6 Purification of Cap Analogs 530
3 Synthetic Cap Analogs in Search for Cap-Regulated Cellular Processes 531
3.1 Cap Analogs in mRNA Translation and Translation Inhibition Studies (Early Studies) 531
3.2 Cap Analogs in Search for Cap Affinities Toward Canonical eIF4Es and Their Isoforms 534
3.3 Alternatively Capped mRNAs with High Translational Efficiency 537
3.4 Synthetic Cap Analogs in RNA Maturation (Splicing) and RNA Intracellular Transport 540
3.5 Cap Analogs in mRNA Metabolism 542

J. Stepinski
Department of Biophysics, Institute of Experimental Physics, Faculty of Physics, University of Warsaw, Zwirki i Wigury 93, 02-089 Warsaw, Poland
e-mail: jastep@biogeo.uw.edu.pl

E. Darzynkiewicz (✉)
Department of Biophysics, Institute of Experimental Physics, Faculty of Physics, University of Warsaw, Zwirki i Wigury 93, 02-089 Warsaw, Poland

Centre of New Technologies, University of Warsaw, Zwirki i Wigury 93, 02-089 Warsaw, Poland
e-mail: edek@biogeo.uw.edu.pl

V.A. Erdmann et al. (eds.), *Chemical Biology of Nucleic Acids*, RNA Technologies, DOI 10.1007/978-3-642-54452-1_28, © Springer-Verlag Berlin Heidelberg 2014

3.6 Affinity Resins 544
3.7 Translation Inhibition by miRNAs 548
4 Synthetic Cap Analogs: Relevance to Medicine 549
5 Future Directions 550
References 551

Abstract Almost all eukaryotic mRNAs have a monomethylguanosine cap structure consisting of 7-methyl guanosine that is connected *via* 5′–5′ triphosphate bond to the next nucleoside (m^7GpppN; MMG-cap) on their 5′ termini. In unicellular kinetoplastida, including *Leishmanias* (responsible for a wide spectrum of diseases), the cap is unusually highly methylated ($m^7Gpppm_3^{6,6,2'}Apm^{2'}Apm^{2'}Cpm_2^{3,2'}U$), known as cap-4, while in nematodes (e.g., *C. elegans* or *Ascaris*), the mRNA cap is ended with trimethylguanosine ($m_3^{2,2,7}$GpppN; TMG-cap). A large class of uridine-rich small nuclear RNAs (U snRNAs) on their 5′ termini have also TMG-cap.

Over the last three decades several classes of 5′ mRNA cap analogs, including the natural ones (MMG-cap, TMG-cap, cap-4), have been synthesized in our lab and by other groups. They were serving as valuable tools in elucidating molecular mechanisms of such cap-regulated cellular processes as protein translation initiation, pre-mRNA splicing, RNA intracellular transport, mRNA turnover, and cap-dependent translation inhibition by microRNAs. Some of the synthetic cap dinucleotides (anti-reverse cap analogs; ARCAs), adopted to construct mRNA transcripts with the increased translational efficiency, have found commercial application in production of proteins. In this chapter, we describe the strategies and technical approaches in the synthesis of natural and modified cap analogs. Their application in biology and more recently, in medical studies is also reviewed.

Keywords 5′ mRNA cap • Trimethylguanosine cap • cap-4 • Capped oligonucleotides • Anti-reverse cap analogs • Non-hydrolyzable synthetic cap analogs • Cap-binding proteins • Capping enzymes • Decapping enzymes • Translation inhibitors • Antitumor mRNA vaccines

1 Introduction: Cap Structures, Their Occurrence, and Functions

In contrast to their prokaryotic (bacterial) counterparts, the vast majority of eukaryotic mRNAs bear a cap structure on 5′ ends; the cap consists of 7-methyguanosine linked to the next nucleoside (N) *via* 5′–5′ triphosphate bond (m^7GpppN). This cap structure is called a monomethyl guanosine cap, known also as MMG cap. Moreover, the two adjacent nucleosides can be 2′-*O*-methylated. The cap structure with a single methyl group on the terminal guanosine is called cap-0, whereas the 2′-*O*-methylations on the ribose moiety(ies) provide, subsequently, cap-1 (m^7GpppN^m) and cap-2 ($m^7GpppN^mN^m$). The MMG cap is added to the RNA in the nucleus co-transcriptionally by a specific complex of capping enzymes while further methylations, carried out by specific methyltransferases, occur in the nucleus, leading to cap-1, and further, in cytoplasm, leading to cap-2. In unicellular Kinetoplastida, including *Leishmanias*, the organisms responsible for a wide spectrum of diseases, the cap is unusually highly methylated: $m^7Gpppm_3{}^{6,6,2'}Apm^{2'}Apm^{2'}Cpm_2{}^{3,2'}U$, known as cap-4, and all mRNAs acquire this cap by trans-splicing. In nematodes (e.g., *C. elegans* or *Ascaris suum*), except mRNAs capped with MMG cap, 30–50 % of the mRNAs are terminated by a trimethylguanosine cap ($m_3{}^{2,2,7}GpppN$; TMG cap), which is provided by a short RNA spliced leader. Also many small nuclear RNAs (snRNAs) that catalyze splicing of pre mRNAs bear the TMG cap structure on their 5′ end. All these three cap structures found in Nature are shown in Fig. 1.

The identification of MMG cap in cellular and viral mRNAs dates to the mid-1970s. The historical perspective of the events leading to determination of its structure was described in detail by their principal discoverers Furuichi and Shatkin (2000). As early as in 1975, along with the discovery of the cap structure, Shatkin's laboratory proved its crucial role in translation initiation and in stabilization of capped mRNA against exonucleotic degradation. Over the years, it was shown that the MMG cap plays a pivotal regulatory role in many other cellular processes, including promotion of mRNA transcription, mRNA polyadenylation, splicing, nuclear export, and inhibition of capped mRNA translation by microRNAs (miRNAs). In all those events, the cap-related regulatory proteins are involved as they interact with the cap structure directly or indirectly *via* other proteins. We refer the readers to an excellent review by Topisirovic et al. (2011) that covers practically the entirety of the current knowledge on regulatory functions of MMG cap and cap-binding proteins in the control of gene expression.

Much less is known about the role of the TMG cap. It is assumed that it promotes the protein translation of TMG-capped mRNAs in nematodes, but the full evidence is still not conclusive. On the other hand, its promoting role as a component of a nuclear targeting signal in nuclear import of trimethylguanosine-capped U snRNAs has been well documented (see Sect. 3.4.2). The role of cap-4 in Trypanosomatides remains unknown, and it is still to be determined.

Yet another issue is the metabolism of all three caps, which should be considered in the context of their biosynthesis that engages the capping enzymes, as well as their degradation that involves the decapping enzymes [covered briefly in the review by Topisirovic et al. (2011), and in more detail in the reviews cited therein].

Over nearly four decades, many cap-related papers dealing with caps' structures, their biological functions, as well as their interactions with a broad number of cap-binding proteins have been published. It should be pointed out that in many studies that led to important discoveries in this field, the synthetic cap analogs proved to be the tool of extreme value. In this chapter, we highlight the close link between the development of biological chemistry in the field of mRNA cap analogs and the use of these analogs in multiplexed biological and medical studies.

The specific nature of the cap structure is attributed to great extent to the presence of 7-methylguanosine, the chemical properties of which differ significantly from those of non-methylated nucleosides. The main feature of

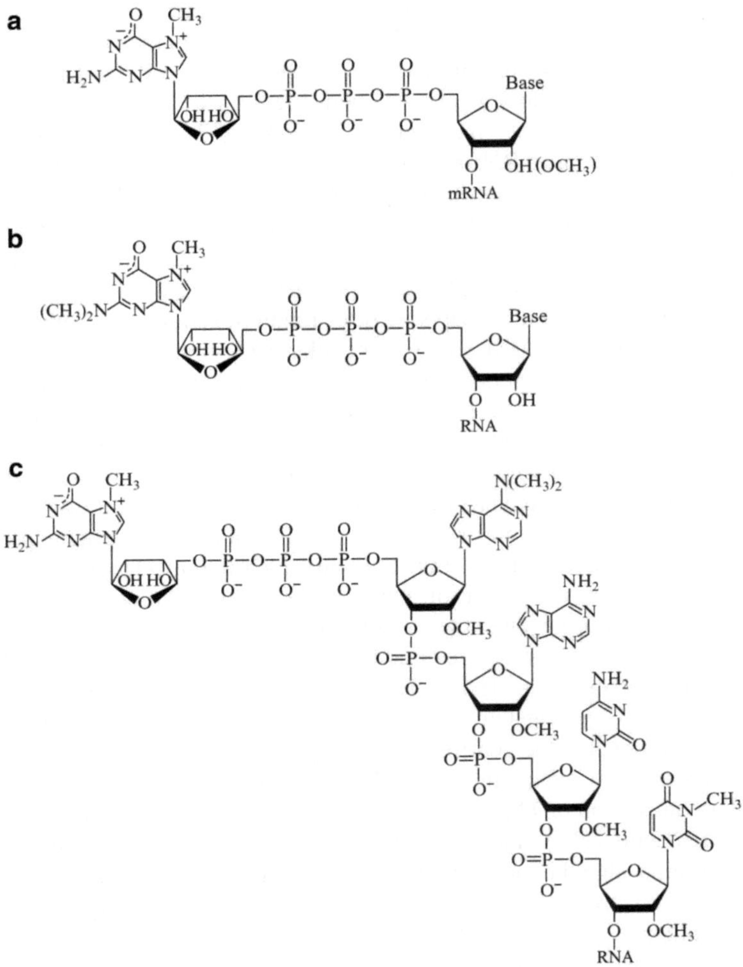

Fig. 1 Structures of natural caps: (**a**) MMG-cap, m^7GpppN, N denotes any nucleoside, (**b**) TMG-cap, $m_3^{2,2,7}GpppN$, (**c**) cap-4, $m^7Gpppm_3^{6,6,2'}Apm^{2'}Apm^{2'}Cpm_2^{3,2'}U$

7-methylguanosine, as well as other 7-alkylguanosine moieties, is the positive charge at 7-nitrogen atom of the imidazole ring. Whereas under neutral and slightly acidic conditions a monocation is the predominant form of 7-methylguanosine, at $pH > 7$ deprotonation occurs at N1 and a zwitterion is formed (Fig. 2).

The first consequence of this specific structure is susceptibility to binding with other aromatic moieties by π–π interaction called "stacking". 7-methylguanine has strong polarizing properties and interacts preferably with easily polarizable molecules, such as aromatic amino acids: tryptophan, phenylalanine, and tyrosine. Since

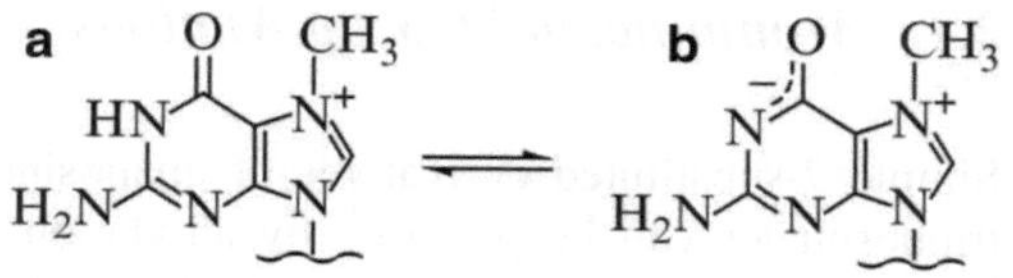

Fig. 2 Ionization of 7-methylguanosine. (**a**) Monocation (keto) form, (**b**) zwitterion (enolate) form

7-methylguanine is fluorescent, fluorescence methods are widely used to study binding processes of the cap and proteins. It is worth to note that in dinucleotide cap analogs, intramolecular stacking between 7-methylguanine and second base, for instance, guanine in m^{7}GpppG, also can be observed. Another consequence of the unique structure of 7-methylguanosine is a different conformation of the ribose ring. Unlike other purine nucleosides and nucleotides, 7-methylguanosine and its 5′-phosphorylated derivatives prefer 3′-*endo* D-ribose ring conformation. Moreover, 7-methylguanosine nucleotides prefer an *anti* conformation around the *N*-glycosidic bond due to electrostatic interaction between the positively charged base and the negatively charged phosphate and show preference for synclinical arrangement of C4′–C5′, which places the 5′-*O* above the sugar ring and stabilizes the conformation. Conformation of the cap structure is one of the factors that influence the ability of the cap to bind to proteins. Therefore, much attention has been dedicated to ^{1}H NMR and X-ray techniques, which are the main physicochemical methods for studying three-dimensional structures of cap-protein complexes. The positive charge at the 7-nitrogen atom of 7-methylguanosine is also responsible for low chemical stability of cap analogs in alkaline conditions. It is well known (Townsend and Robins 1963; Darzynkiewicz et al. 1990) that at pH > 8 the imidazole ring of 7-methylguanosine undergoes cleavage. Accounting for the fact that nucleotides, in general, can be degraded by depurination or phosphate hydrolysis, this additional instability factor of 7-methylguanosine moiety makes the chemistry of cap analogs even more challenging. Moreover, nucleotides are poorly soluble in organic solvents due to the ionic nature of phosphate moieties, and therefore, prior to any reaction, they must be usually converted into the so-called organic salts such as tributylammonium, triethylammonium, or pyridinium salt. Also, due to the ionic nature of the phosphate, isolation of pure cap analogs requires special and extensive efforts.

2 Chemical Synthesis of 5′ mRNA Cap Analogs

The "cap analog" can be defined as any chemical compound capable of mimicking some structural and biological features of the mRNA's 5′-end. Therefore, from the chemical point of view, the term is not unequivocal. Generally, cap analogs can be divided into three groups: mononucleotides, dinucleotides, and capped oligonucleotides.

2.1 Mononucleotide Cap Analogs

Simple 7-substituted derivatives of guanosine and guanosine 5′-mono-, di-, and triphosphates can be prepared by alkylation of the guanine ring at N7 in mild conditions using appropriate alkyl iodide, bromide, or chloride (Adams et al. 1978; Darzynkiewicz et al. 1985; Jankowska et al. 1993b). Other alkylating reagents such as dimethyl sulfate (Hendler et al. 1970) and methyl methanesulfonate (Fukuoka et al. 1994a) were also used.

In case of other analogs, which contain modified base or/and sugar moieties, 5′-phosphorylation of a particular nucleoside is usually the first step of the chemical synthesis. This can be achieved by means of pyrophosphoryl chloride [$Cl_2P(O)OP(O)Cl_2$] (Imai et al. 1969), but in most cases Yoshikawa's method (Yoshikawa et al. 1967; Burgess and Cook 2000) is preferred; it includes reaction of the starting nucleoside with phosphorus oxytrichloride in trimethyl or triethyl phosphate. The nucleotide 5′-phosphate residue can be elongated into di-, tri-, tetra-, or even longer oligophosphate chain by means of appropriate coupling reactions. To obtain a diphosphate analog, a monophosphate is converted to an active derivative, most often into *P*-imidazolide, using 1,1′-carbonyldiimidazole (Hoard and Ott 1965; Adams et al. 1978) or imidazole, triphenylphosphine, and 2,2-dithiodipyridine (Mukaiyama and Hashimoto 1971; Lohrmann and Orgel 1978), and then coupled with mono-(tri-*n*-butylammonium) phosphate (Adams et al. 1978; Peng et al. 2002) or mono-(triethylammonium) phosphate (Stepinski et al. 2001) in DMF (dimethylformamide), preferably in the presence of anhydrous zinc chloride, which proved to be a good promoter of the pyrophosphate bond formation (Kadokura et al. 1997; Stepinski et al. 2001). For synthesis of nucleotides that bear a bridging or non-bridging modifications in the oligophosphate chain properly substituted ortophosphate *P*-imidazolides may be employed as an alternative phosphate group donors (Kadokura et al. 1997; Strenkowska et al. 2012).

An attractive feature of Yoshikawa's 5′-monophosphorylation reaction is that it can be adapted to one-pot synthesis of nucleotide triphosphates. The procedure involves the reaction of chlorophosphate intermediate with bis-(tri-*n*-butylammonium) pyrophosphate (Hoard and Ott 1965; Ludwig 1981; Ruth and Cheng 1981). Many aspects of nucleoside 5′-triphosphate syntheses were described earlier in a comprehensive review of Burgess and Cook (2000).

Also, the method of Ludwig and Eckstein (1989) has numerous examples of use; this method is based on nucleoside 5′-*O* phosphorylation with 2-chloro-4*H*-1,3,2-benzodioxaphosphorin-4-one (salicylchlorophosphite) followed by sequential one-pot reactions to prepare nucleoside 5′-di- or triphosphates or their derivatives with non-bridging substitutions at α-phosphate (He et al. 1998; Li and Shaw 2004; Li et al. 2005). However, in contrast to Yoshikawa's procedure, 2′-*O* and 3′-*O* positions need to be protected in this method. Extension of such preparative methods allowed also the synthesis of 5′-tetra- and 5′-pentaphosphate of 7-methylguanosine (m^7Gp_4 and m^7Gp_5, respectively) (Zuberek et al. 2004).

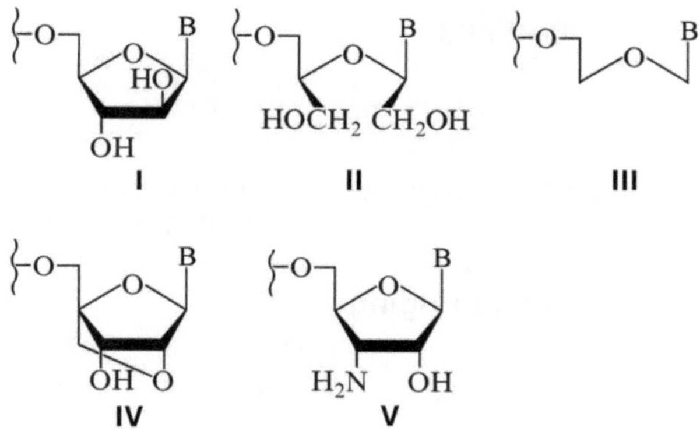

Fig. 3 Examples of sugar modifications in cap analogs: **I**—D-arabinose, **II**—*seco*-ribose, **III**—*acyclo*-ribose, **IV**—2′-*O*,4′-*C*-methylene-D-ribose, **V**—3′-amino-3′-deoxy-D-ribose; **B**—7-methylguanine moiety

However, not only simple cap analogs like derivatives of nucleosides alkylated at any position can be synthesized. Among cap analogs modified within the base and/or sugar moiety, there are also several examples of more profound structural changes, which may include for instance phosphates of 7-methylguanine β-D-arabinofuranoside that changes the configuration at C2′ of the sugar moiety (Darzynkiewicz et al. 1985), 7-methyl-*acyclo*- and 7-methyl-*seco*-guanosine (Darzynkiewicz et al. 1985, 1987; Stepinski et al. 1990), 7-methylguanine LNA-analog (7-methyl-2′-*O*,4′-*C*-methyleneguanosine) (Kore et al. 2009), or 7-methyl-3′-amino-3′-deoxyguanosine (Jemielity et al. 2012a). The modified sugar moiety structures are shown in Fig. 3.

Labeled cap analogs may serve as valuable tools for some biophysical and biochemical studies. These can include photoaffinity probes with phenyl azide- and benzophenone-substituted phosphonamides of 7-methylguanosine 5′-triphosphates (Chavan et al. 1990), 2′-*O*- and 3′-*O*-anthraniloyl derivatives of 7-methylguanosine 5′-triphosphate (Ren and Goss 1996), 3′-*O*-anthraniloyl and 3′-*O*-(*N*-methylanthraniloyl) mono- and dinucleotide cap analogs (Ziemniak et al. 2013b), fluorescein-containing 7-methylguanosine 5′-monophosphate (Natarajan et al. 2004), spin-labeled (i.e., consisting of free radical residue) compounds (Stepinski et al. 2007; Jankowska-Anyszka et al. 2011), and biotin-containing analogs (Moreno et al. 2009; Jankowska-Anyszka et al. 2011; Jemielity et al. 2012a).

Compounds modified within the oligophosphate residue are another important class of cap analogs. Although attempts to introduce an unnatural structural feature into nucleotide oligophosphate backbone have been made for a few decades (Myers et al. 1963; Yount et al. 1971; Eckstein 1970; Engel 1977; Stock 1979; Sood et al. 1990; Ma et al. 1992; Blackburn et al. 1992), analogs consisting of 7-methylguanosine moiety constitute a kind of strict cap analogs. Basically, the analogs were prepared as intermediates for 5′,5′-dinucleotides syntheses (see Sect. 2.2); nevertheless, some of them were independently characterized in many biochemical tests. Among them, there are simple methyl esters at phosphate group of the nucleotide (Darzynkiewicz et al. 1981, 1985, 1987), as well as compounds, in

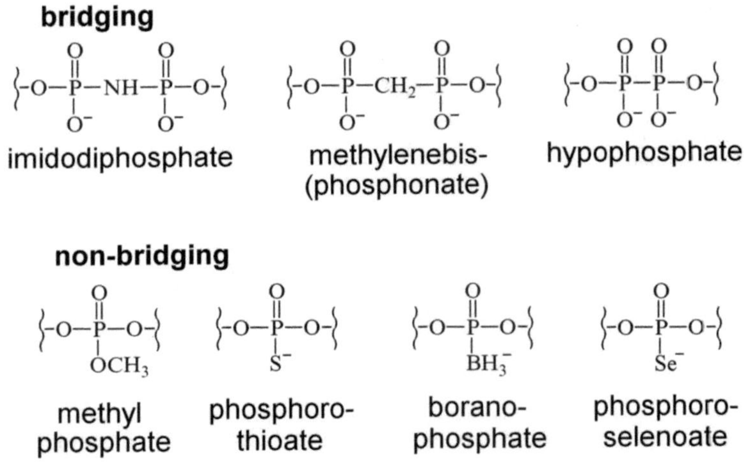

Fig. 4 Possible modifications within the phosphate chain of the cap analogs

which the bridging (anhydride) oxygen is replaced with a methylene group (Kalek et al. 2005a, b, 2006; Jemielity et al. 2005a; Rydzik et al. 2009) or with imidophosphate (P–NH–P) (Rydzik et al. 2012) and other compounds, in which one of the phosphorus atoms bears sulfur (Kowalska et al. 2005, 2007a, b), selene (Glass et al. 1993; Kowalska et al. 2009), or borane (BH_3^-) (Kowalska et al. 2008a; Su et al. 2011) instead any external (non-bridging) oxygen atom. Figure 4 summarizes schematically the modifications within the phosphate chain of the analogs.

Introduction of a methylene moiety into a nucleotide was achieved by modification of Yoshikawa's method using commercially available methylene (bisphosphonic dichloride) as a phosphorylating agent. The reaction proceeds regioseletively into the 5′ position of an unprotected nucleoside. The obtained nucleoside 5′-bisphosphonate is susceptible to activation with imidazole and also, on the other hand, to reaction with other imidazole-activated phosphates or pyrophosphates. Thus, it is possible to prepare nucleoside 5′-phosphono-phosphates bearing methylene bridge in any oligophosphate position (Table 3). Of course, it is possible to introduce 5′-oligophosphate moiety with two (or theoretically even more) methylene modifications. The example, although involving a different way of synthesis, consists of nucleophilic displacement of nucleoside 5′-*O*-tosyl group by phosphonic–bisphosphonic acid (Stock 1979).

In the same manner, imidodiphosphate moiety can be introduced into the nucleoside 5′ position using dichlorophosphorylphosphorimidoyl trichloride (Tomasz et al. 1988; Rydzik et al. 2012). Unfortunately, in contrast to methylene-phosphonate analogs, some nucleoside 5′-imidodiphosphate-phosphate compounds (for instance, GppNHp) are unstable (Tomasz et al. 1988; Rydzik et al. 2012), and therefore the imidodiphosphate series are limited (Table 3). However, the synthesis of a nucleoside 5′-triphosphate analog with two imidodiphosphate modifications was reported (Ma et al. 1992).

5′-Thiophosphate moiety can be also introduced into a nucleoside according to the modified Yoshikawa procedure using $PSCl_3$ in trimethylphosphate with 2,6-dimethylpyridine as promoter (Murray and Atkinson 1968; Moran and Whitesides 1984; Kowalska et al. 2005, 2008b; Strenkowska et al. 2010). An introduction of a thiophosphate residue into position other than α (position β, γ) can be performed by treatment of the appropriate nucleoside *P*-imidazolide with excess of thiophosphate triethylammonium salt in dimethylformamide (DMF) in the presence of $ZnCl_2$. Each thiophosphate compound exists in a form of two diastereoisomers, which can usually be separated by a preparative HPLC (high-pressure liquid chromatography). From the chemical point of view, it is important that in case when 7-alkylguanosine containing thiophosphate analog is prepared, the alkylation step should be performed before the introduction of the thiophosphate moiety; otherwise the alkylation proceeds preferably on the sulfur atom (Kowalska et al. 2005). It is worth to note that an analog with a terminal thiophosphate residue could not be activated with imidazole (Strenkowska et al. 2010). A few syntheses of closely related analogs with a selenophosphate residue were also reported (Glass et al. 1993; Kowalska et al. 2009).

Chemistry can offer an even much wider spectrum of different modifications of the oligophosphate chain, including for instance inserting dihalogenomethylenebis-(phosphonates) (Blackburn et al. 1992), hypophosphates (Setondji et al. 1970; Guranowski et al. 2010a), and phosphosulfates (Guranowski et al. 2010b; Kowalska et al. 2012).

2.2 *Dinucleotide Cap Analogs*

Among the dinucleotide cap analogs, P^1-guanosine-5′ P^3-(7-methylguanosine-5′) triphosphate (m^7GpppG) is a standard compound. A similar structure has m^7GpppA, compound that is also very often exploited in biochemical and structural studies. And, similarly to the mononucleotide compounds referred above, the dinucleotide analogs can be modified in one or both bases, in the ribose ring(s), and/or in the triphosphate bridge, offering a broad range of tools for specific biochemical, biophysical, biological, biomedical, and physicochemical studies.

Generally, it is possible to prepare dinucleoside 5′,5′-oligophosphates by means of condensation of a nucleoside mono-, di-, or triphosphate with a second nucleoside 5′-oligophosphate activated at the terminal phosphate with an easily leaving group (Blackburn et al. 1992; Mikkola et al. 2005; Kore et al. 2008a, 2010b; Jemielity et al. 2010). Imidazole derivatives, imidazolidephosphates (phosphorimidazolidates), are most widely used as leaving groups (Cramer et al. 1961; Hoard and Ott 1965; Sawai et al. 1991, 1992; Stepinski et al. 1995, 2001; Kadokura et al. 1997; Jemielity et al. 2010). Other leaving groups of choice include phenylthio- (Nakagawa et al. 1980), 4-methoxyphenylthio- (Kohno et al. 1985), 4-chlorophenylthio- (Fukuoka et al. 1994a, b), morpholine- (Adam and Moffat 1966), or 5-chloro-8-quinolyl- (Fukuoka et al. 1994a, b) nucleoside 5′-phosphate

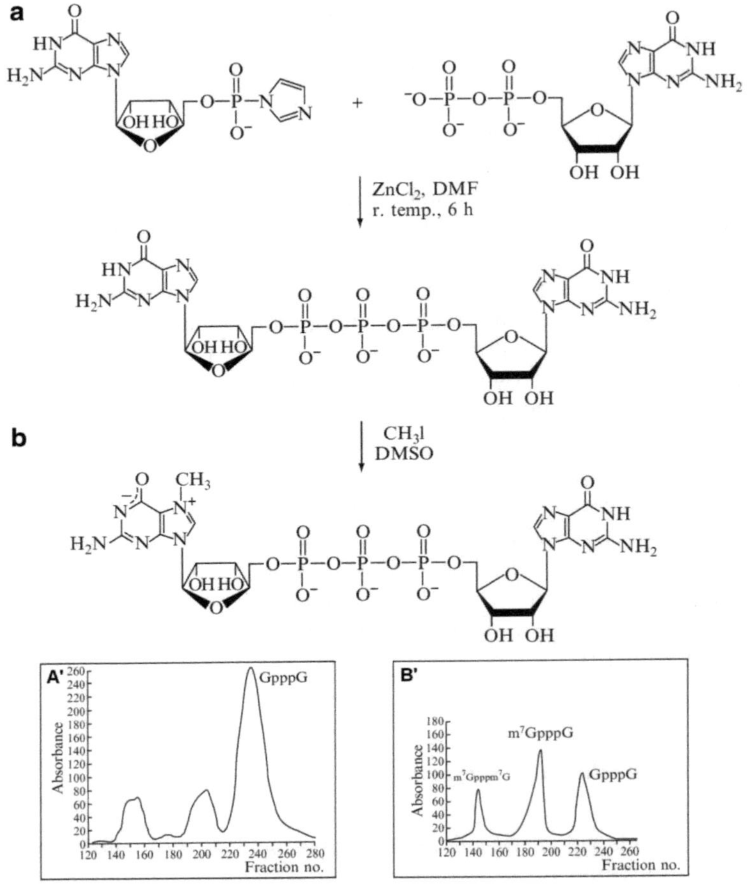

Fig. 5 Two-stage synthesis of m^{7}GpppG: (**a**) coupling reaction to GpppG; (**b**) methylation to the final product. *Inserts* show clear separation of GpppG (**a′**) and m^{7}GpppG (**b′**) from the reaction mixtures by DEAE-Sephadex A-25 (HCO_3^-) ion-exchange chromatography with linear gradient of triethylammonium bicarbonate aqueous buffer (pH 7.4) as mobile phase, absorption at $\lambda = 260$ nm, arbitrary units (Niedzwiecka et al. 2007)

derivatives. The new pyrophosphate bond is formed by means of a nucleophilic attack of the terminal phosphoryl hydroxyl group on the activated phosphate. Usually, the coupling reactions are carried on in anhydrous solutions, DMF, pyridine, 1-methyl-2-pyrrolidinone, dimethylsulfoxide (DMSO), or other polar aprotic solvent, most often with an appropriate catalyst. However, in case when imidazolides are used as substrates, the coupling can be also performed in aqueous buffer at pH 7 (Shimazu et al. 1990; Sawai et al. 1991, 1992).

Small-scale chemical syntheses of m^{7}GpppG and m^{7}GpppA were first accomplished using di-*n*-butylphosphinothioyl- (Hata et al. 1976) or phenylthio- (Nakagawa et al. 1980) activating groups. Independently, tetraphosphate analogs,

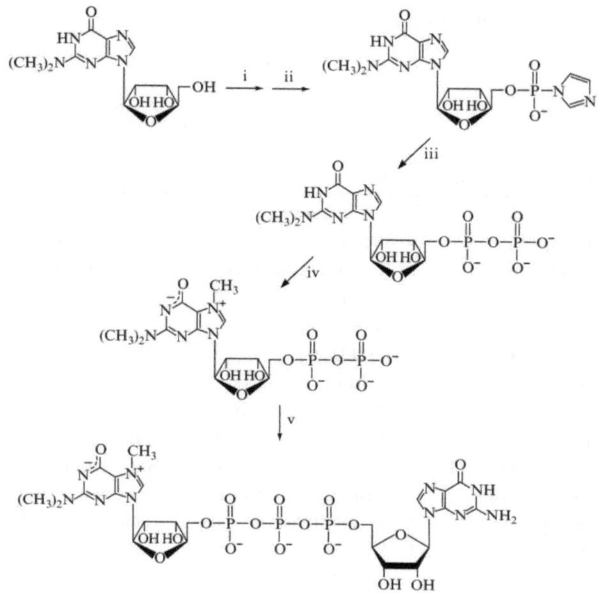

Fig. 6 Synthesis of $m_3{}^{7,2,7}$GpppG: (i) $POCl_3$, $(CH_3O)_3PO$; (ii) imidazole, 2,2′-dithiodipyridine, triphenylphosphine, DMF; (iii) $ZnCl_2$, triethylammonium dihydrogenphosphate, DMF; (iv) CH_3I, DMSO; (v) $ZnCl_2$, guanosine 5′-(*P*-imidazolido) monophosphate (triethylammonium salt), DMF (Niedzwiecka et al. 2007).

m^7GppppG and m^7Gppppm^{7}G, were prepared starting from GppppG, which was isolated from natural sources and methylated using dimethyl sulfate (Sasavage et al. 1979). A new, two-step synthesis of m^7GpppG on a larger scale proceeded by a chemical formation of GpppG, and its kinetically controlled methylation at N7 (Fukuoka et al. 1994a, b; Stepinski et al. 1995). This approach is superior to alternative methods due to efficient isolation of GpppG and m^7GpppG from the appropriate reaction mixtures (Fig. 5). The improved procedure for GpppG, m^7GpppG, and m^7Gpppm^{7}G synthesis with use of $ZnCl_2$ as a promoter during formation of the new pyrophosphate bond was described in details by Niedzwiecka et al. (2007).

Modifications of the bases or ribose ring(s) require more laborious synthetic procedures. They should be introduced at the nucleoside (or even base) level and the intermediates can be successively converted to the required nucleotides and dinucleotides. A representative example of $m_3{}^{2,2,7}$GpppG synthesis is shown below on Fig. 6. First syntheses of this TMG-cap analog were accomplished using phenylthio groups as the activating agents (Darzynkiewicz et al. 1988, 1990; Iwase et al. 1989). An alternative procedure exploited coupling of N^2,N^2,7-trimethylguanosine 5′-monophosphate imidazolide with GDP in aqueous HCl/N-ethylmorpholine buffer, pH 7, in the presence of Mg(II) ions (Stepinski et al. 1995) or in an organic solvent (DMF) in the presence of anhydrous zinc chloride (Niedzwiecka et al. 2007).

Similarly, other dinucleotide 5′,5′-di-, tri-, tetra-, penta-, and heksaphosphates modified in the base(s) and/or in the ribose(s) can be synthesized (Stepinski

Table 1 Selected cap analogs with 7-methylguanosine connected to various nucleosides

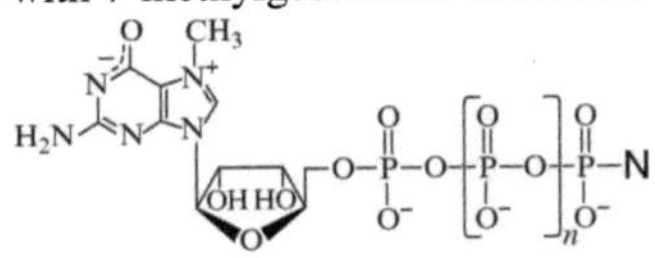

Cap abbreviation	n	N	Ref.
m^7GppG	0	Guanosine	Sawai et al. (1992), Stepinski et al. (1995)
m^7GppI	0	Inosine	Adams et al. (1978)
$m^7Gpp(m^7G)$	0	7-methylguanosine	Stepinski et al. (1995)
m^7GpppG	1	Guanosine	[a]
m^7GpppA	1	Adenosine	[b]
m^7GpppC	1	Cytidine	Sawai et al. (1991), Jankowska et al. (1996)
m^7GpppU	1	Uridine	Sawai et al. (1991), Jankowska et al. (1996)
$m^7Gppp(2'dG)$	1	2′-deoxyguanosine	Jankowska et al. (1996)
$m^7Gppp(m^{2'O}G)$	1	2′-*O*-methylguanosine	Jankowska et al. (1996)
$m^7Gppp(m^{3'O}G)$	1	3′-*O*-methylguanosine	Stepinski et al. (2005)
$m^7Gppp(m^7G)$	1	7-methylguanosine	Fukuoka et al. (1994a, b), Stepinski et al. (1995)
$m^7Gppp(m^6A)$	1	N^6-methyladenosine	Jankowska et al. (1996)
$m^7Gppp(m_3^{6,6,2'}A)$	1	$N^6,N^6,O^{2'}$-trimethyladenosine	Lewdorowicz et al. (2004)
$m^7Gppp(araG)$	1	Arabinoguanosine	Stepinski et al. (2005)
$m^7Gppp(isoG)$	1	Isoguanosine	Stepinski et al. (2005)
m^7Gppp (formycin)	1	Formycin A	Stepinski et al. (2005)
m^7Gppp (2-aminoA)	1	2,6-diaminopurine	Kore and Charles (2010a)
m^7Gppp (7-dazaG)	1	7-deazaguanosine	Kore et al. (2010a)
$m^7GppppG$	2	Guanosine	Sasavage et al. (1979), Stepinski et al. (1995), Jemielity et al. (2003a, b)
$m^7Gpppp(m^7G)$	2	7-methylguanosine	Sasavage et al. (1979), Stepinski et al. (1995)
$m^7GpppppG$	3	Guanosine	Jemielity et al. (2003a, b)
$m^7GppppppG$	4	Guanosine	Zdanowicz et al. (2009)

[a]Hata et al. (1976), Nakagawa et al. (1980), Fukuoka et al. (1994a, b), Sawai et al. (1991), Stepinski et al. (1995), Niedzwiecka et al. (2007)
[b]Hata et al. (1976), Nakagawa et al. (1980), Fukuoka et al. (1994a, b), Sawai et al. (1991), Jankowska et al. (1996), Matsuo et al. (2000)

et al. 2001; Jemielity et al. 2003a, b, 2010; Grudzien-Nogalska et al. 2007; Kore et al. 2008a, 2010b).

Among modified dinucleotide cap analogs, anti-reverse cap analogs (ARCAs), which do not possess free 3′-OH in 7-methylguanosine moiety and therefore can be incorporated in in vitro tests with the use of polymerases into RNA transcripts exclusively into the proper orientation (see Sect. 3.3), play a special role. Firstly

Table 2 Selected anti-reverse cap analogs (ARCAs)

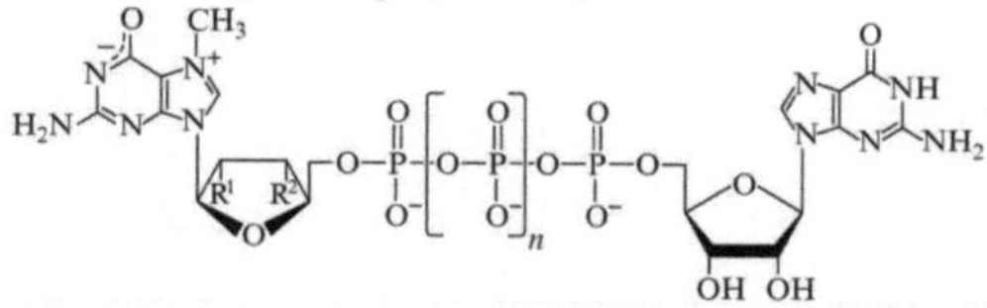

Cap abbreviation	n	R^1	R^2	Ref.
$m_2^{7,3'O}$GpppG	1	OH	OCH_3	Stepinski et al. (2001), Peng et al. (2002)
m^7(3′dG)pppG	1	OH	H	Stepinski et al. (2001)
$m_2^{7,2'O}$GpppG	1	OCH_3	OH	Jemielity et al. (2003a, b)
m^7(2′dG)pppG	1	H	OH	Jemielity et al. (2003a, b)
m^7(2′FG)pppG	1	F	OH	Kore et al. (2007)
m^7(2′allylG)pppG	1	$OCH_2CH{=}CH_2$	OH	Kore and Charles (2010b)
m^7(iPrG)pppG	1	2′,3′-*O*-isopropylidene		Kore et al. (2008b), Warminski et al. (2013)
$m_2^{7,3'O}$GppppG	2	OH	OCH_3	Jemielity et al. (2003a, b)
$m_2^{7,2'O}$GppppG	2	OCH_3	OH	Jemielity et al. (2003a, b)
m^7(2′dG)ppppG	2	H	OH	Jemielity et al. (2003a, b)
$m_2^{7,3'O}$GpppppG	3	OH	OCH_3	Jemielity et al. (2003a, b)
$m_2^{7,2'O}$GppppppG	4	OCH_3	OH	Zdanowicz et al. (2009)

(Stepinski et al. 2001) described synthesis of two analogs including 7-methyl-3′-deoxy- and 7,3′*O*-dimethylguanosine moieties. The key reaction of pyrophosphate bond formation was achieved in DMF solution employing catalytic properties of zinc chloride. Independently, Peng et al. (2002) reported similar preparation route leading to 7,3′*O*-dimethylguanosine cap, but the final coupling reaction was carried out with activated GMP-morpholidate in the presence of 1*H*-tetrazole as catalyst. Later it was determined that also 2′-substituted (Jemielity et al. 2003b; Kore et al. 2010b) as well as 2′,3′-*O*-isopropylidene-7-methylguanosine caps (Kore et al. 2008b; Warminski et al. 2013) exhibit ARCA properties.

Modifications of the triphosphate bridge (other than the length) have been already listed in the previous section (Sect. 2.1). Theoretically, in each mono- or diphosphate compound mentioned there may be incorporated into a dinucleotide cap analog. Limitations of the methods noted in individual instances result from instability of certain substrates, from poor performance of some activation reactions, sometimes from low yield of final couplings, and finally from difficulties with isolation of the desired product of a satisfactory purity grade (Stepinski et al. 2002; Jemielity et al. 2010).

Over the last few decades, more than hundred structures of cap dinucleotides have been published. Representative examples are shown below (Tables 1, 2, 3, 4, 5, and 6), sorted with respect to specific structural features of the cap analogs.

Table 3 Selected cap analogs modified in bridging position of 5′,5′-oligophosphate chain

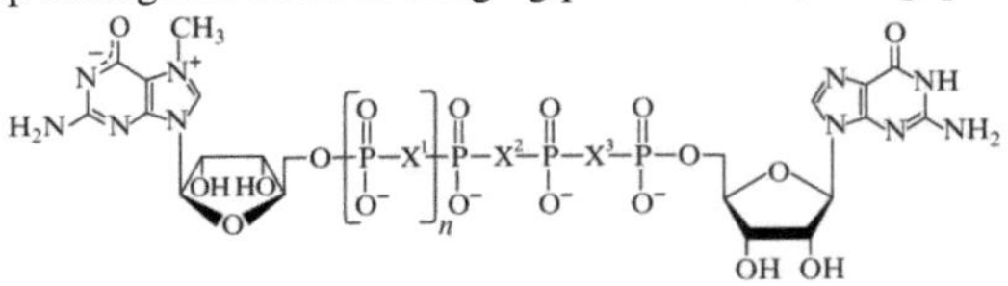

Cap abbreviation	*n*	X^1	X^2	X^3	Ref.
m^{7}Gp**CH$_2$**ppG	0	–	**CH$_2$**	O	Kalek et al. (2006)
m^{7}Gpp**CH$_2$**pG	0	–	O	**CH$_2$**	Kalek et al. (2006)
m^{7}Gp**CH$_2$**pppG	1	**CH$_2$**	O	O	Rydzik et al. (2009)
m^{7}Gpp**CH$_2$**ppG	1	O	**CH$_2$**	O	Rydzik et al. (2009)
m^{7}Gppp**CH$_2$**pG	1	O	O	**CH$_2$**	Rydzik et al. (2009)
m^{7}Gp**NH**ppG	0	–	**NH**	O	Rydzik et al. (2012)
m^{7}Gpp**NH**pG	0	–	O	**NH**	Rydzik et al. (2012)
m^{7}Gp**NH**pppG	1	**NH**	O	O	Rydzik et al. (2012)
m^{7}Gppp**NH**pG	1	O	O	**NH**	Rydzik et al. (2012)

To summarize, although different coupling reactions that lead to the production of dinucleotide cap analogs have never been compared directly, the following conclusions may be formulated:

- Methods that include phenylthio (or similar) leaving group require strict anhydrous conditions, are onerous due to hazardous properties of thiols and pyridine used, and give numerous side products, which are problematic during isolation of pure products
- Methods including nucleoside imidazolide or morpholide as activating group are less sensitive to moisture and usually give a smaller number of side products, and high yields can be obtained if reactions are carried out in the presence of a promoter
- It appears that $ZnCl_2$ in DMF is an especially effective promoter of the pyrophosphate bond formation, but in some cases, use of anhydrous $MgCl_2$ can be advantageous; promoters improve solubility of reactants, serve as a Lewis acid catalyst activating the leaving group, and act as a template coordinating both nucleotide subunits
- When the above conditions are used, nucleotide imidazolides undergo coupling with the broad spectrum of nucleophiles including phosphate, pyrophosphate, nucleoside mono-, di-, and triphosphate and with analogous compounds, such as methylenebisphosphonates, thiophosphates, and imidodiphosphates
- Imidazole derivatives of nucleotides are stable enough to perform the coupling reaction in aqueous medium at pH ~7, and when the water conditions are applied, Mn^{2+} ions show the best catalytic properties
- There are few examples of alternative coupling with the use of 5-chloro-8-hydroxyquinoline as a leaving group in an organic solvent (a mixture of 1-methyl-2-pyrtrolidinone and hexamethylphosphoroamide) with $CuCl_2$ as a catalyst (Fukuoka et al. 1994a, b; Koukhareva and Lebedev 2004)

Table 4 Selected cap analogs modified in non-bridging position of 5′,5′-oligophosphate chain[a]

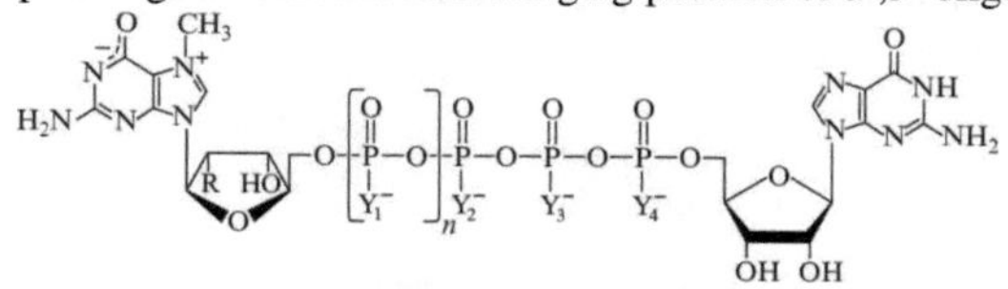

Cap abbreviation	n	R	Y_1	Y_2	Y_3	Y_4	Ref.
m^7Gp_SppG	0	OH	–	S	O	O	Kowalska et al. (2008b)
m^7Gpp_SpG	0	OH	–	O	S	O	Kowalska et al. (2008b)
m^7Gppp_SG	0	OH	–	O	O	S	Kowalska et al. (2008b)
$m_2{}^{7,2'O}Gp_SppG$	0	OCH_3	–	S	O	O	Kowalska et al. (2008b)
$m_2{}^{7,2'O}Gpp_SpG$	0	OCH_3	–	O	S	O	Kowalska et al. (2008b)
$m_2{}^{7,2'O}Gppp_SG$	0	OCH_3	–	O	O	S	Kowalska et al. (2008b)
$m_2{}^{7,2'O}Gp_SpppG$	1	OCH_3	S	O	O	O	Strenkowska et al. (2010)
$m_2{}^{7,2'O}Gpp_SppG$	1	OCH_3	O	S	O	O	Strenkowska et al. (2010)
$m_2{}^{7,2'O}Gppp_SpG$	1	OCH_3	O	O	S	O	Strenkowska et al. (2010)
$m_2{}^{7,2'O}Gpppp_SG$	1	OCH_3	O	O	O	S	Strenkowska et al. (2010)
$m_2{}^{7,2'O}Gpp_{Se}pG$	0	OCH_3	–	O	Se	O	Kowalska et al. (2009)
$m_2{}^{7,2'O}Gpp_{BH3}pG$	0	OCH_3	–	O	BH_3	O	Su et al. (2011)
$m^7(iPrG)pp_SpG$	0	[b]	–	O	S	O	Warminski et al. (2013)

[a]Each presented analog was obtained in form of two diastereoisomers. In most cases the diastereoisomers (depicted as D1 and D2) were isolated by means of preparative RP HPLC
[b]iPr = 2′,3′-*O*-isopropylidene

– In case of tetra- and pentaphosphate dinucleotides synthesis, quite different chemical route may be applied; it is based on a one-pot procedure using phosphitylation of a protected nucleoside with 2-chloro-4*H*-1,3,2-benzodioxaphosphorin-4-one (salicylchlorophosphite), followed by a sequential reaction with inorganic pyrophosphate and a nucleoside 5′- mono- or diphosphate (Han et al. 2006)

It is worth to note that the presented Tables 1, 2, 3, 4, 5, and 6 do not include all known cap analogs, especially the ones with multiple modifications. More chemical information can be found in the cited reviews (Mikkola et al. 2005; Kore et al. 2008a, 2010b; Jemielity et al. 2010).

2.3 *Capped Oligonucleotides*

For nearly three decades attempts have been made to synthesize the analogs of mRNA's 5′-end that contain not only the 5′–5′ oligophosphate bridge, but also a fragment of a RNA strand of different length and composition.. Most often, syntheses of such compounds are performed in a stepwise manner. Firstly, mRNA oligomer with a planned sequence is obtained using chemical (conventional

Table 5 Selected cap analogs consisting of a nucleoside other than 7-methylguanosine and N^2, N^2,7-trimethylguanosine

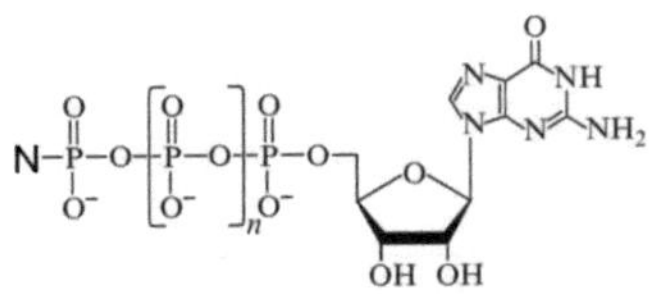

Cap abbreviation	n	N	Ref.
GppG	0	Guanosine	Stepinski et al. (1995)
m^3RBppG	0	3-methylbenzimidazole β-D-riboside	Chlebicka et al. (1995)
GpppG	1	Guanosine	Adam and Moffat (1966), Fukuoka et al. (1994a, b), Stepinski et al. (1995), Niedzwiecka et al. (2007)
et^7GpppG	1	7-ethylguanosine	Darzynkiewicz et al. (1990)
bn^7GpppG	1	7-benzylguanosine	Darzynkiewicz et al. (1990)
RVpppG	1	Ribavirin	Westman et al. (2005)
m^7*acyclo*GpppG	1	7-methyl-*acyclo*-guanosine	Stepinski et al. (1990)
m^7*seco*GpppG	1	7-methyl-*seco*-guanosine	Stepinski et al. (1990)
m^3RBpppG	1	3-methylbenzimidazole β-D-riboside	Chlebicka et al. (1995)
$m_2^{2,7}$GpppG	1	N^2,7-dimethylguanosine	Darzynkiewicz et al. (1988, 1990), Stepinski et al. (1995)
GppppG	2	Guanosine	Adam and Moffat (1966), Stepinski et al. (1995)
m^3RBppppG	2	3-methylbenzimidazole β-D-riboside	Chlebicka et al. (1995)
$m_2^{2,7}$GppppG	2	N^2,7-dimethylguanosine	Stepinski et al. (1995)

or in solid phase) or biochemical (in an appropriate transcription system) methods. Then, a cap structure is attached thereto using chemical or enzymatic approach.

The enzymatic approaches, not covered in this review, are suitable for a small-scale synthesis of capped oligoribonucleotides, while the chemical routes seem to be more effective in case of preparation of larger product amounts (at a milligram scale).

The first chemical synthesis of capped mRNA fragments was performed in Hata's laboratory (Yamaguchi et al. 1984) using phenylthio activation method. Short capped oligonucleotides (m^7GpppA^mGU and m^7GpppA^mG) were obtained by means of a reaction between P^2-*S*-phenyl-7-methylguanosine 5′-diphosphate and 5′-phosphorylated oligonucleotides in a pyridine solution with iodine or silver nitrate added as activators. Later, some improvements in the yield and performance of the methods were introduced by the same group (Iwase et al. 1988, 1992; Sekine et al. 1985, 1989).

Table 6 Selected dinucleotide TMG-cap analogs

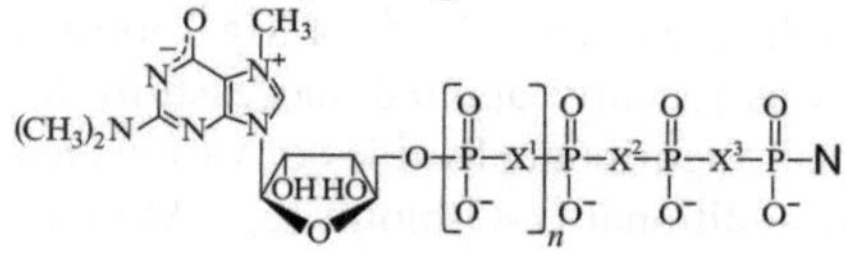

Cap abbreviation	*n*	N	X^1	X^2	X^3	Ref.
$m_3^{2,2,7}$GpppG	0	Guanosine	–	O	O	Darzynkiewicz et al. (1988, 1990), Iwase et al. (1989), Stepinski et al. (1995), Niedzwiecka et al. (2007)
$m_3^{2,2,7}$GpppA	0	Adenosine	–	O	O	Iwase et al. (1989)
$m_3^{2,2,7}$GpppA^m	0	2′-*O*-methyl-adenosine	–	O	O	Honcharenko et al. (2013)
$m_3^{2,2,7}$GpppA~N_3[a]	0	Adenosine[a]	–	O	O	Honcharenko et al. (2012)
$m_3^{2,2,7}$GpppA^m~N_3[b]	0	2′-*O*-methyl-adenosine[b]	–	O	O	Honcharenko et al. (2012)
$m_3^{2,2,7}$Gp**CH₂**ppA^m	0	2′-*O*-methyl-adenosine	–	**CH_2**	O	Honcharenko et al. (2013)
$m_3^{2,2,7}$Gpp**CH₂**pA^m	0	2′-*O*-methyl-adenosine	–	O	**CH_2**	Honcharenko et al. (2013)
$m_3^{2,2,7}$Gpp**CH₂CH₂**pA^m	0	2′-*O*-methyl-adenosine	–	O	**CH_2CH_2**	Ohkubo et al. (2013)
$m_3^{2,2,7}$GppppG	1	Guanosine	O	O	O	Stepinski et al. (1995)

[a]2′-*O*-substituted adenosine with a linker having terminal "clickable" azido residue
[b]3′-*O*-substituted 2′-*O*-methyladenosine with a linker having terminal "clickable" azido residue

Sekine et al. (1996) used imidazolidephosphate method of activation for synthesis of TMG-capped triribonucleotide, $m_3^{2,2,7}$GpppA^{m}U^{m}A. The target tetramer, a short 5′ end fragment of U1 snRNA which plays an important role in pre-mRNA splicing, was synthesized by the condensation of a partially protected triribonucleotide 5′-terminal diphosphate with imidazole-activated N^2,N^2,7-trimethylguanosine 5′-monophosphate. Later, a similar final coupling was performed by the same group (Kadokura et al. 2001; Ohkubo et al. 2013) for solid-phase synthesis of the trinucleotide (see Sect. 2.4).

Sawai et al. (1999) also utilized the imidazole activation method in the synthesis of two model capped oligoribonucleotides: m^7GpppAAAAA and m^7GpppACACUUGCUUU. However, the capping reaction of appropriate

5′-monophosphorylated oligoribonucleotides was carried out with imidazole-activated diphosphate in the presence of Mn^{2+} or Mg^{2+} ion as catalyst in an aqueous solution. This method was recently applied successfully to the synthesis of G-, MMG-, and TMG-capped oligoribonucleotides of various lengths (up to 22-mers), some of them with an additional 3′-O-biotin tag (Moreno et al. 2009; Piecyk et al. 2012; Honcharenko et al. 2012).

An alternative method was described by Zuberek et al. (2002, 2003). In this approach, a triribonucleotide, $pA^mU^mA^m$, was activated at 5′-phosphate with imidazole and then coupled with 7-methylguanosine 5′-diphosphate in DMF solution in the presence of $ZnCl_2$ to prepare $m^7GpppA^mU^mA^m$. However, further experiments (Worch et al. 2005a, b) proved that the method is less successful than yet another approach, in which imidazole-activated 7-methylguanosine 5′-diphosphate was coupled with a 5′-phosphorylated triribonucleotide under the same conditions. The latter approach was found to be more reproducible and gave higher yield (48.0 % vs. 18.6 %). The improved approach was also applied to the chemical synthesis of a unique cap-4 characteristic for trypanosomatid species. Thus, the multistep synthesis of $m^7Gpppm_3{}^{6,6,2'}Apm^{2'}Apm^{2'}Cpm_2{}^{3,2'}U$ (Lewdorowicz et al. 2004), and of shorter analogs: $m^7Gpppm_3{}^{6,6,2'}Apm^{2'}A$ and $m^7Gpppm_3{}^{6,6,2'}Apm^{2'}Apm^{2'}C$, the "cap-2," and "cap-3," respectively (Lewdorowicz et al. 2007b), was described including the preparation of nucleoside synthons, phosphoramidite solid-phase synthesis of the 5′-phosphorylated tetramer and trimer, conventional "in solution" synthesis of the dimer ($pm_3{}^{6,6,2'}Apm^{2'}A$), and their final couplings with imidazole-activated 7-methylguanosine 5′-diphosphate in DMF in the presence of $ZnCl_2$. Moreover, non-methylated counterpart of the cap-4, $m^7GpppAACU$, as an essential reference compound, was prepared (Yoffe et al. 2004). However, the coupling reaction in anhydrous conditions failed and the synthesis was carried out in a water solution with manganese salt as a catalyst. Here, it can be stated that in case of 2′-*O*-methylated oligoribonucleotide fragments the coupling reactions with imidazole-activated 7-methylguanosine 5′-diphosphate proceed preferably in an organic solvent in the presence of anhydrous $ZnCl_2$, while in case of more hydrophilic oligoribonucleotides, the aqueous conditions and the presence of manganese ions give better results (Piecyk et al. 2012; Smietanski et al. 2014).

Koukhareva and Lebedev (2004) presented another coupling procedure, which was based on a reaction between (5-chloro-8-hydroxyquinoline)-activated 7-methylguanosine 5′-diphosphate and 5′-phosphorylated tetramer (pGACU) in an organic solvent (a mixture of hexamethylphosphoramide and 1-methyl-2-pyrrolidinone) with $CuCl_2$ as a catalyst.

2.4 *Capping on Solid Support*

Synthesis of capped oligoribonucleotides would be significantly simplified if the last step of MMG or TMG incorporation was performed still on a solid support resin. The use of strong basic conditions when the oligonucleotide is being released from the solid support in a standard solid-phase synthesis makes the strategy useless because the cap structure is extremely vulnerable to such treatment. To overcome the problem, a modification should be introduced at an early stage of the synthesis in order to change the chemical nature of the linker between the first nucleoside at 3′-*O*- position and the solid support. Kadokura et al. (2001) used 2-cyanoethyl protected phosphoroamidate linker. This linker is sufficiently stable during the chain elongation, and at the end of the synthesis, just before the capping step, the linker becomes acid labile. Thus, the capped oligoribonucleotide ($m_3^{2,2,7}GpppA^{m}U^{m}Ap$) can be released in acidic conditions without damaging the cap structure. A novel approach with the use of disulfide-tethered solid support was described by Jemielity et al. (2005b). The crucial step of the synthesis was a $ZnCl_2$-promoted coupling of m^7GDP imidazolide to a fully deprotected 5′-phosphorylated oligoribonucleotide still on solid support in DMF. The disulfide bond was cleaved with DTT in aqueous methanol containing 0.01 % triethylamine. In this manner, $m^7GpppA^{m}U^{m}A^{m}p$- (Jemielity et al. 2005b) and $m^7Gpppm_3^{6,6,2'}Apm^{2'}Apm^{2'}Cpm_2^{3,2'}Up$- (Lewdorowicz et al. 2007a) capped oligoribonucleotides were obtained. However, surprisingly, in the latter example the coupling reaction performed in water solution in the presence of manganese salt gave better results (Lewdorowicz et al. 2007a).

2.5 *Mixed Chemical–Biochemical Capping*

Since both in vitro enzymatic syntheses of capped mRNAs and the chemical approaches that were briefly characterized above have several limitations and still remain challenging, methods combining chemical and enzymatic processes are alternatively explored. Brownlee et al. (1995) synthesized oligoribonuclotides (11–13 mers) by a solid-phase automated procedure, then introduced 5′-diphosphate group chemically, and after deprotection and removal of the oligonucleotides from the solid support, the cap structure was added enzymatically with guanylyltransferase, GTP (guanosine 5′-triphosphate), and S-adenosylmethionine. Another possibility was presented by Thillier et al. (2012), who synthesized short RNAs on a solid support by phosphoramidite chemistry. The cap structure was then coupled by the addition of GDP (guanosine 5′-diphosphate) after phosphorylation of the terminal 5′-OH and activation by imidazole. After deprotection and release from the support, the obtained GpppN-RNAs were purified, and the N^7-methyl group was introduced by enzymatic means using the human (guanine-N^7)-methyl transferase with *S*-adenosylmethionine.

In some in vitro systems, synthetic dinucleotide cap analogs can be elongated at 3′ position forming short capped oligoribonucleotides. Using this approach, Hodel et al. (1998) obtained m^{7}GpppGAAAAA during polimerase T7 transcription in the presence of m^{7}GpppG and ATP-αS (adenosine 5′-α-thiotriphosphate). Matsuo et al. (2000) and Peyrane et al. (2007) synthesized m^{7}GpppAC$_n$ ($n = 1$–9) using T7 DNA primase with synthetic m^{7}GpppA and CTP (cytidine 5′-triphosphate). The T7 primase has substrate specificity for adenosine at the first position of the primer preventing other analogs (non-adenosine, for instance, m^{7}GpppG) from being incorporated and excluding formation of "reverse" cap structures.

2.6 Purification of Cap Analogs

Chemical syntheses leading to nucleotides are usually multi-reagent, and extensive effort is required in order to isolate the expected product. In case of small-scale synthesis (order of micrograms), semi-preparative techniques of electrophoresis, paper chromatography, or HPLC may be used. For purification of crude large-scale reaction mixtures (order of multi-milligrams and more), ion-exchange liquid column chromatography (IE LC) is preferred. DEAE-Cellulose, QAE-Sephadex, or DEAE-Sephadex A-25 has been utilized as a stationary phase, but recently the latter is definitely preferred because of the best separation results. The preferred mobile phase consists of increasing gradient of triethylammonium bicarbonate buffer in water (TEAB, pH ~7.4), but the use of aqueous ammonium bicarbonate was also reported. However, ion-exchange chromatography is often insufficient for full separation and further purification is needed. In such case, preparative reverse phase HPLC (RP HPLC) approach can be an effective isolation method. Columns with C-18 as a stationary phase and various gradients of methanol or acetonitrile in ammonium acetate or triethylammonium acetate water buffers as eluants are commonly used. Generally, eluants for preparative chromatography need to be volatile and can be removed with decomposition under vacuum to yield appropriate nucleotide (triethyl)ammonium salt. It is very important to have an efficient rotary evaporator system (vacuum < 10 hPa, cooling medium ~0 °C) at disposal for concentration of the collected IE-LC fractions, especially if the sample includes base-labile 7-methylguanosine (or similar) moiety; otherwise partial cleavage of the imidazole ring may occur. Successive additions of ethanol to the sample also significantly facilitate evaporation and decomposition of volatile buffer.

Detection of fractions (at $\lambda = 250$–280 nm) as well as UV absorption spectra can be performed on easily accessible UV–Vis spectrometers. Figure 5 in Sect. 2.2 shows IE-LC elution profiles for a two-step synthesis of m^{7}GpppG (Niedzwiecka et al. 2007) as an example of efficient separation of the products.

The structure identity and purity of a cap analog can be checked routinely by means of HPLC, mass spectrometry, ^{1}H and ^{31}P NMR, UV-absorption, and fluorescence spectroscopy. Sometimes, analytical TLC in several solvent systems and other special techniques has also to be employed.

Analytical HPLC can be performed on RP and/or IE columns. As for the mobile phase, the best results are achieved with increasing gradient of methanol or acetonitrile in potassium phosphate aqueous buffers of various pHs and concentrations depending on the method used and nature of analyzed sample.

Often, structures of cap analogs are confirmed by ^{1}H NMR spectra analysis. Moreover, the recent enormous development of various NMR techniques allows study conformation and interaction with other chemical objects on the level comparable with X-ray spectroscopy. ^{31}P NMR has a special value for determination of cap analog structures. It offers a direct determination of a number and structure of phosphorus moieties in a given compound.

The majority of cap analogs as the final product can be converted into sodium salts, and in general, they can be safely stored at −20 to −80 °C for months or even years.

3 Synthetic Cap Analogs in Search for Cap-Regulated Cellular Processes

Synthetic cap analogs played a significant role in research that led to understanding of several cellular processes which engage the 5′ terminus of mRNA or snRNA. Availability of these analogs and their use in numerous studies greatly contributed toward comprehension of the regulatory role of the cap structure. In this section of the chapter, we briefly review the most evident of their applications. Due to space limitation it was impossible to cite all published articles or to list all synthesized cap analogs. We refer therefore the reader to more specialized reviews in this field that were recently published. Figure 7 displays the key directions of biological and medical research in which the synthetic cap analogs have been, or might be in future, employed.

3.1 Cap Analogs in mRNA Translation and Translation Inhibition Studies (Early Studies)

Protein synthesis of MMG-capped mRNAs was the first cellular process shown to be strongly dependent on the presence of the cap structure. Historically, simple cap analogs, such as m^{7}GMP, m^{7}GDP, m^{7}GTP, and m^{7}GpppA, shown to inhibit protein translation or ribosome binding to the capped mRNAs (Hickey et al. 1976, 1977; Canaani et al. 1976; Weber et al. 1976), were the main tools used to elucidate the role of the cap during translation. In the mid-1970s, our knowledge on molecular mechanism of translation initiation involving mRNA cap was in its infancy. It was then believed that cap analogs compete with the cap on mRNA for the very recently discovered cap-binding protein (CBP) suspected to be engaged in protein

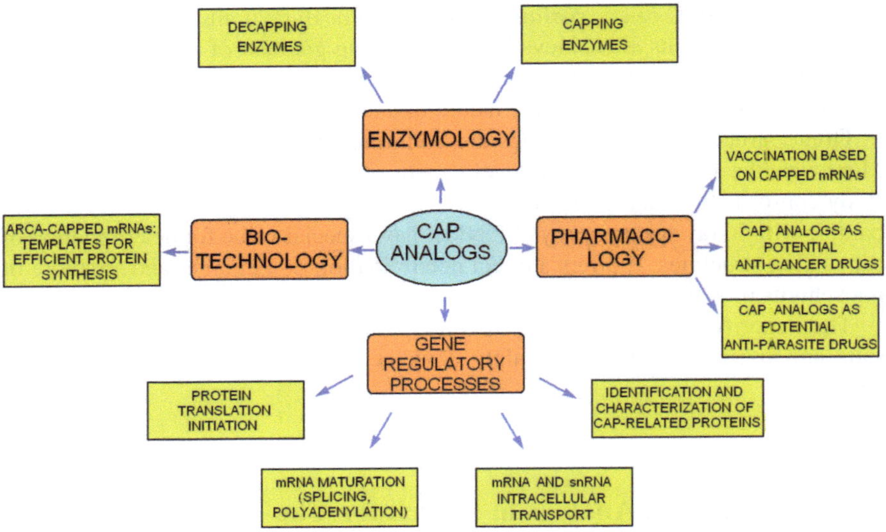

Fig. 7 Biological and medical applications of RNA cap analogs

translation along with the cap. Although at that time the structure of CBP was unknown, several labs, including our, became interested in identifying which structural elements of the MMG cap are fundamental for cap recognition by the protein, since it was suspected that the formation of cap-CBP complex was an essential in early step in the protein synthesis. Using guanosine 5′ nucleotides with different alkyl and aryl modifications, it was shown that the presence of N-7 substituent in the guanine moiety, introducing positive charge in to the imidazole ring, was absolutely necessary, however, with relatively broad tolerance towards its nature; e.g., 7-benzylated guanosine nucleotides were as active as their 7-methylated counterparts (Adams et al. 1978). Modifications in the guanine moiety made by Adams et al. (1978) and Darzynkiewicz et al. (1987), leading to the introduction of a methyl substituent on N-1, C-8, or NH_2 group in m^7GMP or m^7GDP, proved the importance of maintaining unsubstituted N-1 position and the absolute necessity of the presence of at least one free hydrogen atom in the amino group. On the other hand, 2′ and 3′ methylations or replacing 2′-OH with 2′-H (2′ deoxy analog) in the ribose ring of 7-methylguanosine have either minor or no effect on the inhibitory activity. However, the disruption (opening) of the ribose ring leading to 2′,3′-*seco* (diol) or *acyclo* m^7GMP derivatives significantly abolished the inhibitory activity of 7-methylguanosine 5′-monophosphates modified in such a manner (Darzynkiewicz et al. 1985, 1987). The in vitro translational studies with aberrantly capped synthetic mRNA transcripts obtained in SP6 or T7 RNA polymerase reaction confirmed the data from translation inhibition: RNAs containing cap structure elements restoring inhibitory activity in parent cap analogs were also active as translators, and vice versa, those inactive as inhibitors delivered translationally inert mRNA transcripts (Darzynkiewicz et al. 1988, 1989). As to the

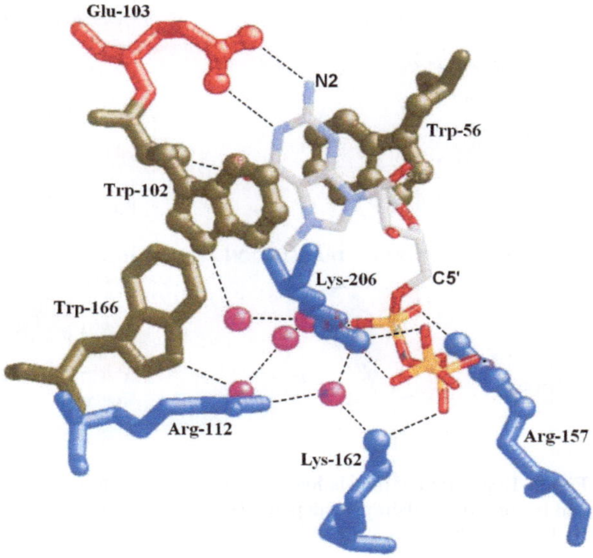

Fig. 8 Interatomic contacts in m^{7}GpppG–eIF4E complex. Hydrogen bonds, salt bridges, and van der Waals interactions are indicated with *dotted lines*. Trapped water molecules supporting hydrogen-bonded network are shown as *magenta spheres*. Taken from Niedzwiecka et al. (2002)

phosphate moiety in 7-methylated guanosine nucleotides, studies with 7-methylguanosine mono-, di-, and triphosphates and their alpha, beta, and gamma methyl esters showed a strong dependence of the inhibitory activity on the number of negatively charged phosphate groups, with m^{7}GTP being the strongest and m^{7}GMP being the weakest inhibitor (Darzynkiewicz et al. 1981, 1989). In later studies, Cai et al. (1999) investigated 58 cap analogs for their ability to inhibit in vitro protein synthesis. Interestingly, there is a broad tolerance toward the kind of the second nucleoside (N) within a series of m^{7}GpppN dinucleotide cap analogs where N was guanosine (G), 2′-*O*-methylguanosine (G^m), 2′-deoxyguanosine (dG), adenosine (A), N^6-methyladenosine (m^{6}A), cytidine (C), and uridine (U). The use of algorithm developed in this work enabled the authors to attain dissociation constants (K_I) for the complexes of cap analogs with cap-binding protein CBP that was renamed in 1990s as the eukaryotic initiation factor 4E (eIF4E).

To sum up, long before the structure of cap-eIF4E complex was resolved by NMR (Matsuo et al. 1997) and crystallography (Marcotrigiano et al. 1997; Niedzwiecka et al. 2002), the early works performed on the in vitro translational system with taking an advantage of the use of synthetic cap analogs obtained valuable information on the structural requirements of the MMG cap for its translational activity and, hence, (although indirectly) for its recognition by eIF4E protein. Figure 8 shows the crystallographic structure of m^{7}GpppG–eIF4E complex (Niedzwiecka et al. 2002).

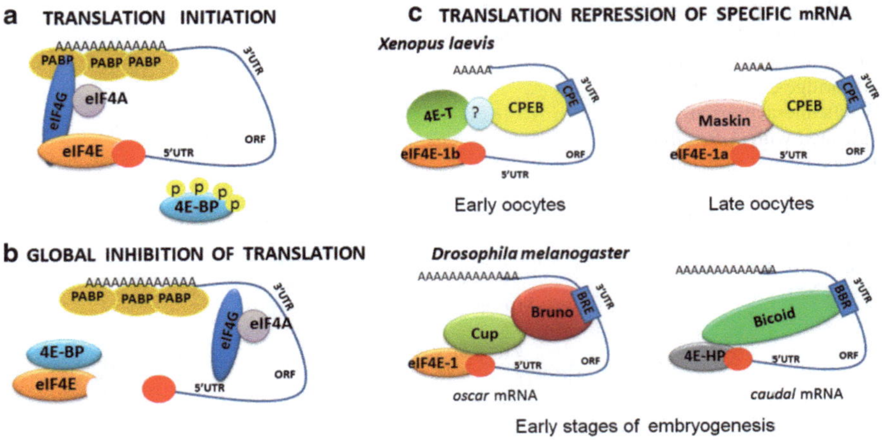

Fig. 9 Regulation of translation initiation and translation repression, involving eIF4Es (canonical and isoforms) and their protein partners (eIF4G, 4E-BPs, other binding proteins). Figure provided by Joanna Zuberek

3.2 *Cap Analogs in Search for Cap Affinities Toward Canonical eIF4Es and Their Isoforms*

eIF4E canonical form plays a fundamental role in assembling translation initiation complex and therefore it regulates the cap-dependent initiation of protein synthesis. In the complex with MMG-cap of mRNA with the large scaffold protein eIF4G, which binds to few other initiation proteins within eIF4F complex, it participates in recruitment of mRNA to small ribosomal subunits (Gingras and Raught 1999; von der Haar et al. 2004; Sonenberg and Hinnebusch 2009), see Fig. 9.

Assembly of the cap-binding complex can be regulated by the eIF4E-binding proteins (4E-BPs), which interfere with interaction between eIF4E and eIF4G by competing for a shared binding surface on eIF4E (Fig. 9). Due to rapidly growing availability of genome sequences, hundreds of proteins from different species, homologous to the human canonical translational factor eIF4E, have been discovered over the last two decades. Evolutionarily, it seems that a single early eIF4E gene underwent a series of gene duplications, and at present, eIF4E and its relatives comprise a family of structurally related proteins within a given organism (Joshi et al. 2004). To distinguish canonical eIF4E from its relatives, it has been renamed as eIF4E-1 or eIF4E-1a. With the exception of some eIF4Es from protozoa, all eIF4Es can be grouped into one of three structural classes (Joshi et al. 2004). Class I members (where the canonical eIF4E belongs) carry Trp residues equivalent to Trp 43 and Trp 46 of *H. sapiens* eIF4E-1a and appear to be present in all eukaryotes. Class II members have Trp 43 substituted by Tyr/Phe/Leu and Trp 56 by Tyr/Phe. Class III members have a Trp residue equivalent to Trp 43 but carry Cys/Tyr substitution relative to Trp56.

The present knowledge on the physiological role of eIF4E family members is rather modest. It is suggested that expression of specific eIF4Es may occur at different developmental stages of an organism, and acts as a repressor by affecting the recruitment of a subset of mRNAs to eIF4F complex as it is observed also in case of eIF4E (Fig. 9). Such role is well known for Class II isoform (4E-HP) in *Drosophila melanogaster*. During the early stage of embryogenesis when the embryo's axis is formed, the translation of mRNA encoding two morphogens *caudal* and *hunchback* is repressed in 3′UTR-specific manner with participation of 4E-HP in a certain part of the embryo. In case of *caudal* mRNA repression, Bicoid protein present at the anterior part of embryo binds simultaneously to the 3′UTR of caudal mRNA and to 4E-HP, which interacts with cap structure. Similarly, 4E-HP binds to a protein complex containing Nanos protein at the 3′UTR of *hunchback* mRNA. On the other hand, similar mechanism of recruitment of Class I eIF4E-1b isoform is observed during a process aimed at silencing the translation of maternal mRNA in the early stage of oogenesis in *X. laevis* (Fig. 9). In this case, 4E-T is a binding partner for eIF4E-1b (Minshall et al. 2007).

The discovery of multiple eIF4E family members within a single organism, which was rather unexpected, has opened a new path of investigation on the physiological roles of these different family members. In line with the above, it is of importance to know the cap recognition specificities of all eIF4E isoforms from different species, beginning with the lower ones (protozoa, yeast) up to the highest (mammalian).

Niedzwiecka et al. (2002, 2007) developed an unambiguous time-synchronized fluorescence titration method for the characterization of cap-binding interactions and measured the association constants between recombinant murine canonical eIF4E with 20 MMG- or TMG-cap analogs, including mono- and dinucleotides, differing in number of phosphate groups and in substituent (alkyl *vs.* aryl) on N-7 in guanine base (Table 7). The obtained quantitative data showed a high specificity of this canonical mammalian protein exclusively towards MMG-cap analogs with a broad tolerance towards the nature of the first transcribed nucleotide. Generally, the association constants between synthetic cap analogs and eIF4Es obtained by means of fluorescence titration, together with the binding properties of these proteins to the cap-affinity resins (see Sect. 3.7), allowed to discriminate the canonical and isoformic eIF4E proteins, within a certain organism as well between different organisms. It has to be pointed out that sometimes relatively minor differences in cap-binding affinities may be responsible for their different biological function. Certain findings from the author's laboratory, relevant to the above, are presented as follows:

(i) Phosphorylation of canonical mammalian eIF4E attenuates its interaction with a broad series of mono- and dinucleotide MMG-cap analogs, differing in number of phosphate groups, by electrostatic repulsion (Zuberek et al. 2003, 2004). The obtained results point to the significance of electrostatic forces for the processes of formation, stabilization, and dissociation of the eIF4E-cap complex.

Table 7 Apparent K_I values[a] for inhibition of in vitro protein translation by representative cap analogs and equilibrium association constants (K_{as})[b] for murine eIF4E(26-217)-cap analog complexes

Compound	Apparent K_I (μM)[a]	$K_{as} \times 10^{-6}$ (M^{-1})[b]
m^7GMP	382	0.806
bn^7GMP	113	–
$m_2^{2,7}$GMP	151	–
m^7GDP	7.5	20.4
m^7GTP	4.39	108.7
bn^7GTP	8.42	17.6
$m_2^{2,7}$GTP	4.31	29.2
$m_3^{2,2,7}$GTP	75.8	0.143
m^7GppG	128	–
m^7Gpp(m^7G)	371	–
GpppG	246	0.005
m^7GpppG	17.1	7.4
m^7Gppp(m^7G)	31.0	1.9
m^7GpppG^m	20.8	8.0
m^7GpppA	31.5	4.7
$m_2^{2,7}$GpppG	14.7	–
GppppG	181	–
m^7GppppG	3.62	102.8
m^7Gpppp(m^7G)	2.67	23.5
$m_2^{2,7}$GppppG	2.15	–

[a]According to Cai et al. (1999)
[b]According to Niedzwiecka et al. (2002)

(ii) The human 4EHP (h4EHP) isoform binds cap analogs: m^7GpppG and m^7GTP, with respectively 30-fold and 100-fold lower affinity than the canonical eIF4E; hence, it cannot compete with eIF4E for binding to the cap structure of some mRNAs to inhibit translation (Zuberek et al. 2007). However, the translation inhibition may take place after binding of h4EHP to yet unidentified protein(s), analogously as it was shown for translational inhibition of *caudal* mRNA from *Drosophila melanogaster* due to association of d4EHP with Bicoid protein (Cho et al. 2005).

(iii) The use of synthetic cap-4, its analogs, as well as MMG- and TMG-cap analogs allowed determination of cap-binding specificities of four eIF4E isoforms (LeishIF4E-1 to LeishIF4E-4) found in the parasite *Leishmania* genome database (Lewdorowicz et al. 2004; Yoffe et al. 2004, 2006, 2009). Based on the obtained data, LeishIF4E-1 was hypothesized as translation initiation factor. The search for the role of eIF4E orthologs in trypanosomatides is still in its infancy; however, their current biophysical characterization makes it possible to distinguish them from their host (human, animal) counterparts. The interaction between specific ligands, synthetic caps, and cap-binding proteins could serve as a potential target for drug development.

(iv) Fluorescence titration of five eIF4E isoforms from nematode *Caenorhabditis elegans* with MMG- and TMG-cap analogs made it possible to distinguish the binding affinities of the proteins toward both natural caps present in mRNAs of these parasites. The experiments have shown two isoforms primarily specific for MMG-cap, whereas the remaining three interact with both caps, which was in agreement with their behavior on cap-affinity columns (see Sect. 3.7) (Jankowska-Anyszka et al. 1998; Miyoshi et al. 2002; Keiper et al. 2000; Stachelska et al. 2002). In another nematode, human parasite *Schistosoma mansoni*, only a simple isoform eIF4E present in this metazoa binds to both MMG- and TMG-cap with approximately the same strength (Liu et al. 2009). The large difference in affinity for TMG cap between *Schistosoma* and mammalian eIF4E (Niedzwiecka et al. 2002; Yoffe et al. 2006) suggests that this protein may represent a potential target for rational drug design against the parasite. More recent biophysical studies employing NMR spectroscopy, isothermal titration calorimetry, and fluorescence titration on *Ascaris suum* eIF4E-3 binding with synthetic MMG- and TMG-cap analogs, including their capped 22-oligonucleotide SL (Splice Leaders) counterparts, delivered insight on conformational changes of the protein upon binding with both types of cap analogs. The findings suggested a key interaction between the *trans*-spliced nematode 5′-UTR and eIF4E that leads to efficient translation of mRNAs in nematodes (Liu et al. 2011; Piecyk et al. 2012).

3.3 Alternatively Capped mRNAs with High Translational Efficiency

3.3.1 Anti-Reverse Cap Analogs

Synthetic capped mRNAs, obtained by in vitro transcription in the presence of all four 5′ triphosphate nucleotides, with excess of m^7Gp_3G natural cap dinucleotide, cDNA template, and bacterial (Contreras et al. 1982) or bacteriophage (Konarska et al. 1984; Yisraeli and Melton 1989) RNA polymerase, proved to be useful tools to study such cellular processes as mRNA translation, pre-mRNA splicing, mRNA turnover, and intracellular transport of mRNAs and snRNAs. However, due to the presence of free 3′-OH groups on both 7-methylguanosine and guanosine moieties of the MMG-cap structure, up to one-half of mRNAs acquire a cap incorporated in the unwanted reverse, Gp_3Gm^7, orientation (Pasquinelli et al. 1995), which results in reduction of translational efficiency of such mRNAs. We (Stepinski et al. 2001) and others (Peng et al. 2002) solved this problem by substituting the 3′-OH of 7-methylguanosine in a conventional m^7Gp_3G cap dinucleotide by an *O*-methyl group or a hydrogen atom, thus obtaining $m_2{}^{7,3'\text{-}O}Gp_3G$ or $m^{7,3'\text{-deoxy}}Gp_3G$, respectively. These cap dinucleotides were incorporated into RNA transcripts exclusively in the correct orientation and therefore we named them "anti-reverse cap analogs" (ARCAs). The idea of ARCA was supported by our earlier observation (Darzynkiewicz et al. 1985) that the modification of a simple cap analog, m^7GMP, by 3′- or 2′-O-methylation ($m_2{}^{7,3'\text{-}O}GMP$ and $m_2{}^{7,2'\text{-}O}GMP$), or by replacing 2′-OH with a hydrogen atom ($m^{7,2'\text{-deoxy}}GMP$), practically does not decrease their translational inhibitory properties as compared to the parent m^7GMP. This suggested that such modifications in ARCA-capped mRNAs might be tolerated by the translational machinery (recognition by eIF4E cap-binding protein). And in fact, $m_2^{7,3'-O}Gp_3G$-capped mRNAs showed twofold higher translational efficiency in vitro as compared to RNAs capped with the conventional m^7Gp_3G (Stepinski et al. 2001). Therefore, this analog, known as "ARCA," was patented (Darzynkiewicz et al. 2006) and found a wide biotechnological application in obtaining ARCA-capped mRNA transcripts as templates for preparative production of proteins. It is commercially available, both separately and within transcription and translation kits.

We have also synthesized ARCAs with tetra- and pentaphosphate bridges, which, after incorporation into the mRNA, lead to even higher translational efficiency (Jemielity et al. 2003b). In agreement with our expectations (see above), analogs with 2′-*O*-methyl or 2′-deoxy modifications also behave as ARCAs (Jemielity et al. 2003b). It has been shown that mRNAs capped with various ARCAs are also translated 1.2- to 2.5-fold more efficiently in cultured mammalian cells than mRNAs capped with m^7Gp_3G (Grudzien et al. 2006). Similar biological evaluations were obtained using other 2′-substituted dinucleotide cap analogs, including 2′-deoxy-2′-fluoro- (Kore et al. 2007), 2′,3′-*O*-isopropylidene- (Kore et al. 2008b), and 2′-*O*-allyl-7-methylguanosine moieties (Kore and Charles 2010b).

Analogs with a general formula of $m^7Gp_nm^7G$ ($n = 3$ or 4), the so-called two-headed caps, may be a special solution for the problem of reverse incorporation of the dinucleotide cap into mRNA. Due to their symmetrical structure, these analogs form the same transcripts irrespective of incorporation direction but with 7-methylguanosine at the first transcribed nucleoside. Moreover, such in vitro prepared mRNAs have high translational efficiency. In the study by Grudzien et al. (2004), $m^7Gp_3m^7G$-RNA and $m^7Gp_4m^7G$-RNA were translated more effectively by a factor of 2.6- and 3.1-fold, respectively, than the standard m^7Gp_3G-RNA. Some advantages of two-headed analogs in biological laboratory tests were also noted by Kore et al. (Kore et al. 2007; Kore and Shanmugasundaram 2008).

3.3.2 ARCAs with Modified Phosphate Bridge

Also, series of cap dinucleotides, ARCAs and non-ARCAs, containing bridging and non-bridging phosphate(s) modifications, were synthesized. In order to improve the stability of mRNA transcripts against the decapping enzymes Dcp1/Dcp2, Nudt16, and DcpS, we replaced the bridging oxygen atoms by methylene groups (Grudzien et al. 2006; Kalek et al. 2006). In eukaryotic cells, two major pathways of degradation of capped and polyadenylated mRNAs exist (Liu et al. 2008; Song et al. 2013), both initiated by shortening the 3′ poly(A) tail (Fig. 10).

In the 5′→3′ pathway, right after poly(A) track shortening, decapping enzyme Dcp1/Dcp2 or/and Nudt16 hydrolyzes the cap structure between α and β phosphates, releasing m^7GDP and then the remaining 5′-phosphate-terminated mRNA is

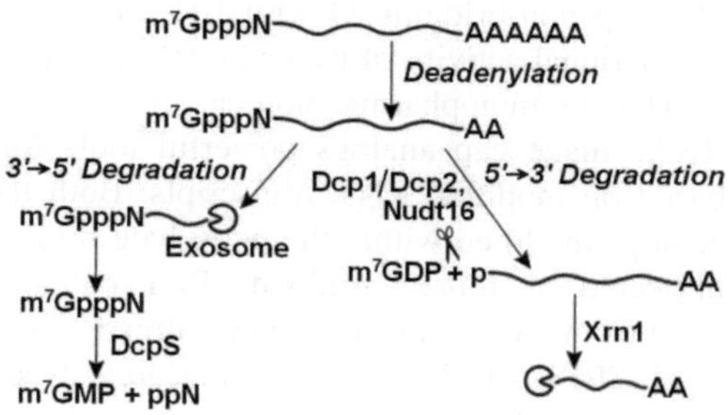

Fig. 10 General eukaryotic mRNA decay pathways

hydrolyzed by exonuclease Xrn1. In 3′→5′ pathway, deadenylated mRNA is degraded processively by the exosome, whereas the obtained short capped oligonucleotides are decapped by a scavenger decapping enzyme DcpS, with release of m^7GMP as a result of cleavage between β and γ phosphates. In order to protect mRNA against 5′→3′ degradation, we introduced a methylene bridging group between α and β phosphate moieties. To avoid the reverse incorporation into the RNA transcript, the 7-methylguanosine moiety was additionally modified by 3′-*O*-methylation, leading to $m_2^{7,3'O}GppCH_2pG$. Due to its selective stabilization of mRNA, this type of ARCA was used to investigate the relative contribution of the 5′→3′ *versus* 3′→5′ pathways of mRNA degradation in mammalian cells, suggesting the predomination of the first pathway (Grudzien et al. 2006). The same type of ARCA served as a useful tool to construct a capped Dcp1/Dcp2-resistant oligonucleotide that was subsequently used in enzyme–ligand binding studies by means of NMR spectroscopy and kinetic analysis. It was shown that the yeast Dcp2 subunit of the enzyme resolves interactions with the cap and RNA body using a bipartite surface forming a channel intersecting the catalytic and regulatory Dcp1-binding domains (Deshmukh et al. 2008).

Another series of cap dinucleotides synthesized in our laboratory during the last decade include both ARCAs and non-ARCAs, with bridging and non-bridging modifications within the phosphate chain (Tables 3 and 4). Among many others, after transcriptional incorporation into mRNAs, non-bridging ARCAs with a thiophosphate instead of phosphate group in β position produced transcripts more resistant to cleavage by Dcp1/Dcp2 and 4–5-fold translationally more active in cultured cells than their counterparts capped with natural m^7GpppG (Kowalska et al. 2008b; Grudzien-Nogalska et al. 2013). Importantly, mRNAs modified with D1 diastereomer of β-S-ARCA ($m_2^{7,2'-O}Gpp_SpG$ D1) have superior properties not only in cultured cells but also in vivo in animals (Kuhn et al. 2010). Injection of luciferase mRNA capped with β-S-ARCA (D1) into lymph nodes of mice, where the RNA transcript is taken up by immature dendritic cells, results in much higher production of luciferase than injection with ARCA-capped mRNA (Kuhn et al. 2010). Moreover, the in vivo delivery of the antigen as β-S-ARCA (D1)-capped mRNA species is highly efficient in protein expression and in priming and expansion of naive antigen-specific T cells in mice (Kuhn et al. 2010). Substitution of the β non-bridging O with BH_3 also enhances significantly the stability and translational activity of capped mRNAs (Su et al. 2011).

The immunopharmacological properties of recombinant antigen-encoding RNAs make cap analogs powerful tools for production of anticancer vaccines based on modified RNA transcripts. Both the synthesis of series of ARCA cap analogs modified within the phosphate bridge, and their medical applications, are protected by three worldwide PCT patent applications. The first two patents (US) from these series have been already issued (Jemielity et al. 2012b; Kowalska et al. 2013). In 2012 the pharmaceutical company BionTech AG (Mainz, Germany), which obtained the license for the production of cap modified mRNAs, initiated the first phase of clinical trials of the new-generation anticancer mRNA-based vaccine. The mRNA capped with D1 diastereomer of ARCA, containing phosphothioate group instead of phosphate group in β position ($m_2^{7,2'-O}Gpp_SpG$ D1), already passed the preclinical tests and is currently being in clinical trials assessed for treatment of malignant melanoma.

We refer the reader to the reviews by Jemielity et al. (2010) and Ziemniak et al. (2013a) on the synthesis and applications of several series of cap dinucleotides with different modifications within the phosphate bridge.

3.4 *Synthetic Cap Analogs in RNA Maturation (Splicing) and RNA Intracellular Transport*

3.4.1 The Role of MMG-Cap and Cap-Binding Complex in Pre-mRNA Splicing and Nuclear Export of Polymerase II-Transcribed RNAs

In the biosynthesis of mRNAs and some of uridine-rich small nuclear RNAs (U snRNAs), addition of a monomethylated guanosine cap (m^7GpppN; MMG-cap) to their 5′ ends is one of the early events. Synthesis of MMG-capped RNAs takes place in nucleus and is carried on by RNA polymerase II (reviewed in Topisirovic et al. 2011). As early as in 1984, Konarska et al. and then Inoue et al. (1989) showed that the MMG-cap is necessary for efficient pre-mRNA splicing. Later studies in Mattaj's lab employed a broad class of synthetic cap analogs to investigate the involvement of the cap structure and a specific nuclear cap-binding protein in pre-mRNA splicing as well as in nuclear export of mature mRNAs and U snRNAs. Originally, Hamm and Mattaj (1990b) showed that m^7GpppG proved to be an effective competitive inhibitor for export of polymerase II-transcribed mRNAs from *Xenopus* oocyte nucleus. The use of a set of modified synthetic cap analogs, including GpppG, m^7GpppG, $m_2^{2,7}$GpppG, $m_3^{2,2,7}$GpppG, and et^7GpppG, to analyze their ability to inhibit export of U1 snRNA from *Xenopus* oocytes nucleus, enabled the researchers to define recognition specificity of a protein factor involved in RNA export (originally referred to as a cap-binding activity, CBA), different from the cytoplasmic translation initiation factor, eIF4E

(Izaurralde et al. 1992). In subsequent studies the same series of compounds was used to investigate the role of this cap-binding activity (later renamed and now known as a cap-binding complex, CBC) from HeLa nuclear extract in pre-mRNA splicing (Izaurralde et al. 1994). Analyses performed within the scope of these studies on in vitro and in vivo inhibition of pre-mRNA splicing by chemically modified cap analogs and on in vitro binding of these analogs to CBC suggested engagement of the CBC in pre-mRNA splicing. In further investigations, with considerable involvement of synthetic cap analogs, the role of cap-binding complex in m^7G-capped RNA export and in pre-mRNA splicing was investigated in detail (e.g., Jarmolowski et al. 1994; Izaurralde et al. 1995; Lewis et al. 1996).

CBC consists of two proteins: CBP20 that interacts with the cap and CBP80 that facilitates this binding (Izaurralde et al. 1994). The three-dimensional structure of human CBC has been solved, both for the *apo*-protein (Mazza et al. 2001; Calero et al. 2002) and for the m^7GpppG-bound protein (Calero et al. 2002; Mazza et al. 2002). Worch et al. (2005b) adapted the fluorescence titration spectroscopy method (Niedzwiecka et al. 2002) for characterization of the human CBC cap-binding specificity by measuring association constants for 17 synthetic mono-, di-, and tetranucleotide cap analogs. The data from fluorescence studies showed that human CBC and cytoplasmic eIF4E exhibit different recognition specificity for the cap, pointing, e.g., to much higher affinity of CBC to dinucleotides than to mononucleotides with the same number of phosphate groups, which is in evident contrast to eIF4E cap affinity (see Sect. 3.2).

3.4.2 Involvement of TMG-Cap and Snurportin1 in snRNA Nuclear Import

In 1986, Mattaj showed that U2 snRNA synthesized in nucleus and terminated with a monomethylguanosine cap (MMG-cap) during its transient stay in cytoplasm undergoes hypermethylation to its trimethylguanosine-capped counterpart. Subsequently, it forms a complex with a set of proteins and returns to the nucleus as a ribonucleoprotein U2 snRNP; there, it participates in pre-mRNA splicing. The synthetic trimethylated cap analogs proved to be extremely helpful in showing the role of TMG-cap as a component of a bipartite nuclear targeting signal (Fischer and Lührmann 1990; Hamm et al. 1990). Obtained synthetically with the use of T7 RNA polymerase, U snRNAs with a variety of cap structures were micro-injected into the cytoplasm of *Xenopus* oocytes; only $m_3^{2,2,7}$G-capped RNAs migrated to the nucleus therein (Hamm et al. 1990; Fischer et al. 1991). Moreover, only TMG-capped analogs were able to inhibit this import competitively. Further studies (e.g., Fischer et al. 1993; Marshallsay and Lührmann 1994) provided an insight into some cytosolic protein factors mediating TMG-cap directed transport. The use of TMG-capped oligonucleotide ($m_3^{2,2,7}$GpppA^{m}pU^{m}pA) and a biotinylated affinity complex based on it [$m_3^{2,2,7}$GpppA^{m}pU^{m}pAp-$(CH_2)_6$-biotin] in Lührmann's lab (Huber et al. 1998) enabled to identify and isolate a 45 kDa protein, called Snurportin1, which recognizes specifically TMG-cap but not MMG-cap and facilitates the nuclear import of the U snRNPs by binding to importin-β. The crystal structure of the protein complexed with $m_3^{2,2,7}$GpppG and fluorescence titration spectroscopy studies with $m_3^{2,2,7}$GpppG, $m_3^{2,2,7}$GpppA, and $m_3^{2,2,7}$GpppA^{m}pU^{m}pA revealed a distinct mode of cap-Snurportin1 binding and specificity from that for eIF4E or CBC, pointing out to the significance of the presence of additional nucleotides with 2′*O*-methylated ribose moieties (Strasser et al. 2005).

3.5 *Cap Analogs in mRNA Metabolism*

3.5.1 Capping Enzymes

Several capping enzymes, including guanylyltransferases, triphosphatases, and methyltransferases, participate in the synthesis of all three natural caps: MMG-cap, TMG-cap, and cap-4 found in mRNAs and/or snRNAs. Below, we present a few examples that illustrate the benefit of using cap analogs, synthesized in our lab, in characterization of a few specific methyltransferases from different species.

(i) Belanger et al. (2010) identified and characterized the methyltransferase 1 (hMTr1) responsible for cap-1 formation in human cells. The authors have shown in vitro that hMTr1 catalyzes the specific methylation on the 2′-*O*-ribose of the first nucleotide within a capped RNA transcript. As a control compound in the search for hMTr1 activity and for identification of the enzymatic product they used the synthetic m^7GpppG^m (G^m = 2′-*O*-methyl guanosine) dinucleotide. Recently Smietanski et al. (2014) solved crystal structure of hMTr1 in its *apo* form as well as in complex with m^7GpppG and with synthetic capped tetranucleotide m^7GpppGAUC.

(ii) The biochemical properties of a recombinant methyltransferase from *Trypanosoma brucei* (TbMTr1) responsible for 2′-*O*-ribose methylation of the first transcribed nucleotide in cap-4 were investigated in the Campbell's laboratory (Mittra et al. 2008). In these studies several synthetic cap dinucleotides, e.g., m^7GpppG, m^7GpppA, m^7Gppp(m^7G), and $m_3^{2,2,7}$GpppG, were used for enzymatic competition assays. Interestingly, TbMTr1 was shown to exhibit some similarities to human cap-1 methyltransferase (hMTr1) activity.

(iii) In search for three exemplary Tgs enzymes, guanine-N2 methyltransferases (from mammalian, protozoan, and viral sources), responsible for the synthesis of 2,2,7-trimethylguanosine (TMG) RNA cap, a broad series of synthetic mono- and dinucleotide cap analogs with a different number of phosphate groups (n = 1–4) and different number of guanosine methylation (7-methyl-, N^2,7-dimethyl-, and N^2,N^2,7-trimethylguanosine) have been employed by Benarroch et al.(2010). The authors observed that hTgs1 utilized all tested 7-methylated guanosine monucleotides, including mono-, di-, tri-, and tetraphosphates, as well as cap tri- and tetraphosphate dinucleotides, converting them to their trimethylguanosine counterparts. However, it was unable to methylate non 7-methylated analogs. In contrast, the protozoan *Giardia lamblia* Tgs (GlaTgs2) was shown to exhibit different guanine-N2 methyltransferase specificity as it makes only a 2,7-dimethylguanosine (DMG) cap from 7-methylguanosine cap and is unable to convert it further into the TMG cap. Mimivirus Tgs (MimiTgs) catalyzed only one step of a single methyl addition at guanine-N2, but could be distinguished by its ability to methylate guanine-N2 in the absence of earlier N7 methylation. It has to be pointed out that the availability of over a dozen of strictly designed cap analogs enabled the researchers to provide the presented above substrate specificity of all those three enzymes.

3.5.2 Decapping Enzymes

In Sect. 3.3.2 we introduced two pathways of mRNA degradation in eukaryotic cells, pointing out to the significance of ARCA cap dinucleotides, modified in phosphate bridge, resistant to hydrolysis by Dcp1/Dcp2 and Nudt16 decapping enzymes, for the synthesis of stable and translationally highly active RNA transcripts. We focus here mainly on the DcpS, a scavenger enzyme, participating in the 3′→5′ pathway of mRNA degradation and responsible for hydrolysis of short capped oligonucleotides or capped dinucleotides with release of m^7GMP. Synthetic cap analogs also proved to be valuable tools for characterization of some of the enzymes.

(i) A series of MMG- and TMG-dinucleotide triphosphates was used in Davis lab to monitor their degradation by scavenger DcpS from nematode *Ascaris* embryos (Cohen et al. 2004). In contrast to human DcpS, which practically did not hydrolyze TMG-caps, the nematode enzyme exhibited a comparable activity towards both types of caps. The kinetic data obtained by means of fluorescence and HPLC methods, with the use of a broad class of cap analogs, allowed determination of structural requirements for the nematode *Caenorhabditis elegans* DcpS (Wypijewska et al. 2010). Interestingly, TMG-cap dinucleotides were hydrolyzed with higher rates, but their MMG counterparts exhibited greater specificity towards the enzyme, which was reflected by lower K_M values. Modification of the first transcribed nucleotide did not affect the activity. The data obtained in this study suggest a flexibility in the first transcribed nucleoside-binding pocket in *C. elegans* DcpS.

(ii) In another paper (Wypijewska et al. 2012), it was shown that DcpS recombinant enzymes from human, yeast, and nematode (*C. elegans*, *A. suum*) were completely inert towards m^7GDP and $m_3{}^{2,2,7}$GDP, which was in a discord with some earlier published data. As the enzymes from all three sources exhibited high affinity for m^7GDP, we postulated that m^7GDP—a natural product of Dcp1/Dcp2 mRNA degradation pathway—may function as a DcpS inhibitor. Its concurrent high affinity towards eIF4E translation initiation factor sets it in a role of a potential regulatory molecule in mRNA metabolic pathways and also in regulation of mRNA translation.

(iii) A collection of 23 synthetic cap analogs, including 17 cap dinucleotides with modifications in the phosphate bridge (different number of phosphate groups and bridging or non-bridging modifications: methylenebisphosphonates, imidodiphosphates, phoshorothioates), was investigated for their inhibitory properties for DcpS from *C. elegans* (Wypijewska del Nogal et al. 2013). Structure-affinity relationship studies revealed molecular determinants for efficient cap binding: a strong dependence on the type of substituents in the phosphate bridge and weak binding affinity for extended triphosphate chain. The comparison of CeDcpS with human DcpS (hDcpS) provided a general insight into the molecular mechanisms of DcpS-cap interactions.

(iv) The availability of the synthetic cap-4 ($m^7Gpppm_3{}^{6,6,2'}Apm^{2'}Apm^{2'}Cpm^{3,2'}U$) concurrently with its shorter counterparts and other synthetic cap analogs was fundamental for characterization of a new member of the FHIT protein family, a HIT-45 protein from *Trypanosoma brucei* (Banerjee et al. 2009). It hydrolyzes cap dinucleotides, yielding, similarly like scavenger decapping enzymes, terminal nucleoside 5′-monophosphate (e.g., m^7GMP from m^7GpppG) with the highest specificity for diadenosine triphosphate (ApppA) and for the shortest version of cap-4, hipermethylated dinucleotide $m^7Gpppm_3{}^{6,6,2'}Apm^{2'}$. Interestingly, HIT-45 is practically inert towards any compounds longer than dinucleotide cap analogs.

3.6 *Affinity Resins*

Cap analogs covalently bound to affinity resins constitute a special group of biochemical tools which are used for purification of proteins that recognize the 5′ end of mRNA specifically. Sonenberg et al. (1979) coupled Sepharose to a mononucleotide cap analog by 2′,3′-*O*-ethylidene bond obtaining a resin with anchored 2′,3′-*O*-[1-(2-carboxyethyl)ethylidene]-7-methylguanosine 5′-diphosphate residues (Panel **A**). Rupprecht et al. (1981) placed the connection to the resin at position 7 of the guanosine moiety, preparing Sepharose modified with 7-(5-carboxypentyl) guanosine 5′-diphosphate residues (Panel **B**). Webb et al. (1984) anchored cap analog at the phosphate, thus synthesizing Sepharose modified with *p*-aminophenyl γ-ester of 7-methylguanosine 5′-triphosphate residues (Panel **C**). The latter approach was then used to prepare affinity resin with N^2,7-di-, and N^2,N^2,7-trimethylguanosine 5′-triphosphate residues (Jankowska et al. 1993a, Panel **D**) and with 7-methylguanosine 5′-(P^1,P^2-methylenebisphosphonate)-phosphate (Szczepaniak et al. 2008, 2012, Panel **E**).

As a continuation of the synthesis of cap-based affinity resins, new products have recently been reported. In these resins, the whole dinucleotide cap moieties are used as anchors; this potentially permits tighter binding of some cap-binding proteins to the affinity medium (Jankowska-Anyszka and Piecyk 2011, Panel **F**;

Szczepaniak et al. 2012, Panel **G**). The Panels **A–G** illustrate the cap-including structures of the known Sepharose-based affinity resins and briefly characterize their applications:

A

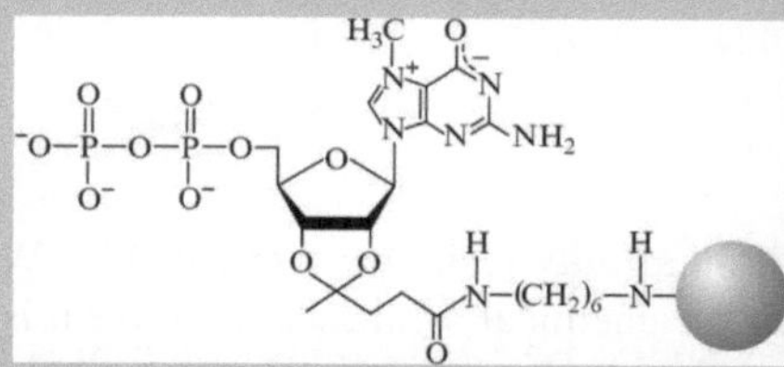

The above structure was the first example of an affinity resin, which was used for purification of CBP (renamed later as eIF4E) from rabbit reticulocytes by one chromatographic passage. The affinity resin was prepared by coupling of levulinic acid $O^{2',3'}$-acetal of m^7GDP to AH-Sepharose 4B with the use of water-soluble carbodiimide (Sonenberg et al. 1979). However, this resin showed a relatively low capacity for CBP retention.

B

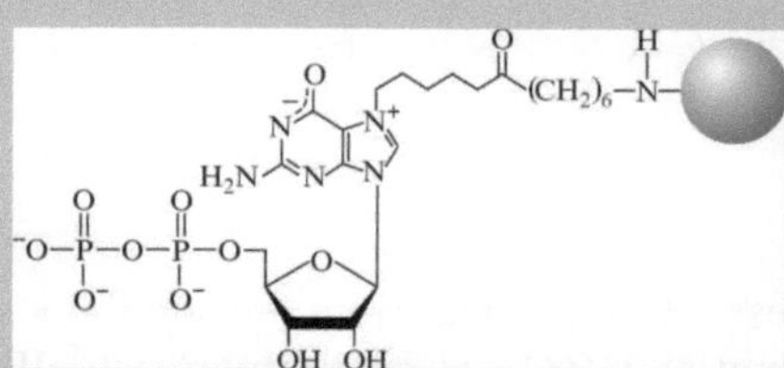

In addition to the previously published affinity resin (**A**), the same group (Rupprecht et al. 1981) by means of similar chemical coupling methods synthesized a new resin, which retained the protein essentially fully attached during the purification of CBP (eIF4E).

(continued)

(continued)

C

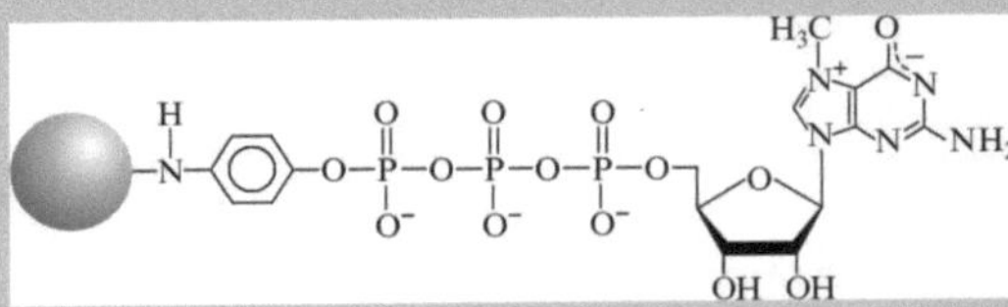

This new kind of an affinity resin was presented by Webb et al. in 1984. Authors showed that the chemical synthesis of the resin is much simpler and that CBP (eIF4E) can be obtained from crude ribosomal wash without prior purification. The initial stage of coupling the m^7GTP ligand to Sepharose included activation of the resin by cyanogen bromide (BrCN). This affinity resin is commercially available.

D

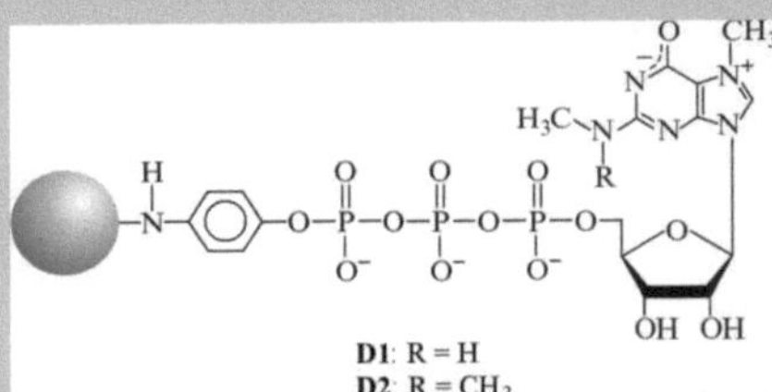

These affinity resins (**D1** and **D2**) were prepared in a similar way as was described by Webb et al. (1984). Human erythrocyte eIF4E can be isolated with higher purity on **D1** ($m_2^{2,7}$GTP-Sepharose) resin than on the commercially available m^7GTP-Sepharose (Jankowska et al. 1993a).

Both resins were used for isolation and characterization of all five eIF4E isoforms from *C. elegans*, among which only IFE-1, IFE-2, and IFE-5 can bind TMG-cap (Jankowska-Anyszka et al. 1998; Keiper et al. 2000; see Sect. 3.2).

(continued)

(continued)

E

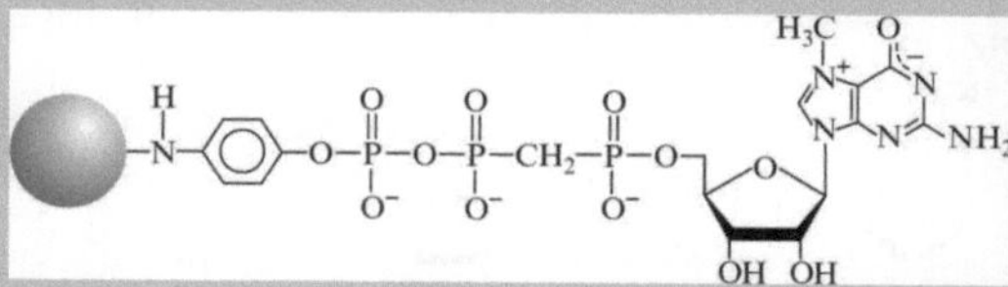

The above structure shows a new affinity resin containing a mononucleotide cap analog resistant to hydrolysis by DcpS. The resin has been designed to identify *Arabidopsis thaliana* DcpS decapping enzyme, but it may be also adopted for purification of other pyrophosphatases (Szczepaniak et al. 2008, 2012).

F

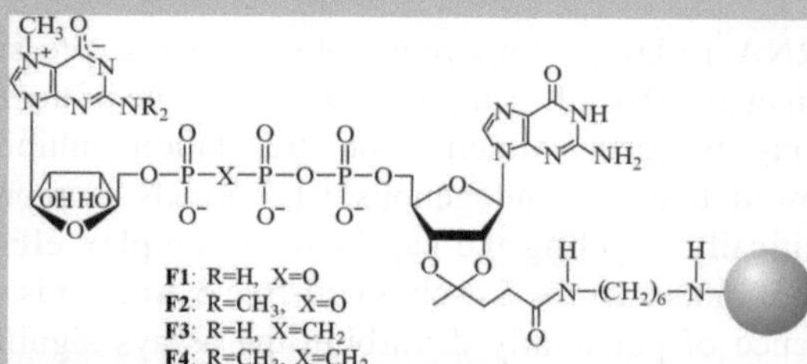

The structures show the first cap-dinucleotide affinity resins. The resins have either a naturally occurring mono- or trimethylated cap or their analogs resistant against enzymatic degradation, bearing CH_2 bridge between β and γ position of the 5′,5′-triphosphate chain (Jankowska-Anyszka and Piecyk 2011). The resins could serve as powerful tools for isolation of already known cap-binding proteins and those to be still identified. However, some limitations may occur for those proteins that need a free 2′-hydroxyl group of the second nucleoside to properly recognize and bind the 5′-end of mRNA.

(continued)

(continued)

G

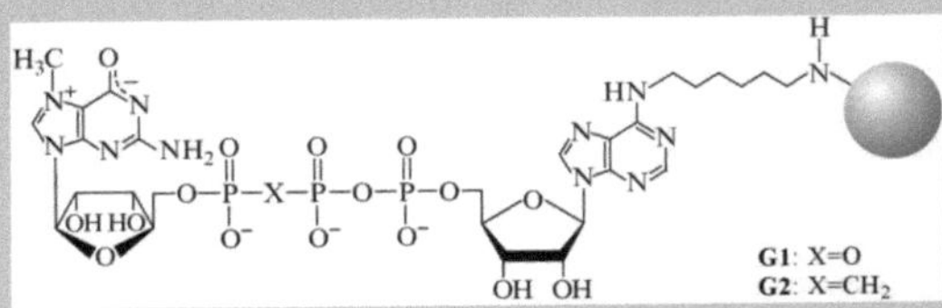

Both matrices (**G1** and **G2**) can be used for isolation and identification of cap-binding proteins from different organisms (yeast, mouse, human). Resin **G2** (m^7GpCH_2ppA-Sepharose), due to the resistance of the attached ligand against DcpS enzyme, might be of especial use for purification and characterization of scavenger-type enzymes (Szczepaniak et al. 2012).

3.7 *Translation Inhibition by miRNAs*

MicroRNAs (miRNAs) play an important role in gene expression regulation in eukaryotes (Fabian et al. 2010). During the last decade one of the most investigated miRNA's functions is their influence on translation inhibition. Mathonnet et al. (2007) showed that the endogenous let-7 miRNAs represses translation initiation by specifically targeting the cap-binding complex eIF4F in an in vitro translation system of mouse Krebs-2 ascites cell-free extract. It is worth to note that to avoid the presence of potentially disturbing the assays significant amounts of transcripts capped incorrectly (reversly), the synthetic mRNAs employed for testing their translational ability upon treatment with miRNAs were capped with ARCA ($m_2^{3'O,7}$GpppG), instead of using conventional m^7GpppG cap dinucleotide (see Sect. 3.3.1).

Kiriakidou et al. (2007), Boland et al. (2010), and Djuranovic et al. (2010) suggested that translation repression by miRNAs is due to the direct interaction of 5′-mRNA cap with Argonaute (AGO) protein from miRISC complex, thereby competing with cap recognition by translation initiation complex eIF4F. However, the crystallographic and NMR data showed that synthetic cap analogs (m^7GpppG, m^7GpppA, m^7GTP) do not bind, either significantly or specifically, to the MID domain of human Argonaute 2 (hAGO2), which was supported by the in vitro pull-down experiments with full-length hAGO2 (Frank et al. 2011). Although the authors proved unequivocally nonexistence of direct cap-AGO interaction, they did not discount the possibility that there might be other cellular factors in vivo allowing to accommodate capped mRNA by the miRISC complex.

Another approach to search the mechanism of translation repression by miRNAs with the involvement of cap structure was undertaken in Henzte's lab by using developed in his lab the cell-free system from *Drosophila* embryos (Zdanowicz et al. 2009). By screening a broad library (27 compounds all together) of synthetic dinucleotide cap analogs, mainly ARCA type, modified additionally in phosphate bridge, the authors managed to identify the analogs that augment miRNA (miR2)-mediated translation initiation inhibition, being "neutral" toward general cap-dependent translation. Interestingly, these analogs augmented also translation inhibition by 4E-BP, a specific protein factor, disrupting eIF4E interaction with eIF4G within the translation initiation complex (Fig. 9). Presented data clearly demonstrate a primary role of the m^{7}GpppN cap structure in miRNA-mediated translational inhibition, supporting also previous observations by Mathonnet et al. (2007). Moreover, the findings implicate structural determinants outside the core eIF4E-binding region in this process and suggest that miRNAs may target cap-dependent translation with similar mode to that otherwise performed by 4E-BP factors in the regulation of translation initiation/inhibition.

All the above three examples of utility of the synthetic cap analogs in studies related to RNA interference demonstrate their potential in expanding this relatively new area of research so important in field of biology and medicine.

4 Synthetic Cap Analogs: Relevance to Medicine

Engagement of cap structures in several pivotal cell regulatory processes provokes to investigate their synthetic analogs and cap-related proteins as potential drugs, or drug targets, in medicine. Below we preset briefly two, most promising in our opinion, potential applications. We refer to an exhaustive recent review dealing with this topic (Ziemniak et al. 2013a).

Inhibitors Targeting eIF4E in Cancer Therapy The cellular mRNAs can be subdivided into two categories: (1) the strong mRNAs, with short unstructuralized 5′ untranslating region (5′UTR), with low sensitivity to the presence of eIF4E, and (2) the week mRNAs, characteristic for malignancy-related (oncogenic) mRNAs, with long, highly structuralized 5′UTRs, much more susceptible to the presence of eIF4E and translated preferably in the presence of abundance of eIF4E. In many human cancers the eukaryotic translation initiation factor 4E is frequently overexpressed. At present three approaches are being intensively explored which lead to the reduction of either eIF4E expression or its activity (Blagden and Willis 2011; Konicek et al. 2008). (1) The first one is based on targeting the *eIF4E*-mRNA template by specific antisense oligonucleotides (ASOs) (Graff et al. 2007). (2) Search for inhibitors disrupting the eIF4E–eIF4G complex, by competing with eIF4E for eIF4G-binding site, might be another way of silencing translation initiation, providing an attractive potential therapy for some tumors (Graff et al. 2008; Moerke et al. 2007). (3) The third option would be disrupting the eIF4E–mRNA cap complex by using the synthetic cap analogs competing with mRNA cap for the cap-binding site in eIF4E (see e.g., Jia et al. 2010; Li et al. 2013). However, the main difficulty is intracellular delivery of negatively charged cap analogs due to their very poor permeability across the plasma membrane. There are several trials to increase the permeability to cap analogs, for example, by reducing their negative charge *via* replacing the phosphate moiety with phosphoroamidate (Kim et al. 2004) or by encapsulation of cap analogs into magnetic-nanoparticle polypyrrole microvessels (Kijewska et al. 2013). It has to be pointed out, however, that certain approaches listed above (e.g., with the use o ASOs) meet already preclinical or clinical trials with positive perspectives for being advanced towards clinical applications.

Alternatively Capped mRNAs in Gene Therapy: Antitumor Vaccines The most explored approach in gene therapy over the past decades had been focused on methods involving the use of DNA. This approach however carries the risks related to introduction of nonreversible, sometimes dangerous changes into the genome as well as has technical problems in delivering the genetic material into the nucleus across the nuclear envelope. The alternative and in many respects safer and easier way is the use of synthetic mRNA coding for a given therapeutic protein. However, the main limitation in using mRNA is its relatively short lifetime in the cell due to nucleolytic degradation. This has been noticeably overcome by terminating the RNAs with anti-reverse cap dinucleotides (ARCAs) modified in phosphate bridge, thereby protecting them against degradation by decapping enzymes and significantly enhancing their translational efficiency (see Sect. 3.3.2).

It has been observed that the efficiency of translation of mRNAs capped with the same cap analog may differ when used in different cell lines (Kuhn et al. 2010; Grudzien-Nogalska et al. 2007; Su et al. 2011). Extensive efforts therefore are being carried on to develop alternatively capped mRNAs designed for particular antigens, specifically, to maximize their translational efficiencies in dendritic cells in hope to obtain highly active vaccines for particular cancers.

5 Future Directions

In this review we attempted to illustrate the impact of the progress in synthesis of cap analogs on our knowledge pertaining to cellular events in which the interactions of different cell constituents with cap play a role in the regulation of translation itself and the associated processes. The experience gained during over 30 years of our engagement in "capology" leads us to recognize that we are still in the middle of the road and we may expect new discoveries and surprises as it happened in the past. For instance, the selective cap-affinity resins, including magnetic beads, may help to identify and characterize the new cap-related proteins such as capping and decapping enzymes or cap-binding regulatory factors. This refers especially to the proteins involved in the metabolism and functions of still poorly explored hypermethylated cap-4 and TMG-cap. Fluorescently tagged cap analogs open new possibilities to investigate the cap-protein molecular interactions, monitor capped mRNA cellular pathways, etc. Development of new generations of pharmaceuticals based on cap analogs is another exciting area to be widely explored. Such pharmaceuticals, designed to regulate the rate of translation of particular proteins, could be of special value in treatment of different diseases. Likewise, modulation of the rate of translation *via* altered cap analogs can be used to efficiently in vitro generate specific proteins for clinical or industrial application.

Acknowledgements This work was supported by grants from the Ministry of Science and Higher Education (Poland) N 301 096 339, National Science Centre (Poland) UMO-2012/07/B/NZ1/00118 and UMO-2013/08/A/NZ1/00866, and National Center of Research and Development (02/EuroNanoMed/2011).

References

Adam A, Moffat JG (1966) Dismutation reactions of nucleoside polyphosphates. V. Syntheses of P1, P4-di(guanosine-5′) tetraphosphate and P1, P3-di(guanosine-5′) triphosphate. J Am Chem Soc 88:838–842

Adams BL, Morgan M, Muthukrishnan S et al (1978) The effect of "cap" analogs on reovirus mRNA binding to wheat germ ribosomes. J Biol Chem 253:2589–2595

Banerjee H, Palenchar JB, Lukaszewicz M et al (2009) Identification of the HIT-45 protein from *Trypanosoma brucei* as an FHIT protein/dinucleoside triphosphatase: Substrate specificity studies on the recombinant and endogenous proteins. RNA 15:1554–1564

Belanger F, Stepinski J, Darzynkiewicz E et al (2010) Characterization of hMTr1, a human Cap1 2′-O-ribose methyltransferase. J Biol Chem 285:33037–33044

Benarroch D, Jankowska-Anyszka M, Stepinski J et al (2010) Cap analog reveal three clades of cap guanine-N2 methyltransferases with distinct methyl acceptor specificities. RNA 16:211–220

Blackburn GM, Guo M, McLennan AG (1992) Synthetic structural analogues of dinucleoside polyphosphates. In: McLennan AG (ed) Ap4A and other dinuleoside polyphosphates. CRC, Boca Raton, FL, pp 305–342

Blagden SP, Willis AE (2011) The biological and therapeutic relevance of mRNA translation in cancer. Nat Rev Clin Oncol 8:280–291

Boland A, Tritschler F, Heimstaedt S et al (2010) Crystal structure and ligand binding of the MID domain of a eukaryotic Argonaute protein. EMBO Rep 11:522–527

Brownlee GG, Fodor E, Pritlove DC et al (1995) Solid phase synthesis of 5′-diphosphorylated oligoribonucleotides and their conversion to capped m^7Gppp-oligoribonucleotides for use as primers for influenza A virus RNA polymerase *in vitro*. Nucleic Acids Res 23:2641–2647

Burgess K, Cook D (2000) Syntheses of nucleoside triphosphates. Chem Rev 100:2047–2059

Cai A, Jankowska-Anyszka M, Centers A et al (1999) Quantitative assessment of mRNA cap analogues as inhibitors of in vitro translation. Biochemistry 38:8538–8547

Calero G, Wilson K, Ly T et al (2002) Structural basis of m^7GpppG binding to the nuclear cap-binding complex. Nat Struct Biol 9:912–917

Canaani D, Revel M, Groner Y (1976) Translational discrimination of "capped" and "non-capped" mRNAs: Inhibition by a series of chemical analogs of m^7GpppX. FEBS Lett 64:326–331

Chavan AJ, Rychlik W, Blaas D et al (1990) Phenyl azide substituted and benzophenone-substituted phosphonamides of 7-methylguanosine 5′-triphosphate as photoaffinity probes for protein-synthesis initiation factor eIF-4E and a proteolytic fragment containing the cap-binding site. Biochemistry 29:5521–5529

Chlebicka L, Wieczorek Z, Stolarski R et al (1995) Synthesis and properties of mRNA 5′-cap analogues with 7-methylguanine replaced by benzimidazole or 3-methylbenzimidazole. Nucleosides Nucleotides 14:771–775

Cho PF, Poulin F, Cho-Park YA et al (2005) A new paradigm for translational control: inhibition via 5′-3′ mRNA tethering by Bicoid and the eIF4E cognate 4EHP. Cell 121:411–423

Cohen LS, Mikhli C, Friedman C et al (2004) Nematode m^7GpppG – and $m_3{}^{2,2,7}$GpppG – RNA decapping: Activities in *Ascaris* embryos and characterization of *C. elegans* scavenger DcpS. RNA 10:1609–1624

Contreras R, Cheroutre H, Degrave W et al (1982) Simple, efficient in vitro synthesis of capped RNA useful for direct expression of cloned eukaryotic genes. Nucleic Acids Res 10:6353–6362

Cramer F, Schaller H, Staab HA (1961) Zur Chemie der "Energiereichen Phosphate" XI. Darstellung von Imidazoliden der Phosphorsäure. Chem Ber 94:1612–1621

Darzynkiewicz E, Antosiewicz J, Ekiel I et al (1981) Methyl esterification of m^7Gp reversibly blocks its activity as an analog of eukaryotic mRNA 5′-caps. J Mol Biol 153:451–453

Darzynkiewicz E, Ekiel I, Tahara SM et al (1985) Chemical synthesis and characterization of 7-methylguanosine cap analogues. Biochemistry 24:1701–1707

Darzynkiewicz E, Ekiel I, Lassota P et al (1987) Inhibition of eukaryotic translation by analogues of messenger RNA 5′-cap: chemical and biological consequences of 5′-phosphate modifications of 7-methylguanosine 5′-monophosphate. Biochemistry 26:4372–4380

Darzynkiewicz E, Stepinski J, Ekiel I et al (1988) ß-Globin mRNAs capped with m^7G, $m_2{}^{2,7}$G or $m_3{}^{2,2,7}$G differ in intrinsic translation efficiency. Nucleic Acids Res 16:8953–8962

Darzynkiewicz E, Stepinski J, Ekiel I et al (1989) Inhibition of eukaryotic translation by nucleoside 5′-monophosphate analogues of mRNA 5′-cap: Changes in N7 substituent affect analogue activity. Biochemistry 28:4771–4778

Darzynkiewicz E, Stepinski J, Tahara SM et al (1990) Synthesis, conformation and hydrolytic stability of P1, P3-dinucleoside triphosphates related to mRNA 5′-cap, and comparative kinetic studies on their nucleoside and nucleoside monophosphate analogs. Nucleosides Nucleotides 9:599–618

Darzynkiewicz E, Rhoads RE, Stepinski J (2006) Synthesis and use of anti-reverse mRNA cap analogues. US Patent 7,074,596, 11 July 2006

Deshmukh MV, Jones BN, Quang-Dang D-U et al (2008) mRNA decapping is promoted by an RNA binding channel in Dcp2. Mol Cell 29:324–336

Djuranovic S, Zinchenko MK, Hur JK et al (2010) Allosteric regulation of Argonaute proteins by miRNAs. Nat Struct Mol Biol 17:144–150

Eckstein F (1970) Nucleoside phosphorothioates. J Am Chem Soc 92:4718–4723

Engel R (1977) Phosphonates as analogs of natural phosphates. Chem Rev 77:349–367

Fabian MR, Sonenberg N, Filipowicz W (2010) Regulation of mRNA translation and stability by microRNAs. Annu Rev Biochem 79:351–379

Fischer U, Lührmann R (1990) An essential signaling role for the m_3G cap in the transport of U1 snRNP to the nucleus. Science 249:786–790

Fischer U, Darzynkiewicz E, Tahara SM et al (1991) Diversity in the signals required for nuclear accumulation of U snRNPs and variety in the pathways of nuclear transport. J Cell Biol 113:705–714

Fischer U, Sumpter V, Sekine M et al (1993) Nucleocytoplasmic transport of u snRNPs – definition of a nuclear location signal in the Sm core domain that binds a transport receptor independently of the m_3G cap. EMBO J 12:573–583

Frank F, Fabian MR, Stepinski J et al (2011) Structural analysis of 5′-mRNA-cap interactions with the human AGO2 MID domain. EMBO Rep 12:415–420

Fukuoka K, Suda F, Suzuki R et al (1994a) One-pot reaction for the synthesis of $m^7G^{5'}$pppG and $m^7G^{5'}$pppA by using an activatable bifunctional phoshorylating reagent. Tetrahedron Lett 35:1063–1066

Fukuoka K, Suda F, Suzuki R et al (1994b) Large scale synthesis of the cap part in messenger RNA using new type of bifunctional phoshorylating reagent. Nucleosides Nucleotides 13:1557–1567

Furuichi Y, Shatkin AJ (2000) Viral and cellular mRNA capping: Past and prospects. Adv Virus Res 55:135–184

Gingras AC, Raught B (1999) eIF4E initiation factors: Effectors of mRNA recruitment to ribosomes and regulators of translation. Annu Rev Biochem 68:913–963

Glass RS, Singh WP, Jung W et al (1993) Monoselenophosphate: Synthesis, characterization, and identity with the prokaryotic biological selenium donor, compound SePX. Biochemistry 32:12555–12559

Graff JR, Konicek BW, Vincent TM et al (2007) Therapeutic suppression of translation initiation factor eIF4E expression reduces tumor growth without toxicity. J Clin Invest 117:2638–2648

Graff JR, Konicek BW, Carter JH et al (2008) Targeting the eukaryotic translation initiation factor 4E for cancer therapy. Cancer Res 68:631–634

Grudzien E, Stepinski J, Jankowska-Anyszka M et al (2004) Novel cap analogs for in vitro synthesis of mRNAs with high translational efficiency. RNA 10:1479–1487

Grudzien E, Kalek M, Jemielity J et al (2006) Differential inhibition of mRNA degradation pathways by novel cap analogs. J Biol Chem 281:1857–1867

Grudzien-Nogalska E, Stepinski J, Jemielity J et al (2007) Synthesis of anti-reverse cap analogs (ARCAs) and their application in protein translation and stability. Methods Enzymol 431:203–227

Grudzien-Nogalska E, Kowalska J, Su W et al (2013) Synthetic mRNAs with superior translation and stability properties. Methods Mol Biol 969:55–72

Guranowski A, Wojdyla AM, Zimny J et al (2010a) Recognition of different nucleotidyl-derivatives as substrates of reactions catalyzed by various HIT-proteins. New J Chem 34:888–893

Guranowski A, Wojdyla AM, Zimny J et al (2010b) Dual activity of certain HIT-proteins: A. thaliana Hint4 and C-elegans DcpS act on adenosine 5′-phosphosulfate as hydrolases (forming AMP) and as phosphorylases (forming ADP). FEBS Lett 584:93–98

Hamm J, Mattaj IW (1990) Monomethylated cap structures facilitate RNA export from the nucleus. Cell 63:109–118

Hamm J, Darzynkiewicz E, Tahara SM et al (1990) The trimethylguanosine cap structure of U1 snRNA Is a component of a bipartite nuclear targeting signal. Cell 62:569–577

Han QW, Gaffney BL, Jones RA (2006) One-flask synthesis of dinucleoside tetra- and pentaphosphates. Org Lett 8:2075–2077

Hata T, Nakagawa I, Shimotohno K et al (1976) The synthesis of α, γ-dinucleoside triphosphates. The confronted nucleotide structure found at the 5′-terminus of eukaryote messenger ribonucleic acid. Chem Lett 1976:987–990

He KH, Hasan A, Krzyzanowska B et al (1998) Synthesis and separation of diastereomers of ribonucleoside 5′-(α-P-borano)triphosphates. J Org Chem 63:5769–5773

Hendler SS, Fürer E, Srinivasan PR (1970) Synthesis and chemical properties of monomers and polymers containing 7-methylguanine and an investigation of their substrate or template properties for bacterial deoxyribonucleic acid or ribonucleic acid polymerases. Biochemistry 9:4141–4153

Hickey ED, Weber LA, Baglioni C (1976) Inhibition of initiation of protein synthesis by 7-methylguanosine-5′-monophosphate. Proc Natl Acad Sci U S A 73:19–23

Hickey ED, Weber LA, Baglioni C et al (1977) Relation between inhibition of protein synthesis and conformation of 5′-phosphorylated 7-methylguanosine derivatives. J Mol Biol 109:173–183

Hoard DE, Ott DG (1965) Conversion of mono- and oligodeoxyribonucleotides to 5′-triphosphates. J Am Chem Soc 87:1785–1788

Hodel AE, Gershon PD, Quiocho FA (1998) Structural basis for sequence-nonspecific recognition of 5′-capped mRNA by a cap-modifying enzyme. Mol Cell 1:443–447

Honcharenko M, Romanowska J, Alvira M et al (2012) Capping of oligonucleotides with "clickable" m_3G-CAPs. RSC Adv 2:12949–12962
Honcharenko M, Zytek M, Bestas B et al (2013) Synthesis and evaluation of stability of m_3G-CAP analogues in serum-supplemented medium and cytosolic extract. Bioorg Med Chem 21 (24):7921–7928, http://dx.doi.org/10.1016/j.bmc.2013.10.002
Huber J, Cronshagen U, Kadokura M et al (1998) Snurportin1, an m_3G-cap-specific nuclear import receptor with a novel domain structure. EMBO J 17:4114–4126
Imai KI, Fujii S, Takanohashi K et al (1969) Studies on phosphorylation. 4. Selective phosphorylation of primary hydroxyl group in nucleosides. J Org Chem 34:1547–1550
Inoue K, Ohno M, Sakamoto H et al (1989) Effect of the cap structure on pre-messenger-RNA splicing in Xenopus oocyte nuclei. Genes Dev 3:1472–1479
Iwase R, Sekine M, Hata T et al (1988) A new method for the synthesis of capped oligoribonucleotides by use of an appropriately protected 7-methylguanosine diphosphate derivative as a donor for the triphosphate bond formation. Tetrahedron Lett 29:2969–2972
Iwase R, Sekine M, Tokumoto Y et al (1989) Synthesis of N2, N2,7-trimethylguanosine cap derivatives. Nucleic Acids Res 17:8979–8989
Iwase R, Maeda M, Fujiwara T et al (1992) Molecular design of eukaryotic messenger RNA and its chemical synthesis. Nucleic Acids Res 20:1643–1648
Izaurralde E, Stepinski J, Darzynkiewicz E et al (1992) A cap binding protein that may mediate nuclear export of RNA polymerase II-transcribed RNAs. J Cell Biol 118:1287–1295
Izaurralde E, Lewis J, McGuigan C et al (1994) A nuclear cap binding protein complex involved in pre-mRNA splicing. Cell 78:657–668
Izaurralde E, Lewis J, Gamberi C et al (1995) A cap-binding protein complex mediating U snRNA export. Nature 376:709–712
Jankowska M, Temeriusz A, Stolarski R et al (1993a) Synthesis of $m^{2,7}$GTP- and $m^{2,2,7}$GTP-Sepharose 4B: New affinity resins for isolation of cap binding proteins. Collect Czech Chem Commun 58(Special issue):132–137
Jankowska M, Stepinski J, Stolarski R et al (1993b) Synthesis and properties of new NH_2 and N7 substituted GMP and GTP 5′-mRNA cap analogues. Collect Czech Chem Commun 58(Special issue):138–141
Jankowska M, Stepinski J, Stolarski R et al (1996) ^{1}H NMR and fluorescence studies of new mRNA 5′-cap analogues. Collect Czech Chem Commun 61(Special issue):S197–S202
Jankowska-Anyszka M, Piecyk K (2011) Dinucleotide cap analogue affinity resins for purification of proteins that specifically recognize the 5′ end of mRNA. Bioorg Med Chem Lett 21:6131–6134
Jankowska-Anyszka M, Lamphear BJ, Aamodt EJ et al (1998) Multiple isoforms of eukaryotic protein synthesis initiation factor 4E in *Caenorhabditis elegans* can distinguish between mono- and trimethylated mRNA cap structures. J Biol Chem 273:10538–10542
Jankowska-Anyszka M, Piecyk K, Samonina-Kosicka J (2011) Synthesis of a new class of ribose functionalized dinucleotide cap analogues for biophysical studies on interaction of cap-binding proteins with the 5′ end of mRNA. Org Biomol Chem 9:5564–5572
Jarmolowski A, Boelens WC, Izaurralde E et al (1994) Nuclear export of different classes of RNA is mediated by specific factors. J Cell Biol 124:627–635
Jemielity J, Stepinski J, Jaremko M et al (2003a) Synthesis of novel mRNA 5′ cap-analogues: Dinucleoside P1, P3-tri- P1, P4-tetra- and P1, P5-pentaphosphates. Nucleosides Nucleotides Nucleic Acids 22:691–694
Jemielity J, Fowler T, Zuberek J et al (2003b) Novel 'anti-reverse' cap analogs with superior translational properties. RNA 9:1108–1122
Jemielity J, Pietrowska-Borek M, Starzynska E et al (2005a) Synthesis and enzymatic characterization of methylene analogs of adenosine 5′-tetraphosphate (p_4A). Nucleosides Nucleotides Nucleic Acids 24:589–593

Jemielity J, Heinonen P, Lönnberg H et al (2005b) A novel approach to solid phase chemical synthesis of oligonucleotide mRNA cap analogs. Nucleosides Nucleotides Nucleic Acids 24:601–605

Jemielity J, Kowalska J, Rydzik AM et al (2010) Synthetic mRNA cap analogs with a modified triphosphate bridge – synthesis, applications and prospects. New J Chem 34:829–844

Jemielity J, Lukaszewicz M, Kowalska J et al (2012a) Synthesis of biotin labelled cap analogue – incorporable into mRNA transcripts and promoting cap-dependent translation. Org Biomol Chem 10:8570–8574

Jemielity J, Grudzien-Nogalska E, Kowalska J et al (2012b) Synthesis and use of anti-reverse phosphorothioate analogs of the messenger RNA cap. US Patent 08,153,773, 10 Apr 2012

Jia Y, Chiu T-L, Amin EA et al (2010) Design, synthesis and evaluation of analogs of initiation factor 4E (eIF4E) cap-binding antagonist Bn-7-GMP. Eur J Med Chem 45:1304–1313

Joshi B, Cameron A, Jagus R (2004) Characterization of mammalian eIF4E-family members. Eur J Biochem 271:2189–2203

Kadokura M, Wada T, Urashima C et al (1997) Efficient synthesis of γ-methyl-capped guanosine 5′-triphosphate as a 5′-terminal unique structure of U6 RNA *via* a new triphosphate bond formation involving activation of methyl phosphorimidazolidate using $ZnCl_2$ as a catalyst in DMF under anhydrous conditions. Tetrahedron Lett 38:8359–8362

Kadokura M, Wada T, Seio K et al (2001) Solid-phase synthesis of 5′-terminal TMG-capped trinucleotide block of U1 snRNA. Tetrahedron Lett 42:8853–8856

Kalek M, Jemielity J, Grudzien E et al (2005a) Synthesis and biochemical properties of novel mRNA 5′ cap analogs resistant to enzymatic hydrolysis. Nucleosides Nucleotides Nucleic Acids 24:615–621

Kalek M, Jemielity J, Stepinski J et al (2005b) A direct method for the synthesis of nucleoside 5′-methylenebis(phosphonate)s from nucleosides. Tetrahedron Lett 46:2417–2421

Kalek M, Jemielity J, Darzynkiewicz ZM et al (2006) Enzymatically stable 5′ mRNA cap analogs: synthesis and binding studies with human DcpS decapping enzyme. Bioorg Med Chem 14:3223–3230

Keiper BD, Lamphear BJ, Deshpande AM et al (2000) Functional characterization of five eIF4E isoforms in Caenorhabditis elegans. J Biol Chem 275:10590–10596

Kijewska K, Jarzebinska A, Kowalska J et al (2013) Magnetic-nanoparticle-decorated polypyrrole microvessels: Toward encapsulation of mRNA cap analogues. Biomacromolecules 14:1867–1876

Kim J, Chou T-F, Griesgraber GW et al (2004) Direct measurement of nucleoside monophosphate delivery from a phosphoramidate pronucleotide by stable isotope labeling and LC-ESI$^-$-MS/MS. Mol Pharm 1:102–111

Kiriakidou M, Tan GS, Lamprinaki S et al (2007) An mRNA m^7G cap binding-like motif within human Ago2 represses translation. Cell 129:1141–1151

Kohno K, Nishiyama S, Kamimura T et al (1985) Chemical synthesis of capped RNA fragments and their ability to complex with eukaryotic ribosomes. Nucleic Acids Res Symp Ser 16:233–236

Konarska MM, Padgett RA, Sharp PA (1984) Recognition of cap structure in splicing in vitro of messenger-RNA precursors. Cell 38:731–736

Konicek BW, Dumstorf CA, Graff JR (2008) Targeting the eIF4E translation initiation complex for cancer therapy. Cell Cycle 7:2466–2471

Kore AR, Charles I (2010a) Synthesis of new dinucleotide mRNA cap analogs containing 2,6-diaminopurine moiety. Lett Org Chem 7:200–202

Kore AR, Charles I (2010b) Synthesis and evaluation of 2′-*O*-allyl substituted dinucleotide cap analog for mRNA translation. Bioorg Med Chem 18:8061–8065

Kore AR, Shanmugasundaram M (2008) Synthesis and biological evaluation of trimethyl-substituted cap analogs. Bioorg Med Chem Lett 18:880–884

Kore AR, Shanmugasundaram M, Charles I et al (2007) Synthesis and application of 2′-fluoro-substituted cap analogs. Bioorg Med Chem Lett 17:5295–5299

Kore AR, Charles I, Shanmugasundaram M et al (2008a) Recent developments in 5′-terminal cap analogs: synthesis and biological ramifications. Mini Rev Org Chem 5:179–192

Kore AR, Shanmugasundaram M, Vlassov AV (2008b) Synthesis and application of a new 2′,3′-isopropylidene guanosine substituted cap analog. Bioorg Med Chem Lett 18:4828–4832

Kore AR, Shanmugasundaram M, Charles I et al (2009) Locked nucleic acid (LNA)-modified dinucleotide mRNA cap analogue: Synthesis, enzymatic incorporation, and utilization. J Am Chem Soc 131:6364–6365

Kore AR, Shanmugasundaram M, Barta TJ (2010a) Synthesis and substrate validation of cap analogs containing 7-deazaguanosine moiety by RNA polymerase. Nucleosides Nucleotides Nucleic Acids 29:821–830

Kore AR, Charles I, Shanmugasundaram M (2010b) Organic synthesis and improved biological properties of modified mRNA cap analogs. Curr Org Chem 14:1083–1098

Koukhareva II, Lebedev AV (2004) Chemical route to the capped RNAs. Nucleosides Nucleotides Nucleic Acids 23:1667–1680

Kowalska J, Lewdorowicz M, Zuberek J et al (2005) Synthesis and properties of mRNA cap analogs containing phosphorothioate moiety in 5′,5′-triphosphate chain. Nucleosides Nucleotides Nucleic Acids 24:595–600

Kowalska J, Lewdorowicz M, Zuberek J et al (2007a) Assignment of the absolute configuration of P-chiral 5′ mRNA cap analogues containing phosphorothioate moiety. Nucleosides Nucleotides Nucleic Acids 26:1301–1305

Kowalska J, Lewdorowicz M, Darzynkiewicz E et al (2007b) A simple and rapid synthesis of nucleotide analogues containing a phosphorothioate moiety at the terminal position of the phosphate chain. Tetrahedron Lett 48:5475–5479

Kowalska J, Zuberek J, Darzynkiewicz ZM et al (2008a) Synthesis and properties of boranophosphate mRNA cap analogues. In: Hocek M (ed) Collection symposium series, vol 10. Institute of Organic Chemistry and Biochemistry, Academy of Sciences of Czech Republic, Prague, pp 383–385

Kowalska J, Lewdorowicz M, Zuberek J et al (2008b) Synthesis and characterization of mRNA cap analogs containing phosphorothioate substitutions that bind tightly to eIF4E and are resistant to the decapping pyrophosphatase DcpS. RNA 14:1119–1131

Kowalska J, Lukaszewicz M, Zuberek J et al (2009) Phosphoroselenoate dinucleotides for modification of mRNA 5′ end. Chembiochem 10:2469–2473

Kowalska J, Osowniak A, Zuberek J et al (2012) Synthesis of nucleoside phosphosulfates. Bioorg Med Chem Lett 22:3661–3664

Kowalska J, Jemielity J, Darzynkiewicz E et al (2013) mRNA cap analogs. US Patent 08,519,110, 27 Aug 2013

Kuhn AN, Diken M, Kreiter S et al (2010) Phosphorothioate cap analogs increase stability and translational efficiency of RNA vaccines in immature dendritic cells and induce superior immune responses in vivo. Gene Ther 17:961–971

Lewdorowicz M, Yoffe Y, Zuberek J et al (2004) Chemical synthesis and binding activity of the trypanosomatid cap-4 structure. RNA 10:1469–1478

Lewdorowicz M, Jemielity J, Kierzek R et al (2007a) Solid-supported synthesis of 5′-mRNA cap-4 from trypanosomatides. Nucleosides Nucleotides Nucleic Acids 26:1329–1333

Lewdorowicz M, Stepinski J, Kierzek R et al (2007b) Synthesis of *Leishmania* cap-4 intermediates, cap-2 and cap-3. Nucleosides Nucleotides Nucleic Acids 26:1339–1348

Lewis J, Izaurralde E, Jarmolowski A et al (1996) A nuclear cap-binding complex facilitates association of U1 snRNP with the cap-proximal 5′ splice site. Genes Dev 10:1683–1698

Li P, Shaw BR (2004) Convenient synthesis of nucleoside borane diphosphate analogues: Deoxy- and ribonucleoside 5′-P$^{\alpha}$-boranodiphosphates. J Org Chem 69:7051–7057

Li P, Xu ZH, Liu HY et al (2005) Synthesis of α-P-modified nucleoside diphosphates with ethylenediamine. J Am Chem Soc 127:16782–16783

Li S, Jia Y, Jacobson B, McCauley J et al (2013) Treatment of breast and lung cancer cells with a N-7 benzyl guanosine monophosphate tryptamine phosphoramidate pronucleotide (4Ei-1) results in chemosensitization to gemcitabine and induced eIF4E proteasomal degradation. Mol Pharm 10:523–531

Liu SW, Jiao X, Welch S et al (2008) Analysis of mRNA decapping. Methods Enzymol 448:3–21

Liu W, Zhao R, McFarland C et al (2009) Structural insights into parasite eIF4E binding specificity for m^7G and $m^{2,2,7}G$ mRNA caps. J Biol Chem 284:31336–31349

Liu W, Jankowska-Anyszka M, Piecyk K et al (2011) Structural basis for nematode eIF4E binding an $m^{2,2,7}G$-cap and its implications for translation initiation. Nucleic Acids Res 39:8820–8832

Lohrmann R, Orgel LE (1978) Preferential formation of (2′-5′)-linked internucleotide bonds in non-enzymatic reactions. Tetrahedron 34:853–855

Ludwig J (1981) A new route to nucleoside 5′-triphosphates. Acta Biochim Biophys Hung 16:131–133

Ludwig J, Eckstein F (1989) Rapid and efficient synthesis of nucleoside 5′-*O*-(1-thiotriphosphates), 5′-triphosphates and 2′,3′-cyclophosphorothioates using 2-chloro-4*H*-1,3,2-benzodioxaphosphorin-4-one. J Org Chem 54:631–635

Ma QF, Bathurst IC, Barr PJ et al (1992) New thymidine triphosphate analog inhibitors of human immunodeficiency virus-1 reverse-transcriptase. J Med Chem 35:1938–1941

Marcotrigiano J, Gingras AC, Sonenberg N et al (1997) Cocrystal structure of the messenger RNA 5′ cap-binding protein (eIF4E) bound to 7-methyl-GDP. Cell 89:951–961

Marshallsay C, Lührmann R (1994) In-vitro nuclear import of snRNPs – cytosolic factors mediate m_3G-cap dependence of U1 and U2 snRNP transport. EMBO J 13:222–231

Mathonnet G, Fabian MR, Svitkin YV et al (2007) MicroRNA function in vitro: inhibition of translational initiation by targeting eIF4F. Science 317:1764–1767

Matsuo H, Li H, McGuire AM et al (1997) Structure of translation factor eIF4E bound to m^7GDP and interaction with 4E-binding protein. Nat Struct Biol 4:717–724

Matsuo H, Moriguchi T, Takagi T et al (2000) Efficient synthesis of ^{13}C,^{15}N-labeled RNA containing the cap structure m^7GpppA. J Am Chem Soc 122:2417–2421

Mattaj IW (1986) Cap trimethylation of U-snRNA is cytoplasmic and dependent on U-snRNP protein-binding. Cell 46:905–911

Mazza C, Ohno M, Segref A et al (2001) Crystal structure of the human nuclear cap binding complex. Mol Cell 8:383–396

Mazza C, Segref A, Mattaj IW et al (2002) Large-scale induced fit recognition of m^7GpppG cap analogue by the human nuclear cap binding complex. EMBO J 21:5548–5557

Mikkola S, Salomaki S, Zhang Z et al (2005) Preparation and properties of mRNA 5′-cap structure. Curr Org Chem 9:999–1022

Minshall N, Reiter MH, Weil D et al (2007) CPEB interacts with an ovary-specific eIF4E and 4E-T in early Xenopus oocytes. J Biol Chem 282:37389–37401

Mittra B, Zamudio JR, Bujnicki JM et al (2008) The TbMTr1 spliced leader RNA cap-1 2′-*O*-ribose methyltransferase from *Trypanosoma brucei* acts with substrate specificity. J Biol Chem 283:3161–3172

Miyoshi H, Dwyer DS, Keiper BD et al (2002) Discrimination between mono- and trimethylated cap structures by two isoforms of *Caenorhabditis elegans* eIF4E. EMBO J 21:4680–4690

Moerke NJ, Aktas H, Chen H et al (2007) Small-molecule inhibition of the interaction between the translation initiation factors eIF4E and eIF4G. Cell 128:257–267

Moran JR, Whitesides GM (1984) A practical enzymatic-synthesis of (S_P)-adenosine 5′-*O*-(1-thiotriphosphate) ((S_P)-ATP-α-S). J Org Chem 49:704–706

Moreno PMD, Wenska M, Lundin KE et al (2009) A synthetic snRNA m_3G-CAP enhances nuclear delivery of exogenous proteins and nucleic acids. Nucleic Acids Res 37:1925–1935

Mukaiyama T, Hashimoto M (1971) Phosphorylation by oxidation-reduction condensation. Preparation of active phosphorylating reagents. Bull Chem Soc Jpn 44:2284

Murray AW, Atkinson MR (1968) Adenosine 5′-phosphorothioate. A nucleotide analog that is a substrate competitive inhibitor or regulator of some enzymes that interact with adenosine 5′-phosphate. Biochemistry 7:4023–4029

Myers TC, Nakamura K, Flesher JW (1963) Phosphonic acid analogs of nucleoside phosphates.1. Synthesis of 5′-adenylyl methylenediphosphonate, a phosphonic acid analog of ATP. J Am Chem Soc 85:3292–3295

Nakagawa I, Konya S, Ohtani S et al (1980) A "capping" agent: P^1-S-phenyl P^2-7-methylguanosine-5′ pyrophosphorothioate. Synthesis 1980:556–557

Natarajan A, Moerke N, Fan YH et al (2004) Synthesis of fluorescein labeled 7-methylguanosinemonophosphate. Bioorg Med Chem Lett 14:2657–2660

Niedzwiecka A, Marcotrigiano J, Stepinski J et al (2002) Biophysical studies of eIF4E cap-binding protein: Recognition of mRNA 5′ cap structure and synthetic fragments of eIF4G and 4E-BP1 proteins. J Mol Biol 319:615–635

Niedzwiecka A, Stepinski J, Antosiewicz JM et al (2007) Biophysical approach to studies of cap-eIF4E interaction by synthetic cap analogues. Methods Enzymol 430:209–246

Ohkubo A, Kondo Y, Suzuki M et al (2013) Chemical synthesis of U1 snRNA derivatives. Org Lett 15:4386–4389

Pasquinelli AE, Dahlberg JE, Lund E (1995) Reverse 5′ caps in RNAs made in vitro by phage RNA polymerases. RNA 1:957–967

Peng ZH, Sharma V, Singleton SF et al (2002) Synthesis and application of a chain-terminating dinucleotide mRNA cap analog. Org Lett 4:161–164

Peyrane F, Selisko B, Decroly E et al (2007) High-yield production of short GpppA- and ^{7Me}GpppA-capped RNAs and HPLC-monitoring of methyltransfer reactions at the guanine-N7 and adenosine-2′O positions. Nucleic Acids Res 35:e26

Piecyk K, Davis RE, Jankowska-Anyszka M (2012) 5′-Terminal chemical capping of spliced leader RNAs. Tetrahedron Lett 53:4843–4847

Ren JH, Goss DJ (1996) Synthesis of a fluorescent 7-methylguanosine analog and a fluorescence spectroscopic study of its reaction with wheatgerm cap binding proteins. Nucleic Acids Res 24:3629–3634

Rupprecht KM, Sonenberg N, Shatkin AJ et al (1981) Design and preparation of affinity columns for the purification of eukaryotic messenger ribonucleic-acid cap binding-protein. Biochemistry 20:6570–6577

Ruth JL, Cheng YC (1981) Nucleoside analogs with clinical potential in antivirus chemotherapy – the effect of several thymidine and 2′-deoxycytidine analog 5′-triphosphates on purified human (alpha, beta) and herpes-simplex virus (type-1, type-2) DNA-polymerases. Mol Pharmacol 20:415–422

Rydzik AM, Lukaszewicz M, Zuberek J et al (2009) Synthetic dinucleotide mRNA cap analogs with tetraphosphate 5′,5′ bridge containing methylenebis(phosphonate) modification. Org Biomol Chem 7:4763–4776

Rydzik AM, Kulis M, Lukaszewicz M et al (2012) Synthesis and properties of mRNA cap analogs containing imidodiphosphate moiety-fairly mimicking natural cap structure, yet resistant to enzymatic hydrolysis. Bioorg Med Chem 20:1699–1710

Sasavage NL, Friderici K, Rottman FM (1979) Specific-inhibition of capped messenger-RNA translation in vitro by $m^7G^{5'}pppp^{5'}G$ and $m^7G^{5'}pppp^{5'}$-m^7G. Nucleic Acids Res 6:3613–3624

Sawai H, Wakai H, Shimazu M (1991) Facile synthesis of cap portion of messenger RNA by Mn (II) ion catalyzed pyrophosphate formation in aqueous solution. Tetrahedron Lett 32:6905–6906

Sawai H, Shimazu M, Wakai H et al (1992) Divalent metal ion-catalyzed pyrophosphate bond formation in aqueous solution. Synthesis of nucleotides containing polyphosphate. Nucleosides Nucleotides 11:773–785

Sawai H, Wakai H, Nakamura-Ozaki A (1999) Synthesis and reactions of nucleoside 5′-diphosphate imidazolide. A nonenzymatic capping agent for 5′-monophosphorylated oligoribonucleotides in aqueous solution. J Org Chem 64:5836–5840

Sekine M, Nishiyama S, Kamimura T et al (1985) Chemical synthesis of capped oligoribonucleotides, $m^7G^{5'}$pppAUG and $m^7g^{5'}$pppAUGACC. Bull Chem Soc Jpn 58:850–860

Sekine M, Iwase R, Hata T et al (1989) Synthesis of capped oligoribonucleotides by use of protected 7-methylguanosine 5′-diphosphate derivatives. J Chem Soc Perkin Trans I 1989:969–978

Sekine M, Kadokura M, Satoh T et al (1996) Chemical synthesis of a 5′-terminal TMG-capped triribonucleotide $m_3^{2,2,7}G^{5'}$pppAmpUmpA of U1 RNA. J Org Chem 61:4412–4422

Setondji J, Remy P, Dirheime G et al (1970) Analogues of nucleoside polyphosphates. 4. synthesis of adenosine 5′-hypophosphate – a structural analogue of ADP. Biochim Biophys Acta 224:136–143

Shimazu M, Shinozuka K, Sawai H (1990) Facile synthesis of nucleotides containing polyphosphates by Mn(II) and Cd(II) ion-catalyzed pyrophosphate bond formation in aqueous solution. Tetrahedron Lett 31:235–238

Smietanski M, Werner M, Purta E et al (2014) Structural analysis of human 2′–O-ribose methyltransferases involved in mRNA cap structure formation. Nat Commun 5:3004. doi:10.1038/ncomms4004

Sonenberg N, Hinnebusch AG (2009) Regulation of translation initiation in eukaryotes: mechanisms and biological targets. Cell 136:731–745

Sonenberg N, Rupprecht KM, Hecht SM et al (1979) Eukaryotic messenger-RNA cap binding-protein – purification by affinity chromatography on Sepharose-coupled m^7GDP. Proc Natl Acad Sci USA 76:4345–4349

Song M-G, Bail S, Kiledjian M (2013) Multiple Nudix family proteins possess mRNA decapping activity. RNA 19:390–399

Sood A, Shaw BR, Spielvogel BF (1990) Boron-containing nucleic-acids. 2. Synthesis of oligodeoxynucleoside boranophosphates. J Am Chem Soc 112:9000–9001

Stachelska A, Wieczorek Z, Ruszczynska K et al (2002) Interaction of three Caenorhabditis elegans isoforms of translation initiation factor eIF4E with mono- and trimethylated mRNA 5′ cap analogues. Acta Biochim Pol 49:671–682

Stepinski J, Grabowska L, Darzynkiewicz E et al (1990) Synthesis, conformation and hydrolytic stability of modified mRNA 5′-cap structures: P1,P3-dinucleoside triphosphates derived from guanosine and acyclic analogues of 7-methyl-, N2,7-dimethyl- and N2,N2,7-trimethylguanosines. Collect Czech Chem Commun 55(Special Issue):117–120

Stepinski J, Bretner M, Jankowska M et al (1995) Synthesis and properties of P1, P2-, P1, P3- and P1, P4-dinucleoside di-, tri- and tetraphosphate mRNA 5′-cap analogues. Nucleosides Nucleotides 14:717–721

Stepinski J, Waddell C, Stolarski R et al (2001) Synthesis and properties of mRNAs containing the novel "anti-reverse" cap analogues 7-methyl-(3′-O-methyl)GpppG and 7-methyl-(3′-deoxy) GpppG. RNA 7:1486–1495

Stepinski J, Jemielity J, Lewdorowicz M et al (2002) Catalytic efficiency of divalent metal salts in dinucleoside 5′,5′-triphosphate bond formation. In: Točik Z, Hocek M (eds) Collection symposium series, vol 5. Institute of Organic Chemistry and Biochemistry, Academy of Sciences of Czech Republic, Prague, pp 154–158

Stepinski J, Zuberek J, Jemielity J et al (2005) Novel dinucleoside 5′,5′-triphosphate cap analogues. Synthesis and affinity for murine translation factor eiF4E. Nucleosides Nucleotides Nucleic Acids 24:629–633

Stepinski J, Wojcik J, Sienkiewicz A et al (2007) Synthesis and NMR spectral properties of spin labelled mRNA 5′ cap analogue, a new tool for biochemical studies of cap binding proteins. J Phys Condens Matter 19:285202 (10 pp)

Stock JA (1979) Synthesis of phosphonate analogs of thymidine diphosphate and triphosphate from 5′-*O*-toluenesulfonylthymidine. J Org Chem 44:3997–4000

Strasser A, Dickmanns A, Lührmann R et al (2005) Structural basis for m_3G-cap-mediated nuclear import of spliceosomal UsnRNPs by snurportin1. EMBO J 24:2235–2243

Strenkowska M, Kowalska J, Lukaszewicz M et al (2010) Towards mRNA with superior translational activity: synthesis and properties of ARCA tetraphosphates with single phosphorothioate modifications. New J Chem 34:993–1007

Strenkowska M, Wanat P, Ziemniak M et al (2012) Preparation of synthetically challenging nucleotides using cyanoethyl P-imidazolides and microwaves. Org Lett 14:4782–4785

Su W, Slepenkov S, Grudzien-Nogalska E et al (2011) Translation, stability, and resistance to decapping of mRNAs containing caps substituted in the triphosphate chain with BH_3, Se, and NH. RNA 17:978–988

Szczepaniak SA, Jemielity J, Zuberek J et al (2008) Bisphosphonate mRNA cap analog attached to Sepharose for affinity chromatography of decapping enzymes. Nucleic Acids Symp Ser 52:295–296

Szczepaniak SA, Zuberek J, Darzynkiewicz E et al (2012) Affinity resins containing enzymatically resistant mRNA cap analogs – a new tool for the analysis of cap-binding proteins. RNA 18:1421–1432

Thillier Y, Decroly E, Morvan F et al (2012) Synthesis of 5′ cap-0 and cap-1 RNAs using solid-phase chemistry coupled with enzymatic methylation by human (guanine-N^7)-methyl transferase. RNA 18:856–868

Tomasz J, Vaghefi MM, Ratsep PC et al (1988) Nucleoside imidodiphosphates synthesis and biological-activities. Nucleic Acids Res 16:8645–8664

Topisirovic I, Svitkin YV, Sonenberg N et al (2011) Cap and cap-binding proteins in the control of gene expression. Wiley Interdiscip Rev RNA 2:277–298

Townsend LB, Robins RK (1963) Ring cleavage of purine nucleosides to yield possible biogenic precursors of pteridines and riboflavin. J Am Chem Soc 85:242–243

von der Haar T, Gross JD, Wagner G et al (2004) The mRNA cap-binding protein eIF4E in post-transcriptional gene. Nat Struct Mol Biol 11:503–511

Warminski M, Kowalska J, Buck J et al (2013) The synthesis of isopropylidene mRNA cap analogs modified with phosphorothioate moiety and their evaluation as promoters of mRNA translation. Bioorg Med Chem Lett 23:3753–3758

Webb NR, Chari RVJ, DePillis G et al (1984) Purification of the messenger RNA cap-binding protein using a new affinity medium. Biochemistry 23:177–181

Weber LA, Feman ER, Hickey ED et al (1976) Inhibition of HeLa cell messenger RNA translation by 7-methylguanosine 5′-monophosphate. J Biol Chem 251:5657–5662

Westman B, Beeren L, Grudzien E et al (2005) The antiviral drug ribavirin does not mimic the 7-methylguanosine moiety of the mRNA cap structure *in vitro*. RNA 11:1505–1513

Worch R, Stepinski J, Niedzwiecka A et al (2005a) Novel way of capping mRNA trimer and studies of its interaction with human nuclear cap-binding complex. Nucleosides Nucleotides Nucleic Acids 24:1131–1134

Worch R, Niedzwiecka A, Stepinski J et al (2005b) Specificity of recognition of mRNA cap by human nuclear cap-binding complex. RNA 11:1355–1363

Wypijewska del Nogal A, Surleac MD, Kowalska J et al (2013) Analysis of decapping scavenger cap complex using cap analogs reveals molecular determinants for efficient cap binding. FEBS J 280(24):6508–6527. doi:10.1111/febs.12553

Wypijewska A, Bojarska E, Stepinski J et al (2010) Structural requirements for Caenorhabditis elegans DcpS substrates based on fluorescence and HPLC enzyme kinetic studies. FEBS J 277:3003–3013

Wypijewska A, Bojarska E, Lukaszewicz M et al (2012) 7-Methylguanosine diphosphate (m^7GDP) is not hydrolyzed but strongly bound by decapping scavenger (DcpS) enzymes and potently inhibits their activity. Biochemistry 51:8003–8013

Yamaguchi K, Nakagawa I, Sekine M et al (1984) Chemical synthesis of the 5′-terminal part bearing cap structure of messenger RNA of cytoplasmic polyhedrosis virus (CPV): $m^7G^{5'}$pppAmpG and $m^7G^{5'}$pppAmpGpU. Nucleic Acids Res 12:2939–2954

Yisraeli JK, Melton DA (1989) Synthesis of long, capped transcripts in vitro by SP6 and T7 RNA-polymerases. Methods Enzymol 180:42–50

Yoffe Y, Zuberek J, Lewdorowicz M et al (2004) Cap-binding activity of an eIF4E homolog from *Leishmania*. RNA 10:1764–1775

Yoffe Y, Zuberek J, Lerer A et al (2006) Binding specificities and potential roles of isoforms of eukaryotic initiation factor 4E in Leishmania. Eukaryot Cell 5:1969–1979
Yoffe Y, Léger M, Zinoviev A et al (2009) Evolutionary changes in the Leishmania eIF4F complex involve variations in the eIF4E-eIF4G interactions. Nucleic Acids Res 37:3243–3253
Yoshikawa M, Kato T, Takenishi T (1967) A novel method for phosphorylation of nucleosides to 5′-nucleotides. Tetrahedron Lett 8:5065–5068
Yount RG, Babcock D, Ballanty W et al (1971) Adenylyl imidodiphosphate, an adenosine triphosphate analog containing a P-N-P linkage. Biochemistry 10:2484–2489
Zdanowicz A, Thermann R, Kowalska J et al (2009) Drosophila miR2 Primarily Targets the m^7GpppN Cap Structure for Translational Repression. Mol Cell 35:881–888
Ziemniak M, Strenkowska M, Kowalska J et al (2013a) Potential therapeutic applications of RNA cap analogs. Future Med Chem 5:1141–1172
Ziemniak M, Szabelski M, Lukaszewicz M et al (2013b) Evaluation of fluorescent cap analogues for mRNA labeling. RSC Adv 3:20943–20958
Zuberek J, Stepinski J, Niedzwiecka A et al (2002) Synthesis of tetraribonucleotide cap analogue m7GpppAm2′pUm2′pAm2′ and its interaction with eukaryotic initiation factor eIF4E. In: Točik Z, Hocek M (eds) Collection symposium series, vol 5. Institute of Organic Chemistry and Biochemistry, Academy of Sciences of Czech Republic, Prague, pp 399–403
Zuberek J, Wyslouch-Cieszynska A, Niedzwiecka A et al (2003) Phosphorylation of eIF4E attenuates its interaction with mRNA 5′ cap analogs by electrostatic repulsion: Intein-mediated protein ligation strategy to obtain phosphorylated protein. RNA 9:52–61
Zuberek J, Jemielity J, Jablonowska A et al (2004) Influence of electric charge variation at residues 209 and 159 on interaction of eIF4E with the mRNA 5′ terminus. Biochemistry 43:5370–5379
Zuberek J, Kubacka D, Jablonowska A et al (2007) Weak binding affinity of human 4EHP for mRNA cap analogs. RNA 13:691–697

Innovative Chemistry for Synthesis of Regular RNA, 5′-Triphosphate RNA, or 5′-Capped RNA

Yann Thillier, François Morvan, Jean-Jacques Vasseur, and Françoise Debart

Contents

1 Introduction 564
2 Novel Chemical Synthesis of Regular RNA 565
2.1 Progress in RNA Synthesis Development 565
2.2 Original PivOM Technology 567
2.3 Benefits of PivOM Technology 570
3 Chemical Synthesis of 5′-Triphosphate RNA on Solid Support 573
3.1 Role and Applications of 5′-Triphosphate RNA 573
3.2 Existing Chemical Methods Based on the Use of Salicylphosphochloridite 573
3.3 Synthesis of 5′-Triphosphate RNA via H-Phosphonate Chemistry 574
4 Chemical Synthesis of 5′-Capped RNA on Solid Support 576
4.1 Roles and Significance of the Cap Structure at the 5′-End of mRNA 576
4.2 State of the Art on Methods for 5′-Capped RNA Synthesis 577
4.3 Chemical Synthesis on Solid-Support of GpppRNA and Their Enzymatic Methylation in Solution to ^{7m}GpppRNA 581
5 Conclusion and Prospects 585
References 586

Abstract A great renewal in the chemical synthesis of RNA has grown up since the advent of RNA interference (RNAi) with the crucial need of short RNA molecules for biological research and therapeutic applications. RNA synthesis requires a suitable protection for 2′-OH to ensure successful RNA assembly and to avoid RNA damage during the deprotection process. Of special interest is our all-base-labile strategy which provides highly pure RNA sequences efficiently and rapidly. This method consists in protecting 2′-OH with base-labile pivaloyloxymethyl (PivOM) protection compatible with standard protections for 5′-OH, phosphates and nucleobases. The main advantage of this RNA synthesis is a

Y. Thillier • F. Morvan • J.-J. Vasseur • F. Debart (✉)
IBMM UMR 5247 CNRS-UM1-UM2, Université Montpellier 2, Place Eugène Bataillon, 34095, Montpellier cedex 05, France
e-mail: debart@univ-montp2.fr

V.A. Erdmann et al. (eds.), *Chemical Biology of Nucleic Acids*, RNA Technologies,
DOI 10.1007/978-3-642-54452-1_29, © Springer-Verlag Berlin Heidelberg 2014

simple and straightforward two-step deprotection in mild basic conditions without any concomitant RNA degradation or 2′-*O*-migration. In this chapter this original method will be described. Besides development of innovative RNA synthesis, the chemical production of 5′-functionalized RNA with triphosphate (TP/ppp) or cap moieties is also attractive for biologists since these molecules are important substrates in many biological processes such as ligation of RNA molecules, detection of viral responses via activation of the RIG-1 protein, induction of antiviral immunity, enzymatic synthesis of 5′-capped (^{7m}Gppp)-RNA, recognition signal for translation, RNA splicing, and nucleocytoplasmic exportation. To face the high demand of such molecules, we have developed a new synthetic approach on solid support using *H*-phosphonate intermediate for the large-scale synthesis of the 5′-triphosphate or 5′-GpppRNA where RNA assembly was performed with PivOM technology. We will report on the synthetic procedures which represent a highly efficient alternative to the methods described in the literature to date.

Keywords RNA chemistry • Solid-phase synthesis • Triphosphate RNA • Capped RNA • Phosphoroimidazolidate

1 Introduction

A renewed interest in the rapid and efficient chemical synthesis of RNA has grown up since the advent of RNA interference (RNAi) (Elbashir et al. 2001; Fire et al. 1998) with the high demand for synthetic RNA molecules as powerful tools for biological research and therapeutic applications. Compared to DNA synthesis, RNA production is more complex due to the presence of 2′-OH functions which require suitable protection to enable successful RNA assembly and to avoid RNA damage during the deprotection process. The current and most popular approach to the automated synthesis of oligoribonucleotides on solid support (Beaucage 2008) is based on phosphoramidite chemistry which was initially developed for the synthesis of oligodeoxynucleotides by Beaucage and Caruthers (1981). Besides lower coupling yields in chain assembly compared to DNA elongation, the main difficulty in RNA chemistry results from the RNA instability in basic or acidic media leading to strand cleavage and/or phosphate migration with formation of unnatural 2′–5′ internucleotide linkages. The success of solid-phase RNA synthesis largely depends on the right combination of orthogonal protecting groups for the reactive functions in ribonucleotide amidites. It is noteworthy that as for DNA synthesis most of RNA synthesis methods regularly use acid-labile 4,4′-dimethoxytrityl group (DMTr) for blocking 5′-OH function, base-labile acyl groups to protect exocyclic amino functions on nucleobases (C, G and A) and 2-cyanoethyl (CNE) group for phosphate protection. Ammonolytic conditions are commonly applied for nucleobases and phosphates deprotection and release of oligoribonucleotides from the solid support. Finally, while the choice of a suitable 2′-protection has been the main concern for the chemist in RNA synthesis field for many years, it was universally stated that the 2′-protecting group should be stable under ammonia treatment to avoid RNA degradation.

Actually for these last 10 years, numerous 2′-*O*-protecting groups (Aralov and Chakhmakhcheva 2013) were developed to find an alternative to the commonly used *tert*-butyldimethylsilyl (TBDMS) group (Usman et al. 1987) but none of them could really supplant it despite its drawbacks. Of special interest is our recent all base-labile protections strategy which provides highly pure RNA sequences efficiently and rapidly. This improved synthetic method consists in protecting 2′-OH with base-labile pivaloyloxymethyl (PivOM) protections compatible with standard protections for 5′-OH (DMTr), phosphates (2-cyanoethyl), and nucleobases (acyl groups) (Debart et al. 2009; Lavergne et al. 2008). The main advantage of this RNA chemistry strategy is a straightforward two-step deprotection in mild conditions during a short period of time at room temperature without any concomitant RNA degradation or migration. This strategy was challenging and overcome the well-known RNA instability in basic media. In this chapter we will describe this original method of RNA synthesis currently on the market.

Besides RNAi field where efficient RNA synthesis is required, the chemical production of 5′-functionalized RNA with triphosphate (TP) moieties is of great interest for many biological applications. For example, 5′-TP RNA are used for the ligation of RNA molecules, the detection of viral responses via activation of the RIG-1 protein, the induction of antiviral immunity, and in the enzymatic synthesis of 5′-capped (^{7m}Gppp)-RNA. In the same way, 5′-capped RNA of defined and appropriate length in sufficient quantity and purity are useful tools for crystallography studies of their complex with enzymes involved in the viral capping system to identify new potent antiviral targets. We developed a new synthetic approach on solid support using *H*-phosphonate intermediate for the synthesis of the 5′-triphosphate or 5′-GpppRNA where chain assembly was performed with PivOM technology. We will report on the synthetic procedure which represents a highly efficient alternative to the methods described in the literature to date.

2 Novel Chemical Synthesis of Regular RNA

2.1 Progress in RNA Synthesis Development

The automated RNA chemical synthesis mimics DNA synthesis and consists in the assembly of suitably protected ribonucleoside phosphoramidite-building blocks on solid support (Beaucage 2008). The standard approach for orthogonal protection of 5′-OH and 2′-OH of ribonucleoside is based on 5′-*O*-DMTr-3′-cyanoethyl diisopropylphosphoramidites and for 2′-OH protection, a large variety of groups have been proposed in combination with the DMTr group. The most commonly used were

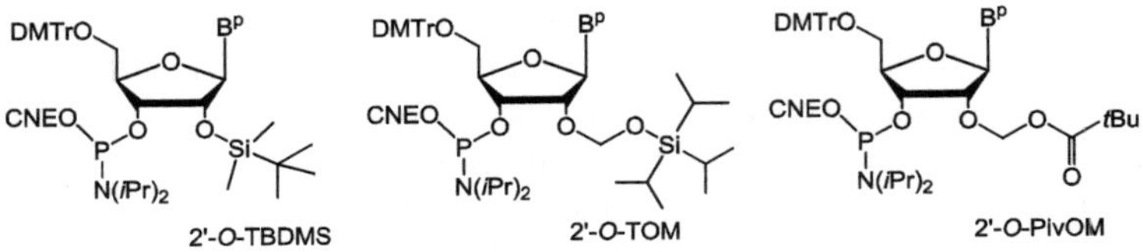

Fig. 1 Some ribonucleoside phosphoramidite building blocks for RNA synthesis bearing common TBDMS and TOM groups and the original PivOM group for 2′-OH protection

certainly the two fluoride-labile silyl groups: the early *tert*-butyldimethylsilyl (TBDMS) (Usman et al. 1987) and the more recent triisopropylsilyloxymethyl (TOM) (Pitsch et al. 2001) groups (Fig. 1). Although some improvements in the coupling time and efficiency of TBDMS phosphoramidites were achieved thanks to powerful activators such as 5-ethylthio-*1H*-tetrazole (ETT) (Sproat et al. 1995) and 5-benzylthio-*1H*-tetrazole (BTT) (Welz and Müller 2002), the major drawback with TBDMS is the high potential of the alkylsilyl ether to migrate from the 2′- to the 3′-position during the preparation of the phosphoramidite (Reese 2005). Of this migration results the presence of unnatural 2′–5′-phophodiester internucleosidic linkages in the final RNA.

The development of the TOM group with the acetal moiety avoids first the 2′–3′ isomerization during phosphitylation and therefore the risk of unnatural linkages is excluded. Secondly, the lower steric hindrance of the TOM group compared with TBDMS allows higher coupling efficiency (>99 %) in shorter coupling times (2–3 min). In the same way, many recent 2′-protecting groups exploit the advantage of a flexible formacetal moiety in reducing the steric hindrance next to the 3′-phosphoramidite function (Fig. 2). Thus, the 2-cyanoethoxymethyl (CEM) (Ohgi et al. 2005; Shiba et al. 2007), the 2-cyanoethoxyethyl (CEE) (Umemoto and Wada 2004), the 2-(4-toluylsulfonyl)ethoxymethyl (TEM) (Zhou et al. 2007), the 2-nitrobenzyloxymethyl (2-NBOM) (Miller et al. 2001),the 4-nitrobenzyloxymethyl (4-NBOM) (Gough et al. 1996),the 4-(*N*-dichloroacetyl-*N*-methylamino)benzyloxymethyl (4-DCA-MABOM) (Cieslak et al. 2007, 2008), the 2-cyano-2,2-dimethylethanimine-*N*-oxymethyl (Cieslak et al. 2013), the levulinylacetal ester (ALE) (Lackey et al. 2009), and the 2-*tert*-butyldithiomethyl (DTM) (Semenyuk et al. 2006) were developed to be competitive to the most widely used RNA synthesis strategies and allowed successful RNA chemical synthesis. Nevertheless, some weaknesses still persist. Particularly, the fluoride ions treatment after ammonolysis performed for removing most of these groups like TBDMS and TOM represents an extra step in the deprotection process. Moreover, this treatment requires a further desalting step by precipitation or cartridge purification leading to additional time-consuming work-up procedures and a loss of material to obtain pure oligoribonucleotides.

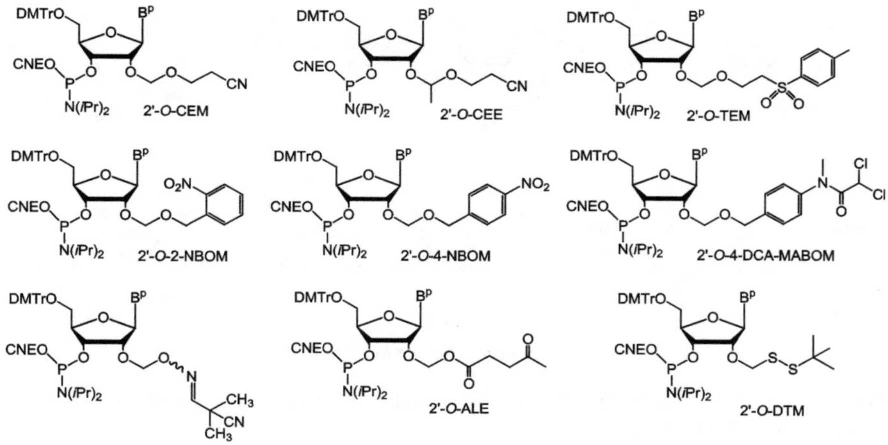

Fig. 2 Various 5′-*O*-DMTr-protected phosphoramidite ribonucleotides with 2′-protecting groups bearing a formacetal moiety

2.2 *Original PivOM Technology*

With the aim of reducing the two steps for RNA deprotection found in most of RNA synthesis methods in a simple ammonia treatment as for DNA deprotection, we designed base-labile protecting groups for 2′-OH. This was a major challenge due to the well-known RNA instability in basic media. Furthermore, the use of acyl (acetyl or benzoyl) groups for 2′-protection reported in an early work resulted in very poor yields (Kempe et al. 1982). In a preliminary work, with the goal for searching 2′-*O*-biolabile protecting groups which would increase the RNA stability in biological media and the cell penetration due to the lipophilic nature, we tested several acyloxymethyl groups (Parey et al. 2006). Pivaloyloxymethyl (PivOM), acetyloxymethyl (AcOM), pivaloylthiomethyl (PivSM), and acetylthiomethyl (AcSM) groups were first evaluated in oligouridylates as biolabile protections and it was found that the best properties were obtained with the PivOM group in terms of prodrug criteria. In parallel, we used this group as 2′-protection for the chemical synthesis of natural oligoribonucleotides with the objective to remove it under basic conditions simultaneously with the acyl groups from nucleobases, the CNE from phosphates, and the hydrolysis of the linker between RNA chain and solid support.

2.2.1 Preparation of 2′-*O*-PivOM Phosphoramidites

The 2′-*O*-PivOM-building blocks were prepared following two routes (Lavergne et al. 2010). The first one was based on the site-specific introduction of the PivOM group in which the Markiewicz reagent ($TIPSiCl_2$) simultaneously blocks 5′-OH and 3′-OH and leaves the 2′-OH free to accept PivOM (Fig. 3). This method avoids

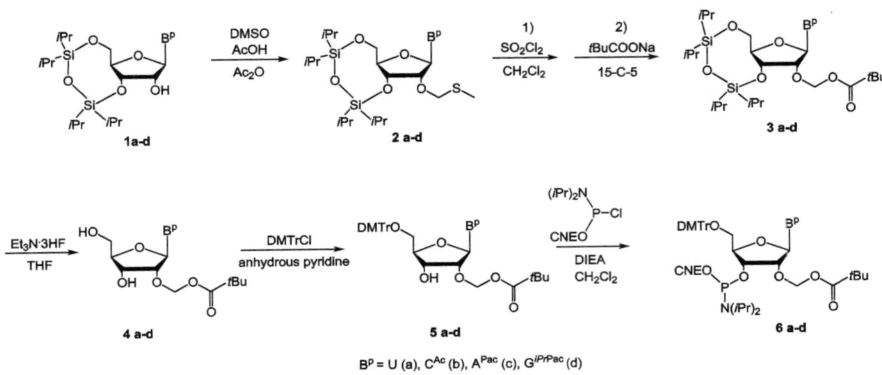

Fig. 3 Synthesis of 2′-*O*-PivOM 3′-*O*-phosphoramidite ribonucleotides with site-specific introduction of the PivOM group

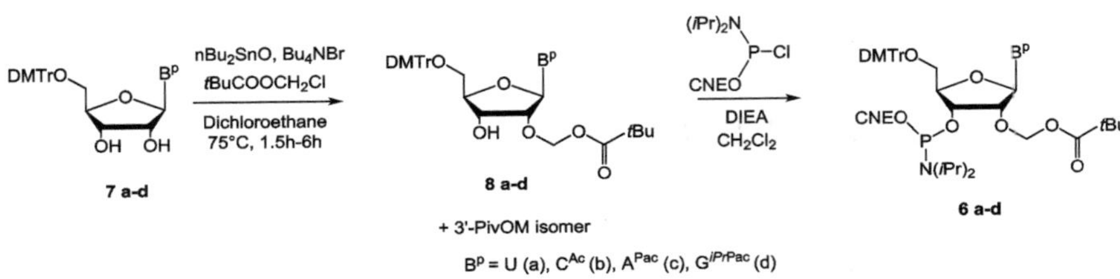

Fig. 4 Synthesis of 2′-*O*-PivOM 3′-*O*-phosphoramidite ribonucleotides with direct unspecific introduction of the PivOM group

separation of 2′- and 3′-isomers with several tedious purifications by column chromatography but it requires additional steps for the whole phosphoramidite synthesis in comparison to the alternative method.

This alternative synthesis consists in derivatizing 2′-OH of 5′-DMTr base-protected ribonucleotides **7a–d** with PivOM via a 2′,3′-*O*-dibutylstannylidene intermediate which was treated with the alkylating agent pivaloyloxymethyl chloride to give a mixture of 2′-PivOM and 3′-PivOM derivatives (Fig. 4) (Lavergne et al. 2008). After separation by silica gel chromatography, the desired fast-eluting 2′-PivOM compounds **8a–d** were isolated with higher yield (34–49 %) from **7a–d** than the undesired 3′-isomers. The tritylated 2′-PivOM compounds **8a–d** were converted to the corresponding amidites **6a–d** with 75–82 % yield by using 2-cyanoethyl *N,N*-diisopropylchlorophosphoramidite. It is noteworthy that the four suitably protected 2′-PivOM phophoramidites **6a–d** are currently available at Chemgenes Corporation (USA) (Debart et al. 2009).

2.2.2 Automated RNA Assembly on Solid Support

The suitably protected 2′-*O*-PivOM phosphoramidites were assembled on a DNA synthesizer by using commercially available controlled-pore glass (LCAA-CPG) linked to 5′-*O*-DMTr 2′-*O*-Ac ribonucleotides through a 3′-*O*-succinyl linker (Link Technologies, UK). Dichloroacetic acid in dichoromethane was used as detritylation agent. Condensation of the 2′-PivOM amidites (0.1 M in CH_3CN) with the 5′-OH of the supported growing RNA chain was performed with 180 s coupling time in the presence of BTT as activator (Welz and Müller 2002). The average stepwise yield was 99.7 %, which is comparable with DNA or other efficient RNA amidites. The capping step was achieved with phenoxyacetic anhydride to avoid the transacylation on the amino functions of purine residues which would occur if acetic anhydride had been used as capping reagent. Indeed the resulting acetyl groups could not be removed from guanosine residues upon mild basic conditions (room temperature, short time reaction) which are applied for RNA deprotection.

2.2.3 2′-PivOM RNA Deprotection

After chain elongation, RNA sequences anchored to CPG solid support were either treated for 3 min with 1 M DBU in acetonitrile (Fig. 5) or piperidine (10 %) in acetonitrile for 15 min to eliminate cyanoethyl groups from phosphates. Then a 28 % aqueous ammonia solution was applied to the column for 3–4 h at room temperature for simultaneous removal of fast-labile acyl groups from nucleobases, PivOM groups, and cleavage of the succinyl linker to release unprotected RNA. Before evaporation of the ammonia solution, isopropylamine should be added to trap formaldehyde resulting from acetal moiety hydrolysis in order to prevent adduct formation between formaldehyde and guanine residues in the presence of ammonia (Fig. 6) (Lavergne et al. 2008). Nevertheless, it is noteworthy that guanosine adducts are unstable and decomposed after several hours in water or in triethylammonium acetate (TEAAc) buffer. If mass spectrometry analysis of the crude material revealed the presence of such adducts (peaks at +41), after reverse-phase HPLC purification and evaporation of the TEAAc eluents no adducts would be detected any more (Fig. 7).

2.2.4 Quality Control of RNA Synthesized with PivOM Technology

The HPLC profiles of crude RNA sequences illustrate the efficiency of the chain assembly and the deprotection protocol. For comparison, the same RNA 21-mer was synthesized with PivOM technology and with TBDMS chemistry according to standard conditions. In our hands, the average stepwise coupling yield of TBDMS phosphoramidites was 96.8 % affording the 21-mer with a lower purity than that obtained with the PivOM methodology.

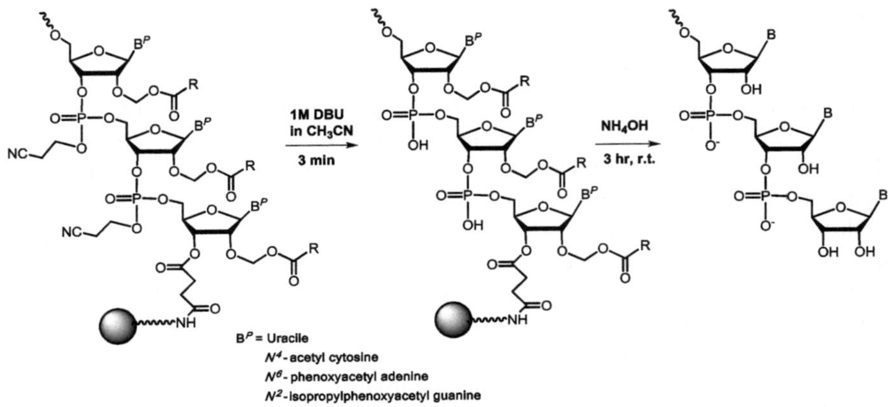

Fig. 5 Two-step basic deprotection of RNA sequences assembled on solid support from 2′-PivOM phosphoramidites

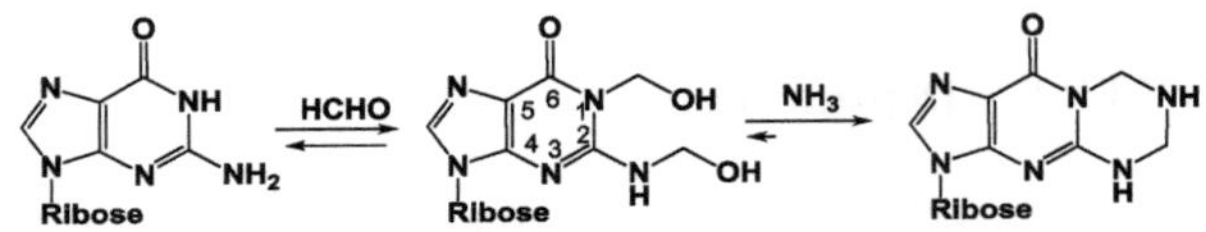

Fig. 6 Reversible side reaction in guanine residues during ammonia evaporation

The digestion of a 21-mer mixed-base sequence by nuclease P1 for specific cleavage of the 3′–5′-internucleotide linkages and alkaline phosphatase gave the four natural ribonucleosides as attested by HPLC analysis of the digestion mixture (Fig. 8). This data proved the absence of any nucleobase modification or unnatural 2′–5′-internucleotide linkages.

The activity of a siRNA 21-mer duplex prepared with PivOM technology was evaluated in an RNAi assay that targets the Ret/PTC1 junction oncogene involved in papillary thyroid carcinoma (Fig. 9). The siRNA duplex obtained with PivOM chemistry had a similar gene silencing activity (60 % inhibition) than a commercial siRNA duplex with the same sequence (40 % inhibition) prepared with standard TBDMS chemistry. This result attested the integrity and purity of the siRNA synthesized by the PivOM technology.

2.3 *Benefits of PivOM Technology*

The main advantage of the RNA strategy with an acetal ester group as PivOM is first the high coupling in a few minutes and secondly the straightforward two-step basic deprotection applied for a short period of time at room temperature which represents mild conditions for RNA. Indeed, the PivOM technology which makes

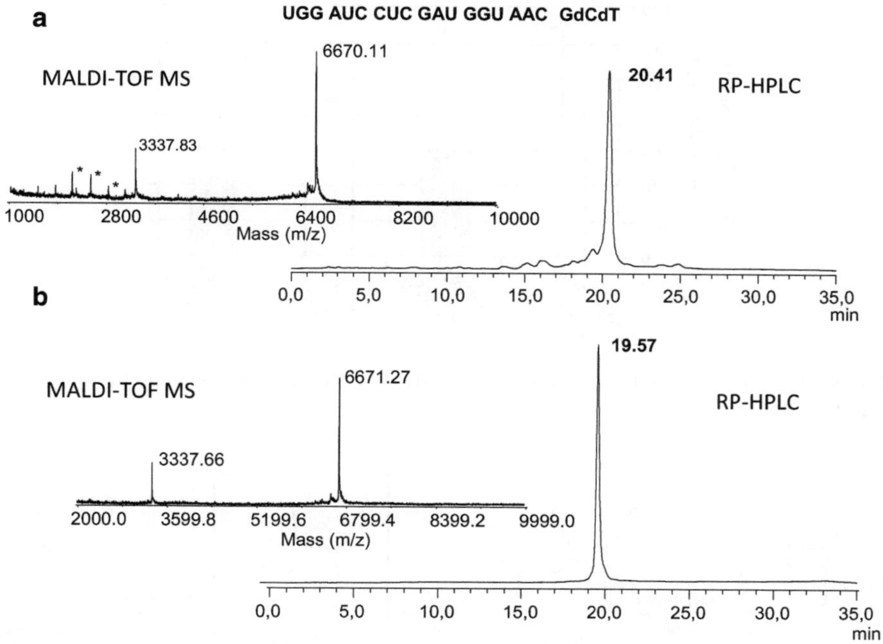

Fig. 7 RP-HPLC profiles and MALDI-TOF mass spectrometry analysis of a 21-mer RNA: (**a**) crude material and (**b**) purified material

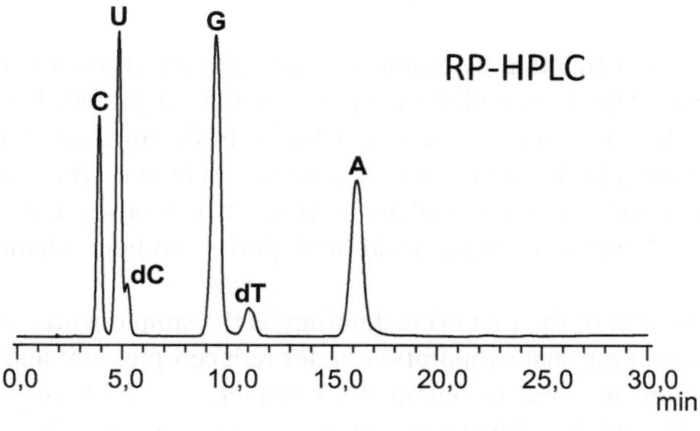

Fig. 8 RP-HPLC analysis of the ribonucleosides mixture obtained after nuclease P1 and phosphatase alkaline digestion of the 21-mer RNA sequence: UGG AUC CUC GAU GGU AAC GdCdT

exclusive use of base-labile protecting groups is challenging because it disrupts the dogma "RNA synthesis is incompatible with a whole base-labile protections strategy" related to the well-known RNA instability in basic media. The proposed mechanism of base-mediated hydrolysis of the acetal ester group consists of

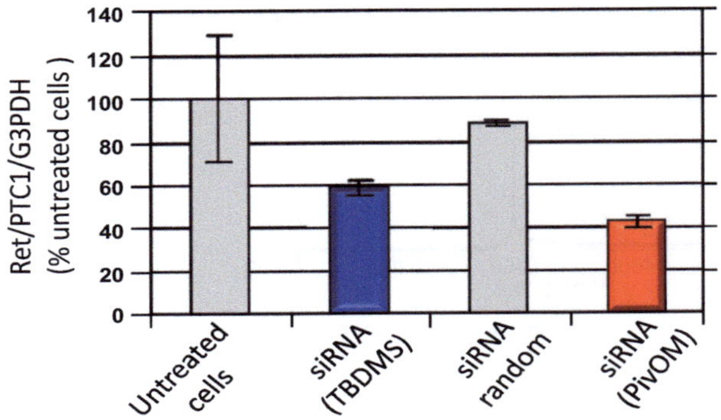

Fig. 9 Analysis after RT-PCR quantification of Ret/PTC1 expression inhibition by cytofectin-transfected siRNA in vitro

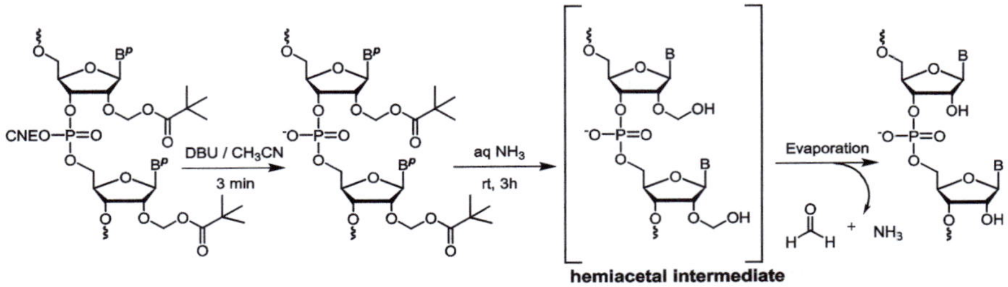

Fig. 10 Mechanism of 2′-PivOM RNA deprotection under ammonia conditions

cleavage of the ester function by ammonia leading to the formation of a formaldehyde hemiacetal. This intermediate is stable enough to protect RNA in aqueous ammonia and thus to avoid its degradation in basic medium. Upon ammonia evaporation, while pH decreases, the hemiacetal undergoes fragmentation to the 2′-OH ribonucleoside and formaldehyde (Fig. 10). Consequently crude RNA sequences are obtained in high yield and purity without chain rupture and migration.

A further strength of the PivOM technology is the simple work-up process after ammonia deprotection: just evaporation under reduced pressure and no additional desalting steps are required to obtain RNA sequences with high purity. Furthermore, the PivOM-building blocks are commercially available. This original technology is a powerful and efficient method for fast RNA synthesis.

3 Chemical Synthesis of 5′-Triphosphate RNA on Solid Support

3.1 Role and Applications of 5′-Triphosphate RNA

Triphosphate (TP) is a key entity for nucleosides and nucleic acids and confers crucial properties to them which have motivated research and applications with such derivatives for these last years. 5′-TP nucleosides are the natural substrates for enzymatic DNA or RNA synthesis whereas modified TP nucleosides are used as antiviral drugs and diagnostic reagents. A TP entity is initially present at the 5′-end of any viral or self-RNA molecules generated by RNA polymerases. In bacteria, the 5′-TP of primary transcripts is converted to a 5′-monophosphate by a RNA pyrophosphohydrolase (RppH), allowing access to both endo- and 5′-exoribonucleases which ensure mRNA degradation. In eukaryotes a maturation process of mRNA cleaves the TP to DP, caps the 5′-OH with a guanosine triphosphate, and modifies RNA in the nucleus before its release in the cytoplasm. In addition, 5′-TP DNA or RNA are used as substrates for ligation in vitro (Joyce 2007), the detection of viral responses via activation of the RIG-1 helicase (Hornung et al. 2006) and the induction of antiviral immunity (Allam et al. 2008). More recently, it has been discovered a synergic effect of the immune response triggered by the binding of 5′-TP RNA to RIG-1 with gene silencing mediated by small interfering RNA (siRNA) (Poeck et al. 2008). Thus the use of 5′-TP siRNA against melanoma provoked a massive apoptosis of tumor cells in lung metastases in vivo. Therefore, this data emphasizes that 5′-TP RNA play an important role in biochemical reactions and have a promising future for many medical applications. The demand for synthetic 5′-TP RNA is obviously increasing and cannot be satisfied by enzymatic methods but chemical methods on solid-support afford highly pure RNA with reproducible yield independent of RNA sequence. Other advantages of the chemical synthesis are the possibility to introduce chemical modifications in the sequence and finally to produce larger quantities.

3.2 Existing Chemical Methods Based on the Use of Salicylphosphochloridite

Despite numerous methods reported on the synthesis of 5′-TP nucleosides (Burgess and Cook 2000), very few approaches were described to produce efficiently 5′-TP DNA or RNA. Of particular interest, the well-known versatile salicylphosphochloridite method of Ludwig and Eckstein (1989) was adapted to the solid-supported synthesis of several 5′-DP or 5′-TP DNA or RNA (Fig. 11).

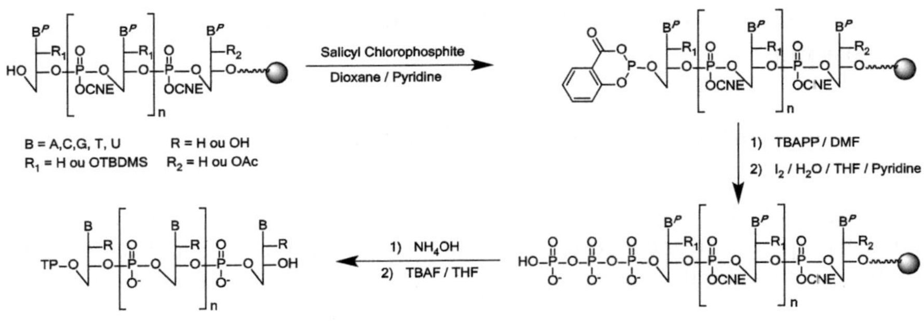

Fig. 11 Solid-supported synthesis of 5′-TP RNA via Ludwig and Eckstein method

A first synthesis of 5′-TP 11-mer or 13-mer RNA was tentatively performed but only characterization of 5′-DP RNA was reported despite the use of pyrophosphate (Brownlee et al. 1995). Later, several 5′-TP 1- to 5-mer DNA were obtained with 15–30 % yield, longer sequences could not be isolated (Lebedev et al. 2001). More recently, a 5′-TP 24-mer RNA was synthesized via the same method but neither global yield nor reaction efficiency was mentioned (Schlee et al. 2009). Lately the Ludwig–Eckstein method was improved by the use of the highly reactive pyrophosphorylating reagent tris(tetra-*n*-butylammonium) hydrogen pyrophosphate instead of the conventional tri-*n*-butylammonium salt (Schmidt et al. 2009) for the nucleophilic substitution reaction. This improved method was reproducible and gave a 5′-TP 25-mer RNA with acceptable yield (12 % isolated yield) (Nagata et al. 2012). Our contribution to the field was to develop a robust, reproducible, scalable, and simple procedure for the solid-phase synthesis of 5′-TP DNA and RNA of various lengths, sequences, and chemical modifications that provides reasonably good yields and satisfactory purity of the crude products (Zlatev et al. 2010).

3.3 Synthesis of 5′-Triphosphate RNA via H-Phosphonate Chemistry

Most of reported strategies involve the coupling of inorganic pyrophosphate with an activated monophosphorylated derivative for triphosphate synthesis. The efficacy of the substitution reaction depends on the choice of the activated entity. Based on the use of a phosphoroimidazolidate (*ImP*) intermediate, synthesis of 5′-TP nucleosides and oligonucleotides have been originally obtained with yield up to 80 % (Hoard and Ott 1965). Several methods were described to give rise this activated phosphate through the reaction of a phosphate monoester with 1,1′-carbonyldiimidazole (Hoard and Ott 1965) or a mixture of triphenylphosphine, 4,4′-dithiopyridine, and imidazole (Mukaiyama and Hashimoto 1971). In our case, we opted for the amidative oxidation of an *H*-phosphonate (Hp) monoester

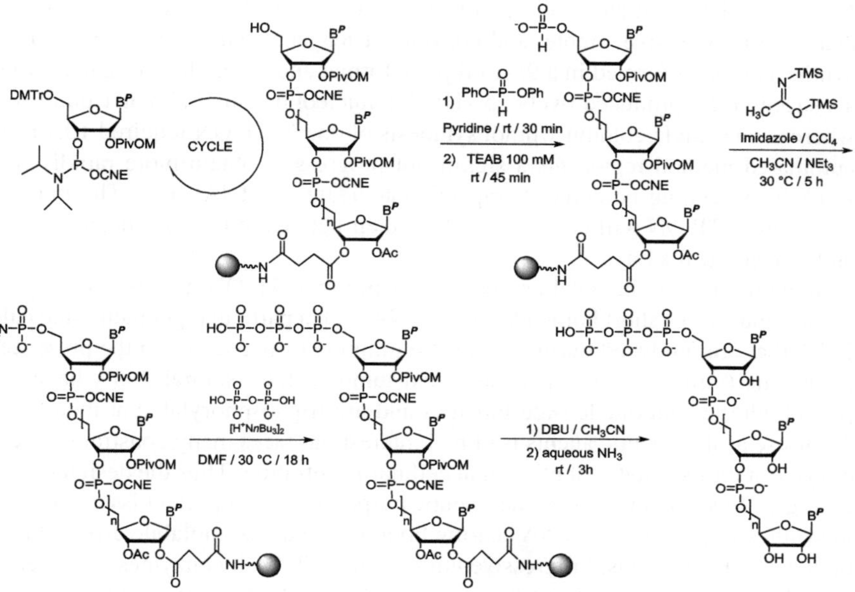

Fig. 12 General synthesis of 5′-TP RNA on solid support using 2′-PivOM chemistry

derivative firstly introduced at the 5′-end of solid-supported oligonucleotides (ON) as a stable and easily accessible intermediate. The 5′-*Hp*-ON was obtained by treatment with a solution of diphenylphosphite in pyridine, followed by hydrolysis, and it was further oxidized to an activated 5′-phosphoroimidazolidate using CCl_4 in the presence of imidazole and *N,O*-bis-trimethylsilylacetamide (Fig. 12). Substitution of the imidazole by pyrophosphate tri-*n*-butylammonium salt was performed overnight at 30 °C. This approach was successfully applied to the synthesis of 5′-TP DNA, RNA, and modified 2′-OMe RNA (Zlatev et al. 2010, 2012). When 5′-TP RNA are prepared through standard TBDMS chemistry, the long desilylation reaction times with fluoride ions cause slight damage to the triphosphate moiety and furthermore subsequent tedious desalting or purification is a major drawback for this method. As alternative synthesis for 5′-TP RNA, PivOM chemistry was rather preferred for RNA assembly and 5′-end solid-supported RNA was then functionalized with TP with better yield providing 5′-TP RNA with higher purity (crude purity: 80 % of 5′-TP ON). Indeed compared to the TBDMS approach, PivOM are removed by ammonia treatment which preserves TP backbone. This efficient solid-phase method is surely versatile, reproducible to provide 5′-TP RNA of various lengths and sequences with different 5′-terminal nucleotides and sugar and backbone chemical modifications. More recently a fully automated chemical method for the parallel and high-throughput solid-phase synthesis of 5′-TP and 5′-DP ON was reported (Zlatev et al. 2013). This automated synthetic approach is universal since it can be applied to oligonucleotides of different length, sequence, and nature of the 5′-terminal nucleotide.

Moreover, the automated solid-phase synthesis involved the use of commercially available inexpensive, stable, and convenient reagents. The procedure for the TP synthesis was performed in a 96-well plate format and the high-throughput synthesis successfully produced twelve 5′-TP oligonucleotides in 16.7 h at 1 μmol scale, whereas one single column 1 μmol synthesis of one 5′-TP ON required 10.9 h. The automated method represents a significant progress to obtain more rapidly 5′-TP RNA for screening potential therapeutic candidates in RNAi field. The improved access to 5′-TP ON will be also useful for their applications in mechanistic studies and biochemical assays.

In addition, our chemical process for obtaining 5′-TP ON has also been applied to the synthesis of short homoadenosine RNA with a particular geometry and called 2–5A that are produced during the innate immune response due to the presence of pathogens (virus, bacteria, parasites, and tumor cells). Natural 2–5A have 2′–5′ phosphodiester internucleoside linkages and are triphosphorylated at their 5′ end. Their use as therapeutic agents is of real interest but faces many constraints such as enzyme stability, biodistribution, and cellular penetration. Thus the design of 2–5A biologically stable in vivo for therapeutic purposes is an important issue. We chose to synthesize prodrugs of 2–5A bearing lipophilic and enzymolabile groups PivOM on the 3′-OH functions. Progressive unmasking of 2–5A by enzymes would ensure a prolonged release with an increased half-life of these messengers to make interferon response longer. To obtain these 2–5A 3′-*O*-PivOM 5′-TP, we have developed a new strategy for RNA prodrugs synthesis which consisted in retaining conventional protecting groups for RNA and applying a new deprotection treatment with an amine in anhydrous organic medium. These conditions also guarantee the integrity of 2–5A sequence with a 5′-TP backbone and 3′-*O*-PivOM groups. Thus two 2–5A analogues (trimer and tetramer) were synthesized with an efficiency of 70 % and proved to be good activators of RNase L, a key player in the innate immune response. The most promising 2–5A trimer was produced at several milligrams scale and its activity was evaluated in cell cultures infected with influenza and respiratory syncytial virus (RSV). The modified trimer exhibited a noteworthy antiviral activity in a Vero cell line infected by a VRS strain (Thillier et al. 2013).

4 Chemical Synthesis of 5′-Capped RNA on Solid Support

4.1 Roles and Significance of the Cap Structure at the 5′-End of mRNA

The cap structure found at the 5′-end of eukaryotic messenger RNA and many viral RNA consists of a N^7-methylguanosine (^{7m}G) linked to the 5′-terminal nucleotide (N) of the pre-mRNA via a 5′–5′-triphosphate linkage (^{7m}GpppN-RNA) (Decroly et al. 2011; Shatkin 1976; Shuman 2001). This cap modification is essential for efficient RNA translation (Hodel et al. 1998) and for its protection from degradation by 5′-exonucleases. Moreover, the cap plays an important role in RNA splicing and export from the nucleus and avoids recognition of mRNA by the innate immunity machinery (Daffis et al. 2010; Zust et al. 2011). In eukaryotes, the capping reaction was achieved by four enzymes: an RNA triphosphatase (RTPase) which converts the 5′-TP RNA in 5′-DP RNA to allow the transfer of a GMP molecule onto the mRNA by a guanylyltransferase (GTase) giving rise to the capped RNA:GpppN-RNA (Fig. 13). Finally a guanine N^7-methyltransferase (N^7-MTase) adds a methyl group at N^7 position of the guanine residue in the presence of S-adenosyl-L-methionine (AdoMet) and the cap-0 structure (^{7m}GpppN-RNA) is obtained. This methylation step makes the GTase reaction irreversible. A further methylation at the 2′-*O*-position of the first nucleotide can be achieved by a 2′-*O*-methyltransferase (2′-*O*-MTase) leading to the cap-1 structure (^{7m}GpppN$_{2'\text{-Om}}$-RNA). Most RNA viruses also add a similar cap moiety which is essential for virus replication. However recent studies have shown that viral RNA capping pathways are different in enzymes, reaction sequences, and mechanisms (Decroly et al. 2012) which render viral enzymes potential novel targets for antiviral drug design. To speed up drug design, biologists need a good definition of these viral enzymes at the molecular and atomic level. Nevertheless the characterization and use of the viral capping system as antiviral targets have been greatly limited by the low availability of suitable amount of pure capped RNA substrates of defined and appropriate length required for structural studies by crystallography. Indeed due to this important bottleneck very few crystal structures of binary complexes with capping enzymes and RNA virus exist. Moreover, for the same reason it is noteworthy that capped RNA have never been considered as therapeutic tools for gene silencing.

4.2 State of the Art on Methods for 5′-Capped RNA Synthesis

Due to the instability of the positively charged N^7-methylguanosine under acidic and basic conditions, the chemical methods for synthesis of 5′-capped RNA has remained a real challenge (Mikkola et al. 2005). Consequently numerous approaches have been developed to produce such molecules: in vitro enzymatic synthesis, chemical methods, and finally a combination of chemical and biochemical processes.

4.2.1 In vitro Enzymatic Methods

It is known that viral enzymes cap defined and specific RNA sequences present at 5′-end which are difficult to synthesize with T3, T7, or SP6 DNA-dependent RNA polymerases (Bollati et al. 2010). Actually the bacteriophagic promoter initiates polymerization with GTP and synthesizes RNA beginning with pppGG (T3 and T7) or GA (SP6), whereas the specific T7 promotor φ2.5 has been used for RNA

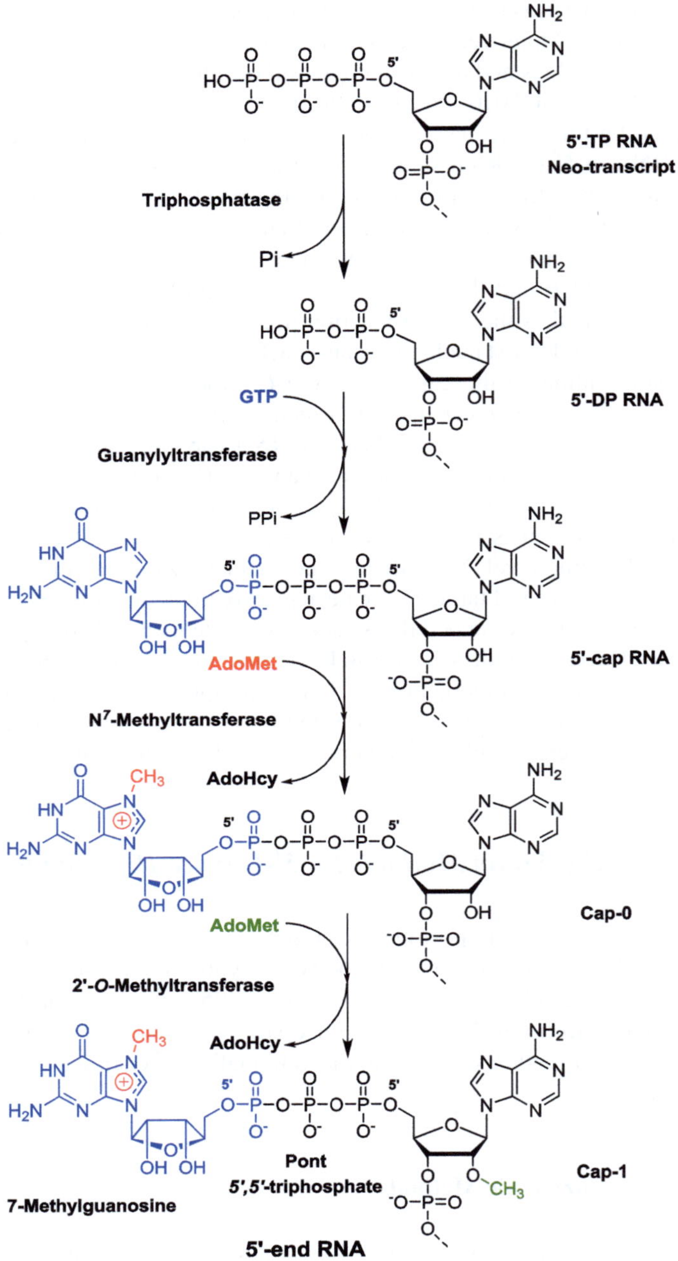

Fig. 13 Eukaryotic capping pathways

synthesis initiated by pppAG. This specificity of sequence is the major limitation. After RNA transcription, vaccinia virus-capping system that contains RTPase, GTase, and N^7-MTase activities is able to add a cap structure to the 5′-end of RNA (Brownlee et al. 1995).

Alternatively short-capped RNA can be prepared by the use of a DNA primase that initiates the synthesis from a specific codon on a given DNA template and RNA contains two nucleotides complementary to the initiation codon. Particularly, primase of the bacteriophage T7 could produce ^{7m}GpppAC$_n$ and GpppAC$_n$ ($1 \leq n \leq 7$) with a 50 % average yield but one major limitation of this method was the required pppAC RNA sequence at the 5′-end (Peyrane et al. 2007).

4.2.2 Combination of Enzymatic Methods with Chemical Synthesis

To overcome the weakness of enzymatic methods depending of defined sequences, chemical synthesis can be combined to biochemical methods to afford 5′-capped RNA. One consists in first synthesizing the dinucleotide cap^{7m}GpppN or some analogues which can be incorporated by RNA viral polymerases (T7 polymerase). Various cap analogues have been synthesized and added successfully by in vitro transcription to prevent the wrong incorporation of the cap structure in reverse mode (Jemielity et al. 2003; Kore et al. 2009). The second method consists of first preparing a 5′-DP oligonucleotide on solid support then the cap structure is added in solution with the enzymatic complex of vaccinia virus (RTPase, GTase, and MTase) (Nagata et al. 2010). In the last combined approach, RNA fragments are prepared by chemical synthesis and then are assembled by enzymatic ligation. For example, a capped 42-mer RNA was obtained after enzymatic ligation of three fragments (14-mer, 18-mer, and capped 9-mer) by T4 RNA ligase (Iwase et al. 1992). The capped RNA fragment resulted from the coupling in solution of ^{7m}GDP with a 5′-phosphoroimidazolidate 9-mer RNA previously synthesized in solution. The capped 42-mer was isolated with 5.4 % overall yield resulting of the low coupling efficacy of ^{7m}GDP with *ImP*-RNA. More recently, an efficient method was reported to generate a capped 74-mer RNA by ligation of two fragments (capped 13-mer and 5′-MP 62-mer) which were chemically synthesized on solid support. Ligation was performed with T4 RNA ligase 2 with 92 % efficacy (Barral et al. 2013).

4.2.3 Existing Chemical Methods

For a long time, the lack of GDP solubility and reactivity in solution was a major hurdle for chemical synthesis of 5′-capped RNA. The addition of metal chlorides to the capping solution enhanced the solubility of the diphosphate entity and promoted the phosphoric anhydride production. This improvement allowed the development of two chemical strategies: the coupling of an activated ^{7m}GDP derivative either in solution or on solid-supported RNA.

In fact, due to pH instability of 7-methylguanosine under acidic conditions where depurination occurs, whereas the opening of the imidazole ring of the 7-methylguanine nucleobase is observed in basic conditions, the chemical solid-phase synthesis of 5′-capped RNAs still remains challenging. Nevertheless, the capping reaction can be achieved in solution after the release of 5′-MP RNA sequence from solid support. For the coupling reaction, the β-phosphate of ^{7m}GDP was first activated as either an imidazolide derivative (^{7m}Gpp-Im) (Lewdorowicz et al. 2004; Sawai et al. 1991, 1999; Shimazu et al. 1990; Zuberek et al. 2003) or a chloroquinolylester (^{7m}Gpp-Q) (Koukhareva and Lebedev 2004). In both cases, the capped RNA was obtained with a moderate 35 % yield. Although this method was successful to give rise capped RNA in small amounts, the procedure required tedious purification steps to obtain pure 5′-MP RNA sequences before capping, to remove excess reagents, and to afford highly pure ^{7m}GpppRNA.

A solid-phase strategy could overcome these drawbacks and three chemical routes have been investigated for the synthesis of natural 5′-capped RNA. All routes consisted in coupling of an activated phosphoroimidazolidate derivative of N^7-methylguanosine to a solid-supported unprotected 5′-phosphate RNA obtained after RNA deprotection by a standard basic treatment before the coupling with the capping agent. Then the 5′-capped RNA was released from the solid support in a final step using soft media conditions compatible with the cap structure stability. Then, the key step of such process resided in the design of an unconventional linker between the solid support and the oligonucleotide. Three special linkers have been studied to synthesize natural 5′-capped RNA. First a phosphoramidate linker developed by Kadokura et al. was cleaved in soft acidic conditions not detrimental to the ^{7m}G stability and capped RNA was isolated with 20 % yield (Kadokura et al. 2001). Later Jemielity et al. investigated an alternative disulfide linker removed under slight basic pH conditions and could formed a capped trimer 2′-OMe RNA with 50 % efficacy (Jemielity et al. 2005). Lately Ohkubo et al. exploited a fluoride labile silyl-linker to anchor RNA to solid support and the final treatment (TBAF, AcOH, and THF) released 5′-capped RNA with 20–27 % yield after purification (Ohkubo et al. 2009, 2012). However if reported yields of the capping step were over 50 % in these chemical pathways, overall yields were low and purity of 5′-capped RNA were not always satisfactory. Although these chemical methods were successful, most of them produce a very limited amount of capped RNA. Therefore, our aim was to synthesize RNA with various cap structures (cap, cap-0, and cap-1) in great quantities using the combination of chemical synthesis of GpppRNA followed by enzymatic methylation (Fig. 14).

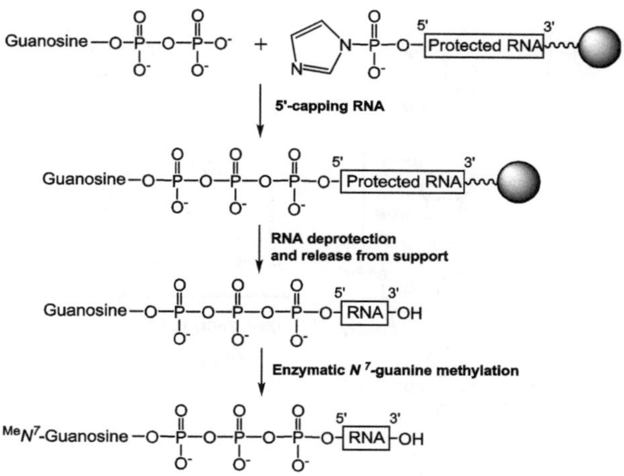

Fig. 14 General scheme for chemical synthesis on solid support of 5′-GpppRNA and their enzymatic methylation in solution

4.3 Chemical Synthesis on Solid-Support of GpppRNA and Their Enzymatic Methylation in Solution to ^{7m}GpppRNA

The originality of our strategy over the existing reported methods is based upon the automated RNA assembly and the capping reaction fully achieved on solid support with commercially available materials, which makes it easy to perform by other researchers without tedious and time-consuming preparation of reagents (Thillier et al. 2012).

4.3.1 Chemical Synthesis of GpppRNA

Automated elongation of RNA on solid support was accomplished at 1 μmol scale by the 2′-*O*-PivOM phosphoramidite method described in the first part of this chapter (Lavergne et al. 2008). The PivOM technology has the advantage of the deprotection basic conditions over the standard TBDMS chemistry for RNA synthesis because the mild ammonia treatment (3 h, rt) prevents degradation of the TP moiety in the cap structure. Actually the standard fluoride treatment (TBAF overnight) applied to remove TBDMS groups resulted in a partial hydrolysis of the TP moiety. Moreover, the ammonia deprotection is more convenient for RNA recovery since desalting procedures necessary after fluoride treatment are avoided and replaced by a simple evaporation. To obtain RNA with a cap-1, the introduction of a 2′-*O*-methyl ribonucleoside at the 5′-end of the sequence could be performed under the same assembly conditions via the corresponding commercially available 2′-*O*-methyl phosphoramidites.

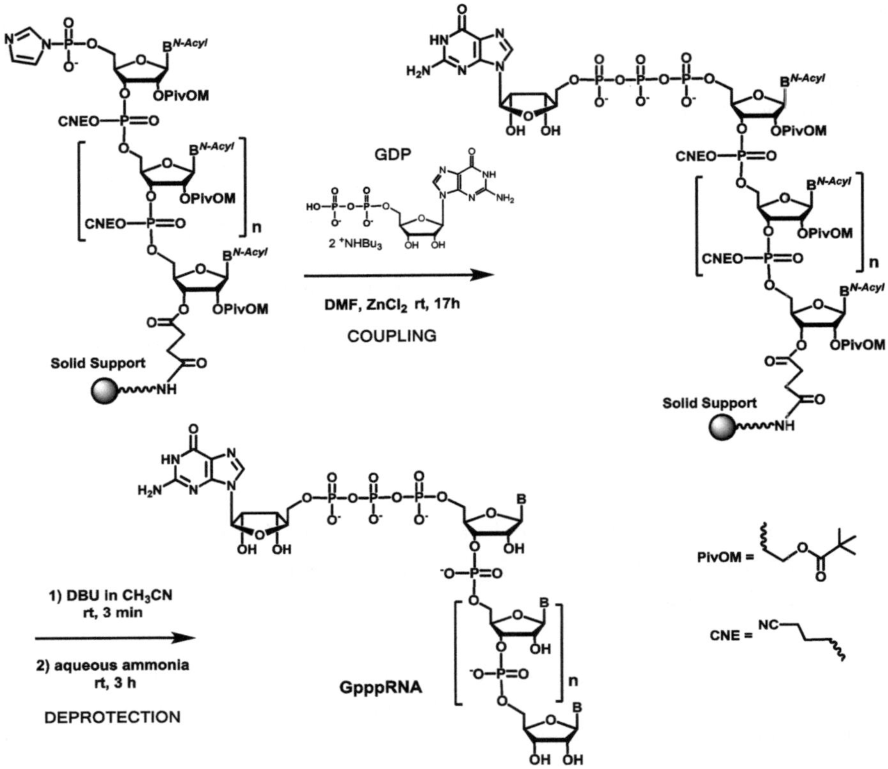

Fig. 15 Solid-phase synthesis of 5′-GpppRNA

As for 5′-TP RNA synthesis, we have chosen to form the triphosphate motif by the reaction of GDP with an activated phosphate group at the 5′-end of the solid-supported and protected RNA (Fig. 15). We have shown that the 5′-phosphoroimidazolidate RNA was efficient to give TP derivatives in good yield (Zlatev et al. 2010). Another strategy would have been to activate GDP with imidazole and to react with 5′-phosphate RNA on solid support. This approach requires more steps for the preparation of GDP *Im* and 5′-phosphate RNA instead of just modifying the 5′-OH RNA. Imidazole activation in organic solvent for the synthesis of 5′-capped oligonucleotides initially reported by Sekine's group (Sekine et al. 1989) seems to be the most powerful activation method for the phosphate chain formation (Jemielity et al. 2010). As described above the 5′-*ImP* RNA was prepared from the *H*-phosphonate intermediate with quantitative yield by amidative oxidation.

Before coupling with GDP in anhydrous DMF, due to its poor solubility in organic solvents, commercially available GDP sodium salt was first converted into its bis-(tri-*n*-butylammonium) salt by passing through a DOWEW-H^+ resin in the presence of tri-*n*-butylamine. Moreover to overcome the low solubility and poor reactivity of GDP, the addition of metal chlorides was reported to make a clear homogenous solution and to speed up the coupling reaction (Kadokura et al. 1997).

Among several metal chlorides tested in this capping reaction, $ZnCl_2$ was demonstrated to be the best catalyst in anhydrous organic solvent such as DMF due to the high affinity of the divalent cation Zn^{2+} for the phosphate oxygen and for the nitrogen atom of the imidazolidate entity. Thus it would act as a promoter that would bring the reagents closer to each other to enable the substitution. Therefore the triphosphate bond formation between GDP bis-(tri-*n*-butylammonium) salt and supported 5′-*ImP* RNA was performed at 30 °C overnight in the presence of 400 mM $ZnCl_2$ in dry DMF (Fig. 15). After removal of the capping solution from the synthesis reactor and several washings, deprotection procedure was applied with DBU for CNE elimination followed by ammonia treatment for PivOM and acyl group removal and linker cleavage.

Analysis by ion-exchange HPLC of the crude materials indicated the presence of GpppRNA with 40–50 % acceptable purity in the chromatograms. The other noticeable peaks essentially correspond to the 5′-*ImP* RNA resulting from the incomplete coupling reaction and leading to the 5′-phosphate RNA after hydrolysis. It is noteworthy that the capping yield would not be dependent either on the length or on the 5′-terminal nucleotide (A, C, G, or U) of RNA sequence since the yields were quite similar. Additionally, when the capping reaction was performed on a sequence containing a 5′-end 2′-OMe ribonucleotide, it was similarly efficient with yields reaching 50 %. After purification GpppRNA were isolated with satisfactory yields (40 %) and with high purity (>97 %) (Fig. 16). It should be emphasized that this chemical synthesis, in contrast to enzymatic methods produces great amounts of GpppRNA with a defined length and without any limitation regarding 5′-end RNA sequence.

4.3.2 Enzymatic Methylation of Cap Structure by Human N^7-MTase

GpppRNA obtained by solid-phase chemistry correspond to the product of the RTPase and GTase reactions and may be used as substrates for N^7-MTase in biochemical studies or structural investigations by crystallography. To our knowledge no N^7-MTase was commercially available except vaccinia virus N^7-MTase but it is coupled with GTase activity. Therefore, we had to prepare the human N^7-MTase which should specifically methylate the N^7 position of guanosine (Peyrane et al. 2007; Thillier et al. 2012). First the optimized concentration of N^7-MTase (250 nM) was determined to get 100 % methylation over 5 h at 30 °C in the presence of the methyl donor (0.4 mM AdoMet). The reaction products were also digested by P1 nuclease, and TLC analysis confirmed that the cap structure was methylated at its N^7 position as expected. N^7-Methylation was performed at 70 nmol scale of GpppRNA. The progress of the reaction was followed by IEX-HPLC analysis after 1 h and 5 h incubation, time when the methylation was complete since no trace of the starting material was detected (Fig. 17). No further HPLC purification was necessary and ^{7m}GpppRNA were next separated from the enzyme using a RP-C_{18} cartridge and from the remaining AdoMet and the formed AdoHcy by gel exclusion chromatography on a Sephadex G-25 column. It is noteworthy that hMTase-mediated N^7-methylation was also efficient irrespective

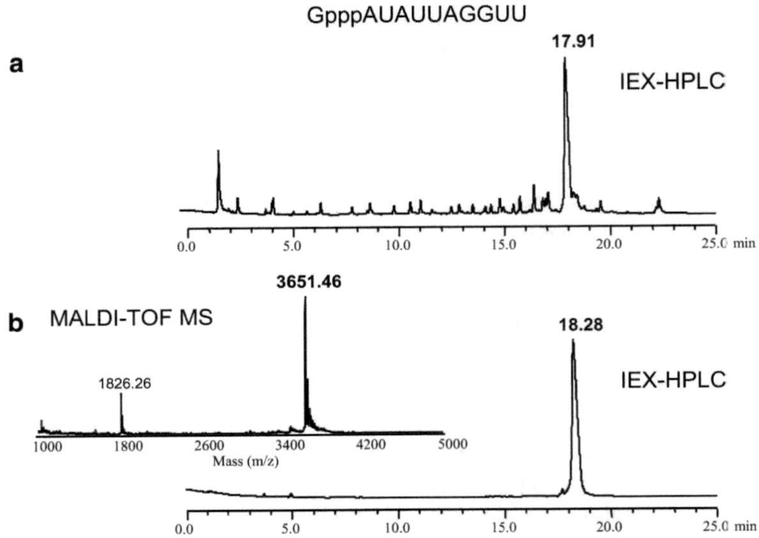

Fig. 16 IEX-HPLC profiles and MALDI-TOF mass spectrometry analysis of a 10-mer capped GpppRNA: (**a**) crude material and (**b**) purified material

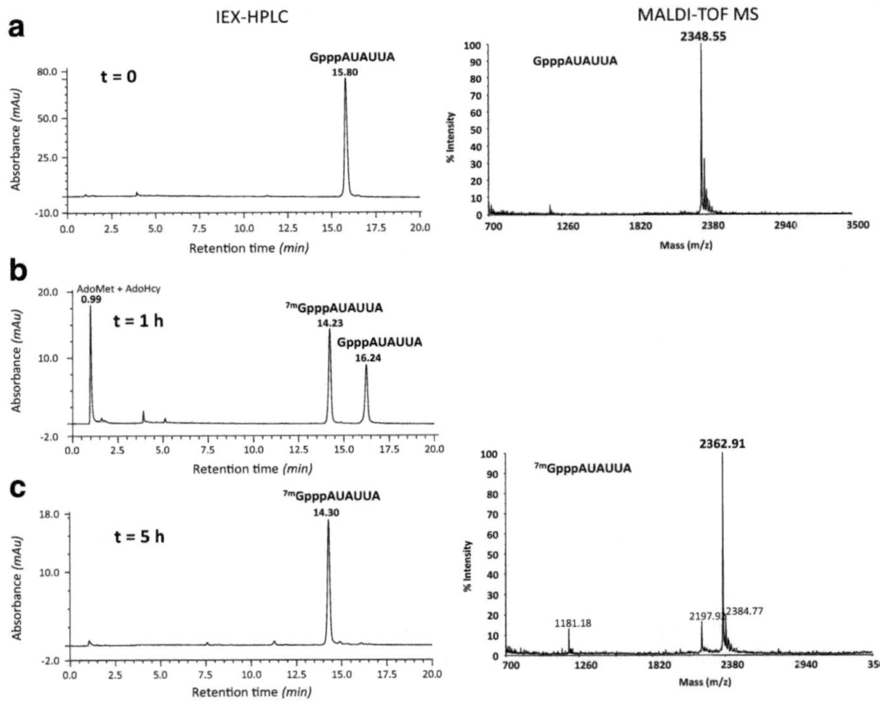

Fig. 17 Guanine N^7-methylation reaction of 5′-GpppAUAUUA followed by IEX-HPLC

of the nucleotide at 5′-end (2′-OH or 2′-OMe) and this observation means that N^7-methylation can occur even if the 5′-ribonucleotide carries already a 2′-OMe group. Large amounts of various dengue or SRAS virus GpppRNA of 4–18 nucleotides in length were successfully methylated with human N^7-MTase to give rise ^{7m}GpppRNA with cap-0 or cap-1 structure and were isolated with 80 % average yield.

5 Conclusion and Prospects

RNA with a triphosphate entity or a cap structure at their 5′-end play a key role in RNA maturation process or in RNA recognition in many biological pathways. This significance makes them very attractive for biologists as enzyme substrates for structural studies by crystallography or as tools for stimulation of innate immunity response. Enzymatic methods are limiting and not satisfactory to produce large amount of such molecules. Therefore face to the high demand chemists were urged to develop original and efficient synthesis to overcome this bottleneck.

In this chapter, we have summarized some existing chemical methods for regular RNA synthesis and for functionalization of 5′-end RNA with a TP or a cap moiety. We have highlighted recent developments from our research in RNA synthesis with the use of exclusively base-labile protections (even in 2′-OH) and in solid-support synthesis of RNA carrying triphosphate moiety at their 5′-end. The originality of this chemical synthesis is based on the use of an activated intermediate: the 5′-phosphoroimidazolide RNA that remains protected and anchored to solid support during the functionalization. This monophosphorylated RNA is coupled with a diphosphate entity such as pyrophosphate or GDP to afford after deprotection 5′-TP RNA or capped RNA, respectively (Fig. 18). Our strategy is advantageous over existing ones since the easier access to these molecules and their availability in sufficient amounts are very attractive and promising for detailed insights into structural and mechanistic studies of 5′-TP RNA and capped RNA involved in various biochemical processes. Future research efforts are in progress to yield RNA with modified triphosphate bridges more resistant to enzyme hydrolysis (Piton et al. 2013).

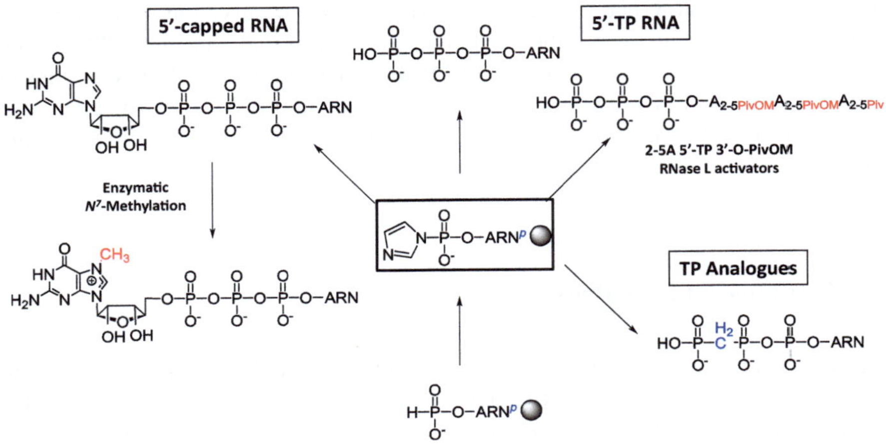

Fig. 18 Synthetic scheme for chemical synthesis of 5′-functionalized RNA on solid support

Acknowledgments The financial support of CNRS funding through the Programme Interdisciplinaire: Maladies Infectieuses Emergentes, of the Ministère de l'EducationNationale de la Recherche et Technologie, of the project SILVER (Health-F3-2010-260644) of the European Union 7th Framework Programme is gratefully acknowledged.

References

Allam R, Pawar RD, Kulkarni OP et al (2008) Viral 5′-triphosphate RNA and non-CpG DNA aggravate autoimmunity and lupus nephritis via distinct TLR-independent immune responses. Eur J Immunol 38:3487–3498

Aralov AV, Chakhmakhcheva OG (2013) Protective groups in the chemical synthesis of oligoribonucleotides. Russ J Bioorg Chem 39:1–21

Barral K, Sallamand C, Petzold C et al (2013) Development of specific dengue virus 2′-O- and N7-methyltransferase assays for antiviral drug screening. Antiviral Res 99:292–300

Beaucage SL (2008) Solid-phase synthesis of siRNA oligonucleotides. Curr Opin Drug Discov Dev 11:203–216

Beaucage SL, Caruthers MH (1981) Deoxynucleoside phosphoramidites - a new class of key intermediates for deoxypolynucleotide synthesis. Tetrahedron Lett 22:1859–1862

Bollati M, Alvarez K, Assenberg R et al (2010) Structure and functionality in flavivirus NS-proteins: perspectives for drug design. Antiviral Res 87:125–148

Brownlee GG, Fodor E, Pritlove DC et al (1995) Solid-phase synthesis of 5′-diphosphorylated oligoribonucleotides and their conversion to capped ^{7Me}Gppp-oligoribonucleotides for use as primers for influenza A virus RNA polymerase in vitro. Nucleic Acids Res 23:2641–2647

Burgess K, Cook D (2000) Syntheses of nucleoside triphosphates. Chem Rev 100:2047–2060

Cieslak J, Kauffman JS, Kolodziejski J et al (2007) Assessment of 4-Nitrogenated benzyloxymethyl groups for 2′-hydroxyl protection in solid-phase RNA synthesis. Org Lett 9:671–674

Cieslak J, Grajkowski A, Kauffman JS et al (2008) The 4-(*N*-dichloroacetyl-*N*-methylamino) benzyloxymethyl group for 2′-hydroxyl protection of ribonucleosides in the solid-phase synthesis of oligoribonucleotides. J Org Chem 73:2774–2783

Cieslak J, Ausin C, Grajkowski A et al (2013) The 2-cyano-2,2-dimethylethanimine-N-oxymethyl group for the 2′-hydroxyl protection of ribonucleosides in the solid-phase synthesis of RNA sequences. Chem Eur J 19:4623–4632

Daffis S, Szretter KJ, Schriewer J et al (2010) 2′-*O* methylation of the viral mRNA cap evades host restriction by IFIT family members. Nature 468:452–456

Debart F, Vasseur J-J, Lavergne T (2009) Chemical RNA synthesis method. Publication date: 3 Décembre 2010, PCT International Application, WO 2009144418 (A1)

Decroly E, Debarnot C, Ferron F et al (2011) Crystal structure and functional analysis of the SARS-coronavirus RNA cap 2′-O-methyltransferase nsp10/nsp16 complex. PLoS Pathog 7: e1002059

Decroly E, Ferron F, Lescar J et al (2012) Conventional and unconventional mechanisms for capping of virus mRNA. Nat Rev Microbiol 10:51–65

Elbashir SM, Harborth J, Lendeckel W et al (2001) Duplexes of 21-nucleotide RNAs mediate RNA interference in cultured mammalian cells. Nature 411:494–498

Fire A, Xu S, Montgomery MK et al (1998) Potent and specific genetic interference by double-stranded RNA in *Caenorhabditis elegans*. Nature 391:806–811

Gough GR, Miller TJ, Mantick NA (1996) p-nitrobenzyloxymethyl: a new fluoride-removable protecting group for ribonucleoside 2′-hydroxyls. Tetrahedron Lett 37:981–982

Hoard DE, Ott DG (1965) Conversion of mono- and oligodeoxyribonucleotides to 5′-triphosphates. J Am Chem Soc 87:1785–1788

Hodel AE, Gershon PD, Quiocho FA (1998) Structural basis for sequence-non-specific recognition of 5′-capped mRNA by a cap-modifying enzyme. Mol Cell 1:443–447

Hornung V, Ellegast J, Kim S et al (2006) 5′-triphosphate RNA is the ligand for RIG-I. Science 314:994–997

Iwase R, Maeda M, Fujiwara T et al (1992) Molecular design of a eukaryotic messenger RNA and its chemical synthesis. Nucleic Acids Res 20:1643–1648

Jemielity J, Fowler T, Zuberek J et al (2003) Novel "anti-reverse" cap analogs with superior translational properties. RNA 9:1108–1122

Jemielity J, Heinonen P, Lonnberg H et al (2005) A novel approach to solid phase chemical synthesis of oligonucleotide mRNA cap analogs. Nucleosides Nucleotides Nucleic Acids 24:601–605

Jemielity J, Kowalska J, Rydzik AM et al (2010) Synthetic mRNA cap analogues with a modified triphosphate bridge synthesis, applications and prospects. New J Chem 34:829–844

Joyce GF (2007) Forty years of in vitro evolution. Angew Chem Int Ed 46:6420–6436

Kadokura M, Wada T, Urashima C et al (1997) Efficient synthesis of γ-methyl-capped guanosine 5′-triphosphate as a 5′-terminal unique structure of U6 RNA via a new triphosphate bond formation involving activation of methyl phosphoroimidazolidate using $ZnCl_2$ as a catalyst in DMF under anhydrous conditions. Tetrahedron Lett 38:8359–8362

Kadokura M, Wada T, Seio K et al (2001) Solid-phase synthesis of a 5′-terminal TMG-capped trinucleotide block of U1 snRNA. Tetrahedron Lett 42:8853–8856

Kempe T, Chow F, Sundquist WI et al (1982) Selective 2′-benzoylation at the *cis* 2′,3′-diols of protected ribonucleosides. New solid phase synthesis of RNA and DNA-RNA mixtures. Nucleic Acids Res 10:6695–6714

Kore AR, Shanmugasundaram M, Charles I et al (2009) Locked nucleic acid (LNA)-modified dinucleotide mRNA cap analogue: synthesis, enzymatic incorporation, and utilization. J Am Chem Soc 131:6364–6365

Koukhareva II, Lebedev AV (2004) Chemical route to the capped RNAs. Nucleosides Nucleotides Nucleic Acids 23:1667–1680

Lackey JG, Mitra D, Somoza MM et al (2009) Acetal levulinyl ester (ALE) groups for 2′-hydroxyl protection of ribonucleosides in the synthesis of oligoribonucleotides on glass and microarrays. J Am Chem Soc 131:8496–8502

Lavergne T, Bertrand J-R, Vasseur J-J et al (2008) A base-labile group for 2′-OH protection of ribonucleosides: a major challenge for RNA synthesis. Chem Eur J 14:9135–9138

Lavergne T, Janin M, Dupouy C et al (2010) Chemical synthesis of RNA with base-labile 2′-O-(pivaloyloxymethyl)-protected ribonucleoside phosphoramidites. In: Beaucage SL, Bergstrom DE, Glick GD, Jones RA (eds) Current protocols in nucleic acid chemistry, vol 43. John Wiley, New York, NY, pp 3.19.11–13.19.27

Lebedev AV, Koukhareva II, Beck T et al (2001) Preparation of oligodeoxynucleotide 5′-triphosphates using solid support approach. Nucleosides Nucleotides Nucleic Acids 20:1403–1409

Lewdorowicz M, Yoffe Y, Zuberek J et al (2004) Chemical synthesis and binding activity of the trypanosomatid cap-4 structure. RNA 10:1469–1478

Ludwig J, Eckstein F (1989) Rapid and efficient synthesis of nucleoside 5′-*O*-(1-thiotriphosphates), 5′-triphosphates and 2′,3′-cyclophosphorothioates using 2-chloro-4*H*-1,3,2-benzodioxaphosphorin-4-one. J Org Chem 54:631–635

Mikkola S, Zhang Z, Mäki E et al (2005) Preparation and properties of mRNA 5′-cap structure. Curr Org Chem 9:1–23

Miller TJ, Schwartz ME, Gough GR (2001) Synthesis of oligoribonucleotides using the 2-nitrobenzyloxymethyl group for 2′-hydroxyl protection. In: Current protocols in nucleic acid chemistry, vol Unit 3.7. Wiley, New York, NY, pp 3.7.1–3.7.8

Mukaiyama T, Hashimoto M (1971) Phosphorylation by oxidation-reduction condensation. Preparation of active phosphorylating reagents. Bull Chem Soc Jpn 44:2284

Nagata S, Hamasaki T, Uetake K et al (2010) Synthesis and biological activity of artificial mRNA prepared with novel phosphorylating reagents. Nucleic Acids Res 38:7845–7857

Nagata S, Takagaki K, Wada T (2012) Improved method for the solid-phase synthesis of oligoribonucleotide 5′-triphosphates. Chem Pharm Bull 60:1212–1215

Ohgi T, Masutomi Y, Ishiyama K et al (2005) A new RNA synthetic method with a 2′-*O*-(2-cyanothoxymethyl) protecting group. Org Lett 7:3477–3480

Ohkubo A, Sasaki K, Noma Y et al (2009) Efficient solid-phase synthesis of oligodeoxynucleotides having a 5′-terminal 2,2,7-trimethylguanosine pyrophosphate linkage. Bioorg Med Chem 17:4819–4824

Ohkubo A, Tago N, Yokouchi A et al (2012) Synthesis of 5′-terminal capped oligonucleotides using O-N phosphoryl migration of phosphoramidite derivatives. Org Lett 14:10–13

Parey N, Baraguey C, Vasseur J-J et al (2006) First evaluation of acyloxymethyl or acylthiomethyl groups as biolabile 2′-*O*-protections of RNA. Org Lett 8:3869–3872

Peyrane F, Selisko B, Decroly E et al (2007) High-yield production of short GpppA- and ^{7Me}GpppA-capped RNAs and HPLC-monitoring of methyltransfer reactions at the guanine-N^7 and adenosine-2′-*O*-positions. Nucleic Acids Res 35:e26

Piton J, Larue V, Thillier Y et al (2013) The B. subtilis RNA deprotection enzyme RppH recognizes guanosine in the second position of its substrates. Proc Natl Acad Sci U S A 110:8858–8863

Pitsch S, Weiss PA, Jenny L et al (2001) Reliable chemical synthesis of oligoribonucleotides (RNA) with 2′-*O*-[(triisopropylsilyl)oxy]methyl(2′-*O*-TOM)-protected phosphoramidites. Helv Chim Acta 84:3773–3795

Poeck H, Besch R, Maihoefer C et al (2008) 5′-triphosphate-siRNA: turning gene silencing and Rig-I activation against melanoma. Nat Med 14:1256–1263

Reese CB (2005) Oligo- and poly-nucleotides: 50 years of chemical synthesis. Org Biomol Chem 3:3851–3868

Sawai H, Wakai H, Shimazu M (1991) Facile synthesis of cap portion of messenger RNA by Mn (II) ion-catalyzed pyrophosphate formation in aqueous solution. Tetrahedron Lett 32:6905–6906

Sawai H, Wakai H, Nakamura-Ozaki A (1999) Synthesis and reactions of nucleoside 5′-diphosphate imidazolide. A nonenzymatic capping agent for 5′-monophosphorylated oligoribonucleotides in aqueous solution. J Org Chem 64:5836–5840

Schlee M, Roth A, Hornung V et al (2009) Recognition of 5′-triphosphate by RIG-I Helicase requires short blunt double-stranded RNA as contained in panhandle of negative-strand virus. Immunity 31:25–34

Schmidt A, Schwerd T, Hamm W et al (2009) 5′-triphosphate RNA requires base-paired structures to activate antiviral signaling via RIG-I. Proc Natl Acad Sci U S A 106:12067–12072

Sekine M, Iwase R, Hata T et al (1989) Synthesis of capped oligoribonucleotides by use of protected 7-methylguanosine 5′-diphosphate derivatives. J Chem Soc Perkin Trans 1:969–978

Semenyuk A, Földesi A, Johansson T et al (2006) Synthesis of RNA using 2′-*O*-DTM protection. J Am Chem Soc 128:12356–12357

Shatkin AJ (1976) Capping of eukaryotic messenger-RNAs. Cell 9:645–653

Shiba Y, Masuda H, Watanabe N et al (2007) Chemical synthesis of a very long oligoribonucleotide with 2-cyanoethoxymethyl (CEM) as the 2′-*O*-protecting group: structural identification and biological activity of a synthetic 110 mer precursor-microRNA candidate. Nucleic Acids Res 35:3287–3296

Shimazu M, Shinozuka K, Sawai H (1990) Facile synthesis of nucleotides containing polyphosphates by Mn(II) and Cd(II) ion-catalyzed pyrophosphate bond formation in aqueous solution. Tetrahedron Lett 31:235–238

Shuman S (2001) Structure, mechanism, and evolution of the mRNA capping apparatus. Prog Nucleic Acid Res Mol Biol 66:1–40

Sproat BS, Calonna F, Mullah B et al (1995) An efficient method for the isolation and purification of oligoribonucleotides. Nucleosides Nucleotides 14:255–273

Thillier Y, Decroly E, Morvan F et al (2012) Synthesis of 5′ cap-0 and cap-1 RNAs using solid-phase chemistry coupled with enzymatic methylation by human (guanine-N^7)-methyltransferase. RNA 18:856–868

Thillier Y, Stevens SK, Moy C et al (2013) Solid-phase synthesis of 5′-triphosphate 2′-5′-oligoadenylates analogs with a 3′-O-biolabile group and their evaluation as RNase L activators and antiviral drugs. Bioorg Med Chem 21:5461–5469

Umemoto T, Wada T (2004) Oligoribonucleotide synthesis by the use of 1-(2-cyanoethoxy)ethyl (CEE) as a 2′-hydroxy protecting group. Tetrahedron Lett 45:9529–9531

Usman N, Ogilvie KK, Jiang MY et al (1987) The automated chemical synthesis of long oligoribuncleotides using 2′-*O*-silylated ribonucleoside 3′-*O*-phosphoramidites on a controlled-pore glass support: synthesis of a 43-nucleotide sequence similar to the 3′-half molecule of an *Escherichia coli* formylmethionine tRNA. J Am Chem Soc 109:7845–7854

Welz R, Müller S (2002) 5-(Benzylmercapto)-1H-tetrazole as activator for 2′-*O*-TBDMS phosphoramidite building blocks in RNA synthesis. Tetrahedron Lett 43:795–797

Zhou C, Honcharenko D, Chattopadhyaya J (2007) 2-(4-Tolylsulfonyl)ethoxymethyl (TEM)- a new 2′-OH protecting group for solid-supported RNA synthesis. Org Biomol Chem 5:333–343

Zlatev I, Lavergne T, Debart F et al (2010) Efficient solid-phase chemical synthesis of 5′-triphosphates of DNA, RNA, and their analogues. Org Lett 12:2190–2193. doi:10.1021/ol1004214

Zlatev I, Manoharan M, Vasseur J-J et al (2012) Solid-phase chemical synthesis of 5′-triphosphate DNA, RNA, and chemically modified oligonucleotides. In: Beaucage SL, Bergstrom DE, Glick GD, Jones RA (eds) Current protocols in nucleic acid chemistry, vol 1.28. Wiley, New York, NY, pp 1.28.21–21.28.16

Zlatev I, Lackey JG, Zhang L et al (2013) Automated parallel synthesis of 5′-triphosphate oligonucleotides and preparation of chemically modified 5′-triphosphate small interfering RNA. Bioorg Med Chem 21:722–732

Zuberek J, Wyslouch-Cieszynska A, Niedzweicka A et al (2003) Phosphorylation of eIF4E attenuates its interaction with mRNA 5′ cap analogs by electrostatic repulsion: intein-mediated protein ligation strategy to obtain phosphorylated protein. RNA 9:52–61

Zust R, Cervantes-Barragan L, Habjan M et al (2011) Ribose 2′-*O*-methylation provides a molecular signature for the distinction of self and non-self mRNA dependent on the RNA sensor Mda5. Nat Immunol 12:137–143

Index

A
2–5A, 576
Abasic sites, 88
Acridine, 238, 464, 465
Acridone, 464
Actinomycin, 466
Adenosine triphospate (ATP), 252, 255, 256, 258, 262, 263
Affinity resins, 535, 544–548, 550
  cap-based, 544
Affinity, native DNA, 214–215
2-Agmatinylcytidine, 36
Aldehyde reactive probe (ARP), 88
Alphaproteobacteria, 36
Ambiguous intermediate hypothesis, 32
Ammonia deprotection, 581
Animate, alive, 7
Anthraquinone, 465, 466
Anticancer vaccine, mRNA-based vaccine, 540
Anti-inflammatory, 230
Anti-mRNA strategies
  antisense approach, 200
  DNA enzyme, 200
  micro-RNAs, 200
  ribozymes, 200
  RNA interference, 200
  triple helix-forming oligonucleotides (TFOs), 200
Antisense, 228, 367–382
Antisense oligonucleotide (AON), 200, 498, 499, 501, 502, 503, 506–507
Apoptosis, 238
Aptamer(s), 228, 272, 401, 470
  domain, riboswitches, 424
Arginine, 236
Argonaute (AGO)
  cap-AGO, 548
  human Argonaute 2 (hAGO2), 548
  MID domain, 548
Artificial base pairs, 114
Ascidian, 28
Asialoglycoprotein (ASGP), 238
A site (ribosomal), 33
Association constants, 535, 536, 541
Avastin, 261

B
Bacterial ribosomal A-site, 177
Base-pairing properties, 81–84
Base pair stack, 381–382
Base stacking, 195
Basic fibroblast growth factor (bFGF), 255
Bevacizumab, 261
Bilayers, 237
Binding affinity, 209–211
Biomacromolecules, 386
Blood serum stabilities, 200
BNA, 200, 501
Boot strapping approach, 194
Bowel, 230
Bowen–Conradi syndrome, 402
Budding yeast, *Saccharomyces cerevisiae*, 31

C
Cancer, 228
Cap. *See also* mRNA cap
  affinity, 536
  cap-0, 513
  cap-1, 513
  cap-2, 513
  cap-3, 513
  cap-4, 513
  two-headed, 538

V.A. Erdmann et al. (eds.), *Chemical Biology of Nucleic Acids*, RNA Technologies,
DOI 10.1007/978-3-642-54452-1,

Cap analogs
anti-reverse (ARCA), 512, 522, 523, 537–538, 550
dinucleotides, 512, 515, 517, 521, 535, 537–540, 542, 543, 544, 550
synthesis, 519–525
labeled, 517
modified within
base, 517
oligophosphate, 516, 517, 524, 525
sugar moiety, 517
mononucleotides, 515, 541
purification, 530–531
synthesis on solid support, 529
Cap binding complex (CBC), 540–541
CPB20, 541
CPB80, 541
human, 541
Cap binding interactions, 535
Cap binding protein (CBP), 513, 531, 533, 536, 538, 540, 544, 547, 548
cap-CBP complex, 532
Capped oligonucleotides, 515, 525–528, 539, 543, 582
Capped RNA, 577, 579, 580, 585
(^{7m}Gppp)-RNA, 565
Capping enzymes
Giardia lamblia (GlaTgs2), 543
guanine-N2 methyltransferase (Tgs), 542, 543
methyltransferase 1 (hMTr1), 542
Mimivirus Tgs (MimiTgs), 543
Trypanosoma brucei methyltransferase (TbMTr1), 542
Cap structure, 576, 577, 579, 581, 583
Carbocyanine, 464
Carbohydrates, 229
5-Carboxymethyluridine, 29
Carl–Purcell–Meiboom–Gill (CPMG), 185
Cationic peptides, 176
Cell penetrating peptides (CPP), 175
Cellular penetration, 462, 464
C2′-endo, 195
C3′-endo, 195
Chemical exchange, 184
Chemical ligation
acceptor, 484–487
alkaline, 480, 488, 489
azide, 480, 488–490
click ligation, 489, 490
donor, 484–487
splint, 485, 487, 490
Chemical shifts, 195
Chemo-enzymatic, in vitro, 409–419
Chemo-enzymatic modification
AdoMet-analogs, 417
mRNAs, 418
sequence-specific labeling, 418
tRNAs, 417
Chemotherapeutics, 283
3′-Cholesterol, 235
Choline dhp, 69, 70
Hoogsteen base pairs, 68, 70
Hoogsteen pair, 68, 70
mismatches, 68, 70
sensing system, 68–70
triple helix, 68
Choline ions, 61
Circular dichroism, 401
β Clamp, 104
Clamp loader, 104
Cluster of differentiation 4 (CD4), 255
Co $(NH3)_6^{3+}$can substitute for Mg2+, 430
Codon
capture hypothesis, 32
reassignment, 32
Compartmentalized self-tagging (CST), 254
Control of gene expression, 367–382
Coupling reaction, 516, 520, 523, 524, 528, 529
Crystallization constructs, tetraloops, 433
Cyclobutane pyrimidine dimer (CPD), 88
Cytomegalovirus, 228

D

Decapping enzymes
Arabidopsis thaliana DcpS, 547
C. elegans DcpS (CeDcpS), 543, 544
Dcp1/Dcp2, 538, 539
DcpS, 539, 543, 544, 547, 548
HIT-45, 544
human DcpS (hDcpS), 543, 544
Nudt16, 538, 543
scavenger, 539, 543, 544, 548
Decoding centre, ribosome, 27
Decoding system
early decoding system, 26
vertebrate mitochondrial, 25
Decoy molecule, 116, 125
Decreased stacking, 195
Dendritic cells, 550
immature, 539
Deprotection, 564, 565, 567, 569, 585
2D heteronuclear single quantum coherence (HSQC), 190
1,3-Diaza-2-oxophenothiazine (tC), 100

Differential scanning calorimetry (DSC), 371
Dilution heat, 371, 372, 378
Dinucleotide, 229
Directional mutation pressure
GC pressure, 35
AT pressure, 27
Distamycin, 466, 467
DMD, 507
DNA
aptamer, 244–249, 254–263
duplex, 114
polymerase, 96
polymerase DinB (Pol IV), 106
polymerase III α, 106
secondary structure, 367–382
DNA/RNA hybrid, 368, 398–400
Dogma 13, 4–5, 8, 17, 18, 21
Double-stranded RNA, 168
RNA triple helices, 169
sequence-selective recognition, 167
Downfield-shifted C10 chemical shift, 195
Dpo4, 102
Drisapersen, 499, 500
Drosophila melanogaster, 29, 30, 32, 535, 536
D. embryos, 549
Duchenne muscular dystrophy (DMD), 498
Duplex structure
amplitude of pseudorotation, 221
pseudorotational phase angle, 221
Shift, Slide, Rise, Tilt, Roll, Twist, 221
X-disp, Y-disp, h-Rise, Inclination, Tip, h-Twist), 221
Dystrophin, 498, 501, 506, 507

E
Electron paramagnetic resonance (EPR), 387
Electro-osmotic flow (EOF), 257
ENA, 501–502
End-labeling
cis-diol, 482
hydrazide, 482
periodate, 482–484
primary amine, 482
semicarbazide, 482
Endocytosis, 238
Endosomal escape, 275–277
Enthalpy, 371
driven reactions, 367–382
Entropy, 371
Enzymatic ligation
splint, 485
T4 DNA ligase, 484, 487
T4 RNA ligase, 484–487
T4 Rnl 1, 484–486
T4 Rnl 2, 484, 485
Equilibrium association constant, 427
Escherichia coli, 27, 28, 29, 36
Eteplirsen, 500
Eukaryotic initiation factor 4E (eIF4E)
binding proteins (4E-BPs), 534
canonical, 534–537
cap-eIF4E complex, 533
4E-HP, 535
eIF4G complex, 549
isoforms, 535, 536, 546
LeishIF4E, 536
mammalian, 535, 537–539
murine, 535, 536
orthologs, 536
Evolution of replication, 3–7, 17, 20, 21
Evolution of translation, 5–7, 21
Evolvability, 1–22
Excited states (ESs), 183
Exon-skipping, 498–502, 506, 507
Exonuclease, 101, 200
3′-Exonucleases, 232
stability, 215–216
Exonuclease Xrn1, 539
Expression platform, riboswitches, 424

F
Family box(es), 27, 29, 31, 33, 35, 36
FANA, 251
ANA, 254
CeNA, 254
HNA, 254
TNA, 254
Fenton-type reactions, 78
Fibrosarcoma, 238
Fidelity, 97
Fleximers, 151
Flex-GTP, 154
reverse, 155
9-Fluorenylmethoxycarbonyl, 235
Fluorescence-based pH sensors, 282
Fluorescence titration, 535–537, 541, 542
Fluorescent bioprobes, 160–161
Fluorophores, 281–282
FMRP. *See* Fragile X mental retardation protein (FMRP)
FNM, 261
Foerster resonance energy transfer (FRET), 388
5-Formylcytidine, 29
Fragile X mental retardation protein (FMRP), 469

Free-radical cyclization, 202
Free-radical ring closure, 203–208
F-test, 193
Furan, 467

G
Galactose, 239
GDP, 582, 583, 585
Genetic alphabet expansion, 143, 145, 146
Genetic code
  early genetic code, 25, 26, 33, 34, 35, 37
  universal genetic code, 26, 27, 37
  vertebrate genetic code, 25
  vertebrate mitochondrial genetic code, 33
Genomic economisation, 27
Gibbs free energy, equation, 371
Glucose, 239
Glycosidic angles, 195
gp43 (DNA polymerase), 105
gp44/62 (clamp loader), 105
gp45 (sliding clamp), 105
5′-GpppRNA, 565
G-quadruplex(es), 442
  AS1411, 451–452
  backbone modifications, 445
  biomembranes, 59, 60, 64
  biosensors, 449
  cautions, 444
  crowded environment, 59
  drug-delivery, 453–454
  drug targets, 444
  enthalpic contributions, 61
  heterogeneous, 59
  ion sensors, 450
  liposomes, 59, 60, 63
  loop length, 444
  molecular crowding, 444
  nuclear confinement, 64
  oncogens, 449
  promotors, 449
  telomeres, 447–449
  therapeutic agents, 452–455
  thermodynamic parameters, 61, 62
G-quartet, 460, 466, 467, 470
Guanidinium, 235
Guanine, 515, 532
  alkylation, 516
Guanine (G)-quadruplex, 460–461
Guanine N^7-methyltransferase, 577

H
Hairpin, 460, 466
Hartman–Hahn matching condition, 192, 193
Heat, 367–382
  endothermic, 373, 375, 377, 378, 379
  exothermic, 372, 377, 379, 380, 381
Heat compensation block, 191
Heavy atom
  adaption into LIOAc, 432
  Cs^+ anomalous-diffraction, 434
  Cs^+ coordination, 434
  Cs^+ RNA phasing, 434
  G•U binding wobble base, 432
  halogenated pyrimidines, 431
  $Ir(NH3)_6^{3+}$ binding in Mg^{2+}, 431
  mode of osmium pentaammine binding, 432
Heavyn atom Cs^+, 431
Helix stability, 157–160
Hemicyanine, 465
Hen egg lysozyme (HEL), 257
Hepatocytes, 238
Hess cycle, 375, 378, 379
High-throughput solid-phase synthesis of 5′-TP, 575
(HIV) Rev protein, 255
hMTase, 583
^{1}H NMR studies, 117
H-phosphonate, 574
5′-Hp-ON, 575
Human blood serum, 219–220
Human neutrophil elastase (HNE), 254, 258
Human N^7-MTase, 583, 585
Hydration contributions, 381–382
  electrostricted water, 375, 378, 380, 381
  structural water, 375, 378
Hydrogen bonding (H-bonding), 114
Hypercholesterolemia, 228

I
ILPR. *See* Insulin-linked polymorphic region (ILPR)
Imidazolidate, 583
Immunoglobulin E (IgE), 255
I-motifs, 442, 460
  biosensors, 449
  drug delivery, 454–455
  glucose sensors, 451
  molecules, 446–447
  pH, 446–447
  pH sensor, 451–452
  therapeutic agents, 452–455

5′-ImP RNA, 582, 583
Inanimate, not yet alive, 7
In cell EPR, 404
Incorporation of non-natural nucleotides
  chemical synthesis, 412–415
  enzymatic synthesis, 415
  nucleotidyltransferases, 416
  RNA polymerases, 415
  T7-polymerase, 415, 416
Increased stacking, 195
Insulin, 470
Insulin-like growth factor 2 (IGF-2), 470
Insulin-linked polymorphic region (ILPR), 470
Intercalating, 229
Interferon-γ (IFN-γ), 255, 256
Interleaved experiments, 193
Ionic liquids (ILs)
  Hoogsteen base pairs, 65
  Watson–Crick base pairs, 65
Isothermal titration calorimetry (ITC), 170, 171, 370–371, 426
  enthalpy change ($\Delta H°$), 427
  ligand binding, 430
  metal depletion by EDTA, 430
Isotope patterns, 391

K
Keratinocyte growth factor (KGF), 255, 258
Kinetoplastida
  Leishmania, 536
  Trypanosoma brucei, 536
  trypanosomatides, 513, 536
K^+ ions, RNA binding, 426
Klenow, 232
Klenow fragment (KF), 100
Klentaq1, 99

L
Larmor frequency, 184
Leukemia, 238
Levenberg–Marquardt minimization algorithm, 193
Ligand, 462, 465, 468, 469
Ligation, 234
Line broadening, 186
Lipids, 229
Lipolexes, 236
Liposomes, 234
Little finger (LF), 102
Locked nucleic acids (LNA), 200, 231, 462
Loligo bleekeri, 30, 32
Ludwig–Eckstein method, 574
Lysidine, 2-lysylcytidine, 36

M
Macrocycle, 467
Macugen, 246
Malignant melanoma, 540
MCE, 499
4M choline dhp, 70
Membrane, 237
Messenger RNA, 228
Metaphor, 4
Method of Ludwig and Eckstein, 573
7-Methylguanine, 514, 515
7-Methylguanosine, 513–517, 519, 522, 523, 526, 528, 530, 532, 537–539, 544
Mg^+ ions, RNA binding, 426
Mg^{2+} ions, RNA binding, 426
Michael reaction, 503, 504, 505
MicroRNA (miRNA), 177, 229, 512, 513, 548
  let-7 miRNA, 548
  miRISC complex, 548
Mitochondria, 25
Mitochondrial respiratory chain, 77
Modification enzyme, 32
Modified oligos
  2′, 4′-locked 7′S or 7′R-Me-cLNA-A,-G,-^{Me}C' and-T, 199
Modified triphosphate, 585
Molecular dynamics, 182
Mono-exponential decay, 193
Mono-layers, 237
MonoLEX, 249
Monte-Carlo approach, 193
Mother liquor
  high-salt complications, 431
  poly(ethylene) glycol, 431
mRNA
  ARCA-capped, 538, 539
  capped, 526, 529, 531
  degradation, 513, 539, 543
  intracellular transport, 512, 537, 540–542
  nuclear export, 513, 540–541
  polyadenylation, 513
  splicing, 513, 527, 537, 540–541
  synthetic capped, 537
  transcription, 513, 539
  translation, 548

mRNA (*cont.*)
turnover, 537
mRNA cap, 513, 515–531, 539, 540, 548, 549, 550
MMG-cap, 531, 534, 537, 540–542
MMG-capped oligoribonucleotides, 528
monomethylguanosine, 541
TMG-cap, 513, 514, 521, 527, 528, 535, 536, 537, 541–543, 546, 551
TMG-capped oligoribonucleotides, 528
trimethylguanosine, 513
Mutagenesis, 97
Mutagenicity, oxidized RNA bases, 84
Mutate-and chemical-shift-fingerprint (MCSF), 195
Mycoplasma capricolum, 27, 28, 31

N
N-acetylgalactosamine, 234
Naive antigen-specific T cells, 539
Nanocarriers, 286
Nanoparticles, 234
Nearest-neighbor parameters, 375
NECEEM, 257
Nematodes, 29, 512, 513, 536, 537, 543
Ascaris suum, 31, 513, 537
Caenorhabditis elegans, 536, 543
Radopholus similis, 28
Schistosoma mansoni, 537
Neomycin riboswitch, 401–402
Nep1, 402
Neurodegenerative diseases, 75
NF-κB, 126
Nitroxide radicals, 390
N^7-methylation, 585
NMR-spectroscopy, 389
N^7-MTase, 579, 583
Non-universal codons, 26
North-type (N-type) conformation, 201
North-type conformationally constrained nucleoside analogues
amino-LNA, 201
aza-ENA, 201
1′, 2′-azetidine, 201
2′, 4′-BNA^{COC}, 201
carba-ENA, 201
2′–*C*,4′–*C*-bridged carba-LNA, 201
α-L-carba-LNAs analogues, 201
2′–*O*, 4′–*C*-ethylene-bridged nucleic acid (ENA), 201
1′, 2′-oxetane, 201
6′-substituted LNA, 201

Nucleases, 230
Nucleic acid-based therapeutics, 286
Nucleic acids, 228
heterogeneous conditions, 58
homogenous conditions, 58
ionic liquid (IL), 58
liposomes, 58
molecular crowding conditions, 58
Nucleobase
imidazo[5′, 4′, 4.5]pyrido[2,3-*d*] pyrimidine, 116
modifications, 150–154
1,8-naphthyridine, 116
Nucleotides
diphosphates, 516
oligophosphate, 515
5′-phosphates, 516
triphosphates, 516
Nucleotides analogs, 200

O
Off-equilibrium, 3, 4, 5, 8, 20
Off-resonance $R_{1\rho}$ profile, 192
Off-target, 229
Oligodeoxynucleotides, 116
Oligofectamine, 238
Oligonucleotides, 228, 369
O-methylisourea, 235
One bond coupling constant, 192
One-dimensional (1D) acquisition scheme, 189
On-resonance $R_{1\rho}$ profile, 192
Origin of life, 3, 6
Ornithine, 236
Oxazole, 467
Oxidative damage, 76
Oxidative lesions
e-rA, 80
e-rC, 80
FapyA, 80
FapyG, 80
5-hydroxycytidine, 80
5-hydroxyuridine, 80
8-oxo-7,8-dihydroguanosine, 80
Oxidative stress, 77
8-Oxo-7,8-dihydro-2′-deoxyguanine (8-oxoG), 103
Oxyphosphorane, 42, 43

P
Pegaptanib, 246
Peptide library, 465, 470

Peptide nucleic acids (PNAs), 169–172, 468–469, 498, 501
  cellular uptake, 175, 176
  nucleobase-modified, 173
Peptides, 229
Peptidyl transferase centre, 27, 86
Phasing module
  general G•U Cs^+ binding, 435
  G•U advantages, 435
  minor groove Cs^+ binding, 437
  single-site G•U wobble, 435
  tetraloop Cs^+ binding, 436
Phosphodiesterase, 232
Phosphonate analogues, 48
  uridine 2′/3′-methylphosphonates, 51
Phosphoramidite, 236
Phosphorodiamidate morpholino oligomers (PMO), 498, 500, 501, 502, 505, 506
Phosphoroimidazolidate (ImP), 574, 575, 582
5′-Phosphoroimidazolide RNA, 585
Phosphorothioate, 45, 233
  phosphorodithioate, 53
  3′, 5′-phosphoro-monothiates, 52
  triester, 49
  uridylyl-3′-5′-thymidine, 49
Phosphotriesters, 47
Photosensitizers, 277
Phthalocyanine, 462, 467
Pivaloyloxymethyl (PivOM), 565, 567, 569, 570, 572, 575, 581
Plasma, 236
PNPase, 87
Polymerase, 232
Polymerase chain reaction (PCR), 133–146, 245, 251, 252
  DNA sequencing, 135, 136
  efficiency, 133, 134
  fidelity, 134, 135, 142
  misincorporation, 133, 135, 136, 138, 140, 141, 142
  replication, 133, 135, 137, 138, 140, 141, 142
  selectivity, 133, 135, 137, 138, 141, 142
  transcription, 133, 134, 146
  triphosphate, 133, 137
Polyplexes, 236
Porphyrin, 462, 467
Postsynthetic, 393
  derivatization, 395
Pre-mRNA splicing, 512, 527, 537, 540–541
  trans-splicing, 513
PreQ1-I riboswitches, 424
PreQ1-II riboswitches, 426
PreQ1 metabolite, 424
Proliferating cellular nuclear antigen (PCNA), 104
Promoter, 461, 466, 470
2′-Protecting group, 565
2′-Protectiom, 567
Proto-oncogene, 461, 467, 470
Pseudorotation barrier(s), 44, 45, 52
Pseudosugars, 229
Pseudouridine, 29
Pulsed electron–electron double resonance (PELDOR), 387
Pulse sequences, 189
Pyrophosphate bond, 516, 520, 521, 523, 524

Q
Queuosine, 29
  tRNA nucleotide, 424
Quindoline, 238

R
Radionuclides, 282
RALyase, 88
RB69, 101
Reactive nitrogen species
  nitric oxide, 77
  nitrogen dioxide, 77
  peroxynitrite, 77
Reactive oxygen species
  hydrogen peroxide, 77
  hydroxyl radical, 77
  singlet oxygen, 77
  superoxide, 77
RecA, 106
Receptor, 229
Receptor-mediated endocytosis, 273–275
Relaxation delays, 191
Relaxation dispersion NMR spectroscopy, 184
Release factor (translation), 28
Replication factor C (RFC), 104
Respiratory syncytial virus (RSV), 255
Retrogression, 27
Reverse transcriptases
  AMV-RT, 84, 85
  HIV-RT, 85
  HIV1-RT, 84
  MMLV-RT, 84, 85
  RAV2-RT, 84, 85
Rho-independent transcription, 425

Ribonucleic acid (RNA), 228
affinity, 211–215
aptamer, 245–248, 254–263
functionalization
ACE, 472
biotin, 482, 483, 492
dye, 478, 480, 481, 492, 493
isotopes, 478, 482
linker, 480, 481, 483, 486
post-synthetic, 478–482, 489, 490, 493
reporter, 478, 480, 482, 483
spin probes, 478, 480
TBDMS, 472
TOM, 472
heavy atom binding, 431
modification
CuAAC, 410
Diels–Alder reaction, 412
SPAAC, 411
Staudinger ligation, 411
modifying systems, 32
phasing, 431
phosphodiester, 42
3′-alkylphosphates, 45
ribonucleoside 3′-phosphodiesters, 52
UpU, 51, 52
3′,5′-UpU, 45
uridine 3′-(3-nitrobenzyl phosphate), 44
uridine 3′-phosphodiesters, 48
selectivity, 215, 222
synthesis, 392, 564, 565, 567, 571, 572, 573, 581, 582, 585
monomer building blocks, 480, 481, 489, 493
phosphoramidite, 479, 480, 482, 489
phosphoramidite-building blocks, 478–480
soid phase, 483
TBDMS, 479
world, 5–7
abiotic, 6
prebiotic, 6, 8, 9, 16–20
primordial, 3, 6, 7, 8, 9, 10, 21
Ribosome-binding site (RBS), 426
Riboswitches gene regulation, 424
Ribozyme(s), 244
hairpin, 491
hammerhead, 489
HDV, 489
twin, 478, 491–493
RNA-induced silencing complex (RISC), 228
RPA, 105
$R_1\rho$ relaxation dispersion, 185
Ruler for DNA and RNA, 397–398

S
Secondary interaction, 124
Secondary structures, 195
Secondary/tertiary structures of mRNA, 369
SELEX, 244, 245
aptamer, 132, 135, 143–146
CE-SELEX, 257, 258
library, 143, 144, 145
non-selection, 249
post-SELEX modifications, 246–248
sublibrary, 143, 144
Sequence context effect
effect of configuration, 218–219
effect of nucleobase, 217–218
Serum, 232
Shape complementarity, 121
Single-molecule FRET (smFRET), 100
siRNA, 228
Site directed spin labelling (SDSL), 386
Slow-off-rate, 262
Snake venom phosphodiesterase exonuclease stability, 200
snRNA, 511–551
nuclear import, 513, 541–542
U snRNAs, 512, 513, 540, 541
U snRNP, 542
Snurportin1, 541–542
cap-snurportin 1, 542
SO_42-ion, 431
Solid phase synthesis, 393
SOMAmers, 254, 256
Sonogashira cross-coupling, 395
South–North sugar conformational equilibrium, 201
Spiegelmer, 248
Spin labels, 390
Spinlock power, 191, 192
ssDNA-binding protein (SSB), 105
Stability, 219–220
Stacking, 514, 515
interaction, 119
Stoichiometry (n) of binding, ITC, 427
Systems chemistry, 1–22
animate, 2–4, 6
synthetic, 2–5, 8, 9, 10, 18, 20, 21, 22

T
Targeting reactions, 372
5-Taurinomethyluridine, 29
TcDNA, 501
Telomerase, 461, 462, 464, 465
Telomere, 461, 465
TEMPA, 391

TEMPO, 391
Tenascin-C (TN-C), 247, 248
Tert-butoxycarbonyl, 235
Tetraloop, 400–401
Tetramethyl-pyrrolin-1-oxyl-3-acetylene (TPA), 391
Tetrathiafulvalene (TTF), 238
Therapeutic, 232
Thermal stability, 117
  melting temperatures, 118
Thermodynamic profiles, 367–382
Threoninol, 237
Thrombin, 255, 256, 257, 258, 259, 470
Thyroid stimulating hormone (TSH), 255
Tikhonov regularization, 387
T_M, 375
Toxins, 285
5′-TP, 573
5′-TP DNA or RNA, 573
5′-TP 25-mer RNA, 574
5′-TP ON, 576
5′-TP RNA, 573, 575–577, 585
5′-TP siRNA, 573
Trans-activating response RNA (TAR), 257
Transcriptional activators
  SAHA-PIPs, 358–360
  TALENs, 355
  ZFNs, 355
Transcription priming
  initiator molecule, 483
  T7 RNA polymerase, 483, 485
Transfection, 236
Transient structures, 182
Translation, 513, 534, 540
  cap dependent, 512, 534, 549
  inhibition, 531–534, 536, 548–551
  initiation, 512, 513, 531, 534, 540, 543, 548, 549
  initiation complex, 534, 548, 549
  regulation, 425
  repression, 534, 548, 549
Translational studies, 532
Translesion synthesis, 97
Transverse relaxation delays, 192
Tricyclo-DNA (TcDNA), 498
Trifluoroacetyl, 235
Triphosphate (TP), 565, 573–575, 582, 585
5′-Triphosphate, 565
5′-5′-Triphosphate bond, 512
Triple helix
  2-aminopyridine, 174
  Hoogsteen hydrogen bonding, 169
  3-oxo-2,3-dihydropyridazine **E,** 172
  physiological conditions, 174
  PNA:RNA, 170
  protonated cytosine, 174
  pyrimidine inversions, 172
  2-pyrimidinone **P,** 172
Tumor
  antigens, 280
  necrosis, 230
Two-codon sets, 27
Two-dimensional (2D) relaxation dispersion, 189

U
UmuD, 106
Unassigned codon, 32
Unmodified anticodon, 25
Unnatural base pair
  Ds-Px pair, 134–146
  isoG-isoC pair, 132
  P-Z pair, 134, 135
  5SICS-MMO2 pair, 134
Uptake, 229, 234
5′-UTR, 537, 549
UV melting curves, 371

V
Vascular endothelial growth factor (VEGF), 228, 246, 255, 256, 258, 260, 261
Vesicles, 237

W
Watson–Crick base pairs, 114

X
Xenopus laevis, 535
Xenopus oocytes, 540, 541
XNAs, 254
  CeNA, 257
  HNA, 257
  TNA, 257

Y
YB-1 protein, 87
Y-family DNA polymerases, 102
Yoshikawa's method, 516, 518

MIX
Papier aus verantwortungsvollen Quellen
Paper from responsible sources
FSC® C105338

If you have any concerns about our products,
you can contact us on
ProductSafety@springernature.com

In case Publisher is established outside the EU,
the EU authorized representative is:
Springer Nature Customer Service Center GmbH
Europaplatz 3, 69115 Heidelberg, Germany

Printed by Libri Plureos GmbH
in Hamburg, Germany